VOLUME ONE HUNDRED AND THIRTY SIX

# Advances in IMMUNOLOGY

## G Protein-Coupled Receptors in Immune Response and Regulation

# ASSOCIATE EDITORS

VOLUME ONE HUNDRED AND THIRTY SIX

# ADVANCES IN IMMUNOLOGY

## G Protein-Coupled Receptors in Immune Response and Regulation

Edited by

**ARUN K. SHUKLA**

*Department of Biological Sciences,*
*Indian Institute of Technology,*
*Kanpur, India*

ELSEVIER

ACADEMIC PRESS

An imprint of Elsevier

Academic Press is an imprint of Elsevier
50 Hampshire Street, 5th Floor, Cambridge, MA 02139, United States
525 B Street, Suite 1800, San Diego, CA 92101-4495, United States
The Boulevard, Langford Lane, Kidlington, Oxford OX5 1GB, United Kingdom
125 London Wall, London, EC2Y 5AS, United Kingdom

First edition 2017



ISBN: 978-0-12-812403-1
ISSN: 0065-2776

For information on all Academic Press publications
visit our website at https://www.elsevier.com/books-and-journals



*Publisher:* Zoe Kruze
*Acquisition Editor:* Ashlie M. Jackman
*Senior Editorial Project Manager:* Helene Kabes
*Production Project Manager:* James Selvam
*Cover Designer:* Greg Harris

Typeset by SPi Global, India

# CONTENTS

*Contributors* *ix*
*Preface* *xi*

**1. β2-Adrenoceptor Function in Asthma** **1**

Yassine Amrani and Peter Bradding

1. Therapeutic Actions of β2-Adrenoceptor Agonists in Asthma 2
2. β2-Adrenoceptor Dysfunction in Asthmatics 5
3. Cross-Talk Between Mast Cells and β2-Receptor Function in Asthma 13
4. Bitter Taste Receptor Agonists: An Alternative Bronchodilatory Strategy 16
5. Concluding Remarks 18

Acknowledgments 19
References 19

**2. G Protein-Coupled Kinin Receptors and Immunity Against Pathogens** **29**

Julio Scharfstein, Pablo I.P. Ramos, and Manoel Barral-Netto

1. Introduction 30
2. The KKS and Kinin Receptors 31
3. Kinins as Effectors of Immunity 41
4. KKS in Infection Immunity 47
5. Concluding Remarks 70

Acknowledgments 70
References 70

**3. P2Y Receptors in Immune Response and Inflammation** **85**

Diana Le Duc, Angela Schulz, Vera Lede, Annelie Schulze, Doreen Thor, Antje Brüser, and Torsten Schöneberg

1. Introduction 86
2. Molecular Structure and Pharmacology of P2Y Receptors 87
3. Expression and Physiological Function of P2Y Receptors in Immune Cells 99
4. Animal Models of P2Y Deficiency 100
5. Human Phenotypes and Diseases Involving P2Y Receptors 103
6. Clinical Perspectives: P2Y Receptors as Drug Targets in Immune Diseases 105
7. Conclusions/Perspective 109

Acknowledgments 109
References 109
Further Reading 121

**4. Emerging Roles for MAS-Related G Protein-Coupled Receptor-X2 in Host Defense Peptide, Opioid, and Neuropeptide-Mediated Inflammatory Reactions** **123**

Hydar Ali

1. Introduction 125
2. Expression Profile and Functions of MRGPRX1–MRGPRX4 126
3. MRGPRX2 Is a Novel GPCR for HDPs on MCs 130
4. MRGPRX2 Is a Novel Tachykinin Receptor in Human but Not Murine MCs 134
5. MRGPRX2 Is a Novel GPCR for Cationic Drugs and Opioids 137
6. Ligand-Binding Sites for Opioids and ZINC-3573 on MRGPRX2 and Signaling 143
7. Regulation of MRGPRX2 Function in MCs 146
8. Concluding Remarks 149

Acknowledgments 150
References 151

**5. Adhesion GPCRs in Regulating Immune Responses and Inflammation** **163**

Hsi-Hsien Lin, Cheng-Chih Hsiao, Caroline Pabst, Josée Hébert, Torsten Schöneberg, and Jörg Hamann

1. Introduction 164
2. Origin and Evolution 165
3. Molecular Structure 168
4. Mechanisms of Activation and Signal Transduction 170
5. Cellular Distribution 174
6. Involvement in Immune Cell Development 176
7. Involvement in Innate Immunity 180
8. Involvement in Adaptive Immunity 187
9. Outlook 189

Acknowledgments 190
References 191

**6. GPCR Signaling in *C. elegans* and Its Implications in Immune Response** **203**

Anjali Gupta and Varsha Singh

1. Introduction 204
2. Regulation of Innate Immune Response by the Nervous System 206

3. GPCR Signaling at the Site of Infection Regulates Innate Immune Responses 213
4. Behavioral Response of *C. elegans* to Pathogens 217
5. Concluding Remarks 221
Acknowledgments 223
References 223

**7. G Protein-Coupled Receptor Kinases in the Inflammatory Response and Signaling 227**

Michael D. Steury, Laura R. McCabe, and Narayanan Parameswaran

1. Introduction 228
2. GRKs in Inflammatory Signaling and Disease 238
3. Conclusion 262
References 262

**8. Canonical and Noncanonical Signaling Roles of β-Arrestins in Inflammation and Immunity 279**

Mohammad M. Ahmadzai, David Broadbent, Christopher Occhiuto, Canchai Yang, Rupali Das, and Hariharan Subramanian

1. Introduction 280
2. Properties of β-Arrestins 281
3. Canonical Roles of β-Arrestins in the GPCR Life-Cycle 284
4. β-Arrestins Modulate Immune Cell Activity via Their Canonical Interactions 288
5. β-Arrestins Modulate Immune Cell Functions via Their Noncanonical Interactions 295
6. β-Arrestins Are Implicated in the Pathogenesis of Several Immunological Disorders 300
7. Future Perspectives and Conclusions 304
Acknowledgments 306
References 306

**9. Emerging Roles of Regulators of G Protein Signaling (RGS) Proteins in the Immune System 315**

Kirk M. Druey

1. Introduction 316
2. The RGS Protein Family 320
3. Mechanisms by Which RGS Proteins Control GPCR Signaling 322
4. RGS Proteins and Cell Migration Signaling 325

5. RGS Proteins in Innate Immune Cells 328
6. RGS Functions in Adaptive Immune Responses 333
7. Noncanonical Functions of RGS Proteins in the Immune System 339
8. RGS Association With Diseases of the Immune System 341
9. Conclusions 343
Acknowledgment 344
References 344

**10. Roles of NHERF Family of PDZ-Binding Proteins in Regulating GPCR Functions 353**

**David Broadbent, Mohammad M. Ahmadzai, Ananth K. Kammala, Canchai Yang, Christopher Occhiuto, Rupali Das, and Hariharan Subramanian**

1. Introduction 354
2. PDZ Proteins 355
3. NHERF Family of PDZ-Binding Proteins 356
4. NHERFs Regulate Phosphate Reabsorption 358
5. NHERF-Mediated Regulation of Ion Transport and Disease 361
6. NHERFs and Cancer 369
7. NHERFs in Immunity and Inflammation 373
8. Clinical Relevance and Future Perspectives 377
Acknowledgments 377
References 378

# CONTRIBUTORS

**Mohammad M. Ahmadzai**
Michigan State University, East Lansing, MI, United States

**Hydar Ali**
University of Pennsylvania School of Dental Medicine, Philadelphia, PA, United States

**Yassine Amrani**
University of Leicester, Leicester, United Kingdom

**Manoel Barral-Netto**
Instituto Gonçalo Moniz, Fundação Oswaldo Cruz (FIOCRUZ), Salvador, Brazil

**Peter Bradding**
University of Leicester, Leicester, United Kingdom

**David Broadbent**
Michigan State University, East Lansing, MI, United States

**Antje Brüser**
Rudolf Schönheimer Institute of Biochemistry, Molecular Biochemistry, University of Leipzig, Leipzig, Germany

**Rupali Das**
Michigan State University, East Lansing, MI, United States

**Kirk M. Druey**
Molecular Signal Transduction Section, Laboratory of Allergic Diseases, NIAID/NIH, Bethesda, MD, United States

**Anjali Gupta**
Department of Molecular Reproduction, Development and Genetics, Indian Institute of Science, Bangalore, India

**Jörg Hamann**
Academic Medical Center, University of Amsterdam, Amsterdam, The Netherlands

**Josée Hébert**
Maisonneuve-Rosemont Hospital; Medical Faculty, University of Montreal, Montreal, QC, Canada

**Cheng-Chih Hsiao**
Academic Medical Center, University of Amsterdam, Amsterdam, The Netherlands

**Ananth K. Kammala**
Michigan State University, East Lansing, MI, United States

**Vera Lede**
Rudolf Schönheimer Institute of Biochemistry, Molecular Biochemistry, University of Leipzig, Leipzig, Germany

**Diana Le Duc**
Rudolf Schönheimer Institute of Biochemistry, Molecular Biochemistry, University of Leipzig, Leipzig, Germany

**Hsi-Hsien Lin**
College of Medicine, Chang Gung University; Chang Gung Memorial Hospital-Linkou, Tao-Yuan, Taiwan

**Laura R. McCabe**
Michigan State University, East Lansing, MI, United States

**Christopher Occhiuto**
Michigan State University, East Lansing, MI, United States

**Caroline Pabst**
Medical Faculty, Martin-Luther-University Halle-Wittenberg, Halle (Saale), Germany

**Narayanan Parameswaran**
Michigan State University, East Lansing, MI, United States

**Pablo I.P. Ramos**
Instituto Gonçalo Moniz, Fundação Oswaldo Cruz (FIOCRUZ), Salvador, Brazil

**Julio Scharfstein**
Institute of Biophysics Carlos Chagas Filho, Federal University of Rio de Janeiro, Center of Health Sciences (CCS), Cidade Universitária, Rio de Janeiro, Brazil

**Torsten Schöneberg**
Medical Faculty, University of Leipzig, Leipzig, Germany

**Angela Schulz**
Rudolf Schönheimer Institute of Biochemistry, Molecular Biochemistry, University of Leipzig, Leipzig, Germany

**Annelie Schulze**
Rudolf Schönheimer Institute of Biochemistry, Molecular Biochemistry, University of Leipzig, Leipzig, Germany

**Varsha Singh**
Department of Molecular Reproduction, Development and Genetics, Indian Institute of Science, Bangalore, India

**Michael D. Steury**
Michigan State University, East Lansing, MI, United States

**Hariharan Subramanian**
Michigan State University, East Lansing, MI, United States

**Doreen Thor**
Rudolf Schönheimer Institute of Biochemistry, Molecular Biochemistry, University of Leipzig, Leipzig, Germany

**Canchai Yang**
Michigan State University, East Lansing, MI, United States

# PREFACE

The largest class of cell surface receptors in our body, referred to as G protein-coupled receptors (GPCRs), play critical roles in many different aspects of immune response and regulation. For example, chemokine receptors are key players in chemotaxis, histamine and complement receptors are crucially involved in inflammatory responses, and emerging data suggest important roles of adhesion GPCRs in innate effector functions. However, a concerted discussion of these roles and mechanisms on one platform is currently lacking. Therefore, we have put together this volume of *Advances in Immunology* focused on GPCRs and their regulatory partners in various aspects of the immune system.

Considering that it is difficult to provide an extensive coverage of every receptor system involved in the process of immune response, we have chosen a handful of receptors and regulatory proteins as examples to highlight their involvement in various processes related to the immune system. In the first part, we have chosen five different receptor systems, namely, the β2-adrenergic receptor, kinin receptors, the MAS-related receptor-X2, nucleotide-binding P2Y receptors, and adhesion receptors, to showcase the diverse roles of GPCRs in mediating and regulating the process of immunity and inflammation. In the second part, we have included a set of chapters that underline the roles of GPCR-interacting proteins such as GRKs (GPCR kinases), β-arrestins, RGS proteins (regulator of G protein signaling), and PDZ–GPCR interactions in various inflammatory and immune responses. These chapters not only discuss the well-known aspects of GPCR signaling in immunology but also present many emerging paradigms that have not yet been reported in classical textbooks. In addition, each chapter presents a forward-looking discussion, providing a glimpse of the tremendous potential associated with the specific system that is being discussed.

I take this opportunity to sincerely thank all the contributors who have spent their valuable time to make this volume possible. I sincerely hope that the chapters included in this volume will benefit many researchers working in the broad area of GPCR signaling in immune system. My sincere thanks to our editorial staff and production team who worked on this project relentlessly to complete it in time.

On behalf of the entire *Advances in Immunology* team, I am delighted to present this special volume entitled "G Protein-Coupled Receptors in Immune Response and Regulation" to you, and very much look forward to your feedback and suggestions.

Arun K. Shukla, PhD
Department of Biological Sciences and Bioengineering
Indian Institute of Technology, Kanpur, India

CHAPTER ONE

# β2-Adrenoceptor Function in Asthma

**Yassine Amrani[1], Peter Bradding**
University of Leicester, Leicester, United Kingdom
[1]Corresponding author: e-mail address: ya26@le.ac.uk

## Contents

1. Therapeutic Actions of β2-Adrenoceptor Agonists in Asthma 2
2. β2-Adrenoceptor Dysfunction in Asthmatics 5
   2.1 Evidence From Asthmatic Patients 5
   2.2 Proasthmatic Factors Affecting $\beta_2$-Adrenergic Function 7
3. Cross-Talk Between Mast Cells and β2-Receptor Function in Asthma 13
   3.1 Mast Cell Infiltration of ASM Is a Key Determinant of the Asthmatic Phenotype 13
   3.2 Bidirectional Interactions Between Mast Cells and ASM 14
   3.3 Cross-Talk Between Mast Cell–ASM Leads to β2-Adrenoceptor Dysfunction 15
4. Bitter Taste Receptor Agonists: An Alternative Bronchodilatory Strategy 16
   4.1 Bronchodilator Activity 17
   4.2 Other Therapeutic Benefits of Bitter Taste Receptors 18
5. Concluding Remarks 18
Acknowledgments 19
References 19

## Abstract

β2-adrenoceptor agonists, often used in combination with corticosteroids, have been extensively used for the treatment of asthma. However, concerns have been raised regarding their adverse effects and safety including poor asthma control, life-threatening exacerbations, exacerbations that often require hospitalization, and asthma-related deaths. The question as to whether these adverse effects relate to the loss of their bronchoprotective action remains an interesting possibility. In the chapter, we will review the experimental evidence that describes the different potential factors and associated mechanisms that can blunt the therapeutic action of β2-adrenoceptor agonists in asthma. We show here evidence that various key inflammatory cytokines, growth factors, some respiratory viruses, certain allergens, unknown factors present in serum from atopic asthmatics have the capacity to impair β2-adrenoceptor function in airway smooth muscle, the main target of these drugs. More importantly, we present our latest research describing the role played by mast cells in impairing β2-adrenoceptor function. Although no definitive conclusion could be made regarding the implication of one single

*Advances in Immunology*, Volume 136
ISSN 0065-2776
http://dx.doi.org/10.1016/bs.ai.2017.06.003

mechanism, receptor uncoupling, or receptor desensitization due to phosphorylation represents the main inhibitory pathways associated with a loss of β2-adrenoceptor function in airway smooth muscle. Targeting the pathways leading to β2-adrenoceptor dysfunction will likely provide novel therapies to improve the efficacy of β2-agonists in asthma.

## 1. THERAPEUTIC ACTIONS OF β2-ADRENOCEPTOR AGONISTS IN ASTHMA

β2-adrenoceptor agonists are highly effective bronchodilator drugs used in the management of pulmonary diseases such as asthma and chronic obstructive pulmonary diseases. They provide rapid relief of symptoms via their potent bronchodilatory action by specifically promoting the relaxation of airway smooth muscle (ASM) (Giembycz & Newton, 2006). $\beta_2$-agonists act on the $\beta_2$-adrenoceptor which activates adenylyl cyclase via the stimulatory G-protein Gs, resulting in cyclic adenosine monophosphosphate (cAMP) production and protein kinase A (PKA) activation which has been described as the main mechanism driving relaxation of ASM. The relaxant pathways are multiple and include a inhibition of the myosin light chain kinase activity leading to a reduction of calcium sensitivity of the contractile proteins, modulation of intracellular calcium levels through the blockade of calcium release from intracellular stores, and induction of membrane hyperpolarization through activation of large-conductance calcium-activated potassium channels ($BK_{ca}$) (nicely reviewed in Giembycz & Newton, 2006; Pera & Penn, 2016). Inhibition of Rho pathways via the activation of cAMP-mediated exchange protein (Epac) has been also described as a mechanism for the beneficial effect of $\beta_2$-agonists in ASM cells (Fogli et al., 2015). Fig. 1 summarizes the mechanisms leading to bronchodilation in response to $\beta_2$-agonists.

The therapeutic value of combining long-acting $\beta_2$-agonists with corticosteroids in the treatment of asthmatic patients is undeniable as evidenced by numerous reports showing better disease control, pulmonary function, and reduced frequency of exacerbations (Newton & Giembycz, 2016). The mechanisms underlying the superior clinical benefits provided by the combination therapies have not been completely elucidated, but several in vivo and in vitro studies have suggested that $\beta_2$-agonists might possess antiinflammatory properties in addition to their bronchoprotective effect. In vivo studies investigated whether several key markers of airway

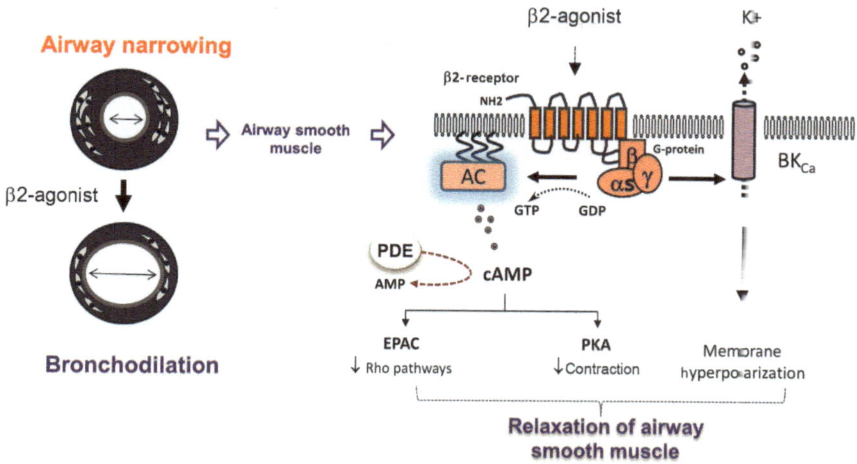


**Fig. 1** Molecular mechanisms driving the bronchodilator action of β2-agonists. β2-agonists directly activate β2-receptor coupled to Gs on airway smooth muscle to augment the activity of adenylyl cyclase (AC) leading to cAMP generation, activation of cAMP-dependent protein kinase A (PKA) and ASM relaxation and bronchodilation. Other pathways may also be involved including the activation of cAMP-mediated exchange protein (Epac) which inhibits the activation of the Rho pathways or the opening of the large conductance calcium-activated potassium channels ($BK_{Ca}$) which causes membrane hyperpolarization. It is believed that cAMP-dependent pathways represent the main mechanism driving ASM relaxation by β2-agonists.

mucosal inflammation in asthmatic patients can be modulated by long-acting $\beta_2$-adrenoceptor agonists. Levels of IL-8 and neutrophil number in the bronchoalveolar lavage fluids were reduced in patients with persistent asthma treated for 12 weeks with salmeterol (Reid et al., 2003). Patients with mild asthma treated with formoterol for 8 weeks had a significant reduction of number of eosinophils in both the submucosa and epithelium (EG2+) (Wallin, Sandstrom, Cioppa, Holgate, & Wilson, 2002), as well as number of mast cells and eosinophils in the lung mucosa (Wallin et al., 1999). Salmeterol was also reported to decrease neutrophils number in the submucosa in mild asthmatics following a 6 weeks of treatment (Jeffery et al., 2002), while in children with asthma, short-term treatment with formoterol (4 weeks) decreased the number of eosinophil in the blood as well as levels of eosinophil cationic protein (ECP) and IL-4 (Stelmach et al., 2002). The antiinflammatory efficacy of salmeterol was also demonstrated when given as an add-on therapy for 3 weeks to inhaled beclomethasone dipropionate which led to a reduction of levels of sputum ECP in mild and moderate asthmatics (Pinto Pereira et al., 2003). However, Roberts et al. (1999) did not find any evidence of antiinflammatory activity in bronchial biopsies or BAL when salmeterol was administered to steroid-naïve subjects for 6 weeks (Roberts et al., 1999). Furthermore, unlike budesonide, formoterol had no effect on the sputum eosinophil count in patients with poorly controlled asthma (Green et al., 2006). It is therefore unclear whether long-acting $\beta_2$-agonists exert inhibitory actions on the accumulation of inflammatory cells and production of pro-asthmatic cytokines in vivo.

Parallel in vitro studies were also performed to investigate whether $\beta_2$-agonists exert antiinflammatory actions on different lung cells by studying their effect on the production of inflammatory mediators in response to various pro-asthmatic stimuli. In airway epithelial cells, for example, salmeterol and formoterol inhibited production of CXCL-8 and VEGF induced by LPS (Chiu, Hsu, Fu, Chu, & Chi, 2007) and thymus- and activation-regulated chemokine (TARC/CCL17) in response to TNF$\alpha$, IL-4, and IFN$\gamma$ (Hung et al., 2008), while albuterol suppressed IL-1$\beta$/IFN$\gamma$-induced GM-CSF production via iNOS pathways (Chorley, Li, Fang, Park, & Adler, 2006). IgE-mediated histamine release can be suppressed by SABAs and LABAs not only in basophils (Kleine-Tebbe, Frank, Josties, & Kunkel, 1994) but also in human lung mast cells (HLMC) via various mechanisms involving $\beta_2$-adrenoceptor-dependent and -independent pathways (Chong, Cooper, Vardey, & Peachell, 1998; Lewis, Chachi, Newby, Amrani, & Bradding, 2016) or via the inhibition of KCa3.1 channels, which regulate cell membrane hyperpolarization and potentiation of both $Ca^{2+}$ influx and mast cell degranulation (Duffy, Cruse, Lawley, & Bradding, 2005). Interesting observations were also made in studies using human ASM cells where combining short-acting (salbutamol) or long-acting (salmeterol, formoterol) $\beta_2$-agonists with dexamethasone (or budesonide) results in additive or synergistic inhibition of the production of various proinflammatory proteins (Ammit et al., 2002; Hallsworth, Twort, Lee, & Hirst, 2001; Kaur, Chivers, Giembycz, & Newton, 2008; Newton, 2014; Nie, Knox, & Pang, 2005; Pang & Knox, 2000, 2001). Among the inflammatory cytokines strongly repressed by the drug combination include various key players in asthma pathogenesis including the chemotactic proteins CCL11, CCL5, and CXCL8 known for the recruitment of eosinophils, T lymphocytes, and neutrophils, respectively (Castan, Magnan, & Bouchaud, 2017). Activation of $\beta_2$-adrenoceptor, cAMP-dependent pathways leading to the induction of steroid-inducible MAP kinase phosphatase-1 (MKP-1) has been described as one mechanism explaining the augmented antiinflammatory action of combining

$\beta_2$-agonists/corticosteroids in ASM cells (Kaur et al., 2008; Manetsch, Che, Seidel, Chen, & Ammit, 2012; Manetsch et al., 2013; Quante et al., 2008). The synergistic induction of the regulator of G-protein signaling 2 (RGS2), a GTPase-activating protein that specifically switch-off $G_q$ pathways, represent another mechanism of bronchoprotection that is induced by combining $\beta_2$-agonists and corticosteroids (Holden et al., 2011). This hypothesis was later confirmed by the same authors who used knockout animals to demonstrate the bronchoprotective role of RGS2 in a murine model of house dust mite-induced allergic inflammation. Interestingly, the authors also found that the presence of RGS2 was required to suppress allergen-associated pulmonary inflammation including production of different chemokines such as CCL3, CCL11, CXCL9, and CXCL10 (George et al., 2017). Combined therapy also inhibits features of ASM remodeling where low doses of fluticasone synergized with fenoterol to suppress PDGF-induced proliferation and hypocontractility in bovine tracheal smooth muscle cells (Dekkers et al., 2012). Another study using precision cut lung slices (PCLS) showed that combining dexamethasone and albuterol prevented homologous desensitization of $\beta_2$-receptor typically seen with chronic stimulation of the receptor (Cooper & Panettieri, 2008). The overall beneficial effects of $\beta_2$-adrenoceptor agonists in asthma likely result from their bronchodilator effect by relaxing ASM but also on their ability to potentiate the antiinflammatory and bronchoprotector actions of corticosteroids through their direct action on ASM cells and other lung cells. Fig. 2 summarizes the therapeutic actions of $\beta_2$-agonists acting in ASM in asthma.

The following section will address the challenging question regarding the mechanisms behind the loss of bronchoprotection induced by $\beta_2$-agonists often seen in patients with asthma.

## 2. β2-ADRENOCEPTOR DYSFUNCTION IN ASTHMATICS

### 2.1 Evidence From Asthmatic Patients

The hypothesis of a reduced $\beta_2$-adrenergic function in asthma was first demonstrated in animal models of allergic asthma and nonspecific airway hyperresponsiveness (Barnes, Dollery, & MacDermot, 1980; Emala, Black, Curry, Levine, & Hirshman, 1993). In the Basenji-Greyhound dog model of airway hyperresponsiveness, Emala and colleagues found a close correlation between impaired $\beta_2$-adrenoceptor responses and the decreased production of cAMP in response to isoproterenol in isolated tracheal muscle

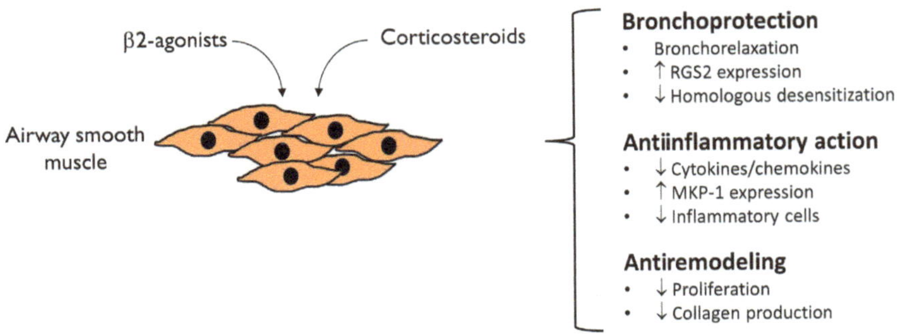


**Fig. 2** Therapeutic actions of β2-agonists in ASM that are enhanced by corticosteroids. β2-agonists stimulate different therapeutic responses in ASM that have been shown to be enhanced by corticosteroids. Corticosteroids have been shown to enhance bronchodilator action of β2-agonists by (i) increasing levels of regulator of G-protein signaling 2 (RGS2) which interferes with agonist-evoked ASM contraction or (ii) reducing receptor homologous desensitization. Corticosteroids also enhance the inhibitory action of β2-agonists on production of various inflammatory mediators which may explain, at least in part, their impact on influx of inflammatory cells within the airways seen in some studies. Finally, β2-agonists have been shown to reduce key features of ASM remodeling.

preparations of these animals (Emala et al., 1993). The existence of altered $\beta_2$-receptor function has been suggested by the observed decreased receptor expression and cAMP production seen in mononuclear cells in patients with asthma (Sato, Bewtra, Hopp, Nair, & Townley, 1990). Allergen exposure was found to be a critical factor involved in driving by more than 21% the reduction of $\beta_2$-receptor number on lymphocytes in patients with allergic asthma (Meurs, Koeter, de Vries, & Kauffman, 1982). Trian and colleagues found the cAMP responses to both short- and long-acting $\beta_2$-agonists (i.e., isoproterenol and formoterol) were reduced by more than 50% in ASM cells derived from asthmatic subjects compared to non-asthmatics (Trian et al., 2011). A recent report also found that the coupling of adenylyl cyclase and cAMP production to $\beta_2$-receptor was markedly reduced in ASM cells derived from patients who had died from fatal asthma. Increased activity of phosphatidylinositol-3-kinase (PI3K) leading to the prevention of $\beta_2$-receptor resensitization was found to be the main mechanism of $\beta_2$-receptor dysfunction in these patients (Gupta et al., 2015). The relationship between the loss of $\beta_2$-adrenergic function and the severity of asthma was also suggested with the observed reduced response to $\beta_2$-agonists therapy experienced by asthma patients during exacerbations (Reddel et al., 1999). Bronchial smooth muscle tissues derived from patients who died from fatal asthma also demonstrate reduced bronchodilator responses to $\beta_2$-adrenergic agonists (Bai, 1991; Goldie, Spina, Henry, Lulich, & Paterson, 1986).

There is also evidence that in addition to impaired $\beta_2$-adrenergic responses in asthma, $\beta_2$-adrenergic agonists may make asthma worse, particularly in the absence of corticosteroid therapy. For example, regular $\beta_2$-adrenoceptor agonist use may enhance airway hyperresponsiveness (Cockcroft, McParland, Britto, Swystun, & Rutherford, 1993; Swystun, Gordon, Davis, Zhang, & Cockcroft, 2000; Taylor et al., 1992; van Schayck et al., 1990) and eosinophilic airway inflammation (Aldridge et al., 2000; Green et al., 2006), accelerate lung function decline (van Schayck et al., 1991), reduce asthma control (Nelson et al., 2006; Taylor et al., 1993, 1998), and potentially contribute to asthma deaths in both short-acting and long-acting form (Johnston & Edwards, 2009; Nelson et al., 2006; Taylor et al., 1993).

The mechanism(s) behind the reduced response to therapy and potential adverse effects on asthma pathophysiology are poorly understood, and mechanisms such as agonist-evoked receptor desensitization (tachyphylaxis) and downregulation have been suggested as possible causes (Johnston & Edwards, 2009). It is likely that targeting the pathways leading to poor $\beta_2$-agonist response would have the potential to enhance both their efficacy and safety in many patients. A large body of evidence demonstrates that numerous factors known to be been associated with airway inflammation and remodeling in asthma can alter ASM cells responsiveness to $\beta_2$-agonists through different mechanisms (summarized in Table 1).

## 2.2 Proasthmatic Factors Affecting $\beta_2$-Adrenergic Function

A major clue for the mechanisms explaining the poor bronchodilator response to $\beta_2$-adrenoceptor agonists seen in asthma has derived from ex vivo studies that have used isolated airway preparations isolated from different animal species. The bronchorelaxant responses of isolated rabbit tracheal smooth muscle to isoproterenol (both maximum ($E_{max}$) and potency ($EC_{50}$)) were dramatically reduced in tissues preincubated for 24 h with serum derived from atopic asthmatic patients (Hakonarson, Herrick, Serrano, & Grunstein, 1997; Hakonarson, Maskeri, Carter, Chuang, & Grunstein, 1999). Interestingly, the authors found that this loss of $\beta_2$-adrenoceptor function by atopic serum was due to an increased activation of the muscarinic M2 receptor/$G_i$ protein-coupled pathways, leading to the attenuation of cAMP production. These ex vivo studies were among the first to highlight the existence of soluble mediators in the serum of

**Table 1** Factors and Associated Mechanisms Blunting β2-Receptor Function in ASM

| **Proasthmatic Factors** | **Mechanisms of β2-Receptor Dysfunction** |
|---|---|
| Cytokines | |
| TNFα | Receptor uncoupling to Gs |
| IL-1β | Receptor uncoupling to Gs |
| | Receptor desensitization via COX-2-derived prostanoids |
| IL-13 | Autocrine action of IL-5 |
| | GRK2-dependent receptor phosphorylation |
| IL-5 | Not investigated |
| Asthmatic serum | Inhibitory cross-talk with M2 receptor/$G_i$ activity |
| | IgE activation (?) |
| | CD23 activation |
| Allergen exposure | |
| Dermatophagoides | Indirect via cysteinyl-leukotrienes |
| Derp 1 activity | Direct activation of MAPK pathways |
| Respiratory viruses | |
| Rhinovirus | Indirect via autocrine action of IL-1β |
| | Receptor heterologous desensitization via autocrine action of prostanoids |
| Respiratory syncyal virus (RSV) | Receptor downregulation |
| Extracellular matrix proteins | |
| Fibronection | Augmented receptor function |
| Collagen | Reduced receptor function |
| Laminin | Reduced receptor function |
| Growth factors | |
| TGF, FGF | Receptor phosphorylation at tyrosine residues (Tyr340, Tyr141) |

allergic patients that could interfere with the responsiveness of ASM to $\beta_2$-receptor agonists. Follow-up studies have identified a number of mediators that are capable of modulating $\beta_2$-receptor function in different in ASM models including isolated tracheal/bronchial tissues or in cultured cells. The section later summarizes all known factors present in asthma that have the capacity of blunting $\beta_2$-receptor function in ASM.

### *2.2.1 Proasthmatic Cytokines*

A number of reports have shown that TNFα is an important player in the pathogenesis of asthma by acting on different lung structural cells including ASM (Brightling, Berry, & Amrani, 2008). Our group provided the initial evidence that murine tracheal rings treated for 24h with TNFα have a 60% reduction in relaxant responses to isoproterenol (Chen, Tliba, Van Besien, Panettieri, & Amrani, 2003), an effect not due to a decreased β-adrenoceptor affinity as no change in $EC_{50}$ could be demonstrated. A similar observation of reduced isoproterenol response was observed in guinea-pig tracheal tissues treated with TNFα (Wills-Karp et al., 1993), suggesting that the effect of TNFα can be seen in different animal species. This attenuated relaxant response could be due to reduced $\beta_2$-receptor activity as studies performed using cultured human ASM cells clearly demonstrated that a decreased $\beta_2$-receptor coupling in TNFα-treated cells as evidenced by the reduced adenylyl cyclase activity (Emala, Kuhl, Hungerford, & Hirshman, 1997) and cAMP-dependent gene expression (Lahiri et al., 2002) in response to isoproterenol stimulation. Interestingly, the relaxant response induced by other stimuli including PGE2 or forskolin, a nonreceptor-dependent relaxant agent that directly activates adenylate cyclase, was not affected by TNFα. These studies suggest a specific effect of TNFα on the bronchorelaxant responses driven by the $\beta_2$-receptor via mechanisms involving a receptor uncoupling to downstream signaling pathways (Chen et al., 2003).

Extensive work has also been done around the basic mechanisms by which another proasthmatic cytokine, IL-1β, impaired $\beta_2$-receptor responsiveness in ASM. This hypothesis was initially described by studies performed in both isolated tracheal rings from guinea-pigs and rabbits reporting that IL-1β caused a reduction of the maximal bronchorelaxant responses to isoproterenol (Wills-Karp et al., 1993) or right shift of the dose–response relaxation curve (Hakonarson, Herrick, Serrano, & Grunstein, 1996). In the latter study, the authors demonstrated this effect was mediated via the negative cross-talk between muscarinic M2 receptors and $\beta_2$-receptors, resulting in reduced cAMP accumulation. The impact of IL-1β has been also demonstrated in studies assessing ASM cell responsiveness to $\beta_2$-agonists by their ability to block cell stiffness measured by magnetic twisting cytometry as readout of cell responsiveness. It was reported that ASM responsiveness to isoproterenol was markedly suppressed by IL-1β in a concentration- and time-dependent manner, via a mechanism involving $\beta_2$-receptor uncoupling to its downstream signaling pathways as revealed by the impaired Gs-adenylyl cyclase activation (Shore, Schwartzman, Le Blanc, Murthy, & Doerschuk, 2001), and decreased CRE-inducible gene expression (Lahiri et al., 2002). Several groups found that IL-1β-induced $\beta_2$-receptor hyporesponsiveness resulted in part from the heterologous desensitization of the receptor caused by the autocrine action of COX-2-derived prostanoids including $PGE_2$ (Laporte et al., 1998; Pang, Holland, & Knox, 1998). Interestingly, this synergistic induction COX-2-dependent mechanism appears to explain the augmented $\beta_2$-receptor hypo-responsiveness in cells treated with a combination of TNFα and IL-1β (Moore et al., 2001) or TNFα and IL-17A (Rumzhum, Patel, et al., 2016). It is interesting to note that IL-1β-induced $\beta_2$-receptor dysfunction could be prevented by the presence of dexamethasone which inhibited COX-2 induction and $PGE_2$ production (Moore et al., 1999). The restoration of the $\beta_2$-agonist-induced relaxation by corticosteroids was suggested as one potential mechanism explaining their beneficial action in asthma.

Like TNFα, IL-13 has been also involved in the pathogenesis of allergic asthma in part by acting on different lung structural cells including ASM cells and airway epithelial cells (Wills-Karp, 2001). In different reports, IL-13 was shown to alter $\beta_2$-receptor function although the underlying mechanisms are still controversial due to the use of different experimental models. In cultured human ASM cells, overnight incubation with IL-13 can blunt the ability of isoproterenol to reduce cell stiffness by involving mechanisms that are dependent on ERK pathways (Laporte et al., 2001). In isolated rabbit tracheal rings, the reduced bronchochorelaxant response of isoproterenol induced by IL-13 occurred indirectly via the autocrine action of another TH2 cytokine IL-5 (Grunstein et al., 2002). An interesting recent report confirmed that IL-13 could also impair $\beta_2$-agonist-induced cAMP generation in human airway epithelial cells via a mechanism involving G protein-coupled receptor kinase (GRK2)-dependent phosphorylation of the $\beta_2$-receptor (Albano et al., 2015). A similar mechanism involving both GRK2/PI3K pathways seems to be involved in the suppression isoproterenol-induced alveolar epithelial fluid transport induced by the chemokine IL-8 and TGFβ (Roux et al., 2013; Wagener, Roux, Carles, & Pittet, 2015). Together, these studies show that TH2 cytokines are capable of blunting $\beta_2$-receptor responsiveness in key lung structural cells via multiple direct and indirect mechanisms affecting β2-receptor phosphorylation status, and coupling to downstream signaling pathways.

### *2.2.2 Respiratory Viruses*

There is clinical evidence showing that asthmatic patients with evidence of respiratory viral infection poorly respond to $\beta_2$-agonists. Isolated leukocytes from two different cohorts of patients (mild and severe asthma) exhibit a reduced response to isoproterenol which was further decreased by the presence of respiratory infections believed to viral in origin based on the symptoms (Busse, 1977). Adult asthmatic patients experiencing virally induced exacerbations have a reduced bronchodilator response to $\beta_2$-agonists (Reddel et al., 1999). Similarly, asthmatic children presenting for acute asthma associated with symptoms of respiratory infection required more treatment with $\beta_2$-agonists compared to children without any virus infection (Rueter et al., 2012). Whether respiratory viruses including rhinovirus (RV) directly interfere with the bronchoprotective of $\beta_2$-agonists has been suggested by multiple experimental studies. Hakonarson and colleagues reported that RV-exposed tracheal smooth muscle tissues present a reduced bronchodilatory response to isoproterenol, an effect mediated via IL-1β (Hakonarson, Carter, Maskeri, Hodinka, & Grunstein, 1999). A direct effect of respiratory viruses on $\beta_2$-receptor function was demonstrated by the failure of sulfonterol to inhibit antigen-induced contraction of isolated guinea pig ASM derived from animals exposed intranasally with to the parainfluenza 3 (Buckner et al., 1981). It was shown that loss of β2-receptor function could be induced in cultured human ASM following exposure to conditioned media collected from lung epithelial cells infected with RV (Trian et al., 2010). The main mechanism was later shown to involve an autocrine effect of COX-2-dependent production of prostaglandins ($PGE_2$, $PGF_{2\alpha}$) by ASM cells causing a heterologous desensitization of $\beta_2$-receptors (Van Ly et al., 2013).

The ability of $PGE_2$ to heterologously desensitize $\beta_2$-receptor in human ASM has been well documented as a main mechanism driving poor responses to $\beta_2$-agonists (Penn, Panettieri, & Benovic, 1998). Recently, two novel proasthmatic stimuli (TLR2 agonists or sphingosine 1-phosphate (S1-P)) were reported to cause a $PGE_2$-dependent heterologous desensitization of the $\beta_2$-receptor responses (assessed by cAMP accumulation to salbutamol and formoterol) in human ASM cells (Alkhouri et al., 2014; Rumzhum, Rahman, Oliver, & Ammit, 2016). Of interest, there is evidence that certain leukotrienes can also desensitize $\beta_2$-receptor responses. $LTD_4$, for example, was described to inhibit isoproterenol-induced responses in human ASM cells via CysLT1R and PKC pathways (Rovati et al., 2006). This is an interesting hypothesis as some animal reports have described the therapeutic benefit of cysteine leukotriene antagonists in respiratory syncytial virus bronchiolitis (Fullmer et al., 2005). Moore and colleagues demonstrated that direct infection of ASM with respiratory syncytial virus was associated with a profound inhibition of isoproterenol-induced cAMP production. The inhibitory effect was time and concentration dependent and was not seen with cells infected with ultraviolet-inactivated virus. This specific effect of the virus on isoproterenol response was possibly due to a downregulation of $\beta_2$-receptor expression (Moore et al., 2006). These studies show the capacity of different respiratory viruses to interfere with the bronchodilator response to $\beta_2$-agonist therapy via multiple mechanisms resulting in receptor desensitization.

### 2.2.3 Other Proasthmatic Stimuli

Because airway remodeling is an important feature in asthma (Brightling, Gupta, Gonem, & Siddiqui, 2012), some groups have questioned whether remodeled ASM had an altered response to $\beta_2$-agonists. Indeed, stimuli including the nature of extracellular matrix (ECM) environment on which the cells grow on were described to dramatically affect the response to $\beta_2$-agonists. For example, ASM cells grown on fibronectin significantly increased isoproterenol-induced cAMP production, while exposure to collagen V or laminin decreased response to isoproterenol (Freyer, Billington, Penn, & Hall, 2004). These studies show that the ECM environment trigger phenotypic changes to ASM cells that could dramatically affect $\beta_2$-receptor function. This is an interesting observation as accumulation of collagen V is a defining feature of airway remodeling in asthma (Bergeron, Tulic, & Hamid, 2010), an observation that could explain why severe asthma patients with features of airway remodeling display a poor bronchodilator response (Goleva, Hauk, Boguniewicz, Martin, & Leung, 2007).

Human bronchial tissues sensitized with asthmatic serum challenged with the house dust mite allergen displayed a rightward shift of the dose–response curve to salbutamol. This effect was blocked by montelukast, suggesting a key role played by cysteinyl-leukotrienes in the desensitizing action induced by the allergen (Rovati et al., 2006). Whether this effect results from a direct activation of IgE receptor expressed on ASM cells remains a possibility. Indeed, several reports demonstrated that activation of the low affinity receptor FCεRII (CD23) has the capacity to alter the bronchorelaxant response of isoproterenol in isolated rabbit tracheal rings (Hakonarson & Grunstein, 1998). The presence of functional IgE receptors on ASM has been confirmed by different groups. IgE receptor activation was reported to activate several signaling pathways including $ERK_{1/2}$, p38, JNK MAPK, and Akt kinases and to regulate ASM cell proliferation via the transcription factor STAT3 (Redhu et al., 2013). CD23 upregulation by IL-4/GM-CSF was found to be closely associated with markers of ASM cells hypertrophy (Belleau, Gandhi, McPherson, & Lew, 2005), while CD23 was reported to be constitutively increased in cells from asthmatic compared to nonasthmatic subjects and to induce the production of IL-1β (Hakonarson, Carter, Kim, & Grunstein, 1999), known to inhibit $\beta_2$-receptor relaxation in ASM tissues (Hakonarson et al., 1996; Wills-Karp et al., 1993). Finally, another study concluded that the sensitization phase was not a requirement for the aeroallergen Derp 1 to blunt $\beta_2$-agonist bronchorelaxant responses in rabbit bronchial rings. This nice study suggested that allergen, through its intrinsic cysteine protease activity, had a direct effect on $\beta_2$-receptor function via the activation of MAPK pathways (Grunstein et al., 2005).

Finally, antigens extracted from *Candida albicans* can impair $\beta_2$-receptor responses in peripheral human T cells. While isoproterenol suppressed IL-2 secretion induced by concavalin, it failed to suppress the IL-2 response by *C. albicans* antigens (Aihara et al., 1999). Although this observation was not tested in the lungs, one can speculate that the reason why some fungal allergens are closely associated with asthma severity in some patients (Castanhinha et al., 2015) may derive from their ability to directly or indirectly cause $\beta_2$-receptor dysfunction in bronchial smooth muscle.

# 3. CROSS-TALK BETWEEN MAST CELLS AND β2-RECEPTOR FUNCTION IN ASTHMA

## 3.1 Mast Cell Infiltration of ASM Is a Key Determinant of the Asthmatic Phenotype

Mast cells play a major role in asthma pathophysiology through their activation by both allergen and diverse nonimmunological stimuli. Secretion of the autacoid mediators histamine, $PGD_2$ and $LTC_4$, induces bronchoconstriction, mucus secretion, and mucosal oedema, and thus contributes to the symptoms of asthma. Mast cells also secrete proinflammatory cytokines including IL-4, IL-5, and IL-13, which regulate IgE synthesis and the development of eosinophilic inflammation (Cruse & Bradding, 2016). In addition, the mast cell-derived growth factors and neutral proteases including TGFβ, basic FGF (FGF2), tryptase, and chymase potentially contribute to airway wall remodeling (Zuyderduyn, Sukkar, Fust, Dhaliwal, & Burgess, 2008).

Mast cells infiltrate several important structures within asthmatic airways including the bronchial epithelium (Birring, Berry, Brightling, & Pavord, 2003), mucous glands (Carroll, Mutavdzic, & James, 2002), and ASM (Brightling et al., 2002). We provided the first demonstration that mast cell infiltration of the ASM is a key determinant of the asthmatic phenotype (Brightling et al., 2002). This important observation was later confirmed by other groups (Amin, Janson, Boman, & Venge, 2005; Begueret et al., 2007; El-Shazly et al., 2006). The number of mast cells within the ASM bundles correlates with the severity of airway hyperresponsiveness suggesting that their presence is functionally relevant (Brightling et al., 2002; Siddiqui et al., 2008). Importantly, mast cells within the ASM demonstrate ongoing cytokine secretion (Brightling et al., 2003) and ultrastructural features of activation, with piecemeal degranulation evident using electron microscopy (Begueret et al., 2007). Since numerous mast cell mediators affect ASM function and vice versa (Arthur & Bradding, 2016), there is likely to be an intimate interaction between mast cells and ASM cells in vivo.

## 3.2 Bidirectional Interactions Between Mast Cells and ASM

There is no animal model which recapitulates the infiltration of ASM by mast cells that is present in patients with asthma. However, *our* coculture studies have shown several important bidirectional interactions which occur in the absence of exogenous cell activators. HLMC adhere to ASM cells, in part through the adhesion molecule, cell adhesion molecule-1 (CADM1) expressed on HLMC (Yang et al., 2006). Stem cell factor (SCF) which is an essential survival factor for HLMC is expressed by both ASM cells and lung fibroblasts in both soluble- and membrane-bound forms (Hollins et al., 2008; Kassel et al., 1999). Therefore mast cell–ASM interactions also occur via the SCF receptor Kit (CD117) expressed on mast cells and membrane SCF on ASM cells (Lewis et al., 2016). When mast cells are adherent to ASM cells, they survive in the absence of exogenous growth factors/cytokines, and proliferate rapidly. This is mediated through a cooperative interaction between CADM1, SCF, and soluble IL-6 (Hollins et al., 2008). CADM1 and c-kit coprecipitate and colocalize within HLMC, and we have proposed that CADM1-dependent adhesion presents Kit to membrane-bound SCF on the ASM membrane (Hollins et al., 2008).

HLMC in coculture with ASM cells also become activated with evidence of increased constitutive histamine release and reduced histamine content (Hollins et al., 2008). They nevertheless maintain their ability to respond to FcεRI cross-linking. This is not an artifact of cell culture as the opposite is seen when HLMC are cocultured with airway epithelial cells (Yang et al., 2006). The mechanism behind this requires further investigation but the findings are compatible with the in vivo observations of mast cell piecemeal degranulation within the ASM in asthma (Begueret et al., 2007).

## 3.3 Cross-Talk Between Mast Cell–ASM Leads to β2-Adrenoceptor Dysfunction

We have shown that within 2 min of SCF exposure, there is uncoupling of the $\beta_2$-receptor in HLMC leading to impaired inhibition of both histamine release and KCa3.1 ion channel modulation, followed by later $\beta_2$-receptor internalization (Cruse et al., 2010). This effect was associated with phosphorylation of the $\beta_2$-receptor at specific tyrosine residues including $Tyr_{350}$ but not $Tyr_{141}$. Remarkably, in the presence of IgE and SCF, salbutamol actually increased histamine release in a dose-dependent manner (Cruse et al., 2010), although the reasons for the enhanced mast cell responses by salbutamol remain to be determined. This receptor cross-talk between receptor tyrosine kinases (RTKs) and $\beta_2$-receptor has also been reported in other cell types including pancreatic β cells, DDT1MF-2 smooth muscle cells, or Chinese hamster ovary cells stably overexpressing $\beta_2$-receptor. It was found that phosphorylation of the $\beta_2$-receptor at $Tyr_{350/354}$, $Tyr_{141}$, and $Tyr_{360}$ can be induced by insulin or insulin-like growth factor and was associated with β2-receptor uncoupling and internalization (Doronin, Wang Hy, & Malbon, 2002; Karoor, Baltensperger, Paul, Czech, & Malbon, 1995; Karoor & Malbon, 1996; Karoor, Wang, Wang, & Malbon, 1998). In addition the $\beta_2$-receptor can also be phosphorylated by insulin at $Ser_{345/346}$ residues through the PI3K-dependent activation of Akt (Doronin, Shumay, Wang, & Malbon, 2002). Our studies show that phosphorylation of the $\beta_2$-receptor at Tyr350 residues by the Kit RTK can result in receptor desensitization and internalization and loss in cell response to $\beta_2$-agonists.

In keeping with the findings that exogenously added SCF leads to β2-receptor dysfunction in HLMCs, coculture of HLMCs with human ASM cells produced very similar results. Thus coculture increased "constitutive" HLMC histamine release, and this was increased further by the addition of $\beta_2$-adrenoceptor agonists (Lewis et al., 2016). Inhibition of FcεRI-dependent HLMC mediator release by β2-agonists was also greatly reduced in HLMC–HASMC coculture. These effects were again associated with phosphorylation of Tyr350 on the β2-receptor and were reversed by neutralization of SCF or CADM1.

We also made the interesting finding that formoterol failed to inhibit ASM cell contraction when cells were cocultured with HLMC (Lewis et al., 2016). Another study also found that culturing ASM with activated T cells led to an attenuation of β-adrenoceptor-mediated airway relaxation (Hakonarson, Kim, Whelan, Campbell, & Grunstein, 2001), although the hypothesis of a change in receptor phosphorylation was not investigated. Our study showed that the loss of ASM responsiveness to formoterol in coculture was also associated with the rapid phosphorylation of the $\beta_2$-receptor at tyrosine 350 residues (Lewis et al., 2016). This effect was again dependent on both SCF and CADM1 suggesting that cell–cell interactions played an important role in driving $\beta_2$-receptor dysfunction. Whether ASM $\beta_2$-receptor phosphorylation also occurred indirectly via the release of a mast cell mediator with RTK activity remains a strong possibility. Indeed a number of RTK mediators such as VEGF, FGF2, TGF, and NGF have been shown to trigger various cellular responses in ASM cells (Bosse & Rola-Pleszczynski, 2008; Freund-Michel & Frossard, 2008; Kazi et al., 2004; Kumawat et al., 2016). We have now evidence to suggest that at least FGF and TGF can desensitize $\beta_2$-receptors in ASM cells via receptor phosphorylation on tyrosine residues (manuscript in preparation) (all mechanisms are summarized in Fig. 3).

## 4. BITTER TASTE RECEPTOR AGONISTS: AN ALTERNATIVE BRONCHODILATORY STRATEGY

Because of the loss of the therapeutic benefits of $\beta_2$-agonists in asthma, there is an unmet need for a better understanding of the mechanisms that prevent the drugs from working properly in patients. Recently, an alternative therapeutic strategy to provide bronchodilation has also come under scrutiny.

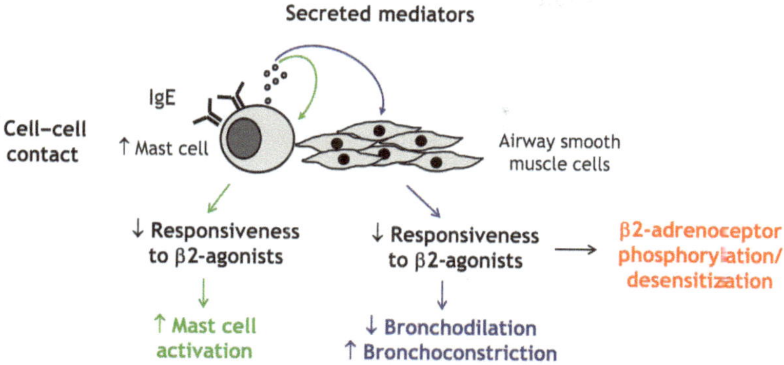


**Fig. 3** Mechanisms of β2-receptor dysfunction in ASM by mast cells. Mast cells can impair ASM responsiveness to β2-agonists via heterologous receptor desensitization which occurs via two main mechanisms. The first one is a direct mechanism where cell–cell physical interactions requiring SCF on ASM and c-kit on mast cells (and CADM1) drive β2-receptor phosphorylation at tyrosine residues. The second mechanism of β2-receptor dysfunction is indirect and mediated via the paracrine action of mediators released by activated mast cells such as growth factors which also stimulate β2-receptor phosphorylation at tyrosine residues. These mast cell mediators could also act on mast cells in autocrine manner and desensitize the antiinflammatory action of β2-agonists via β2-receptor phosphorylation thus preventing mast cell activation and degranulation.

## 4.1 Bronchodilator Activity

In 2010 Deshpande and colleagues made the surprising discovery that activation of bitter taste receptors (TAS2Rs) by using the different agonists (saccharin, chloroquine, and denatonium) evoked marked relaxation in isolated murine and human bronchial tissues despite triggering calcium responses (Deshpande et al., 2010). This finding has fuelled a lot of enthusiasm as TAS2R signaling and effects were different from those initiated by $\beta_2$-receptors. TAS2R activation leads to a stronger bronchodilatory action when compared to isoproterenol, a response that was not dependent on the classical cAMP-dependent manner. Of strong clinical interest was the observation that activation of TAS2Rs showed a relatively modest degree of agonist-promoted receptor desensitization ranging from 31% (Robinett, Deshpande, Malone, & Liggett, 2011) to a nonsignificant 11% (An et al., 2012), and did not affect the therapeutic responses to $\beta_2$-agonists (An et al., 2012). Although the precise mechanisms by which TAS2R agonists trigger bronchorelaxation have not been completely elucidated, it was reported that chloroquine or quinine can both inhibit contractile agonist-evoked calcium oscillations and sensitivity in murine PCLS

(Tan & Sanderson, 2014). It is interesting to note that the relaxant responses of TAS2Rs were still preserved in asthmatic ASM cells or was not affected by IL-13 (Robinett et al., 2014). These observations describe the interesting therapeutic profile of TAS2Rs with their unique mechanism of relaxation and lack of promoting heterologous desensitization of $\beta_2$-receptors.

### 4.2 Other Therapeutic Benefits of Bitter Taste Receptors

Emerging studies have been performed to determine the preclinical value of TAS2R agonists in animal models of allergic asthma. Recently, it was reported that aerosol administration of TAS2R agonists chloroquine or quinine can reverse house dust mite-associated key allergic responses including features of airway remodeling (i.e., ASM mass, lung fibrosis, and mucus accumulation), airway inflammation, and airway hyperresponsiveness (Sharma et al., 2017). The striking effect of TAS2R agonists on ASM remodeling seen in the animal study could result from their direct cytotoxic action through changes in mitochondrial structure and function recently reported in cultured ASM cells (Pan, Sharma, Shah, & Deshpande, 2017). Combined together, these exciting studies suggest that TAS2R agonists represent a potential novel bronchodilatory strategy with additional antiallergic properties.

Future clinical trials are needed to determine whether TAS2R agonists represent an additional bronchodilator therapy in the treatment of lung diseases, especially in patients who are not adequately responding to traditional drugs.

## 5. CONCLUDING REMARKS

A wealth of clinical and experimental evidence has suggested that $\beta_2$-receptors may be dysfunctional in certain groups of patients with asthma who lose their bronchoprotection to $\beta_2$-agonists, and who may experience decreased asthma control with chronic β2-agonist exposure. These patients include those suffering from a severe form of the disease, those experiencing exacerbations caused by a viral infection, and those who died from asthma. Both "constitutive" and "inducible" alterations of $\beta_2$-receptors have been reported in asthmatics not only in the ASM tissues/cells, the main tissue target of agonists, but also in various key inflammatory cells (mast cells, lymphocytes) and lung structural cells (epithelial cells). Different studies including those using different experimental models to study bronchorelaxation have identified some of the key proasthmatic factors and associated mechanisms that promote this status of hypo-responsiveness to $\beta_2$-agonists seen in asthma. Soluble factors including proasthmatic cytokines (both Th1 and Th2), growth factors, respiratory viruses, certain allergens, and activation of IgE receptors can interfere with $\beta_2$-receptors at different levels. The changes in $\beta_2$-receptor function susceptible of blunting cell responsiveness to agonists include a reduction receptor density, uncoupling of the receptor to downstream signaling pathways, heterologous desensitization of the receptor (growth factors, PGE2 receptors), and receptor phosphorylation by RTKs or GRK2. Our group has shown that cell–cell interactions between mast cells and ASM could play a key role in driving $\beta_2$-receptor dysfunction in asthma. Additional studies are required to determine whether targeting these different pathways could restore the bronchoprotective action of $\beta_2$-agonists. A promising bronchodilator strategy was recently uncovered with the discovery of functional bitter taste receptors on ASM. Activation of TAS2Rs led to strong bronchorelaxant responses in both mouse- and human-isolated airway preparations, was not associated with marked agonist-evoked receptor desensitization, and was not perturbed in asthmatic cells. The pathways inhibiting $\beta_2$-receptor function and TAS2Rs represents therefore novel therapeutic targets for the design of novel bronchodilator drugs to treat lung diseases.

## ACKNOWLEDGMENTS

The research was supported by the NIHR Leicester Biomedical Research Centre. The views expressed are those of the authors and do not necessarily reflect those of the National Health Service, the National Institute for Health Research, or the U.K. Department of Health.

## REFERENCES

Aihara, M., Dobashi, K., Horie, T., Iizuka, K., Nakazawa, T., & Mori, M. (1999). Effects of isoproterenol on IL-2 and cAMP production in peripheral T cells from asthmatic and non-asthmatic subjects sensitive to Candida. *Biology of the Cell*, *91*(7), 525–531.

Albano, G. D., Zhao, J., Etling, E. B., Park, S. Y., Hu, H., Trudeau, J. B., et al (2015). IL-13 desensitizes beta2-adrenergic receptors in human airway epithelial cells through a 15-lipoxygenase/G protein receptor kinase 2 mechanism. *The Journal of Allergy and Clinical Immunology*, *135*(5), 1144–1153. e1141–1149.

Aldridge, R. E., Hancox, R. J., Robin Taylor, D., Cowan, J. O., Winn, M. C., Frampton, C. M., et al. (2000). Effects of terbutaline and budesonide on sputum cells and bronchial hyperresponsiveness in asthma. *American Journal of Respiratory and Critical Care Medicine*, *161*(5), 1459–1464.

Alkhouri, H., Rumzhum, N. N., Rahman, M. M., FitzPatrick, M., de Pedro, M., Oliver, B. G., et al. (2014). TLR2 activation causes tachyphylaxis to beta2-agonists in vitro and ex vivo: Modelling bacterial exacerbation. *Allergy*, *69*(9), 1215–1222.

Amin, K., Janson, C., Boman, G., & Venge, P. (2005). The extracellular deposition of mast cell products is increased in hypertrophic airways smooth muscles in allergic asthma but not in nonallergic asthma. *Allergy*, *60*(10), 1241–1247.

Ammit, A. J., Lazaar, A. L., Irani, C., O'Neill, G. M., Gordon, N. D., Amrani, Y., et al. (2002). Tumor necrosis factor-alpha-induced secretion of RANTES and interleukin-6 from human airway smooth muscle cells. Modulation by glucocorticoids and beta-agonists. *American Journal of Respiratory Cell and Molecular Biology*, *26*(4), 465–474.

An, S. S., Wang, W. C., Koziol-White, C. J., Ahn, K., Lee, D. Y., Kurten, R. C., et al. (2012). TAS2R activation promotes airway smooth muscle relaxation despite beta(2)-adrenergic receptor tachyphylaxis. *American Journal of Physiology. Lung Cellular and Molecular Physiology*, *303*(4), L304–L311.

Arthur, G., & Bradding, P. (2016). New developments in mast cell biology: Clinical implications. *Chest*, *150*(3), 680–693.

Bai, T. R. (1991). Abnormalities in airway smooth muscle in fatal asthma. A comparison between trachea and bronchus. *The American Review of Respiratory Disease*, *143*(2), 441–443.

Barnes, P. J., Dollery, C. T., & MacDermot, J. (1980). Increased pulmonary alpha-adrenergic and reduced beta-adrenergic receptors in experimental asthma. *Nature*, *285*(5766), 569–571.

Begueret, H., Berger, P., Vernejoux, J. M., Dubuisson, L., Marthan, R., & Tunon-de-Lara, J. M. (2007). Inflammation of bronchial smooth muscle in allergic asthma. *Thorax*, *62*(1), 8–15.

Belleau, J. T., Gandhi, R. K., McPherson, H. M., & Lew, D. B. (2005). Research upregulation of CD23 (FcepsilonRII) expression in human airway smooth muscle cells (huASMC) in response to IL-4, GM-CSF, and IL-4/GM-CSF. *Clinical and Molecular Allergy*, *3*, 6.

Bergeron, C., Tulic, M. K., & Hamid, Q. (2010). Airway remodelling in asthma: From benchside to clinical practice. *Canadian Respiratory Journal*, *17*(4), e85–e93.

Birring, S. S., Berry, M., Brightling, C. E., & Pavord, I. D. (2003). Eosinophilic bronchitis: Clinical features, management and pathogenesis. *American Journal of Respiratory Medicine*, *2*(2), 169–173.

Bosse, Y., & Rola-Pleszczynski, M. (2008). FGF2 in asthmatic airway-smooth-muscle-cell hyperplasia. *Trends in Molecular Medicine*, *14*(1), 3–11.

Brightling, C., Berry, M., & Amrani, Y. (2008). Targeting TNF-alpha: A novel therapeutic approach for asthma. *The Journal of Allergy and Clinical Immunology*, *121*(1), 5–10 [quiz 11–12].

Brightling, C. E., Bradding, P., Symon, F. A., Holgate, S. T., Wardlaw, A. J., & Pavord, I. D. (2002). Mast-cell infiltration of airway smooth muscle in asthma. *The New England Journal of Medicine*, *346*(22), 1699–1705.

Brightling, C. E., Gupta, S., Gonem, S., & Siddiqui, S. (2012). Lung damage and airway remodelling in severe asthma. *Clinical and Experimental Allergy*, *42*(5), 638–649. http://dx.doi.org/10.1111/j.1365-2222.2011.03917.x.

Brightling, C. E., Symon, F. A., Holgate, S. T., Wardlaw, A. J., Pavord, I. D., & Bradding, P. (2003). Interleukin-4 and -13 expression is co-localized to mast cells within the airway smooth muscle in asthma. *Clinical and Experimental Allergy*, *33*(12), 1711–1716.

Buckner, C. K., Clayton, D. E., Ain-Shoka, A. A., Busse, W. W., Dick, E. C., & Shult, P. (1981). Parainfluenza 3 infection blocks the ability of a beta adrenergic receptor agonist to inhibit antigen-induced contraction of guinea pig isolated airway smooth muscle. *The Journal of Clinical Investigation*, *67*(2), 376–384.

Busse, W. W. (1977). Decreased granulocyte response to isoproterenol in asthma during upper respiratory infections. *The American Review of Respiratory Disease*, *115*(5), 783–791.

Carroll, N. G., Mutavdzic, S., & James, A. L. (2002). Increased mast cells and neutrophils in submucosal mucous glands and mucus plugging in patients with asthma. *Thorax*, *57*(8), 677–682.

Castan, L., Magnan, A., & Bouchaud, G. (2017). Chemokine receptors in allergic diseases. *Allergy*, *72*(5), 682–690.

Castanhinha, S., Sherburn, R., Walker, S., Gupta, A., Bossley, C. J., Buckley, J., et al. (2015). Pediatric severe asthma with fungal sensitization is mediated by steroid-resistant IL-33. *The Journal of Allergy and Clinical Immunology*, *136*(2), 312–322. e317.

Chen, H., Tliba, O., Van Besien, C. R., Panettieri, R. A., Jr., & Amrani, Y. (2003). Selected contribution: TNF-{alpha} modulates murine tracheal rings responsiveness to G-protein-coupled receptor agonists and KCl. *Journal of Applied Physiology*, *95*(2), 864–872.

Chiu, J. C., Hsu, J. Y., Fu, L. S., Chu, J. J., & Chi, C. S. (2007). Comparison of the effects of two long-acting beta2-agonists on cytokine secretion by human airway epithelial cells. *Journal of Microbiology, Immunology, and Infection*, *40*(5), 388–394.

Chong, L. K., Cooper, E., Vardey, C. J., & Peachell, P. T. (1998). Salmeterol inhibition of mediator release from human lung mast cells by beta-adrenoceptor-dependent and independent mechanisms. *British Journal of Pharmacology*, *123*(5), 1009–1015.

Chorley, B. N., Li, Y., Fang, S., Park, J. A., & Adler, K. B. (2006). (R)-albuterol elicits antiinflammatory effects in human airway epithelial cells via iNOS. *American Journal of Respiratory Cell and Molecular Biology*, *34*(1), 119–127.

Cockcroft, D. W., McParland, C. P., Britto, S. A., Swystun, V. A., & Rutherford, B. C. (1993). Regular inhaled salbutamol and airway responsiveness to allergen. *Lancet*, *342*(8875), 833–837.

Cooper, P. R., & Panettieri, R. A., Jr. (2008). Steroids completely reverse albuterol-induced beta(2)-adrenergic receptor tolerance in human small airways. *The Journal of Allergy and Clinical Immunology*, *122*(4), 734–740. http://dx.doi.org/10.1016/j.jaci.2008.07.040.

Cruse, G., & Bradding, P. (2016). Mast cells in airway diseases and interstitial lung disease. *European Journal of Pharmacology*, *778*, 125–138.

Cruse, G., Yang, W., Duffy, S. M., Chachi, L., Leyland, M., Amrani, Y., et al. (2010). Counterregulation of beta(2)-adrenoceptor function in human mast cells by stem cell factor. *The Journal of Allergy and Clinical Immunology*, *125*(1), 257–263. e251–255.

Dekkers, B. G., Pehlic, A., Mariani, R., Bos, I. S., Meurs, H., & Zaagsma, J. (2012). Glucocorticosteroids and beta(2)-adrenoceptor agonists synergize to inhibit airway smooth muscle remodeling. *The Journal of Pharmacology and Experimental Therapeutics*, *342*(3), 780–787.

Deshpande, D. A., Wang, W. C., McIlmoyle, E. L., Robinett, K. S., Schillinger, R. M., An, S. S., et al. (2010). Bitter taste receptors on airway smooth muscle bronchodilate by localized calcium signaling and reverse obstruction. *Nature Medicine*, *16*(11), 1299–1304.

Doronin, S., Shumay, E., Wang, H. Y., & Malbon, C. C. (2002a). Akt mediates sequestration of the beta(2)-adrenergic receptor in response to insulin. *The Journal of Biological Chemistry*, *277*(17), 15124–15131.

Doronin, S., Wang Hy, H. Y., & Malbon, C. C. (2002b). Insulin stimulates phosphorylation of the beta 2-adrenergic receptor by the insulin receptor, creating a potent feedback inhibitor of its tyrosine kinase. *The Journal of Biological Chemistry*, *277*(12), 10698–10703.

Duffy, S. M., Cruse, G., Lawley, W. J., & Bradding, P. (2005). Beta2-adrenoceptor regulation of the K+ channel iKCa1 in human mast cells. *The FASEB Journal, 19*(8), 1006–1008.

El-Shazly, A., Berger, P., Girodet, P. O., Ousova, O., Fayon, M., Vernejoux, J. M., et al. (2006). Fraktalkine produced by airway smooth muscle cells contributes to mast cell recruitment in asthma. *Journal of Immunology, 176*(3), 1860–1868.

Emala, C., Black, C., Curry, C., Levine, M. A., & Hirshman, C. A. (1993). Impaired beta-adrenergic receptor activation of adenylyl cyclase in airway smooth muscle in the basenji-greyhound dog model of airway hyperresponsiveness. *American Journal of Respiratory Cell and Molecular Biology, 8*(6), 668–675.

Emala, C. W., Kuhl, J., Hungerford, C. L., & Hirshman, C. A. (1997). TNF-alpha inhibits isoproterenol-stimulated adenylyl cyclase activity in cultured airway smooth muscle cells. *The American Journal of Physiology, 272*(4 Pt. 1), L644–L650.

Fogli, S., Stefanelli, F., Battolla, B., Bianchi, F., Breschi, M. C., & Mattii, L. (2015). Salbutamol inhibits RhoA activation in normal but not in desensitized bronchial smooth muscle cells. *The Journal of Pharmacy and Pharmacology, 67*(10), 1416–1420.

Freund-Michel, V., & Frossard, N. (2008). Overexpression of functional TrkA receptors after internalisation in human airway smooth muscle cells. *Biochimica et Biophysica Acta, 1783*(10), 1964–1971.

Freyer, A. M., Billington, C. K., Penn, R. B., & Hall, I. P. (2004). Extracellular matrix modulates beta2-adrenergic receptor signaling in human airway smooth muscle cells. *American Journal of Respiratory Cell and Molecular Biology, 31*(4), 440–445.

Fullmer, J. J., Khan, A. M., Elidemir, O., Chiappetta, C., Stark, J. M., & Colasurdo, G. N. (2005). Role of cysteinyl leukotrienes in airway inflammation and responsiveness following RSV infection in BALB/c mice. *Pediatric Allergy and Immunology, 16*(7), 593–601.

George, T., Bell, M., Chakraborty, M., Siderovski, D. P., Giembycz, M. A., & Newton, R. (2017). Protective roles for RGS2 in a mouse model of house dust mite-induced airway inflammation. *PloS One, 12*(1), e0170269.

Giembycz, M. A., & Newton, R. (2006). Beyond the dogma: Novel beta2-adrenoceptor signalling in the airways. *The European Respiratory Journal, 27*(6), 1286–1306.

Goldie, R. G., Spina, D., Henry, P. J., Lulich, K. M., & Paterson, J. W. (1986). In vitro responsiveness of human asthmatic bronchus to carbachol, histamine, beta-adrenoceptor agonists and theophylline. *British Journal of Clinical Pharmacology, 22*(6), 669–676.

Goleva, E., Hauk, P. J., Boguniewicz, J., Martin, R. J., & Leung, D. Y. (2007). Airway remodeling and lack of bronchodilator response in steroid-resistant asthma. *The Journal of Allergy and Clinical Immunology, 120*(5), 1065–1072.

Green, R. H., Brightling, C. E., McKenna, S., Hargadon, B., Neale, N., Parker, D., et al. (2006). Comparison of asthma treatment given in addition to inhaled corticosteroids on airway inflammation and responsiveness. *The European Respiratory Journal, 27*(6), 1144–1151.

Grunstein, M. M., Hakonarson, H., Leiter, J., Chen, M., Whelan, R., Grunstein, J. S., et al. (2002). IL-13-dependent autocrine signaling mediates altered responsiveness of IgE-sensitized airway smooth muscle. *American Journal of Physiology. Lung Cellular and Molecular Physiology, 282*(3), L520–L528.

Grunstein, M. M., Veler, H., Shan, X., Larson, J., Grunstein, J. S., & Chuang, S. (2005). Proasthmatic effects and mechanisms of action of the dust mite allergen, Der p 1, in airway smooth muscle. *The Journal of Allergy and Clinical Immunology, 116*(1), 94–101.

Gupta, M. K., Asosingh, K., Aronica, M., Comhair, S., Cao, G., Erzurum, S., et al. (2015). Defective resensitization in human airway smooth muscle cells evokes beta-adrenergic receptor dysfunction in severe asthma. *PloS One, 10*(5), e0125803.

Hakonarson, H., Carter, C., Kim, C., & Grunstein, M. M. (1999a). Altered expression and action of the low-affinity IgE receptor FcepsilonRII (CD23) in asthmatic airway smooth muscle. *The Journal of Allergy and Clinical Immunology, 104*(3 Pt. 1), 575–584.
Hakonarson, H., Carter, C., Maskeri, N., Hodinka, R., & Grunstein, M. M. (1999b). Rhinovirus-mediated changes in airway smooth muscle responsiveness: Induced autocrine role of interleukin-1beta. *The American Journal of Physiology, 277*(1 Pt. 1), L13–L21.
Hakonarson, H., & Grunstein, M. M. (1998). Autologously up-regulated Fc receptor expression and action in airway smooth muscle mediates its altered responsiveness in the atopic asthmatic sensitized state. *Proceedings of the National Academy of Sciences of the United States of America, 95*(9), 5257–5262.
Hakonarson, H., Herrick, D. J., Serrano, P. G., & Grunstein, M. M. (1996). Mechanism of cytokine-induced modulation of beta-adrenoceptor responsiveness in airway smooth muscle. *The Journal of Clinical Investigation, 97*(11), 2593–2600.
Hakonarson, H., Herrick, D. J., Serrano, P. G., & Grunstein, M. M. (1997). Autocrine role of interleukin 1beta in altered responsiveness of atopic asthmatic sensitized airway smooth muscle. *The Journal of Clinical Investigation, 99*(1), 117–124.
Hakonarson, H., Kim, C., Whelan, R., Campbell, D., & Grunstein, M. (2001). Bi-directional activation between human airway smooth muscle cells and T lymphocytes: Role in induction of altered airway responsiveness. *Journal of Immunology, 166*(1), 293–303.
Hakonarson, H., Maskeri, N., Carter, C., Chuang, S., & Grunstein, M. M. (1999c). Autocrine interaction between IL-5 and IL-1beta mediates altered responsiveness of atopic asthmatic sensitized airway smooth muscle. *The Journal of Clinical Investigation, 104*(5), 657–667.
Hallsworth, M. P., Twort, C. H., Lee, T. H., & Hirst, S. J. (2001). Beta(2)-adrenoceptor agonists inhibit release of eosinophil-activating cytokines from human airway smooth muscle cells. *British Journal of Pharmacology, 132*(3), 729–741.
Holden, N. S., Bell, M. J., Rider, C. F., King, E. M., Gaunt, D. D., Leigh, R., et al. (2011). Beta2-adrenoceptor agonist-induced RGS2 expression is a genomic mechanism of bronchoprotection that is enhanced by glucocorticoids. *Proceedings of the National Academy of Sciences of the United States of America, 108*(49), 19713–19718.
Hollins, F., Kaur, D., Yang, W., Cruse, G., Saunders, R., Sutcliffe, A., et al. (2008). Human airway smooth muscle promotes human lung mast cell survival, proliferation, and constitutive activation: Cooperative roles for CADM1, stem cell factor, and IL-6. *Journal of Immunology, 181*(4), 2772–2780.
Hung, C. H., Chu, Y. T., Hua, Y. M., Hsu, S. H., Lin, C. S., Chang, H. C. et al. (2008). Effects of formoterol and salmeterol on the production of Th1- and Th2-related chemokines by monocytes and bronchial epithelial cells. *The European Respiratory Journal, 31*(6), 1313–1321.
Jeffery, P. K., Venge, P., Gizycki, M. J., Egerod, I., Dahl, R., & Faurschou, P. (2002). Effects of salmeterol on mucosal inflammation in asthma: A placebo-controlled study. *The European Respiratory Journal, 20*(6), 1378–1385.
Johnston, S. L., & Edwards, M. R. (2009). Mechanisms of adverse effects of {beta}-agonists in asthma. *Thorax, 64*(9), 739–741.
Karoor, V., Baltensperger, K., Paul, H., Czech, M. P., & Malbon, C. C. (1995). Phosphorylation of tyrosyl residues 350/354 of the beta-adrenergic receptor is obligatory for counterregulatory effects of insulin. *The Journal of Biological Chemistry, 270*(43), 25305–25308.
Karoor, V., & Malbon, C. C. (1996). Insulin-like growth factor receptor-1 stimulates phosphorylation of the beta2-adrenergic receptor in vivo on sites distinct from those phosphorylated in response to insulin. *The Journal of Biological Chemistry, 271*(46), 29347–29352.

Karoor, V., Wang, L., Wang, H. Y., & Malbon, C. C. (1998). Insulin stimulates sequestration of beta-adrenergic receptors and enhanced association of beta-adrenergic receptors with Grb 2 via tyrosine 350. *The Journal of Biological Chemistry*, *273*(49), 33035–33041.

Kassel, O., Schmidlin, F., Duvernelle, C., Gasser, B., Massard, G., & Frossard, N. (1999). Human bronchial smooth muscle cells in culture produce stem cell factor. *The European Respiratory Journal*, *13*(5), 951–954.

Kaur, M., Chivers, J. E., Giembycz, M. A., & Newton, R. (2008). Long-acting beta2-adrenoceptor agonists synergistically enhance glucocorticoid-dependent transcription in human airway epithelial and smooth muscle cells. *Molecular Pharmacology*, *73*(1), 203–214.

Kazi, A. S., Lotfi, S., Goncharova, E. A., Tliba, O., Amrani, Y., Krymskaya, V. P., et al. (2004). Vascular endothelial growth factor-induced secretion of fibronectin is ERK dependent. *American Journal of Physiology. Lung Cellular and Molecular Physiology*, *286*(3), L539–L545.

Kleine-Tebbe, J., Frank, G., Josties, C., & Kunkel, G. (1994). Influence of salmeterol, a long-acting beta 2-adrenoceptor agonist, on IgE-mediated histamine release from human basophils. *Journal of Investigational Allergology & Clinical Immunology*, *4*(1), 12–17.

Kumawat, K., Koopmans, T., Menzen, M. H., Prins, A., Smit, M., Halayko, A. J., et al. (2016). Cooperative signaling by TGF-beta1 and WNT-11 drives sm-alpha-actin expression in smooth muscle via Rho kinase-actin-MRTF-A signaling. *American Journal of Physiology. Lung Cellular and Molecular Physiology*, *311*(3), L529–L537.

Lahiri, T., Moore, P. E., Baraldo, S., Whitehead, T. R., McKenna, M. D., Panettieri, R. A., Jr., et al. (2002). Effect of IL-1beta on CRE-dependent gene expression in human airway smooth muscle cells. *American Journal of Physiology. Lung Cellular and Molecular Physiology*, *283*(6), L1239–L1246.

Laporte, J. C., Moore, P. E., Baraldo, S., Jouvin, M. H., Church, T. L., Schwartzman, I. N., et al. (2001). Direct effects of interleukin-13 on signaling pathways for physiological responses in cultured human airway smooth muscle cells. *American Journal of Respiratory and Critical Care Medicine*, *164*(1), 141–148.

Laporte, J. D., Moore, P. E., Panettieri, R. A., Moeller, W., Heyder, J., & Shore, S. A. (1998). Prostanoids mediate IL-1beta-induced beta-adrenergic hyporesponsiveness in human airway smooth muscle cells. *The American Journal of Physiology*, *275*(3 Pt. 1), L491–L501.

Lewis, R. J., Chachi, L., Newby, C., Amrani, Y., & Bradding, P. (2016). Bidirectional counterregulation of human lung mast cell and airway smooth muscle beta2 adrenoceptors. *Journal of Immunology*, *196*(1), 55–63.

Manetsch, M., Che, W., Seidel, P., Chen, Y., & Ammit, A. J. (2012). MKP-1: A negative feedback effector that represses MAPK-mediated pro-inflammatory signaling pathways and cytokine secretion in human airway smooth muscle cells. *Cellular Signalling*, *24*(4), 907–913.

Manetsch, M., Rahman, M. M., Patel, B. S., Ramsay, E. E., Rumzhum, N. N., Alkhouri, H., et al. (2013). Long-acting beta2-agonists increase fluticasone propionate-induced mitogen-activated protein kinase phosphatase 1 (MKP-1) in airway smooth muscle cells. *PloS One*, *8*(3), e59635.

Meurs, H., Koeter, G. H., de Vries, K., & Kauffman, H. F. (1982). The beta-adrenergic system and allergic bronchial asthma: Changes in lymphocyte beta-adrenergic receptor number and adenylate cyclase activity after an allergen-induced asthmatic attack. *The Journal of Allergy and Clinical Immunology*, *70*(4), 272–280.

Moore, P. E., Cunningham, G., Calder, M. M., DeMatteo, A. D., Jr., Peeples, M. E., Summar, M. L., et al. (2006). Respiratory syncytial virus infection reduces beta2-adrenergic responses in human airway smooth muscle. *American Journal of Respiratory Cell and Molecular Biology, 35*(5), 559–564.

Moore, P. E., Lahiri, T., Laporte, J. D., Church, T., Panettieri, R. A., Jr., & Shore, S. A. (2001). Selected contribution: Synergism between TNF-alpha and IL-1beta in airway smooth muscle cells: Implications for beta-adrenergic responsiveness. *Journal of Applied Physiology, 91*(3), 1467–1474.

Moore, P. E., Laporte, J. D., Gonzalez, S., Moller, W., Heyder, J., Panettieri, R. A., Jr., et al. (1999). Glucocorticoids ablate IL-1beta-induced beta-adrenergic hyporesponsiveness in human airway smooth muscle cells. *The American Journal of Physiology, 277*(5 Pt. 1), L932–L942.

Nelson, H. S., Weiss, S. T., Bleecker, E. R., Yancey, S. W., Dorinsky, P. M., & SMART Study Group. (2006). The Salmeterol Multicenter Asthma Research Trial: A comparison of usual pharmacotherapy for asthma or usual pharmacotherapy plus salmeterol. *Chest, 129*(1), 15–26.

Newton, R. (2014). Anti-inflammatory glucocorticoids: Changing concepts. *European Journal of Pharmacology, 724*, 231–236.

Newton, R., & Giembycz, M. A. (2016). Understanding how long-acting beta2-adrenoceptor agonists enhance the clinical efficacy of inhaled corticosteroids in asthma—An update. *British Journal of Pharmacology, 173*(24), 3405–3430.

Nie, M., Knox, A. J., & Pang, L. (2005). Beta2-adrenoceptor agonists, like glucocorticoids, repress eotaxin gene transcription by selective inhibition of histone H4 acetylation. *Journal of Immunology, 175*(1), 478–486.

Pan, S., Sharma, P., Shah, S. D., & Deshpande, D. A. (2017). Bitter taste receptor agonists alter mitochondrial function and induce autophagy in airway smooth muscle cells. *American Journal of Physiology. Lung Cellular and Molecular Physiology, 313*(1), L154–L165. http://dx.doi.org/10.1152/ajplung.00106.2017.

Pang, L., Holland, E., & Knox, A. J. (1998). Role of cyclo-oxygenase-2 induction in interleukin-1beta induced attenuation of cultured human airway smooth muscle cell cyclic AMP generation in response to isoprenaline. *British Journal of Pharmacology, 125*(6), 1320–1328.

Pang, L., & Knox, A. J. (2000). Synergistic inhibition by beta(2)-agonists and corticosteroids on tumor necrosis factor-alpha-induced interleukin-8 release from cultured human airway smooth-muscle cells. *American Journal of Respiratory Cell and Molecular Biology, 23*(1), 79–85.

Pang, L., & Knox, A. (2001). Regulation of TNF-alpha-induced eotaxin release from cultured human airway smooth muscle cells by beta2-agonists and corticosteroids. *The FASEB Journal, 15*(1), 261–269.

Penn, R. B., Panettieri, R. A., Jr., & Benovic, J. L. (1998). Mechanisms of acute desensitization of the beta 2AR-adenylyl cyclase pathway in human airway smooth muscle. *American Journal of Respiratory Cell and Molecular Biology, 19*(2), 338–348.

Pera, T., & Penn, R. B. (2016). Bronchoprotection and bronchorelaxation in asthma: New targets, and new ways to target the old ones. *Pharmacology & Therapeutics, 164*, 82–96.

Pinto Pereira, L. M., Clement, Y., Rouse, J., Matthew, J., Asgarali, Z., Ramcutar, D., et al. (2003). Salmeterol decreases eosinophilic cationic protein and rescue medication in patients inhaling beclomethasone dipropionate: Preliminary study in mild and moderate asthma in Trinidad, West Indies. *International Journal of Clinical Pharmacology Research, 23*(2–3), 69–74.

Quante, T., Ng, Y. C., Ramsay, E. E., Henness, S., Allen, J. C., Parmentier, J., et al. (2008). Corticosteroids reduce IL-6 in ASM cells via up-regulation of MKP-1. *American Journal of Respiratory Cell and Molecular Biology, 39*(2), 208–217.

Reddel, H., Ware, S., Marks, G., Salome, C., Jenkins, C., & Woolcock, A. (1999). Differences between asthma exacerbations and poor asthma control. *Lancet*, *353*(9150), 364–369.

Redhu, N. S., Shan, L., Al-Subait, D., Ashdown, H. L., Movassagh, H., Lamkhioued, B., et al. (2013). IgE induces proliferation in human airway smooth muscle cells: Role of MAPK and STAT3 pathways. *Allergy, Asthma and Clinical Immunology*, *9*(1), 41.

Reid, D. W., Ward, C., Wang, N., Zheng, L., Bish, R., Orsida, B., et al. (2003). Possible anti-inflammatory effect of salmeterol against interleukin-8 and neutrophil activation in asthma in vivo. *The European Respiratory Journal*, *21*(6), 994–999.

Roberts, J. A., Bradding, P., Britten, K. M., Walls, A. F., Wilson, S., Gratziou, C., et al. (1999). The long-acting beta2-agonist salmeterol xinafoate: Effects on airway inflammation in asthma. *The European Respiratory Journal*, *14*(2), 275–282.

Robinett, K. S., Deshpande, D. A., Malone, M. M., & Liggett, S. B. (2011). Agonist-promoted homologous desensitization of human airway smooth muscle bitter taste receptors. *American Journal of Respiratory Cell and Molecular Biology*, *45*(5), 1069–1074.

Robinett, K. S., Koziol-White, C. J., Akoluk, A., An, S. S., Panettieri, R. A., Jr., & Liggett, S. B. (2014). Bitter taste receptor function in asthmatic and nonasthmatic human airway smooth muscle cells. *American Journal of Respiratory Cell and Molecular Biology*, *50*(4), 678–683.

Roux, J., McNicholas, C. M., Carles, M., Goolaerts, A., Houseman, B. T., Dickinson, D. A., et al. (2013). IL-8 inhibits cAMP-stimulated alveolar epithelial fluid transport via a GRK2/PI3K-dependent mechanism. *The FASEB Journal*, *27*(3), 1095–1106.

Rovati, G. E., Baroffio, M., Citro, S., Brichetto, L., Ravasi, S., Milanese, M., et al. (2006). Cysteinyl-leukotrienes in the regulation of beta2-adrenoceptor function: An in vitro model of asthma. *Respiratory Research*, 7, 103. http://dx.doi.org/10.1186/1465-9921-7-103.

Rueter, K., Bizzintino, J., Martin, A. C., Zhang, G., Hayden, C. M., Geelhoed, G. C., et al. (2012). Symptomatic viral infection is associated with impaired response to treatment in children with acute asthma. *The Journal of Pediatrics*, *160*(1), 82–87.

Rumzhum, N. N., Patel, B. S., Prabhala, P., Gelissen, I. C., Oliver, B. G., & Ammit, A. J. (2016a). IL-17A increases TNF-alpha-induced COX-2 protein stability and augments PGE2 secretion from airway smooth muscle cells: Impact on beta2-adrenergic receptor desensitization. *Allergy*, *71*(3), 387–396.

Rumzhum, N. N., Rahman, M. M., Oliver, B. G., & Ammit, A. J. (2016b). Effect of sphingosine 1-phosphate on cyclo-oxygenase-2 expression, prostaglandin E2 secretion, and beta2-adrenergic receptor desensitization. *American Journal of Respiratory Cell and Molecular Biology*, *54*(1), 128–135.

Sato, T., Bewtra, A. K., Hopp, R. J., Nair, N., & Townley, R. G. (1990). Alpha- and beta-adrenergic-receptor systems in bronchial asthma and in subjects without asthma: Reduced mononuclear cell beta-receptors in bronchial asthma. *The Journal of Allergy and Clinical Immunology*, *86*(6 Pt. 1), 839–850.

Sharma, P., Yi, R., Nayak, A. P., Wang, N., Tang, F., Knight, M. J., et al. (2017). Bitter taste receptor agonists mitigate features of allergic asthma in mice. *Scientific Reports*, 7, 46166.

Shore, S. A., Schwartzman, I. N., Le Blanc, B., Murthy, G. G., & Doerschuk, C. M. (2001). Tumor necrosis factor receptor 2 contributes to ozone-induced airway hyperresponsiveness in mice. *American Journal of Respiratory and Critical Care Medicine*, *164*(4), 602–607.

Siddiqui, S., Mistry, V., Doe, C., Roach, K., Morgan, A., Wardlaw, A., et al. (2008). Airway hyperresponsiveness is dissociated from airway wall structural remodeling. *The Journal of Allergy and Clinical Immunology*, *122*(2), 335–341.

Stelmach, I., Gorski, P., Jerzynska, J., Stelmach, W., Majak, P., & Kuna, P. (2002). A randomized, double-blind trial of the effect of treatment with formotero on clinical and inflammatory parameters of asthma in children. *Annals of Allergy, Asthma & Immunology, 89*(1), 67–73.

Swystun, V. A., Gordon, J. R., Davis, E. B., Zhang, X., & Cockcroft, D. W. (2000). Mast cell tryptase release and asthmatic responses to allergen increase with regular use of salbutamol. *The Journal of Allergy and Clinical Immunology, 106*(1 Pt. 1), 57–64.

Tan, X., & Sanderson, M. J. (2014). Bitter tasting compounds dilate airways by inhibiting airway smooth muscle calcium oscillations and calcium sensitivity. *British Journal of Pharmacology, 171*(3), 646–662.

Taylor, I. K., O'Shaughnessy, K. M., Choudry, N. B., Adachi, M., Palmer, J. B., & Fuller, R. W. (1992). A comparative study in atopic subjects with asthma of the effects of salmeterol and salbutamol on allergen-induced bronchoconstriction, increase in airway reactivity, and increase in urinary leukotriene E4 excretion. *The Journal of Allergy and Clinical Immunology, 89*(2), 575–583.

Taylor, D. R., Sears, M. R., Herbison, G. P., Flannery, E. M., Print, C. G., Lake, D. C., et al. (1993). Regular inhaled beta agonist in asthma: Effects on exacerbations and lung function. *Thorax, 48*(2), 134–138.

Taylor, D. R., Town, G. I., Herbison, G. P., Boothman-Burrell, D., Flannery, E. M., Hancox, B., et al. (1998). Asthma control during long-term treatment with regular inhaled salbutamol and salmeterol. *Thorax, 53*(9), 744–752.

Trian, T., Burgess, J. K., Niimi, K., Moir, L. M., Ge, Q., Berger, P., et al. (2011). Beta2-agonist induced cAMP is decreased in asthmatic airway smooth muscle due to increased PDE4D. *PloS One, 6*(5), e20000. http://dx.doi.org/10.1371/journal.pone.0020000.

Trian, T., Moir, L. M., Ge, Q., Burgess, J. K., Kuo, C., King, N. J., et al. (2010) Rhinovirus-induced exacerbations of asthma: How is the {beta}2-adrenoceptor implicated? *American Journal of Respiratory Cell and Molecular Biology, 43*(2), 227–233.

Van Ly, D., Faiz, A., Jenkins, C., Crossett, B., Black, J. L., McParland, B., et al. (2013). Characterising the mechanism of airway smooth muscle beta2 adrenoceptor desensitization by rhinovirus infected bronchial epithelial cells. *PloS One, 8*(2), e56058.

van Schayck, C. P., Dompeling, E., van Herwaarden, C. L., Folgering, H., Verbeek, A. L., van der Hoogen, H. J., et al. (1991). Bronchodilator treatment in moderate asthma or chronic bronchitis: Continuous or on demand? A randomised controlled study. *BMJ, 303*(6815), 1426–1431.

van Schayck, C. P., Graafsma, S. J., Visch, M. B., Dompeling, E., van Weel, C., & van Herwaarden, C. L. (1990). Increased bronchial hyperresponsiveness after inhaling salbutamol during 1 year is not caused by subsensitization to salbutamol. *The Journal of Allergy and Clinical Immunology, 86*(5), 793–800.

Wagener, B. M., Roux, J., Carles, M., & Pittet, J. F. (2015). Synergistic inhibition of beta2-adrenergic receptor-mediated alveolar epithelial fluid transport by interleukin-8 and transforming growth factor-beta. *Anesthesiology, 122*(5), 1084–1092.

Wallin, A., Sandstrom, T., Cioppa, G. D., Holgate, S., & Wilson, S. (2002) The effects of regular inhaled formoterol and budesonide on preformed Th-2 cytokines in mild asthmatics. *Respiratory Medicine, 96*(12), 1021–1025.

Wallin, A., Sandstrom, T., Soderberg, M., Howarth, P., Lundback, B., De la-Cioppa, G., et al. (1999). The effects of regular inhaled formoterol, budesonide, and placebo on mucosal inflammation and clinical indices in mild asthma. *American Journal of Respiratory and Critical Care Medicine, 159*(1), 79–86.

Wills-Karp, M. (2001). IL-12/IL-13 axis in allergic asthma. *The Journal of Allergy and Clinical Immunology, 107*(1), 9–18.

Wills-Karp, M., Uchida, Y., Lee, J. Y., Jinot, J., Hirata, A., & Hirata, F. (1993). Organ culture with proinflammatory cytokines reproduces impairment of the beta-adrenoceptor-mediated relaxation in tracheas of a guinea pig antigen model. *American Journal of Respiratory Cell and Molecular Biology*, *8*(2), 153–159.

Yang, W., Kaur, D., Okayama, Y., Ito, A., Wardlaw, A. J., Brightling, C. E., et al. (2006). Human lung mast cells adhere to human airway smooth muscle, in part, via tumor suppressor in lung cancer-1. *Journal of Immunology*, *176*(2), 1238–1243.

Zuyderduyn, S., Sukkar, M. B., Fust, A., Dhaliwal, S., & Burgess, J. K. (2008). Treating asthma means treating airway smooth muscle cells. *The European Respiratory Journal*, *32*(2), 265–274.

CHAPTER TWO

# G Protein-Coupled Kinin Receptors and Immunity Against Pathogens

**Julio Scharfstein*[,1], Pablo I.P. Ramos[†], Manoel Barral-Netto[†]**
*Institute of Biophysics Carlos Chagas Filho, Federal University of Rio de Janeiro, Center of Health Sciences (CCS), Cidade Universitária, Rio de Janeiro, Brazil
[†]Instituto Gonçalo Moniz, Fundação Oswaldo Cruz (FIOCRUZ), Salvador, Brazil
[1]Corresponding author: e-mail address: jscharf2@gmail.com

## Contents

1. Introduction 30
2. The KKS and Kinin Receptors 31
   2.1 Intravascular Release of Kinins and GPCR Regulation of Cell Signaling 31
   2.2 Functional Links Between Thrombogenesis and the KKS 34
   2.3 Multiple Kininogenases Promote Kinin Release Extravascularly 36
   2.4 Pharmacological Specificity of G Protein-Coupled B1 and B2 Kinin Receptors 39
   2.5 Regulation of B2R and B1R Signaling by Metallopeptidases 40
3. Kinins as Effectors of Immunity 41
   3.1 Role of B1R in Neutrophil Transmigration to Inflamed Tissues 41
   3.2 KKS Expression During Macrophage Development 42
   3.3 Dendritic Cells: Functional Roles of B2R and B1R 43
   3.4 Mast Cell Coupling to the KKS: Lessons From Allergic Inflammation 46
   3.5 KKS and CD4$^+$T Lymphocytes: Controversial Roles in EAE 46
4. KKS in Infection Immunity 47
   4.1 Viral Infections 47
   4.2 KKS- and B2R-Dependent Immunity Against Bacterial Infection 48
   4.3 KKS and Parasitic Infections Caused by Intracellular Protozoa 51
5. Concluding Remarks 70
Acknowledgments 70
References 70

## Abstract

For decades, immunologists have considered the complement system as a paradigm of a proteolytic cascade that, acting cooperatively with the immune system, enhances host defense against infectious organisms. In recent years, advances made in thrombosis research disclosed a functional link between activated neutrophils, monocytes, and platelet-driven thrombogenesis. Forging a physical barrier, the fibrin scaffolds generated by synergism between the extrinsic and intrinsic (contact) pathways of coagulation

*Advances in Immunology*, Volume 136
ISSN 0065-2776
http://dx.doi.org/10.1016/bs.ai.2017.05.007

entrap microbes within microvessels, limiting the systemic spread of infection while enhancing the clearance of pathogens by activated leukocytes. Insight from mice models of thrombosis linked fibrin formation via the intrinsic pathway to the auto-activation of factor XII (FXII) by negatively charged "contact" substances, such as platelet-derived polyphosphates and DNA from neutrophil extracellular traps. Following cleavage by FXIIa, activated plasma kallikrein (PK) initiates inflammation by liberating the nonapeptide bradykinin (BK) from an internal domain of high molecular weight kininogen (HK). Acting as a paracrine mediator, BK induces vasodilation and increases microvascular permeability via activation of endothelial B2R, a constitutively expressed subtype of kinin receptor. During infection, neutrophil-driven extravasation of plasma fuels inflammation via extravascular activation of the kallikrein-kinin system (KKS). Whether liberated by plasma-borne PK, tissue kallikrein, and/or microbial-derived proteases, the short-lived kinins activate immature dendritic cells via B2R, thus linking the infection-associated innate immunity/inflammation to the adaptive arm of immunity. As inflammation persists, a GPI-linked carboxypeptidase M removes the C-terminal arginine from the primary kinin, converting the B2R agonist into a high-affinity ligand for B1R, a GPCR subtype that is transcriptionally upregulated in injured/inflamed tissues. As reviewed here, lessons taken from studies of kinin receptor function in experimental infections have shed light on the complex proteolytic circuits that, acting at the endothelial interface, reciprocally couple immunity to the proinflammatory KKS.

## 1. INTRODUCTION

The discovery of the kallikrein-kinin system (KKS) traces back to 1909, when Abelous and Bardier (1909), two French physiologists from Toulouse, observed a reduction of blood pressure in dogs injected with human urine. Several years later, Frey isolated and characterized this non-dialyzable hypotensive substance and, together with Kraut, Frey, and Werle (1930), named it as "kallikrein"—under the premise that it was produced in the pancreas. Subsequent investigations identified a second kallikrein in the plasma (Kraut, Frey, & Werle, 1933), both of which able to release the hypotensive substance(s) from a precursor molecule circulating in the blood (Werle, Götze, & Kappler, 1937). Ten years later, Brazilian pharmacologists (Rocha e Silva, Beraldo, & Rosenfeld, 1949) discovered that "bradykinin"—a hypotensive substance that slowly contracted guinea pig ileum, was released from a precursor circulating in the blood, following administration of snake venoms or trypsin. Shortly afterward, Werle, Kehl, and Koebke (1950) reported that kallidin and bradykinin originated from the cleavage of the same plasma precursor proteins, termed as "kininogens." In the early 1960s, the sequences of the decapeptide kallidin (lysyl-BK) and BK (Arg-Pro-Pro-Gly-Phe-Ser-Pro-Phe-Arg) were elucidated. In parallel, Erdos and Sloane (1962) characterized a kinin-metabolizing peptidase from plasma, termed as kininase I (carboxypeptidase N) and discovered that kininase II, a peptidase that degrades the hypotensive BK, is dipeptidylcarboxypeptidase identical to angiotensin-converting enzyme (ACE) (Yang, Erdös, & Levin, 1970). While these studies were in progress, research in Brazil revealed that the biological function of BK was potentiated by a polypeptide isolated from the snake *Bothrops jararaca* (Ferreira, 1965). Subsequently characterized as ACE inhibitor, this discovery paved way for the development of captopril, the first synthetic antihypertensive drug. In the mid-1950s, the discovery of an asymptomatic patient that had prolonged clotting time in vitro due to deficiency of FXII (Hageman factor) was the start point for the biochemical studies that characterized the contact system as a procoagulative cascade (Ratnoff & Colopy, 1955).

In the early 1980s, Regoli and Barabé (1980) distinguished the existence of two kinin receptors based on differences of agonist potencies and synthesis of subtype-specific synthetic antagonists further validated their predictions (Vavrek & Stewart, 1985). Gene cloning in the 1990s led to the characterization of kinin receptors as members of the large rhodopsin family of GPCRs containing seven-membrane-spanning domains (Hess, Borkowski, Young, Strader, & Ransom, 1992; Kammerer, Braun, Arnold, & Roscher, 1995; McEachern et al., 1991; McIntyre, Phillips, Skidmore, Brown, & Webb, 1993; Pesquero et al., 1994, 1996). Insight on the functional role of B2R and B1R came with the development of knockout mice for each subtype of kinin receptors (Borkowski et al., 1995; Cayla et al., 2007; Pesquero et al., 2000). As covered by Bader (2011), kinin receptors and the KKS are involved in an impressive number of physiological and pathophysiological processes. In this chapter, we will present an outline of progress in our knowledge about the roles of kinin receptors as functional linkers between coagulation/thrombosis inflammation and immunity in infection.

## 2. THE KKS AND KININ RECEPTORS

### 2.1 Intravascular Release of Kinins and GPCR Regulation of Cell Signaling

Integrated by two serine proteases (FXII and PK) and by the cofactor HK, the contact system is an enzymatic mechanism that (i) initiates the intrinsic pathway of coagulation (ii) triggers bradykinin (BK)-induced inflammatory cascades (Kaplan & Joseph, 2014; Renné et al., 2005; Scharfstein, 2016; Schmaier, 2008). Activation of the contact system usually starts when the blood is exposed to "contact" substances, a structurally diverse group of substances of endogenous or exogenous origin that triggers the autocatalytic activation of zymogen FXII (Kaplan & Joseph, 2014). Included in the group of contact substances of endogenous origin are polyphosphates (polyP) released and retained at the surface of activated platelets (Müller et al., 2009; Verhoef et al., 2017) and DNA-nucleosomes integrating neutrophil extracellular traps (NETs) (Oehmcke, Mörgelin, & Herwald, 2009). Although misfolded proteins induce autoactivation of FXII (Maas et al., 2008), leading to BK release by PK, it is intriguing that structural constraints impede the activation of the intrinsic pathway of coagulation.

Although PK may be directly generated by prolylcarboxypeptidase, a matrix-associated serine protease that is expressed constitutively by endothelial cells (Moreira et al., 2002; Shariat-Madar, Mahdi, & Schmaier, 2002, 2004), it is currently believed that trace levels of FXIIa generated in vivo by autocatalysis (triggered by endogenous "contact" substances) are sufficient to convert the plasma prekallikrein zymogen into active PK (Fig. 1). Since PK reciprocally activates circulating forms of FXII, this feedback amplification loop eventually overcomes the regulatory constraints exerted by C1 inhibitor (Kaplan & Joseph, 2014). Acting further downstream, FXIIa converts FXI into FXIa, which then initiates FXIa, which then initiates the intrinsic pathway of coagulation by converting FIX into FIXa (Fig. 1). Alternatively, FXIa can be generated (independently of FXIIa) by thrombin (Gailani & Broze, 1991).

While the procoagulative cascade progresses, PKa liberates BK nanopeptide from HK (Fig. 1). Acting as a paracrine mediator, the short-lived BK increases microvascular permeability by triggering B2R, a high-affinity GPCR expressed constitutively by vascular endothelial cells. Beyond its role in vascular homeostasis, endothelial activation via the BK/B2R pathway is thought to promote fibrinolysis and regulate thrombus development by (i) upregulating plasminogen activator (tPA) expression by endothelial cells (Brown, Gainer, Murphey, & Vaughan, 2000), (ii) inhibiting tissue factor expression via the eNOS/Akt pathway (Dong et al., 2015), and (iii) upregulating prostacyclin and nitric oxide (NO) production via B2R cross talk with angiotensin 1–7 receptors (Schmaier, 2016). PK may also drive fibrinolysis by generating plasmin, a serine protease that cleaves native C3 and C5, hence linking activation of the contact system/KKS to fibrinolytic and complement cascades (Amara et al., 2010).

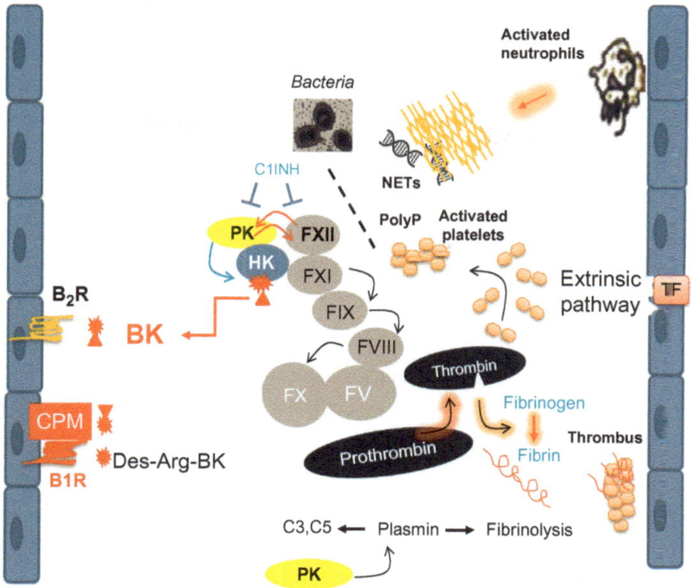


**Fig. 1** Intravascular activation of the KKS. Although there are major differences in the activation requirements and regulation of the intrinsic and extrinsic pathways, they converge in thrombin activation—the effector of fibrin formation. Beyond its role in the initiation of coagulation, the contact pathway triggers the KKS, a proinflammatory/profibrinolytic cascade. The intravascular activation of the contact system (*upper part* of panel, *center*) is initiated when FXII (zymogen) undergoes autocatalytic activation upon binding to a broad group of "contact" molecules, some of which are displayed at bacterial surfaces while others (endogenous origin) are negatively charged polymers that are either released/exposed at the surface of activated platelets or associated (DNA) to neutrophil extracellular traps (NETS) (*top, center*). Misfolded proteins, such as amyloids, also activate FXII and trigger the KKS without initiating the intrinsic pathway of coagulation. Whichever the origin of the contact substance, the formation of trace levels of FXIIa is sufficient to promote the cleavage of plasma prekallikrein (bound to HK) into PK, which in turn, reciprocally cleaves FXII zymogens. Although this feedback mechanism is counter-regulated by the serpin C1INH, intravascular levels of PK and FXIIa may rapidly rise, as regulation is overcome. Acting as a prothrombogenic factor, FXIIa initiates coagulation by converting FXI into FXIa, which then generates FIXa, an effector of fibrin formation via the intrinsic pathway. While FXIIa initates the procoagulative cascade, PK triggers inflammation by liberating the nanopeptide BK from an internal moiety of HK. Acting as a paracrine signal, the short-lived BK activates B2R, a GPCR subtype that is constitutively expressed by endothelial cells (*left side* of panel) as well as by other cell types, including innate sentinel cells (not represented in the scheme). As the inflammation continues, cytokines and other proxidative signals upregulate endothelial B1R expression, rendering the microvascular beds sensitive to kinin metabolites (des-Arg-BK). PK may also promote plasmin-dependent fibrinolysis and activate complement C5 and C3, thus linking the KKS to fibrinolytic and complement cascades.

Although C1 inhibitor was originally characterized as a serpin that inactivates C1 serine esterases, it is now well established that the multiple swelling episodes afflicting patients with hereditary angioedema are due to dysregulated contact activation, whether caused by mutations that generate a dysfunctional protein (type 1) or result in deficient plasma levels (type 2) of this serpin (Kaplan & Joseph, 2014). Irrespective of the origin of the initial "contact" stimuli, C1 inhibitor targets activated forms of the contact factors FXIIa and PK, indirectly protecting endothelial B2R from excessive activation by BK. Analysis of the dynamics of cutaneous inflammation in mice revealed that the edema response is amplified extravascularly as result of cleavage/inactivation of C1 inhibitor by matrix metalloprotease 2 (MMP2) (auf dem Keller, Prudova, Eckhard, Fingleton, & Overall, 2013). It is also known that patients undergoing antihypertensive medication with ACE inhibitors may develop acquired angioedema due to excessive formation of BK (Nielsen & Gramstad, 2006). Clinical studies showed that the PK inhibitor Ecallantide (DX-88, Kalbitor) or B2R antagonists (Icatibant) (Cugno, Zanichelli, Bellatorre, Griffini, & Cicardi, 2009; Longhurst et al., 2015) efficiently reduce the frequency of swelling reactions in these patients. As opposed to type 1 and type 2 hereditary angioedema, the expression and function of C1 inhibitor is fully preserved in patients with type 3 hereditary angioedema. Notably, these patients display a FXII mutation that affects a critical glycosylation site, rendering the zymogen sensitive to cleavage/activation by plasmin (de Maat et al., 2016).

## 2.2 Functional Links Between Thrombogenesis and the KKS

Although the contact system does not significantly contribute to hemostasis (the bleeding time is not increased in patients with genetic deficiency of FXII, HK, and FXI), the findings that mice deficient of these key plasma factors (FXII, FXI, HK) are resistant to experimental thrombosis (Kleinschnitz et al., 2006; Merkulov et al., 2008; Rosen, Gailani, & Castellino, 2002) represented an important advance in this field (Renné et al., 2005). Inspired by lessons from arthropod evolution, Engelmann and Massberg (2013) proposed that physiological thrombogenesis, an innate response termed "immunothrombosis," is reminiscent of a primitive of the primitive procoagulative mechanism that protects insects from microbial infection (Fig. 2). Using a mouse models of systemic infection with Gram-negative bacteria, these workers provided compelling evidences that blood coagulation/thrombogenesis (in microvessels) promotes the retention of these pathogens within fibrin scaffolds that are formed via cooperation

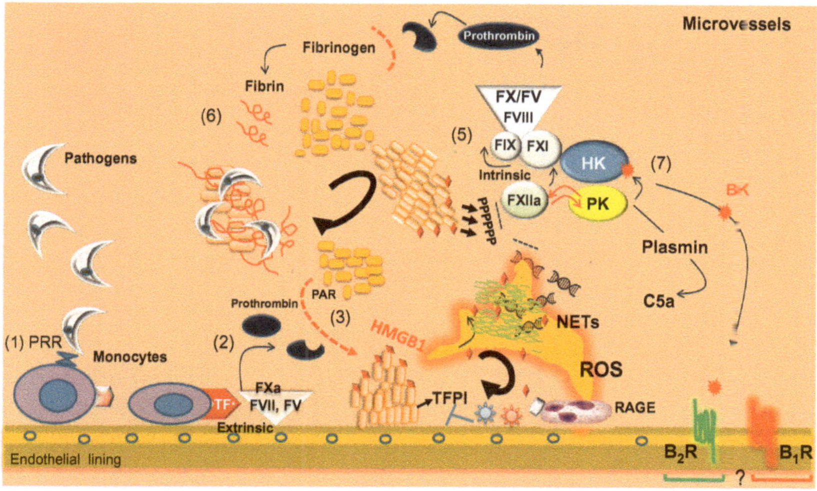


**Fig. 2** Functional coupling between innate immunity and thrombogenesis. Differently from arterial or venous thrombosis, i.e., conditions in which tissue factor is only exposed to the blood after tissue injury, this paradigm predicts that PRRs trigger upregulated expression of tissue factor expression in circulating myeloid leukocytes. At the *left side* of the panel, monocytes rely on PRRs to sense the threat posed by a blood-borne pathogen. Following innate recognition, the procoagulative process is unleashed by active TF factor exposed at the surface of activated monocytes (*bottom* of panel) and by TF delivered within microvesicles (not represented in the scheme). Upon binding to FVII, thrombin is generated by the FXa/FVa-dependent prothrombinase complex. After activating platelets via PAR1/4, the release of tissue factor protein inhibitor (TFPI) dampens the extrinsic pathway by inactivating FVIIa and FXa. However, the formation of fibrin networks is nonetheless sustained via activation of the FXII-dependent contact pathway by long-chain polyP exposed at the surface of activated platelets (*center* of panel). In parallel, HK is cleaved by PK, liberating the proinflammatory and profibrinolytic BK in the lumen of the microvessel. Responsive to adhesive interactions with activated platelets (P-Selectin and PSGL), neutrophils expel their extracellular traps (NETs), providing the lattices that subsequently incorporate neutrophil elastase (NE), and platelet adhesive molecules, such as Von Willebrand factor. Recapitulating the contact activation responses elicited by long-chain PolyP, NETs/DNA activates the FXII-dependent contact pathway. An additional mechanism of amplification depends on NE-dependent cleavage of TFPI. Apart from promoting fibrin formation via the extrinsic pathway, thrombin also activates FXI (bypassing FXII), hence further amplifying the coagulation process. Studies of pathological thrombosis-in models of sterile inflammation-revealed that thrombus formation is further accelerated by HMGB1, a prototypic damage-associated motif pattern (DAMP) displayed (reduced form) at the surface of activated platelets. Once exposed to reactive oxygen species, the disulfide-linked HMGB1 is incorporated into NETs, reinforcing its prothrombogenic role. Future studies may clarify whether HGMB1/RAGE optimizes intravascular immunity against the fibrin-entrapped pathogens. Although PK and BK are thought to counter-modulate thrombogenesis in large vessels (arterial or venous), a more rapid and subtle procoagulative response may suffice to retain pathogens within tiny capillary beds. Under these circumstances, local release of BK might optimize the phagocytic uptake and immune clearance of fibrin-entrapped microbes via activation of kinin receptors (B2R and/or B1R). *Adapted from Engelmann, B., & Massberg, S. (2013). Thrombosis as an intravascular effector of innate immunity.* Nature Reviews. Immunology, 13*(1), 34–45. https://doi.org/10.1038/nri3345.*

between the tissue factor-driven extrinsic and FXII/PK-driven intrinsic pathways of coagulation (Gaertner & Massberg, 2016; Massberg et al., 2010). Based on models of deep venous thrombosis, it was recently reported that severity of thrombogenesis is linked to platelet surface exposure of high mobility group box 1 (HMGB1), a nonhistone architectural DNA-binding nuclear protein that is subsequently exported to the extracellular space (Stark et al., 2016; Vogel et al., 2015). Acting as a typical damage-associated molecular pattern (DAMP), HMGB1 activates circulating monocytes through the binding of receptors for advanced glycation end products (RAGE) and classical innate sensor receptors, such as TLR2, TLR4, and TLR9 (Park et al., 2005). In their study, Stark et al. (2016) also demonstrated that pathological outcome is critically influenced by the redox status of HMGB1. Acting at the endothelial interface, HMGB1 propagates venous thrombosis by (i) upregulating the extrinsic pathway of coagulation via stimulation of monocyte recruitment and upregulated expression of tissue factor; (ii) increasing platelet accumulation and aggregation; and (iii) modifying the DNA/nucleosome scaffold formed by NETs, which then reinforces thrombogenesis via activation of the contact system by platelet-derived polyP and/or NETs/DNA expelled by activated neutrophils (Fig. 2). Although this knowledge provides a novel mechanistic framework to understand the functional connectivity between platelets and circulating neutrophils/monocytes (Ali, Wuescher, & Worth, 2015), it is still unclear whether HGMB1, acting synergistically with platelet-derived polyP and/or DNA/NETs might harness intravascular immunity via cooperation between PRRs (RAGE/TLR pathway) and kinin-driven proteolytic cascades (B2R and/or B1R pathway) (Fig. 2). Interestingly, in this context, it has been recently proposed that the coupling between PK/kinin/B2R and eNOS/phosphoinositide 3-kinase/Akt signaling pathways might negatively regulate tissue factor-induced arterial thrombosis (Dong et al., 2015; Schmaier, 2016). Future studies may clarify whether subtle changes in prostacyclin/NO-dependent regulation of thrombogenesis may underlie the immune dysfunctions described in infected $B2R^{-/-}$ mice (discussed further later in this chapter).

## 2.3 Multiple Kininogenases Promote Kinin Release Extravascularly

While PK is the main BK-liberating protease acting in the bloodstream, the extravascular generation of kinins may involve the participation of several kininogenases. In normal conditions, this task is executed by tissue kallikrein

(KLK1), a constitutively activated serine protease that releases lysyl-BK (LBK) mostly from low-molecular-weight kininogen (LK) diffusing through interstitial spaces (LK is present in the bloodstream at threefold higher proportion than HK). During inflammation, however, oxidized kininogens are not efficiently cleaved by tissue kallikrein due to the oxidation of a methionine residue localized upstream of the N-terminal flanking site of the internal kinin moiety (Kozik et al., 1998). Biochemical studies suggest that neutrophil elastase (NE) and mast cell (MC)-derived tryptase, acting in synergy, might release bioactive kinins from oxidized kininogens. In a study involving patients with vasculitis, Kahn et al. (2009) reported that neutrophil proteinase 3 (PR3) generates an extended kinin peptide (PR3-kinin) containing two extra amino acids on both sides of the kinin moiety (Met-Lys-BK-Ser-Ser).

Plasma leakage through postcapillary venules, a common manifestation of acute inflammation, occurs within a few minutes after endothelium exposure to potent inducers of microvascular permeability, such as histamine and bradykinin. Following the opening of endothelial junctions, the subendothelial matrix and interstitial spaces are exposed to a myriad of plasma molecules, including the zymogens that integrate the procoagulative contact system. Recent studies of MC function in mice models of allergic inflammation demonstrated that MC degranulation leads to histamine release along with the discharge of heparin and long-chained polyP, both of which characterized as endogenous triggers of the contact pathway of coagulation (Moreno-Sanchez, Hernandez-Ruiz, Ruiz, & Docampo, 2012; Oschatz et al., 2011). Following histamine-induced leakage of plasma contact factors, the plasma-borne FXII is activated by heparin/polyP (likewise released from MC granules), leading to PK-dependent BK release from HK (Fig. 3).

Circulating neutrophils comprise the first line of defense against infection. Although neutrophil-induced microvascular leakage is thought to occur independently of endothelium crossing, studies in mouse models of inflammation have implicated TNF-α as a key effector of neutrophil-evoked plasma extravasation induced by C5a and LTB4 (Finsterbusch, Voisin, Beyrau, Williams, & Nourshargh, 2014). As the inflammation progresses, DNA/nucleosomes associated to NETs may function as "contact" substances, promptly activating the plasma factors (FXII, HK-prekallikrein complexes) that diffuse through injured/inflamed tissues.

Although not mutually exclusive, microvascular leakage may lead to retention of plasma contact factors at the cell surface of a broad range of tissue cells, including resident macrophages and infiltrating neutrophils—as documented

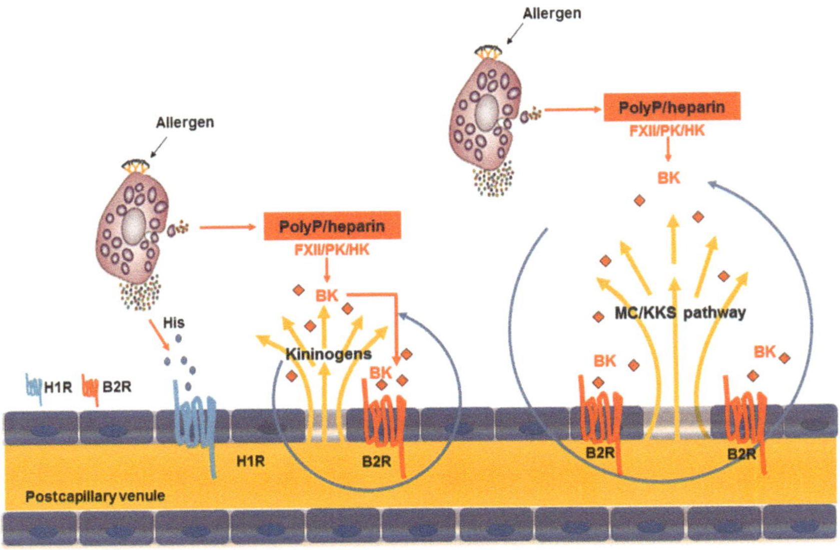


**Fig. 3** Role of the MC/KKS axis in anaphylaxis. In the *upper part* of the scheme, allergen binding to surface-bound IgE (via FcεRI) triggers MC release of (i) histamine and (ii) heparin and polyP. Following histamine-induced leakage of plasma, the "contact" of FXII with the negatively charged polymers leads to autocatalytic generation of FXIIa. Following reciprocal cycles of activation between PK and FXIIa, BK is liberated plasma-borne HK. Further downstream, iterative cycles of endothelial B2R activation, plasma leakage, MC degranulation, and heparin-induced activation of contact plasma factors may account for the severe hypotension and systemic plasma leakage associated to anaphylactic shock (Sala-Cunill et al., 2015).

in studies performed with human neutrophils (Henderson, Figueroa, Müller-Esterl, & Bhoola, 1994) and monocytes/macrophages (Barbasz & Kozik, 2009). Motivated by the findings that contact system factors assemble at the surface of platelets (Gustafson, Schutsky, Knight, & Schmaier, 1986), studies with endothelial cells revealed that heparan sulfate proteoglycans retain HK molecules via interaction with the internal histidine-rich (positively charged) D5 domain (Herwald, Dedio, Kellner, Loos, & Müller-Esterl, 1996; Renne, Dedio, David, & Muller-Esterl, 2000). Also noteworthy, there is biochemical evidence that HSP-90 binding to the HK promotes the autocatalytic activation of plasma prekallikrein independently of FXIIa (Kaplan & Joseph, 2014). In addition, studies with endothelial cells suggested that prolylcarboxypeptidase, a lysosomal enzyme, promotes prekallikrein conversion into the BK-releasing PK independently of FXIIa (Shariat-Madar et al., 2004). During infection, the intertwined nature of the proteolytic circuitries underlying the KKS network is further complicated when inflamed tissues are exposed to pathogens that express their own kinin-releasing proteases (Petempa & Herwald, 2011; Scharfstein, 2016).

## 2.4 Pharmacological Specificity of G Protein-Coupled B1 and B2 Kinin Receptors

Belonging to the rhodopsin superfamily of GPCRs, B1R and B2R are single-chain polypeptide receptors in which the seven-transmembrane (7-TM) sequences are interspersed by three intracellular and three extracellular loops—spatially separating the external amino-terminal domain from the intracellular carboxy-terminal domain (Leeb-Lundberg, 2011). Mapped to human chromosome 14q32 and rat, the genes coding for B1R (BDKRB1) and B2R (BDKRB2) consist of 353 and 364 amino acids, respectively, showing 36% of homology in the protein sequence (Leeb-Lundberg, Marceau, Müller-Esterl, Pettibone, & Zuraw, 2005). As already stated, the natural ligands of B2R are BK or LBK, i.e., the nona/decapeptides that are, respectively, released from an internal moiety of HK or LK by plasma and tissue kallikrein (KLK1). The high-affinity binding of BK to B2R is dependent on both ionic interactions provided by the N-terminal side chain of BK (Arg1) and the B2 receptor N-terminus as well as by other nonionic with BK residues reaching two helical turns along the interior face of TM-VI (Leeb-Lundberg, 2011). In contrast, the binding of des-Arg-LBK to B1R involves primarily ionic interactions between Lys1 and Arg2 and the negatively charged C-terminus (Regoli & Barabé, 1980).

Whether involving formation of homo- or heterooligomers, the activation signals that B2R transduce in vascular cells are either coupled to Gαq, Gαi, Gαs, or Gα12/13 (Leeb-Lundberg et al., 2005). BK induces B2R sequestration in caveolae of vascular endothelial cells where it associates to NO synthase (eNOS) (Haasemann, Cartaud, Muller-Esterl, & Dunia, 1998). Follow-up studies performed with bovine aortic endothelial cells and human umbilical vein endothelial cells linked B2R-dependent production of nitric oxide (NO) to calcium and phosphorylation-dependent activation of eNOS (Venema, 2002). In another in vitro study, Petrovic et al. (2007) reported that BK-dependent internalization of B2R promotes the downstream activation of Rho family GTPase (Cdc42), which then induced actin polymerization and the formation of filopodia in cultured human endothelial cells. As previously mentioned, BK is thought to stimulate B2R-dependent fibrinolysis by upregulating the release of tissue plasminogen activator from human vascular cells via cyclooxygenase-dependent pathways, independently of NOS (Brown et al., 2000).

Although B1R is hardly detectable in most tissues, activation of NF-κB proinflammatory cytokines (IL-1β, TNF-α, and IFN-γ) upregulates the transcription of this GPCR subtype in injured tissues in injured tissues (McLean, Ahluwalia, & Perretti, 2000; Medeiros et al., 2004; Sabourin, Morissette, Bouthillier, Levesque, & Marceau, 2002; Schanstra et al., 1998). While B2R is activated by intact kinins, the agonists of B1R are kinin metabolites (des-Arg-kinins) that lose their C-terminal Arginine residue upon cleavage by "kininase I," a general designation for carboxypeptidase M (CPM) or carboxypeptidase N (CPN). Once generated, des-Arg-kinins are no longer able to trigger B2R. Studies in mammalian cell lines transfected with mutated proteins yielded a wealth of information about the structure–activity relationship of B2R and B1R. The role of B1R in NO production was studied in human microvascular cells activated by proinflammatory cytokines (Ignjatovic, Stanisavljevic, Brovkovych, Skidgel, & Erdös, 2004; Zhang, Tan, Zhang, & Skidgel, 2008; Zhang et al., 2007). Once stimulated by B1R ligands, the activated microvascular cells upregulated NO production via activation of iNOS, in contrast to the B2R/eNOS-dependent responses that BK induced in normal endothelial cells. Although the bulk of these studies focused on B2R or B1R, the biological responses relayed by these GPCRs are sometimes modulated by formation of complexes with other surface receptors. For example, studies performed with smooth muscle cells demonstrated that (i) B2R forms heterodimers with angiotensin 2 receptors (AT2) independently of B2R binding to BK and (ii) formation of the heterodimers increased the efficiency of AT2 signaling by angiotensin II (AbdAlla, Lother, & Quitterer, 2000).

## 2.5 Regulation of B2R and B1R Signaling by Metallopeptidases

Overlapping regulatory mechanisms limit the long-range signaling activity of intact kinins (BK/LKK). Typically acting as paracrine signals, the intact kinins are swiftly degraded by dipeptidyl-carboxypeptidases (kininase II), such as ACE, a membrane-bound metallopeptidase displayed at endothelial surfaces, and highly expressed in the pulmonary microcirculation (Pesquero, Jubilut, Lindsey, & Paiva, 1992). Intact kinins are also degraded by endopeptidase, another well-characterized membrane-bound kininase II. Desensitization of the constitutively expressed B2R is another fundamental regulatory mechanism that protect tissues from excessive inflammation induced by BK. Studies with mammalian cells transfected with individual G protein-coupled receptor kinases (GRK) revealed that agonist-induced activation of B2R drives GRK4α-dependent drives serine phosphorylation at the C-terminal tail B2R. The differential sensitivity to agonist-induced desensitization is considered an important distinction between agonist-induced regulation of B2R and B1R. In contrast to B2R, B1R is not efficiently phosphorylated nor desensitized because it lacks the conserved serine and threonines that are present at the C-terminal tail of B2R (Blaukat & Dikic, 2001; Leeb-Lundberg, Kang, Lamb, & Fathy, 2001). Notably, although B1R (in the absence of agonist binding) undergoes clathrin-mediated endocytosis and GAP-mediated lysosomal degradation by endothelial cells, its surface display is stabilized upon binding to its cognate ligand (des-Arg-BK/LBK) (Enquist Skröder, Whistler, & Leeb-Lundberg, 2007). Further optimizing B1R signaling, Zhang et al. (2008) demonstrated that CPM, the glycosylphosphatidylinositol (GPI)-anchored carboxypeptidase that generates des-Arg-BL/LBK, is also upregulated during inflammation. Remarkably, catalytically inactive CMP mutants (bound the intact substrate, i.e., BK/LBK) induce B1R signaling independently of the ability to convert the intact kinins into B1R ligands (Zhang, Tan, & Skidgel, 2013). These findings suggested that surface CPM (upon binding to kinin substrate) has a dual role in B1R signaling: acting as a B1R cofactor, it induces a conformational cross talk that signals B1R while the catalytic site releases the cognate agonist. It was proposed that the integration between these molecular events might render B1R hypersensitive to the low levels of kinin metabolites released by CPM.

# 3. KININS AS EFFECTORS OF IMMUNITY

## 3.1 Role of B1R in Neutrophil Transmigration to Inflamed Tissues

Although the upregulated kinin/B1R pathway is clearly implicated in chronic inflammation and tissue remodeling, it has been difficult to predict the functional outcome of B1R engagement in different diseases. For example, B1R-deficient mice were protected from experimentally induced diabetic cardiomyopathy (Westermann et al., 2009). On the other hand, Merino et al. (2009) reported that high-fat diet aggravated atherosclerosis and angiotensin II-induced aortic aneurysms in transgenic mice with double deficiency of apolipoprotein E and B1R.

Using the classical air pouch model of inflammation, Ahluwalia and Perretti (1996) demonstrated that kinins induce IL-1β-dependent leukocyte transmigration via activation of the inducible B1R pathway. Consistent with these results, Vianna and Calixto (1998) reported that selective B1R blockers impaired neutrophil infiltration in the mouse model of lung inflammation, a finding that was later substantiated by the analysis of the phenotypic properties of neutrophils from B1R$^{-/-}$ mice (Pesquero et al., 2000). While not mutually exclusive, B1R agonists induce leukocyte transmigration by upregulating endothelial expression of the chemokine CXCL5 (Duchene et al., 2007). Although the role of kinin receptors in sepsis was not explored in the clinical settings, it has been reported that kinins induce chemotaxis of human neutrophils (Paegelow, Trzeczak, Böckmann, & Vietinghoff, 2002) via activation of the B1R pathway (Ehrenfeld et al., 2006). Follow-up studies in human neutrophils linked activation of the B1R pathway to (i) upregulated surface expression of LFA-1 and Mac-1 and (ii) ERK1/2- and p38 MAPK-dependent release of matrix metalloproteinase 9 (MMP9) and myeloperoxidase (Ehrenfeld et al., 2009). Of further interest, clinical studies with asthma patients demonstrated that KKS components were increased in the bronchoalveolar lavage fluid of asthmatic patients, and the levels of these components were further increased upon allergen challenge (Christiansen, Proud, & Cochrane, 1987). While examining the expression profile of B2R and B1R in peripheral blood neutrophils of asthma patients, Bertram, Baltic, et al. (2007) did not find significant changes in their cohorts, as compared to normal donor. However, these authors verified that B1R transcription (NF-κB and AP-1 dependent) was decreased in asthmatic patients that inhaled high doses of corticosteroid.

## 3.2 KKS Expression During Macrophage Development

Although the study of infection-associated plasma leakage provided a wealth of information about the functional coupling between the KKS and microbial immunity (Scharfstein, 2016), it is unclear whether kinins and their GPCRs are important players in the development of innate immunity. More recently, advances in high-throughput RNA sequencing studies shed new light about the origin and functional diversity of tissue macrophages in adult mice (Perdiguero & Geissmann, 2016). Challenging the classical view that tissue macrophages depends on polarization of the infiltrating monocytes that originate from bone marrow hematopoietic stem cells, it is now firmly established that this subgroup of innate sentinel cells only represent a minor subset of "passenger" or "transitory" myeloid cells. In contrast, the majority of tissue-resident macrophages, such as Kupffer cells in the liver, Langerhans cells in the epidermis, alveolar macrophages, and microglia in the brain, are self-renewing phagocytes that originated from yolk sac-derived erythromyeloid progenitors, being genetically and developmentally distinct from hematopoietic stem cells. Using high-throughput RNA sequencing, Mass et al. (2016) analyzed the dynamic changes of gene expression during the process of macrophage differentiating into tissue-resident F4/80$^+$ macrophages. The results from these fundamental studies revealed that macrophages precursors derived from erythromyeloid progenitors (major subset) presented a core differentiation program that differed from those (minor subset) originating from bone marrow-derived myeloid cells like monocytes and neutrophils. Apart from the differential regulation of well-defined transcription factors, these authors observed that macrophage precursors originating from yolk sac-derived liver upregulate several genes coding for innate immunity functions, including pattern-recognition receptors, phagocytic receptors, Fc gamma receptors, lysosomal cathepsins, complement components, and CX3CR1, a chemokine receptor required for tissue colonization. Intriguingly, the expression of bradykinin B2 and B1 receptors genes during embryonic development or postnatally was not significantly different (Mass & Geissmann, personal communication). However, tissue-resident F4/80$^+$ macrophages that colonize the liver showed transcriptional upregulation of the contact factors FXII and HK-1, without influencing expression in tissue macrophages residing in the limbs/skin, lung, kidney, and brain. As discussed further later, it will be worth exploring whether the increased expression of contact factors by Kupffer cells lining the liver sinusoids may influence macrophage migration and granuloma development in the mouse model of visceral leishmaniasis.

## 3.3 Dendritic Cells: Functional Roles of B2R and B1R

Consisting of highly heterogeneous populations, DCs have evolved the ability to sample antigens from normal or injured/infected tissues and present processed peptides in MHC-restricted fashion to T cells in specialized anatomical areas of secondary lymphoid tissues. Although antigen-bearing DCs normally execute tolerogenic functions in the steady state their conversion into immunogenic antigen-presenting cells usually requires the signaling of innate receptors (pattern-recognition receptors) by pathogen-associated motif patterns (PAMPs). Apart from the activation signals elicited by pathogens, the antigen-bearing DCs express receptors for endogenous inflammatory cues, whose levels may vary as a function of microbial virulence and pathogenicity. The cross talk forged between activated PRRs and receptors for endogenous danger signals coordinates DC reprogramming: the antigen-bearing DCs are attracted to lymphatics, from where these antigen-presenting cells immigrate to T cell-rich areas of draining lymph nodes. Earlier in this decade, we hypothesized that immature DCs may be activated by BK, whether generated systemically or released in inflamed peripheral tissues. To address this question, in a systemic model, a sublethal dose of LBK (B2R agonist) was injected i.p. in WT- and B2R-deficient mice (Aliberti et al., 2003). By using IL-12 serum levels as the readout of systemic inflammation, we verified that BK induced a steep rise in IL-12p70 levels in WT mice, whereas B2R$^{-/-}$ mice showed a blunted response. Peaking at 1 h postinjection and declining precipitously thereafter, the short-lived IL-12 response induced by LBK was not unanticipated because ACE swiftly degrades intact kinins (B2R agonists). As predicted, LBK induced IL-12 responses of higher magnitude and duration in WT mice pretreated with the ACE inhibitor captopril. Furthermore, captopril upregulated innate immunity in B2R$^{+/+}$ mice challenged by carrageenin, a classic activator of the KKS—the phenotype being expressed as (i) increased migration of CD11c$^{+}$ DC to the periarteriolar lymphoid sheath area of the spleen and (ii) increased IL-12 and IL-10 production ex vivo by WT splenic CD11c$^{+}$ DCs, but not by DCs from B2R-deficient mice. Along similar lines, the injection of LBK or carrageenin upregulated CD40, CD80, and CD86 in WT (spleen) CD11c$^{+}$ DCs. Extending this approach to the in vitro settings, we verified that LBK induced dose-dependent secretion of both IL-12 and IL-10 in WT CD11c$^{+}$ DCs, but not in B2R$^{-/-}$ DCs. FACS analysis revealed higher frequencies of IL-12$^{+}$ DCs in the subsets of CD11cCD8$^{+}$ CD4$^{-}$and CD11cCD8$^{-}$ CD4$^{-}$(DN), but not in the CD11cCD8$^{-}$ CD4$^{+}$ of WT splenic DCs. Noteworthy, immature DCs, HOE-140 (icatibant; B2R antagonist) but not a B1R antagonist (des-Arg-HOE-140) blocked IL-12p70 and p40 production by WT splenic DC activated by LBK. Consistent with the in vivo findings, LBK also stimulated B2R-dependent induction of IL-10-producing DCs. Of note, WT DCs (but not B2R$^{-/-}$ DCs) secreted higher levels of IL-10 upon addition of captopril, i.e., in the absence of exogenous LBK. Combined, these results suggested that the low levels of endogenous kinins generated in cultures of B2R$^{+/+}$ DCs are sufficient to induce IL-10 production, but not IL-12. Of further interest, BK induced a more intense IL-12 response splenic DCs derived from IL-10$^{-/-}$ mice. These results suggested that IL-10, acting as a counter regulator of IL-12, fine-tunes B2R-driven IL-12 production elicited by intact kinins.

Since the systemic injection of LBK inevitably induces B2R-dependent hypotension, we then asked whether the exogenous LBK could upregulate adaptive immunity in mice immunized with ovalbumin (OVA). Using the classical Th2-dependent pleurisy model established in BALB/c mice immunized with alum/OVA, we evaluated the impact of the addition of LBK (LPS free) to the alum/OVA emulsions. Recall assays performed with T cells isolated from draining lymph nodes showed that the inclusion of LBK in alum/OVA emulsions reduced IL-4 production by OVA-specific Th2 cells. Conversely, LBK promoted an increase in OVA-specific effector Th1 cells, and the phenotype switch (Th2 > Th1) was accompanied by decreased eosinophil infiltration in the pleural cavity of BALB/c mice. The evidence that exogenous B2R agonists drive Th1 responses via IL-12 was further validated by the negative results observed in IL-12p40$^{-/-}$ mice immunized with OVA/alum/BK.

While the study of the Th1-directing function of BK were still in progress, Bertram, Baltic, et al. (2007) and Bertram, Misso, et al. (2007) analyzed the functional properties of B2R and B1R expression in monocyte-derived human DCs (hMo-DCs) during their differentiation in vitro. According to these studies, the transcription of B2R and B1R was upregulated (3rd and 4th day of culture) independently of the presence of IL-4 or GM-CSF. Of note, the transcriptional expression of both GPCRs was upregulated by immature hMo-DCs stimulated with the corresponding ligands. Measurements of intracellular $[Ca^{2+}]_i$ fluxes and cellular migration in vitro revealed that hMo-DCs responded to LBK via B2R signaling pathways. However, in contrast to mouse (splenic) DCs, immature hMo-DCs did not upregulate the expression of costimulatory molecules, not upon stimulation with BK nor des-Arg-LBK. Interestingly, addition of BK to cultures of mature hMo-DCs upregulated the expression of the type 1-polarizing IL-12p70 (Kaman, Wolterink, Bader, Boele, & van der Kleij, 2009). Surprisingly, although B1R agonists induced chemotaxis of immature hMo-DCs, Gulliver et al. (2011) observed that upon maturation, these activated dendritic cells became unresponsive to both des-Arg-LBK and the chemokine CCL19—in spite of the fact that CCR7 expression remained unchanged. Interestingly, des-Arg-LBK enhanced IL-12p70 production by mature hMo-DCs, whereas IL-12p40 was conversely inhibited, indicating that B1R signaling function was preserved upon DC maturation. Complementing these studies, Gulliver et al. (2011) compared the profile of kinin receptor expression of hMo-DCs isolated from the blood of asthmatic patients vs normal patients. Although DC differentiation requires several days of cultivation, the expression levels of B1R in hMo-DCs from asthmatic patients were higher as compared to those of normal donors. These results, which mimic the B1R phenotypes of eosinophils isolated from the blood of asthmatic patients (Bertram et al., 2008), suggest that these peripheral blood leukocytes are responsive to asthma-associated inflammatory cytokines and/or other circulating mediators.

## 3.4 Mast Cell Coupling to the KKS: Lessons From Allergic Inflammation

Triggered by systemic degranulation of mast cells and basophils, anaphylactic shock is a life-threatening systemic disorder usually (but not always) triggered by allergens. Following the recognition of allergen by IgE bound to mast cells/basophils via FcεRI, signal transduction leads to the secretion of mast cell granules containing a myriad of preformed inflammatory mediators. Based on vitro studies, MC-derived heparin and polyphosphates were recently characterized as the negatively charged polymers that, upon exposure to plasma, might trigger the release of BK via activation of the contact system (Fig. 3). Using the classical model of passive anaphylaxis, Sala-Cunill et al. (2015) have recently demonstrated that mice deficient in FXII, PK, HK-1, and B2R$^{-/-}$ are protected from allergen-induced hypotension, which was preserved in B1R or FXI. Reconstitution experiments in FXII$^{-/-}$ mice confirmed that purified human FXII rescued allergen-induced hypotension. In a complementary study involving analysis of contact phase factors in plasma from patients that suffered anaphylactic shock, the authors found increased heparin levels in plasma, correlating with increased formation of cleaved/activated form of FXII, PK, and HK. Collectively, these studies suggest that severity of allergic inflammation is worsened as result of MC coupling to BK-induced inflammatory cascades (Fig. 3). As a corollary, it has been suggested that therapeutic drugs targeting contact phase enzymes (e.g., FXII/PK) or B2R (e.g., iticabant) might reduce risk of anaphylactic shock in highly allergic patients.

## 3.5 KKS and CD4$^{+}$T Lymphocytes: Controversial Roles in EAE

In a study involving patients with multiple sclerosis, Prat et al. (1999) reported that T cell migration is negatively regulated by DABK. Using an experimental model of EAE established in the SJL mouse strain,

Schulze-Topphoff et al. (2009) demonstrated that mice treated with a B1R ligand showed reduced CNS infiltration by encephalitogenic Th17 effector cells and the EAE symptoms were ameliorated. While SJL mice benefited from treatment with B1R agonists, studies of EAE in C57BL/6 mice revealed an opposite phenotype: CNS pathology and spatial reference memory deficits were both attenuated in B1R-knockout mice or in wild-type mice treated with B1R antagonists (Dutra et al., 2011, 2013; Göbel et al., 2011). The reasons for the contrasting phenotypes of B1R in EAE established in different mice strains were not elucidated.

# 4. KKS IN INFECTION IMMUNITY

## 4.1 Viral Infections

Sindbis virus (SINV), a prototype member of the Alphavirus genus, is an enveloped, single-strand positive-sense RNA virus that causes murine encephalitis. Mice infected with SINV have been used as a model to study the pathogenesis of human arthropod-borne encephalitis, rash, and arthritis caused by other alphaviruses. Although the role of the KKS in SINV-induced encephalitis is still unknown, in vitro studies revealed that the expression of B2R is upregulated in SIV-infected human brain microvascular cells (Rust et al., 2012). Using BK as the exogenous agonist for B2R, activation of PI3 kinase and ERK pathways blunted SINV-induced host cell apoptosis and increased virus replication in the human brain microvascular cells. Extending the breadth of these culture studies to an intracerebral model of infection in mice, Rust et al. (2012) demonstrated that the B2R antagonist HOE-140 reduced the local viral load. Based on these results, it was proposed that SINV might usurp the KKS to stimulate endothelial cell survival and fuel its own replication in the infected endothelial cells through upregulated expression of B2R.

Although the KKS has not been systematically explored as a therapeutic target in patients infected with hemorrhagic viruses, there is a punctual report indicating that iticabant (B2R antagonist) led to clinical recovery (hemorrhage and kidney syndrome) of a patient infected with Puumala virus (Antonen et al., 2013). Decades ago, it has been reported that the plasma levels of activated PK is increased in patients with moderate or severe hemorrhagic fever associated to renal dysfunction (Sidel'nikov & Obukhova, 1990). Hantavirus has been recently used as a model to study the impact of the KKS in infection by hemorrhagic viruses, which commonly induce edema, hypotension, shock, and coagulation abnormalities (Taylor,

Wahl-Jensen, Copeland, Jahrling, & Schmaljohn, 2013). Using reconstituted capillary blood vessels exposed to Hantavirus in combination to models of endothelial barrier using microvascular cells, these authors demonstrated that Hantavirus infection drives the activation of the KKS (contact pathway). Given the absence of cytopathic effects on Hantavirus-infected microvascular cells, it is unlikely that viral RNA released from injured microvascular cells have acted as "contact" factors in this experimental setting. Considering that PK is directly generated (FXIIa independently) by prolylcarboxypeptidase (Shariat-Madar et al., 2004), it is possible that Hantavirus-infected endothelial cells might upregulate the expression of this matrix-associated serine protease, perhaps accounting for BK release and ensuing breakdown of the endothelial barrier. Alternatively, Hantavirus infection may induce misfolded proteins or exposure of HSP-90, previously implicated in FXII and prekallikrein cleavage, respectively (Joseph, Tholanikunnel, & Kaplan, 2002; Maas et al., 2008). Finally, as pointed out by Taylor et al. (2013), Hantavirus infection may trigger the KKS by enhancing the endothelial expression of the multiple cell surface docking sites for HK, such as heparan sulfate proteoglycans (Renne et al., 2000) or the multimolecular receptor complex consisting of urokinase plasminogen activator receptor (uPAR), cytokeratin 1 (CK1), and the globular heads of the complement C1q receptor (gC1qR) (Kaplan & Joseph, 2014; Mahdi, Madar, Figueroa, & Schmaier, 2002).

## 4.2 KKS- and B2R-Dependent Immunity Against Bacterial Infection

In the late 1990s, meningococcal septic shock and streptococcal toxic shock syndrome were associated to the activation of the contact system (Sriskandan et al., 2000; Wuillemin et al., 1995). Although Gram-positive or Gram-negative bacteria are able to induce hypotension and inflammation through the release of kinins (Potempa & Herwald, 2011), an early work by Fauve and Hévin (1979) provided the first evidences that exogenous kinins, acting as effectors of immunity, reduce bacterial burden in mice infected by *Listeria monocytogenes* and *Salmonella typhimurium*. Years later, Kaman et al. (2009) reported that $B2R^{-/-}$ mice challenged by *L. monocytogenes* displayed a susceptible phenotype, manifested as body weight loss and increased hepatic burden of *Listeria*-correlating with deficient production of IL-12p70 and IFN-γ. Unfortunately, it is not known whether the susceptible phenotype of Listeria-infected $B2R^{-/-}$ mice was caused by deficient generation of TNF-α/NO-producing DCs (Serbina, Salazar-Mather, Biron, Kuziel, &

Pamer, 2003) and/or resulted from impaired formation of effector type-1 $CD8^+$ T cells, as previously described in $B2R^{-/-}$ mice challenged systemically by a different intracellular pathogen, i.e., *Trypanosoma cruzi* (Monteiro et al., 2007). Several years later, Fang et al. (2013) reported that arterial thrombosis is delayed in B2R-deficient mice and ascribed this dysfunction to increased production of prostacyclin and NO. Of further interest, these authors documented that tissue factor expression in vessels was inhibited due to compensatory overstimulation of the angiotensin 1–7/MAS pathway. Based on these findings, it is possible that the impaired thrombogenesis in $B2R^{-/-}$-infected mice might translate into deficient intravascular immunity. Moreover, it is not known whether the upregulated levels of prostacyclin, NO and/or other mediators produced in B2R-deficient mice might have impaired the differentiation of bone marrow-derived monocytes into the immune-protective splenic DCs (Serbina et al., 2003).

It is well established that some hypotensive Gram-positive bacteria strains activate the contact system via LPS or teichoic acid (Potempa & Herwald, 2011). Early biochemical studies identified M-proteins of *Streptococcus pyogenes* as the docking sites for HK (Ben Nasr et al., 1997). In addition, *S. pyogenes* liberates kinins from kininogens via a bacterial cysteine protease designated streptopain (Herwald et al., 1996). Subsequent studies revealed that *Staphylococcus aureus* releases kinins via concerted action between two staphopains (A and B) (Potempa & Herwald, 2011); although staphopain B by itself does not generate a vasoactive peptide from intact kininogens, staphopain A generates a kininogen fragment that is subsequently processed by staphopain B, releasing a slightly longer kinin peptide (Leu-Met-Lys-BK) (Imamura et al., 2005).

During the progression of periodontal diseases, the composition of the subgingival biofilm is gradually enriched with anaerobic Gram-negative bacteria belonging to the red complex group of periodontal pathogens, whereas the facultative Gram-positive bacteria dominate the normal flora. Multiple lines of evidences suggest that chronic infection by red complex bacteria favors bone erosion by osteoclasts as result of aberrant immune responses to bacterial antigens exposed in the inflamed periodontal tissues (Cekici, Kantarci, Hasturk, & Van Dyke, 2014). Efforts to characterize osteoclastogenic T cell subsets have recently converged at Th17, a proinflammatory helper T cell subset whose differentiation depends on the cooperative action of IL-6, TGF-β, and IL-23. Beyond the involvement in periodontitis, systemic infection with *Porphyromonas gingivalis* is also considered a risk factor in the development of atherosclerosis (Cai, Kobayashi, Hashizume-Takizawa, & Kurita-Ochiai, 2014).

Molecular studies have linked *P. gingivalis* virulence and pathogenicity to the expression of several biochemically characterized molecules, such as *Fimbriae*, LPS, cysteine proteases and hemagglutinin, all of which required for bacterial colonization and survival in the inflamed gingival tissues. Innate sentinel cells recognize two major TLR2 ligands of *P. gingivalis*: an atypical LPS and Fimbriae, a multiunit complex composed of polymerized fimbrillin A combined to other accessory proteins. Analysis of NFκ-B-responsive genes in macrophages infected by *P. gingivalis* or treated with purified wall components (*Fimbriae* and LPS) linked the responses to TLR2/MyD88-p38 MAPK pathways (Zhou & Amar, 2007). In another in vitro study, it was shown that *Fimbriae* provide adhesive filamentous appendages for gingival epithelial cells without triggering TLR2-mediated proinflammatory responses, because these particular host cell types do not express the CD14 cofactor required for TLR2 signaling (Eskan, Hajishengallis, & Kinane, 2007). As for *Fimbriae*, this bacterial protein subverts immunity by stimulating TLR2/PI3K-dependent upregulation of CR3, a surface receptor that downmodulates intracellular killing by macrophages (Hajishengallis, Shakhatreh, Wang, & Liang, 2007; Wang et al., 2007). Cysteine proteases from the gingipain family are responsible for 85% of the general proteolytic activity generated by *P. gingivalis* (Potempa & Herwald, 2011). Consisting of arginine-specific gingipains (Rgps; RgpA and RgpB) and a lysine-specific gingipain (Kgp), these proteases promote bacterial virulence through a wide range of molecular mechanisms, such as disruption of plasma clotting, subversion of complement system and neutrophil functions, degradation of antibacterial peptides and host cell surface receptors (Imamura et al., 1995; Travis, Pike, Imamura, & Potempa, 1997). Rather than strictly acting as a virulence factor, gingipains reciprocally stimulate host defenses by inducing inflammation via cleavage of protease-activated receptors (Lourbakos et al., 2001). Evidence of synergism between the atypical *P. gingivalis* LPS (TLR2 agonist) and gingipains was obtained in mice infected intramucosally (oral route) with wild-type *P. gingivalis* vs gingipain-R-deficient mutants (Monteiro et al., 2009). Analysis of the functional profile of Fimbriae-specific CD4+ T responses in submandibular lymph nodes linked Th1 responses (C57BL/6 mice) or mixed Th1/Th17 responses (BALB/c mice) to cross-talk between TLR2/B2R, a transcellular activation pathway initiated by LPS (an atypical TLR2 ligand) and propagated by gingipain R (acting as kinin-releasing enzyme). Using a mouse model of ligature-induced experimental periodontitis, i.e., a procedure that does not involve deliberate injection of periodontal pathogens,

Gonçalves-Zillo et al. (2013) reported that B1R-deficient mice displayed a susceptible phenotype, characterized by increased bone loss and upregulated secretion of the osteoclastogenic IL-17 cytokine. Consistent with their in vivo results, analysis of osteoclast differentiation linked absence of B1R in bone marrow progenitor cells to upregulated development of osteoclasts, suggesting that activation of the kinin/B1R pathway may limit the extent of bone resorption by regulating osteoclast formation and inhibiting IL-17 production.

## 4.3 KKS and Parasitic Infections Caused by Intracellular Protozoa

### 4.3.1 *Chagas Disease*

Afflicting 6–8 million people living in poverty-stricken rural areas of South and Central America, Chagas disease is a pleomorphic clinical entity caused by chronic infection with *T. cruzi*, a parasitic protozoan that displays remarkable intraspecific genetic/phenotypic heterogeneity, whether translated as differences in parasite virulence, tissue tropism, and/or pathogenicity (Tibayrenc, 2010). Following vector transmission of *T. cruzi* by the blood-seeking triatomine insects, this obligate intracellular pathogen undergoes successive cycles of invasion, intracellular replication, and morphogenetic development in susceptible mammalian cells. In the classical mode of disease transmission, the insect-derived infective forms (metacyclic trypomastigotes) initiate the infection by invading susceptible cells present in tissues lacerated by insect proboscis. Following phagocytosis by tissue-resident macrophages and/or endocytic internalization by other cell types (e.g., epithelial cells, fibroblasts, cardiomyocytes, adipocytes, endothelial cells), the parasites rapidly escape from the phagolysosomes, reaching the host cell cytoplasm, where they transform into the oval-shaped amastigotes (replicating forms). After coopting host cell metabolism to support their intracellular replication for several days (Caradonna, Engel, Jacobi, Lee, & Burleigh, 2013), the amastigotes transform into infectious trypomastigotes, which upon host cell death, are released in interstitial spaces, from where the parasites reach the lymphatics or fall in the bloodstream. Although acute Chagas disease is characterized by high blood parasitism, fever, and hepatosplenomegaly, numerous cases of facial swelling have been recently reported in patients (Amazonian region) that were accidentally infected orally, following ingestion of stocks of fresh (nonpasteurized) fruit juices that had been previously macerated with sylvatic bugs contaminated with *T. cruzi* (Pinto, Ferreira, Valente, Harada, & Valente, 2009).

During their navigation through host tissues, the trypomastigotes initiate innate immunity through the shedding of lipid vesicles that are enriched with a glycosylphophatidylinositol-linked anchor (tGPI), a potent TLR2 agonist (Almeida & Gazzinelli, 2001; Campos et al., 2001) (Fig. 4). With the onset of acquired immunity, the blood parasitemia and acute symptoms subside due to combined action of protective antiparasite antibodies and type-1 effector T cells, but the parasites are not eliminated from the immune host. Although the majority of chronically infected patients (~70%) remain asymptomatic, susceptible individuals develop a progressive chronic "chagasic" cardiomyopathy (CCC) (Dutra, Menezes, Magalhães, & Gollob, 2014; Marin-Neto, Cunha-Neto, Maciel, & Simões, 2007) that remains incurable to this date. Although heart parasitism is hardly detectable during the chronic stage of the disease, studies in cohorts of CCC patients and mice models suggest that chronic myocarditis/fibrosis is worsened as result of an immunoregulatory imbalance that translates into upregulated heart infiltration by TNF-α-producing $CD8^+$ T cells (Silverio et al., 2012). Beyond the collateral damage inflicted by these pathogenic subsets of effector T cells, CCC is aggravated by arrhythmias, thromboembolism, and poorly understood microvascular abnormalities (Higuchi et al., 1999). Years later, Tanowitz et al. (2005) demonstrated that ET-1, a potent vasoconstrictor and profibrosing polypeptide upregulated in *T. cruzi*-infected cardiomyocytes, fuels myocardial fibrosis in chronically infected mice. Although the functional interplay between ET-1 and the KKS was not investigated in the context of CCC, there is precedent that the prooxidative ET-1 and angiotensin II upregulate B1R expression by cardiovascular cells (Morand-Contant, Anand-Srivastava, & Couture, 2010). The possibility that the ET-1/kinin/B1R axis worsens heart fibrosis is worth exploring in view of recent evidences that chronic myocarditis and heart fibrosis are markedly attenuated in $B1R^{-/-}$ mice (unpublished observations), somewhat reminiscent of the cardioprotective phenotype which these mutants displayed in streptozotocin-induced diabetic cardiomyopathy (Westermann et al., 2009). Among several other potential explanations for the pathogenic role of B1R, it is possible that formation of des-Arg-kinins in the infiltrated heart tissues might induce a low-grade (but sustained) extravasation of plasma, which in turn drives interstitial fibrosis via the mast cell/MMP2-dependent pathway. Albeit not exclusively, B1R may also support the transendothelial migration of TNF-α-producing infiltrating effector $CD8^+$ T cells into the parasitized heart.

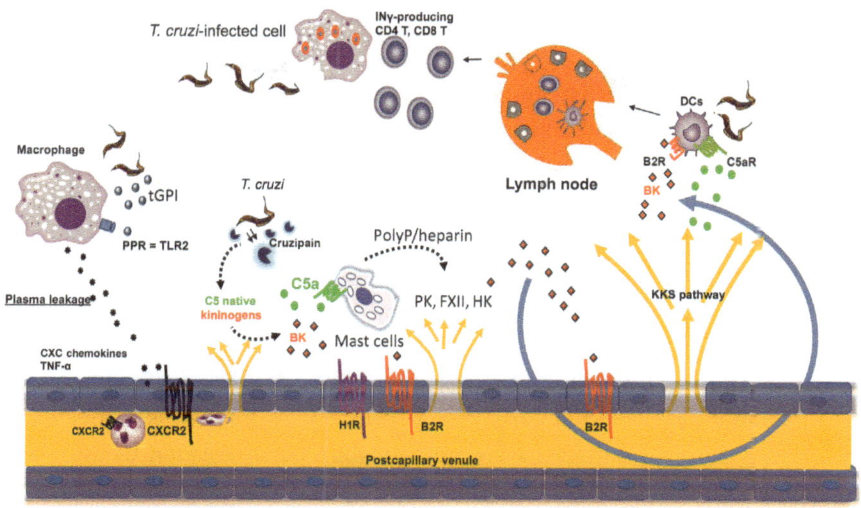


**Fig. 4** Microvascular leakage is coupled to type-1 adaptive immunity in experimental *T. cruzi* infection. The panel illustrates the dynamics of inflammation evoked by *T. cruzi* infective forms (trypomastigotes). Upon rupture of parasitized host cells (*upper side/left*), the interstitial spaces are swarmed with the flagellated parasites. During their navigation, innate sentinel cells (e.g., macrophages) are activated by tGPI (TLR2 ligand) contained in lipid vesicles that are continuously shed by the parasites. Following secretion of chemokines/cytokines (e.g., TNF-α), activation of CXCR2-expressing neutrophils/endothelium leads to a subtle destabilization of the endothelial barrier, allowing for the diffusion/accumulation of plasma proteins (including kininogens and complement components) in the parasite-laden tissues. Upon association to sulphated proteoglycans, HK becomes sensitive to the kinin-releasing activity of cruzipain, the major lysosomal-like cysteine protease of *T. cruzi*. Tightly regulated by the kinin-degrading activity of ACE (not represented in this scheme), the inflammatory response initiated by PRR (TLR2) is amplified by the vasoactive activity of kinins (via activation of endothelial B2R) and C5a anaphylatoxin (Schmitz et al., 2014). Perivascular MCs, acting at the endothelial interface, fuel the proinflammatory phenotype of *T. cruzi* via activation of the KKS (contact) pathway (Nascimento et al., 2017), reminiscent of mechanisms described in severe allergy. As the inflammation is propagated, the released kinins (acting jointly with C5a) activate antigen-bearing DCs via the B2R/C5aRI pathway. After reaching the lymphatics, these APCs coordinate IL-12-dependent development of type 1 effector T cells (Monteiro et al., 2006; Schmitz et al., 2014).

#### 4.3.1.1 *T. cruzi*-Evoked Microvascular Leakage and Th1 Development

In a previous section, we have briefly mentioned that Fimbriae-specific Th1 effector development in *P. gingivalis*-infected C57BL/6 mice is preceded by the activation of TLR2/B2R, a transcellular cross talk forged by the cooperation between LPS and gingipain, respectively, acting as an atypical TLR2 ligand- and kinin-releasing protease (Monteiro et al., 2009). Prior to this report, studies in peripheral (subcutaneous) models of *T. cruzi* infection provided the first precedent that endogenously released kinins, acting as a Th1-directing adjuvant, upregulate the production of *T. cruzi*-specific effector T cells (type-1) in regional lymph nodes (Monteiro et al., 2006; Schmitz et al., 2014). Insight on the underlying mechanisms emerged from intravital microscopy studies in the hamster cheek pouch, used in combination to mouse models of *T. cruzi* infection (footpad). Briefly, we found that lipid anchors (TLR2 agonists) shed by tissue culture trypomastigotes initiate neutrophil-dependent inflammation via joint activation of CXCR2 (Schmitz, Svensjö, Serra, Teixeira, & Scharfstein, 2009) and endothelin receptors (ETaR/ETbR) (Andrade et al., 2012). A subtle leakage induced by neutrophils activated at early stages of infection (Fig. 4) is sufficient to enable the diffusion of plasma proteins through parasite-laden tissues. Further downstream, the major cysteine protease of *T. cruzi* (cruzipain) cleaves kininogens and native C5, generating kinins and C5a (Monteiro et al., 2006; Schmitz et al., 2014). Following MC degranulation, plasma leakage is drastically enhanced due to the extravascular activation of the KKS (Nascimento et al., 2017). (Fig. 4) whereas ACE, acting as a regulator of BK-induced cascades, blunts *T. cruzi*-induced inflammation (Scharfstein et al., 2013). Interestingly in this context, studies in *T. cruzi*-infected mice revealed that ACE inhibitors potentiated Th1 responses in regional lymph nodes (Monteiro et al., 2006; Schmitz et al., 2014) by inhibiting the connection between innate immunity (TLR2 sensing of *T. cruzi*) and the proteolytic cascades (KKS/Complement) that, acting further downstream, propagate the inflammatory response (Fig. 4).

#### 4.3.1.2 The Immunological Role of B2R in Systemic *T. cruzi* Infection

Differently from the studies in models involving peripheral infection, which required inflammation/endothelial barrier to enable the influx of plasma contact factors and complement components through extracellular spaces, the sinusoids of liver and spleen are lined by fenestrated endothelium. As such, circulating parasites navigating through the hepatic and spleen stroma might be faced with cell-bound HK, whether displayed by tissue-resident macrophages, DCs, and/or other cell types dispersed in the tissue stroma. Interestingly in this context, in vitro studies demonstrated that *T. cruzi* trypomastigotes upregulated IL-12p70 and costimulatory molecules (CD40, CD86, CD80) in $CD11c^{+}$ $B2R^{+/+}$ DCs (splenic) via the cruzipain/kinin/B2R pathway while failing to induce maturation of $B2R^{-/-}$ DCs (Monteiro et al., 2007). Consistent with the maturation dysfunction of

$B2R^{-/-}$ DCs, we found that $B2R^{-/-}$ mice acutely infected by *T. cruzi* succumbed, the susceptible phenotype being ascribed to the progressive decline in the formation of INF-γ-producing (immunoprotective) effector T cells in the spleen, which was coupled to a sharp rise of antigen-specific IL-17-producing T cells. Importantly, the adaptive transfer of immature (splenic) $CD11c^{+}$ $B2R^{+/+}$ DCs restored the type-1 responses of the recipient $B2R^{-/-}$ mice, while reciprocally suppressing the formation of IL-17-producing T cells (Monteiro et al., 2007). Recently, it has been reported that patients with chronic Chagas disease display a hypercoagulative phenotype (Pinazo et al., 2016). Although the levels of circulating parasites are extremely low during the chronic phase of human Chagas disease, it is conceivable that monocytes activated by type-1-expressing $CD4^{+}$ and $CD8^{+}$ T cells acquire a prothrombogenic phenotype, which in the long run may account for the hypercoagulative phenotype described in chronically infected patients. During the acute phase, however, it is well known that the blood parasitemia rises before subsiding at the face of antimicrobial immunity. Although not explored in our published studies, it will be interesting to know if circulating trypomastigotes trigger immunothrombosis, being transiently retained in the delicate fibrin scaffolds generated intravascularly. At the time of the publication of our study (Monteiro et al., 2007), we were not aware that the global ablation of B2R or PK genes resulted in delayed thrombosis in mice (Fang et al., 2013; Schmaier, 2016). As previously discussed, these authors reported that B2R-deficient mice upregulate the angiotensin 1–7/MAS pathway, the dysfunction being associated to increased formation of prostacyclin and inhibition of tissue factor expression in arterial vessels. Considering that $B2R^{-/-}$ mice is increasingly susceptible to acute systemic challenge by at least three pathogens (*T. cruzi*, *L. chagasi*, and *L. monocytogenes*), it is plausible that the primitive fibrin-dependent barrier forged by immunothrombosis was somewhat compromised in the mutant mice, consequently facilitating the systemic dissemination of these pathogens to the liver and spleen. Noteworthy, the finding that adoptive transfer of $B2R^{+/+}$ DCs conferred resistance to the $B2R^{-/-}$ mice and rescued development of *T. cruzi*-specific (immunoprotective) type-1 effector T cells, while conversely suppressing Th17 responses (Monteiro et al., 2007) implies that the APC functions of the adoptively transferred splenic $B2R^{+/+}$ DCs (i.e., sampling of *T. cruzi* antigens, antigen processing/presentation of MHC–peptide complexes and upregulated production of Th1 polarizing cytokines) were not impaired by the systemic alterations originally described in normal

B2R$^{-/-}$ mice. This notwithstanding, we cannot rule out the possibility that prostacyclin, NO, and/or any other systemic compensatory responses observed in normal B2R$^{-/-}$ mice might have impaired the development of bone marrow-derived B2R$^{-/-}$ monocyte precursors into splenic CD11c$^{+}$ B2R$^{-/-}$ DCs (Serbina et al., 2003).

#### 4.3.1.3 GPCRs as Cellular Gateways for *T. cruzi* Invasion

Subjected to intense selective pressure by antiparasite immunity, *T. cruzi* has evolved multiple strategies to survive in the hostile inflammatory environment. Although the studies in *T. cruzi*-infected WT and B2R$^{-/-}$ mice have linked the proinflammatory KKS/B2R pathway to the generation of immunoprotective type-1 effector T cells, in a long-lasting infection such as Chagas disease, patients may not experience drastic/prolonged changes in KKS homeostasis, such as observed in mice in which B2R was globally ablated. Instead, subtle changes in KKS function may occur as the infection progresses from the asymptomatic form to clinically active chronic heart disease. If we accept at face value the premise that a few parasitized heart cells sporadically rupture, releasing large numbers of trypomastigotes to interstitial spaces, we may predict that inflammatory cues (of microbial or endogenous origin) may increase the microvascular permeability of the surrounding capillaries, leading to the extravasation of plasma. The concept that *T. cruzi* trypomastigotes may increment their infectivity at expense of the formation of inflammatory edema (Scharfstein, Andrade, Svensjö, Oliveira, & Nascimento, 2013; Scharfstein et al., 2000) was initially inspired by a biochemical finding: the characterization of cruzipain as kinin-releasing cysteine protease (Del Nery, Juliano, Lima, Scharfstein, & Juliano, 1997; Lima et al., 2002). Differently from gingipains—bacterial cysteine proteases that release kinins from soluble forms of kininogens—the *T. cruzi* cysteine protease (a member of the papain-like superfamily) is rapidly inactivated by the cystatin-like domains present at the terminal end of soluble kininogens (Stoka et al., 1995). However, cruzipain efficiently release kinins when the parasites interact with HK molecules that are tethered on docking platforms formed by heparan sulfate proteoglycans (Lima et al., 2002). Once bound to sulfated proteoglycans, the cystatin-like inhibitory sites of HK are masked, rendering HK sensitive to the kinin-releasing activity of cruzipain (Lima et al., 2002). Consistent with this interpretation, studies of the outcome of *T. cruzi* interaction with endothelial cells or B2R-transfected cells revealed that cruzipain potentiated parasite invasion via activation of the HK/kinin/B2R pathway (Scharfstein et al., 2000). Of note, natural cysteine protease inhibitors, such as cystatins and $\alpha_2$-macroglobulin, did not prevent parasite invasion of host cells via the kinin/B2R pathway. Based on these results, it was suggested that cruzipain cleaves surface-bound HK, releasing kinins (B2R agonists) within a "synapse"-containing B2R, presumably formed by the juxtaposition of host cells and the parasites' plasma membranes. In a series of complementary studies adding trypomastigotes to LPS-activated endothelial cells and neonatal mouse cardiomyocytes, we documented that parasite invasion was potentiated via cross talk between the upregulated B1R and B1R (Todorov et al., 2003). More recently, the high-order complexity of protein/protein interactions modulating GPCR signaling during host cell invasion was highlighted in culture studies performed with human smooth muscle cells (steady state) and cardiomyocytes: in both systems, *T. cruzi* potentiated host cell invasion via mechanisms involving interdependent activation of B2R and endothelin receptors (ETaR/ETbR) (Andrade et al., 2012; Scharfstein et al., 2013). Using an intracardiac model of infection, we recently demonstrated that mice pretreated with specific antagonists of B2R, or ETaR/ETbR displayed markedly reduced heart parasitism and ameliorated severity of myocarditis/fibrosis (Nascimento et al., 2017). In the same study, we provided evidences that *T. cruzi* trypomastigotes fueled their infectivity by generating kinins and endothelins intracardiacally via activation of the MC/KKS (contact) pathway (Nascimento et al., 2017).

### 4.3.2 KKS in Experimental Models of Leishmaniasis

#### 4.3.2.1 Immunity at the Interface Between the Sandfly Vector and *Leishmania*

Transmitted through the bite of infected phlebotomine sandflies, the parasitic protozoan of genus *Leishmania* are the etiologic agents of cutaneous and visceral Leishmaniasis, a group of Neglected tropical diseases prevalent in 88 countries, afflicting approximately 12 million people and responsible for 2 million annual cases (http://www.who.int/leishmaniasis/burden/en/). Involving more than 20 species and subspecies of *Leishmania*, the geographic distribution of these parasites is determined by the availability of competent sandfly vectors. Pathological manifestations are also influenced by the diversity of sandfly saliva (both composition and exposure) and quality/intensity of host immune response. Additional factors, such as the microbiome in the local of sandfly bites, may also influence the outcome of *Leishmania* infection (Andrade, Carvalho, & Rocha, 1990; Lopes, Carneiro, dos Santos, & Vieira, 2016; Salgado et al., 2016).

The life cycle of all *Leishmania* species involves two morphological forms, the amastigotes and the promastigotes. Mammalian infection is initiated when the sand flies regurgitate following blood feeding, releasing infectious forms (metacyclic promastigotes) in superficial layers of the dermis. At this time, the mechanical action of the sandfly proboscis provokes lacerations of blood capillaries in the host. Owing to the combined action of multiple antihemostatic factors and vasodilating substances present in the sandfly saliva, the hemorrhages are for a while intensified at the site of proboscis (Ribeiro, 1987). Current knowledge of sandfly transmission of *Leishmania* suggests that promastigotes deposited in the dermis are either engulfed by neutrophils, dermal macrophages, or dermal DCs (Kaye & Scott, 2011). Sandfly salivary components also have a direct capacity to modulate local and perhaps systemic immune responses. Considerable progress has been obtained at defining the mechanisms by which *Leishmania* subverts immunity. It is currently believed that, soon after sandfly proboscis, neutrophils recruited to primary site of infection internalize *Leishmania major* promastigotes (Peters & Sacks, 2009). Establishment of productive infection depends on the parasite ability to escape from neutrophil killing by blocking oxidative burst and entering a nonlytic compartment that does not fuse with lysosomes (Laufs et al., 2002). Although some species of *Leishmania* are sensitive to the microbicidal function of NETs (Guimaraes-Costa et al., 2009), there are examples of parasites that subvert this innate defense response, such as *Leishmania donovani (L. donovani)* (Gabriel, McMaster, Girard, & Descoteaux, 2010)—etiologic agent of visceral leishmaniasis. Studies of the outcome of intracellular parasitism in macrophages revealed that efferocytosis of apoptotic neutrophils critically influences the fate of *L. major* (Filardy et al., 2010).

As the *Leishmania* infection progresses, immunity is gradually reinforced by the recruitment of IFN-γ-producing Th1 effector T cells to infected tissues. Exemplifying the complexity of the interplay between the selective forces that operate at the primary site of infection shortly after sandfly proboscis, several studies demonstrated that immunity against sandfly saliva antigens has impact on the establishment of *Leishmania* parasitism. For example, the increased blow flow associated to delayed-type hypersensitivity (DTH) responses to sandfly saliva antigens facilitate blood feeding by the sandfly *Phlebotomus papatasi* (Belkaid et al., 2000). Reciprocally, Aquino et al. (2010) reported that individuals that underwent seroconversion to sandfly saliva antigens were more prone to develop efficient DTH against the parasite. Although the composition of sandfly saliva varies as a function of age and feeding pattern of the insects (Prates et al., 2008), maxadilan (a potent vasodilator) and hyaluronidase are among the first of such molecules capable of interfering with *Leishmania* survival at the early stages of parasite interaction with the vertebrate host (Brodie, Smith, Morris, & Titus, 2007; Lerner, Ribeiro, Nelson, & Lerner, 1991; Volfova, Hostomska, Cerny, Votypka, & Volf, 2008). Gradually, additional mediators from the sandfly saliva intervene, inhibiting coagulation and inflammatory responses (Fig. 5). The existence of a complex interplay between *Leishmania*, sandfly saliva, and the immunocompetent host was illustrated by studies in mice showing that immunity against sandfly saliva extracts modulate the outcome of *Leishmania* infection, either enhancing disease progression (Belkaid et al., 1998) or protecting against *Leishmania* infection (de Moura et al., 2013; Gomes et al., 2016). Studies with *Leishmania*-infected neutrophils revealed that the infection-promoting activity of the sandfly saliva extract is neutralized by antibodies raised against saliva antigens (Prates et al., 2011). As a corollary, we may predict that neutrophils from individuals that produce high-titers of serum antibodies to the sandfly saliva antigens might develop more potent leishmanicidal functions, hence reinforcing immune resistance against cutaneous leishmaniasis (Fig. 5).

Although NETs can efficiently kill several extracellular pathogens in vitro, *Streptococcus pneumoniae* expresses a DNase-1 that subverts the microbicidal activity of NETs through the inhibition of reactive oxygen species (ROS) formation independently of phagocytosis (Beiter et al., 2006; Munafo et al., 2009). For similar reasons, *Leishmania* promastigotes subvert the microbicidal function of NETs/DNA (Oehmcke et al., 2009) through Lundep, a DNase expressed in the salivary glands of *Lu. longipalpis* Chagas et al. (2014). Using a mouse model, these workers showed that passively transferred antibodies to Lundep decreased blood feeding by the sandfly. This implies that in normal animals (i.e., in the absence of anti-Lundep antibodies) the viscosity of the blood is decreased as result of Lundep-dependent degradation of DNA/NETs (hence blunting fibrin formation via activation of the contact system). Notably, the authors also showed that at 9 weeks p.i., there was a marked reduction of tissue parasitism in naïve mice that were coinjected with *L. major* promastigotes and Lundep (or saliva extracts). Since the half-life of rLundep in vivo is fairly brief, it is conceivable that the insect mediator has fueled tissue parasitism by suppressing and/or down modulating Th1 induction at early stages of *L. major* infection. Given indications that BK acts as Th1-directing endogenous adjuvant (Aliberti et al., 2003), it is possible that rLundep has

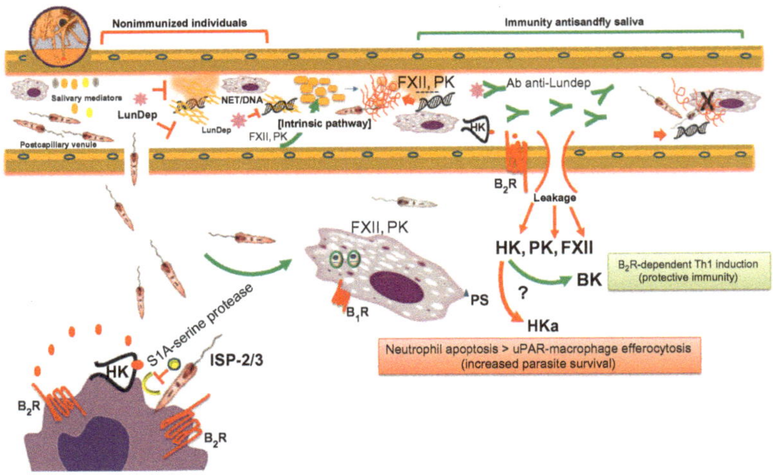


**Fig. 5** KKS sandfly saliva and immunity in Leishmaniasis. Following sandfly inoculation of Leishmania (*upper left*), saliva mediators diffuse through the parasite-laden dermal tissues. The scheme shows that *L. major* promastigotes might take advantage of the fact that Lundep, a DNAse that inhibits the contact system of coagulation, degrades DNA associated to NETs (expelled by activated neutrophils) (Chagas et al., 2014). Other saliva factors might inhibit platelet activation, hence reducing blood exposure to long-chain polyP, another potent prothrombogenic trigger of the contact system (not shown in the scheme). By inhibiting the intravascular formation of fibrin scaffolds (otherwise induced by DNA/NETs), Lundep might facilitate Leishmania spread through damaged tissues. At the right side of the panel (*upper*), the prothrombogenic activity of DNA/NETs (contact system dependent) is enhanced in individuals that produce high titers of Lundep-specific antibodies following natural exposure to insect saliva or after immunization with Lundep (*center/right side* of panel). Leishmania retention within fibrin scaffolds may limit pathogen spread via the bloodstream and/or proximal dermal tissues. After reaching deeper areas of the dermis (*center*), the insect-derived promastigotes are either phagocytosed by tissue-infiltrating neutrophils (*center of panel*) or by resident macrophages (*bottom, left*). During their internalization by macrophages, *L. major* promastigotes rely on the membrane-bound ecotin-like proteínas (ISP2/3) to limit extent of phagocytosis. Studies with ISP2/3 mutants suggest that wild-type *L. major* blunts B2R-driven innate immunity by inhibiting the kinin-releasing activity of S1 serine proteases. In the absence of B2R activation, the phagocytic uptake of *L. major* promastigotes is dominantly dependent on CR3, a receptor that facilitates the intracellular survival/replication of *L. major*.

indirectly blunted DC activation via B2R. Admittedly, it is also possible that long-term immunoregulatory effects resulting from rLundep-mediated inhibition of the contact pathway at early stages of *L. major* infection (Chagas et al., 2014) might have blunted the activation of the RAGE/TLR pathway by inhibiting the formation of oxidized HGMB1, a prototypic "danger-associated motif pattern" (Fig. 5). Future studies in animal models of infections may clarify whether activation of contact system by NETs (DNA) might stimulate DC-driven Th1 responses in *L. major* infection by forging the cross-talk between the proinflammatory HGMB1/TLR/RAGE and the BK/B2R pathways.

#### 4.3.2.2 KKS and Microvascular Leakage Distinguishes *Leishmania* Species

Although saliva mediators transiently inhibit hemostasis in the site of sandfly transmission, the flagellated promastigotes navigate away from the site of inoculation before being engulfed by neutrophils. Intravital microscopy in the hamster cheek pouch revealed marked intra-species differences in the proinflammatory phenotypes of promastigotes. Whereas the topically applied *L. major* promastigotes (etiologic agent of cutaneous leishmaniasis) elicited a robust kinin/B2R-driven plasma leakage in the hamster cheek pouch (Svensjö, Almeida, Vellasco, Juliano, & Scharfstein, 2014), fairly milder responses were observed when the microvascular beds were exposed to *L. donovani* or *L. chagasi* promastigotes (Svensjö et al., 2006). Interestingly, upon addition of ACE inhibitors to the superfusate, there was a B2R-dependent potentiation of microvascular leakage in tissues exposed to visceralizing species of *Leishmania* (Svensjö et al., 2006), suggesting that the released kinins (B2R agonists) were swiftly degraded by ACE. Involving activation of B2R, the kinin responses elicited by *L. donovani* peaked 15 min after topical application of the pathogen and decreased sharply thereafter. The extended timing of *Leishmania*-evoked leakage contrasts with the fast response induced by topical application of BK or histamine, which typically peaked after 2 min, but not later than 5 min and also differed from the vascular leakage observed after ischemia/reperfusion (peak value at 10 min). *L. chagasi* promastigotes elicited a somewhat delayed B2R-dependent leakage response (peaking at 18 min) as compared to *L. donovani*. Of note, the *Leishmania* promastigotes consistently induced rolling/adhesion of leukocytes in postcapillary venules soon after delivery to the tissue stroma, suggesting that mast cells or other innate immunity cells rapidly sensed these pathogens, releasing inflammatory cues that destabilize the endothelial barrier. As explained earlier in this chapter, tissue culture-derived *T. cruzi* trypomastigotes initiate neutrophil-dependent plasma leakage via activation of the tGPI/TLR2 pathway (Monteiro et al., 2006; Schmitz et al., 2009). Based on findings obtained in other experimental systems, innate sensing of visceralizing *L. infantum* is either mediated by C-type lectins SIGNR3, dectin1, or mannose receptor (Kaye & Beattie, 2016). Whereas SIGNR3 favored parasite survival through selective modulation of the LTB4/IL-1β axis, recognition by dectin1, and mannose receptor resulted in upregulated ROS production (Lefèvre et al., 2013). Irrespective of the nature of mechanisms implicated in the initial destabilization of the endothelial barrier, it is interesting that inflammation evoked by *L. chagasi* promastigotes (B2R dependent) was blunted upon parasite treatment with N-Pip-F-hF-VSPh (K11777) (Svensjö et al., 2006), an irreversible cysteine proteinase inhibitor that targets cruzipain (Doyle, Zhou, Engel, & McKerrow, 2007). So far, it is unknown whether *L. chagasi* promastigotes are empowered with cruzipain-like kininogenases (Del Nery et al., 1997; Lima et al., 2002). Generation of null mutants in *L. donovani* was so far restricted to cathepsin B, a lysosomal-like cysteine protease implicated in the processing of a wide range of secreted proteins, including those involved in oxidation–reduction, cell death, and protein turnover (Gerbaba, Gedamu, Isermann, Bruchhaus, & Tomás, 2013).

#### 4.3.2.3 Kinins Drive Macrophage Phagocytosis and Leishmanicidal Responses

Although neutrophils are known to rapidly engulf sandfly *Leishmania* promastigotes deposited in the dermis, a fraction of these flagellated parasites might target tissue-resident macrophages. In a study of the role of kinins in the outcome of *Leishmania* interaction with resident macrophages, Svensjö et al. (2006) observed that the phagocytic uptake via the kinin/B2R pathway was conditioned by the presence of ACE (CD143) inhibitors in the cultures. Consistent with B2R-dependent phagocytic uptake, addition of exogenous BK or HK to serum-free cultures supplemented with ACE inhibitors led to dose-dependent increase of macrophage uptake of *L. chagasi*. Mirroring the anti-inflammatory effects mediated by inhibitors of parasite cysteine proteases, the active site targeting of these enzymes inhibited B2R-dependent uptake of *L. chagasi* by the macrophages. Intriguingly, although HOE-140 decreased the number of promastigotes internalized by the macrophages, analysis of the fate of the parasites (72 h p.i.) revealed that increased outgrowth/survival in spite of the fact that fewer amastigotes were initially internalized by the macrophages. These results suggested that activation of the kinin/B2R pathway at early stages of the phagocytic response harness the microbicidal power of macrophages. Consistent with this, addition of HOE-140 (in the absence of ACE inhibitor) to cultures of thioglycollate (TG)-elicited inflammatory macrophages revealed that the B2R antagonist robustly increased the intracellular survival/replication of *L. chagasi*. Based on these results, we concluded that the activation threshold of B2R (microbicidal response) is lowered in TG-induced macrophages (microbicidal responses) as compared to normal macrophages (Svensjö et al., 2006). Extending the breadth of these studies to *T. cruzi*, we subsequently demonstrated that activation of B2R links phagocytic uptake of the pathogen to intracellular activation of trypanocidal responses. Further highlighting the complexity of protein/protein interactions during macrophage internalization, we demonstrated that *T. cruzi* phagocytosis involves cross talk between at least two GPCRs, i.e., B2R and C5aR1 (Schmitz et al., 2014).

The examples described earlier came from studies of macrophage interaction with two wild-type species of pathogenic trypanosomatids (*L. chagasi* and *T. cruzi*), both of which promote visceral infection (liver and spleen). Surprisingly, a markedly different phenotype was observed when we analyzed the role of the kinin/B2R pathway in macrophage interaction with *L. major*, a species that causes cutaneous leishmaniasis (Svensjö et al., 2014). Unlike the phenotype of the visceralizing parasites, the phagocytic uptake of *L. major* promastigotes by macrophages was not potentiated by ACE inhibitors nor was it modulated by B2R antagonists (Svensjö et al., 2014). Remarkably, however, a clear-cut phagocytic dependence of B2R was observed when mouse macrophages were challenged in vitro with *L. major* promastigotes mutants that lacked ecotin-like inhibitors of serine peptidases (ISPs) (Svensjö et al., 2014) (Fig. 5). To understand the rationale behind these experiments, it is important to briefly recapitulate the phenotypic characteristics of *L. major* ($\Delta$*isp2/3*), the genetically modified lines of parasites deficient of ISP-2/3 originally developed by Mottram, Lima, and coworkers (Eschenlauer et al., 2009; Faria et al., 2011). Previous functional studies with the archetype of the family, *Escherichia coli* ecotin (Lee et al., 1991), revealed that this inhibitor targeted NE, a member of the trypsin-fold S1A-family serine peptidases. Since leishmania do not express S1A-family serine proteases, Faria et al. (2011) sought to determine whether ISPs could facilitate parasite survival in macrophages by targeting elastase or other S1A-family proteases. Interestingly, in vitro studies performed with ISP-2/3 leishmania revealed that macrophages internalize these parasite mutants in vitro (via CR3) far more efficiently than wild-type leishmania (Faria et al., 2011). Furthermore, the enhanced phagocytosis of ISP-2/3 deficient leishmania was associated to enhanced killing. As reported by Faria et al. (2011), molecular studies linked the enhanced leishmanicidal response of macrophages to cross-talk between leukocyte elastase (NE) and the microbicidal TLR4/PKR/TNF-α/IFN-β axis.

While not excluding leukocyte elastase as a potenical target of ISP-2/3, Svensjö et al. (2014) reported that the kinin/B2R pathway also contributes to the intriguing phenotype of ISP2/3 mutants this intriguing phenotype. Using classical pharmacological tools (HOE-140 and ACE inhibitors) to manipulate the kinin/B2R pathway at early stages of *L. major* interaction with mouse macrophages, these experiments showed that B2R blockade reverted (to normal levels) the increased phagocytic uptake of the ISP-2/3-deficient parasites (Fig. 5). Based on these results, we inferred that wild-type *L. major* promastigotes exploits the membrane-bound ISP-2/3 as a strategy to inhibit the pericellular release of kinins (B2R agonists) by S1A-serine proteases. As an attempt to investigate the in vivo role of ISP-2/3, Svensjö et al. (2014) compared the proinflammatory phenotype of wild-type *L. major* promastigotes vs Δ*isp2/3* mutants in the hamster cheek pouch model of topical infection. Intriguingly, the intravital microscopy studies revealed that both parasite lines elicited equivalent (potent) microvascular leakage via activation of kinin/B2R pathway (in the absence of ACE inhibitors). This implies that the membrane-bound ISP2/3 do not interfere with the propagation of the kinin/B2R-dependent proteolytic cascade (Fig. 5). Although speculative, we may predict that the robust plasma leakage induced by the *L. major* promastigotes (wild-type or ISP2/3 mutants) might have a dual role on host/parasite balance: on one hand, the released kinins may harness innate immunity by stimulating the microbicidal function of neutrophils and macrophages and/or by reprogramming $B2R^{+/+}$ DCs. Conversely, the diffusion of plasma proteins through the extravascular spaces may propitiate the binding of HK to phosphatidylserine (PS) residues exposed on apoptotic cells. Following PK-dependent cleavage, the two chain glycoprotein HKa (free of BK) may act as an opsonin (Fig. 5), promoting urokinase plasminogen receptor-dependent efferocytosis by macrophages (Yang et al., 2014). Interestingly in this context, it has been reported that clearance of apoptotic neutrophils induces an $IL\text{-}12^{low}IL\text{-}10^{high}$ regulatory phenotype in macrophages (Filardy et al., 2010). In summary, although *L. major* promastigotes induced robust plasma leakage, the expression of the membrane-bound ISP-2/3 did not interfere with the fluid-phase propagation of inflammation via the kinin/B2R pathway.

Instead, the membrane-bound ISP2/3 of *L. major* promastigotes target the kinin-releasing activity of S1a-family proteases displayed at the macrophage surfaces (Svensjö et al., 2014). Using this strategy, the vector-transmitted *L. major* may limit paracrine signaling of the prooxidative kinin//B2R pathway without interfering with the fluid-phase activation of the KKS (Fig. 5).

#### 4.3.2.4 The Role of B2R in the Mouse Model of Visceral Leishmaniasis

As recently reviewed by Kaye and Beattie (2016), the histopathological profile of human visceral leishmaniasis (VL) has a peculiar organ-specific dichotomy that is also manifested in mice models: in the liver the presence of inflammatory granuloma correlates fairly well with the immune status and tissue parasite burden; however, this relationship is less clearly defined in the disorganized spleen structure. For example, in canine VL, symptomatic dogs show the absence of hepatic granuloma associated to higher parasite burdens whereas asymptomatic dogs conversely display robust hepatic granuloma formation and low parasite burden enriched by effector T cells and activated macrophages/DCs. The presence of epitheliod granulomas in asymptomatic patients is associated to low tissue parasitism, a feature that was also observed in the mouse model of VL (Kaye & Beattie, 2016).

Intravital microscopy provided a wealth of information about the relationship between infection, inflammation/immunity, and formation of hepatic granuloma (Beattie et al., 2010). Although the mouse model of VL has a serious drawback-requiring intravenous injection of massive numbers of visceralizing parasites, important clues emerged from analysis of the dynamics of granuloma formation in the liver. As mentioned earlier in this chapter, Kupffer cells originating from macrophage precursors derived from embryonic hematopoiesis in the yolk sac (Perdiguero & Geissmann, 2016). Strategically lined in the liver sinusoids, Kupffer cells continuously extend and retract lamellipodia, and upon contact with amastigotes, passively internalize these parasites into vacuoles (Beattie et al., 2013, 2010). Interestingly, this early phagocytic event does not result in massive recruitment of neutrophils to the sinusoids, suggesting that liver infection starts rather silently, most likely reflecting the paucity of PAMPs expressed by the *Leishmania* (Kaye & Beattie, 2016). As the liver infection progresses, a nascent granuloma is formed, forged by mononuclear cell recruitment, repositioning of resident Kupffer cells (infected or not) and fusion/multinucleation. Studies involving antibody-dependent neutrophil depletion in mice revealed that these leukocytes coordinate different organ-specific functions, such as promotion of parasite clearance from the spleen/bone marrow and fueling formation of the inflammatory granuloma in the liver (McFarlane et al., 2008). Hepatic granuloma limits the extent of Kupffer cell presentation of MHC-class I/peptide complexes to immunoprotective effector $CD8^+$ T cells (Beattie et al., 2010; Kaye & Beattie, 2016). In addition, it has been reported that targeting checkpoint regulators reinforces protective immunity. For example, antibodies that interfere with CD86–CTLA4 interactions enhanced Th1 responses and granuloma formation (Murphy, Cotterell, Gorak, Engwerda, & Kaye, 1998). More recently, the control of intracellular parasitism was associated to specific transcriptomic profiles of uninfected macrophages activated by inflammatory cues (Beattie et al., 2013). Considering that the liver and spleen sinusoids are lined by nonfenestrated endothelium, we may visualize a scenario in which circulating leukocytes, including activated neutrophils/monocytes recruited to sinusoid, are constantly exposed to plasma proteins, including contact system factors. Viewed from this perspective, it is interesting that $B2R^{-/-}$ mice challenged intravenously with a high dose of *L. chagasi* amastigotes displayed a susceptible VL phenotype Nico et al. (2012). At 30-day postinfection, the mutant mice displayed increased spleen and liver weight relative to whole-body weight and showed elevated parasite burden in the liver. Importantly, the worsening of pathogenic outcome was correlated with weakened DTH responses, a classic surrogate marker of protective immunity in this model, In addition, *L. chagasi*-infected $B2R^{-/-}$ mice decreased production of IFN-γ-producing T cells in the spleen, and presented lower serum titers of IgG2B antibodies to *Leishmania* antigen. More recently, a report indicating that $B2R^{-/-}$ mice (naïve) have delayed arterial thrombosis due to upregulated production of prostacyclin and reduced tissue factor expression (Schmaier, 2016) raises the possibility that the susceptible phenotype of the mutant mice is (at least in part) reflecting deficient retention of visceralizing parasites within fibrin scaffolds, reminiscent of intravascular immunity against bacterial pathogens (Engelmann & Massberg, 2013).

#### 4.3.2.5 Data Mining: Preliminary Search for Transcriptional Changes in KKS Components in Human Leishmaniasis

Visceralization of *Leishmania* in humans is more likely dependent on the migration of infected myeloid cells to the liver and spleen than free-living parasites, and the timing of granuloma formation in the liver of patients is hard to predict. In *Rhesus* monkeys infected with a highly virulent strain of *L. infantum* strain, the formation of poorly differentiated hepatic granulomas containing aggregated macrophages and lymphocytes was only observed at 6 weeks p.i., shortly before visualization of amastigotes within a granuloma formed by multinucleated cells (Kaye & Beattie, 2016). Mirroring the histopathological features of the spleen pathology in human VL, the congested splenic sinusoids of Rhesus monkeys displayed disorganized follicular structures and heavily infected macrophages in the subcapsular region and red pulp.

Literature search on human or canine VL revealed only scattered reports of coagulopathies, including a fatal case (Budd-Chiari syndrome) that was linked to procoagulative factor V Leiden mutation (Gürkan, Ünsal, Başlamışlı, & Arslan, 2004), three patients displaying disseminated intravascular coagulation associated to hemophagocytosis (Boukhris et al., 2015), and one unusual case of disseminated intravascular coagulation in dogs (Honse et al., 2013). Except for punctual reports showing reduced plasma levels of fibrinogens (Bruno et al., 2015) and platelet abnormalities in symptomatic dogs (Ciaramella et al., 2005), there is no obvious indication that low-grade immunothrombosis and BK-induced inflammation participate in human VL, whether modulating the inflammatory granuloma in the liver and/or influencing the outcome of splenic pathology (e.g., congestion of splenic sinusoids). So far, very limited information has emerged from our preliminary efforts at identifying transcriptional changes in B2R/B1R or other KKS components in tissue lesions and immune cells infected by different species of *Leishmania* (Table 1). Although several members of the KKS are selectively upregulated as the yolk sac-derived macrophage precursors develop in the mouse liver (Perdigueiro & Gleissmann, personal communication), there is no information on the expression profile of these components in spleen/liver sections from symptomatic VL patients. In a limited in vitro study conducted with *L. donovani*, B1R was upregulated (2.8-fold) in human DCs and to a lower extent (1.78-fold) in *L. major*-infected DCs; irrespective of the parasite species examined so far, *Leishmania* did not upregulate B1R in human macrophages. A summary of these preliminary results are provided in the legend of Table 1.

**Table 1** Mining of Transcriptomic Data Related to the KKS in Datasets of Leishmaniasis and Other Diseases

| Gene | Synonyms | Protein | Log$_2$FC (Disease/Control), Positive Value Meaning Higher Expression in the Infected/Disease Sample | | | | | | | | | | | | |
|---|---|---|---|---|---|---|---|---|---|---|---|---|---|---|---|
| | | | Leishmaniasis | | | | | | | | | | | | |
| | | | GSE42088 (Dendritic Cells, *L. major*) | GSE69252 (NK Cells, *L. mexicana* LCL) | GSE69252 (NK Cells, *L. mexicana* DCL) | GSE61211 (Macrophages, *L. braziliensis*) | GSE63931 (Lesion) | GSE55664 (Lesion) | GSE43880 (Lesion) | GSE25819 (Macrophages, *L. panamensis*) | GSE360 (Dendritic Cell, *L. donovani*) | GSE360 (Macrophages, *L. donovani*) | GSE360 (Dendritic Cell, *L. major*) | GSE360 (Macrophages, *L. major*) |
| ***G protein-coupled receptors*** | | | | | | | | | | | | | | | |
| BDKRB1 | B1R, BRADYB1 | B1 bradykinin receptor | 0.05 | 0.16 | 0.18 | −0.30 | −0.24 | −0.24 | −0.69 | 0.43 | 2.98 | −0.43 | 1.79 | 0.66 |
| BDKRB2 | BKR2 | B2 bradykinin receptor | 0.02 | 0.16 | 0.25 | 0.20 | 0.26 | 0.14 | −0.16 | 0.25 | −0.01 | 0.99 | −0.19 | 0.67 |
| ***Serine proteases and cofactors that liberate kinins from kininogens*** | | | | | | | | | | | | | | | |
| KLK1 | — | Kallikrein-1 | 0.21 | 0.09 | −0.04 | 0.13 | −4.07 | −1.44 | −1.44 | −1.06 | — | — | — | 3.11 |
| KLKB1 | KLK3 | Plasma kallikrein | 0.05 | 0.40 | 0.27 | 0.10 | 0.31 | −0.02 | 0.12 | −0.50 | 0.11 | −0.03 | −0.20 | 0.88 |
| F12 | HAF, HAE3, HAEX | Coagulation factor XII | 0.03 | 0.08 | 0.00 | 0.14 | −0.58 | −0.56 | −0.69 | −1.00 | 0.01 | −1.86 | 0.12 | −0.43 |
| KNG1 | BDK, KNG | Kininogen-1 | 0.04 | 0.31 | −0.01 | 0.14 | −0.16 | −0.03 | 0.01 | −0.53 | 0.96 | 1.50 | 2.26 | 1.59 |
| ***Kinin-degrading peptidases*** | | | | | | | | | | | | | | | |
| ACE | DCP, ACE1, DCP1, C | Angiotensin I-converting enzyme | −0.34 | −0.22 | −0.22 | 0.04 | 1.40 | 0.12 | 0.14 | −0.06 | −0.16 | −0.14 | −0.40 | −0.33 |
| ACE2 | — | Angiotensin I-converting enzyme 2 | 0.03 | 0.20 | 0.13 | −0.05 | 0.65 | 0.18 | −0.07 | −0.96 | — | — | — | — |
| MME | EPN | Neprilysin | 0.03 | 0.45 | 0.00 | 0.12 | 3.59 | 2.71 | 2.86 | 0.68 | 0.38 | 0.11 | 0.43 | −2.56 |
| CPM | — | Carboxypeptidase M | −0.61 | −1.30 | −1.65 | −0.26 | — | 0.05 | −0.27 | −0.12 | −0.52 | 0.31 | −0.60 | 0.36 |
| CPN1 | ACBP | Carboxypeptidase N catalytic chain | 0.02 | 0.16 | −0.01 | 0.09 | 0.40 | 0.09 | 0.13 | −0.90 | — | 1.00 | — | 1.50 |
| CPN2 | ACBP | Carboxypeptidase N subunit 2 | 0.09 | 0.27 | 0.20 | 0.18 | −0.44 | 0.07 | 0.09 | 0.38 | 5.43 | 4.25 | 6.57 | — |
| ***C1 inhibitor and other potentially interesting factors*** | | | | | | | | | | | | | | | |
| SERPING1 | C1IN, C1NH | Plasma protease C1 inhibitor | −0.07 | 0.49 | −1.25 | −0.06 | 2.58 | 2.26 | 3.12 | −0.57 | 1.40 | −0.03 | 1.63 | 0.69 |
| PLAUR | CD87, MO3, UPAR | Urokinase plasminogen activator | 0.98 | 0.57 | −1.26 | 0.05 | 3.75 | 3.02 | 2.49 | 0.29 | 1.11 | 1.10 | 1.40 | 1.73 |
| ***Tryptase (mast cell marker) subunits*** | | | | | | | | | | | | | | | |
| TPSAB1 | TPS1, TPS2, TPSB1 | Tryptase alpha/beta-1 | 0.02 | 0.07 | −0.06 | — | −0.16 | 0.06 | −1.22 | — | 0.30 | 1.87 | 0.26 | 1.16 |
| TPSB2 | | Tryptase beta-2 | 0.02 | 0.07 | −0.06 | — | — | 0.11 | −0.59 | — | 0.30 | 1.87 | 0.26 | 1.16 |
| TPSD1 | | Tryptase delta | 0.05 | 0.19 | 0.15 | 0.01 | −0.45 | −0.07 | −0.14 | 0.65 | — | — | — | |
| TPSG1 | PRSS31, TMT | Tryptase gamma | 0.02 | 0.19 | 0.11 | 0.02 | −0.15 | −0.17 | −0.17 | 0.68 | — | — | — | — |

**Caption: —, probe for gene is absent in given array.**
Color caption for sample GSE ID according to sample type
- Skin biopsies
- Cells (NK, macrophages, DCs, etc.)
- Blood (whole, PBMC, monocytes)

Cell heatmap scale according to Log2(fold change)
2.0
1.0
0.0
−1.0
−2.0

*(Continued)*

## Table 1 Mining of Transcriptomic Data Related to the KKS in Datasets of Leishmaniasis and Other Diseases—cont'd

| Gene | Synonyms | Protein | Log$_2$FC (Disease/Control), Positive Value Meaning Higher Expression in the Infected/Disease Sample | | | | | | | | | | | | |
|---|---|---|---|---|---|---|---|---|---|---|---|---|---|---|---|
| | | | Fungal infections | | | Bacterial infections | | | | | Cancer | Other inflammatory | | | |
| | | | GSE7355 (Endothelial cells, Candida) | GSE6965 (DCs, Aspergillus) | GSE78000 (Aspergillus) | GSE65517 (Sepsis) | GSE12108 (Francisella) | GSE54992 (Act ve TB) | GSE13670 (Monocytes, S.aureus, 8h p.i.) | GSE13670 (Monocytes, S.aureus, 24h p.i.) | GSE46517 (Melanoma) | GSE75890 (Atopic Dermatitis) | GSE65832 (Atopic Dermatitis) | GSE30999 (Psoriasis) | GSE54456 (Psoriasis) |
| ***G protein-coupled receptors*** | | | | | | | | | | | | | | | |
| BDKRB1 | B1R, BRADYB1 | B1 bradykinin receptor | −0.59 | 0.08 | 0.15 | — | 0.12 | −0.44 | 0.51 | **−1.83** | −0.68 | 0.38 | 0.83 | 0.93 | 0.77 |
| BDKRB2 | BKR2 | B2 bradykinin receptor | 0.24 | 0.62 | −0.01 | **−3.43** | 0.07 | −0.54 | 0.18 | −0.91 | −1.15 | 0.56 | 0.73 | 1.19 | 0.61 |
| ***Serine proteases and cofactors that liberate kinins from kininogens*** | | | | | | | | | | | | | | | |
| KLK1 | — | Kallikrein-1 | −0.38 | −0.23 | −0.19 | −0.95 | 0.36 | 0.21 | 0.29 | −0.69 | **−2.21** | −0.17 | −0.34 | −0.47 | **−1.56** |
| KLKB1 | KLK3 | Plasma kallikrein | 0.26 | −0.26 | 0.05 | — | 0.09 | −1.21 | 0.71 | −0.45 | 0.22 | −0.01 | −0.43 | −0.07 | −0.51 |
| F12 | HAF, HAE3, HAEX | Coagulation factor XII | −0.36 | 0.06 | 0.88 | 1.01 | 0.08 | −0.13 | −0.56 | −0.13 | −0.09 | 0.06 | 0.31 | **1.90** | 1.47 |
| KNG1 | BDK, KNG | Kininogen-1 | −0.35 | 0.08 | −0.03 | — | 0.04 | 0.49 | 0.44 | −0.44 | 0.28 | 0.06 | — | 0.00 | 0.00 |
| ***Kinin-degrading peptidases*** | | | | | | | | | | | | | | | |
| ACE | DCP, ACE1, DCP1, C | Angiotensin I-converting enzyme | −0.22 | 0.17 | 0.07 | — | 0.04 | −0.46 | 0.15 | −0.65 | −0.19 | 0.38 | 0.30 | 0.00 | −0.41 |
| ACE2 | — | Angiotensin I-converting enzyme 2 | −0.48 | — | 0.04 | **1.66** | 0.07 | 0.36 | 1.07 | 1.05 | 0.47 | 0.06 | 0.73 | 0.61 | 1.40 |
| MME | EPN | Neprilysin | 0.44 | 0.16 | −0.35 | 0.13 | 0.24 | 1.19 | −0.12 | −0.22 | 0.48 | 0.64 | −0.52 | −0.31 | −0.26 |
| CPM | — | Carboxypeptidase M | −0.36 | 0.64 | 0.12 | **2.63** | **2.03** | 1.34 | −0.21 | 0.53 | −0.12 | 0.56 | −0.27 | 0.14 | −0.61 |
| CPN1 | ACBP | Carboxypeptidase N catalytic chain | −0.41 | −0.06 | 0.03 | −0.89 | 0.09 | −0.13 | 1.22 | −0.13 | 0.95 | −0.13 | — | 0.00 | 0.00 |
| CPN2 | ACBP | Carboxypeptidase N subunit 2 | −0.96 | −0.15 | 0.07 | 0.35 | 0.03 | 0.85 | −0.85 | −0.31 | 1.33 | 0.24 | −0.28 | 0.01 | 0.22 |
| ***C1 inhibitor and other potentially interesting factors*** | | | | | | | | | | | | | | | |
| SERPING1 | C1IN, C1NH | Plasma protease C1 inhibitor | −0.67 | −0.32 | **1.80** | 1.00 | **−2.66** | **3.65** | **2.54** | **3.21** | −0.29 | 0.17 | −0.41 | −0.40 | −0.68 |
| PLAUR | CD87, MO3, UPAR | Urokinase plasminogen activator | 0.82 | −0.28 | 0.53 | 0.42 | **2.59** | −1.45 | 1.10 | **1.58** | 1.21 | 0.79 | 1.22 | **1.60** | 0.60 |
| ***Tryptase (mast cell marker) subunits*** | | | | | | | | | | | | | | | |
| TPSAB1 | TPS1, TPS2, TPSB1 | Tryptase alpha/beta-1 | −0.84 | −0.40 | −0.21 | 0.26 | 0.87 | −0.86 | 0.79 | −0.79 | **−2.31** | −0.05 | −0.52 | −0.57 | −0.65 |
| TPSB2 | | Tryptase beta-2 | −0.41 | −0.22 | — | **−3.08** | 0.73 | 0.84 | 0.48 | −0.28 | **−2.12** | 0.55 | −0.52 | −0.22 | −0.75 |
| TPSD1 | | Tryptase delta | 0.70 | −0.13 | 0.06 | — | 0.00 | −0.66 | **1.58** | −0.47 | 0.40 | −0.24 | — | 0.00 | −0.71 |
| TPSG1 | PRSS31, TMT | Tryptase gamma | −0.29 | 0.09 | −0.10 | — | 0.36 | −0.41 | −0.54 | 0.08 | 0.57 | 0.03 | −0.26 | −0.03 | **−1.99** |

**Caption: —, probe for gene is absent in given array.**

Extended legend: Datasets related to leishmaniasis, fungal/bacterial diseases, cancer, and other inflammatory diseases were searched and retrieved from the Gene Expression Omnibus (GEO, http://www.ncbi.nlm.nih.gov/geo) and fold changes (diseased vs healthy) are reported for 18 genes of interest in log$_2$ scale. Absolute fold changes greater than 1.5 were considered significant and are marked in *bold letter*. Missing values (absence of a probe for a gene in a given array platform) are reported as "—." This preliminary survey allows direct comparison of the expression of multiple components of the KKS in different diseases. For leishmaniasis, experiments in cells (DCs, macrophages, NK) with multiple species and skin biopsies of the cutaneous form of the disease caused by *Leishmania braziliensis* are shown. Although a pristine expression signature on the KKS genes alone cannot be inferred, there are consistent alteration patterns for instance within the skin biopsies samples (GSE63931, GSE55664, GSE43880) where upregulation of MME, SERPING1, and PLAUR on the infected group were seem. On the other hand, there is a decrease in the expression of KLK1 in the skin biopsies, although not consistent in all studies. The opposing expression shifts observed for some genes between biopsies and cells studies might punctuate different molecular mechanisms involved in individuals subjected to the natural disease and when performing in vitro experiments, hinting a complex picture of the infection by *Leishmania*. For instance, in one study (GSE360), consisting of several experiments using DCs or macrophages infected by *L. major* and *L. donovani*, upregulation of B1R could be seem in DCs infected by both *Leishmania* species at 16 h postinfection, with fold changes varying from 1.79 (infection by *L. major*) to 2.79 (infection by *L. donovani*), although the small number of replicates (pool of RNA from seven donors independently evaluated twice) limits the interpretation of these findings. In this dataset, *Leishmania*-infected macrophages did not elicit a similar response, arguing for a DC-specific response. Interestingly, downregulation of KLK1 was observed in datasets of melanoma (GSE46517) and psoriasis (GSE54456). On the other hand, in bacterial infections, upregulation of SERPING1 was also observed in the response to Mtb (GSE54992) and *S. aureus* (GSE13670) infections, but not *Francisella* (GSE12108), this gene appearing significantly downregulated in the latter. This initial survey pinpoints the involvement of at least some components of the KKS in leishmaniasis and other infectious and inflammatory diseases, and follow-up studies are warranted to confirm and expand these findings. *Gray horizontal stretches* below some of the datasets indicate that distinct samples from a same study were analyzed, for instance focusing on distinct cell lines.

## 5. CONCLUDING REMARKS

Despite of the vitality of KKS research, there have been limited attempts to systematically analyze the role of G protein-coupled kinin receptors in the immunopathogenesis of infectious diseases. For many years, experts in blood coagulation were intrigued that bleeding was not increased in patients with genetic deficiencies of plasma contact factors. As reviewed in this chapter, it is possible that the convergence between the extrinsic and intrinsic pathway of coagulation fuels host defense against pathogens via two interrelated mechanisms. In the bloodstream, the entrapment of pathogens within fibrin-driven scaffolds might facilitate pathogen clearance by circulating neutrophils and monocytes activated by systemically released kinins. In addition, lessons taken from models of peripheral infection provided readers with a conceptual framework to understand the functional interplay between innate immunity and the proteolytic cascades (KKS and Complement) that shape Th1 effector development in the draining lymph nodes. Development of transgenic mice with conditional expression of B2R and B1R in different subsets of immune cells (neutrophils, monocytes and DCs) is required to evaluate whether in vivo activation of kinin receptors optimize phagocytosis and microbicidal functions of circulating leukocytes—perhaps reminiscent of the phenotypic properties of macrophages infected in vitro with *T. cruzi* and *Leishmania*.

## ACKNOWLEDGMENTS

The authors wish to thank Dr. Elvira G. Perdiguero (CNRS, URA, Paris) and Dr. Frederic Geissmann (Memorial Sloan Kettering Cancer Center) for generously sharing the results of differential transcription (KKS components) during mouse macrophage development. We also wish to express our gratitude to our students, Lucas Vellasco for the valuable help in the edition of the manuscript and Rafaela Serra in the preparation of the illustrations. We also wish to thank Mr. James Selvam and his team for useful suggestions as to how we could improve the presentation of the figures and legends during preparation of the galley proof.

## REFERENCES

AbdAlla, S., Lother, H., & Quitterer, U. (2000). AT1-receptor heterodimers show enhanced G-protein activation and altered receptor sequestration. *Nature*, *407*(6800), 94–98. https://doi.org/10.1038/35024095.

Abelous, J. E., & Bardier, E. (1909). Les substances hypotensives de l'urine humaine normale. *Comptes Rendus Société Biologie*, *66*, 511–512.

Ahluwalia, A., & Perretti, M. (1996). Involvement of bradykinin B1 receptors in the polymorphonuclear leukocyte accumulation induced by IL-1 beta in vivo in the mouse.

*Journal of Immunology (Baltimore, Md.: 1950)*, *156*(1), 269–274. Retrieved from http://www.ncbi.nlm.nih.gov/pubmed/8598472.

Ali, R. A., Wuescher, L. M., & Worth, R. G. (2015). Platelets: Essential components of the immune system. *Current Trends in Immunology*, *16*, 65–78. Retrieved from http://www.ncbi.nlm.nih.gov/pubmed/27818580.

Aliberti, J., Viola, J. P. B., Vieira-de-Abreu, A., Bozza, P. T., Sher, A., & Scharfstein, J. (2003). Cutting edge: Bradykinin induces IL-12 production by dendritic cells: A danger signal that drives Th1 polarization. *Journal of Immunology (Baltimore, Md.: 1950)*, *170*(11), 5349–5353. Retrieved from http://www.ncbi.nlm.nih.gov/pubmed/12759407.

Almeida, I. C., & Gazzinelli, R. T. (2001). Proinflammatory activity of glycosylphosphatidylinositol anchors derived from Trypanosoma cruzi: Structural and functional analyses. *Journal of Leukocyte Biology*, *70*(4), 467–477. Retrieved from http://www.ncbi.nlm.nih.gov/pubmed/11590183.

Amara, U., Flierl, M. A., Rittirsch, D., Klos, A., Chen, H., Acker, B., et al. (2010). Molecular intercommunication between the complement and coagulation systems. *The Journal of Immunology*, *185*(9), 5628–5636. https://doi.org/10.4049/jimmunol.0903678.

Andrade, T. M., Carvalho, E. M., & Rocha, H. (1990). Bacterial infections in patients with visceral leishmaniasis. *The Journal of Infectious Diseases*, *162*(6), 1354–1359. Retrieved from http://www.ncbi.nlm.nih.gov/pubmed/2230265.

Andrade, D., Serra, R., Svensjö, E., Lima, A. P. C., Ramos, E. S., Fortes, F. S., et al. (2012). Trypanosoma cruzi invades host cells through the activation of endothelin and bradykinin receptors: A converging pathway leading to chagasic vasculopathy. *British Journal of Pharmacology*, *165*(5), 1333–1347. https://doi.org/10.1111/j.1476-5381.2011.01609.x.

Antonen, J., Leppänen, I., Tenhunen, J., Arvola, P., Mäkelä, S., Vaheri, A., et al. (2013). A severe case of Puumala hantavirus infection successfully treated with bradykinin receptor antagonist icatibant. *Scandinavian Journal of Infectious Diseases*, *45*(6), 494–496. https://doi.org/10.3109/00365548.2012.755268.

Aquino, D. M. C., Caldas, A. J. M., Miranda, J. C., Silva, A. A. M., Barral-Netto, M., & Barral, A. (2010). Epidemiological study of the association between anti-Lutzomyia longipalpis saliva antibodies and development of delayed-type hypersensitivity to Leishmania antigen. *The American Journal of Tropical Medicine and Hygiene*, *83*(4), 825–827. https://doi.org/10.4269/ajtmh.2010.10-0182.

auf dem Keller, U., Prudova, A., Eckhard, U., Fingleton, B., & Overall, C. M. (2013). Systems-level analysis of proteolytic events in increased vascular permeability and complement activation in skin inflammation. *Science Signaling*, *6*(258), rs2. https://doi.org/10.1126/scisignal.2003512.

Bader, M. (Ed.), (2011). *Kinins*. Berlin, Boston: DE Gruyter. https://doi.org/10.1515/9783110252354.

Barbasz, A., & Kozik, A. (2009). The assembly and activation of kinin-forming systems on the surface of human U-937 macrophage-like cells. *Biological Chemistry*, *390*(3), 269–275. https://doi.org/10.1515/BC.2009.032.

Beattie, L., d'El-Rei Hermida, M., Moore, J. W. J., Maroof, A., Brown, N., Lagos, D., et al. (2013). A transcriptomic network identified in uninfected macrophages responding to inflammation controls intracellular pathogen survival. *Cell Host & Microbe*, *14*(3), 357–368. https://doi.org/10.1016/j.chom.2013.08.004.

Beattie, L., Peltan, A., Maroof, A., Kirby, A., Brown, N., Coles, M., et al. (2010). Dynamic imaging of experimental Leishmania donovani-induced hepatic granulomas detects Kupffer cell-restricted antigen presentation to antigen-specific CD8 T cells. *PLoS Pathogens*, *6*(3), e1000805. https://doi.org/10.1371/journal.ppat.1000805.

Beiter, K., Wartha, F., Albiger, B., Normark, S., Zychlinsky, A., & Henriques-Normark, B. (2006). An endonuclease allows Streptococcus pneumoniae to escape from neutrophil extracellular traps. *Current Biology*, *16*(4), 401–407. https://doi.org/10.1016/j.cub.2006.01.056.

Belkaid, Y., Kamhawi, S., Modi, G., Valenzuela, J., Noben-Trauth, N., Rowton, E., et al. (1998). Development of a natural model of cutaneous leishmaniasis: Powerful effects of vector saliva and saliva preexposure on the long-term outcome of Leishmania major infection in the mouse ear dermis. *The Journal of Experimental Medicine*, *188*(10), 1941–1953. Retrieved from http://www.ncbi.nlm.nih.gov/pubmed/9815271.

Belkaid, Y., Valenzuela, J. G., Kamhawi, S., Rowton, E., Sacks, D. L., & Ribeiro, J. M. (2000). Delayed-type hypersensitivity to Phlebotomus papatasi sand fly bite: An adaptive response induced by the fly? *Proceedings of the National Academy of Sciences of the United States of America*, *97*(12), 6704–6709. Retrieved from http://www.ncbi.nlm.nih.gov/pubmed/10841567.

Ben Nasr, A., Herwald, H., Sjöbring, U., Renné, T., Müller-Esterl, W., & Björck, L. (1997). Absorption of kininogen from human plasma by Streptococcus pyogenes is followed by the release of bradykinin. *Biochemical Journal*, *326*(3), 657–660.

Bertram, C. M., Baltic, S., Misso, N. L., Bhoola, K. D., Foster, P. S., Thompson, P. J., et al. (2007a). Expression of kinin B1 and B2 receptors in immature, monocyte-derived dendritic cells and bradykinin-mediated increase in intracellular Ca2+ and cell migration. *Journal of Leukocyte Biology*, *81*(6), 1445–1454. https://doi.org/10.1189/jlb.0106055.

Bertram, C. M., Misso, N. L., Fogel-Petrovic, M., Figueroa, C. D., Foster, P. S., Thompson, P. J., et al. (2008). Expression of kinin receptors on eosinophils: Comparison of asthmatic patients and healthy subjects. *Journal of Leukocyte Biology*, *85*(3), 544–552. https://doi.org/10.1189/jlb.0508283.

Bertram, C., Misso, N. L., Fogel-Petrovic, M., Figueroa, C., Thompson, P. J., & Bhoola, K. D. (2007b). Comparison of kinin B1 and B2 receptor expression in neutrophils of asthmatic and non-asthmatic subjects. *International Immunopharmacology*, 7(14), 1862–1868. https://doi.org/10.1016/j.intimp.2007.07.012.

Blaukat, A., & Dikic, I. (2001). Activation of sphingosine kinase by the bradykinin B2 receptor and its implication in regulation of the ERK/MAP kinase pathway. *Biological Chemistry*, *382*(1), 135–139. https://doi.org/10.1515/BC.2001.020.

Borkowski, J. A., Ransom, R. W., Seabrook, G. R., Trumbauer, M., Chen, H., Hill, R. G., et al. (1995). Targeted disruption of a B2 bradykinin receptor gene in mice eliminates bradykinin action in smooth muscle and neurons. *The Journal of Biological Chemistry*, *270*(23), 13706–13710. Retrieved from http://www.ncbi.nlm.nih.gov/pubmed/7775424.

Boukhris, I., Azzabi, S., Chérif, E., Kéchaou, I., Mahjoub, S., Kooli, C., et al. (2015). Hémophagocytose et coagulation intravasculaire disséminée au cours de la leishmaniose viscérale de l'adulte: Trois nouveaux cas. *The Pan African Medical Journal*, *22*, 96. https://doi.org/10.11604/pamj.2015.22.96.5662.

Brodie, T. M., Smith, M. C., Morris, R. V., & Titus, R. G. (2007). Immunomodulatory effects of the Lutzomyia longipalpis salivary gland protein Maxadilan on mouse macrophages. *Infection and Immunity*, *75*(5), 2359–2365. https://doi.org/10.1128/IAI.01812-06.

Brown, N. J., Gainer, J. V., Murphey, L. J., & Vaughan, D. E. (2000). Bradykinin stimulates tissue plasminogen activator release from human forearm vasculature through B(2) receptor-dependent, NO synthase-independent, and cyclooxygenase-independent pathway. *Circulation*, *102*(18), 2190–2196. Retrieved from http://www.ncbi.nlm.nih.gov/pubmed/11056091.

Bruno, B., Maurella, C., Falco, S., Tarducci, A., Zanatta, R., D'Angelo, A., et al. (2015). Thromboelastometric evaluation of hemostasis in dogs infected with Leishmania infantum. *Journal of Veterinary Emergency and Critical Care*, *25*(4), 502–511. https://doi.org/10.1111/vec.12325.

Cai, Y., Kobayashi, R., Hashizume-Takizawa, T., & Kurita-Ochiai, T. (2014). Porphyromonas gingivalis infection enhances Th17 responses for development of atherosclerosis. *Archives of Oral Biology*, *59*(11), 1183–1191. https://doi.org/10.1016/j.archoralbio.2014.07.012.

Campos, M. A., Almeida, I. C., Takeuchi, O., Akira, S., Valente, E. P., Procópio, D. O., et al. (2001). Activation of toll-like receptor-2 by glycosylphosphatidylinositol anchors from a protozoan parasite. *Journal of Immunology (Baltimore, Md.: 1950)*, *167*(1), 416–423. Retrieved from http://www.ncbi.nlm.nih.gov/pubmed/11418678.

Caradonna, K. L., Engel, J. C., Jacobi, D., Lee, C.-H., & Burleigh, B. A. (2013). Host metabolism regulates intracellular growth of Trypanosoma cruzi. *Cell Host & Microbe*, *13*(1), 108–117. https://doi.org/10.1016/j.chom.2012.11.011.

Cayla, C., Todiras, M., Iliescu, R., Saul, V. V., Gross, V., Pilz, B., et al. (2007). Mice deficient for both kinin receptors are normotensive and protected from endotoxin-induced hypotension. *FASEB Journal: Official Publication of the Federation of American Societies for Experimental Biology*, *21*(8), 1689–1698. https://doi.org/10.1096/fj.06-7175com.

Cekici, A., Kantarci, A., Hasturk, H., & Van Dyke, T. E. (2014). Inflammatory and immune pathways in the pathogenesis of periodontal disease. *Periodontology 2000*, *64*(1), 57–80. https://doi.org/10.1111/prd.12002.

Chagas, A. C., Oliveira, F., Debrabant, A., Valenzuela, J. G., Ribeiro, J. M. C., & Calvo, E. (2014). Lundep, a sand fly salivary endonuclease increases Leishmania parasite survival in neutrophils and inhibits XIIa contact activation in human plasma. *PLoS Pathogens*, *10*(2), e1003923. https://doi.org/10.1371/journal.ppat.1003923.

Christiansen, S. C., Proud, D., & Cochrane, C. G. (1987). Detection of tissue kallikrein in the bronchoalveolar lavage fluid of asthmatic subjects. *Journal of Clinical Investigation*, *79*(1), 188–197. https://doi.org/10.1172/JCI112782.

Ciaramella, P., Pelagalli, A., Cortese, L., Pero, M. E., Corona, M., Lombardi, P., et al. (2005). Altered platelet aggregation and coagulation disorders related to clinical findings in 30 dogs naturally infected by Leishmania infantum. *The Veterinary Journal*, *169*(3), 465–467. https://doi.org/10.1016/j.tvjl.2004.03.009.

Cugno, M., Zanichelli, A., Bellatorre, A. G., Griffini, S., & Cicardi, M. (2009). Plasma biomarkers of acute attacks in patients with angioedema due to C1-inhibitor deficiency. *Allergy*, *64*(2), 254–257. https://doi.org/10.1111/j.1398-9995.2008.01859.x.

de Maat, S., Björkqvist, J., Suffritti, C., Wiesenekker, C. P., Nagtegaal, W., Koekman, A., et al. (2016). Plasmin is a natural trigger for bradykinin production in patients with hereditary angioedema with factor XII mutations. *Journal of Allergy and Clinical Immunology*, *138*(5), e9 1414–1423. https://doi.org/10.1016/j.jaci.2016.02.021.

de Moura, T. R., Oliveira, F., Carneiro, M. W., Miranda, J. C., Clarêncio, J., Barral-Netto, M., et al. (2013). Functional transcriptomics of wild-caught Lutzomyia Intermedia salivary glands: Identification of a protective salivary protein against Leishmania braziliensis infection. *PLoS Neglected Tropical Diseases*, 7(5), e2242. https://doi.org/10.1371/journal.pntd.0002242.

Del Nery, E., Juliano, M. A., Lima, A. P., Scharfstein, J., & Juliano, L. (1997). Kininogenase activity by the major cysteinyl proteinase (cruzipain) from Trypanosoma cruzi. *The Journal of Biological Chemistry*, *272*(41), 25713–25718. Retrieved from http://www.ncbi.nlm.nih.gov/pubmed/9325296.

Dong, R., Chen, W., Feng, W., Xia, C., Hu, D., Zhang, Y., et al. (2015). Exogenous bradykinin inhibits tissue factor induction and deep vein thrombosis via activating the eNOS/phosphoinositide 3-kinase/Akt signaling pathway. *Cellular Physiology and Biochemistry*, *37*(4), 1592–1606. https://doi.org/10.1159/000438526.

Doyle, P. S., Zhou, Y. M., Engel, J. C., & McKerrow, J. H. (2007). A cysteine protease inhibitor cures Chagas' disease in an Immunodeficient-mouse model of infection.

*Antimicrobial Agents and Chemotherapy*, *51*(11), 3932–3939. https://doi.org/10.1128/AAC.00436-07.

Duchene, J., Lecomte, F., Ahmed, S., Cayla, C., Pesquero, J., Bader, M., et al. (2007). A novel inflammatory pathway involved in leukocyte recruitment: Role for the kinin B1 receptor and the chemokine CXCL5. *Journal of Immunology (Baltimore, Md.: 1950)*, *179*(7), 4849–4856. Retrieved from http://www.ncbi.nlm.nih.gov/pubmed/17878384.

Dutra, R. C., Leite, D. F. P., Bento, A. F., Manjavachi, M. N., Patrício, E. S., Figueiredo, C. P., et al. (2011). The role of Kinin receptors in preventing neuroinflammation and its clinical severity during experimental autoimmune encephalomyelitis in mice. *PloS One*, *6*(11), e27875. https://doi.org/10.1371/journal.pone.0027875.

Dutra, W. O., Menezes, C. A. S., Magalhães, L. M. D., & Gollob, K. J. (2014). Immunoregulatory networks in human Chagas disease. *Parasite Immunology*, *36*(8), 377–387. https://doi.org/10.1111/pim.12107.

Dutra, R. C., Moreira, E. L. G., Alberti, T. B., Marcon, R., Prediger, R. D., & Calixto, J. B. (2013). Spatial reference memory deficits precede motor dysfunction in an experimental autoimmune encephalomyelitis model: The role of kallikrein–kinin system. *Brain, Behavior, and Immunity*, *33*, 90–101. https://doi.org/10.1016/j.bbi.2013.06.002.

Ehrenfeld, P., Matus, C. E., Pavicic, F., Toledo, C., Nualart, F., Gonzalez, C. B., et al. (2009). Kinin B1 receptor activation turns on exocytosis of matrix metalloprotease-9 and myeloperoxidase in human neutrophils: Involvement of mitogen-activated protein kinase family. *Journal of Leukocyte Biology*, *86*(5), 1179–1189. https://doi.org/10.1189/jlb.0109012.

Ehrenfeld, P., Millan, C., Matus, C. E., Figueroa, J. E., Burgos, R. A., Nualart, F., et al. (2006). Activation of kinin B1 receptors induces chemotaxis of human neutrophils. *Journal of Leukocyte Biology*, *80*(1), 117–124. https://doi.org/10.1189/jlb.1205744.

Engelmann, B., & Massberg, S. (2013). Thrombosis as an intravascular effector of innate immunity. *Nature Reviews Immunology*, *13*(1), 34–45. https://doi.org/10.1038/nri3345.

Enquist, J., Skröder, C., Whistler, J. L., & Leeb-Lundberg, L. M. F. (2007). Kinins promote B2 receptor endocytosis and delay constitutive B1 receptor endocytosis. *Molecular Pharmacology*, *71*(2), 494–507. https://doi.org/10.1124/mol.106.030858.

Erdos, E. G., & Sloane, E. M. (1962). An enzyme in human blood plasma that inactivates bradykinin and kallidins. *Biochemical Pharmacology*, *11*, 585–592. Retrieved from http://www.ncbi.nlm.nih.gov/pubmed/13890520.

Eschenlauer, S. C. P., Faria, M. S., Morrison, L. S., Bland, N., Ribeiro-Gomes, F. L., DosReis, G. A., et al. (2009). Influence of parasite encoded inhibitors of serine peptidases in early infection of macrophages with Leishmania major. *Cellular Microbiology*, *11*(1), 106–120. https://doi.org/10.1111/j.1462-5822.2008.01243.x.

Eskan, M. A., Hajishengallis, G., & Kinane, D. F. (2007). Differential activation of human gingival epithelial cells and monocytes by Porphyromonas gingivalis fimbriae. *Infection and Immunity*, *75*(2), 892–898. https://doi.org/10.1128/IAI.01604-06.

Fang, C., Stavrou, E., Schmaier, A. A., Grobe, N., Morris, M., Chen, A., et al. (2013). Angiotensin 1-7 and Mas decrease thrombosis in Bdkrb2−/− mice by increasing NO and prostacyclin to reduce platelet spreading and glycoprotein VI activation. *Blood*, *121*(15), 3023–3032. https://doi.org/10.1182/blood-2012-09-459156.

Faria, M. S., Reis, F. C. G., Azevedo-Pereira, R. L., Morrison, L. S., Mottram, J. C., & Lima, A. P. C. A. (2011). Leishmania inhibitor of serine peptidase 2 prevents TLR4 activation by neutrophil elastase promoting parasite survival in murine macrophages. *The Journal of Immunology*, *186*(1), 411–422. https://doi.org/10.4049/jimmunol.1002175.

Fauve, R. M., & Hévin, M. B. (1979). Inflammation and host resistance against bacteria. I. Increased resistance against *Listeria monocytogenes* and *Salmonella typhimurium* in mice, following their treatment with bradykinin, kallidin and methionyl-lysyl-bradykinin (author's transl.). *Annales d'Immunologie, 130C*(5), 743–748. Retrieved from http://www.ncbi.nlm.nih.gov/pubmed/393161.

Ferreira, S. H. (1965). A bradykinin-potentiating factor (BPF) present in the venom of *Bothrops jararaca. British Journal of Pharmacology and Chemotherapy, 24*(1), 163–169. Retrieved from http://www.ncbi.nlm.nih.gov/pubmed/14302350.

Filardy, A. A., Pires, D. R., Nunes, M. P., Takiya, C. M., Freire-de-Lima, C. G., Ribeiro-Gomes, F. L., et al. (2010). Proinflammatory clearance of apoptotic neutrophils induces an IL-12$^{low}$ IL-10$^{high}$ regulatory phenotype in macrophages. *The Journal of Immunology, 185*(4), 2044–2050. https://doi.org/10.4049/jimmunol.1000017.

Finsterbusch, M., Voisin, M.-B., Beyrau, M., Williams, T. J., & Nourshargh, S. (2014). Neutrophils recruited by chemoattractants in vivo induce microvascular plasma protein leakage through secretion of TNF. *The Journal of Experimental Medicine, 211*(7), 1307–1314. https://doi.org/10.1084/jem.20132413.

Gabriel, C., McMaster, W. R., Girard, D., & Descoteaux, A. (2010). Leishmania donovani promastigotes evade the antimicrobial activity of neutrophil extracellular traps. *The Journal of Immunology, 185*(7), 4319–4327. https://doi.org/10.4049/jimmunol.1000893.

Gaertner, F., & Massberg, S. (2016). Blood coagulation in immunothrombosis—At the frontline of intravascular immunity. *Seminars in Immunology, 28*(6), 561–569. https://doi.org/10.1016/j.smim.2016.10.010.

Gailani, D., & Broze, G. J. (1991). Factor XI activation in a revised model of blood coagulation. *Science (New York, NY), 253*(5022), 909–912. Retrieved from http://www.ncbi.nlm.nih.gov/pubmed/1652157.

Gerbaba, T. K., Gedamu, L., Isermann, K., Bruchhaus, I., & Tomás, A. (2013). Cathepsin B Gene disruption induced Leishmania donovani proteome remodeling implies Cathepsin B role in Secretome regulation. *PloS One, 8*(11), e79951. https://doi.org/10.1371/journal.pone.0079951.

Göbel, K., Pankratz, S., Schneider-Hohendorf, T., Bittner, S., Schuhmann, M. K., Langer, H. F., et al. (2011). Blockade of the kinin receptor B1 protects from autoimmune CNS disease by reducing leukocyte trafficking. *Journal of Autoimmunity, 36*(2), 106–114. https://doi.org/10.1016/j.jaut.2010.11.004.

Gomes, R., Cavalcanti, K., Teixeira, C., Carvalho, A. M., Mattos, P. S., Cristal, J. R., et al. (2016). Immunity to Lutzomyia whitmani saliva protects against experimental Leishmania braziliensis infection. *PLoS Neglected Tropical Diseases, 10*(11), e0005078. https://doi.org/10.1371/journal.pntd.0005078.

Gonçalves-Zillo, T. O., Pugliese, L. S., Sales, V. M. T., da Silva Mori, M. A., Squaiella-Baptistão, C. C., Longo-Maugéri, I. M., et al. (2013). Increased bone loss and amount of osteoclasts in kinin B1 receptor knockout mice. *Journal of Clinical Periodontology, 40*(7), 653–660. https://doi.org/10.1111/jcpe.12097.

Guimaraes-Costa, A. B., Nascimento, M. T. C., Froment, G. S., Soares, R. P. P., Morgado, F. N., Conceicao-Silva, F., et al. (2009). Leishmania amazonensis promastigotes induce and are killed by neutrophil extracellular traps. *Proceedings of the National Academy of Sciences, 106*(16), 6748–6753. https://doi.org/10.1073/pnas.0900226106.

Gulliver, R., Baltic, S., Misso, N. L., Bertram, C. M., Thompson, P. J., & Fogel-Petrovic, M. (2011). Lys-des[Arg9]-bradykinin alters migration and production of interleukin-12 in monocyte-derived dendritic cells. *American Journal of Respiratory Cell and Molecular Biology, 45*(3), 542–549. https://doi.org/10.1165/rcmb.2010-0238OC.

Gürkan, E., Ünsal, Ç., Başlamışlı, F., & Arslan, D. (2004). Budd-Chiari syndrome associated with visceral Leishmaniasis and factor V Leiden mutation. *Journal of Thrombosis and Thrombolysis, 18*(3), 205–207. https://doi.org/10.1007/s11239-005-0347-4.

Gustafson, E. J., Schutsky, D., Knight, L. C., & Schmaier, A. H. (1986). High molecular weight kininogen binds to unstimulated platelets. *The Journal of Clinical Investigation, 78*(1), 310–318. https://doi.org/10.1172/JCI112567.

Haasemann, M., Cartaud, J., Muller-Esterl, W., & Dunia, I. (1998). Agonist-induced redistribution of bradykinin B2 receptor in caveolae. *Journal of Cell Science*, *111*(7), 917–928.

Hajishengallis, G., Shakhatreh, M.-A. K., Wang, M., & Liang, S. (2007). Complement receptor 3 blockade promotes IL-12-mediated clearance of Porphyromonas gingivalis and negates its virulence in vivo. *Journal of Immunology (Baltimore, Md.: 1950)*, *179*(4), 2359–2367. Retrieved from http://www.ncbi.nlm.nih.gov/pubmed/17675497.

Henderson, L. M., Figueroa, C. D., Müller-Esterl, W., & Bhoola, K. D. (1994). Assembly of contact-phase factors on the surface of the human neutrophil membrane. *Blood*, *84*(2), 474–482. Retrieved from http://www.ncbi.nlm.nih.gov/pubmed/8025275.

Herwald, H., Dedio, J., Kellner, R., Loos, M., & Müller-Esterl, W. (1996). Isolation and characterization of the kininogen-binding protein p33 from endothelial cells. Identity with the gC1q receptor. *The Journal of Biological Chemistry*, *271*(22), 13040–13047. https://doi.org/10.1074/JBC.271.22.13040.

Hess, J. F., Borkowski, J. A., Young, G. S., Strader, C. D., & Ransom, R. W. (1992). Cloning and pharmacological characterization of a human bradykinin (BK-2) receptor. *Biochemical and Biophysical Research Communications*, *184*(1), 260–268. Retrieved from http://www.ncbi.nlm.nih.gov/pubmed/1314587.

Higuchi, M. L., Fukasawa, S., De Brito, T., Parzianello, L. C., Bellotti, G., & Ramires, J. A. (1999). Different microcirculatory and interstitial matrix patterns in idiopathic dilated cardiomyopathy and Chagas' disease: A three dimensional confocal microscopy study. *Heart (British Cardiac Society).*, *82*(3), 279–285. Retrieved from http://www.ncbi.nlm.nih.gov/pubmed/10455076.

Honse, C. O., Figueiredo, F. B., de Alencar, N. X., Madeira, M. d. F., Gremião, I. D. F., & Schubach, T. M. P. (2013). Disseminated intravascular coagulation in a dog naturally infected by Leishmania (Leishmania) chagasi from Rio de Janeiro—Brazil. *BMC Veterinary Research*, *9*, 43. https://doi.org/10.1186/1746-6148-9-43.

Ignjatovic, T., Stanisavljevic, S., Brovkovych, V., Skidgel, R. A., & Erdös, E. G. (2004). Kinin B1 receptors stimulate nitric oxide production in endothelial cells: Signaling pathways activated by angiotensin I-converting enzyme inhibitors and peptide ligands. *Molecular Pharmacology*, *66*(5), 1310–1316. https://doi.org/10.1124/mol.104.001990.

Imamura, T., Potempa, J., Pike, R. N., Moore, J. N., Barton, M. H., & Travis, J. (1995). Effect of free and vesicle-bound cysteine proteinases of Porphyromonas gingivalis on plasma clot formation: Implications for bleeding tendency at periodontitis sites. *Infection and Immunity*, *63*(12), 4877–4882. Retrieved from http://www.ncbi.nlm.nih.gov/pubmed/7591149.

Imamura, T., Tanase, S., Szmyd, G., Kozik, A., Travis, J., & Potempa, J. (2005). Induction of vascular leakage through release of bradykinin and a novel kinin by cysteine proteinases from *Staphylococcus aureus*. *The Journal of Experimental Medicine*, *201*(10), 1669–1676. https://doi.org/10.1084/jem.20042041.

Joseph, K., Tholanikunnel, B. G., & Kaplan, A. P. (2002). Heat shock protein 90 catalyzes activation of the prekallikrein-kininogen complex in the absence of factor XII. *Proceedings of the National Academy of Sciences of the United States of America*, *99*(2), 896–900. https://doi.org/10.1073/pnas.022626899.

Kahn, R., Hellmark, T., Leeb-Lundberg, L. M. F., Akbari, N., Todiras, M., Olofsson, T., et al. (2009). Neutrophil-derived proteinase 3 induces Kallikrein-independent release of a novel vasoactive Kinin. *The Journal of Immunology*, *182*(12), 7906–7915. https://doi.org/10.4049/jimmunol.0803624.

Kaman, W. E., Wolterink, A. F. W. M., Bader, M., Boele, L. C. L., & van der Kleij, D. (2009). The bradykinin B2 receptor in the early immune response against Listeria infection. *Medical Microbiology and Immunology*, *198*(1), 39–46. https://doi.org/10.1007/s00430-008-0103-4.

Kammerer, S., Braun, A., Arnold, N., & Roscher, A. A. (1995). The human bradykinin B2 receptor gene: Full-length cDNA, genomic organization and identification of the regulatory region. *Biochemical and Biophysical Research Communications*, *211*(1), 226–233. https://doi.org/10.1006/bbrc.1995.1800.

Kaplan, A. P., & Joseph, K. (2014). Pathogenic mechanisms of bradykinin mediated diseases: Dysregulation of an innate inflammatory pathway. *Advances in Immunology*, *121*, 41–89. https://doi.org/10.1016/B978-0-12-800100-4.00002-7.

Kaye, P. M., & Beattie, L. (2016). Lessons from other diseases: Granulomatous inflammation in leishmaniasis. *Seminars in Immunopathology*, *38*(2), 249–260. https://doi.org/10.1007/s00281-015-0548-7.

Kaye, P., & Scott, P. (2011). Leishmaniasis: Complexity at the host–pathogen interface. *Nature Reviews. Microbiology*, *9*(8), 604–615.

Kleinschnitz, C., Stoll, G., Bendszus, M., Schuh, K., Pauer, H.-U., Burfeind, P., et al. (2006). Targeting coagulation factor XII provides protection from pathological thrombosis in cerebral ischemia without interfering with hemostasis. *The Journal of Experimental Medicine*, *203*(3), 513–518. https://doi.org/10.1084/jem.20052458.

Kozik, A., Moore, R. B., Potempa, J., Imamura, T., Rapala-Kozik, M., & Travis, J. (1998). A novel mechanism for bradykinin production at inflammatory sites. Diverse effects of a mixture of neutrophil elastase and mast cell tryptase versus tissue and plasma kallikreins on native and oxidized kininogens. *The Journal of Biological Chemistry*, *273*(50), 33224–33229. Retrieved from http://www.ncbi.nlm.nih.gov/pubmed/9837892.

Kraut, H., Frey, E. K., & Werle, E. (1930). Der Nachweis eines Kreislaufhormons in der Pankreasdrüse (IV. Mitteilung über dieses Kreislaufhormon.). *Hoppe-Seyler's Zeitschrift für Physiologische Chemie*, *189*(3–4), 97–106. https://doi.org/10.1515/bchm2.1930.189.3-4.97.

Kraut, H., Frey, E. K., & Werle, E. (1933). Über den Nachweis und das Vorkommen des Kallikreins im Blut. VIII. Mitteilung über Kallikrein. *Hoppe-Seyler's Zeitschrift für Physiologische Chemie*, *222*(3–4), 73–99. https://doi.org/10.1515/bchm2.1933.222.3-4.73.

Laufs, H., Müller, K., Fleischer, J., Reiling, N., Jahnke, N., Jensenius, J. C., et al. (2002). Intracellular survival of Leishmania major in neutrophil granulocytes after uptake in the absence of heat-labile serum factors. *Infection and Immunity*, *70*(2), 826–835. Retrieved from http://www.ncbi.nlm.nih.gov/pubmed/11796617.

Lee, H. R., Seo, J. H., Kim, O. M., Lee, C. S., Suh, S. W., Hong, Y. M. et al. (1991). Molecular cloning of the ecotin gene in *Escherichia coli*. *FEBS Letters*, *287*(1–2), 53–56. Retrieved from http://www.ncbi.nlm.nih.gov/pubmed/1879537.

Leeb-Lundberg, L. M. F. (2011). Kinin receptor signaling and regulation. In M. Bader (Ed.), *Kinins* (pp. 33–50). Berlin, Boston: DE Gruyter. chapter 3. https://doi.org/10.1515/9783110252354.33.

Leeb-Lundberg, L. M. F., Kang, D. S., Lamb, M. E., & Fathy, D. B. (2001). The human B1 bradykinin receptor exhibits high ligand-independent, constitutive activity: Roles of residues in the fourth intracellular and third transmembrane domains. *Journal of Biological Chemistry*, *276*(12), 8785–8792. https://doi.org/10.1074/jbc.M007396200.

Leeb-Lundberg, L. M. F., Marceau, F., Müller-Esterl, W., Pettibone, D. J., & Zuraw, B. L. (2005). International union of pharmacology. XLV. Classification of the kinin receptor family: From molecular mechanisms to pathophysiological consequences. *Pharmacological Reviews*, *57*(1), 27–77. https://doi.org/10.1124/pr.57.1.2.

Lefèvre, L., Lugo-Villarino, G., Meunier, E., Valentin, A., Olagnier, D., Authier, H., et al. (2013). The C-type lectin receptors Dectin-1, MR, and SIGNR3 contribute both positively and negatively to the macrophage response to Leishmania infantum. *Immunity*, *38*(5), 1038–1049. https://doi.org/10.1016/j.immuni.2013.04.010.

Lerner, E. A., Ribeiro, J. M., Nelson, R. J., & Lerner, M. R. (1991). Isolation of maxadilan, a potent vasodilatory peptide from the salivary glands of the sand fly Lutzomyia longipalpis.

*The Journal of Biological Chemistry*, *266*(17), 11234–11236. Retrieved from http://www.ncbi.nlm.nih.gov/pubmed/2040631.

Lima, A. P. C. A., Almeida, P. C., Tersariol, I. L. S., Schmitz, V., Schmaier, A. H., Juliano, L., et al. (2002). Heparan sulfate modulates kinin release by Trypanosoma cruzi through the activity of cruzipain. *The Journal of Biological Chemistry*, *277*(8), 5875–5881. https://doi.org/10.1074/jbc.M108518200.

Longhurst, H. J., Aberer, W., Bouillet, L., Caballero, T., Fabien, V., Zanichelli, A., et al. (2015). Analysis of characteristics associated with reinjection of icatibant: Results from the Icatibant outcome survey. *Allergy and Asthma Proceedings*, *36*(5), 399–406. https://doi.org/10.2500/aap.2015.36.3892.

Lopes, M. E. M., Carneiro, M. B. H., dos Santos, L. M., & Vieira, L. Q. (2016). Indigenous microbiota and Leishmaniasis. *Parasite Immunology*, *38*(1), 37–44. https://doi.org/10.1111/pim.12279.

Lourbakos, A., Yuan, Y., Jenkins, A. L., Travis, J., Andrade-Gordon, P., Santulli, R., et al. (2001). Activation of protease-activated receptors by gingipains fromPorphyromonas gingivalis leads to platelet aggregation: A new trait in microbial pathogenicity. *Blood*, *97*(12), 3790–3797.

Maas, C., Govers-riemslag, J. W. P., Bouma, B., Schiks, B., Hazenberg, B. P. C., Lokhorst, H. M., et al. (2008). Misfolded proteins activate Factor XII in humans, leading to kallikrein formation without initiating coagulation. *The Journal of Clinical Investigation*, *118*(9), 3208–3218. https://doi.org/10.1172/JCI35424DS1.

Mahdi, F., Madar, Z. S., Figueroa, C. D., & Schmaier, A. H. (2002). Factor XII interacts with the multiprotein assembly of urokinase plasminogen activator receptor, gC1qR, and cytokeratin 1 on endothelial cell membranes. *Blood*, *99*(10), 3585–3596. Retrieved from http://www.ncbi.nlm.nih.gov/pubmed/11986212.

Marin-Neto, J. A., Cunha-Neto, E., Maciel, B. C., & Simões, M. V. (2007). Pathogenesis of chronic Chagas heart disease. *Circulation*, *115*(9), 1109–1123.

Mass, E., Ballesteros, I., Farlik, M., Halbritter, F., Gunther, P., Crozet, L., et al. (2016). Specification of tissue-resident macrophages during organogenesis. *Science*, *353*(6304), aaf4238. https://doi.org/10.1126/science.aaf4238.

Massberg, S., Grahl, L., von Bruehl, M.-L., Manukyan, D., Pfeiler, S., Goosmann, C., et al. (2010). Reciprocal coupling of coagulation and innate immunity via neutrophil serine proteases. *Nature Medicine*, *16*(8), 887–896. https://doi.org/10.1038/nm.2184.

McEachern, A. E., Shelton, E. R., Bhakta, S., Obernolte, R., Bach, C., Zuppan, P., et al. (1991). Expression cloning of a rat B2 bradykinin receptor. *Proceedings of the National Academy of Sciences of the United States of America*, *88*(17), 7724–7728. Retrieved from http://www.ncbi.nlm.nih.gov/pubmed/1715575.

McFarlane, E., Perez, C., Charmoy, M., Allenbach, C., Carter, K. C., Alexander, J., et al. (2008). Neutrophils contribute to development of a protective immune response during onset of infection with Leishmania donovani. *Infection and Immunity*, *76*(2), 532–541. https://doi.org/10.1128/IAI.01388-07.

McIntyre, P., Phillips, E., Skidmore, E., Brown, M., & Webb, M. (1993). Cloned murine bradykinin receptor exhibits a mixed B1 and B2 pharmacological selectivity. *Molecular Pharmacology*, *44*(2), 346–355. Retrieved from http://www.ncbi.nlm.nih.gov/pubmed/8394991.

McLean, P. G., Ahluwalia, A., & Perretti, M. (2000). Association between kinin B(1) receptor expression and leukocyte trafficking across mouse mesenteric postcapillary venules. *The Journal of Experimental Medicine*, *192*(3), 367–380. Retrieved from http://www.ncbi.nlm.nih.gov/pubmed/10934225.

Medeiros, R., Cabrini, D. A., Ferreira, J., Fernandes, E. S., Mori, M. A. S., Pesquero, J. B., et al. (2004). Bradykinin B1 receptor expression induced by tissue damage in the rat portal vein: A critical role for mitogen-activated protein kinase and nuclear factor-

B signaling pathways. *Circulation Research, 94*(10), 1375–1382. https://doi.org/10.1161/01.RES.0000128404.65887.08.

Merino, V. F., Todiras, M., Mori, M. A., Sales, V. M. T., Fonseca, R. G., Saul, V., et al. (2009). Predisposition to atherosclerosis and aortic aneurysms in mice deficient in kinin B1 receptor and apolipoprotein E. *Journal of Molecular Medicine, 87*(10), 953–963. https://doi.org/10.1007/s00109-009-0501-0.

Merkulov, S., Zhang, W.-M., Komar, A. a., Schmaier, A. H., Barnes, E., Zhou, Y., et al. (2008). Deletion of murine kininogen gene 1 (mKng1) causes loss of plasma kininogen and delays thrombosis. *Blood, 111*(3), 1274–1281. https://doi.org/10.1182/blood-2007-06-092338.

Monteiro, A. C., Schmitz, V., Morrot, A., de Arruda, L. B., Nagajyothi, F., Granato, A., et al. (2007). Bradykinin B2 receptors of dendritic cells, acting as sensors of kinins proteolytically released by Trypanosoma cruzi, are critical for the development of protective type-1 responses. *PLoS Pathogens, 3*(11), e185. https://doi.org/10.1371/journal.ppat.0030185.

Monteiro, A. C., Schmitz, V., Svensjo, E., Gazzinelli, R. T., Almeida, I. C., Todorov, A., et al. (2006). Cooperative activation of TLR2 and bradykinin B2 receptor is required for induction of type 1 immunity in a mouse model of subcutaneous infection by Trypanosoma cruzi. *Journal of Immunology (Baltimore, Md.: 1950), 177*(9), 6325–6335. Retrieved from http://www.ncbi.nlm.nih.gov/pubmed/17056563.

Monteiro, A. C., Scovino, A., Raposo, S., Gaze, V. M., Cruz, C., Svensjö, E. et al. (2009). Kinin danger signals proteolytically released by gingipain induce fimbriae-specific IFN-gamma- and IL-17-producing T cells in mice infected intramucosally with Porphyromonas gingivalis. *Journal of Immunology (Baltimore, Md.: 1950), 183*(6), 3700–3711. https://doi.org/10.4049/jimmunol.0900895.

Morand-Contant, M., Anand-Srivastava, M. B., & Couture, R. (2010). Kinin B1 receptor upregulation by angiotensin II and endothelin-1 in rat vascular smooth muscle cells: Receptors and mechanisms. *American Journal of Physiology—Heart and Circulatory Physiology, 299*(5), H1625–H1632.

Moreira, C. R., Schmaier, A. H., Mahdi, F., da Motta, G., Nader, H. B., & Shariat-Madar, Z. (2002). Identification of prolylcarboxypeptidase as the cell matrix-associated prekallikrein activator. *FEBS Letters, 523*(1–3), 167–170. https://doi.org/10.1016/S0014-5793(02)02980-0.

Moreno-Sanchez, D., Hernandez-Ruiz, L., Ruiz, F. A., & Docampo R. (2012). Polyphosphate is a novel pro-inflammatory regulator of mast cells and is located in acidocalcisomes. *The Journal of Biological Chemistry, 287*(34), 28435–28444. https://doi.org/10.1074/jbc.M112.385823.

Müller, F., Mutch, N. J., Schenk, W. A., Smith, S. A., Esterl, L., Spronk, H. M., et al. (2009). Platelet polyphosphates are proinflammatory and procoagulant mediators in vivo. *Cell, 139*(6), 1143–1156. https://doi.org/10.1016/j.cell.2009.11.001.

Munafo, D. B., Johnson, J. L., Brzezinska, A. A., Ellis, B. A., Wood, M. R., & Catz, S. D. (2009). DNase I inhibits a late phase of reactive oxygen species production in neutrophils. *Journal of Innate Immunity, 1*(6), 527–542. https://doi.org/10.1159/000235860.

Murphy, M. L., Cotterell, S. E., Gorak, P. M., Engwerda, C. R., & Kaye, P. M. (1998). Blockade of CTLA-4 enhances host resistance to the intracellular pathogen, Leishmania donovani. *Journal of Immunology (Baltimore, Md.: 1950), 161*(8), 4153–4160. Retrieved from http://www.ncbi.nlm.nih.gov/pubmed/9780188.

Nascimento, C. R., dos Santos Andrade, D., Carvalho-Pinto, C. E., Serra, R. R., Vellasco, L., Brasil, G., et al. (2017). Mast cell coupling to the kallikrein-kinin system fuels intracardiac parasitism and worsens heart pathology in experimental Chagas disease. *Frontiers of Immunology, 8*, 840. https://doi.org/10.3389/fimmu.2017.00840.

Nico, D., Feijó, D. F., Maran, N., Morrot, A., Scharfstein, J., Palatnik, M., et al. (2012). Resistance to visceral leishmaniasis is severely compromised in mice deficient of bradykinin B2-receptors. *Parasites & Vectors*, *5*(1), 261. https://doi.org/10.1186/1756-3305-5-261.

Nielsen, E. W., & Gramstad, S. (2006). Angioedema from angiotensin-converting enzyme (ACE) inhibitor treated with complement 1 (C1) inhibitor concentrate. *Acta Anaesthesiologica Scandinavica*, *50*(1), 120–122. https://doi.org/10.1111/j.1399-6576.2005.00819.x.

Oehmcke, S., Mörgelin, M., & Herwald, H. (2009). Activation of the human contact system on neutrophil extracellular traps. *Journal of Innate Immunity*, *1*(3), 225–230. https://doi.org/10.1159/000203700.

Oschatz, C., Maas, C., Lecher, B., Jansen, T., Björkqvist, J., Tradler, T., et al. (2011). Mast cells increase vascular permeability by heparin-initiated bradykinin formation in vivo. *Immunity*, *34*(2), 258–268. https://doi.org/10.1016/j.immuni.2011.02.008.

Paegelow, I., Trzeczak, S., Böckmann, S., & Vietinghoff, G. (2002). Migratory responses of polymorphonuclear leukocytes to kinin peptides. *Pharmacology*, *66*(3), 153–161. https://doi.org/63797.

Park, J. S., Gamboni-Robertson, F., He, Q., Svetkauskaite, D., Kim, J.-Y., Strassheim, D., et al. (2005). High mobility group box 1 protein interacts with multiple Toll-like receptors. *American Journal of Physiology—Cell Physiology*, *290*(3), C917–C924. https://doi.org/10.1152/ajpcell.00401.2005.

Perdiguero, E. G., & Geissmann, F. (2016). The development and maintenance of resident macrophages. *Nature Immunology*, *17*(1), 2–8. https://doi.org/10.1038/ni.3341.

Pesquero, J. B., Araujo, R. C., Heppenstall, P. A., Stucky, C. L., Silva, J. A., Walther, T., et al. (2000). Hypoalgesia and altered inflammatory responses in mice lacking kinin B1 receptors. *Proceedings of the National Academy of Sciences of the United States of America*, *97*(14), 8140–8145. https://doi.org/10.1073/pnas.120035997.

Pesquero, J. B., Jubilut, G. N., Lindsey, C. J., & Paiva, A. C. (1992). Bradykinin metabolism pathway in the rat pulmonary circulation. *Journal of Hpertension*, *10*(12), 1471–1478. Retrieved from http://www.ncbi.nlm.nih.gov/pubmed/1363432.

Pesquero, J. B., Lindsey, C. J., Zeh, K., Paiva, A. C., Ganten, D., & Bader, M. (1994). Molecular structure and expression of rat bradykinin B2 receptor gene. Evidence for alternative splicing. *The Journal of Biological Chemistry*, *269*(43), 26920–26925. Retrieved from http://www.ncbi.nlm.nih.gov/pubmed/7929432.

Pesquero, J. B., Pesquero, J. L., Oliveira, S. M., Roscher, A. A., Metzger, R., Ganten, D., et al. (1996). Molecular cloning and functional characterization of a mouse bradykinin B1 receptor Gene. *Biochemical and Biophysical Research Communications*, *220*(1), 219–225. https://doi.org/10.1006/bbrc.1996.0384.

Peters, N. C., & Sacks, D. L. (2009). The impact of vector-mediated neutrophil recruitment on cutaneous leishmaniasis. *Cellular Microbiology*, *11*(9), 1290–1296. https://doi.org/10.1111/j.1462-5822.2009.01348.x.

Petrovic, N., Schacke, W., Gahagan, J. R., O'Conor, C. A., Winnicka, B., Conway, R. E., et al. (2007). CD13/APN regulates endothelial invasion and filopodia formation. *Blood*, *110*(1), 142–150. https://doi.org/10.1182/blood-2006-02-002931.

Pinazo, M.-J., Posada, E. d. J., Izquierdo, L., Tassies, D., Marques, A.-F., de Lazzari, E., et al. (2016). Altered hypercoagulability factors in patients with chronic Chagas disease: Potential biomarkers of therapeutic response. *PLoS Neglected Tropical Diseases*, *10*(1), e0004269. https://doi.org/10.1371/journal.pntd.0004269.

Pinto, A. Y. d. N., Ferreira, A. G., Valente, V. d. C., Harada, G. S., & Valente, S. A. d. S. (2009). Urban outbreak of acute Chagas disease in Amazon region of Brazil: Four-year follow-up after treatment with benznidazole. *Revista Panamericana de Salud Publica = Pan American Journal of Public Health*, *25*(1), 77–83. Retrieved from http://www.ncbi.nlm.nih.gov/pubmed/19341528.

Potempa, J., & Herwald, H. (2011). 19 Kinins in bacterial infections. In *Kinins* (pp. 307–320). Berlin, Boston: DE Gruyter. https://doi.org/10.1515/9783110252354.307.

Prat, A., Weinrib, L., Becher, B., Poirier, J., Duquette, P., Couture, R., et al. (1999). Bradykinin B1 receptor expression and function on T lymphocytes in active multiple sclerosis. *Neurology, 53*(9), 2087–2092. Retrieved from http://www.ncbi.nlm.nih.gov/pubmed/10599786.

Prates, D. B., Araujo-Santos, T., Luz, N. F., Andrade, B. B., Franca-Costa, J., Afonso, L., et al. (2011). Lutzomyia longipalpis saliva drives apoptosis and enhances parasite burden in neutrophils. *Journal of Leukocyte Biology, 90*(3), 575–582. https://doi.org/10.1189/jlb.0211105.

Prates, D. B., Santos, L. D., Miranda, J. C., Souza, A. P. A., Palma, M. S., Barral-Netto, M., et al. (2008). Changes in amounts of total salivary gland proteins of Lutzomyia longipallpis (Diptera: Psychodidae) according to age and diet. *Journal of Medical Entomology, 45*(3), 409–413. Retrieved from http://www.ncbi.nlm.nih.gov/pubmed/18533433.

Ratnoff, O. D., & Colopy, J. E. (1955). A familial hemorrhagic trait associated with a deficiency of a clot-promoting fraction of plasma. *The Journal of Clinical Investigation, 34*(4), 602–613. https://doi.org/10.1172/JCI103109.

Regoli, D., & Barabé, J. (1980). Pharmacology of bradykinin and related kinins. *Pharmacological Reviews, 32*(1), 1–46. Retrieved from http://www.ncbi.nlm.nih.gov/pubmed/7015371.

Renne, T., Dedio, J., David, G., & Muller-Esterl, W. (2000). High molecular weight Kininogen utilizes Heparan sulfate proteoglycans for accumulation on endothelial cells. *Journal of Biological Chemistry, 275*(43), 33688–33696. https://doi.org/10.1074/jbc.M000313200.

Renné, T., Pozgajová, M., Grüner, S., Schuh, K., Pauer, H.-U., Burfeind, P , et al. (2005). Defective thrombus formation in mice lacking coagulation factor XII. *The Journal of Experimental Medicine, 202*(2), 271–281. https://doi.org/10.1084/jem.20050664.

Ribeiro, J. M. C. (1987). Role of saliva in blood-feeding by arthropods. *Annual Review of Entomology, 32*(1), 463–478. https://doi.org/10.1146/annurev.en.32.010187.002335.

Rocha e Silva, M., Beraldo, W. T., & Rosenfeld, G. (1949). Bradykinin, a hypotensive and smooth muscle stimulating factor released from plasma globulin by snake venoms and by trypsin. *The American Journal of Physiology, 156*(2), 261–273. Retrieved from http://www.ncbi.nlm.nih.gov/pubmed/18127230.

Rosen, E. D., Gailani, D., & Castellino, F. J. (2002). FXI is essential for thrombus formation following FeCl3-induced injury of the carotid artery in the mouse. *Thrombosis and Haemostasis, 87*(4), 774–776. Retrieved from http://www.ncbi.nlm.nih.gov/pubmed/12008966.

Rust, N. M., Papa, M. P., Scovino, A. M., da Silva, M. M. C., Calzavara-Silva, C. E., Marques, E. T. d. A., et al. (2012). Bradykinin enhances Sindbis virus infection in human brain microvascular endothelial cells. *Virology, 422*(1), 81–91. https://doi.org/10.1016/j.virol.2011.10.003.

Sabourin, T., Morissette, G., Bouthillier, J., Levesque, L., & Marceau, F. (2002). Expression of kinin B 1 receptor in fresh or cultured rabbit aortic smooth muscle: Role of NF-κB. *American Journal of Physiology—Heart and Circulatory Physiology, 283*(1), H227–H237. https://doi.org/10.1152/ajpheart.00978.2001.

Sala-Cunill, A., Björkqvist, J., Senter, R., Guilarte, M., Cardona, V., Labrador, M., et al. (2015). Plasma contact system activation drives anaphylaxis in severe mast cell-mediated allergic reactions. *Journal of Allergy and Clinical Immunology, 135*(4), 1031.e6–1043.e6. https://doi.org/10.1016/j.jaci.2014.07.057.

Salgado, V. R., Queiroz, A. T. L. d., Sanabani, S. S., Oliveira, C. I. d., Carvalho, E. M., Costa, J. M. L., et al. (2016). The microbiological signature of human cutaneous leishmaniasis lesions exhibits restricted bacterial diversity compared to healthy skin. *Memórias do Instituto Oswaldo Cruz, 111*(4), 241–251. https://doi.org/10.1590/0074-02760150436.

Schanstra, J. P., Bataillé, E., Marin Castaño, M. E., Barascud, Y., Hirtz, C., Pesquero, J. B., et al. (1998). The B1-agonist [des-Arg10]-kallidin activates transcription factor NF-kappaB and induces homologous upregulation of the bradykinin B1-receptor in cultured human lung fibroblasts. *The Journal of Clinical Investigation, 101*(10), 2080–2091. https://doi.org/10.1172/JCI1359.

Scharfstein, J. (2016). Kinins. In M. J. Parnham (Ed.), *Compendium of inflammatory diseases* (pp. 815–836). Basel: Springer Basel. https://doi.org/10.1007/978-3-7643-8550-7_127.

Scharfstein, J., Andrade, D., Svensjö, E., Oliveira, A. C., & Nascimento, C. R. (2013). The kallikrein-kinin system in experimental Chagas disease: A paradigm to investigate the impact of inflammatory edema on GPCR-mediated pathways of host cell invasion by Trypanosoma cruzi. *Frontiers in Immunology, 3*, 396. https://doi.org/10.3389/fimmu.2012.00396.

Scharfstein, J., Schmitz, V., Morandi, V., Capella, M. M., Lima, A. P., Morrot, A., et al. (2000). Host cell invasion by Trypanosoma cruzi is potentiated by activation of bradykinin B(2) receptors. *The Journal of Experimental Medicine, 192*(9), 1289–1300. Retrieved from http://www.ncbi.nlm.nih.gov/pubmed/11067878.

Schmaier, A. H. (2008). Assembly, activation, and physiologic influence of the plasma kallikrein/kinin system. *International Immunopharmacology, 8*(2), 161–165. https://doi.org/10.1016/j.intimp.2007.08.022.

Schmaier, A. H. (2016). The contact activation and kallikrein/kinin systems: Pathophysiologic and physiologic activities. *Journal of Thrombosis and Haemostasis, 14*(1), 28–39. https://doi.org/10.1111/jth.13194.

Schmitz, V., Almeida, L. N., Svensjö, E., Monteiro, A. C., Köhl, J., & Scharfstein, J. (2014). C5a and bradykinin receptor cross-talk regulates innate and adaptive immunity in Trypanosoma cruzi infection. *Journal of Immunology (Baltimore, Md: 1950), 193*(7), 3613–3623. https://doi.org/10.4049/jimmunol.1302417.

Schmitz, V., Svensjö, E., Serra, R. R., Teixeira, M. M., & Scharfstein, J. (2009). Proteolytic generation of kinins in tissues infected by Trypanosoma cruzi depends on CXC chemokine secretion by macrophages activated via Toll-like 2 receptors. *Journal of Leukocyte Biology, 85*(6), 1005–1014. https://doi.org/10.1189/jlb.1108693.

Schulze-Topphoff, U., Prat, A., Prozorovski, T., Siffrin, V., Paterka, M., Herz, J., et al. (2009). Activation of kinin receptor B1 limits encephalitogenic T lymphocyte recruitment to the central nervous system. *Nature Medicine, 15*(7), 788–793. https://doi.org/10.1038/nm.1980.

Serbina, N. V., Salazar-Mather, T. P., Biron, C. A., Kuziel, W. A., & Pamer, E. G. (2003). TNF/iNOS-producing dendritic cells mediate innate immune defense against bacterial infection. *Immunity, 19*(1), 59–70. Retrieved from http://www.ncbi.nlm.nih.gov/pubmed/12871639.

Shariat-Madar, Z., Mahdi, F., & Schmaier, A. H. (2002). Identification and characterization of prolylcarboxypeptidase as an endothelial cell prekallikrein activator. *The Journal of Biological Chemistry, 277*(20), 17962–17969. https://doi.org/10.1074/jbc.M106101200.

Shariat-Madar, Z., Mahdi, F., & Schmaier, A. H. (2004). Recombinant prolylcarboxypeptidase activates plasma prekallikrein. *Blood, 103*(12), 4554–4561. https://doi.org/10.1182/blood-2003-07-2510.

Sidel'nikov, I. N., & Obukhova, G. G. (1990). The role of biologically active substances in permeability changes in the blood vessels of patients with hemorrhagic fever with renal syndrome. *Terapevticheskiĭ Arkhiv, 62*(6), 66–68. Retrieved from http://www.ncbi.nlm.nih.gov/pubmed/1977202.

Silverio, J. C., Pereira, I. R., Cipitelli Mda, C., Vinagre, N. F., Rodrigues, M. M., Gazzinelli, R. T., et al. (2012). CD8+ T-cells expressing interferon gamma or perforin play antagonistic roles in heart injury in experimental Trypanosoma Cruzi-elicited cardiomyopathy. *PLoS Pathogens, 8*(4), e1002645. http://dx.doi.org/10.1371/journal.ppat.1002645.

Sriskandan, S., Kemball-Cook, G., Moyes, D., Canvin, J., Tuddenham, E., & Cohen, J. (2000). Contact activation in shock caused by invasive group A Streptococcus pyogenes. *Critical Care Medicine, 28*(11), 3684–3691. Retrieved from http://www.ncbi.nlm.nih.gov/pubmed/11098974.

Stark, K., Philippi, V., Stockhausen, S., Busse, J., Antonelli, A., Miller, M., et al. (2016). Disulfide HMGB1 derived from platelets coordinates venous thrombosis in mice. *Blood, 128*(20), 2435–2449.

Stoka, V., Nycander, M., Lenarcic, B., Labriola, C., Cazzulo, J. J., Björk, I., et al. (1995). Inhibition of cruzipain, the major cysteine proteinase of the protozoan parasite, *Trypanosoma cruzi*, by proteinase inhibitors of the cystatin superfamily. *FEBS Letters, 370*, 101–104.

Svensjö, E., Almeida, L. N. D., Vellasco, L., Juliano, L., & Scharfstein, J. (2014). Ecotin-like ISP of *L. major* promastigotes fine-tunes macrophage phagocytosis by limiting the pericellular release of bradykinin from surface-bound kininogens: A survival strategy based on the silencing of proinflammatory G-protein coupled Kinin B 2 and B1 receptors. *Mediators of Inflammation, 2014*, 143450. https://doi.org/10.1155/2014/143450.

Svensjö, E., Batista, P. R., Brodskyn, C. I., Silva, R., Lima, A. P. C. A., Schmitz, V., et al. (2006). Interplay between parasite cysteine proteases and the host kinin system modulates microvascular leakage and macrophage infection by promastigotes of the Leishmania donovani complex. *Microbes and Infection, 8*(1), 206–220. https://doi.org/10.1016/j.micinf.2005.06.016.

Tanowitz, H. B., Huang, H., Jelicks, L. A., Chandra, M., Loredo, M. L., Weiss, L. M., et al. (2005). Role of endothelin 1 in the pathogenesis of chronic chagasic heart disease. *Infection and Immunity, 73*(4), 2496–2503. https://doi.org/10.1128/IAI.73.4.2496-2503.2005.

Taylor, S. L., Wahl-Jensen, V., Copeland, A. M., Jahrling, P. B., & Schmaljohn, C. S. (2013). Endothelial cell permeability during hantavirus infection involves factor XII-dependent increased activation of the Kallikrein-Kinin system. *PLoS Pathogens, 9*(7), e1003470. https://doi.org/10.1371/journal.ppat.1003470.

Tibayrenc, M. (2010). Modelling the transmission of Trypanosoma cruzi: The need for an integrated genetic epidemiological and population genomics approach. *Advances in Experimental Medicine and Biology, 673*, 200–211. Retrieved from http://www.ncbi.nlm.nih.gov/pubmed/20632539.

Todorov, A. G., Andrade, D., Pesquero, J. B., Araujo, R. d. C., Bader, M., Stewart, J., et al. (2003). Trypanosoma cruzi induces edematogenic responses in mice and invades cardiomyocytes and endothelial cells in vitro by activating distinct kinin receptor (B1/B2) subtypes. *FASEB Journal: Official Publication of the Federation of American Societies for Experimental Biology, 17*(1), 73–75. https://doi.org/10.1096/fj.02-0477fje.

Travis, J., Pike, R., Imamura, T., & Potempa, J. (1997). Porphyromonas gingivalis proteinases as virulence factors in the development of periodontitis. *Journal of Periodontal Research, 32*(1 Pt. 2), 120–125. Retrieved from http://www.ncbi.nlm.nih.gov/pubmed/9085221.

Vavrek, R. J., & Stewart, J. M. (1985). Competitive antagonists of bradykinin. *Peptides, 6*(2), 161–164. Retrieved from http://www.ncbi.nlm.nih.gov/pubmed/4034408.

Venema, R. C. (2002). Post-translational mechanisms of endothelial nitric oxide synthase regulation by bradykinin. *International Immunopharmacology, 2*(13–14), 1755–1762. Retrieved from http://www.ncbi.nlm.nih.gov/pubmed/12489789.

Verhoef, J. J. F., Barendrecht, A. D., Nickel, K. F., Dijkxhoorn, K., Kenne, E., Labberton, L., et al. (2017). Polyphosphate nanoparticles on the platelet surface trigger contact system activation. *Blood, 129*(12), 1707–1717. https://doi.org/10.1182/blood-2016-08-734988.

Vianna, R. M., & Calixto, J. B. (1998). Characterization of the receptor and the mechanisms underlying the inflammatory response induced by des-Arg9-BK in mouse pleurisy. *British Journal of Pharmacology*, *123*(2), 281–291. https://doi.org/10.1038/sj.bjp.0701590.

Vogel, S., Bodenstein, R., Chen, Q., Feil, S., Feil, R., Rheinlaender, J., et al. (2015). Platelet-derived HMGB1 is a critical mediator of thrombosis. *Journal of Clinical Investigation*, *125*(12), 4638–4654. https://doi.org/10.1172/JCI81660.

Volfova, V., Hostomska, J., Cerny, M., Votypka, J., & Volf, P. (2008). Hyaluronidase of bloodsucking insects and its enhancing effect on Leishmania infection in mice. *PLoS Neglected Tropical Diseases*, *2*(9), e294. https://doi.org/10.1371/journal.pntd.0000294.

Wang, M., Shakhatreh, M.-A. K., James, D., Liang, S., Nishiyama, S.-I., Yoshimura, F., et al. (2007). Fimbrial proteins of porphyromonas gingivalis mediate in vivo virulence and exploit TLR2 and complement receptor 3 to persist in macrophages. *Journal of Immunology (Baltimore, Md.: 1950)*, *179*(4), 2349–2358. Retrieved from http://www.ncbi.nlm.nih.gov/pubmed/17675496.

Werle, E., Götze, W., & Kappler, A. (1937). Über die Wirkung des Kallikreins auf den isolierten Darm und über eine neue darmkontrahierende Substanz. *Biochemische Zeitschrift*, *289*, 217–233.

Werle, E., Kehl, R., & Koebke, K. (1950). Bradykinin, kallidin and hypertension. *Biochemische Zeitschrift*, *320*(4), 372–383. Retrieved from http://www.ncbi.nlm.nih.gov/pubmed/14777780.

Westermann, D., Walther, T., Savvatis, K., Escher, F., Sobirey, M., Riad, A., et al. (2009). Gene deletion of the kinin receptor B1 attenuates cardiac inflammation and fibrosis during the development of experimental diabetic cardiomyopathy. *Diabetes*, *58*(6), 1373–1381.

Wuillemin, W. A., Fijnvandraat, K., Derkx, B. H., Peters, M., Vreede, W., ten Cate, H., et al. (1995). Activation of the intrinsic pathway of coagulation in children with meningococcal septic shock. *Thrombosis and Haemostasis*, *74*(6), 1436–1441. Retrieved from http://www.ncbi.nlm.nih.gov/pubmed/8772216.

Yang, A., Dai, J., Xie, Z., Colman, R. W., Wu, Q., Birge, R. B., et al. (2014). High molecular weight Kininogen binds phosphatidylserine and opsonizes Urokinase plasminogen activator receptor-mediated Efferocytosis. *The Journal of Immunology*, *192*(9), 4398–4408.

Yang, H. Y., Erdös, E. G., & Levin, Y. (1970). A dipeptidyl carboxypeptidase that converts angiotensin I and inactivates bradykinin. *Biochimica et Biophysica Acta*, *214*(2), 374–376. Retrieved from http://www.ncbi.nlm.nih.gov/pubmed/4322742.

Zhang, Y., Brovkovych, V., Brovkovych, S., Tan, F., Lee, B.-S., Sharma, T., et al. (2007). Dynamic receptor-dependent activation of inducible nitric-oxide synthase by ERK-mediated phosphorylation of Ser745. *The Journal of Biological Chemistry*, *282*(44), 32453–32461. https://doi.org/10.1074/jbc.M706242200.

Zhang, X., Tan, F., & Skidgel, R. A. (2013). Carboxypeptidase M is a positive allosteric modulator of the kinin B1 receptor. *The Journal of Biological Chemistry*, *288*(46), 33226–33240. https://doi.org/10.1074/jbc.M113.520791.

Zhang, X., Tan, F., Zhang, Y., & Skidgel, R. A. (2008). Carboxypeptidase M and kinin B1 receptors interact to facilitate efficient b1 signaling from B2 agonists. *The Journal of Biological Chemistry*, *283*(12), 7994–8004. https://doi.org/10.1074/jbc.M709837200.

Zhou, Q., & Amar, S. (2007). Identification of signaling pathways in macrophage exposed to Porphyromonas gingivalis or to its purified cell wall components. *Journal of Immunology (Baltimore, Md.: 1950)*, *179*(11), 7777–7790. Retrieved from http://www.ncbi.nlm.nih.gov/pubmed/18025224.

CHAPTER THREE

# P2Y Receptors in Immune Response and Inflammation

**Diana Le Duc*, Angela Schulz*, Vera Lede*, Annelie Schulze*, Doreen Thor*, Antje Brüser*, Torsten Schöneberg[†,1]**

*Rudolf Schönheimer Institute of Biochemistry, Molecular Biochemistry, University of Leipzig, Leipzig, Germany
[†]Medical Faculty, University of Leipzig, Leipzig, Germany
[1]Corresponding author: e-mail address: schoberg@medizin.uni-leipzig.de

## Contents

1. Introduction 86
2. Molecular Structure and Pharmacology of P2Y Receptors 87
   2.1 Evolution and Genomic Structures of P2Y Receptor Genes 87
   2.2 The Molecular Structure of P2Y Receptors 90
   2.3 Nucleotides as Agonists and Signaling Molecules 91
   2.4 Mechanisms of P2Y Receptor Activation, Signaling, and Termination 98
3. Expression and Physiological Function of P2Y Receptors in Immune Cells 99
4. Animal Models of P2Y Deficiency 100
5. Human Phenotypes and Diseases Involving P2Y Receptors 103
6. Clinical Perspectives: P2Y Receptors as Drug Targets in Immune Diseases 105
   6.1 Drugs Targeting P2Y Receptors 105
   6.2 Immune Modulation by P2Y Receptor Drugs 106
   6.3 P2Y in Transplantation 107
   6.4 P2Y in Inflammatory States 108
7. Conclusions/Perspective 109
Acknowledgments 109
References 109
Further Reading 121

## Abstract

Metabotropic pyrimidine and purine nucleotide receptors (P2Y receptors) are expressed in virtually all cells with implications in very diverse biological functions, including the well-established platelet aggregation (P2Y12), but also immune regulation and inflammation. The classical P2Y receptors bind nucleotides and are encoded by eight genes with limited sequence homology, while phylogenetically related receptors (e.g., P2Y12-like) recognize lipids and peptides, but also nucleotide derivatives. Growing lines of evidence suggest an important function of P2Y receptors in immune cell differentiation and maturation, migration, and cell apoptosis. Here, we give a perspective on the P2Y receptors' molecular structure and physiological importance in

*Advances in Immunology*, Volume 136
ISSN 0065-2776
http://dx.doi.org/10.1016/bs.ai.2017.05.006

immune cells, as well as the related diseases and P2Y-targeting therapies. Extensive research is being undertaken to find modulators of P2Y receptors and uncover their physiological roles. We anticipate the medical applications of P2Y modulators and their immune relevance.

## 1. INTRODUCTION

According to the modern paradigm of life origins, living organisms emerged on the primitive earth through a process of chemical evolution, which resulted in the formation of biomonomers in the primordial seas (Miller, 1974). A recent study reported the synthesis of purine and pyrimidine nucleobases starting from one of the most abundant molecules in the Universe, formamide (Kumar, Sharma, & Kamaluddin, 2014). This suggests the important roles they may have played in the prebiotic chemical evolution. Indeed, purinergic signaling, which uses purines and pyrimidines as chemical transmitters, has a long evolutionary history, being omnipresent across species and tissues, with a high versatility (Verkhratsky & Burnstock, 2014). ATP can act as a "damage signaler" (Burnstock & Boeynaems, 2014; Verkhratsky & Burnstock, 2014), a function that became conserved in evolution, such that ATP is one of the main mediators of numerous systems of biological defense (Verkhratsky & Burnstock, 2014). Hence, cell surface receptors evolved to respond to this signal. The evolutionary oldest receptors are the P2X ATP-gated cation channels, present already in the protozoa (Verkhratsky & Burnstock, 2014), while the metabotropic P2Y receptors are evolutionary much younger with the first representatives of the class present in sharks and rays (Hoyle, 2011). P2Y receptors are a family of G protein-coupled receptors (GPCRs) stimulated by different nucleotides and are present in virtually all tissues, with very diverse functions from neurotransmission (Guzman & Gerevich, 2016) and penile erection (Gur & Hellstrom, 2009) to immunity (Burnstock & Boeynaems, 2014).

Once in extracellular fluids, ATP can be degraded by ectonucleotidases into ADP, which is further degraded to AMP and eventually to adenosine. Purinergic receptors have been progressively characterized and in 1978 were initially subdivided by Burnstock into P1 (adenosine) and P2 (ATP, ADP) (Burnstock, 1978). In 1985, P2 receptors were further subdivided into ionotropic P2X and metabotropic P2Y receptors (Burnstock & Kennedy, 1985). The metabotropic P2Y receptors belong to the rhodopsin-like

GPCR family (family/class A or 1), the largest group of GPCRs in vertebrates. Although functional nucleotide receptors were cloned from nonmammals, like p2y$_3$ in *Gallus gallus* (Webb et al., 1993), p2y$_8$ in *Xenopus laevis* (Bogdanov, Dale, King, Whittock, & Burnstock, 1997), and p2y$_{turkey}$ in *Meleagris gallopavo* (Boyer, Waldo, & Harden, 1997), NC-IUPHAR classifies formally only eight mammalian P2Ys (P2Y1, 2, 4, 6, 11–14) (Burnstock et al., 2016). According to the preferential agonist, the eight receptors can be further classified into adenine nucleotide-activated (P2Y1, P2Y11, P2Y12, P2Y13), pyrimidine nucleotide-activated (P2Y4, P2Y6), ATP/UTP-activated (P2Y2), and UDP-sugar-activated (P2Y14) receptors. In addition to these cloned and functionally characterized P2Y receptors, there may exist additional, yet unidentified ones (Burnstock et al., 2016).

The main difficulty in characterizing this class of receptors comes from the fact that many tissues express several P2Y receptors (Schoneberg et al., 2007). The system is even more difficult to outline, given the lack of specific agonists and antagonists of most P2Y receptors and also the absence of specific antibodies for the extracellular domains (Schwiebert, 2003). Moreover, most tissues release nucleotides which act autocrinally and affect the activity of the receptor under study (Schwiebert, 2003). Also ectonucleotidases metabolize and interconvert nucleotide agonists, making it even more difficult to functionally characterize the receptors (Harden, Lazarowski, & Boucher, 1997).

The presented pitfalls may explain why these receptors remained poorly characterized compared to other GPCRs. The involvement of the P2Y receptors in immune functions has only recently met a better understanding. Mouse models deficient for the individual P2Y receptors helped in dissecting the specific relevance of nucleotide receptors. In this chapter, we will provide a perspective specifically on the role of P2Y receptors in the immune system regulation, their involvement in different immune-related diseases, and the potential of P2Y therapeutic targeting.

## 2. MOLECULAR STRUCTURE AND PHARMACOLOGY OF P2Y RECEPTORS

### 2.1 Evolution and Genomic Structures of P2Y Receptor Genes

With the availability of numerous invertebrate and vertebrate genomes the evolutionary history of genes can be reconstructed. Nucleotides and their derivatives function as signaling molecules already in unicellular organisms via GPCRs, such as cAMP in *Dictyostelium* (Manahan, Iglesias, Long, & Devreotes, 2004). However, prototypical P2Y receptors, as we know them from all vertebrates, seem to be an innovation that occurred about 550 million years ago since none of the invertebrate and hemichordate genomes sequenced to date contain related sequences (Krishnan, Almen, Fredriksson, & Schioth, 2013). This is coincidental with many other innovations during the Cambrian explosion including a complex adaptive immune system (Sunyer, 2013). The general presence in all living organisms makes nucleotides most probably useless as specific signal for an adaptive immune response. However, being present in almost all immune cells one can speculate that P2Y receptors specifically modulate immune functions by sensing nucleotides as "master signal" for all kinds of private and foreign cell damages. P2Y receptor-like sequences can be found in jawless lampreys and are present in all bony and cartilaginous fishes sequenced to date by screening with, e.g., human P2Y1 or P2Y12 receptor sequences (Fig. 1).

In-depth phylogenetic analyses of GPCRs have shown that P2Y receptors cluster when compared with other GPCRs of family A (Cvicek, Goddard, & Abrol, 2016; Krishnan, Almen, Fredriksson, & Schioth, 2012). However, the adenine nucleotide-activated P2Y1 and P2Y12, for example, display only minor structural relation at the protein level and phylogenetic analyses support that nucleotide specificity, e.g., for ADP/ATP evolved independently in P2Y1 and P2Y12 (Schoneberg et al., 2007). This is further supported by the fact that within the crystal structures of P2Y1 and P2Y12, the agonist-binding sites significantly differ between both receptors (Zhang et al., 2015; Zhang, Zhang, Gao, Paoletta, et al., 2014; Zhang, Zhang, Gao, Zhang, et al., 2014). Interestingly, several other structurally related GPCRs, such as leukotriene receptor CysLT1R, the succinate receptor GPR91, the 2-oxoglutarate receptor GPR99, and the orphan GPR34, cluster into one of the both groups. Based on their amino acid sequences P2Y receptors can be subdivided into at least two groups (Fig. 1). One comprises P2Y1, 2, 4, 6, and 11, while the second group contains P2Y12–14. This phylogenetic clustering correlates in parts with the genomic localization of P2Y receptors. Chromosomal clustering in the human genome is found for P2Y1, P2Y12, P2Y13, P2Y14, and the P2Y12-related receptors GPR87 and GPR171 at chromosome 3 and for P2Y2 and P2Y6 at chromosome 11, indicating multiple rounds of gene duplications.

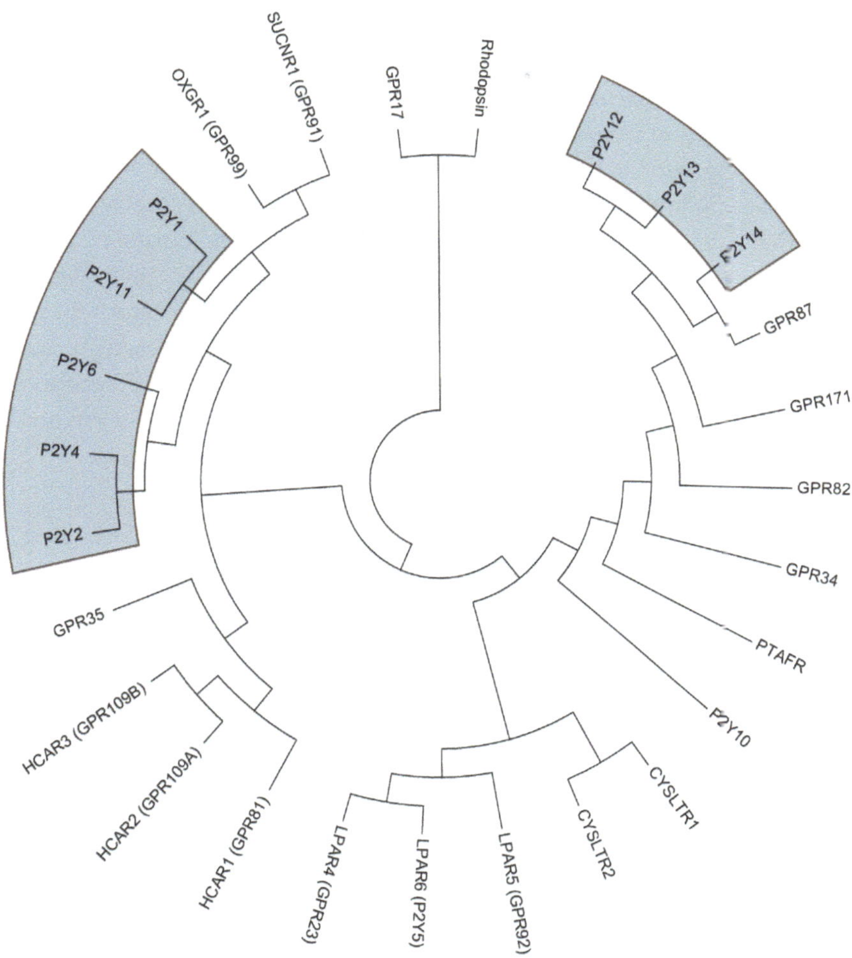


**Fig. 1** Molecular phylogenetic analysis of P2Y receptors and related GPCRs by the Maximum Likelihood method. Classical P2Y receptors (*blue boxed*) cluster in the δ subfamily of rhodopsin-like GPCR (shown) together with cysteinyl leukotriene receptors (CYSLTR#), lysophospholipid receptors (LPAR#), the platelet-activating factor receptor (PTAFR), the hydroxycarboxylic acid receptors (HCAR#), the succinate receptor (SUCNR1), the oxoglutarate receptor (OXGR1), and several orphan GPCRs (GPR#). Human protein sequences were aligned with ClustalW using the PAM matrix and default parameters. The sequence of the human rhodopsin served as outgroup. The evolutionary history was inferred by using the Maximum Likelihood method based on the JTT matrix-based model (Jones, Taylor, & Thornton, 1992). The tree with the highest log likelihood (−15198.2312) is shown. Initial tree(s) for the heuristic search were obtained automatically by applying Neighbor-Join and BioNJ algorithms to a matrix of pairwise distances estimated using a JTT model, and then selecting the topology with superior log likelihood value. The analysis involved 27 amino acid sequences. All positions containing gaps and missing data were eliminated. There was a total of 273 positions in the final dataset. Evolutionary analyses were conducted in MEGA7 (Kumar, Stecher, & Tamura, 2016).

Most coding regions of human P2Y receptors do not contain introns. However, introns in the 5′UTR seem to be common. Some contain cryptic introns in their N terminus-coding regions (P2Y6, GPR87, GPR34). The genomic structure is mainly conserved during evolution. However, introns in the coding regions of several P2Y receptor genes newly appear in some fish species (Schoneberg et al., 2007). The human P2Y11, located at chromosome 19, is an exception in many aspects. P2Y11 orthologs are missing in murine genomes and several human splice variants exist modifying the coding region of the N terminus and even producing a fusion protein with an adjacent gene (PPAN) (Dreisig & Kornum, 2016). Interestingly, these chimeric transcripts are upregulated during granulocyte differentiation; however, the functional features of this fusion protein are largely unknown (Communi, Suarez-Huerta, Dussossoy, Savi, & Boeynaems, 2001).

## 2.2 The Molecular Structure of P2Y Receptors

Functional identification and characterization of P2Y receptors in cell expression systems is problematic because of the abundance of endogenous nucleotide receptors, nucleosidases, and nucleotide release. Therefore, many attempts have been made to identify sequence signatures which may help in annotation and grouping of P2Y receptors (Costanzi, Mamedova, Gao, & Jacobson, 2004; Ivanov, Costanzi, & Jacobson, 2006). Sequence alignments of P2Y receptors do not provide any specific signature for these receptors; moreover, GPCRs having nonnucleotide agonists (leukotrienes, organic acids) cluster within the P2Y-like receptors (Cvicek et al., 2016; Krishnan et al., 2012), lending additional support to the lack of P2Y receptor-specific signatures.

Numerous mutagenesis studies on P2Y receptors, often combined with homology modeling, attempted to identify positions and residues involved in nucleotide binding and specificity. For example, His6.52/Arg6.55 and Lys7.35/Glu7.36/Leu7.39 within transmembrane helix 6 (TM6) and the extracellular loop 3 (ECL3), respectively, were proposed to be involved in nucleotide binding (Abbracchio et al., 2003; Deflorian & Jacobson, 2011; Hoffmann, Sixel, Di Pasquale, & von Kugelgen, 2008; Ignatovica, Megnis, Lapins, Schioth, & Klovins, 2012; Schmidt et al., 2013). In-depth phylogenetic and mutagenesis studies addressing this region in the ADP/ATP receptor P2Y12 could show that all these residues are 100% conserved among species and that mutation of these positions by any other amino acid abolished receptor function (Coster et al., 2012). However, such residue combination is also present in some GPR87, GPR171, and GPR34 orthologs (all P2Y12-like receptors), which are not activated by ADP or ATP. This does not rule out that these residues are involved in nucleotide binding of, e.g., P2Y12, but it implicates additional positions which determine ligand specificity.

Crystal structures of GPCRs, including also molecular structures of P2Y1 (Zhang et al., 2015) and P2Y12 (Zhang, Zhang, Gao, Paoletta, et al., 2014; Zhang, Zhang, Gao, Zhang, et al., 2014), are a major breakthrough of the last decade. They highlight the relevance of the diphosphate (pyrophosphate) group in binding and orientating the nucleotide within the receptor molecule. For example, in the crystallographic pose of 2-MeSADP at P2Y12 (Zhang, Zhang, Gao, Paoletta, et al., 2014) Lys7.35 directly interacts with the β phosphate moiety of the agonist as proposed in homology models (Schmidt et al., 2013). Although the general positioning of the nucleoside moiety in the models was correct, the specific interacting amino acid partners differ between the models and the crystal structure. The crystal structure was generated with a fusion protein and an additional mutation (Asp7.49 to Asn), which somehow improved protein purification and stabilized the receptor for crystallization. However, Asp7.49 is 100% conserved among P2Y12 orthologs and within the P2Y12-like group, and functional analysis of the Asp7.49Asn mutant revealed partial loss of function (Coster et al., 2012). One can, therefore, speculate that the in vivo fine structure of P2Y12 and the position of the agonists may partially differ from the recent crystallography-based model. However, the overall arrangements of the transmembrane helices and even the positioning of some agonist-binding pocket-relevant key residues, such as Lys7.35, mainly overlap between P2Y1 and P2Y12 in the overlay of the structures (Zhang et al., 2015). Interestingly, the identification of an additional antagonist-binding cavity raises the possibility of a second potential nucleotide-binding site in P2Y1. This may provide a clue how dinucleotides can specifically bind to P2Y receptors (see Section 2.3) (Jacobson et al., 2015).

## 2.3 Nucleotides as Agonists and Signaling Molecules

Nucleotides are key compounds in manifold metabolic pathways of the cell. They are also present extracellularly in high concentrations upon active and passive release. Here, nucleotides and their derivatives can serve as signaling molecules specifically detected by membrane-bound receptors. After considerable resistance, it is now generally accepted that nucleotides are genuine signaling molecules being involved in a variety of physiological functions, e.g., in the nervous and immune systems. Classically, the purine (ATP, ADP) and pyrimidine nucleotides (UTP, UDP, UDP-sugars) are considered as the main signaling molecules (Table 1). However, there is evidence that several P2Y receptors can also be activated by other naturally occurring purine and pyrimidine nucleotides (CTP, GTP, ITP) (Bogdanov et al., 1997; Kennedy, Qi, Herold, Harden, & Nicholas, 2000). Diadenosine polyphosphates are nucleotide metabolites present in secretory vesicles together with other chemical messengers (Schluter, Tepel, & Zidek, 1996). For example, diadenosine tri- and tetraphosphates (AP3A, AP4A) are stored in high concentrations in platelet-dense granules and are released when platelets become activated (Luthje & Ogilvie, 1983). AP4A activates P2Y2 and P2Y4 (Castro, Pintor, & Miras-Portugal, 1992; Communi, Motte, Boeynaems, & Pirotton, 1996; Patel et al., 2001), while it seems to be an antagonist at P2Y13 (Marteau et al., 2003). Furthermore, AP4C has good activity at P2Y1, whereas UP4C is an agonist at P2Y2 (Shaver et al., 2005).

In contrast to specific cAMP receptors, forming a separate group of 7TM receptors in *Dictyostelium discoideum*, in mammals, cyclic nucleotides (cAMP, cGMP) have not been identified as extracellular signaling molecules acting via GPCRs. Further, purine nucleotide derivatives such as NAD+ have been identified as agonists on the P2Y11 (Moreschi et al., 2006).

However, the set of endogenous ligands interacting with P2Y receptor seems to be even more complex, suggesting network-modulating and fine-tuning nucleotide receptors (Volonte, Amadio, D'Ambrosi, Colpi, & Burnstock, 2006). As shown in Fig. 1, P2Y receptors cluster in the δ subfamily of rhodopsin-like GPCR together with leukotriene, lipid, and short fatty acid metabolite receptors (Fredriksson, Lagerstrom, Lundin, & Schioth, 2003). It is therefore not surprising that binding of nonnucleotide ligands to P2Y receptors was anecdotally observed, among them several immune response-relevant metabolites. For example, the binding of cysteinyl leukotriene E4 to the human P2Y12 is controversially discussed (Foster, Fuerst, Lee, Cousins, & Woszczek, 2013; Nonaka, Hiramoto, & Fujita, 2005). In a broad screening approach, we have recently identified the prostaglandin $E_2$ glyceryl ester ($PGE_2$-G) as agonist for P2Y6. Previous work suggests that $PGE_2$-G activates a GPCR in the murine macrophage-like cell line RAW264.7 (Nirodi, Crews, Kozak, Morrow, & Marnett, 2004). Interestingly, we showed that $PGE_2$-G and UDP are both agonists at P2Y6, but they

**Table 1** Expression and Function of the Classical P2Y and P2Y-Like Receptors in Cells of the Immune System

| Receptor Name | Classical Nucleotide Ligand | Cell Type | Outcome of Activation | References |
|---|---|---|---|---|
| P2Y1 | ADP > ATP | Platelets | Platelet aggregation | Nylander, Mattsson, Ramstrom, and Lindahl (2003) |
| | | Eosinophils | Chemotaxis and reactive oxygen metabolites production | Ferrari et al. (2000) and Mohanty, Raible, McDermott, Pelleg, and Schulman (2001) |
| | | Monocyte-derived DCs | Reduced capacity to attract monocytes | Horckmans et al. (2006) |
| | | Neutrophils | Undefined roles | Burnstock and Boeynaems (2014) and Jacob, Perez Novo, Bachert, and Van Crombruggen (2013) |
| | | Macrophages | Intracellular calcium increase | Bowler, Bailey, North, and Surprenant (2003), Coutinho-Silva et al. (2005), and Myrtek et al. (2008) |
| | | Mast cells | Calcium influx | Feng, Mery, Beller, Favot, and Boyce (2004) and Schulman et al. (1999) |
| | | B and T cells, CD34+ stem cells | Undefined roles | Lee, Park, Kong, Kim, and Han (2006) and Wang, Jacobsen, Bengtsson, and Erlinge (2004) |
| P2Y2 | ATP = UTP | Neutrophils | Calcium mobilization, neutrophil activation, chemotaxis, primary granule release | Chen et al. (2006, 2010), Meshki, Tuluc, Bredetean, Ding, and Kunapuli (2004), and Meshki, Tuluc, Bredetean, Garcia, and Kunapuli (2006) |
| | | Macrophages | Intracellular calcium increase, chemotaxis,[a] phagocytosis | Elliott et al. (2009), Bowler et al. (2003), Coutinho-Silva et al. (2005), Myrtek et al. (2008), Isfort et al. (2011), and Kronlage et al. (2010) |
| | | Dendritic cells | Migration | Muller et al. (2010) |
| | | Eosinophils | Chemotaxis and cytokines release | Muller et al. (2010), Kobayashi, Kouzaki, and Kita (2010), and Vanderstocken et al. (2010) |
| | | Mast cells | Histamine release | Feng et al. (2004) and Schulman et al. (1999) |
| | | B and T cells, CD34+ stem cells | Undefined roles | Lee et al. (2006) and Wang et al. (2004) |

*Continued*

**Table 1** Expression and Function of the Classical P2Y and P2Y-Like Receptors in Cells of the Immune System—cont'd

| Receptor Name | Classical Nucleotide Ligand | Cell Type | Outcome of Activation | References |
|---|---|---|---|---|
| P2Y4 | UTP > ATP | Neutrophils | Undefined roles | Chen, Shukla, Namiki, Insel, and Junger (2004) |
| | | Macrophages | Intracellular calcium increase | Coutinho-Silva et al. (2005) and Myrtek et al. (2008) |
| | | Microglia | Microglial pinocytosis | Li et al. (2013) |
| | | Dendritic cells | Undefined roles | Berchtold et al. (1999) |
| | | Eosinophils | Activation | Ferrari et al. (2000) |
| | | B and T cells, CD34+ stem cells | Undefined roles | Lee et al. (2006) and Wang et al. (2004) |
| P2Y6 | UDP ≫ UTP $PGE_2$-glycerylester | Neutrophils | Possible involvement in chemotaxis | Kukulski et al. (2009) |
| | | Macrophages | Chemotaxis | Myrtek et al. (2008), Bar et al. (2008), and Kimura et al. (2014) |
| | | Dendritic cells | Undefined roles | Jacob et al. (2013) and Berchtold et al. (1999) |
| | | Eosinophils | Chemokines release | Idzko et al. (2003) |
| | | B and T cells, CD34+ stem cells | Undefined roles | Lee et al. (2006) and Wang et al. (2004) |
| P2Y11 | ATP > UTP and NAADP | Granulocytes | Activation | Moreschi et al. (2008) |
| | | Neutrophils | Inhibition of constitutive neutrophil apoptosis | Vaughan et al. (2007) |
| | | Macrophages | Intracellular calcium release | Coutinho-Silva et al. (2005) and Myrtek et al. (2008) |
| | | Dendritic cells | Dendritic cells maturation, cytokines release, migration | Schnurr et al. (2003), Wilkin et al. (2001), and Wilkin, Stordeur, Goldman, Boeynaems, and Robaye (2002) |
| | | Natural killer cells | Regulation of chemotaxis and cytotoxicity | Gorini et al. (2010) |
| | | Eosinophils | Intracellular calcium increase | Ferrari et al. (2000) |
| | | B and T cells, CD34+ stem cells | Undefined roles | Lee et al. (2006) and Wang et al. (2004) |

| | | | | |
|---|---|---|---|---|
| P2Y12 | ADP > ATP | Platelets | Platelet aggregation | Nylander et al. (2003) |
| | | Monocytes | Monocyte activation | Wang et al. (2004) |
| | | Macrophages | Chemotaxis and phagocytosis | Isfort et al. (2011) and Kronlage et al. (2010) |
| | | Dendritic cells | Dendritic cell maturation and endocytosis | Ben Addi, Cammarata, Conley, Boeynaems, and Robaye (2010) and Marteau, Communi, Boeynaems, and Suarez Gonzalez (2004) |
| | | Mast cells | Possible cytokine release inhibition | Feng et al. (2004) |
| | | B and T cells, CD34+ stem cells | Undefined roles | Lee et al. (2006) and Wang et al. (2004) |
| P2Y13 | ADP ≫ ATP | Red blood cells | Reduces ATP release | Wang et al. (2005) |
| | | Monocytes | Activation | Wang et al. (2004) |
| | | Macrophages | Possible intracellular calcium increase | Myrtek et al. (2008) |
| | | Mast cells | Degranulation | Feng et al. (2004) and Gao, Ding, and Jacobson (2010a) |
| | | B and T cells, CD34+ stem cells | Undefined roles | Lee et al. (2006) and Wang et al. (2004) |
| P2Y14 | UDP, UDP-glucose | Neutrophils | Neutrophil chemotaxis, reduced cAMP accumulation | Barrett et al. (2013) and Scrivens and Dickenson (2006) |
| | | Macrophages | Possible intracellular calcium increase | Myrtek et al. (2008) |
| | | Dendritic cells | Dendritic cell maturation | Skelton, Cooper, Murphy, and Platt (2003) |
| | | Eosinophils | Undefined roles | Jacob et al. (2013) |
| | | Mast cells | Degranulation | Gao, Ding, and Jacobson (2010b) and Gao, Wei, Jayasekara, and Jacobson (2013) |
| | | B and T cells, CD34+ stem cells | T cell cAMP inhibition<br>Stem cell chemotaxis, repopulation | Scrivens and Dickenson (2005) |

*Continued*

**Table 1** Expression and Function of the Classical P2Y and P2Y-Like Receptors in Cells of the Immune System—cont'd

| Receptor Name | Classical Nucleotide Ligand | Cell Type | Outcome of Activation | References |
|---|---|---|---|---|
| *P2Y-like receptors* | | | | |
| GPR34 | *Lyso*-PS[b] | Granulocytes | Undefined roles | Liebscher et al. (2011) |
| | | Monocytes | Undefined roles | Liebscher et al. (2011) |
| | | Microglia | Morphology and phagocytosis | Preissler et al. (2015) |
| | | Dendritic cells | Apoptosis | Jager et al. (2016) |
| GPR171 | BigLEN | Lymph node and spleen | Undefined roles | Unpublished own data |
| | | *Spleen*:<br>B cells<br>$T_H$ cells<br>$T_C$ cells<br>Natural killer cells | Undefined roles | Unpublished own data |
| | | *Lymphocytes*:<br>B cells<br>T cells<br>Natural killer cells | Downregulation of myeloid differentiation | Rossi, Lemoli, and Goodell (2013) |
| | | Dendritic cells | Undefined roles | Unpublished own data |
| GPR82 | Orphan | Mononuclear cells | Undefined roles | Krug et al. (2012) and Teles et al. (2013) |
| | | Dendritic cells | Undefined roles | Ricciardi et al. (2008) |
| | | T cells | Undefined roles | Wang, Ciuffi, Leipzig, Berry, and Bushman (2007) |
| GPR87 | LPA[c] | Dendritic cells | Undefined roles | Njau, Geffers, Thalmann, Haller, and Wagner (2009) |
| | | Macrophages | Undefined roles | Kazeros et al. (2008) |

| | | | | |
|---|---|---|---|---|
| GPR17 | UDP, UDP-galactose, UDP-glucose, leukotrienes[d] | Oligodendrocytes | Differentiation, myelination | Boccazzi et al. (2016), Marucci et al. (2016), and Ou et al. (2016) |

[a]Results are controversial regarding chemotaxis signals: initially considered long-range "find-me" signals (Elliott et al., 2009); subsequent studies using real-time chemotaxis assays proposed that ATP signaling is strictly autocrine or paracrine acting as short-range "touch me" signals (Isfort et al., 2011).
[b]The agonist *lyso*-phosphatidylserine (*lyso*-PS) is controversial since the human GPR34 is activated by *lyso*-PS in vitro (Sugo et al., 2006), but other studies using GPR34-deficient mice do not support *lyso*-PS as the endogenous agonist at GPR34 (Liebscher et al., 2011). *Lyso*-PS was suggested to have only random agonistic activity at some GPR34 orthologs (Ritscher et al., 2012). Modified nucleotides can bind to GPR34 (Ritscher et al., 2012).
[c]Lysophosphatidic acid (LPA) is still under debate whether it is indeed an agonist for GPR87 (Tabata, Baba, Shiraishi, Ito, & Fujita, 2007) or not (Niss Arfelt et al., 2016).
[d]It is still under debate whether UDP nucleotides and leukotrienes are agonists at GPR17 (see text).

activate the receptor with extremely different $EC_{50}$ values of ~1 pM and ~50 nM, respectively (Brüser et al., 2017). One can speculate that P2Y6 integrates the two different chemical signals related to cell damage into a common intracellular response.

Using in silico docking studies and functional assays 5-phosphoribosyl 1-pyrophosphate has been identified as agonist at the P2Y1 and P2Y12 (Hiramoto et al., 2004; Nonaka et al., 2005). The cholesterol biosynthesis intermediate farnesyl pyrophosphate is an antagonist at the P2Y12 (Hogberg et al., 2012). These findings may highlight the importance of the pyrophosphate-binding site in P2Y receptors (see above).

The P2Y12-like group contains a number of orphan GPCRs (GPR34, GPR82, GPR87, GPR171) where the endogenous agonist is not known or still under debate. The human GPR34 was shown to be activated by *lyso*-phosphatidylserine (*lyso*-PS) in vitro (Sugo et al., 2006). *Lyso*-PS is generated by hydrolysis of membrane lipids through phospholipases A1 and A2. *Lyso*-PS is a potent activator of histamine release from mast cells (Bellini, Viola, Menegus, Toffano, & Bruni, 1990). However, studies with GPR34-deficient mice did not support *lyso*-PS as the endogenous *agonist* at GPR34 (Liebscher et al., 2011; Preissler et al., 2015), and in vitro studies suggest that *lyso*-PS has only a random agonistic activity at some GPR34 orthologs (Ritscher et al., 2012). Interestingly, modified nucleotides can bind to GPR34 (Ritscher et al., 2012). No agonists are reported for GPR82, and the finding that lysophosphatidic acid (LPA) is an agonist at GPR87 (Tabata et al., 2007) is controversially discussed (Niss Arfelt et al., 2016). Recently, GPR171 has been deorphanized by showing that a peptide, BigLEN, activates the receptor and regulates appetite in mice (Gomes et al., 2013). GPR17 was primary deorphanized as a receptor for UDP, UDP-glucose, and cysteinyl leukotrienes LTC4 and LTD4 (Ciana et al., 2006). GPR17 couples to Gi/o protein (Simon et al., 2016). However, the action of nucleotides and cysteinyl leukotrienes as agonists at GPR17 is controversially discussed (Benned-Jensen & Rosenkilde, 2010; Hennen et al., 2013; Qi, Harden, & Nicholas, 2013; Simon et al., 2017).

## 2.4 Mechanisms of P2Y Receptor Activation, Signaling, and Termination

P2Y receptors as classical GPCRs activate G protein-dependent signaling cascades upon agonist binding. Thus, the members of the P2Y1 receptor subgroup (P2Y1, P2Y2, P2Y4, P2Y6, and P2Y11) all couple to Gαq/11 proteins activating phospholipase C. P2Y11 also couples to the Gαs protein/adenylyl cyclase pathway (Qi, Kennedy, Harden, & Nicholas, 2001). The members of the P2Y12 receptor subgroup (GPR34, GPR171, GPR87, P2Y12, P2Y13, P2Y14) couple to Gαi/o proteins. Overexpresssion of GPR87 in transiently transfected HEK293 cells revealed ligand-independent coupling to Gαi, Gαq, and Gα12/13 proteins (Niss Arfelt et al., 2016). Agonist-induced internalization seems to be common also for P2Y receptors. For example, the P2Y1 and P2Y12 are internalized by arrestin recruitment (Nisar et al., 2012; Reiner et al., 2009).

Not only receptor internalization and desensitization switch off P2Y signaling (Cunningham, Nisar, & Mundell, 2013) but also degradation of nucleotides by ectonucleotidases. Adenosine is the main product of the sequential hydrolysis of extracellular ATP catalyzed by CD39 and CD73. Adenosine itself acts as agonist on adenosine receptors, which are also GPCRs, and has mainly antiinflammatory properties on immune cells (Linden & Cekic, 2012).

## 3. EXPRESSION AND PHYSIOLOGICAL FUNCTION OF P2Y RECEPTORS IN IMMUNE CELLS

P2Y receptors are expressed in virtually all cells including immune cells. Since GPCRs escape usually from immunological detections in vivo because of low protein levels (Michel, Wieland, & Tsujimoto, 2009), the wealth of expression information comes from RNA-sequencing data. In macrophages (highest expression: P2Y2, P2Y14, P2Y6), dendritic cells (P2Y14, P2Y6, GPR171, P2Y13), and microglia (P2Y12, GPR34, P2Y13, P2Y6) P2Y receptors are among the most abundant GPCRs as demonstrated by RNA-sequencing studies (Hickman et al., 2013; Jager et al., 2016; Preissler et al., 2015). P2Y12 and GPR34 are even considered as a molecular signature for microglia compared to other neuronal cells (Hickman et al., 2013; Preissler et al., 2015). Therefore, P2Y receptors are involved in a number of functions related to immune responses, among them chemotaxis, phagocytosis, and granule release (Table 1). Nucleotides are actively released from synaptic vesicles (neurons) and granules (mononuclear cells, endocrine cells) and may act as (co-)transmitters of immune synapses and auto/paracrine transmitters. Further, ATP and other nucleotides are released by damaged or stressed cells as a "find-me signal" (Elliott et al., 2009). Indeed, P2Y receptor-deficient mouse models showed that several P2Y receptors (P2Y2, P2Y6, P2Y12) and P2Y12-like receptors

(GPR34) are involved in properly orchestrating chemotaxis and phagocytosis of peripheral macrophages (Elliott et al., 2009) and microglia (Haynes et al., 2006; Koizumi et al., 2007; Preissler et al., 2015). However, it is under debate whether the released nucleotides and their degradation products indeed "guide" phagocytes toward inflammatory sites. P2Y receptor activation increases random migration, which may also promote clearance of damaged and apoptotic cells (Junger, 2011). This is supported by a recent study showing that P2Y12 is linked to toll-like receptor 2-mediated random migration of microglia (Ifuku, Buonfiglioli, Jordan, Lehnardt, & Kettenmann, 2016).

The induction of vesicle degranulation is another important function of P2Y receptors. P2Y1 and P2Y12 are major components of the autocrine purinergic signaling system that amplifies platelet aggregation by auto- and paracrine release of ATP and ADP from platelet granules (Burnstock, 2015). P2Y2-induced and pannexin 1-based ATP release has been shown to promote chemotaxis of neutrophils (Junger, 2011). Degranulation and release of histamine from mast cells has been linked to the activation of P2Y2, P2Y13, and P2Y14 (see Table 1).

Besides the well-established role of nucleotide signaling in migration and degranulation of immune and neural cells, activation of P2Y receptors is highly relevant for stem cell proliferation and differentiation (Kaebisch, Schipper, Babczyk, & Tobiasch, 2015). These functions on stem cells seem to be important also for immune cells and their progenitors. For example, GPR171 function negatively regulates myeloid differentiation (Rossi et al., 2013). P2Y14 has a restricted expression in primitive cells in the hematopoietic lineage (Lee et al., 2003), and it is among the most prominent regulators of murine hematopoietic stem cell repopulation (Holmfeldt et al., 2016).

## 4. ANIMAL MODELS OF P2Y DEFICIENCY

For most P2Y receptors (except for P2Y11 because there is no ortholog in rodents and GPR171, for which there is an in vivo knock-down) gene-deficient mouse strains have been generated and characterized. All P2Y-deficient mouse models are vital under standard animal housing conditions but most present distinct phenotypes among them altered immune functions (see Table 2).

As expected from the well-established effect of ATP and ADP and P2Y receptor antagonists on platelets, P2Y1- and P2Y12-deficient mouse strains

**Table 2** P2Y Receptor-Deficient Mouse Models and Their Phenotypes

| Receptor Name | Main Phenotype | Immune Phenotype | References |
|---|---|---|---|
| P2Y1 | Increased bleeding time<br>Decreased platelet aggregation | Microglia-triggered IL-6 secretion from astrocytes | Fabre et al. (1999), Leon et al. (1999), and Shinozaki et al. (2014) |
| P2Y2 | Abnormal neuron differentiation<br>Decreased vasodilation | Dysregulated airway epithelial ion transport<br>Altered apoptotic cell clearance by macrophages | Elliott et al. (2009), Arthur, Akassoglou, and Insel (2005), Cressman et al. (1999), and Wang et al. (2015) |
| P2Y4 | Abnormal digestive secretion | | Robaye et al. (2003) |
| P2Y6 | Altered UDP-induced aorta contraction | Macrophage dysfunction<br>Altered allergen-induced pulmonary inflammation<br>Microglial dysfunction<br>Role in vascular inflammation | Bar et al. (2008), Giannattasio et al. (2011), and Riegel et al. (2011) |
| P2Y12 | Decreased platelet aggregation<br>Increased bleeding time | Microglia migration/activation deficiency<br>Involved in sepsis-induced lung injury<br>Changed cytokine profile of dendritic cells | Haynes et al. (2006), Foster et al. (2001), Andre et al. (2003), Liverani et al. (2016), and Zhang et al. (2017) |
| P2Y13 | Altered plasma and hepatic cholesterol and lipid levels<br>Altered bone development | | Blom et al. (2010), Fabre et al. (2010), and Wang, Robaye, Gossiel, Boeynaems, and Gartland (2014) |

*Continued*

**Table 2** P2Y Receptor-Deficient Mouse Models and Their Phenotypes—cont'd

| Receptor Name | Main Phenotype | Immune Phenotype | References |
|---|---|---|---|
| P2Y14 | Decreased insulin secretion and glucose tolerance<br>Impaired gastric peristalsis<br>Increased mean systemic arterial blood pressure<br>Altered smooth muscle function | Decreased airway responsiveness<br>Increased spleen weight | Bassil et al. (2009) and Meister et al. (2014) |
| GPR17 | Altered oligodendrocyte myelination | Increased vascular permeability | Maekawa, Balestrieri, Austen, and Kanaoka (2009) and Chen et al. (2009) |
| GPR34 | Immune impairment upon pathogen challenge | Macrophage dysfunction<br>Microglial dysfunction | Liebscher et al. (2011) and Preissler et al. (2015) |
| GPR82 | Reduced food intake and body weight | Impaired immune response in DTH test (own observation) | Engel et al. (2011) |
| GPR87 | Abnormal femur morphology<br>Small vertebrae | | The European Mouse Mutant Archive (EMMA) (2017) |
| GPR171 | Altered hypothalamic appetite regulation (in vivo knockdown) | | Gomes et al. (2013) |

displayed altered platelet aggregation upon ADP administration (Foster et al., 2001; Leon et al., 1999). These investigations highlight that both P2Y1 and P2Y12 play essential roles in controlling vascular occlusion and thrombotic states. The immune function of P2Y12 was previously underappreciated until studying its microglial function (see above). Similarly, P2Y6 seems to function as a sensor for diffusible UDP from damaged cells to recruit microglia (Koizumi et al., 2007). In the peripheral immune system, P2Y6 protects the lung against exuberant allergen-induced pulmonary inflammation by inhibiting activation of effector T cells (Giannattasio et al., 2011).

## 5. HUMAN PHENOTYPES AND DISEASES INVOLVING P2Y RECEPTORS

Next-generation sequencing (NGS) data from population genetic studies exposed an unexpected variability of genes in humans. Most variants, so-called single-nucleotide polymorphisms (SNPs), are without an obvious relevance on gene function. However, some SNPs can have functional impact with phenotypical consequences. Identification of genetic variants contributing to phenotypes and even susceptibility to diseases is crucial for understanding the complexity of human health and disease. *Genome*-wide linkage analyses were successfully applied for a number of other single-gene Mendelian diseases, and with the advantage of high-throughput genotyping and sequencing technologies, genome-wide association studies (GWAS) became feasible and made a major contribution to understand the genetics of complex traits and diseases better. Most of the SNPs identified in GWAS are located in inter- or noncoding intragenic regions, suggesting their regulatory role in gene expression. Therefore, one of the crucial steps following GWAS is to identify and validate target genes potentially affected by the associated SNP. Expression quantitative trait locus (eQTL) analyses allow for correlating genotype and tissue-specific gene expression levels. The integrative approach combining GWAS and eQTL studies allows clarifying whether the genotype–phenotype association uncovered in a GWAS might be mediated through the regulation of a gene transcription by the corresponding SNP. With respect to P2Y receptors, several loci carrying the respective P2Y receptor genes have been shown to cosegregate with complex traits and diseases (Table 3). However, only a few are related to immune system dysfunctions.

There is a possible association between variation in the P2Y11 gene (noncoding region) and susceptibility to narcolepsy. The disease-associated

**Table 3** Association of Genetic Variants in P2Y Receptor Genes Identified in GWAS

| Genetic Locus | | | | |
|---|---|---|---|---|
| **Gene Name** | **GPCR** | **SNP** | **Association of Genetic Variants in GWAS-Associated Trait** | **References** |
| *P2RY1* | P2Y1 | | Platelet aggregation | Hetherington et al. (2005) |
| *P2RY2* | P2Y2 | rs10898909 | Glucose homeostasis | Palmer et al. (2015) |
| | | rs1791933 | Insomnia | Byrne et al. (2012) |
| | | rs11603160 | Body mass index in asthmatics | Melen et al. (2013) |
| | | rs4944831 | Hypertension | Wang et al. (2010) |
| *P2RY11* | P2Y11 | rs1551570 | Narcolepsy | Kornum et al. (2011) |
| *P2RY12* | P2Y12 | rs2046934 | Clinical outcome of extracranial or intracranial stenosis | Li et al. (2016) |
| *P2RY13* | P2RY13 | | Asthma (eQTL) | |
| *P2RY14* | P2Y14 | rs1554120 | Heschl's gyrus morphology<br>Asthma (eQTL) | Cai et al. (2014) |
| *GPR17* | GPR17 | | Schizophrenia | Jia et al. (2012) |

GWAS signals according to the GWAS Catalog (Welter et al., 2014).

allele was correlated with reduced expression of P2Y11 in CD8+ T lymphocytes and natural killer cells. P2Y11, as an important regulator of immune cell survival, may have implications in narcolepsy and other autoimmune diseases (Kornum et al., 2011). Expressed QTL analyses detected an association of P2Y13 and P2Y14 expression with asthma (Ferreira et al., 2017). There is evidence that upregulation of GPR34 plays a role in the development of cancers, such as lymphomas (Wlodarska et al., 2009) and gastric adenocarcinoma (Yu et al., 2013). GPR34 was identified as an alternative trigger for p185Bcr-Abl-induced leukemogenesis: knock-down of GPR34 was sufficient to suppress proliferation and survival in vitro and lead to an attenuated leukemogenesis in vivo (Zuo et al., 2014).

In principle, sequence variants can modulate the activity of a given GPCR in two directions—activation (gain of function) and inactivation (loss of function). Activating and inactivating mutations in GPCR can cause human diseases with contrasting phenotypes but also convergent phenotypes (Schoneberg et al., 2004). Patients with platelet-type bleeding disorder due to a 2-bp deletion and compound heterozygous missense mutations in the P2Y12 gene (both are inactivating mutations) were reported (Cattaneo et al., 2003; Hollopeter et al., 2001). As the P2Y1 is also involved in platelet aggregation, it is reasonable that variants of P2Y1 have an impact on platelet function as well (Hetherington et al., 2005). Other monogenic diseases based on functionally relevant mutations in P2Y receptors have not been reported yet, although mouse models are suggestive for further pathologic phenotypes also in humans under certain conditions (see Table 2).

# 6. CLINICAL PERSPECTIVES: P2Y RECEPTORS AS DRUG TARGETS IN IMMUNE DISEASES

## 6.1 Drugs Targeting P2Y Receptors

Although there are numerous studies indicating a potential use of synthetic P2Y receptors agonists/antagonists in the treatment of selected diseases, only two P2Y receptors are target of approved drug therapies. P2Y12 antagonists are broadly used as antithrombotics in the treatment of acute coronary syndrome. The prototypical drug, clopidogrel (thienopyridine), is a prodrug that has as active metabolite a thiol derivative, which irreversibly binds to and inhibits P2Y12. Because of its delayed onset of action and interindividual variability in platelet response, e.g., due to genetic polymorphisms of the metabolizing enzyme, new P2Y12 antagonists (prasugrel, cangrelor, ticagrelor, elinogrel) have been developed to improve the pharmacological and clinical profile of clopidogrel (Ferri, Corsini, & Bellosta, 2013). There is some evidence that besides their antithrombotic action P2Y12 inhibitors may have beneficial effects by blocking microglia function. It was shown in an animal model that ticagrelor can reduce stroke damage by inhibition of P2Y12-mediated microglia activation and chemotaxis (Gelosa et al., 2014).

The P2Y2 dinucleotide agonist diquafosol is approved in treatment of the dry eye syndrome including dry eyes occurring in the Sjögren's syndrome (Bremond-Gignac, Gicquel, & Chiambaretta, 2014; Lau, Samarawickrama, & Skalicky, 2014). The benefit of activation of chloride-ion secretion by P2Y2 agonist, denufosol, was clinically tested in cystic fibrosis (Kellerman et al., 2008), but eventually failed in the late phases of clinical trials

(Ratjen et al., 2012). The proinflammatory role of P2Y2 signaling may have been responsible for the disappointing results in the long-term treatment of cystic fibrosis.

## 6.2 Immune Modulation by P2Y Receptor Drugs

Activation of pyrimidinergic P2Y receptors are involved in controlling *Toxoplasma gondii* infection in macrophages (Moreira-Souza et al., 2015). There is also evidence linking *T. gondii* infection to P2Y2-induced chloride secretion, which may provide a new target for the treatment of pneumonia caused by this parasite (Guo et al., 2015). It was shown that dendritic cell function and Th2 lymphocyte recruitment during lung inflammation involve P2Y2 activation, and the lack of P2Y2 increases the morbidity and mortality rate in response to pneumonia virus infection of mice (Vanderstocken et al., 2012). P2Y2 activation induces keratinocyte proliferation (Dixon et al., 1999) and neutrophil migration (Chen et al., 2006), suggesting a potential use of P2Y2 antagonists in treating psoriasis (Conroy, Kindon, Kellam, & Stocks, 2016). P2Y2 antagonists have also been suggested as potential antimetastatic therapies, since P2Y2 activation by ATP released from tumor cell-activated platelets opens the endothelial barrier and allows cancer cell extravasation (Schumacher, Strilic, Sivaraj, Wettschureck, & Offermanns, 2013). While the P2Y2 agonist diquafosol has already been approved for use, an antagonist molecule is still being researched (Conroy et al., 2016).

P2Y4 is highly expressed in the intestine (Robaye et al., 2003), and just like in the case of P2Y2, P2Y4 is the main mediator of UTP-stimulated chloride secretion in the small and large intestine (Ghanem et al., 2005). Future P2Y4 antagonists could be thus relevant for treating diarrheas, whether infectious or not.

*Clostridium difficile* is the leading cause of nosocomial diarrhea in the developed countries. Toxins released by *C. difficile* trigger CXCL8/IL-8, which attracts neutrophils and induces an inflammatory response with pseudomembranous colitis. Inhibition of the P2Y6 receptor attenuates the release of CXCL8/IL-8, inflammation, and increased intestinal permeability after *C. difficile* infection (Hansen et al., 2013). Hence, P2Y6 antagonists may be useful for treating such diarrheas. Furthermore, P2Y6 was also shown to mediate inflammatory responses to monosodium urate crystals (Uratsuji et al., 2012). Monosodium urate crystals also induce the release of CXCL8/IL-8, IL-1α, and IL-6 from human keratinocytes, which are reduced after P2Y6 blockade. Thus, it was postulated that monosodium urate crystals-associated inflammatory diseases, such as tophaceous gout, may benefit from a P2Y6 antagonist. Although a couple of molecules are being used as P2Y6 antagonists, their physical properties are not the best premises for oral drug discovery programs (Conroy et al., 2016).

However, the unequivocal role of P2Y antagonists in preventing inflammation is demonstrated by the P2Y12 antagonists. Although their main use is in preventing arterial thrombosis, it has been increasingly accepted that platelets play an important role in inflammation and immune responses. P2Y12 antagonists reduce release of proinflammatory α-granule contents from platelets, the formation of proinflammatory platelet–leukocyte aggregates, and reduce markers of systemic inflammation such as tumor necrosis factor α and C-reactive protein following acute coronary syndromes (Thomas & Storey, 2015). Ticagrelor, the more potent inhibitor of P2Y12, compared to clopidogrel, has better outcomes on pulmonary infections and sepsis (James et al., 2009), suggesting again the involvement of P2Y12 in inflammation and immunity.

## 6.3 P2Y in Transplantation

Conditions such as cellular necrosis, ischemia/reperfusion, and activation of immune inflammatory cells can trigger ATP release (Eltzschig, Sitkovsky, & Robson, 2012). Such conditions are usually met in the context of allotransplantation, and ATP is lately considered a signal of the alloimmune response (Vergani et al., 2014). After the ATP egress from intracellular stores, the entire purinergic system is activated, including P2X, P2Y receptors, and after metabolization to adenosine, the P1 receptors. To date only the P2X7 was suggested to be clinically relevant in transplant survival (Vergani et al., 2013). While upregulation of P2X7 is directly related to the alloimmune response (Vergani et al., 2013), P2Y receptors may also play an indirect role, although P2Y receptors have not yet been specifically assessed in transplantation. To this end, P2Y-mediated vascular inflammation might be important in allograft rejection. Systemic lipopolysaccharide challenge induces P2Y6-dependent vascular inflammation (Riegel et al., 2011), which renders P2Y6 as a possible target for graft vasculopathy (Vergani et al., 2014).

## 6.4 P2Y in Inflammatory States

P2Y receptors contribute to the fine balance of inflammatory states, having both pro- and antiinflammatory effects. Their ambivalent function makes them both friends and foe during inflammation (Idzko, Ferrari, & Eltzschig, 2014).

As previously stated P2Y2 activation was related to enhanced mucociliary clearance, which made denufosol a good candidate for cystic fibrosis treatment. However, the lack of pulmonary function improvement on long-term treatment (Ratjen et al., 2012) may be related to a P2Y2-mediated neutrophil-induced hyperinflammation. To this end, in patients with chronic obstructive pulmonary disease, neutrophils express higher levels of P2Y2, which leads to higher migration and elastolytic activity (Lommatzsch et al., 2010). Similarly, upregulation of P2Y2 induces higher migration and production of reactive oxygen species in eosinophils and monocyte-derived dendritic cells from patients with allergies (Muller et al., 2010). Conversely, P2Y2 agonists promote wound healing (Gendaszewska-Darmach & Kucharska, 2011) by recruitment of leukocytes to the site of tissue damage and supporting the clearance of apoptotic cells and bacteria through macrophages (Elliott et al., 2009) and neutrophil activation (Chen et al., 2006, 2010). Hence, P2Y2 signaling can also contribute to resolution of inflammation.

P2Y6 has also been implicated in promoting inflammation in allergy states like asthma (Vieira et al., 2011). As discussed earlier, P2Y6 activation can determine endothelial inflammation (Riegel et al., 2011). Also, experimentally induced inflammatory bowel disease leads to upregulation of both P2Y2 and P2Y6 (Grbic, Degagne, Langlois, Dupuis, & Gendron, 2008). But just like in the case of P2Y2, P2Y6 may contribute to the clearance of dying cells by enhancing the phagocytic capacity of microglia (Koizumi et al., 2007), thereby reducing the inflammation at sites with injured neurons.

Clinical observations have suggested the role of P2Y12 in inflammation (see above). The observations are supported by murine studies, which demonstrate a P2Y12-dependent release of the proasthmatic leukotriene LTE4 (Paruchuri et al., 2009) and P2Y12-mediated activation of allergen-specific T cells (Ben Addi et al., 2010). However, additional research is necessary to decide whether P2Y12 ligands are useful in the treatment or prevention of allergic states.

## 7. CONCLUSIONS/PERSPECTIVE

After almost 40 years starting from a controversial debate, if nucleotides at all are transmitters, the nucleotide P2Y receptors evolved to a highly appreciated group of rhodopsin-like GPCRs. This group is still expanding by group members and by agonists activating P2Y receptors. Classical P2Y receptors, which bind nucleotides, were found to be activated by dinucleoside polyphosphates and even by endogenous compounds with no relation to nucleotides. Nonclassical but phylogenetically related receptors (e.g., P2Y12-like) recognize lipids and peptides, but also nucleotide derivatives. This situation of related and unrelated receptors with multiple and in parts overlapping agonist specificities will challenge the classical pharmacological dogma that receptors are primarily sorted and grouped by their endogenous agonists. Besides pharmacological properties and phylogenetic relation, structural information will add to this debate. We will learn whether given binding pockets with distinct specificities evolved from one prototype of nucleotide-binding site and/or from convergent evolution which led to new and clearly distinct binding sites for identical nucleotide ligands.

The advent of transgenic animals and modern sequencing technologies disclosed a number of previously unknown physiological functions besides their relevance in neurotransmission. These new data point to a high relevance of P2Y receptors in various immune functions, among them stem cell maintenance, immune cell differentiation and maturation, migration, and cell apoptosis. There is currently a high discrepancy between the obviously high physiological importance and the therapeutically targeting P2Y receptor subtypes. Surely, the current major efforts in developing specific ligands and the ongoing dissection of the physiological function of P2Y receptors will ultimately lead to an increased relevance in medical application.

## ACKNOWLEDGMENTS

We thank Diana Tiesel and Heidi Funke for their contributions to the expression analysis of GPR171 in immune cells. This work was supported by the Deutsche Forschung sgemeinschaft (Scho 624/9-1).

## REFERENCES

Abbracchio, M. P., Boeynaems, J. M., Barnard, E. A., Boyer, J. L., Kennedy, C., Miras-Portugal, M. T., et al. (2003). Characterization of the UDP-glucose receptor (re-named here the P2Y14 receptor) adds diversity to the P2Y receptor family. *Trends in Pharmacological Sciences, 24*, 52–55.

Andre, P., Delaney, S. M., LaRocca, T., Vincent, D., DeGuzman, F., Jurek, M., et al. (2003). P2Y12 regulates platelet adhesion/activation, thrombus growth, and thrombus stability in injured arteries. *The Journal of Clinical Investigation*, *112*, 398–406.

Arthur, D. B., Akassoglou, K., & Insel, P. A. (2005). P2Y2 receptor activates nerve growth factor/TrkA signaling to enhance neuronal differentiation. *Proceedings of the National Academy of Sciences of the United States of America*, *102*, 19138–19143.

Bar, I., Guns, P. J., Metallo, J., Cammarata, D., Wilkin, F., Boeynams, J. M., et al. (2008). Knockout mice reveal a role for P2Y6 receptor in macrophages, endothelial cells, and vascular smooth muscle cells. *Molecular Pharmacology*, *74*, 777–784.

Barrett, M. O., Sesma, J. I., Ball, C. B., Jayasekara, P. S., Jacobson, K. A., Lazarowski, E. R., et al. (2013). A selective high-affinity antagonist of the P2Y14 receptor inhibits UDP-glucose-stimulated chemotaxis of human neutrophils. *Molecular Pharmacology*, *84*, 41–49.

Bassil, A. K., Bourdu, S., Townson, K. A., Wheeldon, A., Jarvie, E. M., Zebda, N., et al. (2009). UDP-glucose modulates gastric function through P2Y14 receptor-dependent and -independent mechanisms. *American Journal of Physiology. Gastrointestinal and Liver Physiology*, *296*, G923–G930.

Bellini, F., Viola, G., Menegus, A. M., Toffano, G., & Bruni, A. (1990). Signalling mechanism in the lysophosphatidylserine-induced activation of mouse mast cells. *Biochimica et Biophysica Acta*, *1052*, 216–220.

Ben Addi, A., Cammarata, D., Conley, P. B., Boeynaems, J. M., & Robaye, B. (2010). Role of the P2Y12 receptor in the modulation of murine dendritic cell function by ADP. *Journal of Immunology*, *185*, 5900–5906.

Benned-Jensen, T., & Rosenkilde, M. M. (2010). Distinct expression and ligand-binding profiles of two constitutively active GPR17 splice variants. *British Journal of Pharmacology*, *159*, 1092–1105.

Berchtold, S., Ogilvie, A. L., Bogdan, C., Muhl-Zurbes, P., Ogilvie, A., Schuler, G., et al. (1999). Human monocyte derived dendritic cells express functional P2X and P2Y receptors as well as ecto-nucleotidases. *FEBS Letters*, *458*, 424–428.

Blom, D., Yamin, T. T., Champy, M. F., Selloum, M., Bedu, E., Carballo-Jane, E., et al. (2010). Altered lipoprotein metabolism in P2Y(13) knockout mice. *Biochimica et Biophysica Acta*, *1801*, 1349–1360.

Boccazzi, M., Lecca, D., Marangon, D., Guagnini, F., Abbracchio, M. P., & Ceruti, S. (2016). A new role for the P2Y-like GPR17 receptor in the modulation of multipotency of oligodendrocyte precursor cells in vitro. *Purinergic Signal*, *12*, 661–672.

Bogdanov, Y. D., Dale, L., King, B. F., Whittock, N., & Burnstock, G. (1997). Early expression of a novel nucleotide receptor in the neural plate of Xenopus embryos. *The Journal of Biological Chemistry*, *272*, 12583–12590.

Bowler, J. W., Bailey, R. J., North, R. A., & Surprenant, A. (2003). P2X4, P2Y1 and P2Y2 receptors on rat alveolar macrophages. *British Journal of Pharmacology*, *140*, 567–575.

Boyer, J. L., Waldo, G. L., & Harden, T. K. (1997). Molecular cloning and expression of an avian G protein-coupled P2Y receptor. *Molecular Pharmacology*, *52*, 928–934.

Bremond-Gignac, D., Gicquel, J. J., & Chiambaretta, F. (2014). Pharmacokinetic evaluation of diquafosol tetrasodium for the treatment of Sjogren's syndrome. *Expert Opinion on Drug Metabolism & Toxicology*, *10*, 905–913.

Brüser, A., Zimmermann, A., Crews, B. C., Sliwoski, G., Meiler, J., König, G. M., et al. (2017). Prostaglandin $E_2$ glyceryl ester is an endogenous agonist of the nucleotide receptor $P2Y_6$. *Scientific Reports,* 7, 2380. http://dx.doi.org/10.1038/s41598-017-02414-8.

Burnstock, G. (1978). A basis for distinguishing two types of purinergic receptor. In L. Bolis & R. W. Straub (Eds.), *Cell membrane receptors for drugs and hormones: A multidisciplinary approach* (pp. 107–118). New York: Raven Press.

Burnstock, G. (2015). Blood cells: An historical account of the roles of purinergic signalling. *Purinergic Signal, 11*, 411–434.
Burnstock, G., Abbracchio, M. P., Boeynaems, J. M., Boyer, J. L., Ceruti, S., Fumagalli, M., et al. (2016). P2Y receptors, introduction. In *IUPHAR/BPS guide to pharmacology*.
Burnstock, G., & Boeynaems, J. M. (2014). Purinergic signalling and immune cells. *Purinergic Signal, 10*, 529–564.
Burnstock, G., & Kennedy, C. (1985). Is there a basis for distinguishing two types of P 2-purinoceptor? *General Pharmacology: The Vascular System, 16*, 433–440.
Byrne, E. M., Johnson, J., McRae, A. F., Nyholt, D. R., Medland, S. E., Gehrman, P. R., et al. (2012). A genome-wide association study of caffeine-related sleep disturbance: Confirmation of a role for a common variant in the adenosine receptor. *Sleep, 35*, 967–975.
Cai, D. C., Fonteijn, H., Guadalupe, T., Zwiers, M., Wittfeld, K., Teumer, A., et al. (2014). A genome-wide search for quantitative trait loci affecting the cortical surface area and thickness of Heschl's gyrus. *Genes, Brain, and Behavior, 13*, 675–685.
Castro, E., Pintor, J., & Miras-Portugal, M. T. (1992). Ca(2+)-stores mobilization by diadenosine tetraphosphate, Ap4A, through a putative P2Y purinoceptor in adrenal chromaffin cells. *British Journal of Pharmacology, 106*, 833–837.
Cattaneo, M., Zighetti, M. L., Lombardi, R., Martinez, C., Lecchi, A., Conley, P. B., et al. (2003). Molecular bases of defective signal transduction in the platelet P2Y12 receptor of a patient with congenital bleeding. *Proceedings of the National Academy of Sciences of the United States of America, 100*, 1978–1983.
Chen, Y., Corriden, R., Inoue, Y., Yip, L., Hashiguchi, N., Zinkernagel, A., et al. (2006). ATP release guides neutrophil chemotaxis via P2Y2 and A3 receptors. *Science, 314*, 1792–1795.
Chen, Y., Shukla, A., Namiki, S., Insel, P. A., & Junger, W. G. (2004). A putative osmoreceptor system that controls neutrophil function through the release of ATP, its conversion to adenosine, and activation of A2 adenosine and P2 receptors. *Journal of Leukocyte Biology, 76*, 245–253.
Chen, Y., Wu, H., Wang, S., Koito, H., Li, J., Ye, F., et al. (2009). The oligodendrocyte-specific G protein-coupled receptor GPR17 is a cell-intrinsic timer of myelination. *Nature Neuroscience, 12*, 1398–1406.
Chen, Y., Yao, Y., Sumi, Y., Li, A., To, U. K., Elkhal, A., et al. (2010). Purinergic signaling: A fundamental mechanism in neutrophil activation. *Science Signaling, 3*, ra45.
Ciana, P., Fumagalli, M., Trincavelli, M. L., Verderio, C., Rosa, P., Lecca, D., et al. (2006). The orphan receptor GPR17 identified as a new dual uracil nucleotides/cysteinyl-leukotrienes receptor. *The EMBO Journal, 25*, 4615–4627.
Communi, D., Motte, S., Boeynaems, J. M., & Pirotton, S. (1996). Pharmacological characterization of the human P2Y4 receptor. *European Journal of Pharmacology, 317*, 383–389.
Communi, D., Suarez-Huerta, N., Dussossoy, D., Savi, P., & Boeynaems, J. M. (2001). Cotranscription and intergenic splicing of human P2Y11 and SSF1 genes. *The Journal of Biological Chemistry, 276*, 16561–16566.
Conroy, S., Kindon, N., Kellam, B., & Stocks, M. J. (2016). Drug-like antagonists of P2Y receptors-from lead identification to drug development. *Journal of Medicinal Chemistry, 59*, 9981–10005.
Costanzi, S., Mamedova, L., Gao, Z. G., & Jacobson, K. A. (2004). Architecture of P2Y nucleotide receptors: Structural comparison based on sequence analysis, mutagenesis, and homology modeling. *Journal of Medicinal Chemistry, 47*, 5393–5404.
Coster, M., Wittkopf, D., Kreuchwig, A., Kleinau, G., Thor, D., Krause, G. et al. (2012). Using ortholog sequence data to predict the functional relevance of mutations in G-protein-coupled receptors. *The FASEB Journal, 26*, 3273–3281.

Coutinho-Silva, R., Ojcius, D. M., Gorecki, D. C., Persechini, P. M., Bisaggio, R. C., Mendes, A. N., et al. (2005). Multiple P2X and P2Y receptor subtypes in mouse J774, spleen and peritoneal macrophages. *Biochemical Pharmacology, 69*, 641–655.
Cressman, V. L., Lazarowski, E., Homolya, L., Boucher, R. C., Koller, B. H., & Grubb, B. R. (1999). Effect of loss of P2Y(2) receptor gene expression on nucleotide regulation of murine epithelial Cl(−) transport. *The Journal of Biological Chemistry, 274*, 26461–26468.
Cunningham, M. R., Nisar, S. P., & Mundell, S. J. (2013). Molecular mechanisms of platelet P2Y(12) receptor regulation. *Biochemical Society Transactions, 41*, 225–230.
Cvicek, V., Goddard, W. A., 3rd, & Abrol, R. (2016). Structure-based sequence alignment of the transmembrane domains of all human GPCRs: Phylogenetic, structural and functional implications. *PLoS Computational Biology, 12*, e1004805.
Deflorian, F., & Jacobson, K. A. (2011). Comparison of three GPCR structural templates for modeling of the P2Y12 nucleotide receptor. *Journal of Computer-Aided Molecular Design, 25*, 329–338.
Dixon, C. J., Bowler, W. B., Littlewood-Evans, A., Dillon, J. P., Bilbe, G., Sharpe, G. R., et al. (1999). Regulation of epidermal homeostasis through P2Y2 receptors. *British Journal of Pharmacology, 127*, 1680–1686.
Dreisig, K., & Kornum, B. R. (2016). A critical look at the function of the P2Y11 receptor. *Purinergic Signal, 12*, 427–437.
Elliott, M. R., Chekeni, F. B., Trampont, P. C., Lazarowski, E. R., Kadl, A., Walk, S. F., et al. (2009). Nucleotides released by apoptotic cells act as a find-me signal to promote phagocytic clearance. *Nature, 461*, 282–286.
Eltzschig, H. K., Sitkovsky, M. V., & Robson, S. C. (2012). Purinergic signaling during inflammation. *The New England Journal of Medicine, 367*, 2322–2333.
Engel, K. M., Schrock, K., Teupser, D., Holdt, L. M., Tonjes, A., Kern, M., et al. (2011). Reduced food intake and body weight in mice deficient for the G protein-coupled receptor GPR82. *PLoS One, 6*, e29400.
Fabre, A. C., Malaval, C., Ben Addi, A., Verdier, C., Pons, V., Serhan, N., et al. (2010). P2Y13 receptor is critical for reverse cholesterol transport. *Hepatology, 52*, 1477–1483.
Fabre, J. E., Nguyen, M., Latour, A., Keifer, J. A., Audoly, L. P., Coffman, T. M., et al. (1999). Decreased platelet aggregation, increased bleeding time and resistance to thromboembolism in P2Y1-deficient mice. *Nature Medicine, 5*, 1199–1202.
Feng, C., Mery, A. G., Beller, E. M., Favot, C., & Boyce, J. A. (2004). Adenine nucleotides inhibit cytokine generation by human mast cells through a Gs-coupled receptor. *Journal of Immunology, 173*, 7539–7547.
Ferrari, D., Idzko, M., Dichmann, S., Purlis, D., Virchow, C., Norgauer, J., et al. (2000). P2 purinergic receptors of human eosinophils: Characterization and coupling to oxygen radical production. *FEBS Letters, 486*, 217–224.
Ferreira, M. A., Jansen, R., Willemsen, G., Penninx, B., Bain, L. M., Vicente, C. T., et al. (2017). Gene-based analysis of regulatory variants identifies 4 putative novel asthma risk genes related to nucleotide synthesis and signaling. *The Journal of Allergy and Clinical Immunology, 139*, 1148–1157.
Ferri, N., Corsini, A., & Bellosta, S. (2013). Pharmacology of the new P2Y12 receptor inhibitors: Insights on pharmacokinetic and pharmacodynamic properties. *Drugs, 73*, 1681–1709.
Foster, H. R., Fuerst, E., Lee, T. H., Cousins, D. J., & Woszczek, G. (2013). Characterisation of P2Y(12) receptor responsiveness to cysteinyl leukotrienes. *PLoS One, 8*, e58305.
Foster, C. J., Prosser, D. M., Agans, J. M., Zhai, Y., Smith, M. D., Lachowicz, J. E., et al. (2001). Molecular identification and characterization of the platelet ADP receptor targeted by thienopyridine antithrombotic drugs. *The Journal of Clinical Investigation, 107*, 1591–1598.

Fredriksson, R., Lagerstrom, M. C., Lundin, L. G., & Schioth, H. B. (2003). The G-protein-coupled receptors in the human genome form five main families. Phylogenetic analysis, paralogon groups, and fingerprints. *Molecular Pharmacology, 63*, 1256–1272.

Gao, Z. G., Ding, Y., & Jacobson, K. A. (2010a). P2Y(13) receptor is responsible for ADP-mediated degranulation in RBL-2H3 rat mast cells. *Pharmacological Research, 62*, 500–505.

Gao, Z. G., Ding, Y., & Jacobson, K. A. (2010b). UDP-glucose acting at P2Y14 receptors is a mediator of mast cell degranulation. *Biochemical Pharmacology, 79*, 873–879.

Gao, Z. G., Wei, Q., Jayasekara, M. P., & Jacobson, K. A. (2013). The role of P2Y(14) and other P2Y receptors in degranulation of human LAD2 mast cells. *Purinergic Signal, 9*, 31–40.

Gelosa, P., Lecca, D., Fumagalli, M., Wypych, D., Pignieri, A., Cimino, M., et al. (2014). Microglia is a key player in the reduction of stroke damage promoted by the new antithrombotic agent ticagrelor. *Journal of Cerebral Blood Flow and Metabolism, 34*, 979–988.

Gendaszewska-Darmach, E., & Kucharska, M. (2011). Nucleotide receptors as targets in the pharmacological enhancement of dermal wound healing. *Purinergic Signal, 7*, 193–206.

Ghanem, E., Robaye, B., Leal, T., Leipziger, J., Van Driessche, W., Beauwens, R., et al. (2005). The role of epithelial P2Y2 and P2Y4 receptors in the regulation of intestinal chloride secretion. *British Journal of Pharmacology, 146*, 364–369.

Giannattasio, G., Ohta, S., Boyce, J. R., Xing, W., Balestrieri, B., & Boyce, J. A. (2011). The purinergic G protein-coupled receptor 6 inhibits effector T cell activation in allergic pulmonary inflammation. *Journal of Immunology, 187*, 1486–1495.

Gomes, I., Aryal, D. K., Wardman, J. H., Gupta, A., Gagnidze, K., Rodriguiz, R. M., et al. (2013). GPR171 is a hypothalamic G protein-coupled receptor for BigLEN, a neuropeptide involved in feeding. *Proceedings of the National Academy of Sciences of the United States of America, 110*, 16211–16216.

Gorini, S., Callegari, G., Romagnoli, G., Mammi, C., Mavilio, D., Rosano, G., et al. (2010). ATP secreted by endothelial cells blocks CX(3)CL 1-elicited natural killer cell chemotaxis and cytotoxicity via P2Y(1)(1) receptor activation. *Blood, 116*, 4492–4500.

Grbic, D. M., Degagne, E., Langlois, C., Dupuis, A. A., & Gendron, F. P. (2008). Intestinal inflammation increases the expression of the P2Y6 receptor on epithelial cells and the release of CXC chemokine ligand 8 by UDP. *Journal of Immunology, 180*, 2659–2668.

Guo, H. M., Gao, J. M., Luo, Y. L., Wen, Y. Z., Zhang, Y. L., Hide, G., et al. (2015). Infection by Toxoplasma gondii, a severe parasite in neonates and AIDS patients, causes impaired anion secretion in airway epithelia. *Proceedings of the National Academy of Sciences of the United States of America, 112*, 4435–4440.

Gur, S., & Hellstrom, W. J. (2009). Activation of P2Y1 and P2Y2 nucleotide receptors by adenosine 5′-triphosphate analogues augmented nerve-mediated relaxation of human corpus cavernosum. *Canadian Urological Association Journal, 3*, 314–318.

Guzman, S. J., & Gerevich, Z. (2016). P2Y receptors in synaptic transmission and plasticity: Therapeutic potential in cognitive dysfunction. *Neural Plasticity, 2016*, 1207393.

Hansen, A., Alston, L., Tulk, S. E., Schenck, L. P., Grassie, M. E., Alhassan, B. F., et al. (2013). The P2Y6 receptor mediates Clostridium difficile toxin-induced CXCL8/IL-8 production and intestinal epithelial barrier dysfunction. *PLoS One, 8*, e81491.

Harden, T. K., Lazarowski, E. R., & Boucher, R. C. (1997). Release, metabolism and interconversion of adenine and uridine nucleotides: Implications for G protein-coupled P2 receptor agonist selectivity. *Trends in Pharmacological Sciences, 18*, 43–46.

Haynes, S. E., Hollopeter, G., Yang, G., Kurpius, D., Dailey, M. E., Gan, W. B., et al. (2006). The P2Y12 receptor regulates microglial activation by extracellular nucleotides. *Nature Neuroscience, 9*, 1512–1519.

Hennen, S., Wang, H., Peters, L., Merten, N., Simon, K., Spinrath, A., et al. (2013). Decoding signaling and function of the orphan G protein-coupled receptor GPR17 with a small-molecule agonist. *Science Signaling*, *6*, ra93.

Hetherington, S. L., Singh, R. K., Lodwick, D., Thompson, J. R., Goodall, A. H., & Samani, N. J. (2005). Dimorphism in the P2Y1 ADP receptor gene is associated with increased platelet activation response to ADP. *Arteriosclerosis, Thrombosis, and Vascular Biology*, *25*, 252–257.

Hickman, S. E., Kingery, N. D., Ohsumi, T. K., Borowsky, M. L., Wang, L. C., Means, T. K., et al. (2013). The microglial sensome revealed by direct RNA sequencing. *Nature Neuroscience*, *16*, 1896–1905.

Hiramoto, T., Nonaka, Y., Inoue, K., Yamamoto, T., Omatsu-Kanbe, M., Matsuura, H., et al. (2004). Identification of endogenous surrogate ligands for human P2Y receptors through an in silico search. *Journal of Pharmacological Sciences*, *95*, 81–93.

Hoffmann, K., Sixel, U., Di Pasquale, F., & von Kugelgen, I. (2008). Involvement of basic amino acid residues in transmembrane regions 6 and 7 in agonist and antagonist recognition of the human platelet P2Y(12)-receptor. *Biochemical Pharmacology*, *76*, 1201–1213.

Hogberg, C., Gidlof, O., Deflorian, F., Jacobson, K. A., Abdelrahman, A., Muller, C. E., et al. (2012). Farnesyl pyrophosphate is an endogenous antagonist to ADP-stimulated P2Y(1)(2) receptor-mediated platelet aggregation. *Thrombosis and Haemostasis*, *108*, 119–132.

Hollopeter, G., Jantzen, H. M., Vincent, D., Li, G., England, L., Ramakrishnan, V., et al. (2001). Identification of the platelet ADP receptor targeted by antithrombotic drugs. *Nature*, *409*, 202–207.

Holmfeldt, P., Ganuza, M., Marathe, H., He, B., Hall, T., Kang, G., et al. (2016). Functional screen identifies regulators of murine hematopoietic stem cell repopulation. *The Journal of Experimental Medicine*, *213*, 433–449.

Horckmans, M., Marcet, B., Marteau, F., Bulte, F., Maho, A., Parmentier, M., et al. (2006). Extracellular adenine nucleotides inhibit the release of major monocyte recruiters by human monocyte-derived dendritic cells. *FEBS Letters*, *580*, 747–754.

Hoyle, C. H. (2011). Evolution of neuronal signalling: Transmitters and receptors. *Autonomic Neuroscience*, *165*, 28–53.

Idzko, M., Ferrari, D., & Eltzschig, H. K. (2014). Nucleotide signalling during inflammation. *Nature*, *509*, 310–317.

Idzko, M., Panther, E., Bremer, H. C., Sorichter, S., Luttmann, W., Virchow, C. J., Jr., et al. (2003). Stimulation of P2 purinergic receptors induces the release of eosinophil cationic protein and interleukin-8 from human eosinophils. *British Journal of Pharmacology*, *138*, 1244–1250.

Ifuku, M., Buonfiglioli, A., Jordan, P., Lehnardt, S., & Kettenmann, H. (2016). TLR2 controls random motility, while TLR7 regulates chemotaxis of microglial cells via distinct pathways. *Brain, Behavior, and Immunity*, *58*, 338–347.

Ignatovica, V., Megnis, K., Lapins, M., Schioth, H. B., & Klovins, J. (2012). Identification and analysis of functionally important amino acids in human purinergic 12 receptor using a Saccharomyces cerevisiae expression system. *The FEBS Journal*, *279*, 180–191.

Isfort, K., Ebert, F., Bornhorst, J., Sargin, S., Kardakaris, R., Pasparakis, M., et al. (2011). Real-time imaging reveals that P2Y2 and P2Y12 receptor agonists are not chemoattractants and macrophage chemotaxis to complement C5a is phosphatidylinositol 3-kinase (PI3K)- and p38 mitogen-activated protein kinase (MAPK)-independent. *The Journal of Biological Chemistry*, *286*, 44776–44787.

Ivanov, A. A., Costanzi, S., & Jacobson, K. A. (2006). Defining the nucleotide binding sites of P2Y receptors using rhodopsin-based homology modeling. *Journal of Computer-Aided Molecular Design*, *20*, 417–426.

Jacob, F., Perez Novo, C., Bachert, C., & Van Crombruggen, K. (2013). Purinergic signaling in inflammatory cells: P2 receptor expression, functional effects, and modulation of inflammatory responses. *Purinergic Signal*, *9*, 285–306.

Jacobson, K. A., Paoletta, S., Katritch, V., Wu, B., Gao, Z. G., Zhao, Q., et al. (2015). Nucleotides acting at P2Y receptors: Connecting structure and function. *Molecular Pharmacology*, *88*, 220–230.

Jager, E., Schulz, A., Lede, V., Lin, C. C., Schoneberg, T., & Le Duc, D. (2016). Dendritic cells regulate GPR34 through mitogenic signals and undergo apoptosis in its absence. *Journal of Immunology*, *196*, 2504–2513.

James, S., Akerblom, A., Cannon, C. P., Emanuelsson, H., Husted, S., Katus, H., et al. (2009). Comparison of ticagrelor, the first reversible oral P2Y(12) receptor antagonist, with clopidogrel in patients with acute coronary syndromes: Rationale, design, and baseline characteristics of the PLATelet inhibition and patient Outcomes (PLATO) trial. *American Heart Journal*, *157*, 599–605.

Jia, P., Wang, L., Fanous, A. H., Pato, C. N., Edwards, T. L., C. International Schizophrenia, et al. (2012). Network-assisted investigation of combined causal signals from genome-wide association studies in schizophrenia. *PLoS Computational Biology*, *8*, e1002587.

Jones, D. T., Taylor, W. R., & Thornton, J. M. (1992). The rapid generation of mutation data matrices from protein sequences. *Computer Applications in the Biosciences*, *8*, 275–282.

Junger, W. G. (2011). Immune cell regulation by autocrine purinergic signalling. *Nature Reviews. Immunology*, *11*, 201–212.

Kaebisch, C., Schipper, D., Babczyk, P., & Tobiasch, E. (2015). The role of purinergic receptors in stem cell differentiation. *Computational and Structural Biotechnology Journal*, *13*, 75–84.

Kazeros, A., Harvey, B. G., Carolan, B. J., Vanni, H., Krause, A., & Crystal, R. G. (2008). Overexpression of apoptotic cell removal receptor MERTK in alveolar macrophages of cigarette smokers. *American Journal of Respiratory Cell and Molecular Biology*, *39*, 747–757.

Kellerman, D., Rossi Mospan, A., Engels, J., Schaberg, A., Gorden, J., & Smiley, L. (2008). Denufosol: A review of studies with inhaled P2Y(2) agonists that led to Phase 3. *Pulmonary Pharmacology & Therapeutics*, *21*, 600–607.

Kennedy, C., Qi, A. D., Herold, C. L., Harden, T. K., & Nicholas, R. A. (2000). ATP, an agonist at the rat P2Y(4) receptor, is an antagonist at the human P2Y(4) receptor. *Molecular Pharmacology*, *57*, 926–931.

Kimura, T., Kobayashi, S., Hanihara-Tatsuzawa, F., Sayama, A., MaruYama, T., & Muta, T. (2014). Responses of macrophages to the danger signals released from necrotic cells. *International Immunology*, *26*, 697–704.

Kobayashi, T., Kouzaki, H., & Kita, H. (2010). Human eosinophils recognize endogenous danger signal crystalline uric acid and produce proinflammatory cytokines mediated by autocrine ATP. *Journal of Immunology*, *184*, 6350–6358.

Koizumi, S., Shigemoto-Mogami, Y., Nasu-Tada, K., Shinozaki, Y., Ohsawa, K., Tsuda, M., et al. (2007). UDP acting at P2Y6 receptors is a mediator of microglial phagocytosis. *Nature*, *446*, 1091–1095.

Kornum, B. R., Kawashima, M., Faraco, J., Lin, L., Rico, T. J., Hesselson, S., et al. (2011). Common variants in P2RY11 are associated with narcolepsy. *Nature Genetics*, *43*, 66–71.

Krishnan, A., Almen, M. S., Fredriksson, R., & Schioth, H. B. (2012). The origin of GPCRs: Identification of mammalian like rhodopsin, adhesion, glutamate and frizzled GPCRs in fungi. *PLoS One*, *7*, e29817.

Krishnan, A., Almen, M. S., Fredriksson, R., & Schioth, H. B. (2013). Remarkable similarities between the hemichordate (Saccoglossus kowalevskii) and vertebrate GPCR repertoire. *Gene*, *526*, 122–133.

Kronlage, M., Song, J., Sorokin, L., Isfort, K., Schwerdtle, T., Leipziger, J., et al. (2010). Autocrine purinergic receptor signaling is essential for macrophage chemotaxis. *Science Signaling*, *3*, ra55.

Krug, T., Gabriel, J. P., Taipa, R., Fonseca, B. V., Domingues-Montanari, S., Fernandez-Cadenas, I., et al. (2012). TTC7B emerges as a novel risk factor for ischemic stroke through the convergence of several genome-wide approaches. *Journal of Cerebral Blood Flow and Metabolism*, *32*, 1061–1072.

Kukulski, F., Ben Yebdri, F., Lecka, J., Kauffenstein, G., Levesque, S. A., Martin-Satue, M., et al. (2009). Extracellular ATP and P2 receptors are required for IL-8 to induce neutrophil migration. *Cytokine*, *46*, 166–170.

Kumar, A., Sharma, R., & Kamaluddin. (2014). Formamide-based synthesis of nucleobases by metal(II) octacyanomolybdate(IV): Implication in prebiotic chemistry. *Astrobiology*, *14*, 769–779.

Kumar, S., Stecher, G., & Tamura, K. (2016). MEGA7: Molecular Evolutionary Genetics Analysis version 7.0 for bigger datasets. *Molecular Biology and Evolution*, *33*, 1870–1874.

Lau, O. C., Samarawickrama, C., & Skalicky, S. E. (2014). P2Y2 receptor agonists for the treatment of dry eye disease: A review. *Clinical Ophthalmology*, *8*, 327–334.

Lee, B. C., Cheng, T., Adams, G. B., Attar, E. C., Miura, N., Lee, S. B., et al. (2003). P2Y-like receptor, GPR105 (P2Y14), identifies and mediates chemotaxis of bone-marrow hematopoietic stem cells. *Genes & Development*, *17*, 1592–1604.

Lee, D. H., Park, K. S., Kong, I. D., Kim, J. W., & Han, B. G. (2006). Expression of P2 receptors in human B cells and Epstein-Barr virus-transformed lymphoblastoid cell lines. *BMC Immunology*, *7*, 22.

Leon, C., Hechler, B., Freund, M., Eckly, A., Vial, C., Ohlmann, P., et al. (1999). Defective platelet aggregation and increased resistance to thrombosis in purinergic P2Y(1) receptor-null mice. *The Journal of Clinical Investigation*, *104*, 1731–1737.

Li, H. Q., Chen, C., Dou, Y., Wu, H. J., Liu, Y. J., Lou, H. F., et al. (2013). P2Y4 receptor-mediated pinocytosis contributes to amyloid beta-induced self-uptake by microglia. *Molecular and Cellular Biology*, *33*, 4282–4293.

Li, X. Q., Ma, N., Li, X. G., Wang, B., Sun, S. S., Gao, F., et al. (2016). Association of PON1, P2Y12 and COX1 with recurrent ischemic events in patients with extracranial or intracranial stenting. *PLoS One*, *11*, e0148891.

Liebscher, I., Muller, U., Teupser, D., Engemaier, E., Engel, K. M., Ritscher, L., et al. (2011). Altered immune response in mice deficient for the G protein-coupled receptor GPR34. *The Journal of Biological Chemistry*, *286*, 2101–2110.

Linden, J., & Cekic, C. (2012). Regulation of lymphocyte function by adenosine. *Arteriosclerosis, Thrombosis, and Vascular Biology*, *32*, 2097–2103.

Liverani, E., Rico, M. C., Tsygankov, A. Y., Kilpatrick, L. E., & Kunapuli, S. P. (2016). P2Y12 receptor modulates sepsis-induced inflammation. *Arteriosclerosis, Thrombosis, and Vascular Biology*, *36*, 961–971.

Lommatzsch, M., Cicko, S., Muller, T., Lucattelli, M., Bratke, K., Stoll, P., et al. (2010). Extracellular adenosine triphosphate and chronic obstructive pulmonary disease. *American Journal of Respiratory and Critical Care Medicine*, *181*, 928–934.

Luthje, J., & Ogilvie, A. (1983). The presence of diadenosine 5′,5‴-P1,P3-triphosphate (Ap3A) in human platelets. *Biochemical and Biophysical Research Communications*, *115*, 253–260.

Maekawa, A., Balestrieri, B., Austen, K. F., & Kanaoka, Y. (2009). GPR17 is a negative regulator of the cysteinyl leukotriene 1 receptor response to leukotriene D4. *Proceedings of the National Academy of Sciences of the United States of America*, *106*, 11685–11690.

Manahan, C. L., Iglesias, P. A., Long, Y., & Devreotes, P. N. (2004). Chemoattractant signaling in dictyostelium discoideum. *Annual Review of Cell and Developmental Biology*, *20*, 223–253.

Marteau, F., Communi, D., Boeynaems, J. M., & Suarez Gonzalez, N. (2004). Involvement of multiple P2Y receptors and signaling pathways in the action of adenine nucleotides diphosphates on human monocyte-derived dendritic cells. *Journal of Leukocyte Biology, 76*, 796–803.

Marteau, F., Le Poul, E., Communi, D., Communi, D., Labouret, C., Savi, P., et al. (2003). Pharmacological characterization of the human P2Y13 receptor. *Molecular Pharmacology, 64*, 104–112.

Marucci, G., Dal Ben, D., Lambertucci, C., Santinelli, C., Spinaci, A., Thomas, A., et al. (2016). The G protein-coupled receptor GPR17: Overview and update. *ChemMedChem, 11*, 2567–2574.

Meister, J., Le Duc, D., Ricken, A., Burkhardt, R., Thiery, J., Pfannkuche, H., et al. (2014). The G protein-coupled receptor P2Y14 influences insulin release and smooth muscle function in mice. *The Journal of Biological Chemistry, 289*, 23353–23366.

Melen, E., Granell, R., Kogevinas, M., Strachan, D., Gonzalez, J. R., Wjst, M., et al. (2013). Genome-wide association study of body mass index in 23 000 individuals with and without asthma. *Clinical and Experimental Allergy, 43*, 463–474.

Meshki, J., Tuluc, F., Bredetean, O., Ding, Z., & Kunapuli, S. P. (2004). Molecular mechanism of nucleotide-induced primary granule release in human neutrophils: Role for the P2Y2 receptor. *American Journal of Physiology. Cell Physiology, 286*, C264–C271.

Meshki, J., Tuluc, F., Bredetean, O., Garcia, A., & Kunapuli, S. P. (2006). Signaling pathways downstream of P2 receptors in human neutrophils. *Purinergic Signal, 2*, 537–544.

Michel, M. C., Wieland, T., & Tsujimoto, G. (2009). How reliable are G-protein-coupled receptor antibodies? *Naunyn-Schmiedeberg's Archives of Pharmacology, 379*, 385–388.

Miller, S. L., & Orgel, L. E. (1974). The origins of life on the earth. In *Prentice-Hall biological science series. Concept of modern biology series* (229 pp.). Englewood Cliffs, New Jersey: Prentice Hall Inc. ISBN 0136420826, 9780136420828.

Mohanty, J. G., Raible, D. G., McDermott, L. J., Pelleg, A., & Schulman, E. S. (2001). Effects of purine and pyrimidine nucleotides on intracellular Ca2+ in human eosinophils: Activation of purinergic P2Y receptors. *The Journal of Allergy and Clinical Immunology, 107*, 849–855.

Moreira-Souza, A. C., Marinho, Y., Correa, G., Santoro, G. F., Coutinho, C. M., Vommaro, R. C., et al. (2015). Pyrimidinergic receptor activation controls Toxoplasma gondii infection in macrophages. *PLoS One, 10*, e0133502.

Moreschi, I., Bruzzone, S., Bodrato, N., Usai, C., Guida, L., Nicholas, R. A., et al. (2008). NAADP+ is an agonist of the human P2Y11 purinergic receptor. *Cell Calcium, 43*, 344–355.

Moreschi, I., Bruzzone, S., Nicholas, R. A., Fruscione, F., Sturla, L., Benvenuto, F., et al. (2006). Extracellular NAD+ is an agonist of the human P2Y11 purinergic receptor in human granulocytes. *The Journal of Biological Chemistry, 281*, 31419–31429.

Muller, T., Robaye, B., Vieira, R. P., Ferrari, D., Grimm, M., Jakob, T., et al. (2010). The purinergic receptor P2Y2 receptor mediates chemotaxis of dendritic cells and eosinophils in allergic lung inflammation. *Allergy, 65*, 1545–1553.

Myrtek, D., Muller, T., Geyer, V., Derr, N., Ferrari, D., Zissel, G., et al. (2008). Activation of human alveolar macrophages via P2 receptors: Coupling to intracellular Ca2+ increases and cytokine secretion. *Journal of Immunology, 181*, 2181–2188.

Nirodi, C. S., Crews, B. C., Kozak, K. R., Morrow, J. D., & Marnett, L. J. (2004). The glyceryl ester of prostaglandin E2 mobilizes calcium and activates signal transduction in RAW264.7 cells. *Proceedings of the National Academy of Sciences of the United States of America, 101*, 1840–1845.

Nisar, S. P., Cunningham, M., Saxena, K., Pope, R. J., Kelly, E., & Mundell, S. J. (2012). Arrestin scaffolds NHERF1 to the P2Y12 receptor to regulate receptor internalization. *The Journal of Biological Chemistry, 287*, 24505–24515.

Niss Arfelt, K., Fares, S., Sparre-Ulrich, A. H., Hjorto, G. M., Gasbjerg, L. S., Molleskov-Jensen, A. S., et al. (2016). Signaling via G proteins mediates tumorigenic effects of GPR87. *Cellular Signalling*, *30*, 9–18.

Njau, F., Geffers, R., Thalmann, J., Haller, H., & Wagner, A. D. (2009). Restriction of Chlamydia pneumoniae replication in human dendritic cell by activation of indoleamine 2,3-dioxygenase. *Microbes and Infection*, *11*, 1002–1010.

Nonaka, Y., Hiramoto, T., & Fujita, N. (2005). Identification of endogenous surrogate ligands for human P2Y12 receptors by in silico and in vitro methods. *Biochemical and Biophysical Research Communications*, *337*, 281–288.

Nylander, S., Mattsson, C., Ramstrom, S., & Lindahl, T. L. (2003). The relative importance of the ADP receptors, P2Y12 and P2Y1, in thrombin-induced platelet activation. *Thrombosis Research*, *111*, 65–73.

Ou, Z., Sun, Y., Lin, L., You, N., Liu, X., Li, H., et al. (2016). Olig2-targeted G-protein-coupled receptor Gpr17 regulates oligodendrocyte survival in response to lysolecithin-induced demyelination. *The Journal of Neuroscience*, *36*, 10560–10573.

Palmer, N. D., Goodarzi, M. O., Langefeld, C. D., Wang, N., Guo, X., Taylor, K. D., et al. (2015). Genetic variants associated with quantitative glucose homeostasis traits translate to type 2 diabetes in Mexican Americans: The GUARDIAN (genetics underlying diabetes in Hispanics) consortium. *Diabetes*, *64*, 1853–1866.

Paruchuri, S., Tashimo, H., Feng, C., Maekawa, A., Xing, W., Jiang, Y., et al. (2009). Leukotriene E4-induced pulmonary inflammation is mediated by the P2Y12 receptor. *The Journal of Experimental Medicine*, *206*, 2543–2555.

Patel, K., Barnes, A., Camacho, J., Paterson, C., Boughtflower, R., Cousens, D., et al. (2001). Activity of diadenosine polyphosphates at P2Y receptors stably expressed in 1321N1 cells. *European Journal of Pharmacology*, *430*, 203–210.

Preissler, J., Grosche, A., Lede, V., Le Duc, D., Krugel, K., Matyash, V., et al. (2015). Altered microglial phagocytosis in GPR34-deficient mice. *Glia*, *63*, 206–215.

Qi, A. D., Harden, T. K., & Nicholas, R. A. (2013). Is GPR17 a P2Y/leukotriene receptor? Examination of uracil nucleotides, nucleotide sugars, and cysteinyl leukotrienes as agonists of GPR17. *The Journal of Pharmacology and Experimental Therapeutics*, *347*, 38–46.

Qi, A. D., Kennedy, C., Harden, T. K., & Nicholas, R. A. (2001). Differential coupling of the human P2Y(11) receptor to phospholipase C and adenylyl cyclase. *British Journal of Pharmacology*, *132*, 318–326.

Ratjen, F., Durham, T., Navratil, T., Schaberg, A., Accurso, F. J., Wainwright, C., et al. (2012). Long term effects of denufosol tetrasodium in patients with cystic fibrosis. *Journal of Cystic Fibrosis*, *11*, 539–549.

Reiner, S., Ziegler, N., Leon, C., Lorenz, K., von Hayn, K., Gachet, C., et al. (2009). Beta-arrestin-2 interaction and internalization of the human P2Y1 receptor are dependent on C-terminal phosphorylation sites. *Molecular Pharmacology*, *76*, 1162–1171.

Ricciardi, A., Elia, A. R., Cappello, P., Puppo, M., Vanni, C., Fardin, P., et al. (2008). Transcriptome of hypoxic immature dendritic cells: Modulation of chemokine/receptor expression. *Molecular Cancer Research*, *6*, 175–185.

Riegel, A. K., Faigle, M., Zug, S., Rosenberger, P., Robaye, B., Boeynaems, J. M., et al. (2011). Selective induction of endothelial P2Y6 nucleotide receptor promotes vascular inflammation. *Blood*, *117*, 2548–2555.

Ritscher, L., Engemaier, E., Staubert, C., Liebscher, I., Schmidt, P., Hermsdorf, T., et al. (2012). The ligand specificity of the G-protein-coupled receptor GPR34. *The Biochemical Journal*, *443*, 841–850.

Robaye, B., Ghanem, E., Wilkin, F., Fokan, D., Van Driessche, W., Schurmans, S., et al. (2003). Loss of nucleotide regulation of epithelial chloride transport in the jejunum of P2Y4-null mice. *Molecular Pharmacology*, *63*, 777–783.

Rossi, L., Lemoli, R. M., & Goodell, M. A. (2013). Gpr171, a putative P2Y-like receptor, negatively regulates myeloid differentiation in murine hematopoietic progenitors. *Experimental Hematology, 41*, 102–112.

Schluter, H., Tepel, M., & Zidek, W. (1996). Vascular actions of diadenosine phosphates. *Journal of Autonomic Pharmacology, 16*, 357–362.

Schmidt, P., Ritscher, L., Dong, E. N., Hermsdorf, T., Coster, M., Wittkopf, D., et al. (2013). Identification of determinants required for agonistic and inverse agonistic ligand properties at the ADP receptor P2Y12. *Molecular Pharmacology, 83*, 256–266.

Schnurr, M., Toy, T., Stoitzner, P., Cameron, P., Shin, A., Beecroft, T., et al. (2003). ATP gradients inhibit the migratory capacity of specific human dendritic cell types: Implications for P2Y11 receptor signaling. *Blood, 102*, 613–620.

Schoneberg, T., Hermsdorf, T., Engemaier, E., Engel, K., Liebscher, I., Thor, D., et al. (2007). Structural and functional evolution of the P2Y(12)-like receptor group. *Purinergic Signal, 3*, 255–268.

Schoneberg, T., Schulz, A., Biebermann, H., Hermsdorf, T., Rompler, H., & Sangkuhl, K. (2004). Mutant G-protein-coupled receptors as a cause of human diseases. *Pharmacology & Therapeutics, 104*, 173–206.

Schulman, E. S., Glaum, M. C., Post, T., Wang, Y., Raible, D. G., Mohanty, J., et al. (1999). ATP modulates anti-IgE-induced release of histamine from human lung mast cells. *American Journal of Respiratory Cell and Molecular Biology, 20*, 530–537.

Schumacher, D., Strilic, B., Sivaraj, K. K., Wettschureck, N., & Offermanns, S. (2013). Platelet-derived nucleotides promote tumor-cell transendothelial migration and metastasis via P2Y2 receptor. *Cancer Cell, 24*, 130–137.

Schwiebert, E. M. (2003). *Extracellular nucleotides and nucleosides: Release, receptors, and physiological & pathophysiological effects.* New York: Academic Press.

Scrivens, M., & Dickenson, J. M. (2005). Functional expression of the P2Y14 receptor in murine T-lymphocytes. *British Journal of Pharmacology, 146*, 435–444.

Scrivens, M., & Dickenson, J. M. (2006). Functional expression of the P2Y14 receptor in human neutrophils. *European Journal of Pharmacology, 543*, 166–173.

Shaver, S. R., Rideout, J. L., Pendergast, W., Douglass, J. G., Brown, E. G., Boyer, J. L., et al. (2005). Structure-activity relationships of dinucleotides: Potent and selective agonists of P2Y receptors. *Purinergic Signal, 1*, 183–191.

Shinozaki, Y., Nomura, M., Iwatsuki, K., Moriyama, Y., Gachet, C., & Koizumi, S. (2014). Microglia trigger astrocyte-mediated neuroprotection via purinergic gliotransmission. *Scientific Reports, 4*, 4329.

Simon, K., Hennen, S., Merten, N., Blattermann, S., Gillard, M., Kostenis, E., et al. (2016). The orphan G protein-coupled receptor GPR17 negatively regulates oligodendrocyte differentiation via Galphai/o and its downstream effector molecules. *The Journal of Biological Chemistry, 291*, 705–718.

Simon, K., Merten, N., Schröder, R., Hennen, S., Preis, P., Schmitt, N. K., et al. (2017). The orphan receptor GPR17 is unresponsive to uracil nucleotides and cysteinyl leukotrienes. *Molecular Pharmacology, 91*(5), 518–532.

Skelton, L., Cooper, M., Murphy, M., & Platt, A. (2003). Human immature monocyte-derived dendritic cells express the G protein-coupled receptor GPR105 (KIAA0001, P2Y14) and increase intracellular calcium in response to its agonist, uridine diphosphoglucose. *Journal of Immunology, 171*, 1941–1949.

Sugo, T., Tachimoto, H., Chikatsu, T., Murakami, Y., Kikukawa, Y., Sato, S., et al. (2006). Identification of a lysophosphatidylserine receptor on mast cells. *Biochemical and Biophysical Research Communications, 341*, 1078–1087.

Sunyer, J. O. (2013). Fishing for mammalian paradigms in the teleost immune system. *Nature Immunology, 14*, 320–326.

Tabata, K., Baba, K., Shiraishi, A., Ito, M., & Fujita, N. (2007). The orphan GPCR GPR87 was deorphanized and shown to be a lysophosphatidic acid receptor. *Biochemical and Biophysical Research Communications*, *363*, 861–866.

Teles, R. M., Graeber, T. G., Krutzik, S. R., Montoya, D., Schenk, M., Lee, D. J., et al. (2013). Type I interferon suppresses type II interferon-triggered human anti-mycobacterial responses. *Science*, *339*, 1448–1453.

The European Mouse Mutant Archive (EMMA), 2017. https://www.infrafrontier.eu.

Thomas, M. R., & Storey, R. F. (2015). Effect of P2Y12 inhibitors on inflammation and immunity. *Thrombosis and Haemostasis*, *114*, 490–497.

Uratsuji, H., Tada, Y., Kawashima, T., Kamata, M., Hau, C. S., Asano, Y., et al. (2012). P2Y6 receptor signaling pathway mediates inflammatory responses induced by monosodium urate crystals. *Journal of Immunology*, *188*, 436–444.

Vanderstocken, G., Bondue, B., Horckmans, M., Di Pietrantonio, L., Robaye, B., Boeynaems, J. M., et al. (2010). P2Y2 receptor regulates VCAM-1 membrane and soluble forms and eosinophil accumulation during lung inflammation. *Journal of Immunology*, *185*, 3702–3707.

Vanderstocken, G., Van de Paar, E., Robaye, B., di Pietrantonio, L., Bondue, B., Boeynaems, J. M., et al. (2012). Protective role of P2Y2 receptor against lung infection induced by pneumonia virus of mice. *PLoS One*, *7*, e50385.

Vaughan, K. R., Stokes, L., Prince, L. R., Marriott, H. M., Meis, S., Kassack, M. U., et al. (2007). Inhibition of neutrophil apoptosis by ATP is mediated by the P2Y11 receptor. *Journal of Immunology*, *179*, 8544–8553.

Vergani, A., Tezza, S., D'Addio, F., Fotino, C., Liu, K., Niewczas, M., et al. (2013). Long-term heart transplant survival by targeting the ionotropic purinergic receptor P2X7. *Circulation*, *127*, 463–475.

Vergani, A., Tezza, S., Fotino, C., Visner, G., Pileggi, A., Chandraker, A., et al. (2014). The purinergic system in allotransplantation. *American Journal of Transplantation*, *14*, 507–514.

Verkhratsky, A., & Burnstock, G. (2014). Biology of purinergic signalling: Its ancient evolutionary roots, its omnipresence and its multiple functional significance. *BioEssays*, *36*, 697–705.

Vieira, R. P., Muller, T., Grimm, M., von Gernler, V., Vetter, B., Durk, T., et al. (2011). Purinergic receptor type 6 contributes to airway inflammation and remodeling in experimental allergic airway inflammation. *American Journal of Respiratory and Critical Care Medicine*, *184*, 215–223.

Volonte, C., Amadio, S., D'Ambrosi, N., Colpi, M., & Burnstock, G. (2006). P2 receptor web: Complexity and fine-tuning. *Pharmacology & Therapeutics*, *112*, 264–280.

Wang, G. P., Ciuffi, A., Leipzig, J., Berry, C. C., & Bushman, F. D. (2007). HIV integration site selection: Analysis by massively parallel pyrosequencing reveals association with epigenetic modifications. *Genome Research*, *17*, 1186–1194.

Wang, S., Iring, A., Strilic, B., Albarran Juarez, J., Kaur, H., Troidl, K., et al. (2015). P2Y(2) and Gq/G(1)(1) control blood pressure by mediating endothelial mechanotransduction. *The Journal of Clinical Investigation*, *125*, 3077–3086.

Wang, L., Jacobsen, S. E., Bengtsson, A., & Erlinge, D. (2004). P2 receptor mRNA expression profiles in human lymphocytes, monocytes and CD34+ stem and progenitor cells. *BMC Immunology*, *5*, 16.

Wang, Z., Nakayama, T., Sato, N., Izumi, Y., Kasamaki, Y., Ohta, M., et al. (2010). The purinergic receptor P2Y, G-protein coupled, 2 (P2RY2) gene associated with essential hypertension in Japanese men. *Journal of Human Hypertension*, *24*, 327–335.

Wang, L., Olivecrona, G., Gotberg, M., Olsson, M. L., Winzell, M. S., & Erlinge, D. (2005). ADP acting on P2Y13 receptors is a negative feedback pathway for ATP release from human red blood cells. *Circulation Research*, *96*, 189–196.

Wang, N., Robaye, B., Gossiel, F., Boeynaems, J. M., & Gartland, A. (2014). The P2Y13 receptor regulates phosphate metabolism and FGF-23 secretion with effects on skeletal development. *The FASEB Journal*, *28*, 2249–2259.

Webb, T. E., Simon, J., Krishek, B. J., Bateson, A. N., Smart, T. G., King, B. F., et al. (1993). Cloning and functional expression of a brain G-protein-coupled ATP receptor. *FEBS Letters*, *324*, 219–225.

Welter, D., MacArthur, J., Morales, J., Burdett, T., Hall, P., Junkins, H., et al. (2014). The NHGRI GWAS Catalog, a curated resource of SNP-trait associations *Nucleic Acids Research*, *42*, D1001–D1006.

Wilkin, F., Duhant, X., Bruyns, C., Suarez-Huerta, N., Boeynaems, J. M., & Robaye, B. (2001). The P2Y11 receptor mediates the ATP-induced maturation of human monocyte-derived dendritic cells. *Journal of Immunology*, *166*, 7172–7177.

Wilkin, F., Stordeur, P., Goldman, M., Boeynaems, J. M., & Robaye, B. (2002). Extracellular adenine nucleotides modulate cytokine production by human monocyte-derived dendritic cells: Dual effect on IL-12 and stimulation of IL-10. *European Journal of Immunology*, *32*, 2409–2417.

Wlodarska, I., Tousseyn, T., De Leval, L., Ferreiro, J., Urbankova, H., Michaux, L., et al. (2009). Novel T(X;14)(p11.4;q32.33) resulting in upregulation of gpr34 and activation of the NFκB pathway is recurrent in malt lymphomas. *Haematologica*, *94*. 271–272.

Yu, W., Ma, S., Wang, L., Zuo, B., Li, M., Qiao, Z., et al. (2013). Upregulation of GPR34 expression affects the progression and prognosis of human gastric adenocarcinoma by PI3K/PDK1/AKT pathway. *Histology and Histopathology*, *28*, 1629–1638.

Zhang, D., Gao, Z. G., Zhang, K., Kiselev, E., Crane, S., Wang, J., et al. (2015). Two disparate ligand-binding sites in the human P2Y1 receptor. *Nature*, *520*, 317–321.

Zhang, J., Li, Z., Hu, X., Su, Q., He, C., Liu, J., et al. (2017). Knockout of P2Y12 aggravates experimental autoimmune encephalomyelitis in mice via increasing of IL-23 production and Th17 cell differentiation by dendritic cells. *Brain, Behavior, and Immunity*, *62*, 245–255.

Zhang, J., Zhang, K., Gao, Z. G., Paoletta, S., Zhang, D., Han, G. W., et al. (2014). Agonist-bound structure of the human P2Y12 receptor. *Nature*, *509*, 119–122.

Zhang, K., Zhang, J., Gao, Z. G., Zhang, D., Zhu, L., Han, G. W., et al. (2014). Structure of the human P2Y12 receptor in complex with an antithrombotic drug. *Nature*, *509*, 115–118.

Zuo, B., Li, M., Liu, Y., Li, K., Ma, S., Cui, M., et al. (2014). G-protein coupled receptor 34 activates Erk and phosphatidylinositol 3-kinase/Akt pathways and functions as alternative pathway to mediate p185Bcr-Abl-induced transformation and leukemogenesis. *Leukemia & Lymphoma*, 1–12.

## FURTHER READING

Daniele, S., Trincavelli, M. L., Gabelloni, P., Lecca, D., Rosa, P., Abbracchio, M. P., et al. (2011). Agonist-induced desensitization/resensitization of human G protein-coupled receptor 17: A functional cross-talk between purinergic and cysteinyl-leukotriene ligands. *The Journal of Pharmacology and Experimental Therapeutics*, *338*, 559–567.

CHAPTER FOUR

# Emerging Roles for MAS-Related G Protein-Coupled Receptor-X2 in Host Defense Peptide, Opioid, and Neuropeptide-Mediated Inflammatory Reactions

**Hydar Ali**[1]
University of Pennsylvania School of Dental Medicine, Philadelphia, PA, United States
[1]Corresponding author: e-mail address: alih@upenn.edu

## Contents

1. Introduction 125
2. Expression Profile and Functions of MRGPRX1–MRGPRX4 126
   2.1 MRGPRX1 Is Expressed on the Neurons of Dorsal Root Ganglia 126
   2.2 MRGPRX2 Is Expressed on $MC_{TC}$ and Is a Receptor for Basic Secretagogues 127
   2.3 MrgprB2 Is the Mouse Ortholog of Human MRGPRX2 128
   2.4 MRGPRX3 and MRGPRX4 Are Expressed on Human Keratinocytes 128
3. MRGPRX2 Is a Novel GPCR for HDPs on MCs 130
   3.1 Activation of MC MRGPRX2 by HDPs 130
   3.2 MRGPRX2 Activation by Retrocyclin and Protegrin Produced in Plant Chloroplast 130
   3.3 Role of LL-37 and MRGPRX2 in Rosacea 131
   3.4 Roles of hBDs, *Staphylococcus aureus* δ-Toxin, and MRGPRX2 in Atopic Dermatitis 133
4. MRGPRX2 Is a Novel Tachykinin Receptor in Human but Not Murine MCs 134
   4.1 NK-1R and MrgprB2 as GPCRs for SP on Experimental Allergic Responses in Mice 134
   4.2 Role of MRGPRX2 in Asthma 135
   4.3 Role of MRGPRX2 in Chronic Idiopathic Urticaria 136
   4.4 Role of MRGPRX2 on Rheumatoid Arthritis 136
5. MRGPRX2 Is a Novel GPCR for Cationic Drugs and Opioids 137
   5.1 Pseudoallergic Reactions by FDA-Approved Cationic Drugs 137
   5.2 Red Man Syndrome 138

*Advances in Immunology*, Volume 136
ISSN 0065-2776
http://dx.doi.org/10.1016/bs.ai.2017.06.002

5.3 Opiates and Opioids Activate MCs and Cause Local and Systemic Allergic Reactions 139
5.4 PRESTO-Tango Approach to Identify MRGPRX2 as a Novel Receptor for Opioids 140
5.5 MRGPRX2 Is an Atypical Receptor for Both Synthetic and Endogenous Opioids 141
5.6 ZINC-3573 Is a Selective Nonopioid Ligand for MRGPRX2 142
6. Ligand-Binding Sites for Opioids and ZINC-3573 on MRGPRX2 and Signaling 143
6.1 Structure-Based Docking Studies to Identify Opioid and ZINC-3573-Binding Sites on MRGPRX2 143
6.2 Activation of MRGPRX2 Leads to Rapid and Transient MC Degranulation 144
6.3 MRGPRX2 Signaling for Degranulation via Gαq and Gαi Family of G Proteins 145
6.4 LL-37 Induces TRPV4 Expression in MCs via MRGPRX2 and Gαi 145
7. Regulation of MRGPRX2 Function in MCs 146
7.1 Role of Constitutive MRGPRX2 Signaling on Cell Proliferation 146
7.2 Retinoic Acid Inhibits Skin $MC_{TC}$ Proliferation but Has Not Effect on Their Survival 147
7.3 Retinoic Acid Causes a Decrease in MRGPRX2 and Chymase but Has No Effect on Tryptase Expression 148
7.4 Receptor Phosphorylation and Desensitization 148
8. Concluding Remarks 149
Acknowledgments 150
References 151

## Abstract

Mast cells (MCs) are tissue-resident immune cells that contribute to host defense but are best known for their roles in allergic and inflammatory diseases. In humans, MCs are divided into two subtypes based on the protease content of their secretory granules. Thus, human lung MCs contain only tryptase and are known as $MC_T$, whereas skin MCs contain both tryptase and chymase and are known as $MC_{TC}$. Patients with severe asthma display elevated MCs in the lung, which undergo phenotypic change from $MC_T$ to $MC_{TC}$. Although the human genome contains four Mas related G protein coupled receptor X (MRGPRX) genes, an important feature of $MC_{TC}$ is that they selectively express MRGPRX2. It is activated by antimicrobial host defense peptides such as human β-defensins and the cathelicidin LL-37 and likely contributes to host defense. MRGPRX2 is also a receptor for the neuropeptide substance P, major basic protein, eosinophil peroxidase, opioids, and many FDA-approved cationic drugs. Increased expression of MRGPRX2 or enhanced downstream signaling likely contributes to chronic inflammatory diseases such as rosacea, atopic dermatitis, chronic urticaria, and severe asthma. In this chapter, I will discuss the expression profile and function of MRGPRX1-4 and review the emerging roles of MRGPRX2 on host defense, chronic inflammatory diseases, and drug-induced pseudoallergic reactions. I will also examine the novel aspects of MRGPRX2 signaling in MCs as it related to degranulation and review the mechanisms of its regulation.

# 1. INTRODUCTION

Mast cells (MCs) are multifunctional immune cells that are found beneath the surface epithelia of all organs and play an important role in host defense (Galli, Maurer, & Lantz, 1999). Their host defense function is mediated, in large part, via the rapid release of granules (degranulation) and the subsequent recruitment of neutrophils (Galli et al., 1999; Malaviya, Ikeda, Ross, & Abraham, 1996; Thakurdas et al., 2007). MCs are, however, best known for their roles in allergic diseases such as anaphylaxis, rhinitis, and asthma (Boyce, 2006; Broide, 2010; Finkelman, Boyce, Vercelli, & Rothenberg, 2010; Patou, Holtappels, Affleck, van Cauwenberge, & Bachert, 2011; Simons, 2009). In these situations, aggregation of cell surface high-affinity IgE receptors (FcεRI) by antigen leads to the rapid degranulation and the generation of newly synthesized lipid-derived mediators and cytokines, which activate effector cells to produce symptoms associated with allergic diseases. However, emerging evidence suggests that FcεRI aggregation on MCs can contribute to resistance toward the lethal effects of honeybee or Russell's viper venom (Galli, 2016). These findings support the notion that in addition to allergic diseases activation of MCs via IgE/FcεRI can help protect the host against noxious substances.

MCs are generally classified into two subtypes depending on the protease content of their secretory granules (Gurish & Austen, 2012). In humans, most MCs that are found in the lung and gut express only tryptase and are known as $MC_T$ (Irani, Bradford, Kepley, Schechter, & Schwartz, 1989; Irani, Schechter, Craig, DeBlois, & Schwartz, 1986). In contrast, most MCs that are found in connective tissues such as the skin contain tryptase and chymase and are known as $MC_{TC}$. In mice, mucosal MCs (MMCs) resemble $MC_T$ and connective tissue MCs (CTMCs) resemble $MC_{TC}$ (Gurish & Austen, 2012; Ruitenberg & Elgersma, 1976). Thus, murine mucosal tissues such as the lung and gut contain MMC and the skin contains CTMCs. MCs also differ in their responses to endogenous and exogenous stimuli that promote degranulation. Thus, while both human MC subtypes are activated via the aggregation of FcεRI, $MC_{TC}$ respond to basic secretagogues such as compound 48/80 and substance P (SP) but $MC_T$ do not (Erdei, Kerekes, & Pecht, 1997; Ogasawara et al., 1988; Schafer et al., 2013).

G protein-coupled receptors (GPCRs) represent the largest family of seven-transmembrane domain receptors that transduce extracellular (EC) signals into biological responses such as cell proliferation, development, survival, metabolism, and neuronal signal transmission via the activation of heterotrimeric G proteins and the recruitment of β-arrestin (Allen & Roth, 2011; Allen et al., 2011; Solinski, Gudermann, & Breit, 2014). The Mas-related gene family, which are newly identified GPCRs, comprises of approximately 50 members in mouse, rat, human, macaque, and rhesus monkey that can be subdivided into several subfamilies (Dong, Han, Zylka, Simon, & Anderson, 2001; Lembo et al., 2002; Solinski et al., 2014). In human, there are four MRGPRX members, MRGPRX1–X4, but the mouse genome contains MrgprA (A1–A10), MrgprB (B1–B5, B8), MrgprC (C11), MrgprD, MrgprE, MrgprF, MrgprG, and MrgprH genes (Dong et al., 2001; Zylka, Dong, Southwell, & Anderson, 2003).

MAS-related G protein-coupled receptor-X2 (MRGPRX2) is expressed at high level in human skin $MC_{TC}$ and but not in lung $MC_{T}$ (Fujisawa et al., 2014; Guhl, Artuc, Neou, Babina, & Zuberbier, 2011; Kajiwara et al., 2010). However, lung $MC_{T}$ in normal individual undergoes a phenotypic change to $MC_{TC}$ in patients with severe asthma (Balzar, Chu, Strand, & Wenzel, 2005). MRGPRX2 is a novel GPCR for antimicrobial host defense peptides (HDPs), the neuropeptide SP, FDA-approved cationic drugs, and opioids (Lansu et al., 2017; Subramanian, Gupta, & Ali, 2016). In this chapter, I will first discuss the expression profiles and possible functions of MRGPRX1–MRGPRX4 and compare the properties of human MRGPRX2 with its murine ortholog, MrgprB2. I will then focus on the emerging roles of MRGPRX2 on host defense, chronic inflammatory diseases, pseudoallergic reactions and discuss the mechanisms of receptor activation and regulation.

## 2. EXPRESSION PROFILE AND FUNCTIONS OF MRGPRX1–MRGPRX4

### 2.1 MRGPRX1 Is Expressed on the Neurons of Dorsal Root Ganglia

MRGPRX1 is expressed mainly in small diameter sensory dorsal root ganglia (DRG) nociceptive neurons and has been implicated in the perception of pain and itch sensation. It was the first human MRGPRX to be assigned to a ligand, bovine adrenal medulla 8–22 (BAM8–22) (Liu et al., 2009). Animal model studies indicated that mouse MrgprC may serve similar function to the human MRGPRX1 as an endogenous pain inhibitor. However, human MRGPRX1 has binding and pharmacologic profiles distinct from mouse MrgprC (Han et al., 2002; Kunapuli et al., 2006; Schmidt, Butterworth,

O'Donnell, Santhakumar, & Tomaszewski, 2009). These findings question feasibility of using conventional animal models to study the role of human MRGPRX1 on pain behavior. Recently, a transgenic mouse line was generated in which mouse Mrgprs were replaced with human MRGPRX1 (Li et al., 2017). This humanized mouse model was used to characterize BAM8–22 and a positive allosteric modulator of MRGPRX1. This model may therefore provide an in vivo tool to screen potential novel therapeutics for the management of persistent pain.

## 2.2 MRGPRX2 Is Expressed on $MC_{TC}$ and Is a Receptor for Basic Secretagogues

Tatemoto et al. (2006) provided the first demonstration that MRGPRX2 is expressed in $MC_{TC}$ but not $MC_T$. Human cord blood-derived MCs can be cultured under different conditions that promote their differentiation to either $MC_T$ or $MC_{TC}$. Quantitative PCR analysis revealed that the copy number of MRGPRX2 in in vitro-differentiated $MC_{TC}$ is 17,565 per 5ng total RNA, whereas it is only 32 per 5ng of total RNA in $MC_T$ (Tatemoto et al., 2006). This is consistent with PCR and microarray data showing that the transcript for MRGPRX2 is expressed at high levels in human skin $MC_{TC}$ but at low levels in lung $MC_T$ (Fujisawa et al., 2014; Kajiwara et al., 2010; Tatemoto et al., 2006). Interestingly, $MC_{TC}$ express MRGPRX2 but not MRGPRX3 and MRGPRX4 (Kashem et al., 2011; Solinski et al., 2013; Tatemoto et al., 2006). Studies with human skin MCs are generally performed with either freshly isolated/purified cells or cells that have been expanded in culture (Guhl et al., 2011; Oskeritzian et al., 2005). It has recently been shown that freshly isolated neonatal human skin MCs express MRGPRX2 ~10-fold higher than cells that were maintained in culture (Motakis et al., 2014). However, it is unknown if freshly isolated MCs differ in their responsiveness to MRGPRX2 ligands when compared to cells that were maintained in culture.

A variety of basic compounds collectively known as basic secretagogues such as MC-degranulating peptide, mastoparan, and compound 48/80 have long been known to induce MC degranulation. Since the 1990s it was generally believed that these basic secretagogues bind to and activate G proteins (Gαi2 and Gαi3) directly to induce downstream signaling for degranulation via a receptor-independent mechanism (Mousli, Bronner, Bockaert, Rouot, & Landry, 1990; Mousli, Bueb, Bronner, Rouot, & Landry, 1990). The first demonstration that basic secretagogues couple to MRGPRX2 was the finding that they activated reporter gene activity in

PC12 cells transiently expressing the receptor (Tatemoto et al., 2006). They also induced dose-dependent $Ca^{2+}$ mobilization in human embryonic kidney (HEK293) expressing MRGPRX2 but not MRGPRX1. The specificity of basic secretagogues for MRGPRX2 over MRGPRX1 for MC degranulation was later confirmed using transfected rat basophilic (RBL-2H3) cells (Kashem et al., 2011; Subramanian, Kashem, et al., 2011).

## 2.3 MrgprB2 Is the Mouse Ortholog of Human MRGPRX2

In contrast to the human genome, which encodes for four MRGPRXs, the mouse genome encodes for 22 potential receptor genes making it difficult to identify the mouse ortholog of human MRGPRX2. However, the utilization of stringent reverse transcriptase-PCR in mouse peritoneal MCs led to the demonstration that they express messenger RNA only for MrgprB2 (McNeil et al., 2015). Generation of a transgenic mouse with td-Tomato reporter under the control of MrgprB2 promoter demonstrated that the expression of MrgprB2 is restricted to connective tissue type MC in the skin, gut, and trachea (McNeil et al., 2015). The same group also generated mice with a 4-base pair deletion in the *MrgprB2* coding region, resulting in a frameshift mutation and early termination of the receptor (shortly after the first transmembrane domain). These MrgprB2 mutant mice have no defect in MC number and respond normally to IgE/FcεRI activation. However, peritoneal MCs from these mice show dramatic reduction in $Ca^{2+}$ mobilization and histamine release in response to compound 48/80 in vitro and reduced paw edema in vivo (McNeil et al., 2015). These findings suggest that MrgprB2 is the mouse ortholog of human MRGPRX2.

## 2.4 MRGPRX3 and MRGPRX4 Are Expressed on Human Keratinocytes

The expression of HDPs such as human β-defensins (hBDs) and the cathelicidin LL-37 is enhanced in inflamed skin and cutaneous wound sites (Harder, Bartels, Christophers, & Schroder, 2001; Harder & Schroder, 2005). In addition to their direct antimicrobial activity, HDPs cause $Ca^{2+}$ mobilization in keratinocytes and promote their migration, proliferation, and induce the production of proinflammatory cytokines/chemokines (Niyonsaba et al., 2007). These findings suggest that HDPs may contribute to skin immunity via the activation of a cell surface receptor on keratinocytes. Unlike human MCs, which selectively express MRGPRX2

(Fujisawa et al., 2014), human keratinocytes express MRGPRX3 and MRGPRX4 at high levels (Kiatsurayanon et al., 2016). Expression of MRGPRX3 and MRGPRX4 is upregulated upon stimulation with toll-like receptor (TLR) ligands. Furthermore, siRNA-mediated knockdown of MRGPRX3 and MRGPRX4 suppresses HDP-induced responses in human keratinocytes. These findings suggest that the role of HDPs on cutaneous host defense, at least in part, reflects MRGPRX3/4 activation in keratinocytes (Fig. 1). It is also possible that overexpression of MRGPRX3/4 or their uncontrolled activation contributes to itch, pain, and other chronic inflammatory skin diseases (Takahashi & Gallo, 2017). In this regards, transgenic overexpression of human MRGPRX3 in rats results in cataract formation and an abnormal skin phenotype (Kaisho et al., 2005).

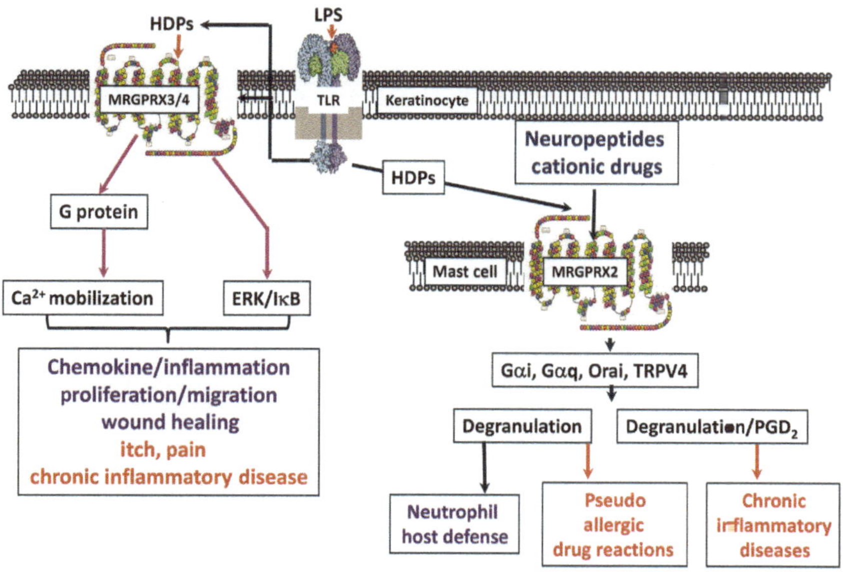


**Fig. 1** Model for an integrated role of MRGPRX3/MRGPRX4 on keratinocytes and MRGPRX2 on MCs on host defense, wound healing, and chronic inflammatory diseases. TLR activation on keratinocytes following microbial infection leads the generation of HDPs, which feedback to activate MRGPRX3/MRGPRX4 on keratinocytes. This activation results in the generation of proinflammatory cytokines, keratinocyte migration, and proliferation. This immunomodulatory effect of HDPs in addition to their antimicrobial activities leads to microbial clearance and would healing. HDPs also activate MCs via MRGPRX2, resulting in their degranulation and subsequent neutrophil recruitment. This is likely to be an early event which is integrated with keratinocyte-derived cytokine induction (late event) to orchestrate host defense. Dysregulation of these pathways may lead to chronic inflammatory diseases (*red text* and see Fig. 2).

# 3. MRGPRX2 IS A NOVEL GPCR FOR HDPs ON MCs

## 3.1 Activation of MC MRGPRX2 by HDPs

MCs play a critical role in host defense and have been referred to as sentinels of innate immunity (Galli et al., 1999; Marshall & Jawdat, 2004; St John & Abraham, 2013). HDPs (LL-37 and hBDs) are potent MC chemoattractants and induce degranulation (Befus et al., 1999; Niyonsaba et al., 2002). HDPs also cause an increase the expression of TLR4 on MCs, which may enhance their ability to detect invading pathogens (Yoshioka et al., 2008). MC activation plays an important role in host defense by causing increased vascular permeability and by initiating the recruitment of neutrophil to the sites of infection (Echtenacher, Mannel, & Hultner, 1996; Malaviya & Abraham, 2000; Thakurdas et al., 2007). It has recently been shown that epithelium and neutrophil-derived HDPs (hBDs and LL-37) activate human MCs via MRGPRX2 (Gupta, Subramanian, & Ali, 2016; Subramanian, Gupta, Guo, Price, & Ali, 2011; Subramanian et al., 2013). LL-37 is produced from activated MCs and neutrophils, and it's antimicrobial activity may reflect, at least in part, the activation of MCs via MRGPRX2 (Di Nardo, Vitiello, & Gallo, 2003; Di Nardo, Yamasaki, Dorschner, Lai, & Gallo, 2008; Subramanian, Gupta, et al., 2011). It is noteworthy that MRGPRX2 activation in MCs results in degranulation but little to no cytokine production (Gaudenzio et al., 2016). However, activation of MRGPRX3/4 in keratinocytes results in the generation of proinflammatory cytokines and the promotion of wound healing (Kiatsurayanon et al., 2016; Niyonsaba et al., 2007). Thus, simultaneous activation of MRGPRX3/4 in keratinocytes and MRGPRX2 in MCs by HDPs likely provides a highly integrated mechanism for cutaneous host defense and wound healing (Fig. 1).

## 3.2 MRGPRX2 Activation by Retrocyclin and Protegrin Produced in Plant Chloroplast

Antibiotics have been used for the treatment of microbial infection since the early 1900s but the emergence of multidrug-resistant strains of microbes poses a tremendous public health concern globally (Laxminarayan et al., 2013). HDPs have been tested in clinical trials with the expectation that they would not develop resistance to microbes and that their mechanism of action would involve harnessing immunomodulatory properties in addition to their direct antimicrobial activities. However, many HDPs cause red blood cell lysis and display cytotoxic activity against immune and nonimmune cells. This explains why the majority of clinical trials with HDPs for microbial infection have involved topically applied peptides (Dawson, McAllister, & Liu, 2010; Westman et al., 2013).

A family of antimicrobial peptides known as θ-defensins is found in leukocytes of rhesus macaques (Tang et al., 1999). Retrocyclin-1 (RC-100) is a cyclic octadecapeptide humanized θ-defensin, which was initially prepared by solid-phase synthesis (Cole et al., 2002). It has broad-spectrum antimicrobial properties and protects human target cells from HIV infection in vitro (Cole et al., 2002). Protegrin-1 (PG-1) is an antimicrobial peptide that was originally isolated from porcine leukocytes (Kokryakov et al., 1993). It shares many structural similarities with θ-defensin; it is a cysteine-rich octadecapeptide with high arginine content but lacks a cyclic backbone (Lehrer, Cole, & Selsted, 2012). Due to the unique structure and broad-spectrum antimicrobial activities, retrocyclins and PG-1 have immense therapeutic potential against infectious diseases.

Unlike hBD3 and LL-37, retrocyclins are nonhemolytic and noncytotoxic (Tran et al., 2008). However, a major limitation of chemically synthesized peptides is that they are prohibitively expensive (~\$600,000–\$700,000/g) and display inappropriate postsynthesis modifications such as lack of folding, disulfide bond formation, and cyclization which render them ineffective as antimicrobial agents (Gupta, Kotian, Subramanian, Daniell, & Ali, 2015). To overcome these limitations, RC-101 and PG-1 have been expressed in transgenic tobacco chloroplasts as green fluorescent protein-fusion proteins (Lee, Li, Jin, & Daniell, 2011). These peptides are efficient in both killing microbes and activating MCs via MRGPRX2 (Gupta et al., 2015; Lee et al., 2011). Thus, HDPs that activate MCs via MRGPRX2 in addition to their direct antimicrobial activity may offer attractive targets for the development of novel antimicrobial therapy (Hilchie, Wuerth, & Hancock, 2013). Low-cost production of HDPs in plants that kill microbes and also harness immunomodulatory properties of MRGPRX2, 3, and 4 (Fig. 1) should facilitate their advancement to the clinic thus overcoming major hurdles in current production systems.

## 3.3 Role of LL-37 and MRGPRX2 in Rosacea

Rosacea is a chronic inflammatory skin condition with different subtypes that include facial flushing, telangiectasia, inflammatory papules, and pustules (Two, Wu, Gallo, & Hata, 2015a; Wilkin et al., 2002). Although

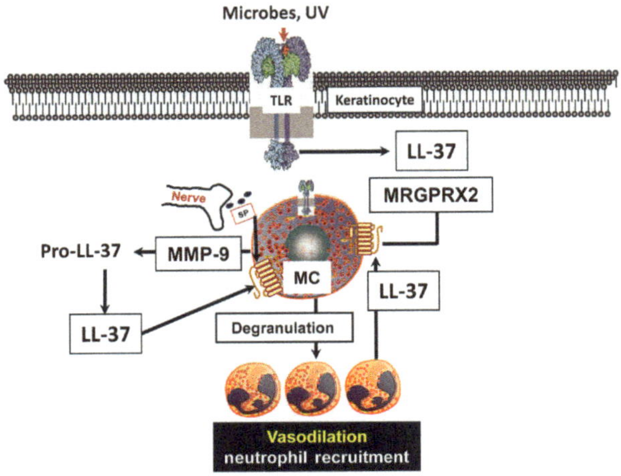


**Fig. 2** Proposed model for the pathogenesis of rosacea. TLR activation on keratinocytes and MCs following microbial infection leads the generation of LL-37. It is also produced by neutrophils which are recruited to the site of infection. LL-37 likely plays an important role in host defense via MC activation through MRGPRX2 and its direct antimicrobial activity. However, dysregulation of this mechanism due to excessive LL-37 production leads to enhanced MC activation, which contributes to the development of rosacea.

the incidence of rosacea is higher among individuals of Celtic and Northern European descent, genomic association studies have yet to identify any causative gene. Instead, as depicted in Fig. 2, substantial evidence points to a role of dysregulated innate immune system in its pathogenesis (Two et al., 2014; Two, Wu, Gallo, & Hata, 2015a, 2015b; Yamasaki et al., 2007; Yamasaki & Gallo, 2009, 2011). The expression of TLR2 is increased in keratinocytes of patients with rosacea and is likely activated by structural molecules present on the cell surface of microbes (Koller, Muller-Wiefel, Rupec, Korting, & Ruzicka, 2011; Lacey, Delaney, Kavanagh, & Powell, 2007; Yamasaki et al., 2011). Patients with rosacea display increased expression and activity of a serine protease, which is responsible for cleaving the cathelicidin Pro-LL-37 into its biologically active form, LL-37 (Yamasaki et al., 2007).

In addition to keratinocytes, LL-37 is produced from MCs and neutrophils (Cowland, Johnsen, & Borregaard, 1995; Di Nardo et al., 2003). Activated MCs also release matrix metalloproteinase-9 (MMP-9), which further contributes to the generation of LL-37. Not surprisingly, LL-37 has been shown to play a central role in the pathogenesis of rosacea (Kanada, Nakatsuji, & Gallo, 2012; Muto et al., 2014; Reinholz, Ruzicka, & Schauber, 2012; Schittek, Paulmann, Senyurek, & Steffen, 2008; Yamasaki et al., 2007). In mice, intradermal injection of LL-37 results in the development of rosacea-like inflammation and this is not observed in MC-deficient $W^{sh}/W^{sh}$ mice (Muto et al., 2014; Yamasaki et al., 2007). Furthermore, reconstitution of MCs in the skin of $W^{sh}/W^{sh}$ mice results in the restoration of LL-37-mediated rosacea-like inflammation (Muto et al., 2014). Neuronal dysregulation and the generation of neuropeptides (NPs) such as pituitary adenylate cyclase-activating peptide (PACAP) and SP are frequently associated with rosacea "flare-ups" (Muto et al., 2014; Schwab et al., 2011; Steinhoff, Schauber, & Leyden, 2013). In addition to MC degranulation, intradermal injection of PACAP results in the expression and release of proinflammatory cytokines which promote neutrophil recruitment in wild type but not in $W^{sh}/W^{sh}$ mice (Muto et al., 2014). These findings strongly suggest that MCs play a critical role in the development and exacerbation of rosacea.

Although rosacea is not a life-threatening disease, there are significant morbidities associated with this condition with regards to the individual's quality of life and the greater risk of being diagnosed with depression (Aksoy, Altaykan-Hapa, Egemen, Karagoz, & Atakan, 2010; Gupta, Gupta, Chen, & Johnson, 2005). Thus, it is important to control signs and symptoms associated with this condition. Indeed, Muto et al. (2014) recently showed that application of cromolyn, a MC stabilizer, reduces erythema associated with rosacea in humans when compared to placebo. Although cromolyn is a general MC stabilizer, its mechanism of action is unknown. Our lab has recently identified MRGPRX2 as a novel GPCR for LL-37 (Subramanian, Gupta, et al., 2011). NPs such as SP were previously thought to activate MCs via neurokinin 1-receptor (NK-1R) (Kulka, Sheen, Tancowny, Grammer, & Schleimer, 2008). However, as described in the Section 4, SP activates human skin $MC_{TC}$ via MRGPRX2. These findings suggest that activation of MRGPRX2 by LL-37 and NPs contributes to the development and exacerbation of rosacea (Fig. 2).

### 3.4 Roles of hBDs, *Staphylococcus aureus* δ-Toxin, and MRGPRX2 in Atopic Dermatitis

Atopic dermatitis is a chronic relapsing inflammatory skin disease that affects 15%–30% of children and ~5% of adults in industrial countries; it is associated genetic predisposition, cutaneous hyperreactivity to environmental triggers, and immune dysregulation (Homey, Steinhoff, Ruzicka, & Leung, 2006; Williams & Flohr, 2006). Itch is one of the most serious clinical symptoms of atopic dermatitis, and repeated itching–scratching episodes may be greatly involved in its acceleration and severity (Boguniewicz, 2005).

MCs are strongly implicated (Kawakami, Ando, Kimura, Wilson, & Kawakami, 2009; Liu, Goodarzi, & Chen, 2011), and it now appears that HDPs (hBDs) and MRGPRX2 are involved in the pathogenesis of atopic dermatitis. Thus, the expression of hBDs is increased in lesional atopic dermatitis skin when compared to nonlesional skin in response to bacteria (Kanda et al., 2011; Kanda & Watanabe, 2012). Furthermore, hBDs cause MC degranulation and induces the pruritogenic cytokine IL-31 via MRGPRX2 (Niyonsaba, Someya, Hirata, Ogawa, & Nagaoka, 2001; Niyonsaba et al., 2010; Subramanian et al., 2013). Moreover, hBDs promote cutaneous vascular permeability through the activation of MCs (Chen et al., 2007).

*Staphylococcus aureus* commonly colonizes or secondarily infects the skin of patients with atopic dermatitis, and the microbe is found in >90% of lesional skin but healthy individuals do not harbor the microbe (Rudikoff & Lebwohl, 1998). *S. aureus* isolates recovered from patients with atopic dermatitis produce large amounts of δ-toxin, which induces MC degranulation ex vivo (Nakamura et al., 2013). Furthermore, δ-toxin causes $Ca^{2+}$ mobilization in HEK293 cells stably expressing MRGPRX2 and induces degranulation a human MC line, LAD2 line (Azimi, Reddy, & Lerner, 2017). These responses are inhibited by the tripeptide QWF, a MRGPRX2 antagonist (Azimi et al., 2017, 2016). These findings suggest that *S. aureus* δ-toxin and hBDs activate MCs via MRGPRX2, which contributes to the pathogenesis of atopic dermatitis.

## 4. MRGPRX2 IS A NOVEL TACHYKININ RECEPTOR IN HUMAN BUT NOT MURINE MCs

### 4.1 NK-1R and MrgprB2 as GPCRs for SP on Experimental Allergic Responses in Mice

It is generally accepted that murine MMCs such as bone marrow-derived cells (BMMCs) do not express NK-1R (van der Kleij et al., 2003). However, BMMCs generated in the presence of stem cell factor and interleukin-4, which favor CTMC phenotype, result in the expression NK-1R. Furthermore, SP causes degranulation in these MCs, which is diminished by an NK-1R antagonist or in cells obtained from NK-1R$^{-}$/$^{-}$ mice (van der Kleij et al., 2003). Subcutaneous injection of SP in mice results in increased vascular permeability and marked infiltration of eosinophils and neutrophils (Matsuda, Kawakita, Kiso, Nakano, & Kitamura, 1989; Yano, Wershil, Arizono, & Galli, 1989). These responses are reduced in MC-deficient mice but could be rescued with local engraftment of cultured MCs. These findings suggest that activation of NK-1R in murine MCs by SP results in increased vascular permeability and granulocyte recruitment in vivo. Nasal allergen provocation induces bronchial hyperresponsiveness in a murine model of asthma via the pulmonary upregulation of SP and NK-1R (Hens et al., 2011). Furthermore, treatment with an NK-1R antagonist abolishes allergen-induced airway hyperresponsiveness, indicating an important role for this GPCR on the development of experimental asthma in mice.

In addition to NK-1R, murine CTMCs express MrgprB2 (McNeil et al., 2015). Furthermore, MrgprB2 mutant mice have no defect in MC number and respond normally to IgE/FcεRI activation. Peritoneal MCs from these mice show dramatic reduction in degranulation in response to SP in vitro and reduced inflammation in vivo (Gaudenzio et al., 2016; McNeil et al., 2015). Azimi et al. (2016) recently showed that conventional NK-1R antagonists have off-target effect on the mouse MrgprB2. It is therefore likely that the effects of NK-1R antagonists on allergen-induced airway hyperresponsiveness mice likely reflect the inhibition of both NK-1R and MrgprB2 (Azimi et al., 2016; Hens et al., 2011).

## 4.2 Role of MRGPRX2 in Asthma

The expression of SP and NK-1R is upregulated in the lungs of human asthmatic when compared to control subjects (Boot et al., 2007; Hens et al., 2011; Nieber et al., 1992; Tomaki et al., 1995). These findings suggest that NK-1R not only participates in experimental allergic responses in mice as discussed above it also plays an important role in the human disease. However, a number of NK-1R antagonists, which are highly effective in animal models of allergic asthma and inflammation, lack efficacy in humans (Boot et al., 2007; Hens et al., 2011; Quartara, Altamura, Evangelista, & Maggi, 2009; Rost, Fleischer, & Nieber, 2006). The striking failure to translate findings from animal models to the clinic suggests that there are important differences in the mechanisms via which NPs activate murine and human MCs. Although normal human lungs MCs are the $MC_T$ type, severe asthma is dominated by the presence of $MC_{TC}$ in the airway submucosa and epithelium (Balzar et al., 2011; Fajt et al., 2013). It is now well documented that $MC_{TC}$-type MCs express MRGPRX2, which is activated by SP (Babina, Artuc, Guhl, & Zuberbier, 2017; Fujisawa et al., 2014; Motakis et al., 2014). Furthermore, while NK-1R antagonists have off-target against MrgprB2, they do not inhibit MRGPRX2-mediated responses (Azimi & Lerner, 2017; Azimi et al., 2016). These findings suggest that while NPs activate murine MCs via NK-1R and MrgprB2, they utilize MRGPRX2 for human MC activation. This difference likely explains why NK-1R antagonists modulate MC-mediated experimental allergic responses in mice but lack efficacy in humans.

## 4.3 Role of MRGPRX2 in Chronic Idiopathic Urticaria

Chronic urticaria refers to the occurrence of daily wheals, which lasts for at least 6 weeks. About 50% of patients have an IgE component and for the other 50%, the cause is unknown and the condition is known as "idiopathic." It is well documented that patients with idiopathic chronic urticaria exhibit stronger wheal reactions to intradermal injection of neuropeptide such as SP when compared to normal subjects (Borici-Mazi, Kouridakis, & Kontou-Fili, 1999; Smith et al., 1992). This effect of SP is not mediated through NK-1R as SP-induced degranulation in skin $MC_{TC}$ is not blocked by an NK-1R antagonist (Fujisawa et al., 2014). MRGPRX2 is expressed in normal human skin MCs and its expression is significantly increased in patients with chronic urticaria (Fujisawa et al., 2014). Furthermore, SP-induced degranulation and prostaglandin $D_2$ ($PGD_2$) generation are substantially attenuated in MRGPRX2-silenced human skin $MC_{TC}$. Eosinophils colocalize with $MC_{TC}$ in the skin of patients with chronic urticaria (Fujisawa et al., 2014; Kaplan & Greaves, 2009). Furthermore, eosinophil-derived major basic protein and eosinophil peroxidase activate human skin $MC_{TC}$ via MRGPRX2. These findings suggest that SP induces MC degranulation in patients with chronic urticaria via MRGPRX2. The subsequent recruitment and activation of eosinophils result in the release of major basic protein and eosinophil peroxidase, which further activate MCs via MRGPRX2 and contribute to the late chronic phase of the response (Fujisawa et al., 2014).

## 4.4 Role of MRGPRX2 on Rheumatoid Arthritis

Rheumatoid arthritis is a chronic systemic inflammatory disease in which cytokines such TNF-$\alpha$, IL-1$\beta$, IL-6, and IL-17A play pivotal roles in the destruction of the joint architecture (Furst & Emery, 2014). In addition, the neuropeptide SP and its receptor NK-1R are implicated in the perception of pain and inflammation in rheumatoid arthritis (Pennefather et al., 2004; Sarzi-Puttini et al., 2014).

Evidence suggests that MCs also participate in the pathogenesis of rheumatoid arthritis. Thus, number of degranulated MCs are increased in synovial tissue of patients with rheumatoid arthritis, and they colocalize with fibroblasts and correlate with disease severity (Bromley, Fisher, & Woolley, 1984; Bromley & Woolley, 1984; Crisp, Chapman, Kirkham, Schiller, & Krane, 1984). Mediators derived from MCs such as histamine and tryptase are also increased in the synovial fluid of patients with rheumatoid arthritis (Malone, Irani, Schwartz, Barrett, & Metcalfe, 1986; Tetlow & Woolley, 1995). Activation of cultured synovial MCs results in the redistribution of scattered SP/tryptase from the cytoplasm to the cell membrane, a pattern of localization similar to that observed in synovial MCs of patients with rheumatoid arthritis. It is interesting to note that activation of synovial MCs via FcεRI results in increased expression of tachykinin (TAC) 1 mRNA and elevated intracellular (IC) SP (Okamura et al., 2017). Human synovial MCs express MRGPRX2 and shRNA-mediated silencing of the receptor results in substantial inhibition of SP-induced degranulation, but a selective NK-1R inhibitor does not have any effect. This suggests that SP induces degranulation in cultured human synovial MCs via MRGPRX2 (Lee et al., 2013; Okamura et al., 2017). Therefore, it is tempting to speculate that upregulation of TAC1 mRNA and the subsequent release of SP following immune complex-mediated MC activation amplify pain and inflammation via the activation of MRGPRX2. However, SP could not be detected in the supernatant following activation partly because it is IC and partly because it is cleaved by proteases released from MCs. Thus, activated MCs may downregulate the SP-mediated activation of synoviocytes in rheumatoid arthritis. By contrast, SP activates MCs via MRGPRX2 to release inflammatory mediators, indicating dual regulation of SP-mediated inflammation by MCs in rheumatoid arthritis (Okamura et al., 2017).

## 5. MRGPRX2 IS A NOVEL GPCR FOR CATIONIC DRUGS AND OPIOIDS

### 5.1 Pseudoallergic Reactions by FDA-Approved Cationic Drugs

Many US Food and Drug Administration (FDA)-approved peptidergic drugs cause injection site pseudoallergic drug reactions in humans. Neuromuscular blocking drugs such as atracurium, mivacurium, tubocurarine, and rocuronium, which are routinely used in surgery to reduce unwanted muscle movement, induce histamine release from skin MCs and cause allergic reactions in surgical settings (Koppert et al., 2001; Mertes et al., 2011). Intravenous administration of sugammadex, a modified γ-cyclodextrin, results in rapid recovery in patients undergoing surgery, who received rocuronium. Sugammadex is well tolerated and thus provides a safe option for the rapid reversal of rocuronium-induced neuromuscular blockade (Keating, 2016). Furthermore, rocuronium-induced injection site reaction is suppressed by sugammadex, indicating that it likely inhibits rocuronium-induced MC degranulation via MRGPRX2 (Spoerl, D'Incau, Roux-Lombard, Harr, & Czarnetzki, 2016).

Icatibant, a bradykinin $B_2$ receptor antagonist, which is used for the treatment of hereditary angioedema, causes injection site erythema and swelling in nearly every patient (Lumry et al., 2011). Furthermore, fluoroquinolone family of antibiotics such as ciprofloxacin and levofloxacin activate MCs to cause histamine release and are associated with allergic reactions (Blanca-Lopez et al., 2013; Kelesidis, Fleisher, & Tsiodras, 2010; Mori, Maru, & Takasuna, 2000). These three drugs stimulate degranulation, $PGD_2$, and TNF-α release in the human MC line, LAD2 and these responses are attenuated in cells with siRNA-mediated knockdown of MRGPRX2 (McNeil et al., 2015).

Peptidergic drugs cause $Ca^{2+}$ mobilization and degranulation in peritoneal MCs from wild type but MrgprB2 mutant mice (McNeil et al., 2015). Additionally, MrgprB2 mutant mice display reduced paw edema in response to icatibant and atracurium and show significantly blunted systemic anaphylaxis response to the fluoroquinolone antibiotics such as ciprofloxacin, when compared to wild-type mice (McNeil et al., 2015). However, the concentrations of ligands required to activate MrgprB2 are significantly higher than those for MRGPRX2 (McNeil et al., 2015). For example, while ciprofloxacin and levofloxacin activate MrgprB2 with $EC_{50}$ values of ~126 and 807 μg/mL, respectively, the same drugs activate MRGPRX2 with $EC_{50}$ values of ~6.8 and 23 μg/mL. These findings suggest that there are important species-specific differences and that MrgprB2 mutant mice may not be an appropriate model to screen drugs for human use.

## 5.2 Red Man Syndrome

Vancomycin is a glycopeptide antibiotic, which has been in used for the past 60 years for the treatment of nosocomial infections by methicillin-resistant staphylococci species (Rubinstein & Keynan, 2014; Sivagnanam & Deleu, 2003). To minimize infusion-related adverse effects vancomycin is normally administered intravenously with a standard infusion time of at least 1 h. The most common adverse effect of vancomycin administration is "red man" or "red neck" syndrome, which is manifested as erythema of the head and upper torso, pruritus, urticaria, and hypotension. It is considered a pseudo-allergic drug reaction mediated via IgE-independent activation of MCs and does not require previous exposure to the drug. Although vancomycin has long been known to induce degranulation of human skin MCs and in vitro-derived MCs the mechanism of its activation was revealed only recently (Veien et al., 2000; Yamaguchi et al., 2013).

It is noteworthy that clinical features of red man syndrome are similar to those frequently observed with ciprofloxacin and atracurium. However, teicoplanin, a glycopeptide antibiotic similar to vancomycin, does not display symptoms associated with red man syndrome (Wilson, 1998). Ciprofloxacin and atracurium active human MCs via MRGPRX2 indicating that vancomycin may also utilize the same pathway to induce MC degranulation. Indeed, it has recently been shown that vancomycin, but not teicoplanin, caused IC $Ca^{2+}$ mobilization in HEK293 cells stably transfected with MRGPRX2 and degranulation in a human MC line endogenously expressing the receptor (Azimi et al., 2017). Furthermore, a peptide inhibitor of MRGPRX2 significantly inhibited vancomycin-induced $Ca^{2+}$ mobilization in transfected HEK293 cells and degranulation in a human MC line (Azimi et al., 2017, 2016).

## 5.3 Opiates and Opioids Activate MCs and Cause Local and Systemic Allergic Reactions

Although opiates and their synthetic counterparts, opioids, are important therapeutic agents for the treatment and management of pain, they can display major side effects, which are mediated via the activation and degranulation of MCs (Casale, Bowman, & Kaliner, 1984; Sheen, Schleimer, & Kulka, 2007). Evidence suggests that many heroin fatalities are caused by MC degranulation and the resulting anaphylactic shock (Edston & van Hage-Hamsten, 1997). The local clinical manifestations of opiates and opioids include urticaria, skin rash, and contact dermatitis at the injection site (Golembiewski, 2002). Although IgE and FcεRI may be involved in rare cases (Harle, Baldo, Coroneos, & Fisher, 1989), most adverse effects of opioids/opiates are mediated via the activation of MCs independently of IgE (Casale et al., 1984).

Morphine induces degranulation in human skin $MC_{TC}$ but it has no effect on intestinal or lung $MC_{T}$ (Tharp et al., 1987). Opiates/opioids induce their biological responses via the activation of three cell surface GPCRs (mu-μ, delta-δ, and kappa-κ) on target cells (Stein, 2016). Human $MC_{TC}$ express transcripts for μ and δ but not the κ-opioid receptor (Sheen et al., 2007). Furthermore, naloxone and naltrexone have no effect on MC degranulation in response to codeine. Based on these findings, it has been suggested that codeine causes MC degranulation independent of its cognate receptors and may involve the direct activation of G proteins similar to that originally proposed for basic secretagogues (Sheen et al., 2007).

## 5.4 PRESTO-Tango Approach to Identify MRGPRX2 as a Novel Receptor for Opioids

In addition to G proteins, nearly all GPCRs including MRGPRX2 signal via β-arrestin (Shenoy & Lefkowitz, 2005; Southern et al., 2013). A number of approaches have been developed to quantify GPCR–β-arrestin interactions, including high-content screening (Barak, Ferguson, Zhang, & Caron, 1997), bioluminescence resonance energy transfer (Bedini, 2015), enzyme complementation, and transcriptional activation following arrestin translocation (Tango) (Barnea et al., 2008). A modified Tango approach named Parallel Receptorome Expression and Screening via transcriptional activation-with Tango (PRESTO-Tango) has recently been developed to provide a platform for rapid, efficient parallel, and simultaneous profiling of biologically active compounds across essentially the entire human druggable GPCRome (Kroeze et al., 2015). The Tango plasmid contains a cleavable signal sequence at the 5′ end of the receptor to promote membrane localization and a Flag epitope tag to allow for the monitoring of cell surface expression. The 3′ end contains the sequences for the tobacco etch virus nuclear inclusion a endopeptidase (TEV)-cleavage site and the tetracycline transactivator (tTA) protein. The sequence from the C-terminus of the $V_2$ vasopressin receptor ($V_2$ tail) was added to GPCR sequence to promote arrestin recruitment (Fig. 3A). Transfection of this Tango plasmid into a HEK293 cell line expressing a tTA-reporter gene (HTLA cells) facilitates high-throughput assay for the activation of β-arrestin-mediated gene expression. In the Tango assay, ligand-dependent recruitment of β-arrestin–TEV protease to the receptor fusion and the subsequent proteolytic cleavage frees tTA to enter the nucleus and activate reporter genes, which can be quantitated by assaying for tTA-dependent luciferase reporter activity (Fig. 3B). This system can also be used to measure IC $Ca^{2+}$ mobilization thus providing a unique platform for assaying both G protein-dependent and -independent signaling (Lansu et al., 2017).

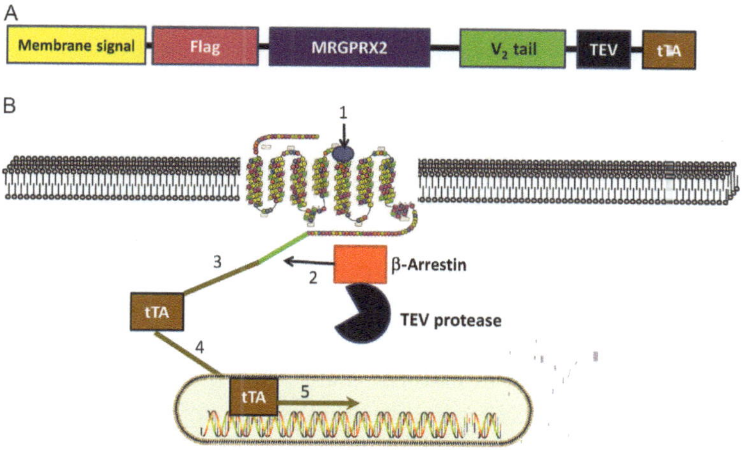


**Fig. 3** Design of the Tango system: (A) Modular design of Tango constructs. (B) Upon activation (1), β-arrestin is recruited to the C-terminus of the $V_2$ tail (2). This is followed by cleavage of the GPCR-fusion protein at the TEV protease–cleavage site (3). Cleavage results in the release of the tTA transcription factor (4), which, after transport to the nucleus, activates transcription of the luciferase reporter gene (5). *Modified from a recent publication by Kroeze, W. K., Sassano, M. F., Huang, X. P., Lansu, K., McCorvy, J. D., Giguere, P. M., et al. (2015). PRESTO-Tango as an open-source resource for interrogation of the druggable human GPCRome.* Nature Structural & Molecular Biology, *22(5), 362–369.*

The PRESTO-Tango system has recently been used to screen ~6000 unique compounds for agonist activity at three members of the MRGPRX family: MRGPRX1, MRGPRX2, and MRGPRX4 (Lansu et al., 2017). This screening revealed that of the 67 compounds that selectively activate MRGPRX2, 5 were opioid-related ligands: ADL-5859, sinomenine, dextromethorphan, dextromethorphan's metabolite dextrorphan, and the previously reported MRGPRX2 ligand, TAN-67 (Lansu et al., 2017; Southern et al., 2013). Other known MRGPRX2 ligands such as cortistatin, SP, and compound 48/80 also activate the β-arrestin-mediated transcription factor in the Tango assay. Furthermore, all five opioid-related MRGPRX2 ligands found in the Tango screen also activated $Ca^{2+}$ mobilization in HEK293 cells transfected with the unmodified receptor (Lansu et al., 2017).

## 5.5 MRGPRX2 Is an Atypical Receptor for Both Synthetic and Endogenous Opioids

Unlike classic opioid receptors, which are activated by levorotary ligands, MRGPRX2 displays ~10-fold higher potency of dextrorotary opioid ligands. In addition, unlike the case for classic opioid receptors (+) and (−) enantiomers of morphine, codeine, thebaine, and TAN-67 activate MRGPRX2 in transfected HEK293 cells with similar potency (Lansu et al., 2017). Thus, it now appears that in addition to δ-opioid receptor, TAN-67 also activates MRGPRX2 but using enantiomers as ligands. Furthermore, naloxone and other canonical opioid antagonists including naltrindole, naltrex and one β-chlornaltrexamine have no effect on TAN-67-induced MRGPRX2 activation. These findings suggest that MRGPRX2, despite low sequence identity with the four classic opioid receptors (~26%), is an atypical opioid receptor with important differences in ligand recognition patterns (Lansu et al., 2017).

Given that MRGPRX2 is activated by synthetic opioids raises an important question regarding endogenous opioid ligands. Of the 20 endogenous opioid peptides tested for activity in PRESTO-Tango and $Ca^{2+}$ mobilization assay, MRGPRX2 was activated by prodynorphin-derived peptides. In addition to transfected HEK293 cells, TAN-67, morphine, and dynorphin A(1–13) also induce robust $Ca^{2+}$ mobilization and degranulation in human MC line (LAD2) that endogenously express MRGPRX2 (Lansu et al., 2017). The effects of these opioids on MC cell degranulation are specific for MRGPRX2 as siRNA-mediated silencing of its expression significantly attenuated degranulation. Furthermore, structurally unrelated κ-, δ-, and μ-opioid receptor agonists (salvinorin A, BW373U86, and DMGO, respectively) do not induce MC degranulation. In addition, naloxone has no effect on the potency or efficacy of MRGPRX2 opioid-induced MC degranulation. These findings suggest that MRGPRX2, not the canonical opioid receptors, mediates opioid-induced degranulation in human MCs. As morphine and structurally similar analgesics induce MC-dependent pruritus (Kumar & Singh, 2013), vasodilation, and hypotension that are poorly reversed by naloxone (Baldo & Pham, 2012; Hutchinson et al., 2011), targeting MRGPRX2 could modulate the side effects of opioid analgesics.

## 5.6 ZINC-3573 Is a Selective Nonopioid Ligand for MRGPRX2

ZINC is a free public resource database containing over 20 million commercially available molecules that can be used for virtual screening studies (Irwin & Shoichet, 2005; Irwin, Sterling, Mysinger, Bolstad, & Coleman, 2012). Virtual screening of the ZINC library with 3.7 million commercially available molecules led to the selection of 20 compounds for MRGPRX2

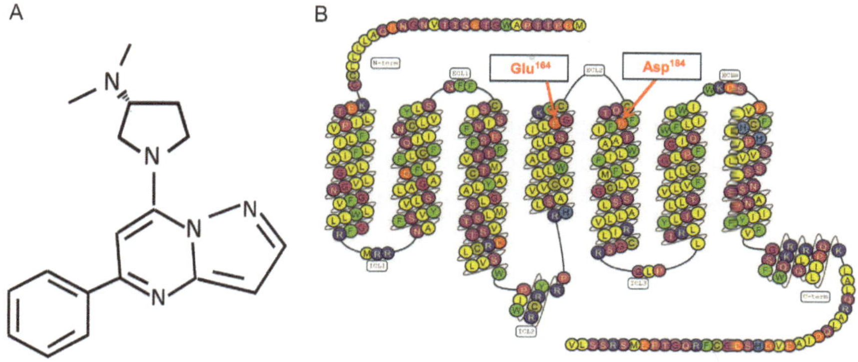


**Fig. 4** (A) Structure of the selective MRGPRX2 agonist, ZINC-3573. (B) Snake-like diagram of MRGPRX2 showing the locations of the potential opioid and ZINC-3537 binding sites (Glu$^{164}$ and Asp$^{184}$).

activity (Lansu et al., 2017). Of the 20 compounds tested one referred to as ZINC-9232 had substantial activity below 10 μM in PRESTO-Tango and calcium release assays. Testing 22 analogs of ZINC-9232 led to the identification of ZINC-3573 ((*R*)-dimethyl-[1-(5-phenyl-pyrazolo[1,5-*a*] pyrimidin-7-yl)-pyrrolidin-3-yl]-amine) with submicromolar potency ($EC_{50}$, 760 nM) (Fig. 4A). The utilization of the PRESTO-Tango system allowed for testing ZINC-3573 at 315 other GPCRs. Interestingly, while *R*-isomer of ZINC-3573 activated MRGPRX2 in both the Tango and $Ca^{2+}$ mobilization assays with an $EC_{50}$ value of $<1$ μM, the *S*-isomer had little activity below 100 μM. Only (R)-ZINC-3573 induced degranulation in MCs, which is significantly reduced in MRGPRX2-silenced MCs. Thus, ZINC-3573 is a highly selective MRGPRX2 agonist, which can be used as a tool to study its activation both in vitro and in vivo.

# 6. LIGAND-BINDING SITES FOR OPIOIDS AND ZINC-3573 ON MRGPRX2 AND SIGNALING

## 6.1 Structure-Based Docking Studies to Identify Opioid and ZINC-3573-Binding Sites on MRGPRX2

Purification of GPCRs has been extremely challenging, and since 1984 crystal structures of only four receptors have been solved (Cherezov et al., 2007; Chien et al., 2010; Deisenhofer, Epp, Miki, Huber, & Michel, 1984; Jaakola et al., 2008; Palczewski et al., 2000; Rasmussen et al., 2007; Warne et al.,

2008; Wu et al., 2010). The NIH's Protein Structure Initiative Program (PSI): Biology GPCR network has been based on the assumption that crystallization of a small number of GPCRs can be used to predict the structural properties on other receptors in the same family. Indeed, the recent crystallography data and comparison of sequence homology have provided prediction regarding the regions of GPCRs that are involved in ligand binding and G protein coupling. The 7TM bundles are connected by three EC loops and three IC loops. It now appears that for most GPCRs, EC, and parts of the transmembrane domains display the greatest structural diversity and are responsible for the binding of diverse ligands, but the IC domains are involved in signaling (Katritch, Cherezov, & Stevens, 2013).

Computer-based modeling and subsequent mutagenesis studies have recently been used to identify bindings sites for a number of GPCRs. Utilization of this approach led to the prediction that the positively charged opioids could bind to $Glu^{164}$ and $Asp^{184}$ on MRGPRX2 (Fig. 4B). The E164Q and D184N substitutions retain the steric properties of the WT residues but have no negative charges. Testing these mutants individually in the PRESTO-Tango system demonstrated loss of activity for opioids and ZINC-3573. These findings suggest that $Glu^{164}$ and $Asp^{184}$ are necessary for activation of MRGPRX2 by opioids and ZINC-3573. However, it is unknown whether HDPs, SP, and other ligands interact with the same site on MRGPRX2 for activation.

## 6.2 Activation of MRGPRX2 Leads to Rapid and Transient MC Degranulation

It has been noted previously that while IgE/FcεRI-mediated cutaneous inflammatory reaction is sustained, injection site reaction to pseudoallergic drugs or SP is transient (Bulfone-Paus, Nilsson, Draber, Blank, & Levi-Schaffer, 2017; Gaudenzio et al., 2016). A possible explanation for difference has recently been described. It was found that while FcεRI-mediated MC degranulation is slow and linked to a sustained $Ca^{2+}$ response, the MRGPRX2-mediated degranulation is very rapid, with a transient $Ca^{2+}$ response. This distinct pattern of $Ca^{2+}$ response evokes a different degranulation dynamics. Thus, FcεRI-mediated degranulation involves the secretion of large granules through limited number of openings presumably due to the fusion of granules. In contrast, MRGPRX2-mediated MC degranulation is characterized by a direct release of individual, spherical granules from all over the MC surface.

Investigation of the signaling events downstream of FcεRI and MRGPRX2 revealed interesting differences (Gaudenzio et al., 2016). Thus, while FcεRI-mediated MC activation results in the phosphorylation of Akt, PKC, and IKK-β, these responses are not activated following the exposure of MCs to MRGPRX2 ligands. Furthermore, in FcεRI-mediated MC activation IKK-β phosphorylation results in the formation of a complex containing the SNARE proteins synaptosomal-associated protein-23 (SNAP23) and syntaxin-4 (STX4). Interestingly, treatment of MCs with an IKK-β antagonist prevents this complex formation and switches the degranulation phenotype toward the rapid, simple degranulation observed in MRGPRX2-mediated MC degranulation. These findings indicate that FcεRI and MRGPRX2 activate different signaling pathways and utilize distinct transport systems for the delivery of granules to the plasma membrane.

## 6.3 MRGPRX2 Signaling for Degranulation via Gαq and Gαi Family of G Proteins

Until the discovery of MRGPRX2, basic secretagogues were thought to directly interact with pertussis toxin (PTx)-sensitive G proteins likely Gαi2 and Gαi3 to propagate signaling for degranulation in a receptor-independent manner (Ferry, Eichwald, Daeffler, & Landry, 2001; Mousli, Bronner, et al., 1990). However, the identification of MRGPRX2 as the receptor for basic secretagogues provides a new opportunity to define the signaling pathways involved in MC degranulation via this receptor. It was found that PTx inhibits MRGPRX2-mediated degranulation in human MCs endogenously expressing MRGPRX2 and transfected RBL-2H3 cells but has no effect on $Ca^{2+}$ mobilization. $La^{3+}$ and 2-aminoethoxydiphenyl borate (2-APB, a dual inhibitor of inositol 1,4,5-triphosphate receptor and transient receptor potential, TRP channels), which cause substantial inhibition of MRGPRX2-mediated $Ca^{2+}$ response, also block degranulation (Subramanian, Gupta, et al., 2011; Subramanian et al., 2013). These findings suggest that MRGPRX2 utilizes two signals for degranulation: one involving Gαq-mediated $Ca^{2+}$ mobilization and the other via a PTx-mediated signaling, which remains to be defined (Subramanian, Gupta, et al., 2011; Subramanian et al., 2013).

## 6.4 LL-37 Induces TRPV4 Expression in MCs via MRGPRX2 and Gαi

Transient receptor potential vanilloid 4 (TRPV4) ion channel is upregulated in rosacea and colocalizes with MCs in the skin (Sulk et al., 2012). In a mouse model of rosacea, injection of LL-37 leads to a dramatic increase in the expression of TRPV4, which colocalizes with MCs (Mascarenhas, Wang, Chang, & Di Nardo, 2017). Exposure of human MCs to LL-37 for 24h results increased expression of TRPV4, which is inhibited by PTx. Inhibition of TRPV4 by HC-067047 or its knockdown using CRISPR/Cas-9 results in substantial inhibition of LL-37-induced MC degranulation. Furthermore, LL-37-induced expression of TRPV4 is ablated in MRGPRX2 knockdown human MCs. These findings suggest that LL-37 activates human MCs via MRGPRX2 to induce the expression of TRPV4, which facilitates MC degranulation. TRPV4 antagonists have been shown to reduce pain-associated behaviors in mouse models and could be utilized to reduce cutaneous inflammation in patients with rosacea and other MRGPRX2-mediated chronic inflammatory diseases (Chen et al., 2017; Kanju et al., 2016; Mascarenhas et al., 2017).

Studies performed with mouse peritoneal MCs isolated from $TRPV4^{-}/^{-}$ mice demonstrated that this channel does not participate in MrgprB2-mediated $Ca^{2+}$ mobilization or degranulation (Solis-Lopez et al., 2017). This could reflect differences in the mechanisms via which murine MrgprB2 and human MRGPRX2 activate MCs. However, another possible explanation is that TRP channels provide physiologically relevant signal for rapid and transient MC degranulation, which could be important for host defense. However, under pathologic conditions, induction of TRPV4 channel via an MRGPRX2/Gαi signaling may allow for a greater influx of cations, which can prime the MC for continuous degranulation (Fig. 5).

## 7. REGULATION OF MRGPRX2 FUNCTION IN MCs

### 7.1 Role of Constitutive MRGPRX2 Signaling on Cell Proliferation

The first demonstration that MRGPRX2 could be involved cell proliferation was seen in transfected NIH-3T3 cells (Burstein et al., 2006). Cells expressing MRGPRX2 undergo proliferation in the absence of any ligand stimulation. This constitutive cell proliferation is partially blocked by PTx indicting the involvement of both Gαq and Gαi family of G proteins. At that time, MRGPRX2 was known to be expressed only in small diameter sensory neurons of DRG. It was proposed that MRGPRX2 is constitutively active in vivo and may be involved in continually monitoring changes in sensory inputs.

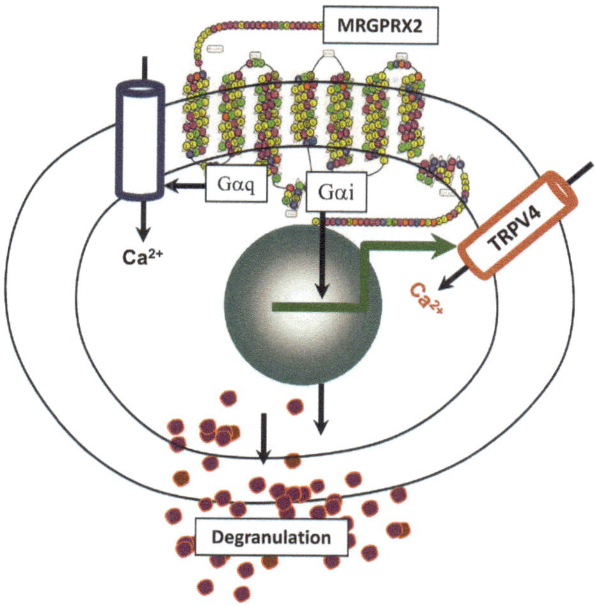


**Fig. 5** Roles of Gαq and Gαi on MRGPRX2-mediated $Ca^{2+}$ mobilization and degranulation. Activation of MRGPRX2 results in $Ca^{2+}$ mobilization via a pathway involving Gαq and possibly via a TRP channel. Chronic exposure of MCs with MRGPRX2 ligands leads to enhanced expression of TRPV4 gene via a Gαi-mediated signaling pathway to facilitate degranulation.

## 7.2 Retinoic Acid Inhibits Skin $MC_{TC}$ Proliferation but Has Not Effect on Their Survival

Although MCs are terminally differentiated cells, they can reenter cell cycle in the presence of stem cell factor. The phenomenon is responsible for the expansion of MCs derived from the human skin (Guhl et al., 2011). Under optimal condition, MCs require up to 2 weeks to complete one cycle of proliferation and during this period express cell cycle regulators E2F1 and proliferating cell nuclear antigen (PCNA) (Guhl et al., 2011; Guhl, Neou, Artuc, Zuberbier, & Babina, 2014; Motakis et al., 2014).

The skin is a rich source of vitamin A metabolite retinoic acid and its precursors such as retinol contribute to the maintenance of skin homeostasis (Mihaly, Gamlieli, Worm, & Ruhl, 2011). Furthermore, retinoids are used for the treatment of a variety of skin disorders including psoriasis, acne, and chronic hand eczema (Kumari, Timm, Kuhl, Heine, & Worm, 2016). It has recently been reported that retinoic acid inhibits stem cell factor-induced increase in MC number. The possible mechanisms for this could involve decreased proliferation or increased apoptosis. However, following retinoic acid treatment, there is no increase in annexin-positive apoptotic cells. Further evidence that retinoic acid does not induce cell death by triggering apoptosis was provided by showing that it slightly increased the anti-apoptotic MCl-1 but had no impact on the expression of Bcl-xl and the proapoptotic Bax (Babina et al., 2017). By contrast, the antiproliferative effect of retinoic acid is accompanied by diminished levels of E2F1 and PCNA expression; genes whose expression correlates with cell cycle progression of skin-derived MCs (Guhl et al., 2014, Motakis et al., 2014). Retinoic acid treatment does not influence granule contents and MC size. This suggests that the effect of retinoic acid on MC number does not reflect reduced viability of noncycling MCs but results from interference with stem cell factor-induced proliferation. This antiproliferative effect of retinoic acid does not appear to be restricted nontransformed skin-derived MCs as it promotes cell cycle arrest in a human MC line, HMC-1, and hematopoietic progenitor-derived MCs (Alexandrakis et al., 2003).

## 7.3 Retinoic Acid Causes a Decrease in MRGPRX2 and Chymase but Has No Effect on Tryptase Expression

In addition to its negative effect on skin MC proliferation, retinoic acid treatment (for 1 week) results in significant reduction in the expression and activity of chymase without affecting histamine or tryptase levels (Babina et al., 2017). Earlier studies showed that human skin exposed to retinoic acid treatment for up to 6 weeks results in increased $MC_T$ with no chance in the $MC_{TC}$ subset (Hjertson et al., 2003). It is therefore possible that appearance of $MC_T$ in retinoic acid-treated skin may result from downregulated chymase expression (Babina et al., 2017). Interestingly, in addition to chymase, retinoic acid causes a decreased expression of MRGPRX2, which results in reduced degranulation in response to compound 48/80 (Babina et al., 2017). The mechanism by which retinoic acid triggers an alteration in the skin MC signature away from $MC_{TC}$ to $MC_T$ is unknown, but patients receiving retinoic acid treatment for certain skin disorder may display reduced sensitivity to drugs that cause pseudoallergic reactions.

## 7.4 Receptor Phosphorylation and Desensitization

Following ligand stimulation, most GPCRs undergo rapid phosphorylation, which promotes β-arrestin-mediated desensitization and internalization (Black, Premont, & Daaka, 2016). Although human MCs express a variety of GPCRs, phosphorylation and desensitization of the anaphylatoxin C3a receptor (C3aR) have been studied in most detail. Following C3a stimulation, C3aR undergoes rapid phosphorylation, desensitization, and internalization (Guo, Subramanian, Gupta, & Ali, 2011; Gupta, Subramanian, Klos, & Ali, 2012; Vibhuti, Gupta, Subramanian, Guo, & Ali, 2011). Furthermore, silencing the expression of G protein-coupled receptor kinase 2 or 3 (GRK2 or GRK3) causes a more sustained $Ca^{2+}$ mobilization and enhanced degranulation (Guo et al., 2011). However, unlike C3aR, MRGPRX2 is resistant to LL-37-induced receptor phosphorylation, desensitization, and internalization (Subramanian, Gupta, et al., 2011). Also, silencing the expression of either GRK2 or GRK3 has no effect on LL-37-induced MC degranulation. Thus, MRGPRX2 appears to be unique among GPCRs expressed in MCs with regards to its resistance to desensitization. It is interesting to note that exposure of MCs to LL-37 results in upregulation of TRPV4 expression resulting in enhanced MRGPRX2-mediated degranulation (Chen et al., 2017; Mascarenhas et al., 2017). Thus, resistance of MRGPRX2 to undergo desensitization (Subramanian, Gupta, et al., 2011) combined with enhanced expression of $Ca^{2+}$ signaling via TRPV4 following chronic exposure to ligands (Mascarenhas et al., 2017) may explain why this receptor contributes to a growing list chronic inflammatory diseases.

## 8. CONCLUDING REMARKS

Although MRGPRX2 was first identified in neurons of the DRG, it has been the subject of intense recent interest because it is highly expressed in human $MC_{TC}$ (Fujisawa et al., 2014; Motakis et al., 2014). Furthermore, it is activated by a diverse group of ligands including HDPs, *S. aureus* δ-toxin, neuropeptides, opioids, and FDA-approved cationic drugs. In addition to the direct killing of microbes the ability of HDPs to activate MRGPRX2 may serve to orchestrate the development of MC-mediated host defense. Thus, harnessing this MC-activating feature of HDPs may provide novel approaches to develop antimicrobial therapy against multidrug-resistant bacteria (Hilchie et al., 2013).

Most studies evaluating clinical potentials of HDPs for antimicrobial activity involve their topical application at the site of infection (Hilchie et al., 2013). At these sites, it is desirable for a therapeutic agent to display antimicrobial activity and to harness MC's host defense properties. Human skin MCs express MRGPRX2 at high levels, and their activation leads to transient degranulation. Therefore, HDP-mediated activation of MCs via MRGPRX2 and the resulting transient inflammatory responses in the context of cutaneous microbial infection likely outweighs the risks of developing adverse reactions.

In addition to host defense MRGPRX2 likely participates in a number of chronic inflammatory diseases. An important feature of MRGPRX2 is that, compared to other GPCRs, it is resistant to desensitization and its prolonged activation leads to the induction of TRPV4 and enhanced degranulation. Thus, it is possible that while transient activation of MRGPRX2 contributes to host defense, its continued activation due to its resistance to desensitization and induction of TRPV4 may lead to chronic inflammatory diseases. Therefore, targeting MRGPRX2 signaling in MCs may provide novel approaches for the development therapy against MC-dependent chronic inflammatory diseases.

MrgprB2 has recently been identified as the mouse ortholog of human MRGPRX2 (McNeil et al., 2015). However, a number of ligands that activate MRGPRX2 have little to no activity against the mouse receptor, indicating that MrgprB2 mutant mice may not be suitable to screen drugs for use in humans (Lansu et al., 2017; McNeil et al., 2015). Recently, a number of humanized mouse models have been developed via the engraftment of human hematopoietic stem cells into NOD-*scid* IL-2R gamma$^{null}$ (NSG) immune-deficient mice with transgenic expression of human cytokines (Bryce et al., 2016; Burton et al., 2016; Pagovich et al., 2016). These mice develop human MCs in the skin, intestine, and lung and have been used to study human IgE/MC-mediated passive cutaneous anaphylaxis (Bryce et al., 2016) and peanut allergy (Burton et al., 2016; Pagovich et al., 2016). As shown in Fig. 6, when immune-deficient NSG mice with transgenic expression of human stem cell factor were engrafted with human hematopoietic stem cells (Burton et al., 2016), they develop human MCs in the lung and that these MCs express MRGPRX2. Thus, humanized mice may serve as a new model to study the role of MC/MRGPRX2-mediated chronic inflammatory diseases.

## ACKNOWLEDGMENTS

This work was supported by National Institutes of Health Grants AI-124182, AI-108585, and OD-017843. I thank Saptarshi Roy for review of the manuscript, Penn Stem cell and Xenograft Core facility for providing humanized mice, and Chizobam Idahosa for performing the experiment shown in Fig. 6.

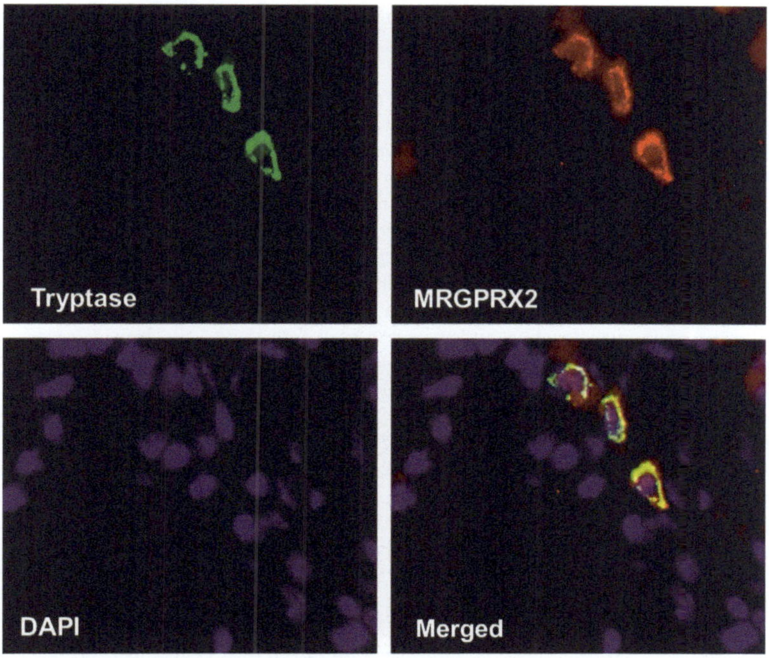


**Fig. 6** MC-specific expression of MRGPRX2 in the lung of humanized mice. NSG mice with transgenic expression of human stem cell factor were engrafted with human $CD34^+$ hematopoietic stem cells (Burton et al., 2016). Paraffin-embedded lung samples were stained with human antitryptase and anti-MRGPRX2 antibodies, and immunofluorescence studies were performed using a procedure similar to the one described recently (Okamura et al., 2017).

## REFERENCES


Aksoy, B., Altaykan-Hapa, A., Egemen, D., Karagoz, F., & Atakan, N. (2010) The impact of rosacea on quality of life: Effects of demographic and clinical characteristics and various treatment modalities. *The British Journal of Dermatology*, *163*(4), 719–725.

Alexandrakis, M. G., Kyriakou, D. S., Seretakis, D., Boucher, W., Letourneau, R., Kempuraj, D., et al. (2003). Inhibitory effect of retinoic acid on proliferation, maturation and tryptase level in human leukemic mast cells (Hmc-1). *International Journal of Immunopathology and Pharmacology*, *16*, 43–47.

Allen, J. A., & Roth, B. L. (2011). Strategies to discover unexpected targets for drugs active at G protein-coupled receptors. *Annual Review of Pharmacology and Toxicology*, *51*, 117–144.

Allen, J. A., Yost, J. M., Setola, V., Chen, X., Sassano, M. F., Chen, M., et al. (2011). Discovery of beta-arrestin-biased dopamine D2 ligands for probing signal transduction pathways essential for antipsychotic efficacy. *Proceedings of the National Academy of Sciences of the United States of America*, *108*(45), 18488–18493.

Azimi, E., & Lerner, E. A. (2017). Implications of MRGPRX2 in human and experimental cardiometabolic diseases. *Nature Reviews. Cardiology*, *14*(2), 124.

Azimi, E., Reddy, V. B., & Lerner, E. A. (2017). Brief communication: MRGPRX2, atopic dermatitis and red man syndrome. *Itch (Philadelphia, Pa)*, *2*(1), pii: e5.

Azimi, E., Reddy, V. B., Shade, K. C., Anthony, R. M., Talbot, S., Pereira, P. J., et al. (2016). Dual action of neurokinin-1 antagonists on Mas-related GPCRs. *JCI Insight*, *1*(16), e89362.

Babina, M., Artuc, M., Guhl, S., & Zuberbier, T. (2017). Retinoic acid negatively impacts proliferation and MCTC specific attributes of human skin derived mast cells, but reinforces allergic stimulability. *International Journal of Molecular Sciences*, *18*(3), pii: e5.

Baldo, B. A., & Pham, N. H. (2012). Histamine-releasing and allergenic properties of opioid analgesic drugs: Resolving the two. *Anaesthesia and Intensive Care*, *40*(2), 216–235.

Balzar, S., Chu, H. W., Strand, M., & Wenzel, S. (2005). Relationship of small airway chymase-positive mast cells and lung function in severe asthma. *American Journal of Respiratory and Critical Care Medicine*, *171*(5), 431–439.

Balzar, S., Fajt, M. L., Comhair, S. A., Erzurum, S. C., Bleecker, E., Busse, W. W., et al. (2011). Mast cell phenotype, location, and activation in severe asthma. Data from the severe asthma research program. *American Journal of Respiratory and Critical Care Medicine*, *183*(3), 299–309.

Barak, L. S., Ferguson, S. S., Zhang, J., & Caron, M. G. (1997). A beta-arrestin/green fluorescent protein biosensor for detecting G protein-coupled receptor activation. *The Journal of Biological Chemistry*, *272*(44), 27497–27500.

Barnea, G., Strapps, W., Herrada, G., Berman, Y., Ong, J., Kloss, B., et al. (2008). The genetic design of signaling cascades to record receptor activation. *Proceedings of the National Academy of Sciences of the United States of America*, *105*(1), 64–69.

Bedini, A. (2015). Bioluminescence resonance energy transfer (BRET) to detect the interactions between kappa opioid receptor and non visual arrestins. *Methods in Molecular Biology*, *1230*, 115–128.

Befus, A. D., Mowat, C., Gilchrist, M., Hu, J., Solomon, S., & Bateman, A. (1999). Neutrophil defensins induce histamine secretion from mast cells: Mechanisms of action. *Journal of Immunology*, *163*(2), 947–953.

Black, J. B., Premont, R. T., & Daaka, Y. (2016). Feedback regulation of G protein-coupled receptor signaling by GRKs and arrestins. *Seminars in Cell & Developmental Biology*, *50*, 95–104.

Blanca-Lopez, N., Ariza, A., Dona, I., Mayorga, C., Montanez, M. I., Garcia-Campos, J., et al. (2013). Hypersensitivity reactions to fluoroquinolones: Analysis of the factors involved. *Clinical and Experimental Allergy*, *43*(5), 560–567.

Boguniewicz, M. (2005). Atopic dermatitis: Beyond the itch that rashes. *Immunology and Allergy Clinics of North America*, *25*(2), 333–351. vii.

Boot, J. D., de Haas, S., Tarasevych, S., Roy, C., Wang, L., Amin, D., et al. (2007). Effect of an NK1/NK2 receptor antagonist on airway responses and inflammation to allergen in asthma. *American Journal of Respiratory and Critical Care Medicine*, *175*(5), 450–457.

Borici-Mazi, R., Kouridakis, S., & Kontou-Fili, K. (1999). Cutaneous responses to substance P and calcitonin gene-related peptide in chronic urticaria: The effect of cetirizine and dimethindene. *Allergy*, *54*(1), 46–56.

Boyce, J. A. (2006). Successful treatment of cold-induced urticaria/anaphylaxis with anti-IgE. *The Journal of Allergy and Clinical Immunology*, *117*(6), 1415–1418.

Broide, D. H. (2010). Allergic rhinitis: Pathophysiology. *Allergy and Asthma Proceedings*, *31*(5), 370–374.

Bromley, M., Fisher, W. D., & Woolley, D. E. (1984). Mast cells at sites of cartilage erosion in the rheumatoid joint. *Annals of the Rheumatic Diseases*, *43*(1), 76–79.

Bromley, M., & Woolley, D. E. (1984). Histopathology of the rheumatoid lesion. Identification of cell types at sites of cartilage erosion. *Arthritis and Rheumatism*, *27*(8), 857–863.

Bryce, P. J., Falahati, R., Kenney, L. L., Leung, J., Bebbington, C., Tomasevic, N., et al. (2016). Humanized mouse model of mast cell-mediated passive cutaneous anaphylaxis and passive systemic anaphylaxis. *The Journal of Allergy and Clinical Immunology*, *138*(3), 769–779.

Bulfone-Paus, S., Nilsson, G., Draber, P., Blank, U., & Levi-Schaffer, F. (2017). Positive and negative signals in mast cell activation. *Trends in Immunology* (in press).
Burstein, E. S., Ott, T. R., Feddock, M., Ma, J. N., Fuhs, S., Wong, S., et al. (2006). Characterization of the Mas-related gene family: Structural and functional conservation of human and rhesus MrgX receptors. *British Journal of Pharmacology, 147*(1), 73–82.
Burton, O. T., Stranks, A. J., Tamayo, J. M., Koleoglou, K. J., Schwartz, L. B., & Oettgen, H. C. (2016). A humanized mouse model of anaphylactic peanut allergy. *The Journal of Allergy and Clinical Immunology, 139*, 314–322.
Casale, T. B., Bowman, S., & Kaliner, M. (1984). Induction of human cutaneous mast cell degranulation by opiates and endogenous opioid peptides: Evidence for opiate and nonopiate receptor participation. *The Journal of Allergy and Clinical Immunology, 73*(6), 775–781.
Chen, Y., Moore, C. D., Zhang, J. Y., Hall, R. P., 3rd, MacLeod, A. S., & Liedtke, W. (2017). TRPV4 moves toward center-fold in rosacea pathogenesis. *The Journal of Investigative Dermatology, 137*(4), 801–804.
Chen, X., Niyonsaba, F., Ushio, H., Hara, M., Yokoi, H., Matsumoto, K., et al. (2007). Antimicrobial peptides human beta-defensin (hBD)-3 and hBD-4 activate mast cells and increase skin vascular permeability. *European Journal of Immunology, 37*(2), 434–444.
Cherezov, V., Rosenbaum, D. M., Hanson, M. A., Rasmussen, S. G., Thian, F. S., Kobilka, T. S., et al. (2007). High-resolution crystal structure of an engineered human beta2-adrenergic G protein-coupled receptor. *Science, 318*(5854), 1258–1265.
Chien, E. Y., Liu, W., Zhao, Q., Katritch, V., Han, G. W., Hanson, M. A., et al. (2010). Structure of the human dopamine D3 receptor in complex with a D2/D3 selective antagonist. *Science, 330*(6007), 1091–1095.
Cole, A. M., Hong, T., Boo, L. M., Nguyen, T., Zhao, C., Bristol, G. et al. (2002). Retrocyclin: A primate peptide that protects cells from infection by T- and M-tropic strains of HIV-1. *Proceedings of the National Academy of Sciences of the United States of America, 99*(4), 1813–1818.
Cowland, J. B., Johnsen, A. H., & Borregaard, N. (1995). hCAP-18, a cathelin/pro-bactenecin-like protein of human neutrophil specific granules. *FEBS Letters, 368*(1), 173–176.
Crisp, A. J., Chapman, C. M., Kirkham, S. E., Schiller, A. L., & Krane, S. M. (1984). Articular mastocytosis in rheumatoid arthritis. *Arthritis and Rheumatism, 27*(8), 845–851.
Dawson, R. M., McAllister, J., & Liu, C. Q. (2010). Characterisation and evaluation of synthetic antimicrobial peptides against Bacillus globigii, Bacillus anthracis and Burkholderia thailandensis. *International Journal of Antimicrobial Agents, 36*(4), 359–363.
Deisenhofer, J., Epp, O., Miki, K., Huber, R., & Michel, H. (1984). X-ray structure analysis of a membrane protein complex. Electron density map at 3 A resolution and a model of the chromophores of the photosynthetic reaction center from Rhodopseudomonas viridis. *Journal of Molecular Biology, 180*(2), 385–398.
Di Nardo, A., Vitiello, A., & Gallo, R. L. (2003). Cutting edge: Mast cell antimicrobial activity is mediated by expression of cathelicidin antimicrobial peptide. *Journal of Immunology, 170*(5), 2274–2278.
Di Nardo, A., Yamasaki, K., Dorschner, R. A., Lai, Y., & Gallo, R. L. (2008). Mast cell cathelicidin antimicrobial peptide prevents invasive group A Streptococcus infection of the skin. *Journal of Immunology, 180*(11), 7565–7573.
Dong, X., Han, S., Zylka, M. J., Simon, M. I., & Anderson, D. J. (2001). A diverse family of GPCRs expressed in specific subsets of nociceptive sensory neurons. *Cell, 106*(5), 619–632.
Echtenacher, B., Mannel, D. N., & Hultner, L. (1996). Critical protective role of mast cells in a model of acute septic peritonitis. *Nature, 381*(6577), 75–77.
Edston, E., & van Hage-Hamsten, M. (1997). Anaphylactoid shock—A common cause of death in heroin addicts? *Allergy, 52*(9), 950–954.

Erdei, A., Kerekes, K., & Pecht, I. (1997). Role of C3a and C5a in the activation of mast cells. *Experimental and Clinical Immunogenetics*, *14*(1), 16–18.

Fajt, M. L., Gelhaus, S. L., Freeman, B., Uvalle, C. E., Trudeau, J. B., Holguin, F., et al. (2013). Prostaglandin D(2) pathway upregulation: Relation to asthma severity, control, and TH2 inflammation. *The Journal of Allergy and Clinical Immunology*, *131*(6), 1504–1512.

Ferry, X., Eichwald, V., Daeffler, L., & Landry, Y. (2001). Activation of betagamma subunits of G(i2) and G(i3) proteins by basic secretagogues induces exocytosis through phospholipase Cbeta and arachidonate release through phospholipase Cgamma in mast cells. *Journal of Immunology*, *167*(9), 4805–4813.

Finkelman, F. D., Boyce, J. A., Vercelli, D., & Rothenberg, M. E. (2010). Key advances in mechanisms of asthma, allergy, and immunology in 2009. *The Journal of Allergy and Clinical Immunology*, *125*(2), 312–318.

Fujisawa, D., Kashiwakura, J., Kita, H., Kikukawa, Y., Fujitani, Y., Sasaki-Sakamoto, T., et al. (2014). Expression of Mas-related gene X2 on mast cells is upregulated in the skin of patients with severe chronic urticaria. *The Journal of Allergy and Clinical Immunology*, *134*(3), 622–633. e629.

Furst, D. E., & Emery, P. (2014). Rheumatoid arthritis pathophysiology: Update on emerging cytokine and cytokine-associated cell targets. *Rheumatology (Oxford)*, *53*(9), 1560–1569.

Galli, S. J. (2016). The mast cell-IgE paradox: From homeostasis to anaphylaxis. *The American Journal of Pathology*, *186*(2), 212–224.

Galli, S. J., Maurer, M., & Lantz, C. S. (1999). Mast cells as sentinels of innate immunity. *Current Opinion in Immunology*, *11*(1), 53–59.

Gaudenzio, N., Sibilano, R., Marichal, T., Starkl, P., Reber, L. L., Cenac, N., et al. (2016). Different activation signals induce distinct mast cell degranulation strategies. *The Journal of Clinical Investigation*, *126*(10), 3981–3998.

Golembiewski, J. A. (2002). Allergic reactions to drugs: Implications for perioperative care. *Journal of Perianesthesia Nursing*, *17*(6), 393–398.

Guhl, S., Artuc, M., Neou, A., Babina, M., & Zuberbier, T. (2011). Long-term cultured human skin mast cells are suitable for pharmacological studies of anti-allergic drugs due to high responsiveness to FcepsilonRI cross-linking. *Bioscience, Biotechnology, and Biochemistry*, *75*(2), 382–384.

Guhl, S., Neou, A., Artuc, M., Zuberbier, T., & Babina, M. (2014). Skin mast cells develop non-synchronized changes in typical lineage characteristics upon culture. *Experimental Dermatology*, *23*, 933–935.

Guo, Q., Subramanian, H., Gupta, K., & Ali, H. (2011). Regulation of C3a receptor signaling in human mast cells by G protein coupled receptor kinases. *PLoS One*, *6*(7), e22559.

Gupta, M. A., Gupta, A. K., Chen, S. J., & Johnson, A. M. (2005). Comorbidity of rosacea and depression: An analysis of the National Ambulatory Medical Care Survey and National Hospital Ambulatory Care Survey—Outpatient Department data collected by the U.S. National Center for Health Statistics from 1995 to 2002. *The British Journal of Dermatology*, *153*(6), 1176–1181.

Gupta, K., Kotian, A., Subramanian, H., Daniell, H., & Ali, H. (2015). Activation of human mast cells by retrocyclin and protegrin highlight their immunomodulatory and antimicrobial properties. *Oncotarget*, *6*(30), 28573–28587.

Gupta, K., Subramanian, H., & Ali, H. (2016). Modulation of host defense peptide-mediated human mast cell activation by LPS. *Innate Immunity*, *22*(1), 21–30.

Gupta, K., Subramanian, H., Klos, A., & Ali, H. (2012). Phosphorylation of C3a receptor at multiple sites mediates desensitization, beta-arrestin-2 recruitment and inhibition of NF-kappaB activity in mast cells. *PLoS One*, 7(10), e46369.

Gurish, M. F., & Austen, K. F. (2012). Developmental origin and functional specialization of mast cell subsets. *Immunity*, *37*(1), 25–33.
Han, S. K., Dong, X., Hwang, J. I., Zylka, M. J., Anderson, D. J., & Simon, M. I. (2002). Orphan G protein-coupled receptors MrgA1 and MrgC11 are distinctively activated by RF-amide-related peptides through the Galpha q/11 pathway. *Proceedings of the National Academy of Sciences of the United States of America*, *99*(23), 14740–14745.
Harder, J., Bartels, J., Christophers, E., & Schroder, J. M. (2001). Isolation and characterization of human beta -defensin-3, a novel human inducible peptide antibiotic. *The Journal of Biological Chemistry*, *276*(8), 5707–5713.
Harder, J., & Schroder, J. M. (2005). Antimicrobial peptides in human skin. *Chemical Immunology and Allergy*, *86*, 22–41.
Harle, D. G., Baldo, B. A., Coroneos, N. J., & Fisher, M. M. (1989). Anaphylaxis following administration of papaveretum. Case report: Implication of IgE antibodies that react with morphine and codeine, and identification of an allergenic determinant. *Anesthesiology*, *71*(4), 489–494.
Hens, G., Raap, U., Vanoirbeek, J., Meyts, I., Callebaut, I., Verbinnen, B., et al. (2011). Selective nasal allergen provocation induces substance P-mediated bronchial hyperresponsiveness. *American Journal of Respiratory Cell and Molecular Biology*, *44*(4), 517–523.
Hilchie, A. L., Wuerth, K., & Hancock, R. E. (2013). Immune modulation by multifaceted cationic host defense (antimicrobial) peptides. *Nature Chemical Biology*, *9*(12), 761–768.
Hjertson, M., Kivinen, P. K., Dimberg, L., Nilsson, K., Harvima, I. T., & Nilsson, G. (2003). Retinoic acid inhibits in vitro development of mast cells but has no marked effect on mature human skin tryptase- and chymase-positive mast cells. *Journal of Investigative Dermatology*, *120*, 239–245.
Homey, B., Steinhoff, M., Ruzicka, T., & Leung, D. Y. (2006). Cytokines and chemokines orchestrate atopic skin inflammation. *The Journal of Allergy and Clinical Immunology*, *118*(1), 178–189.
Hutchinson, M. R., Shavit, Y., Grace, P. M., Rice, K. C., Maier, S. F., & Watkins, L. R. (2011). Exploring the neuroimmunopharmacology of opioids: An integrative review of mechanisms of central immune signaling and their implications for opioid analgesia. *Pharmacological Reviews*, *63*(3), 772–810.
Irani, A. M., Bradford, T. R., Kepley, C. L., Schechter, N. M., & Schwartz, L. B. (1989). Detection of MCT and MCTC types of human mast cells by immunohistochemistry using new monoclonal anti-tryptase and anti-chymase antibodies. *The Journal of Histochemistry and Cytochemistry*, *37*(10), 1509–1515.
Irani, A. A., Schechter, N. M., Craig, S. S., DeBlois, G., & Schwartz, L. B. (1986). Two types of human mast cells that have distinct neutral protease compositions. *Proceedings of the National Academy of Sciences of the United States of America*, *83*(12), 4464–4468.
Irwin, J. J., & Shoichet, B. K. (2005). ZINC—A free database of commercially available compounds for virtual screening. *Journal of Chemical Information and Modeling*, *45*(1), 177–182.
Irwin, J. J., Sterling, T., Mysinger, M. M., Bolstad, E. S., & Coleman, R. G. (2012). ZINC: A free tool to discover chemistry for biology. *Journal of Chemical Information and Modeling*, *52*(7), 1757–1768.
Jaakola, V. P., Griffith, M. T., Hanson, M. A., Cherezov, V., Chien, E. Y., Lane, J. R., et al. (2008). The 2.6 angstrom crystal structure of a human A2A adenosine receptor bound to an antagonist. *Science*, *322*(5905), 1211–1217.
Kaisho, Y., Watanabe, T., Nakata, M., Yano, T., Yasuhara, Y., Shimakawa, K., et al. (2005). Transgenic rats overexpressing the human MrgX3 gene show cataracts and an abnormal skin phenotype. *Biochemical and Biophysical Research Communications*, *330*(3), 653–657.

Kajiwara, N., Sasaki, T., Bradding, P., Cruse, G., Sagara, H., Ohmori, K., et al. (2010). Activation of human mast cells through the platelet-activating factor receptor. *The Journal of Allergy and Clinical Immunology*, *125*(5), 1137–1145. e1136.

Kanada, K. N., Nakatsuji, T., & Gallo, R. L. (2012). Doxycycline indirectly inhibits proteolytic activation of tryptic kallikrein-related peptidases and activation of cathelicidin. *The Journal of Investigative Dermatology*, *132*(5), 1435–1442.

Kanda, N., Kamata, M., Tada, Y., Ishikawa, T., Sato, S., & Watanabe, S. (2011). Human beta-defensin-2 enhances IFN-gamma and IL-10 production and suppresses IL-17 production in T cells. *Journal of Leukocyte Biology*, *89*(6), 935–944.

Kanda, N., & Watanabe, S. (2012). Increased serum human beta-defensin-2 levels in atopic dermatitis: Relationship to IL-22 and oncostatin M. *Immunobiology*, *217*(4), 436–445.

Kanju, P., Chen, Y., Lee, W., Yeo, M., Lee, S. H., Romac, J., et al. (2016). Small molecule dual-inhibitors of TRPV4 and TRPA1 for attenuation of inflammation and pain. *Scientific Reports*, *6*, 26894.

Kaplan, A. P., & Greaves, M. (2009). Pathogenesis of chronic urticaria. *Clinical and Experimental Allergy*, *39*(6), 777–787.

Kashem, S. W., Subramanian, H., Collington, S. J., Magotti, P., Lambris, J. D., & Ali, H. (2011). G protein coupled receptor specificity for C3a and compound 48/80-induced degranulation in human mast cells: Roles of Mas-related genes MrgX1 and MrgX2. *European Journal of Pharmacology*, *668*(1–2), 299–304.

Katritch, V., Cherezov, V., & Stevens, R. C. (2013). Structure-function of the G protein-coupled receptor superfamily. *Annual Review of Pharmacology and Toxicology*, *53*, 531–556.

Kawakami, T., Ando, T., Kimura, M., Wilson, B. S., & Kawakami, Y. (2009). Mast cells in atopic dermatitis. *Current Opinion in Immunology*, *21*(6), 666–678.

Keating, G. M. (2016). Sugammadex: A review of neuromuscular blockade reversal. *Drugs*, *76*(10), 1041–1052.

Kelesidis, T., Fleisher, J., & Tsiodras, S. (2010). Anaphylactoid reaction considered ciprofloxacin related: A case report and literature review. *Clinical Therapeutics*, *32*(3), 515–526.

Kiatsurayanon, C., Niyonsaba, F., Chieosilapatham, P., Okumura, K., Ikeda, S., & Ogawa, H. (2016). Angiogenic peptide (AG)-30/5C activates human keratinocytes to produce cytokines/chemokines and to migrate and proliferate via MrgX receptors. *Journal of Dermatological Science*, *83*(3), 190–199.

Kokryakov, V. N., Harwig, S. S., Panyutich, E. A., Shevchenko, A. A., Aleshina, G. M., Shamova, O. V., et al. (1993). Protegrins: Leukocyte antimicrobial peptides that combine features of corticostatic defensins and tachyplesins. *FEBS Letters*, *327*(2), 231–236.

Koller, B., Muller-Wiefel, A. S., Rupec, R., Korting, H. C., & Ruzicka, T. (2011). Chitin modulates innate immune responses of keratinocytes. *PLoS One*, *6*(2), e16594.

Koppert, W., Blunk, J. A., Petersen, L. J., Skov, P., Rentsch, K., & Schmelz, M. (2001). Different patterns of mast cell activation by muscle relaxants in human skin. *Anesthesiology*, *95*(3), 659–667.

Kroeze, W. K., Sassano, M. F., Huang, X. P., Lansu, K., McCorvy, J. D., Giguere, P. M., et al. (2015). PRESTO-Tango as an open-source resource for interrogation of the druggable human GPCRome. *Nature Structural & Molecular Biology*, *22*(5), 362–369.

Kulka, M., Sheen, C. H., Tancowny, B. P., Grammer, L. C., & Schleimer, R. P. (2008). Neuropeptides activate human mast cell degranulation and chemokine production. *Immunology*, *123*(3), 398–410.

Kumar, K., & Singh, S. I. (2013). Neuraxial opioid-induced pruritus: An update. *Journal of Anaesthesiology Clinical Pharmacology*, *29*(3), 303–307.

Kumari, V., Timm, K., Kuhl, A. A., Heine, G., & Worm, M. (2016). Impact of systemic alitretinoin treatment on skin barrier gene and protein expression in patients with chronic hand eczema. *British Journal of Dermatology*, *175*, 1243–1250.

Kunapuli, P., Lee, S., Zheng, W., Alberts, M., Kornienko, O., Mull, R., et al. (2006). Identification of small molecule antagonists of the human mas-related gene-X1 receptor. *Analytical Biochemistry*, *351*(1), 50–61.

Lacey, N., Delaney, S., Kavanagh, K., & Powell, F. C. (2007). Mite-related bacterial antigens stimulate inflammatory cells in rosacea. *The British Journal of Dermatology*, *157*(3), 474–481.

Lansu, K., Karpiak, J., Liu, J., Huang, X. P., McCorvy, J. D., Kroeze, W. K., et al. (2017). In silico design of novel probes for the atypical opioid receptor MRGPRX2. *Nature Chemical Biology*, *13*(5), 529–536.

Laxminarayan, R., Duse, A., Wattal, C., Zaidi, A. K., Wertheim, H. F., Sumpradit, N., et al. (2013). Antibiotic resistance-the need for global solutions. *The Lancet Infectious Diseases*, *13*(12), 1057–1098.

Lee, H., Kashiwakura, J., Matsuda, A., Watanabe, Y., Sakamoto-Sasaki, T., Matsumoto, K., et al. (2013). Activation of human synovial mast cells from rheumatoid arthritis or osteoarthritis patients in response to aggregated IgG through Fcgamma receptor I and Fcgamma receptor II. *Arthritis and Rheumatism*, *65*(1), 109–119.

Lee, S. B., Li, B., Jin, S., & Daniell, H. (2011). Expression and characterization of antimicrobial peptides Retrocyclin-101 and Protegrin-1 in chloroplasts to control viral and bacterial infections. *Plant Biotechnology Journal*, *9*(1), 100–115.

Lehrer, R. I., Cole, A. M., & Selsted, M. E. (2012). theta-Defensins: Cyclic peptides with endless potential. *The Journal of Biological Chemistry*, *287*(32), 27014–27019.

Lembo, P. M., Grazzini, E., Groblewski, T., O'Donnell, D., Roy, M. O., Zhang, J., et al. (2002). Proenkephalin A gene products activate a new family of sensory neuron—Specific GPCRs. *Nature Neuroscience*, *5*(3), 201–209.

Li, Z., Tseng, P. Y., Tiwari, V., Xu, Q., He, S. Q., Wang, Y., et al. (2017). Targeting human Mas-related G protein-coupled receptor X1 to inhibit persistent pain. *Proceedings of the National Academy of Sciences of the United States of America*, *114*(10), E1996–E2005.

Liu, F. T., Goodarzi, H., & Chen, H. Y. (2011). IgE, mast cells, and eosinophils in atopic dermatitis. *Clinical Reviews in Allergy & Immunology*, *41*(3), 298–310.

Liu, Q., Tang, Z., Surdenikova, L., Kim, S., Patel, K. N., Kim, A., et al. (2009). Sensory neuron-specific GPCR Mrgprs are itch receptors mediating chloroquine-induced pruritus. *Cell*, *139*(7), 1353–1365.

Lumry, W. R., Li, H. H., Levy, R. J., Potter, P. C., Farkas, H., Moldovan, D., et al. (2011). Randomized placebo-controlled trial of the bradykinin B(2) receptor antagonist icatibant for the treatment of acute attacks of hereditary angioedema: The FAST-3 trial. *Annals of Allergy, Asthma & Immunology*, *107*(6), 529–537.

Malaviya, R., & Abraham, S. N. (2000). Role of mast cell leukotrienes in neutrophil recruitment and bacterial clearance in infectious peritonitis. *Journal of Leukocyte Biology*, *67*(6), 841–846.

Malaviya, R., Ikeda, T., Ross, E., & Abraham, S. N. (1996). Mast cell modulation of neutrophil influx and bacterial clearance at sites of infection through TNF-alpha. *Nature*, *381*(6577), 77–80.

Malone, D. G., Irani, A. M., Schwartz, L. B., Barrett, K. E., & Metcalfe, D. D. (1986). Mast cell numbers and histamine levels in synovial fluids from patients with diverse arthritides. *Arthritis and Rheumatism*, *29*(8), 956–963.

Marshall, J. S., & Jawdat, D. M. (2004). Mast cells in innate immunity. *The Journal of Allergy and Clinical Immunology*, *114*(1), 21–27.

Mascarenhas, N. L., Wang, Z., Chang, Y. L., & Di Nardo, A. (2017). TRPV4 mediates mast cell activation in cathelicidin-induced rosacea inflammation. *The Journal of Investigative Dermatology*, *137*(4), 972–975.

Matsuda, H., Kawakita, K., Kiso, Y., Nakano, T., & Kitamura, Y. (1989). Substance P induces granulocyte infiltration through degranulation of mast cells. *Journal of Immunology*, *142*(3), 927–931.

McNeil, B. D., Pundir, P., Meeker, S., Han, L., Undem, B. J., Kulka, M., et al. (2015). Identification of a mast-cell-specific receptor crucial for pseudo-allergic drug reactions. *Nature, 519*(7542), 237–241.

Mertes, P. M., Alla, F., Trechot, P., Auroy, Y., Jougla, E., & Groupe d'Etudes des Reactions Anaphylactoides Peranesthésiques (2011). Anaphylaxis during anesthesia in France: An 8-year national survey. *The Journal of Allergy and Clinical Immunology, 128*(2), 366–373.

Mihaly, J., Gamlieli, A., Worm, M., & Ruhl, R. (2011). Decreased retinoid concentration and retinoid signalling pathways in human atopic dermatitis. *Experimental Dermatology, 20*, 326–330.

Mori, K., Maru, C., & Takasuna, K. (2000). Characterization of histamine release induced by fluoroquinolone antibacterial agents in-vivo and in-vitro. *The Journal of Pharmacy and Pharmacology, 52*(5), 577–584.

Motakis, E., Guhl, S., Ishizu, Y., Itoh, M., Kawaji, H., de Hoon, M., et al. (2014). Redefinition of the human mast cell transcriptome by deep-CAGE sequencing. *Blood, 123*(17), e58–e67.

Mousli, M., Bronner, C., Bockaert, J., Rouot, B., & Landry, Y. (1990). Interaction of substance P, compound 48/80 and mastoparan with the alpha-subunit C-terminus of G protein. *Immunology Letters, 25*(4), 355–357.

Mousli, M., Bueb, J. L., Bronner, C., Rouot, B., & Landry, Y. (1990). G protein activation: A receptor-independent mode of action for cationic amphiphilic neuropeptides and venom peptides. *Trends in Pharmacological Sciences, 11*(9), 358–362.

Muto, Y., Wang, Z., Vanderberghe, M., Two, A., Gallo, R. L., & Di Nardo, A. (2014). Mast cells are key mediators of cathelicidin-initiated skin inflammation in rosacea. *The Journal of Investigative Dermatology, 134*(11), 2728–2736.

Nakamura, Y., Oscherwitz, J., Cease, K. B., Chan, S. M., Munoz-Planillo, R., Hasegawa, M., et al. (2013). Staphylococcus delta-toxin induces allergic skin disease by activating mast cells. *Nature, 503*(7476), 397–401.

Nieber, K., Baumgarten, C. R., Rathsack, R., Furkert, J., Oehme, P., & Kunkel, G. (1992). Substance P and beta-endorphin-like immunoreactivity in lavage fluids of subjects with and without allergic asthma. *The Journal of Allergy and Clinical Immunology, 90*(4 Pt. 1), 646–652.

Niyonsaba, F., Iwabuchi, K., Someya, A., Hirata, M., Matsuda, H., Ogawa, H., et al. (2002). A cathelicidin family of human antibacterial peptide LL-37 induces mast cell chemotaxis. *Immunology, 106*(1), 20–26.

Niyonsaba, F., Someya, A., Hirata, M., Ogawa, H., & Nagaoka, I. (2001). Evaluation of the effects of peptide antibiotics human beta-defensins-1/-2 and LL-37 on histamine release and prostaglandin D(2) production from mast cells. *European Journal of Immunology, 31*(4), 1066–1075.

Niyonsaba, F., Ushio, H., Hara, M., Yokoi, H., Tominaga, M., Takamori, K., et al. (2010). Antimicrobial peptides human beta-defensins and cathelicidin LL-37 induce the secretion of a pruritogenic cytokine IL-31 by human mast cells. *Journal of Immunology, 184*(7), 3526–3534.

Niyonsaba, F., Ushio, H., Nakano, N., Ng, W., Sayama, K., Hashimoto, K., et al. (2007). Antimicrobial peptides human beta-defensins stimulate epidermal keratinocyte migration, proliferation and production of proinflammatory cytokines and chemokines. *The Journal of Investigative Dermatology, 127*(3), 594–604.

Ogasawara, M., Cianciolo, G. J., Snyderman, R., Mitani, M., Good, R. A., & Day, N. K. (1988). Human IFN-gamma production is inhibited by a synthetic peptide homologous to retroviral envelope protein. *Journal of Immunology, 141*(2), 614–619.

Okamura, Y., Mishima, S., Kashiwakura, J. I., Sasaki-Sakamoto, T., Toyoshima, S., Kuroda, K., et al. (2017). The dual regulation of substance P-mediated inflammation via human synovial mast cells in rheumatoid arthritis. *Allergology International*, Mar 30. pii: S1323-8930(17)30022-9.

Oskeritzian, C. A., Zhao, W., Min, H. K., Xia, H. Z., Pozez, A., Kiev, J., et al. (2005). Surface CD88 functionally distinguishes the MCTC from the MCT type of human lung mast cell. *The Journal of Allergy and Clinical Immunology*, *115*(6), 1162–1168.

Pagovich, O. E., Wang, B., Chiuchiolo, M. J., Kaminsky, S. M., Sondhi, D., Jose, C. L., et al. (2016). Anti-hIgE gene therapy of peanut-induced anaphylaxis in a humanized murine model of peanut allergy. *The Journal of Allergy and Clinical Immunology*, *138*, 1652–1662.e7.

Palczewski, K., Kumasaka, T., Hori, T., Behnke, C. A., Motoshima, H., Fox, B. A., et al. (2000). Crystal structure of rhodopsin: A G protein-coupled receptor. *Science*, *289*(5480), 739–745.

Patou, J., Holtappels, G., Affleck, K., van Cauwenberge, P., & Bachert, C. (2011). Syk-kinase inhibition prevents mast cell activation in nasal polyps. *Rhinology*, *49*(1), 100–106.

Pennefather, J. N., Lecci, A., Candenas, M. L., Patak, E., Pinto, F. M., & Maggi, C. A. (2004). Tachykinins and tachykinin receptors: A growing family. *Life Sciences*, *74*(12), 1445–1463.

Quartara, L., Altamura, M., Evangelista, S., & Maggi, C. A. (2009). Tachykinin receptor antagonists in clinical trials. *Expert Opinion on Investigational Drugs*, *18*(12), 1843–1864.

Rasmussen, S. G., Choi, H. J., Rosenbaum, D. M., Kobilka, T. S., Thian, F. S., Edwards, P. C., et al. (2007). Crystal structure of the human beta2 adrenergic G-protein-coupled receptor. *Nature*, *450*(7168), 383–387.

Reinholz, M., Ruzicka, T., & Schauber, J. (2012). Cathelicidin LL-37: An antimicrobial peptide with a role in inflammatory skin disease. *Annals of Dermatology*, *24*(2), 126–135.

Rost, K., Fleischer, F., & Nieber, K. (2006). Neurokinin 1 receptor antagonists—Between hope and disappointment. *Medizinische Monatsschrift für Pharmazeuten*, *29*(6), 200–205.

Rubinstein, E., & Keynan, Y. (2014). Vancomycin revisited—60 Years later. *Frontiers in Public Health*, *2*, 217.

Rudikoff, D., & Lebwohl, M. (1998). Atopic dermatitis. *Lancet*, *351*(9117), 1715–1721.

Ruitenberg, E. J., & Elgersma, A. (1976). Absence of intestinal mast cell response in congenitally athymic mice during Trichinella spiralis infection. *Nature*, *264*(5583), 258–260.

Sarzi-Puttini, P., Salaffi, F., Di Franco, M., Bazzichi, L., Cassisi, G., Casale, R., et al. (2014). Pain in rheumatoid arthritis: A critical review. *Reumatismo*, *66*(1), 18–27.

Schafer, B., Piliponsky, A. M., Oka, T., Song, C. H., Gerard, N. P., Gerard, C., et al. (2013). Mast cell anaphylatoxin receptor expression can enhance IgE-dependent skin inflammation in mice. *The Journal of Allergy and Clinical Immunology*, *131*(2), 541–548. e541–549.

Schittek, B., Paulmann, M., Senyurek, I., & Steffen, H. (2008). The role of antimicrobial peptides in human skin and in skin infectious diseases. *Infectious Disorders Drug Targets*, *8*(3), 135–143.

Schmidt, R., Butterworth, J., O'Donnell, D., Santhakumar, V., & Tomaszewski, M. (2009). Cyclic dimers of C-terminal gamma2-MSH analogs as selective antagonists of the human sensory nerve-specific receptor (SNSR-4). *Advances in Experimental Medicine and Biology*, *611*, 111–112.

Schwab, V. D., Sulk, M., Seeliger, S., Nowak, P., Aubert, J., Mess, C., et al. (2011). Neurovascular and neuroimmune aspects in the pathophysiology of rosacea. *The Journal of Investigative Dermatology. Symposium Proceedings*, *15*(1), 53–62.

Sheen, C. H., Schleimer, R. P., & Kulka, M. (2007). Codeine induces human mast cell chemokine and cytokine production: Involvement of G-protein activation. *Allergy*, *62*(5), 532–538.

Shenoy, S. K., & Lefkowitz, R. J. (2005). Receptor regulation: Beta-arrestin moves up a notch. *Nature Cell Biology*, 7(12), 1159–1161.

Simons, F. E. (2009). Anaphylaxis: Recent advances in assessment and treatment. *The Journal of Allergy and Clinical Immunology*, *124*(4), 625–636. quiz 637-628.

Sivagnanam, S., & Deleu, D. (2003). Red man syndrome. *Critical Care*, 7(2), 119–120.

Smith, C. H., Atkinson, B., Morris, R. W., Hayes, N., Foreman, J. C., & Lee, T. H. (1992). Cutaneous responses to vasoactive intestinal polypeptide in chronic idiopathic urticaria. *Lancet, 339*(8785), 91–93.

Solinski, H. J., Gudermann, T., & Breit, A. (2014). Pharmacology and signaling of MAS-related G protein-coupled receptors. *Pharmacological Reviews, 66*(3), 570–597.

Solinski, H. J., Petermann, F., Rothe, K., Boekhoff, I., Gudermann, T., & Breit, A. (2013). Human Mas-related G protein-coupled receptors-X1 induce chemokine receptor 2 expression in rat dorsal root ganglia neurons and release of chemokine ligand 2 from the human LAD-2 mast cell line. *PLoS One, 8*(3), e58756.

Solis-Lopez, A., Kriebs, U., Marx, A., Mannebach, S., Liedtke, W. B., Caterina, M. J., et al. (2017). Analysis of TRPV channel activation by stimulation of FCepsilonRI and MRGPR receptors in mouse peritoneal mast cells. *PLoS One, 12*, e0171366.

Southern, C., Cook, J. M., Neetoo-Isseljee, Z., Taylor, D. L., Kettleborough, C. A., Merritt, A., et al. (2013). Screening beta-arrestin recruitment for the identification of natural ligands for orphan G-protein-coupled receptors. *Journal of Biomolecular Screening, 18*(5), 599–609.

Spoerl, D., D'Incau, S., Roux-Lombard, P., Harr, T., & Czarnetzki, C. (2016). Non-IgE-dependent hypersensitivity to rocuronium reversed by sugammadex: Report of three cases and hypothesis on the underlying mechanism. *International Archives of Allergy and Immunology, 169*(4), 256–262.

Stein, C. (2016). Opioid receptors. *Annual Review of Medicine, 67*, 433–451.

Steinhoff, M., Schauber, J., & Leyden, J. J. (2013). New insights into rosacea pathophysiology: A review of recent findings. *Journal of the American Academy of Dermatology, 69*(6 Suppl. 1), S15–S26.

St John, A. L., & Abraham, S. N. (2013). Innate immunity and its regulation by mast cells. *Journal of Immunology, 190*(9), 4458–4463.

Subramanian, H., Gupta, K., & Ali, H. (2016). Roles of Mas-related G protein-coupled receptor X2 on mast cell-mediated host defense, pseudoallergic drug reactions, and chronic inflammatory diseases. *The Journal of Allergy and Clinical Immunology, 138*(3), 700–710.

Subramanian, H., Gupta, K., Guo, Q., Price, R., & Ali, H. (2011). Mas-related gene X2 (MrgX2) is a novel G protein-coupled receptor for the antimicrobial peptide LL-37 in human mast cells: Resistance to receptor phosphorylation, desensitization, and internalization. *The Journal of Biological Chemistry, 286*(52), 44739–44749.

Subramanian, H., Gupta, K., Lee, D., Bayir, A. K., Ahn, H., & Ali, H. (2013). beta-Defensins activate human mast cells via Mas-related gene X2. *Journal of Immunology, 191*(1), 345–352.

Subramanian, H., Kashem, S. W., Collington, S. J., Qu, H., Lambris, J. D., & Ali, H. (2011). PMX-53 as a dual CD88 antagonist and an agonist for Mas-related gene 2 (MrgX2) in human mast cells. *Molecular Pharmacology, 79*(6), 1005–1013.

Sulk, M., Seeliger, S., Aubert, J., Schwab, V. D., Cevikbas, F., Rivier, M., et al. (2012). Distribution and expression of non-neuronal transient receptor potential (TRPV) ion channels in rosacea. *The Journal of Investigative Dermatology, 132*(4), 1253–1262.

Takahashi, T., & Gallo, R. L. (2017). The critical and multifunctional roles of antimicrobial peptides in dermatology. *Dermatologic Clinics, 35*(1), 39–50.

Tang, Y. Q., Yuan, J., Osapay, G., Osapay, K., Tran, D., Miller, C. J., et al. (1999). A cyclic antimicrobial peptide produced in primate leukocytes by the ligation of two truncated alpha-defensins. *Science, 286*(5439), 498–502.

Tatemoto, K., Nozaki, Y., Tsuda, R., Konno, S., Tomura, K., Furuno, M., et al. (2006). Immunoglobulin E-independent activation of mast cell is mediated by Mrg receptors. *Biochemical and Biophysical Research Communications, 349*(4), 1322–1328.

Tetlow, L. C., & Woolley, D. E. (1995). Distribution, activation and tryptase/chymase phenotype of mast cells in the rheumatoid lesion. *Annals of the Rheumatic Diseases, 54*(7), 549–555.

Thakurdas, S. M., Melicoff, E., Sansores-Garcia, L., Moreira, D. C., Petrova, Y., Stevens, R. L., et al. (2007). The mast cell-restricted tryptase mMCP-6 has a critical immunoprotective role in bacterial infections. *The Journal of Biological Chemistry, 282*(29), 20809–20815.

Tharp, M. D., Kagey-Sobotka, A., Fox, C. C., Marone, G., Lichtenstein, L. M., & Sullivan, T. J. (1987). Functional heterogeneity of human mast cells from different anatomic sites: In vitro responses to morphine sulfate. *The Journal of Allergy and Clinical Immunology, 79*(4), 646–653.

Tomaki, M., Ichinose, M., Miura, M., Hirayama, Y., Yamauchi, H., Nakajima, N., et al. (1995). Elevated substance P content in induced sputum from patients with asthma and patients with chronic bronchitis. *American Journal of Respiratory and Critical Care Medicine, 151*(3 Pt. 1), 613–617.

Tran, D., Tran, P., Roberts, K., Osapay, G., Schaal, J., Ouellette, A., et al. (2008). Microbicidal properties and cytocidal selectivity of rhesus macaque theta defensins. *Antimicrobial Agents and Chemotherapy, 52*(3), 944–953.

Two, A. M., Hata, T. R., Nakatsuji, T., Coda, A. B., Kotol, P. F., Wu, W. et al. (2014). Reduction in serine protease activity correlates with improved rosacea severity in a small, randomized pilot study of a topical serine protease inhibitor. *The Journal of Investigative Dermatology, 134*(4), 1143–1145.

Two, A. M., Wu, W., Gallo, R. L., & Hata, T. R. (2015a). Rosacea: Part I. Introduction, categorization, histology, pathogenesis, and risk factors. *Journal of the American Academy of Dermatology, 72*(5), 749–758.

Two, A. M., Wu, W., Gallo, R. L., & Hata, T. R. (2015b). Rosacea: Part II. Topical and systemic therapies in the treatment of rosacea. *Journal of the American Academy of Dermatology, 72*(5), 761–770.

van der Kleij, H. P., Ma, D., Redegeld, F. A., Kraneveld, A. D., Nijkamp, F. P., & Bienenstock, J. (2003). Functional expression of neurokinin 1 receptors on mast cells induced by IL-4 and stem cell factor. *Journal of Immunology, 171*(4), 2074–2079.

Veien, M., Szlam, F., Holden, J. T., Yamaguchi, K., Denson, D. D., & Levy, J. H. (2000). Mechanisms of nonimmunological histamine and tryptase release from human cutaneous mast cells. *Anesthesiology, 92*(4), 1074–1081.

Vibhuti, A., Gupta, K., Subramanian, H., Guo, Q., & Ali, H. (2011). Distinct and shared roles of beta-arrestin-1 and beta-arrestin-2 on the regulation of C3a receptor signaling in human mast cells. *PLoS One, 6*(5), e19585.

Warne, T., Serrano-Vega, M. J., Baker, J. G., Moukhametzianov, R., Edwards, P. C., Henderson, R., et al. (2008). Structure of a beta1-adrenergic G-protein-coupled receptor. *Nature, 454*(7203), 486–491.

Westman, J., Hansen, F. C., Olin, A. I., Morgelin, M., Schmidtchen, A., & Herwald, H. (2013). p33 (gC1q receptor) prevents cell damage by blocking the cytolytic activity of antimicrobial peptides. *Journal of Immunology, 191*(11), 5714–5721.

Wilkin, J., Dahl, M., Detmar, M., Drake, L., Feinstein, A., Odom, R., et al. (2002). Standard classification of rosacea: Report of the national rosacea society expert committee on the classification and staging of rosacea. *Journal of the American Academy of Dermatology, 46*(4), 584–587.

Williams, H., & Flohr, C. (2006). How epidemiology has challenged 3 prevailing concepts about atopic dermatitis. *The Journal of Allergy and Clinical Immunology, 118*(1), 209–213.

Wilson, A. P. (1998). Comparative safety of teicoplanin and vancomycin. *International Journal of Antimicrobial Agents, 10*(2), 143–152.

Wu, B., Chien, E. Y., Mol, C. D., Fenalti, G., Liu, W., Katritch, V., et al. (2010). Structures of the CXCR4 chemokine GPCR with small-molecule and cyclic peptide antagonists. *Science*, *330*(6007), 1066–1071.

Yamaguchi, T., Tashiro, K., Tanaka, S., Katayama, S., Ishida, W., Fukuda, K., et al. (2013). Two-step differentiation of mast cells from induced pluripotent stem cells. *Stem Cells and Development*, *22*(5), 726–734.

Yamasaki, K., Di Nardo, A., Bardan, A., Murakami, M., Ohtake, T., Coda, A., et al. (2007). Increased serine protease activity and cathelicidin promotes skin inflammation in rosacea. *Nature Medicine*, *13*(8), 975–980.

Yamasaki, K., & Gallo, R. L. (2009). The molecular pathology of rosacea. *Journal of Dermatological Science*, *55*(2), 77–81.

Yamasaki, K., & Gallo, R. L. (2011). Rosacea as a disease of cathelicidins and skin innate immunity. *The Journal of Investigative Dermatology. Symposium Proceedings*, *15*(1), 12–15.

Yamasaki, K., Kanada, K., Macleod, D. T., Borkowski, A. W., Morizane, S., Nakatsuji, T., et al. (2011). TLR2 expression is increased in rosacea and stimulates enhanced serine protease production by keratinocytes. *The Journal of Investigative Dermatology*, *131*(3), 688–697.

Yano, H., Wershil, B. K., Arizono, N., & Galli, S. J. (1989). Substance P-induced augmentation of cutaneous vascular permeability and granulocyte infiltration in mice is mast cell dependent. *The Journal of Clinical Investigation*, *84*(4), 1276–1286.

Yoshioka, M., Fukuishi, N., Kubo, Y., Yamanobe, H., Ohsaki, K., Kawasoe, Y., et al. (2008). Human cathelicidin CAP18/LL-37 changes mast cell function toward innate immunity. *Biological & Pharmaceutical Bulletin*, *31*(2), 212–216.

Zylka, M. J., Dong, X., Southwell, A. L., & Anderson, D. J. (2003). Atypical expansion in mice of the sensory neuron-specific Mrg G protein-coupled receptor family. *Proceedings of the National Academy of Sciences of the United States of America*, *100*(17), 10043–10048.

CHAPTER FIVE

# Adhesion GPCRs in Regulating Immune Responses and Inflammation

**Hsi-Hsien Lin*,†,1, Cheng-Chih Hsiao‡, Caroline Pabst§, Josée Hébert¶,||, Torsten Schöneberg#, Jörg Hamann‡,1**

*College of Medicine, Chang Gung University, Tao-Yuan, Taiwan
†Chang Gung Memorial Hospital-Linkou, Tao-Yuan, Taiwan
‡Academic Medical Center, University of Amsterdam, Amsterdam, The Netherlands
§Medical Faculty, Martin-Luther-University Halle-Wittenberg, Halle (Saale), Germany
¶Maisonneuve-Rosemont Hospital, Montreal, QC, Canada
||Medical Faculty, University of Montreal, Montreal, QC, Canada
#Medical Faculty, University of Leipzig, Leipzig, Germany
[1]Corresponding authors: e-mail address: hhlin@mail.cgu.edu.tw; j.hamann@amc.uva.nl

## Contents

1. Introduction 164
2. Origin and Evolution 165
3. Molecular Structure 168
4. Mechanisms of Activation and Signal Transduction 170
5. Cellular Distribution 174
6. Involvement in Immune Cell Development 176
   6.1 Hematopoietic Stem and Progenitor Cells 176
   6.2 Leukemia and Lymphoma 179
7. Involvement in Innate Immunity 180
   7.1 Myeloid Cells 180
   7.2 Innate Lymphoid Cells 186
8. Involvement in Adaptive Immunity 187
9. Outlook 189
Acknowledgments 190
References 191

## Abstract

The adhesion family comprises one of the five major clades of G protein-coupled receptors (GPCRs). Unlike conventional GPCRs, adhesion GPCRs (aGPCRs) have extended ectodomains with various protein folds that facilitate protein–protein interactions and, hence, putative cellular adhesive functions. Juxtaposed to the seven-pass transmembrane domain is a GPCR autoproteolysis-inducing domain that enables autoproteolytic cleavage of the receptor, resulting in a bipartite structure of many

*Advances in Immunology*, Volume 136
ISSN 0065-2776
http://dx.doi.org/10.1016/bs.ai.2017.05.005

aGPCRs. aGPCRs are widely distributed and play critical roles in many developmental processes; yet, the underlying mechanisms of activation and signal transduction have emerged only recently. About one-third of the 33 human aGPCRs are expressed in hematopoietic stem, progenitor, or mature cells, where they define distinct cellular populations. Recent studies have demonstrated roles of aGPCR in the control of innate effector functions and the susceptibility for and onset of (auto)inflammatory conditions. We here discuss the current knowledge about aGPCRs in the regulation of immune responses and inflammation.

## 1. INTRODUCTION

The family of adhesion G protein-coupled receptor (aGPCR) was coined and classified by Fredriksson et al. in 2003 after an extensive phylogenetic analysis of the human GPCR superfamily (Fredriksson, Lagerström, Lundin, & Schiöth, 2003). The aGPCR family comprises 33 members in humans and is thought to be the evolutionary prototype of another, secretin-like, GPCR subfamily (Nordström, Lagerström, Wallér, Fredriksson, & Schiöth, 2009). Among the more than 800 distinct GPCRs identified to date, aGPCRs are most structurally complex, comprising a chimeric composition of different functional protein domains, thought to integrate cellular adhesion and intracellular signaling (Lagerström & Schiöth, 2008; Fig. 1).

EMR1 and CD97, expressed on macrophages (Mφs) and T cells, respectively, were the founding members of the aGPCR family (Gordon, Hamann, Lin, & Stacey, 2011), and work on immune cells essentially has enforced the initial understanding of the molecular processing and ligand binding of aGPCRs (Hamann & Petrenko, 2016). In contrast, the functional understanding of aGPCRs in the immune system long lagged behind studies in developmental and neurobiology, and only recently, aGPCRs have been linked to immune responses and inflammation in humans. Notably, vibratory urticaria, a hereditary autoinflammatory condition, has been associated with a missense variant in EMR2, being more sensitive to mechanic activation and leading to an IgE-independent degranulation of mast cells (Boyden et al., 2016). Moreover, GPR56, an aGPCR expressed with great specificity by human cytotoxic lymphocytes, was found to control effector function displayed by natural killer (NK) cells (Chang et al., 2016). Finally, association of variants in aGPCR genes was found in quantitative trait locus analyses and genome-wide genotype–phenotype association studies for rheumatoid

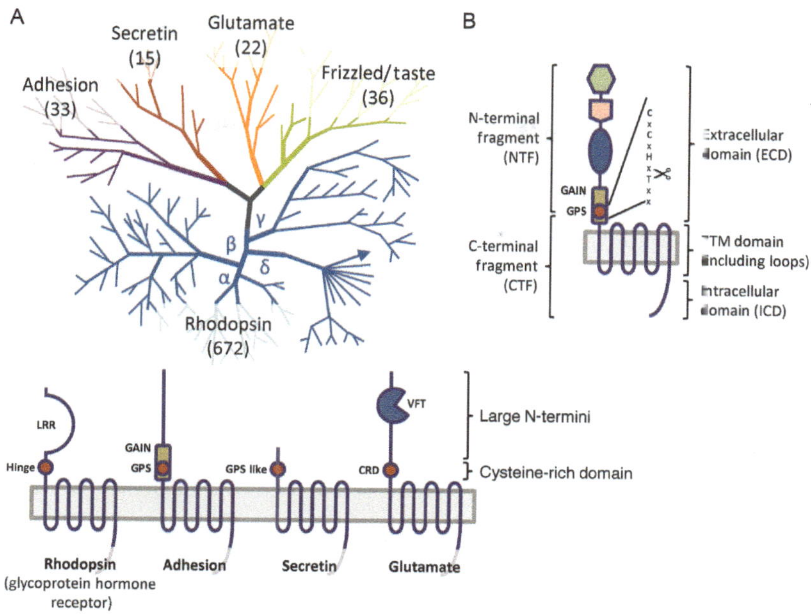


**Fig. 1** Evolution and structure features of aGPCRs. (A) The GRAFS nomenclature classifies GPCRs into five major families, with the number of genes in the human genome indicated between brackets (Lagerström & Schiöth, 2008). Adhesion family members share some structural features with other GPCR families. Cysteine-rich domains (CRDs) and large N-termini are found in rhodopsin (showed is a glycoprotein hormone receptor), secretin, and glutamate GPCRs. *VFT*, Venus flytrap domain. (B) Structural architecture of a prototypical aGPCR, consisting of an ECD, a GAIN domain, a 7TM domain, and an ICD. aGPCR is autocleaved at the GPS into an NTF and a CTF.

arthritis and joint/bone inflammation, multiple sclerosis, periodontitis, amyotrophic lateral sclerosis, response to pathogens, and serum IgE titer (Kovacs & Schöneberg, 2016).

We here review the role of aGPCRs in immune responses and inflammation by discussing their molecular structure, mechanisms of activation and signal transduction, cellular distribution, involvement in immune cell development, involvement in innate and adaptive immunity, and clinical perspectives.

## 2. ORIGIN AND EVOLUTION

Identifying the phylogenetic origin and the structural relation can shed light on the relevance and function of a given protein. In contrast to most other GPCR families, the assignment of aGPCRs to protein classes was difficult because of the very heterogeneous structure of their N-termini. Some of the members of the aGPCR family were even phylogenetically grouped into cadherins until recently (Hulpiau, Gul, & van Roy, 2013). The wealth of sequenced genomes and transcriptomes from numerous species allowed for a more comprehensive classification based on multiple structural features. However, most domains in the N-terminus are variable between aGPCR members, and even the otherwise conserved GPCR proteolytic site (GPS) is absent in some aGPCRs of sea anemones, amoebozoa, and fungi (Krishnan, Almén, Fredriksson, & Schiöth, 2012; Schiöth, Nordström, & Fredriksson, 2010). It turned out that the seven-span transmembrane (7TM) is the only constant domain suitable to track the evolution of aGPCRs (Nordström et al., 2009). Using the amino acid sequence of only the 7TM domain, aGPCRs were identified as receptor family clearly distinguishable from all other GPCR families (Fig. 1A). The branch point with the rhodopsin and glutamate GPCR families traces back over 750 million years ago to fungi and very basal unicellular organisms, such as *Dictyostelium discoideum* (Krishnan et al., 2012; Nordström, Sällman Almén, Edstam, Fredriksson, & Schiöth, 2011). Interestingly, there is strong evidence that the family of secretin-like GPCRs is an evolutionary offspring of aGPCRs (Nordström et al., 2009).

It is almost impossible to fully reconstruct the evolutionary history of aGPCRs from sequence information of recent genomes. However, the modular architecture of aGPCRs, with the highly variable N-terminus and a conserved 7TM domain, suggests that a prototypical 7TM sequence was manifold duplicated and differently equipped with N-terminal modules by chromosomal recombination and genomic exon shuffling. This led to a considerable variability in structure and number of the aGPCRs, set in the different species. The single-cell eukaryote *Capsaspora owczarzaki* contains at least seven aGPCR genes (Krishnan et al., 2012), the acorn worm, a hemichordate, has 18 aGPCR genes (Liebscher, Ackley, et al., 2014), but the worm *Caenorhabditis elegans* has only three aGPCRs (Loveless & Hardin, 2012; Prömel, Frickenhaus, et al., 2012). Mammalian genomes contain about 30 different aGPCRs, and comprehensive phylogenetic analyses on the 7TM of the human receptors classified the aGPCR family into nine subfamilies (I–IX) (Hamann et al., 2015; Krishnan, Nijmeijer, de Graaf, & Schiöth, 2016; Table 1).

The shaping of the aGPCR repertoire in species by gain and loss of genes is still ongoing. One example is the recent history of EMR4. After expansion

**Table 1** Human aGPCR Family Members

| Subfamily (Protein Name) | | Gene Symbol (Protein Names) |
|---|---|---|
| I | ADGRL (latrophilins) | ADGRL1 (LPHN1, CIRL-1, CL1), ADGRL2 (LPHN2, CIRL-2, CL2), ADGRL3 (LPHN3, CIRL-3 CL3), ADGRL4 (ELTD1, ETL) |
| II | ADGRE (EGF-TM7) | ADGRE1 (EMR1, F4/80), ADGRE2 (EMR2), ADGRE3 (EMR3), ADGRE4P (EMR4), ADGRE5 (CD97) |
| III | ADGRA | ADGRA1 (GPR123), ADGRA2 (GPR124), ADGRA3 (GPR125) |
| IV | ADGRC (CELSR) | ADGRC1 (CELSR1), ADGRC2 (CELSR2), ADGRC3 (CELSR3) |
| V | ADGRD | ADGRD1 (GPR133), ADGRD2 (GPR144) |
| VI | ADGRF | ADGRF1 (GPR110), ADGRF2 (GPR111), ADGRF3 (GPR113), ADGRF4 (GPR115), ADGRF5 (GPR116, Ig-Hepta) |
| VII | ADGRB (BAI) | ADGRB1 (BAI1), ADGRB2 (BAI2), ADGRB3 (BAI3) |
| VIII | ADGRG | ADGRG1 (GPR56), ADGRG2 (GPR64, HE6), ADGRG3 (GPR97), ADGRG4 (GPR112), ADGRG5 (GPR114), ADGRG6 (GPR126), ADGRG7 (GPR128) |
| IX | ADGRV | ADGRV1 (VLGR1, GPR98) |

*BAI*, brain-specific angiogenesis inhibitor; *CD*, cluster of differentiation; *CELSR*, cadherin EGF LAG seven-pass G-type receptor; *CIRL*, calcium-independent α-latrotoxin receptor; *CL*, CIRL/latrophilin; *EGF-TM7*, epidermal growth factor–seven-span transmembrane; *ELTD*, EGF, latrophilin, and 7TM domain-containing protein; *EMR*, EGF module-containing, mucin-like hormone receptor; *ETL*, EGF-TM7–latrophilin-related protein; *Ig*, immunoglobulin; *GPR*, GPCR; *HE*, human epididymal gene product; *LPHN*, latrophilin; *P*, pseudogene; *VLGR*, very large GPCR.

of the EMR group in primates up to four members (EMR1–4), e g., in great apes, EMR4 acquired a frameshift mutation in the human linage being a pseudogene in recent (*Homo sapiens*) (Hamann et al., 2003) and extinct (*Homo neanderthalensis, Denisova hominins*) *Hominoidea* (Meyer et al., 2012; Prüfer et al., 2014) and EMR4 is not the only aGPCR showing an adaptive turnover in mammalian genomes. GPR144 is present in human and primate genomes, but murine genomes lack this aGPCR. Similarly, the chromosomally clustered group of the aGPCRs GPR110–116 is highly variable in number among vertebrates (Prömel, Waller-Evans, et al., 2012). These evolutionary dynamics of structure and number are indicative for an involvement of aGPCRs in adaptive processes. Indeed, a recent meta-analysis demonstrated that aGPCRs are significantly involved in adaptation, phenotype variations, and diseases in humans; among them, some aGPCR variants associated with function and phenotypes in the immune system (Kovacs & Schöneberg, 2016).

## 3. MOLECULAR STRUCTURE

The defining feature of aGPCRs is the presence of an unusually large extracellular domain (ECD) of approximately 300–6000 residues, in addition to the signature 7TM moiety, shared by all GPCRs (Hamann et al., 2015; Yona, Lin, Siu, Gordon, & Stacey, 2008). A variety of characteristic protein motifs, arranged in tandem, are normally found at the N-terminal half of ECD. These protein motifs, such as the epidermal growth factor (EGF)-, immunoglobulin (Ig)-, pentraxin (PTX)-, thrombospondin (TSP)-, leucine-rich repeat (LRR)-, and lectin-like domains, are known to facilitate protein–protein interaction and hence a putative cellular adhesion function (Hamann et al., 2015; Yona, Lin, Siu, et al., 2008). Immediately following the adhesive protein motifs is a novel GPCR autoproteolysis-inducing (GAIN) domain that initiates a unique posttranslational proteolytic modification in most aGPCRs (Araç et al., 2012; Nieberler, Kittel, Petrenko, Lin, & Langenhan, 2016; Prömel, Langenhan, & Araç, 2013). The GAIN domain-mediated proteolysis is thought to occur in the lumen of the endoplasmic reticulum during aGPCR biosynthesis and is achieved autocatalytically through a series of nucleophilic reactions at a conserved HL/T(S) sequence of the GPS (Krasnoperov et al., 2002; Lin et al., 2004; Lin, Stacey, Yona, & Chang, 2010). Site-specific N-glycosylation at the GAIN–GPS region controls GPS autoproteolysis (Hsiao et al., 2009).

This novel autocatalytic cleavage was recently depicted by structural analysis of the LPHN1 and BAI3 GAIN domains (Araç et al., 2012). X-ray crystallography revealed a conserved GAIN domain structure comprising a subdomain A of six α-helices, followed by a subdomain B with two small α-helices and 13 β-strands. The GPS cleavage site is located in a unusual kinked turn between the last two β-strands (Araç et al., 2012). It seems this particular domain structure provides a unique configuration, which facilitates the proper arrangement of the scissile bond for the GPS autoproteolytic reaction. Interestingly, a more recent structural characterization of the entire ECD of GPR56 showed a GAIN domain containing a much smaller subdomain A with only three α-helices, but a highly conserved subdomain B (Salzman et al., 2016). These results suggest a more flexible GAIN domain structure than originally thought. Regardless, it is believed that a complete GAIN domain is minimally required and self-sufficient for initiating the autoproteolytic reaction, while the GPS motif, which is located within the last five β-strands, is structurally unsuitable (Araç et al., 2012; Prömel et al., 2013). Following the cleavage, the newly freed last β-strand, which is immediately linked to the first TM region, remains tightly enclosed within the GPS motif. Therefore, a mature aGPCR usually exists on the cell surface as a noncovalent dual subunit complex, composed of an extracellular N-terminal fragment (NTF) and a 7TM C-terminal fragment (CTF) with the putative cellular adhesion and signaling function, respectively (Hamann et al., 2015; Yona, Lin, Siu, et al., 2008; Fig. 1B).

In recent years, deorphanization of aGPCRs has identified multiple endogenous and exogenous binding partners/ligands for many individual aGPCRs (Hamann et al., 2015; Langenhan, Aust, & Hamann, 2013). Most of these ligands are cell membrane and/or extracellular matrix proteins, although polysaccharide, fatty acid, and phospholipid/glycolipid ligands, such as glycosaminoglycans, phosphatidylserine (PtdSer), docosahexaenoic acid, and lipopolysaccharide (LPS), are also able to bind at certain aGPCRs (Chiang et al., 2016; Hamann et al., 2015; Lee, Huang, et al., 2016). The orientation of the ligand–aGPCR interaction could be *trans* (on opposite cells) or *cis* (on the same cell). Interestingly, while a limited number of ligands were capable of inducing aGPCR signaling, the majority of ligands identified to date did not seem able to activate their cognate aGPCRs (Langenhan et al., 2013). This result may suggest that many of the ligands serve merely as the binding partners of aGPCRs without any functional input or simply reflect that proper assay conditions for the analysis of aGPCR activation/signaling were not implemented. Emerging evidence has made the latter explanation more inviting due to the recent revelation of the so-called tethered agonism mechanism of aGPCR activation (Liebscher, Schön, et al., 2014; Schöneberg, Liebscher, Luo, Monk, & Piao, 2015; Stoveken, Hajduczok, Xu, & Tall, 2015). Hence, it is possible that diverse signaling events might be induced by the multiple aGPCR ligands, especially when analyzed in a more in vivo-like condition. We summarize in Table 2 the molecular and functional characteristics of aGPCRs involved in the immune system.

**Table 2** Characteristics of the aGPCRs in the Immune System

| Receptor (Gene Symbol) | Human/ Mouse | Protein Motifs in NTF | Ligands/Interacting Proteins | GPS Proteolysis |
|---|---|---|---|---|
| BAI1 (ADGRB1) | +/+ | RGD, TSR, HRM | αvβ5 Integrin, CD36, PtdSer, LPS | + |
| EMR1 (ADGRE1) | +/+ (F4/80) | EGF (human 6/ mouse 7) | Unknown | − |
| EMR2 (ADGRE2) | +/− | EGF (5) | Dermatan sulfate | + |
| EMR3 (ADGRE3) | +/− | EGF (2) | A putative ligand on Mφ and neutrophils | + |
| EMR4 (ADGRE4) | −/+ | EGF (2) | A putative ligand on A20 cell line | + |
| CD97 (ADGRE5) | +/+ | EGF (human 5/ mouse 4), RGD | CD55, dermatan sulfate, α5β1 integrin, CD90 | + |
| GPR116 (ADGRF5) | +/+ | SEA, Ig | Surfactant protein D | + |
| GPR56 (ADGRG1) | +/+ | PLL | Col III, TG2, heparin, CD9/CD81 | + |
| GPR97 (ADGRG3) | +/+ | Unknown | Unknown | +[a] |

[a]Hsiao et al. (unpublished observation).
*CD*, cluster of differentiation; *Col*, collagen; *EGF*, epidermal growth factor; *GPS*, GPCR proteolysis site; *HRM*, hormone receptor motif; *Ig*, immunoglobulin; *LPS*, lipopolysaccharide; *Mφs*, macrophages; *NTF*, N-terminal fragment; *PLL*, pentraxin/laminin/neurexin/sex hormone-binding globulin like; *PtdSer*, phosphatidylserine; *PTX*, pentraxin; *RGD*, Arg-Gly-Asp; *SEA*, sperm protein, enterokinase, and agrin; *TG*, tissue transglutaminase; *TSR*, thrombospondin type 1 repeat.

# 4. MECHANISMS OF ACTIVATION AND SIGNAL TRANSDUCTION

As shown by phylogenetic studies, the 7TM domain is the most constant structural feature of aGPCRs. It was therefore reasonable to assume that the 7TM domain is, at least in parts, involved in the signal transduction of this receptor family. Although postulated to signal via G proteins and therefore grouped into the superfamily of GPCRs, a clear proof of G-protein coupling was lacking for more than a decade after the discovery of aGPCRs (Hamann et al., 2015).

Today, we know that aGPCRs are able to activate both, G protein-dependent and -independent signaling pathways. The G protein-independent signaling was primarily demonstrated for the aGPCR BAI1. For engulfment of apoptotic cells and Gram-negative bacteria, BAI1 recruits a trimeric complex with ELMO, DOCK1, and Rac via the cytoplasmatic tail of the 7TM domain (Das et al., 2011, 2014; Lee, Penberthy, et al., 2016; Park et al., 2007). However, there are BAI1 functions, such as involvement in cell polarity, which are independent of ELMO/DOCK1/Rac recruitment. Thus, BAI1 association with PAR3/TIAM1 plays a role in synaptogenesis (Duman et al., 2013). G protein-independent signaling has been demonstrated also for other aGPCR family members. Direct cross talk of the aGPCR CELSR1 to Frizzled/Disheveled, DAAM1, and PDZ-RhoGEF leads to actomyosin-dependent contraction in a planar-polarized manner (Nishimura, Honda, & Takeichi, 2012). Involvement of the Wnt/β-catenin pathway and the Wnt/planar cell polarity pathway in aGPCR signaling has been shown for GPR124 (Posokhova et al., 2015; Vanhollebeke et al., 2015; Zhou & Nathans, 2014) and GPR125 (Li et al., 2013). Activation of the Wnt pathway by "classical" GPCRs is not new (de Lau et al., 2011) and seems to be common, e.g., for several members of the LRR-containing GPCR group (LGR group within the rhodopsin-like GPCR family). LGR4–8 have strong homology to the glycoprotein hormone receptors (LGR1–3), and some couple also to the classical G protein-dependent pathways (Kumagai et al., 2002; Luo et al., 2009; Weng et al., 2008).

Evidence for G protein-dependent signaling was long under debate, but taking advantage of increased basal activity of some full lengths and N-terminally truncated receptors helped to solve this open point (Liebscher, Schöneberg, & Prömel, 2013). Most human aGPCRs couple to the $G_s$ protein/adenylyl cyclase pathway, but some to $G_q$, $G_i$, and $G_{12/13}$ proteins (Monk et al., 2015). The involvement of G proteins in aGPCR signaling was further supported by $G_\alpha$ subunit knockdown (Bohnekamp & Schöneberg, 2011; Liebscher, Schön, et al., 2014; Müller et al., 2015), GTPγS-binding studies (Gupte et al., 2012; Stoveken et al., 2015), and different in vivo experiments (Mogha et al., 2013; Müller et al., 2015). There is also evidence that β arrestin-2 binds to activated aGPCRs (Paavola, Stephenson, Ritter, Alter, & Hall, 2011). In sum, aGPCRs signal via well-known pathways of "classical" GPCR; however, the ligands and/or triggers activating the receptors were unknown until recently.

Classically, agonist binding is required to activate a metabotropic receptor. Numerous interaction partners have been identified for aGPCRs, including matrix proteins (e.g., collagens), membrane-bound proteins (e.g., teneurin-2), and small molecules (e.g., PtdSer) (Langenhan et al., 2013) (see Table 2). Most of these molecules did not activate the receptor they bind to. However, there are alternative mechanisms to activate GPCRs. Some GPCRs, such as rhodopsin (retinal), protease-activated receptors (Coughlin, 2001), and glycoprotein hormone receptors (Brüser et al., 2016), have their agonists already covalently bond, or the agonist is an integral part of the receptor protein (tethered agonist), respectively. Recent studies add aGPCRs to the group of GPCRs activated by integral agonists (Liebscher, Schön, et al., 2014). Peptides derived from a conserved sequence C-terminally of the GPS were able to activate the 7TM (Liebscher, Schön, et al., 2014; Stoveken et al., 2015). Based on these finding, it was postulated that, upon structural changes within the ECD, this intramolecular agonist is exposed to the 7TM domain or isomerizes within a 7TM-binding pocket, triggering G-protein activation (Liebscher & Schöneberg, 2016; Fig. 2). This appears to be a common mechanism in aGPCRs since tethered agonistic sequences, called *Stachel* sequences (German word for stinger), have been identified in a number of receptors lately (Demberg, Rothemund,

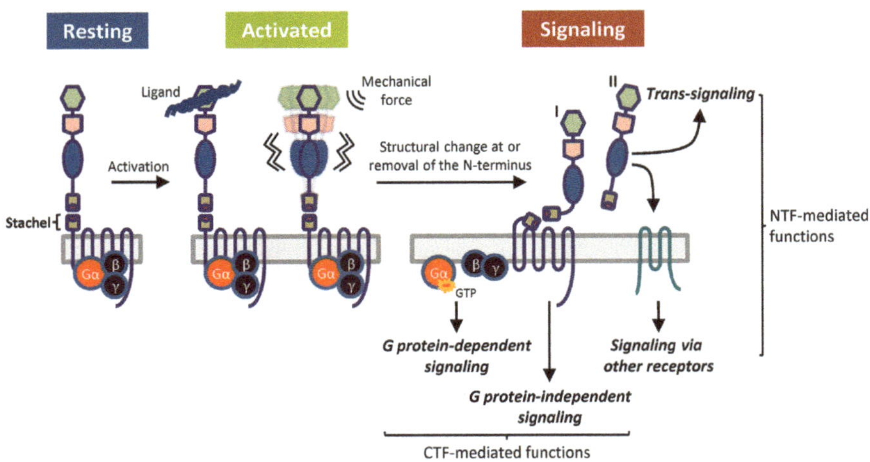


**Fig. 2** Signaling mechanisms of aGPCRs. In the resting state, aGPCRs provide little if any signaling. Ligand binding and/or mechanical forces that structurally change (I) or release (II) the NTF can activate the receptors, resulting in either G protein-dependent or -independent signaling. Binding of a juxtamembranous tethered agonistic sequence, called *Stachel* (German word for stinger), to the CTF has been shown to pose central in these processes. Alternatively, aGPCRs can mediate reverse signaling in *cis* by interacting with other receptors on the same cell, and the liberated NTF can mediate signaling in *trans* by interacting with molecules on opposing cells. See text for further details.

Schöneberg, & Liebscher, 2015; Müller et al., 2015; Wilde et al., 2016). In contrast to protease-activated receptors, cleavage at the GPS seems to be not required for the currently known in vitro and in vivo function of some aGPCRs (Bohnekamp & Schöneberg, 2011; Nieberler et al., 2016; Prömel, Frickenhaus, et al., 2012). The question remained what causes the structural changes of the ECD in order to expose or isomerize the internal agonist. It has been shown that, at least to some extent, binding of extracellular ligands can induce aGPCR-mediated intracellular signaling. For example, collagen IV and the cellular prion protein C can induce (cyclic adenosine monophosphate) cAMP formation upon binding to the N-terminus of GPR126 (Küffer et al., 2016; Paavola, Sidik, Zuchero, Eckart, & Talbot, 2014; Petersen et al., 2015).

A number of recent evidences suggest an additional activation mechanism driven by mechanic forces, implying that aGPCRs may act as metabotropic mechanosensors (Scholz, Monk, Kittel, & Langenhan, 2016). The aGPCR VLGR1 is expressed in inner ear hair cells, and loss of receptor function leads to abnormal stereociliary development and hearing loss, indicating a pivotal role in hearing transduction and/or hair cell development (McGee et al., 2006). Deletion of the *Drosophila* latrophilin dCirl resulted in an impairment of the perception of tactile, proprioceptive, and auditory stimuli through chordotonal neurons, the principal mechanosensors of *Drosophila* (Scholz et al., 2015). Mutation of EMR2 resulted in an increased susceptibility of vibration-mediated degranulation of mast cells in humans (Boyden et al., 2016). First evidence that aGPCRs directly sense mechanical forces came from experiments with GPR126-transfected cells, where shaking of the cells led to receptor-mediated cAMP formation (Petersen et al., 2015). Similar findings with GPR114 revealed that this mechanoactivation is dependent on the presence of the *Stachel* sequence (Wilde et al., 2016). One can speculate that removal of the autoproteolytically cleaved N-terminus and/or relative motion between the N-terminus bound to the extracellular matrix or adjacent cell and the aGPCR-expressing cell result in receptor activation.

Besides providing the surface for ligand binding and mechanotransduction in *cis*-signaling, the large N-terminus of aGPCRs also functions as a ligand for *trans*-signaling. For example, CELSR1 forms a homophilic *trans* interactions required for planar cell polarity (Nishimura et al., 2012). Further, the N-termini of *Drosophila* dCirl and of GPR126 can rescue subphenotypes in receptor-deficient animal models, suggesting 7TM-independent functions (Patra et al., 2013; Prömel, Frickenhaus, et al., 2012).

The existence of activating structures integrated in receptor molecules not only challenges our current pharmacological definition of an agonist but also gives raise to multiple signaling opportunities. The concept of an internal agonist is very straightforward to integrate different extracellular signals, such as ligand binding and mechanical forces, into a uniform intracellular signal (Schöneberg, Kleinau, & Brüser, 2016). Assuming that the *Stachel*-mediated activation is a general mechanism for aGPCRs, one must consider ligands binding to the ECD as allosteric modulators. Different binding partners may promote activation of different signaling pathways (biased signaling). It is also reasonable to assume that a code, meaning the combination of different ligands binding to the N-terminus, is required to fully activate the receptors. In sum, aGPCRs display many features of "classical" GPCRs including multiple G protein-coupling abilities, G protein-independent signaling, and tethered agonism but also novel aspects such as signal integration, mechanosensitivity, and *trans*-signaling via their large ECDs.

## 5. CELLULAR DISTRIBUTION

aGPCRs are found in most cell types and organ systems, including cells of the hematopoietic system (Fig. 3). Three gene clusters, belonging to the ADGRE and the ADGRG subfamilies, are markedly expressed in immune cells.

The five members of the ADGRE subfamily in human, also known as EGF-TM7 receptors (Kwakkenbos et al., 2004), form two gene clusters on chromosome 19p13.3 (EMR1 and EMR4) and chromosome 19p13.1 (EMR2, EMR3, and CD97). The N-terminal EGF-like domains of these receptors are encoded by single exons that, upon alternative splicing, can give rise to isofoms with different numbers of EGF domains in EMR1, EMR2, and CD97. While EMR1–4 are specifically expressed in myeloid cells, CD97 has a much wider distribution, also outside the immune system (Hamann et al., 2015). Their tightly restricted expression makes EGF-TM7/ADGRE receptors suitable markers for subsets of immune cells. For example, EMR1 marks tissue Mφs and eosinophils in the mouse and eosinophils in human (Austyn & Gordon, 1981; Gordon et al., 2011; Hamann et al., 2007), whereas EMR3 indicates maturation of polymorphonuclear granulocytes in human (Matmati et al., 2007).

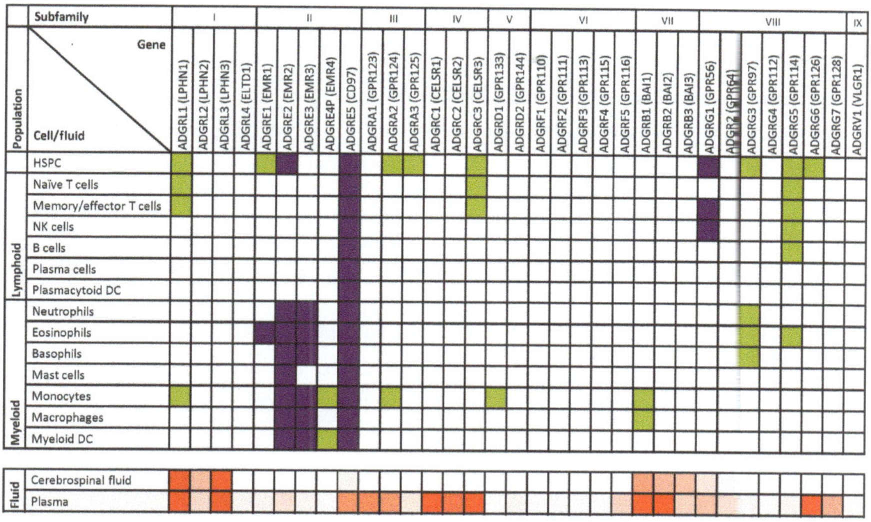


**Fig. 3** Expression of aGPCRs in immune cells. Summary of available published data on the expression of aGPCRs in human hematopoietic cells, obtained by transcriptional profiling (■) and protein analysis techniques (■), and on the presence of soluble aGPCR fragments in plasma (Guldbrandsen et al., 2014; Liu et al., 2007) and cerebrospinal fluid (Guldbrandsen et al., 2014; Liu et al., 2007), obtained by mass spectroscopy (■; *color* gradient indicates the amount of presence). For additional information and references, see the descriptions of individual aGPCRs at www.guidetopharmacology.org/.

Another gene clusters on chromosome 16q21 encode the human ADGRG subfamily members GPR56, GPR97, and GPR114 (Peng et al., 2011). GPR56 is a versatile marker for all human cytotoxic lymphocytes, including NK cells and $CD4^+$ and $CD8^+$ T cells (Peng et al., 2011). During life, the relative abundance of immune cells in whole-blood shifts, with naïve T cells decreasing and highly differentiated effector and memory T cells increasing with chronological age. Of note, in a whole-blood gene expression meta-analysis in 14,983 individuals of European ancestry, GPR56 was the second most highly upregulated gene associated with age (Peters et al., 2015). The cellular distribution of BAI1, GPR97, GPR114, and GPR126, which all show mRNA expression in distinct hematopoietic lineages, still needs to be established at the protein level.

Next to the autoproteolytic processing at the GPS, several aGPCRs undergo secondary cleavage of their extended ECDs. Of note, mass spectrometry of cerebrospinal fluid and blood plasma samples identified fragments of latrophilin/ADGRL, BAI/ADGRB, and CELSR/ADGRC receptors (Guldbrandsen et al., 2014; Liu et al., 2007), which are expressed in the central nervous, but hardly within the immune system (Fig. 3). Whether circulating soluble ECDs of aGPCRs from the central nervous system can signal at distant sites *in trans* is an intriguing question that remains to be studied.

# 6. INVOLVEMENT IN IMMUNE CELL DEVELOPMENT

RNA-sequencing data of sorted human adult bone marrow and cord blood cells now provide detailed insights into the expression patterns of aGPCRs during hematopoietic differentiation. Moreover, several aGPCRs have recently been identified to correlate with survival or response to chemotherapy in different hematologic diseases, hinting at a potential role in malignant transformation.

## 6.1 Hematopoietic Stem and Progenitor Cells

While certain aGPCRs, such as CD97, are expressed in both immature hematopoietic stem and progenitor cells (HSPCs), as well as more differentiated peripheral blood cells (Fig. 3), other aGPCRs show more distinct expression patterns, with either restriction to the HSPC compartment or to differentiated blood cell populations. These very distinct expression profiles point toward specific and nonredundant functional roles of aGPCRs in the hematopoietic system.

Among the aGPCRs expressed in HSC-enriched $CD34^{+}CD45RA^{-}$ cord blood cells (11 of 33 aGPCRs with reads per kilobase per million mapped reads (RPKM) >1), GPR56 is most highly expressed (Fig. 4). Expression decreases with onset of differentiation from HSCs to committed myelomonocytic precursors and is lost at the promyelocyte stage. While GPR56 has been attributed a role in endothelial to HSC transition in mice and zebrafish during the very early steps of hematopoietic tissue development (Solaimani Kartalaei et al., 2015), its functional role and downstream signaling pathways in human primary HSPCs are still under investigation. aGPCRs with similar expression pattern in HSPCs and during granulocytic maturation comprise LPHN1, GPR124, GPR125, CELSR3, GPR113, GPR114, and GPR126 (Fig. 4). Interestingly, the majority of these aGPCRs has been implicated in the Wnt/β-catenin pathway and/or cell polarity, for which a fundamental role in hematopoietic development has unequivocally been shown by several groups (Reya et al., 2003; Scheller

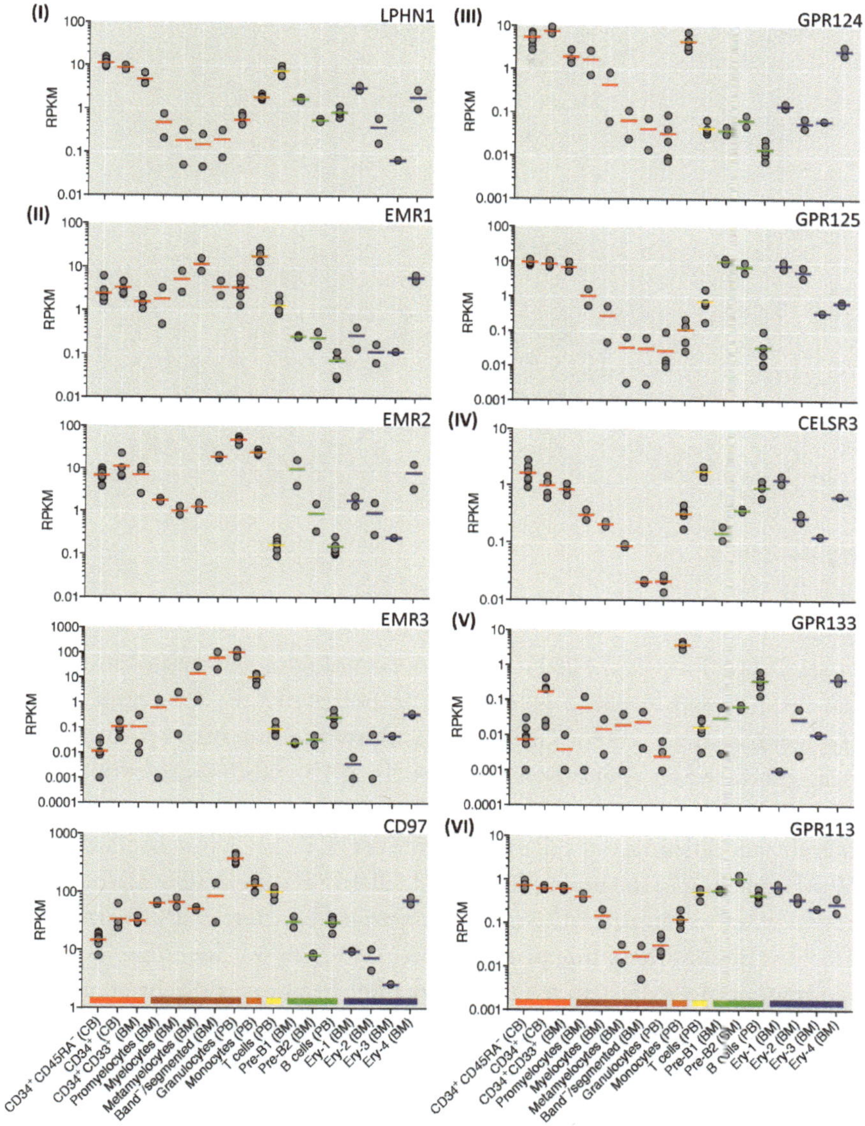


**Fig. 4** Expression of aGPCRs during immune cell development. RNA-sequencing data showing aGPCR expression in sorted cord blood (CB), bone marrow (BM), and peripheral blood (PB) cell populations (Maiga et al., 2016). Gating strategy for subsets was as follows (if not indicated in plot): promyelocytes $CD34^-CD11b^-CD16^-$; myelocytes $CD34^-CD11b^+CD16^-$; metamyelocytes $CD34^-CD11b^+CD16^{med}$; band cells/segmented granulocytes $CD34^-CD11b^+CD16^+$; granulocytes $SSC^{high}CD33^+$; monocytes $FSC^{high}SSC^{low}CD33^{++}CD14^+$; T cells $SSC^{low}CD3^+$; pre-B cells (Pre-B1) $CD34^+$ $CD10^+CD19^+$; pre-B cells (Pre-B2) $CD34^-CD10^+CD19^+$; B cells $SSC^{low}CD19^+$; erythrocytes (Ery-1) $CD34^+CD71^+GlyA^-$; erythrocytes (Ery-2) $CD34^-CD71^+GlyA^-$; erythrocytes (Ery-3) $CD34^-CD71^+GlyA^+$; erythrocytes (Ery-4) $CD34^-CD71^-GlyA^+$. *Roman numerals* indicate aGPCR subfamilies.

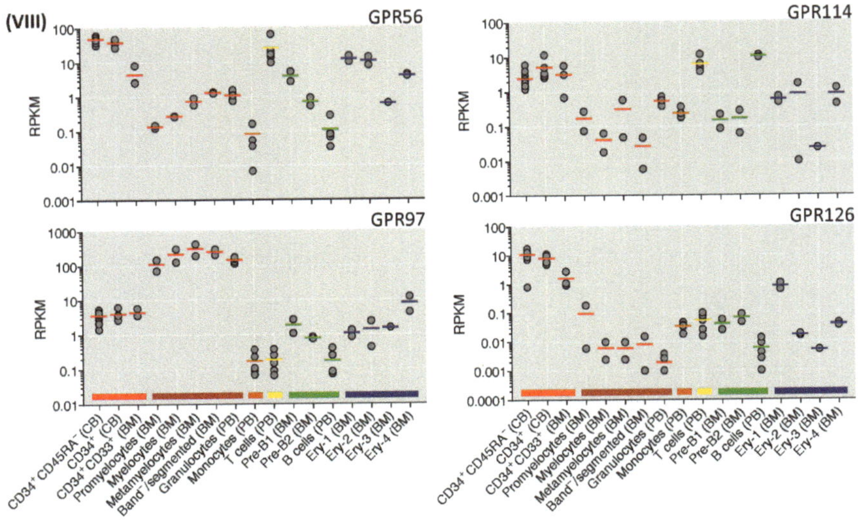


**Fig. 4—Cont'd**

et al., 2006; Ting et al., 2012). Lat-1, a homolog of vertebrate latrophilins, has been shown to play a pivotal role in polarity in the *C. elegans* embryo (Langenhan et al., 2009). GPR124 was shown to potentiate β-catenin signaling induced by Wnt7 (Posokhova et al., 2015), while a direct interaction with disheveled has been shown for GPR125 in zebrafish (Li et al., 2013), thereby modulating cell polarity. Moreover, CELSR3 was associated with Wnt7 signaling in the brain (Wang et al., 2016). Future studies are needed to reveal whether similar pathways are controlled by these aGPCRs in human HSPCs. Besides these functional studies, GPR125 was identified as a novel spermatogonial stem/progenitor cell marker in humans (Sachs et al., 2014; Seandel et al., 2007). Interestingly, GPR125 is not only expressed in noncommitted HSPCs ($CD34^+CD45RA^-$) but also in B cell and erythroid precursors.

In contrast, expression of CD97, EMR2, and EMR3 is low in HSPCs, but gradually increases upon differentiation, reaching maximum expression levels in terminally differentiated mature peripheral blood granulocytes (Fig. 4). GPR97 and EMR1 expression also increases upon granulopoiesis, yet reaches maximum expression levels in metamyelocytes and band form granulocytes, while expression slightly decreases again in terminally differentiated granulocytes. GPR133 shows a unique pattern characterized by expression in peripheral blood monocytes, but no other hematopoietic cell population. Whether these aGPCRs play a functional role in controlling differentiation of HSPCs to the different blood lineages has to be determined in future studies.

## 6.2 Leukemia and Lymphoma

In a study analyzing expression of GPCRs in acute myeloid leukemia (AML) vs normal CD34$^+$ cells, based on RNA-sequencing data, several aGPCRs were identified to be significantly more highly expressed in AML samples compared to normal CD34$^+$ HSPCs, including CD97, EMR1, EMR2, and GPR114 (Maiga et al., 2016). Moreover, we recently identified GPR56 as a surface marker enriching for the leukemia stem cell compartment in AML (Pabst et al., 2016). While GPR56 was expressed on the great majority of CD34$^+$ cord blood cells (>90% by flow-cytometric analysis), AML patient samples displayed distinct gene and protein expression patterns, ranging from complete absence to high expression levels, comparable to those observed in normal HSCs. High GPR56 expression has also been associated with specific high-risk genetic aberrations, such as MECOM/EVI1 rearrangement (Pabst et al., 2016; Saito et al., 2013), complex karyotype, or FLT3/ITD mutations (Daria et al., 2016; Maiga et al., 2016; Pabst et al., 2016), indicating that GPR56 might be crucial for the leukemogenic potential of these oncogenes. GPR56 expression on its own is highly predictable for short survival (Daria et al., 2016; Pabst et al., 2016) and has recently been included in a 17-gene expression signature for rapid risk stratification in AML (Ng et al., 2016). Furthermore, GPR56 overexpression accelerated a HoxA9/Meis1-induced mouse leukemia model (Daria et al., 2016), providing first hints for a functional role of GPR56 in leukemia.

GPR126 was also found to be more highly expressed in AML cells with leukemic stem cell activity (Eppert et al., 2011) and MLL rearrangement (Maiga et al., 2016), but has not been associated with leukemic stem cell function to date. Two other aGPCRs associated with poor survival in AML are GPR97 and GPR114 (Bagger et al., 2016; Cancer Genome Atlas Research Network, 2013; Fig. 5). Interestingly, these three aGPCRs associated with survival in AML (GPR56, GPR97, GPR114) build a gene cluster on human chromosome 16. The pan-leucocyte marker CD97 has recently been shown to be important for AML cell proliferation (Martin et al., 2016) and to be associated with FLT3/ITD-mutated AML (Wobus et al., 2015), and LPHN1 was shown to be inducible by LPS via the mTOR

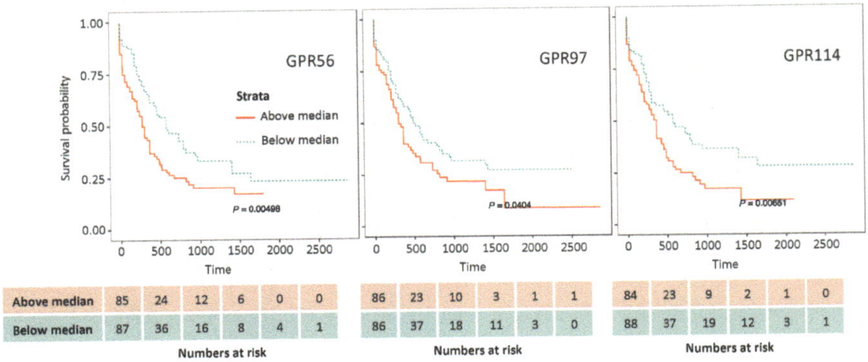


**Fig. 5** Association of aGPCRs expression with survival in AML. High expression of the clustered genes GPR56, GPR97, and GPR114 is associated with poor survival in AML. *Data derived from Cancer Genome Atlas Research Network. (2013). Genomic and epigenomic landscapes of adult de novo acute myeloid leukemia.* The New England Journal of Medicine, 368, *2059–2074 were plotted using BloodSpot Bagger, F. O., Sasivarevic, D., Sohi, S. H., Laursen, L. G., Pundhir, S., Sønderby, C. K., et al. (2016). BloodSpot: A database of gene expression profiles and transcriptional programs for healthy and malignant haematopoiesis.* Nucleic Acids Research, 44, *D917–D924.*

pathway in human AML cell lines (Sumbayev et al., 2016), indicating that LPHN1 is functional in these cells.

Information on aGPCR expression and function in lymphatic diseases is more rare. CD97 was identified as a minimal residual disease marker in acute lymphoblastic leukemia (ALL) (Coustan-Smith et al., 2011), while GPR110 was included in a gene expression signature associated with Philadelphia-like B-ALL (Harvey et al., 2010).

# 7. INVOLVEMENT IN INNATE IMMUNITY

As mentioned, a selected group of distinct aGPCRs are expressed in diverse cells of myeloid and lymphoid lineages in a restricted and temporal fashion; hence, it is not surprising that these aGPCRs are implicated in the immune regulatory functions of both innate and adaptive immunity.

## 7.1 Myeloid Cells

Cells of myeloid lineage include monocytes, Mφs, dendritic cells (DCs), granulocytes (neutrophils, basophil, eosinophil), mast cells, megakaryocytes, and platelets (Janeway, Traver, Walport, & Shlomchik, 2001). In addition to their role in tissue homeostasis, myeloid cells play an essential role in innate immunity as critical phagocytes for pathogen recognition and elimination, as well as cytokine-producing cells for the induction of inflammatory responses. Some of the myeloid cells also work as professional antigen (Ag)-presenting cells (APCs) for initiating the activation of Ag-specific T cells and immune effector cells for tissue hypersensitivity and autoimmunity (Janeway et al., 2001). A subset of aGPCRs is expressed by myeloid cells and known to modulate specific immune functions.

BAI1 was identified originally as an antiangiogenic and antitumor molecule that contains one RGD (Arg-Gly-Asp) motif, five thrombospondin type 1 repeats (TSR), and one hormone receptor motif (HRM) (Kaur, Brat, Devi, & Van Meir, 2005; Nishimori et al., 1997). In addition, BAI1 contains a very long intracellular tail with the PDZ-binding motif QTEV (Gln-Thr-Glu-Val). BAI1 is a multifunctional molecule, which binds numerous endogenous and exogenous ligands. Its TSR domains were shown to interact with the αvβ5 integrin to suppress endothelial cell proliferation (Koh et al., 2004) and with CD36 to inhibit angiogenesis (Cork et al., 2012; Kaur et al., 2009). BAI1 can also bind PtdSer on the surface of dying cells and work as a phagocytic receptor for apoptotic cells in Mφ, microglia, fibroblasts, and epithelial cells (Das et al., 2014; Mazaheri et al., 2014; Park et al., 2007). PtdSer binding to BAI1 leads to its intracellular interaction with ELMO1 and DOCK1, which recruit and activate Rac, enabling the engulfment of apoptotic cells (Park et al., 2007). Indeed, it was shown that microglia in embryonic zebrafish brain use BAI1 to control phagosome formation around and transport of the dying neurons (Mazaheri et al., 2014). BAI1-mediated clearance of apoptotic cells by Mφs was shown to promote antiinflammatory responses via the subsequent production of antiinflammatory mediators and modulation of lipid homeostasis, in part by the upregulation of the ATP-binding cassette transporter 1 (ABCA1) (Fond, Lee, Schulman, Kiss, & Ravichandran, 2015; Hamann, Hsiao, Lee, Ravichandran, & Lin, 2016). Similarly, apoptotic cell removal by BAI1 on colonic epithelial cells reduced the exacerbated inflammation induced by excess apoptotic cells in experimental colitis (Lee, Penberthy, et al., 2016). Interestingly, by binding to the exposed PtdSer on apoptotic myoblast cells, BAI1 was shown to actively promote myoblast fusion (Hochreiter-Hufford et al., 2013).

In addition to the endogenous ligands, BAI1 also acts as a pattern recognition receptor by binding to LPS and mediates the phagocytic uptake of Gram-negative bacteria by Mφs (Das et al., 2011). Again, LPS binding was mediated by the TSR domains of BAI1, which in turn induced Rac activation via an ELMO1/DOCK1-dependent mechanism for bacteria uptake and tumor necrosis factor (TNF) production (Das et al., 2011). It was further revealed that BAI1 can promote reactive oxygen species (ROS) production via Rac1-activated nicotinamide adenine dinucleotide phosphate oxidase, leading to enhanced microbicidal activity in vivo (Billings et al., 2016). More recently, BAI1 was reported to play a role in the antiviral responses of Mφs/microglia by promoting TNF production for the clearance of herpes simplex virus in glioma-bearing mice (Bolyard et al., 2017). The nature of the viral ligand(s) is currently unknown; nevertheless, this capacity might be harmonized with tumor-attacking viruses to be a possible oncolytic viral therapy for brain tumors in the future (Bolyard et al., 2017).

As a prototypic member of the EGF-TM7/ADGRE subfamily, F4/80 (mouse EMR1) contains seven tandem repeats of EGF-like domains in its NTF, a feature shared by all EGF-TM7 receptors (Kwakkenbos et al., 2004; Lin, Stubbs, & Mucenski, 1997; McKnight et al., 1996). F4/80 has been and still is a widely used surface marker of murine tissue Mφs (Austyn & Gordon, 1981; Gordon et al., 2011). Interestingly, however, no obvious phenotypic alteration was found in the Mφ populations of F4/80-deficient animals, indicating that the F4/80 molecule is not required for the normal development and maturation of murine Mφs (Lin et al., 2005; Schaller et al., 2002). Nevertheless, F4/80 was shown to be required for the production of IFN-γ by NK cells during *Listeria monocytogenes* infection, most likely due to its involvement in Mφ–NK cell interaction (Warschkau & Kiderlen, 1999). On the other hand, an unexpected role for F4/80 in the induction of peripheral immune tolerance was established using F4/80-deficient Mφs in two different animal models of peripheral tolerance. It was thought that F4/80 was involved in the interaction of APCs and neighboring immune cells, essential for the generation of Ag-specific efferent regulatory T (Treg) cells (Lin et al., 2005). More recently, the anti-F4/80 monoclonal antibody (mAb) and its derivatives have been employed to track and image specific tissue Mφ subsets in vivo or to inhibit inflammatory pathogenesis by cell type-targeted anticytokine therapy (Efimov et al., 2016; Terry et al., 2015).

Unlike F4/80, the human EMR1 protein was undetectable in monocytes/Mφs and DCs. Instead, EMR1 was expressed restrictedly in human eosinophilic granulocytes (Hamann et al., 2007). Intriguingly, it was demonstrated that specific targeting of eosinophils with a humanized, afucosylated anti-EMR1 mAb leads to enhanced eosinophil killing by

NK cell in vitro and a swift and prolonged eosinopenia in monkeys (Legrand et al., 2014). Hence, EMR1 might be a good candidate target for treating eosinophilic disorders in human.

Activation of EMR2 by the ECD-specific 2A1 mAb primed neutrophils for enhanced adhesion and migration, as well as inflammatory activation by a panel of stimuli (Yona, Lin, Dri, et al., 2008). Furthermore, EMR2 activation modulated cytokine production and survival of LPS-stimulated neutrophils (Chen et al., 2011). Additionally, EMR2 expression was upregulated in neutrophils of systemic inflammatory response syndrome (SIRS) patients, of which a significant association between the percentage of EMR2-expressing neutrophils and the extent of organ failure was identified. Hence, EMR2 was proposed as a potential neutrophil biomarker for SIRS (Lewis et al., 2015; Yona, Lin, Dri, et al., 2008). A more recent report identified upregulated EMR2 expression in neutrophils of cirrhotic patients with infection. Moreover, a strong correlation was noted between the extent of EMR2$^+$ neutrophils in patients with liver cirrhosis and the disease severity and overall mortality (Huang et al., 2016). Similarly, activation of EMR2 by immobilized 2A1 mAb induced the production of proinflammatory cytokines, such as interleukin (IL)-8 and TNF from Mφs (Huang et al., 2012).

Dermatan sulfate was identified as the glycosaminoglycan ligand of EMR2 as well as CD97, and the receptor–ligand pair was highly detected in rheumatoid synovial tissues (Kop et al., 2005; Stacey et al., 2003). Unexpectedly, a missense EMR2 variant (C492Y) was revealed recently as the disease protein solely responsible for autosomal dominant vibratory urticaria, a rare type of dermal allergy caused by vibration-induced mast cell activation (Boyden et al., 2016). Apparently, the EMR2-C492Y variant is structurally less stable and prone to activate mast cells for aberrant histamine release upon vibratory stimulation in the presence of dermatan sulfate or 2A1 mAb. It is thought that vibration somehow enhanced the ligand-induced shedding of the NTF and subsequent activation of the CTF of EMR2 in mast cells (Boyden et al., 2016). The EMR2-mediated signaling pathway that leads to the release of histamine by mast cells has not been determined.

EMR3 is expressed on CD16$^+$ human mature monocytes and certain Mφ subsets, very similar to EMR2 (Matmati et al., 2007; Stacey, Lin, Hilyard, Gordon, & McKnight, 2001). In addition, EMR3 was shown to be a marker of mature granulocytes, though its functional role in granulocytes has not been studied (Matmati et al., 2007). A putative cell surface ligand of EMR3 was identified in monocyte-derived Mφs and activated neutrophils, but the significance of this interaction is not known (Stacey et al., 2001).

Human EMR4 is presumably not a functional receptor due to a one-nucleotide deletion in its exon 8, resulting in a predicted truncated soluble protein (Caminschi et al., 2006; Hamann et al., 2003). By contrast, mouse EMR4, also called F4/80-like receptor (FIRE), is a genuine aGPCR expressed by blood monocytes and certain tissue Mφ subsets (Caminschi et al., 2001; Stacey et al., 2002). $CD8^-$ DCs, but not $CD8^+$ DCs, also express EMR4, which was downregulated following DC activation (Caminschi et al., 2001). It was found recently that EMR4 was also highly expressed in a rare myeloid DC-like subset, termed L-DCs, in the spleen (Hey & O'Neill, 2016). L-DCs are able to activate $CD8^+$ T cells via Ag cross-presentation, but are unable to activate $CD4^+$ T cells due to the lack of major histocompatibility complex (MHC)-II. Interestingly, targeting of the mouse EMR4 molecule by the receptor-specific mAb in vivo significantly enhanced humoral immune responses in the absence of conventional "danger" signal and Fc receptor (FcR) binding. This result suggested a possibility of using the $EMR4^+$ DC-targeting strategy as a substitute for external adjuvants in eliciting immune responses (Corbett et al., 2005). Ligand-binding search identified a potential cellular ligand on a mouse B lymphoma cell line, A20, in a $Ca^{2+}$-independent manner (Stacey et al., 2002). However, the role of EMR4–ligand interaction remains uncharacterized.

CD97 is expressed by nearly all types of myeloid cells, and its expression is further upregulated in certain Mφ subpopulations enriched in autoimmune diseases such as infiltrating Mφs in the rheumatoid synovium, Gaucher spleen Mφ, and foamy Mφs in multiple sclerosis brains (Hamann et al., 1999; van Eijk et al., 2010; Visser et al., 2002). Multiple ligands including CD55, dermatan sulfate, α5β1 integrin, and CD90 were identified for CD97 (Hamann, Vogel, van Schijndel, & van Lier, 1996; Stacey et al., 2003; Wandel, Saalbach, Sittig, Gebhardt, & Aust, 2012; Wang et al., 2005). In rheumatoid synovial tissue, expression of dermatan sulfate, CD55, and α5β1 integrin was readily detected on fibroblast-like synoviocytes, endothelium, and extracellular matrix, suggesting a likely role of CD97 in the pathogenesis of rheumatoid arthritis (Hamann et al., 1999; Kop et al., 2005). Indeed, administration of a mouse CD97 neutralizing mAb ameliorated collagen-induced arthritis in mice, while both CD97-knockout and CD55-knockout animals showed reduced disease activities in two different animal models of rheumatoid arthritis (Hoek et al., 2010; Kop et al., 2006). Importantly, the mAb-mediated CD97 neutralization was reported to have a positive therapeutic potential in mice with established experimental arthritis (de Groot et al., 2009).

Similar to what was found in Mφ, upregulated CD97 expression was detected on neutrophils of patients with psoriasis, and the CD97–CD90 interaction was shown to mediate adhesion of CD97$^+$ myeloid cells to CD90$^+$ activated dermal endothelial cells (Wandel et al., 2012). Hence, CD97–CD90 binding might play a role in regulating granulocyte trafficking to inflammatory sites. Indeed, a role for CD97 in regulating neutrophil migration was first demonstrated in a dextran sulfate sodium-induced colitis model in mice, where the homing of adoptively transferred CD97 mAb-treated neutrophils to the colon was significantly delayed (Leemans et al., 2004). Likewise, impaired neutrophil infiltration to the lung was observed in CD97 mAb-treated mice in a *Streptococcus pneumoniae*-induced pneumonia model. Consequently, an increased bacteria burden in the lungs and a higher death rate were noted in CD97 mAb-treated mice (Leemans et al., 2004). The CD97 mAb-induced granulocytopenia was later found to be mediated via a FcR-dependent mechanism under conditions of acute inflammation, suggesting that CD97 mAbs inhibit the innate immune response most likely by targeting and interfering the recruitment of CD97$^+$ myeloid leukocytes to sites of inflammation (Hamann et al., 2010; Veninga, de Groot, et al., 2011).

Interestingly, however, a mild granulocytosis was detected in CD97-deficient mice at the steady state, and the animals developed an improved antibacterial defense to systemic infection by *L. monocytogenes* due to robust neutrophil accumulation in blood and infected livers (Wang et al., 2007). Hence, a role in maintaining peripheral granulocyte homeostasis was suggested for CD97. This immune modulatory function is very likely mediated in part by the CD97–CD55 interaction, as the CD55-deficient mice were shown to also display a similar phenotype of enhanced granulocytosis and antibacterial activity (Veninga, Hoek, et al., 2011). Intriguingly, it was found recently that CD55–CD97 interaction actively induced down-regulation of surface CD97 on circulating leukocytes in a shear stress-dependent mechanism (Karpus et al., 2013); hence, the in vivo role of CD97 in innate immune response seems to be determined by multiple factors such as the availability of specific ligand(s)/mAbs and conditions of local tissue microenvironment.

GPR116, also called Ig-Hepta, contains in its NTF a sperm protein, enterokinase, and agrin (SEA) domain and two Ig-like domains (Abe, Suzuki, Notoya, Yamamoto, & Hirose, 1999). Interestingly, the SEA domain is also a conserved independent protein fold capable of autoproteolytic cleavage, similar to the GAIN domain (Levitin et al., 2005; Macao, Johansson, Hansson, & Härd, 2006). Hence, GPR116 undergoes two different autoproteolytic modification events during its biosynthesis, generating multiple protein fragments (Abe, Fukuzawa, & Hirose, 2002; Fukuzawa & Hirose, 2006). Early studies showed that GPR116 is highly expressed in alveolar type I and type II epithelial cells. Importantly, GPR116 gene targeting in mice revealed a critical role for GPR116 in maintaining normal lung surfactant levels (Bridges et al., 2013; Fukuzawa et al., 2013; Yang et al., 2013). More recently, analysis of the GPR116-knockout mice showed that GPR116 also played a role in the functional regulation of alveolar Mφs. Indeed, alveolar Mφs of GPR116-deficient mice were foamy and activated, leading to pulmonary local inflammation and emphysema-like symptoms, likely due to the upregulated production of ROS and matrix metalloproteinases (Ariestanti, Ando, Hirose, & Nakamura, 2015).

While little is known of the function of GPR97 in human monocytes/Mφs, a potential role of mouse GPR97 in regulating Mφ inflammation was reported in GPR97-knockout mice (Shi et al., 2016). In a high-fat diet-induced experimental obesity model, the lack of GPR97 leads to increased numbers of infiltrating Mφs and higher TNF levels in liver and kidney, as well as an imbalanced M1 vs M2 Mφ ratio in white adipose tissues. Interestingly, no significant difference in metabolic functions was observed in the GPR97-knockout mice in comparison to the wild-type animals during the high-fat diet-induced obesity (Shi et al., 2016). By contrast, GPR97 was shown to be dispensable for inflammation in an ovalbumin-induced asthma model (Shi et al., 2015).

## 7.2 Innate Lymphoid Cells

Innate lymphoid cells (ILCs) are a group of lymphoid cells with a recently recognized role as regulators of innate immunity, inflammation, and tissue repair at the barrier surfaces (Artis & Spits, 2015; Hepworth & Sonnenberg, 2014; McKenzie, Spits, & Eberl, 2014). As with the innate myeloid cells, ILCs participate in the innate protection of host against pathogens in an Ag nonspecific manner. Hence, ILCs are considered the innate counterparts of Ag-specific T cells (Eberl, Colonna, Di Santo, & McKenzie, 2015). ILCs are divided into subsets, including ILC1s, ILC2s, and ILC3s, representing the innate equivalents of $CD4^+$ T helper 1 ($T_H1$), $T_H2$, and $T_H17$ cells, respectively. While ILC1s, ILC2s, and ILC3s are mostly noncytotoxic, cytotoxic NK cells are considered the innate lymphoid counterparts of cytotoxic CD8$^+$ T cells, although sometimes NK cells are grouped as a part of ILC1 (Eberl et al., 2015; Sonnenberg & Artis, 2015; Spits, Bernink, & Lanier, 2016).

Among all aGPCRs, GPR56 is the only one thus far relevant to ILCs based on its restricted expression in cytotoxic NK- and T-cell populations (Della Chiesa et al., 2010; Peng et al., 2011). It was further shown that forced expression of GPR56 in the human NK-92 cell line suppresses spontaneous and chemokine-induced cell migration (Peng et al., 2011). More recently, GPR56 was shown by us to be a novel inhibitory receptor on human NK cells (Chang et al., 2016). Interestingly, GPR56 expression was mainly detected in mature nondividing NK cells and was tightly regulated by the transcription factor Hobit (homolog of Blimp-1 in T cells). Surface GPR56 protein levels were downregulated upon NK-cell activation via both receptor internalization and shedding. GPR56 formed a receptor complex with the tetraspanin protein CD81, which was identified previously as an inhibitor of NK-cell function (Crotta et al., 2002; Tseng & Klimpel, 2002). It was further shown that dissociation of the GPR56–CD81 complex induced by GPR56 mAbs significantly enhanced NK-cell cytotoxicity and cytokine production. Hence, GPR56 seems to negatively modulated NK-cell functions in a cell-autonomous fashion (Chang et al., 2016). The functional role of GPR56 in ILC1s, ILC2s, and ILC3s is currently not known.

## 8. INVOLVEMENT IN ADAPTIVE IMMUNITY

Adaptive immune responses involve Ag-specific reactions carried out in most part by cells of T and B lineages. To date, only a very selective few aGPCRs are known to play a role in the adaptive immunity. While it is constitutively expressed by myeloid cells, CD97 was identified initially as an early activation marker of T and B cells (Hamann et al., 1995). Interestingly, one of its ligands, CD55, is also expressed by cells of both myeloid and lymphoid lineages, suggesting that the CD97–CD55 interaction might have a role in innate as well as adaptive immunity. Indeed, interaction of this receptor–ligand pair has been shown to provide human T cells a potent costimulatory signal (Spendlove & Sutavani, 2010). Intriguingly, the costimulating effect of CD97–CD55 interaction could be induced either by CD97 on APCs with CD55 on T cells or by CD55 on monocytes and CD97 on T cells (Abbott et al., 2007; Capasso et al., 2006). More specifically, costimulation of freshly isolated $CD4^+$ T cells with plate-bound anti-CD3 mAb and immobilized CD97-Fc fusion protein enhanced cell proliferation and cytokine production. On the other hand, blocking the binding of CD55 on monocytes and CD97 on T cells with specific mAbs to both molecules effectively inhibited T-cell proliferation (Abbott et al., 2007; Capasso et al., 2006).

More recently, it was shown that costimulation of highly pure human naïve $CD4^+$ cells with the CD97 ligand CD55 was able to induce two populations of IL-10-secreting T cells, namely IL-$10^{high}$IFN-$\gamma^-$ T regulatory type 1 (Tr1)-like cells and IL-$10^{low}$IFN-$\gamma^{high}$ $T_H1$-like cells. Interestingly, restimulation of both T-cell populations with recombinant CD55 activated both cell types to produce consistently high levels of IL-10, which suppressed effector T-cell functions. This novel result suggested a possible role for CD97–CD55 interaction in the differentiation and maintenance of a IL-10-producing Tr1-like cell population that could regulate T-effector cell function negatively (Sutavani et al., 2013). In a separate experiment, no significant effects were observed when a CD97 mAb that blocked CD97–CD55 interaction was added in a mixed lymphocyte reaction; however, addition of a mAb against the CD97-GAIN domain was able to inhibit both $CD8^+$ and $CD4^+$ allogeneic T-cell proliferation (Kop et al., 2009). How this specific mAb modulates T-cell activation and proliferation in this reaction is unclear. On the other hand, application of mCD97 mAbs in mouse models of delayed-type hypersensitivity and experimental autoimmune encephalomyelitis did not show any impact on Ag-specific adaptive immune responses (Hamann et al., 2010).

Recently, targeted disruption of the GPR97 gene in mice revealed an unexpected role of mouse GPR97 in B-cell development (Wang et al., 2013). Extensive analyses of the GPR97-deficient mice identified a disorganized splenic architecture, which showed an expanded marginal zone but a decreased follicular B-cell population. Furthermore, the GPR97-mutant animals produced enhanced Ab responses to T-cell-dependent Ags, but reduced Ab responses to T-cell-independent Ags. Interestingly, NF-κB (nuclear factor kappa-light-chain-enhancer of activated B cells) expression and activity were increased in GPR97-deficient spleens, suggesting a critical role of the GPR97–NF-κB axis in regulating follicular vs marginal zone B-cell development in mice (Wang et al., 2013). In addition, a novel GPCR expression profile screening identified GPR97 as the most highly and specifically expressed GPCR in mouse lymphatic endothelium. Functional analysis revealed that GPR97 regulated adhesion and migration of lymphatic endothelial cells via the modulation of RhoA and Cdc42 activity and cytoskeletal reorganization (Valtcheva, Primorac, Jurisic, Hollmén, & Detmar, 2013).

## 9. OUTLOOK

Next-generation sequencing efforts have provided broad insights into gene expression patterns in hematopoietic cells and their malignant counterparts. Evidence was obtained that about one-third of the 33 human aGPCRs are expressed in hematopoietic stem, precursor, or mature cells. It needs to be considered, however, that aGPCRs are shaped by proteolytic processing, initially, during protein synthesis in the endoplasmic reticulum and later, during surface expression at the plasma membrane. Also, aGPCRs molecules can be released form cytoplasmic stores and may have relatively long half-lifes at the cell surface. Quantification of protein expression on cells and in body fluids, using flow cytometry and mass spectrometry, is therefore needed to comprehensively understand the expression dynamics of aGPCRs in the immune system.

Since the seminal finding that F4/80 defines tissue Mφs in the mouse (Austyn & Gordon, 1981; Gordon et al., 2011), expression of aGPCRs has been linked with distinct cell types or populations, e.g., EMR1 with human eosinophils (Hamann et al., 2007), EMR3 with fully mature human granulocytes (Matmati et al., 2007), and, more recently, GPR56 with cytotoxic human lymphocytes (Peng et al., 2011). Protein expression profiling of further aGPCRs, in particular the members of the ADGRG subfamily, may provide further useful cell markers.

Biomarkers that indicate susceptibility to or progression of disease are increasingly important in personalized medicine. Recent studies on leukemia show that several, in part related aGPCRs, regulate HSPC functions and predict survival in AML patients (Martin et al., 2016; Pabst et al., 2016; Solaimani Kartalaei et al., 2015). However, the majority of studies analyzing aGPCRs in leukemia focused on gene expression analyses and statistical association with genetic aberrations or clinical outcome, so that one can only speculate whether these aGPCRs play functional roles in disease initiation or progression.

aGPCRs have been implicated in the development and function of hematopoietic stem, progenitor, or mature cells for quite some time; yet, it has not been clear whether these nonconventional GPCRs regulate cellular homeostasis and activation, or whether they also control specific immune functions. Studies on the role of BAI1 in Mφs (Das et al., 2011; Park et al., 2007), EMR2 in mast cells (Boyden et al., 2016), and GPR56 in NK cells (Chang et al., 2016) imply that aGPCRs indeed regulate innate, and possibly also adaptive, effector functions and paved the way for further exploring their biological activities using novel genetic models and pharmacological tools.

The prominent presence of aGPCRs on polymorphonuclear granulocytes is interesting also from a therapeutic perspective. Legrand et al. showed that an afucosylated mAb directed against EMR1 efficiently eradicates eosinophils through Ab-dependent cell-mediated cytotoxocity, resulting in long-term in vivo depletion of eosinophils in monkeys and, thus, providing an treatment option for eosinophilic disorders (Legrand et al., 2014). Similarly, mAbs to mouse CD97 eliminate neutrophils (and possible also eosinophils) in an FcR-dependent manner, thereby ameliorating various inflammatory conditions (Veninga, de Groot, et al., 2011).

The pharmacological potential of aGPCRs remains to be explored. Screening for natural and synthetic molecules interfering with aGPCR activities has long been hampered by the lack of pharmacological suitable readouts. Only recently, dihydromunduletone has been identified as small-molecule selective aGPCR antagonist with activity against GPR56 and GPR114 (Stoveken et al., 2016). Given the increasing evidence for a functional role of GPR56 in AML, this might represent a novel therapeutic approach for antileukemic therapy. It seems reasonable to assume that further elucidation of the activation mechanisms and signaling pathways as well as the expression profiles and biological functions of aGPCRs will provide novel therapeutic opportunities in the future.

## ACKNOWLEDGMENTS

We thank Guy Sauvageau for access to RNA-sequencing data and the members of the Adhesion GPCR Consortium (www.adhesiongpcr.org) for stimulating discussions. RNA-sequencing data of sorted bone marrow and blood cell populations were generated at the Institute for Research in Immunology and Cancer, University of Montreal, supported by the Government of Canada, through Genome Canada and the Ministère de l'économie, de l'innovation et de l'exportation du Québec, through Génome Québec. We acknowledge support by the Ministry of Science and Technology, Taiwan (MOST-104-2320-B-182-035-MY3), the Chang Gung Memorial Hospital (CMRPD1G0121, CMRPD1D0393, and CORPD1F0061), the Thyssen Foundation (2015-00387), the Industrielle-Alliance Research Chair in leukemia (Université de Montréal), the Max-Eder Program of German Cancer Aid (70111531), the Cole Foundation, and the German Research Foundation (Research Unit 2149).

## REFERENCES

Abbott, R. J. M., Spendlove, I., Roversi, P., Fitzgibbon, H., Knott, V., Teriete, P., et al. (2007). Structural and functional characterization of a novel T cell receptor co-regulatory protein complex, CD97-CD55. *The Journal of Biological Chemistry*, *282*, 22023–22032.

Abe, J., Fukuzawa, T., & Hirose, S. (2002). Cleavage of Ig-Hepta at a "SEA" module and at a conserved G protein-coupled receptor proteolytic site. *The Journal of Biological Chemistry*, *277*, 23391–23398.

Abe, J., Suzuki, H., Notoya, M., Yamamoto, T., & Hirose, S. (1999). Ig-hepta, a novel member of the G protein-coupled hepta-helical receptor (GPCR) family that has immunoglobulin-like repeats in a long N-terminal extracellular domain and defines a new subfamily of GPCRs. *The Journal of Biological Chemistry*, *274*, 19957–19964.

Araç, D., Boucard, A. A., Bolliger, M. F., Nguyen, J., Soltis, S. M., Südhof, T. C., et al. (2012). A novel evolutionarily conserved domain of cell-adhesion GPCRs mediates autoproteolysis. *The EMBO Journal*, *31*, 1364–1378.

Ariestanti, D. M., Ando, H., Hirose, S., & Nakamura, N. (2015). Targeted disruption of Ig-Hepta/Gpr116 causes emphysema-like symptoms that are associated with alveolar macrophage activation. *The Journal of Biological Chemistry*, *290*, 11032–11040.

Artis, D., & Spits, H. (2015). The biology of innate lymphoid cells. *Nature*, *517*, 293–301.

Austyn, J. M., & Gordon, S. (1981). F4/80, a monoclonal antibody directed specifically against the mouse macrophage. *European Journal of Immunology*, *11*, 805–815.

Bagger, F. O., Sasivarevic, D., Sohi, S. H., Laursen, L. G., Pundhir, S., Sønderby, C. K., et al. (2016). BloodSpot: A database of gene expression profiles and transcriptional programs for healthy and malignant haematopoiesis. *Nucleic Acids Research*, *44*, D917–D924.

Billings, E. A., Lee, C. S., Owen, K. A., D'Souza, R. S., Ravichandran, K. S., & Casanova, J. E. (2016). The adhesion GPCR BAI1 mediates macrophage ROS production and microbicidal activity against Gram-negative bacteria. *Science Signaling*, *9*, ra14.

Bohnekamp, J., & Schöneberg, T. (2011). Cell adhesion receptor GPR133 couples to Gs protein. *The Journal of Biological Chemistry*, *286*, 41912–41916.

Bolyard, C. M., Meisen, W. H., Banasavadi-Siddegowda, Y., Hardcastle, J., Yoo, J. Y., Wohleb, E. S., et al. (2017). BAI1 orchestrates macrophage inflammatory response to HSV infection- implications for oncolytic viral therapy. *Clinical Cancer Research*, *23*, 1809–1819. http://dx.doi.org/10.1158/1078-0432.CCR-16-1818.

Boyden, S. E., Desai, A., Cruse, G., Young, M. L., Bolan, H. C., Scott, L. M., et al. (2016). Vibratory urticaria associated with a missense variant in ADGRE2. *The New England Journal of Medicine*, *374*, 656–663.

Bridges, J. P., Ludwig, M.-G., Mueller, M., Kinzel, B., Sato, A., Xu, Y., et al. (2013). Orphan G protein-coupled receptor GPR116 regulates pulmonary surfactant pool size. *American Journal of Respiratory Cell and Molecular Biology*, *49*, 348–357.

Brüser, A., Schulz, A., Rothemund, S., Ricken, A., Calebiro, D., Kleinau, G., et al. (2016). The activation mechanism of glycoprotein hormone receptors with implications in the cause and therapy of endocrine diseases. *The Journal of Biological Chemistry*, *291*, 508–520.

Caminschi, I., Lucas, K. M., O'Keeffe, M. A., Hochrein, H., Laâbi, Y., Köntgen, F., et al. (2001). Molecular cloning of F4/80-like-receptor, a seven-span membrane protein expressed differentially by dendritic cell and monocyte-macrophage subpopulations. *Journal of Immunology*, *167*, 3570–3576.

Caminschi, I., Vandenabeele, S., Sofi, M., McKnight, A. J., Ward, N., Brodnicki, T. C., et al. (2006). Gene structure and transcript analysis of the human and mouse EGF-TM7 molecule, FIRE. *DNA Sequence*, *17*, 8–14.

Cancer Genome Atlas Research Network. (2013). Genomic and epigenomic landscapes of adult de novo acute myeloid leukemia. *The New England Journal of Medicine*, *368*, 2059–2074.

Capasso, M., Durrant, L. G., Stacey, M., Gordon, S., Ramage, J., & Spendlove, I. (2006). Costimulation via CD55 on human CD4+ T cells mediated by CD97. *Journal of Immunology, 177*, 1070–1077.

Chang, G.-W., Hsiao, C.-C., Peng, Y.-M., Vieira Braga, F. A., Kragten, N. A. M., Remmerswaal, E. B. M., et al. (2016). The adhesion G protein-coupled receptor GPR56/ADGRG1 is an inhibitory receptor on human NK cells. *Cell Reports, 15*, 1757–1770.

Chen, T.-Y., Hwang, T.-L., Lin, C.-Y., Lin, T.-N., Lai, H.-Y., Tsai, W.-P., et al. (2011). EMR2 receptor ligation modulates cytokine secretion profiles and cell survival of lipopolysaccharide-treated neutrophils. *Chang Gung Medical Journal, 34*, 468–477.

Chiang, N.-Y., Chang, G.-W., Huang, Y.-S., Peng, Y.-M., Hsiao, C.-C., Kuo, M.-L., et al. (2016). Heparin interacts with the adhesion GPCR GPR56, reduces receptor shedding, and promotes cell adhesion and motility. *Journal of Cell Science, 129*, 2156–2169.

Corbett, A. J., Caminschi, I., McKenzie, B. S., Brady, J. L., Wright, M. D., Mottram, P. L., et al. (2005). Antigen delivery via two molecules on the CD8- dendritic cell subset induces humoral immunity in the absence of conventional "danger" *European Journal of Immunology, 35*, 2815–2825.

Cork, S. M., Kaur, B., Devi, N. S., Cooper, L., Saltz, J. H., Sandberg, E. M., et al. (2012). A proprotein convertase/MMP-14 proteolytic cascade releases a novel 40kDa vasculostatin from tumor suppressor BAI1. *Oncogene, 31*, 5144–5152.

Coughlin, S. R. (2001). Protease-activated receptors in vascular biology. *Thrombosis and Haemostasis, 86*, 298–307.

Coustan-Smith, E., Song, G., Clark, C., Key, L., Liu, P., Mehrpooya, M., et al. (2011). New markers for minimal residual disease detection in acute lymphoblastic leukemia. *Blood, 117*, 6267–6276.

Crotta, S., Stilla, A., Wack, A., D'Andrea, A., Nuti, S., D'Oro, U., et al. (2002). Inhibition of natural killer cells through engagement of CD81 by the major hepatitis C virus envelope protein. *The Journal of Experimental Medicine, 195*, 35–41.

Daria, D., Kirsten, N., Muranyi, A., Mulaw, M., Ihme, S., Kechter, A., et al. (2016). GPR56 contributes to the development of acute myeloid leukemia in mice. *Leukemia, 30*, 1734–1741.

Das, S., Owen, K. A., Ly, K. T., Park, D., Black, S. G., Wilson, J. M., et al. (2011). Brain angiogenesis inhibitor 1 (BAI1) is a pattern recognition receptor that mediates macrophage binding and engulfment of Gram-negative bacteria. *Proceedings of the National Academy of Sciences of the United States of America, 108*, 2136–2141.

Das, S., Sarkar, A., Ryan, K. A., Fox, S., Berger, A. H., Juncadella, I. J., et al. (2014). Brain angiogenesis inhibitor 1 is expressed by gastric phagocytes during infection with Helicobacter pylori and mediates the recognition and engulfment of human apoptotic gastric epithelial cells. *FASEB Journal, 28*, 2214–2224.

de Groot, D. M., Vogel, G., Dulos, J., Teeuwen, L., Stebbins, K., Hamann, J., et al. (2009). Therapeutic antibody targeting of CD97 in experimental arthritis: The role of antigen expression, shedding, and internalization on the pharmacokinetics of anti-CD97 monoclonal antibody 1B2. *Journal of Immunology, 183*, 4127–4134.

de Lau, W., Barker, N., Low, T. Y., Koo, B.-K., Li, V. S. W., Teunissen, H., et al. (2011). Lgr5 homologues associate with Wnt receptors and mediate R-spondin signalling. *Nature, 476*, 293–297.

Della Chiesa, M., Falco, M., Parolini, S., Bellora, F., Petretto, A., Romeo, E., et al. (2010). GPR56 as a novel marker identifying the CD56dull CD16+ NK cell subset both in blood stream and in inflamed peripheral tissues. *International Immunology, 22*, 91–100.

Demberg, L. M., Rothemund, S., Schöneberg, T., & Liebscher, I. (2015). Identification of the tethered peptide agonist of the adhesion G protein-coupled receptor GPR64/ADGRG2. *Biochemical and Biophysical Research Communications, 464*, 743–747.

Duman, J. G., Tzeng, C. P., Tu, Y.-K., Munjal, T., Schwechter, B., Ho, T. S.-Y., et al. (2013). The adhesion-GPCR BAI1 regulates synaptogenesis by controlling the recruitment of the Par3/Tiam1 polarity complex to synaptic sites. *The Journal of Neuroscience*, *33*, 6964–6978.

Eberl, G., Colonna, M., Di Santo, J. P., & McKenzie, A. N. J. (2015). Innate lymphoid cells. Innate lymphoid cells: A new paradigm in immunology. *Science*, *348*, aaa6566.

Efimov, G. A., Kruglov, A. A., Khlopchatnikova, Z. V., Rozov, F. N., Mokhonov, V. V., Rose-John, S., et al. (2016). Cell-type-restricted anti-cytokine therapy: TNF inhibition from one pathogenic source. *Proceedings of the National Academy of Sciences of the United States of America*, *113*, 3006–3011.

Eppert, K., Takenaka, K., Lechman, E. R., Waldron, L., Nilsson, B., van Galen, P., et al. (2011). Stem cell gene expression programs influence clinical outcome in human leukemia. *Nature Medicine*, *17*, 1086–1093.

Fond, A. M., Lee, C. S., Schulman, I. G., Kiss, R. S., & Ravichandran, K. S. (2015). Apoptotic cells trigger a membrane-initiated pathway to increase ABCA1. *The Journal of Clinical Investigation*, *125*, 2748–2758.

Fredriksson, R., Lagerström, M. C., Lundin, L.-G., & Schiöth, H. B. (2003). The G-protein-coupled receptors in the human genome form five main families. Phylogenetic analysis, paralogon groups, and fingerprints. *Molecular Pharmacology*, *63*, 1256–1272.

Fukuzawa, T., & Hirose, S. (2006). Multiple processing of Ig-Hepta/GPR116, a G protein-coupled receptor with immunoglobulin (Ig)-like repeats, and generation of EGF2-like fragment. *Journal of Biochemistry*, *140*, 445–452.

Fukuzawa, T., Ishida, J., Kato, A., Ichinose, T., Ariestanti, D. M., Takahashi, T., et al. (2013). Lung surfactant levels are regulated by Ig-Hepta/GPR116 by monitoring surfactant protein D. *PLoS One*, *8*, e69451.

Gordon, S., Hamann, J., Lin, H.-H., & Stacey, M. (2011). F4/80 and the related adhesion-GPCRs. *European Journal of Immunology*, *41*, 2472–2476.

Guldbrandsen, A., Vethe, H., Farag, Y., Oveland, E., Garberg, H., Berle, M., et al. (2014). In-depth characterization of the cerebrospinal fluid (CSF) proteome displayed through the CSF proteome resource (CSF-PR). *Molecular & Cellular Proteomics*, *13*, 3152–3163.

Gupte, J., Swaminath, G., Danao, J., Tian, H., Li, Y., & Wu, X. (2012). Signaling property study of adhesion G-protein-coupled receptors. *FEBS Letters*, *586*, 1214–1219.

Hamann, J., Aust, G., Araç, D., Engel, F. B., Formstone, C., Fredriksson, R., et al. (2015). International Union of Basic and Clinical Pharmacology. XCIV. Adhesion G protein-coupled receptors. *Pharmacological Reviews*, *67*, 338–367.

Hamann, J., Eichler, W., Hamann, D., Kerstens, H. M., Poddighe, P. J., Hoovers, J. M., et al. (1995). Expression cloning and chromosomal mapping of the leukocyte activation antigen CD97, a new seven-span transmembrane molecule of the secretion receptor superfamily with an unusual extracellular domain. *Journal of Immunology*, *155*, 1942–1950.

Hamann, J., Hsiao, C.-C., Lee, C. S., Ravichandran, K. S., & Lin, H.-H. (2016). Adhesion GPCRs as modulators of immune cell function. *Handbook of Experimental Pharmacology*, *234*, 329–350.

Hamann, J., Koning, N., Pouwels, W., Ulfman, L. H., van Eijk, M., Stacey, M., et al. (2007). EMR1, the human homolog of F4/80, is an eosinophil-specific receptor. *European Journal of Immunology*, *37*, 2797–2802.

Hamann, J., Kwakkenbos, M. J., de Jong, E. C., Heus, H., Olsen, A. S., & van Lier, R. A. W. (2003). Inactivation of the EGF-TM7 receptor EMR4 after the Pan-Homo divergence. *European Journal of Immunology*, *33*, 1365–1371.

Hamann, J., & Petrenko, A. G. (2016). Introduction: History of the adhesion GPCR field. *Handbook of Experimental Pharmacology*, *234*, 1–11.

Hamann, J., Veninga, H., de Groot, D. M., Visser, L., Hofstra, C. L., Tak, P. P., et al. (2010). CD97 in leukocyte trafficking. *Advances in Experimental Medicine and Biology* *706*, 128–137.

Hamann, J., Vogel, B., van Schijndel, G. M., & van Lier, R. A. (1996). The seven-span transmembrane receptor CD97 has a cellular ligand (CD55, DAF). *The Journal of Experimental Medicine*, *184*, 1185–1189.

Hamann, J., Wishaupt, J. O., van Lier, R. A., Smeets, T. J., Breedveld, F. C., & Tak, P. P. (1999). Expression of the activation antigen CD97 and its ligand CD55 in rheumatoid synovial tissue. *Arthritis and Rheumatism*, *42*, 650–658.

Harvey, R. C., Mullighan, C. G., Wang, X., Dobbin, K. K., Davidson, G. S., Bedrick, E. J., et al. (2010). Identification of novel cluster groups in pediatric high-risk B-precursor acute lymphoblastic leukemia with gene expression profiling: Correlation with genome-wide DNA copy number alterations, clinical characteristics, and outcome. *Blood*, *116*, 4874–4884.

Hepworth, M. R., & Sonnenberg, G. F. (2014). Regulation of the adaptive immune system by innate lymphoid cells. *Current Opinion in Immunology*, *27*, 75–82.

Hey, Y.-Y., & O'Neill, H. C. (2016). Antigen presenting properties of a myeloid dendritic-like cell in murine spleen. *PLoS One*, *11*, e0162358.

Hochreiter-Hufford, A. E., Lee, C. S., Kinchen, J. M., Sokolowski, J. D., Arandjelovic, S., Call, J. A., et al. (2013). Phosphatidylserine receptor BAI1 and apoptotic cells as new promoters of myoblast fusion. *Nature*, *497*, 263–267.

Hoek, R. M., de Launay, D., Kop, E. N., Yilmaz-Elis, A. S., Lin, F., Reedquist, K. A., et al. (2010). Deletion of either CD55 or CD97 ameliorates arthritis in mouse models. *Arthritis and Rheumatism*, *62*, 1036–1042.

Hsiao, C.-C., Cheng, K.-F., Chen, H.-Y., Chou, Y.-H., Stacey, M., Chang, G.-W., et al. (2009). Site-specific N-glycosylation regulates the GPS auto-proteolysis of CD97. *FEBS Letters*, *583*, 3285–3290.

Huang, Y.-S., Chiang, N.-Y., Hu, C.-H., Hsiao, C.-C., Cheng, K.-F., Tsai, W.-P., et al. (2012). Activation of myeloid cell-specific adhesion class G protein-coupled receptor EMR2 via ligation-induced translocation and interaction of receptor subunits in lipid raft microdomains. *Molecular and Cellular Biology*, *32*, 1408–1420.

Huang, C.-H., Jeng, W.-J., Ho, Y.-P., Teng, W., Hsieh, Y.-C., Chen, W.-T., et al. (2016). Increased EMR2 expression on neutrophils correlates with disease severity and predicts overall mortality in cirrhotic patients. *Scientific Reports*, *6*, 38250.

Hulpiau, P., Gul, I. S., & van Roy, F. (2013). New insights into the evolution of metazoan cadherins and catenins. *Progress in Molecular Biology and Translational Science*, *116*, 71–94.

Janeway, C. A., Traver, P., Walport, M., & Shlomchik, M. J. (2001). *Immunobiology* (5th ed.). . New York: Garland Science.

Karpus, O. N., Veninga, H., Hoek, R. M., Flierman, D., van Buul, J. D., Vandenakker, C. C., et al. (2013). Shear stress-dependent downregulation of the adhesion-G protein-coupled receptor CD97 on circulating leukocytes upon contact with its ligand CD55. *Journal of Immunology*, *190*, 3740–3748.

Kaur, B., Brat, D. J., Devi, N. S., & Van Meir, E. G. (2005). Vasculostatin, a proteolytic fragment of brain angiogenesis inhibitor 1, is an antiangiogenic and antitumorigenic factor. *Oncogene*, *24*, 3632–3642.

Kaur, B., Cork, S. M., Sandberg, E. M., Devi, N. S., Zhang, Z., Klenotic, P. A., et al. (2009). Vasculostatin inhibits intracranial glioma growth and negatively regulates in vivo angiogenesis through a CD36-dependent mechanism. *Cancer Research*, *69*, 1212–1220.

Koh, J. T., Kook, H., Kee, H. J., Seo, Y.-W., Jeong, B. C., Lee, J. H., et al. (2004). Extracellular fragment of brain-specific angiogenesis inhibitor 1 suppresses endothelial cell proliferation by blocking alphavbeta5 integrin. *Experimental Cell Research*, *294*, 172–184.

Kop, E. N., Adriaansen, J., Smeets, T. J. M., Vervoordeldonk, M. J., van Lier, R. A. W., Hamann, J., et al. (2006). CD97 neutralisation increases resistance to collagen-induced arthritis in mice. *Arthritis Research & Therapy*, *8*, R155.

Kop, E. N., Kwakkenbos, M. J., Teske, G. J. D., Kraan, M. C., Smeets, T. J., Stacey, M., et al. (2005). Identification of the epidermal growth factor-TM7 receptor EMR2 and its ligand dermatan sulfate in rheumatoid synovial tissue. *Arthritis and Rheumatism*, *52*, 442–450.

Kop, E. N., Matmati, M., Pouwels, W., Leclercq, G., Tak, P. P., & Hamann, J. (2009). Differential expression of CD97 on human lymphocyte subsets and limited effect of CD97 antibodies on allogeneic T-cell stimulation. *Immunology Letters*, *123*, 160–168.

Kovacs, P., & Schöneberg, T. (2016). The relevance of genomic signatures at adhesion GPCR loci in humans. *Handbook of Experimental Pharmacology*, *234*, 179–217.

Krasnoperov, V., Lu, Y., Buryanovsky, L., Neubert, T. A., Ichtchenko, K., & Petrenko, A. G. (2002). Post-translational proteolytic processing of the calcium-independent receptor of alpha-latrotoxin (CIRL), a natural chimera of the cell adhesion protein and the G protein-coupled receptor. Role of the G protein-coupled receptor proteolysis site (GPS) motif. *The Journal of Biological Chemistry*, *277*, 46518–46526.

Krishnan, A., Almén, M. S., Fredriksson, R., & Schiöth, H. B. (2012). The origin of GPCRs: Identification of mammalian like rhodopsin, adhesion, glutamate and frizzled GPCRs in fungi. *PLoS One*, *7*, e29817.

Krishnan, A., Nijmeijer, S., de Graaf, C., & Schiöth, H. B. (2016). Classification, nomenclature, and structural aspects of adhesion GPCRs. *Handbook of Experimental Pharmacology*, *234*, 15–41.

Küffer, A., Lakkaraju, A. K. K., Mogha, A., Petersen, S. C., Airich, K., Doucerain, C., et al. (2016). The prion protein is an agonistic ligand of the G protein-coupled receptor Adgrg6. *Nature*, *536*, 464–468.

Kumagai, J., Hsu, S. Y., Matsumi, H., Roh, J.-S., Fu, P., Wade, J. D., et al. (2002). INSL3/Leydig insulin-like peptide activates the LGR8 receptor important in testis descent. *The Journal of Biological Chemistry*, *277*, 31283–31286.

Kwakkenbos, M. J., Kop, E. N., Stacey, M., Matmati, M., Gordon, S., Lin, H.-H., et al. (2004). The EGF-TM7 family: A postgenomic view. *Immunogenetics*, *55*, 655–666.

Lagerström, M. C., & Schiöth, H. B. (2008). Structural diversity of G protein-coupled receptors and significance for drug discovery. *Nature Reviews. Drug Discovery*, *7*, 339–357.

Langenhan, T., Aust, G., & Hamann, J. (2013). Sticky signaling—Adhesion class G protein-coupled receptors take the stage. *Science Signaling*, *6*, re3.

Langenhan, T., Prömel, S., Mestek, L., Esmaeili, B., Waller-Evans, H., Hennig, C., et al. (2009). Latrophilin signaling links anterior-posterior tissue polarity and oriented cell divisions in the C. elegans embryo. *Developmental Cell*, *17*, 494–504.

Lee, J.-W., Huang, B. X., Kwon, H., Rashid, M. A., Kharebava, G., Desai, A., et al. (2016). Orphan GPR110 (ADGRF1) targeted by N-docosahexaenoylethanolamine in development of neurons and cognitive function. *Nature Communications*, *7*, 13123.

Lee, C. S., Penberthy, K. K., Wheeler, K. M., Juncadella, I. J., Vandenabeele, P., Lysiak, J. J., et al. (2016). Boosting apoptotic cell clearance by colonic epithelial cells attenuates inflammation in vivo. *Immunity*, *44*, 807–820.

Leemans, J. C., te Velde, A. A., Florquin, S., Bennink, R. J., de Bruin, K., van Lier, R. A. W., et al. (2004). The epidermal growth factor-seven transmembrane (EGF-TM7) receptor CD97 is required for neutrophil migration and host defense. *Journal of Immunology*, *172*, 1125–1131.

Legrand, F., Tomasevic, N., Simakova, O., Lee, C.-C. R., Wang, Z., Raffeld, M., et al. (2014). The eosinophil surface receptor epidermal growth factor-like module containing mucin-like hormone receptor 1 (EMR1): A novel therapeutic target for eosinophilic disorders. *The Journal of Allergy and Clinical Immunology*, *133*, 1439–1447. 1447.e1–8.

Levitin, F., Stern, O., Weiss, M., Gil-Henn, C., Ziv, R., Prokocimer, Z., et al. (2005). The MUC1 SEA module is a self-cleaving domain. *The Journal of Biological Chemistry*, *280*, 33374–33386.

Lewis, S. M., Treacher, D. F., Edgeworth, J., Mahalingam, G., Brown, C. S., Mare, T. A., et al. (2015). Expression of CD11c and EMR2 on neutrophils: Potential diagnostic biomarkers for sepsis and systemic inflammation. *Clinical and Experimental Immunology, 182,* 184–194.

Li, X., Roszko, I., Sepich, D. S., Ni, M., Hamm, H. E., Marlow, F. L., et al. (2013). Gpr125 modulates Dishevelled distribution and planar cell polarity signaling. *Development, 140,* 3028–3039.

Liebscher, I., Ackley, B., Araç, D., Ariestanti, D. M., Aust, G., Bae, B.-I., et al. (2014). New functions and signaling mechanisms for the class of adhesion G protein-coupled receptors. Presented at the *Annals of the New York Academy of Sciences, 1333,* 43–64.

Liebscher, I., Schön, J., Petersen, S. C., Fischer, L., Auerbach, N., Demberg, L. M., et al. (2014). A tethered agonist within the ectodomain activates the adhesion G protein-coupled receptors GPR126 and GPR133. *Cell Reports, 9,* 2018–2026.

Liebscher, I., & Schöneberg, T. (2016). Tethered Agonism: A common activation mechanism of adhesion GPCRs. *Handbook of Experimental Pharmacology, 234,* 111–125.

Liebscher, I., Schöneberg, T., & Prömel, S. (2013). Progress in demystification of adhesion G protein-coupled receptors. *Biological Chemistry, 394,* 937–950.

Lin, H.-H., Chang, G.-W., Davies, J. Q., Stacey, M., Harris, J., & Gordon, S. (2004). Autocatalytic cleavage of the EMR2 receptor occurs at a conserved G protein-coupled receptor proteolytic site motif. *The Journal of Biological Chemistry, 279,* 31823–31832.

Lin, H.-H., Faunce, D. E., Stacey, M., Terajewicz, A., Nakamura, T., Zhang-Hoover, J., et al. (2005). The macrophage F4/80 receptor is required for the induction of antigen-specific efferent regulatory T cells in peripheral tolerance. *The Journal of Experimental Medicine, 201,* 1615–1625.

Lin, H.-H., Stacey, M., Yona, S., & Chang, G.-W. (2010). GPS proteolytic cleavage of adhesion-GPCRs. *Advances in Experimental Medicine and Biology, 706,* 49–58.

Lin, H. H., Stubbs, L. J., & Mucenski, M. L. (1997). Identification and characterization of a seven transmembrane hormone receptor using differential display. *Genomics, 41,* 301–308.

Liu, X., Valentine, S. J., Plasencia, M. D., Trimpin, S., Naylor, S., & Clemmer, D. E. (2007). Mapping the human plasma proteome by SCX-LC-IMS-MS. *Journal of the American Society for Mass Spectrometry, 18,* 1249–1264.

Loveless, T., & Hardin, J. (2012). Cadherin complexity: Recent insights into cadherin superfamily function in C. elegans. *Current Opinion in Cell Biology, 24,* 695–701.

Luo, J., Zhou, W., Zhou, X., Li, D., Weng, J., Yi, Z., et al. (2009). Regulation of bone formation and remodeling by G-protein-coupled receptor 48. *Development, 136,* 2747–2756.

Macao, B., Johansson, D. G. A., Hansson, G. C., & Härd, T. (2006). Autoproteolysis coupled to protein folding in the SEA domain of the membrane-bound MUC1 mucin. *Nature Structural & Molecular Biology, 13,* 71–76.

Maiga, A., Lemieux, S., Pabst, C., Lavallée, V.-P., Bouvier, M., Sauvageau, G., et al. (2016). Transcriptome analysis of G protein-coupled receptors in distinct genetic subgroups of acute myeloid leukemia: Identification of potential disease-specific targets. *Blood Cancer Journal, 6,* e431.

Martin, G. H., Desrichard, A., Chung, S. S., Woolthuis, C., Hu, W., Garrett-Bakelman, F. E., et al. (2016). CD97 is a critical regulator of acute myeloid leukemia stem cell function. *Blood, 128,* 1077.

Matmati, M., Pouwels, W., van Bruggen, R., Jansen, M., Hoek, R. M., Verhoeven, A. J., et al. (2007). The human EGF-TM7 receptor EMR3 is a marker for mature granulocytes. *Journal of Leukocyte Biology, 81,* 440–448.

Mazaheri, F., Breus, O., Durdu, S., Haas, P., Wittbrodt, J., Gilmour, D., et al. (2014). Distinct roles for BAI1 and TIM-4 in the engulfment of dying neurons by microglia. *Nature Communications, 5,* 4046.

McGee, J., Goodyear, R. J., McMillan, D. R., Stauffer, E. A., Holt, J. R., Locke, K. G., et al. (2006). The very large G-protein-coupled receptor VLGR1: A component of the ankle link complex required for the normal development of auditory hair bundles. *The Journal of Neuroscience*, *26*, 6543–6553.

McKenzie, A. N. J., Spits, H., & Eberl, G. (2014). Innate lymphoid cells in inflammation and immunity. *Immunity*, *41*, 366–374.

McKnight, A. J., Macfarlane, A. J., Dri, P., Turley, L., Willis, A. C., & Gordon, S. (1996). Molecular cloning of F4/80, a murine macrophage-restricted cell surface glycoprotein with homology to the G-protein-linked transmembrane 7 hormone receptor family. *The Journal of Biological Chemistry*, *271*, 486–489.

Meyer, M., Kircher, M., Gansauge, M.-T., Li, H., Racimo, F., Mallick, S., et al. (2012). A high-coverage genome sequence from an archaic Denisovan individual. *Science*, *338*, 222–226.

Mogha, A., Benesh, A. E., Patra, C., Engel, F. B., Schöneberg, T., Liebscher, I., et al. (2013). Gpr126 functions in Schwann cells to control differentiation and myelination via G-protein activation. *The Journal of Neuroscience*, *33*, 17976–17985.

Monk, K. R., Hamann, J., Langenhan, T., Nijmeijer, S., Schöneberg, T., & Liebscher, I. (2015). Adhesion G protein-coupled receptors: From in vitro pharmacology to in vivo mechanisms. *Molecular Pharmacology*, *88*, 617–623.

Müller, A., Winkler, J., Fiedler, F., Sastradihardja, T., Binder, C., Schnabel, R., et al. (2015). Oriented cell division in the C. elegans embryo is coordinated by G-protein signaling dependent on the adhesion GPCR LAT-1. *PLoS Genetics*, *11*, e1005624.

Ng, S. W. K., Mitchell, A., Kennedy, J. A., Chen, W. C., McLeod, J., Ibrahimova, N., et al. (2016). A 17-gene stemness score for rapid determination of risk in acute leukaemia. *Nature*, *540*, 433–437.

Nieberler, M., Kittel, R. J., Petrenko, A. G., Lin, H.-H., & Langenhan, T. (2016). Control of adhesion GPCR function through proteolytic processing. *Handbook of Experimental Pharmacology*, *234*, 83–109.

Nishimori, H., Shiratsuchi, T., Urano, T., Kimura, Y., Kiyono, K., Tatsumi, K., et al. (1997). A novel brain-specific p53-target gene, BAI1, containing thrombospondin type 1 repeats inhibits experimental angiogenesis. *Oncogene*, *15*, 2145–2150.

Nishimura, T., Honda, H., & Takeichi, M. (2012). Planar cell polarity links axes of spatial dynamics in neural-tube closure. *Cell*, *149*, 1084–1097.

Nordström, K. J. V., Lagerström, M. C., Wallér, L. M. J., Fredriksson, R., & Schiöth, H. B. (2009). The secretin GPCRs descended from the family of adhesion GPCRs. *Molecular Biology and Evolution*, *26*, 71–84.

Nordström, K. J. V., Sällman Almén, M., Edstam, M. M., Fredriksson, R., & Schiöth, H. B. (2011). Independent HHsearch, Needleman—Wunsch-based, and motif analyses reveal the overall hierarchy for most of the G protein-coupled receptor families. *Molecular Biology and Evolution*, *28*, 2471–2480.

Paavola, K. J., Sidik, H., Zuchero, J. B., Eckart, M., & Talbot, W. S. (2014) Type IV collagen is an activating ligand for the adhesion G protein-coupled receptor GPR126. *Science Signaling*, 7, ra76.

Paavola, K. J., Stephenson, J. R., Ritter, S. L., Alter, S. P., & Hall, R. A. (2011). The N terminus of the adhesion G protein-coupled receptor GPR56 controls receptor signaling activity. *The Journal of Biological Chemistry*, *286*, 28914–28921.

Pabst, C., Bergeron, A., Lavallée, V.-P., Yeh, J., Gendron, P., Norddahl, G. L., et al. (2016). GPR56 identifies primary human acute myeloid leukemia cells with high repopulating potential in vivo. *Blood*, *127*, 2018–2027.

Park, D., Tosello-Trampont, A.-C., Elliott, M. R., Lu, M., Haney, L. B., Ma, Z., et al. (2007). BAI1 is an engulfment receptor for apoptotic cells upstream of the ELMO/Dock180/Rac module. *Nature*, *450*, 430–434.

Patra, C., van Amerongen, M. J., Ghosh, S., Ricciardi, F., Sajjad, A., Novoyatleva, T., et al. (2013). Organ-specific function of adhesion G protein-coupled receptor GPR126 is domain-dependent. *Proceedings of the National Academy of Sciences of the United States of America*, *110*, 16898–16903.

Peng, Y.-M., van de Garde, M. D. B., Cheng, K.-F., Baars, P. A., Remmerswaal, E. B. M., van Lier, R. A. W., et al. (2011). Specific expression of GPR56 by human cytotoxic lymphocytes. *Journal of Leukocyte Biology*, *90*, 735–740.

Peters, M. J., Joehanes, R., Pilling, L. C., Schurmann, C., Conneely, K. N., Powell, J., et al. (2015). The transcriptional landscape of age in human peripheral blood. *Nature Communications*, *6*, 8570.

Petersen, S. C., Luo, R., Liebscher, I., Giera, S., Jeong, S.-J., Mogha, A., et al. (2015). The adhesion GPCR GPR126 has distinct, domain-dependent functions in Schwann cell development mediated by interaction with Laminin-211. *Neuron*, *85*, 755–769.

Posokhova, E., Shukla, A., Seaman, S., Volate, S., Hilton, M. B., Wu, B., et al. (2015). GPR124 functions as a WNT7-specific coactivator of canonical β-catenin signaling. *Cell Reports*, *10*, 123–130.

Prömel, S., Frickenhaus, M., Hughes, S., Mestek, L., Staunton, D., Woollard, A., et al. (2012a). The GPS motif is a molecular switch for bimodal activities of adhesion class G protein-coupled receptors. *Cell Reports*, *2*, 321–331.

Prömel, S., Langenhan, T., & Araç, D. (2013). Matching structure with function: The GAIN domain of adhesion-GPCR and PKD1-like proteins. *Trends in Pharmacological Sciences*, *34*, 470–478.

Prömel, S., Waller-Evans, H., Dixon, J., Zahn, D., Colledge, W. H., Doran, J., et al. (2012b). Characterization and functional study of a cluster of four highly conserved orphan adhesion-GPCR in mouse. *Developmental Dynamics*, *241*, 1591–1602.

Prüfer, K., Racimo, F., Patterson, N., Jay, F., Sankararaman, S., Sawyer, S., et al. (2014). The complete genome sequence of a Neanderthal from the Altai Mountains. *Nature*, *505*, 43–49.

Reya, T., Duncan, A. W., Ailles, L., Domen, J., Scherer, D. C., Willert, K., et al. (2003). A role for Wnt signalling in self-renewal of haematopoietic stem cells. *Nature*, *423*, 409–414.

Sachs, C., Robinson, B. D., Andres Martin, L., Webster, T., Gilbert, M., Lo, H.-Y., et al. (2014). Evaluation of candidate spermatogonial markers ID4 and GPR125 in testes of adult human cadaveric organ donors. *Andrology*, *2*, 607–614.

Saito, Y., Kaneda, K., Suekane, A., Ichihara, E., Nakahata, S., Yamakawa, N., et al. (2013). Maintenance of the hematopoietic stem cell pool in bone marrow niches by EVI1-regulated GPR56. *Leukemia*, *27*, 1637–1649.

Salzman, G. S., Ackerman, S. D., Ding, C., Koide, A., Leon, K., Luo, R., et al. (2016). Structural basis for regulation of GPR56/ADGRG1 by its alternatively spliced extracellular domains. *Neuron*, *91*, 1292–1304.

Schaller, E., Macfarlane, A. J., Rupec, R. A., Gordon, S., McKnight, A. J., & Pfeffer, K. (2002). Inactivation of the F4/80 glycoprotein in the mouse germ line. *Molecular and Cellular Biology*, *22*, 8035–8043.

Scheller, M., Huelsken, J., Rosenbauer, F., Taketo, M. M., Birchmeier, W., Tenen, D. G., et al. (2006). Hematopoietic stem cell and multilineage defects generated by constitutive beta-catenin activation. *Nature Immunology*, 7, 1037–1047.

Schiöth, H. B., Nordström, K. J. V., & Fredriksson, R. (2010). The adhesion GPCRs; gene repertoire, phylogeny and evolution. *Advances in Experimental Medicine and Biology*, *706*, 1–13.

Scholz, N., Gehring, J., Guan, C., Ljaschenko, D., Fischer, R., Lakshmanan, V., et al. (2015). The adhesion GPCR latrophilin/CIRL shapes mechanosensation. *Cell Reports*, *11*, 866–874.

Scholz, N., Monk, K. R., Kittel, R. J., & Langenhan, T. (2016). Adhesion GPCRs as a putative class of metabotropic mechanosensors. *Handbook of Experimental Pharmacology*, *234*, 221–247.

Schöneberg, T., Kleinau, G., & Brüser, A. (2016). What are they waiting for?—Tethered agonism in G protein-coupled receptors. *Pharmacological Research*, *108*, 9–15.

Schöneberg, T., Liebscher, I., Luo, R., Monk, K. R., & Piao, X. (2015). Tethered agonists: A new mechanism underlying adhesion G protein-coupled receptor activation. *Journal of Receptor and Signal Transduction Research*, *35*, 220–223.

Seandel, M., James, D., Shmelkov, S. V., Falciatori, I., Kim, J., Chavala, S., et al. (2007). Generation of functional multipotent adult stem cells from GPR125+ germline progenitors. *Nature*, *449*, 346–350.

Shi, J.-P., Li, X.-N., Zhang, X.-Y., Du, B., Jiang, W.-Z., Liu, M.-Y., et al. (2015). Gpr97 is dispensable for inflammation in OVA-induced asthmatic mice. *PLoS One*, *10*, e0131461.

Shi, J., Zhang, X., Wang, S., Wang, J., Du, B., Wang, Z., et al. (2016). Gpr97 is dispensable for metabolic syndrome but is involved in macrophage inflammation in high-fat diet-induced obesity in mice. *Scientific Reports*, *6*, 24649.

Solaimani Kartalaei, P., Yamada-Inagawa, T., Vink, C. S., de Pater, E., van der Linden, R., Marks-Bluth, J., et al. (2015). Whole-transcriptome analysis of endothelial to hematopoietic stem cell transition reveals a requirement for Gpr56 in HSC generation. *The Journal of Experimental Medicine*, *212*, 93–106.

Sonnenberg, G. F., & Artis, D. (2015). Innate lymphoid cells in the initiation, regulation and resolution of inflammation. *Nature Medicine*, *21*, 698–708.

Spendlove, I., & Sutavani, R. (2010). The role of CD97 in regulating adaptive T-cell responses. *Advances in Experimental Medicine and Biology*, *706*, 138–148.

Spits, H., Bernink, J. H., & Lanier, L. (2016). NK cells and type 1 innate lymphoid cells: Partners in host defense. *Nature Immunology*, *17*, 758–764.

Stacey, M., Chang, G.-W., Davies, J. Q., Kwakkenbos, M. J., Sanderson, R. D., Hamann, J., et al. (2003). The epidermal growth factor-like domains of the human EMR2 receptor mediate cell attachment through chondroitin sulfate glycosaminoglycans. *Blood*, *102*, 2916–2924.

Stacey, M., Chang, G.-W., Sanos, S. L., Chittenden, L. R., Stubbs, L., Gordon, S., et al. (2002). EMR4, a novel epidermal growth factor (EGF)-TM7 molecule up-regulated in activated mouse macrophages, binds to a putative cellular ligand on B lymphoma cell line A20. *The Journal of Biological Chemistry*, *277*, 29283–29293.

Stacey, M., Lin, H. H., Hilyard, K. L., Gordon, S., & McKnight, A. J. (2001). Human epidermal growth factor (EGF) module-containing mucin-like hormone receptor 3 is a new member of the EGF-TM7 family that recognizes a ligand on human macrophages and activated neutrophils. *The Journal of Biological Chemistry*, *276*, 18863–18870.

Stoveken, H. M., Bahr, L. L., Anders, M. W., Wojtovich, A. P., Smrcka, A. V., & Tall, G. G. (2016). Dihydromunduletone is a small-molecule selective adhesion G protein-coupled receptor antagonist. *Molecular Pharmacology*, *90*, 214–224.

Stoveken, H. M., Hajduczok, A. G., Xu, L., & Tall, G. G. (2015). Adhesion G protein-coupled receptors are activated by exposure of a cryptic tethered agonist. *Proceedings of the National Academy of Sciences of the United States of America*, *112*, 6194–6199.

Sumbayev, V. V., Gonçalves Silva, I., Blackburn, J., Gibbs, B. F., Yasinska, I. M., Garrett, M. D., et al. (2016). Expression of functional neuronal receptor latrophilin 1 in human acute myeloid leukaemia cells. *Oncotarget*, *7*, 45575–45583.

Sutavani, R. V., Bradley, R. G., Ramage, J. M., Jackson, A. M., Durrant, L. G., & Spendlove, I. (2013). CD55 costimulation induces differentiation of a discrete T regulatory type 1 cell population with a stable phenotype. *Journal of Immunology*, *191*, 5895–5903.

Terry, S. Y. A., Boerman, O. C., Gerrits, D., Franssen, G. M., Metselaar, J. M., Lehmann, S., et al. (2015). $^{111}$In-anti-F4/80-A3-1 antibody: A novel tracer to image macrophages. *European Journal of Nuclear Medicine and Molecular Imaging*, *42*, 1430–1438.

Ting, S. B., Deneault, E., Hope, K., Cellot, S., Chagraoui, J., Mayotte, N., et al. (2012). Asymmetric segregation and self-renewal of hematopoietic stem and progenitor cells with endocytic Ap2a2. *Blood*, *119*, 2510–2522.

Tseng, C.-T. K., & Klimpel, G. R. (2002). Binding of the hepatitis C virus envelope protein E2 to CD81 inhibits natural killer cell functions. *The Journal of Experimental Medicine*, *195*, 43–49.

Valtcheva, N., Primorac, A., Jurisic, G., Hollmén, M., & Detmar, M. (2013). The orphan adhesion G protein-coupled receptor GPR97 regulates migration of lymphatic endothelial cells via the small GTPases RhoA and Cdc42. *The Journal of Biological Chemistry*, *288*, 35736–35748.

van Eijk, M., Aust, G., Brouwer, M. S. M., van Meurs, M., Voerman, J. S. A., Dijke, I. E., et al. (2010). Differential expression of the EGF-TM7 family members CD97 and EMR2 in lipid-laden macrophages in atherosclerosis, multiple sclerosis and Gaucher disease. *Immunology Letters*, *129*, 64–71.

Vanhollebeke, B., Stone, O. A., Bostaille, N., Cho, C., Zhou, Y., Maquet, E., et al. (2015). Tip cell-specific requirement for an atypical Gpr124- and Reck-dependent Wnt/β-catenin pathway during brain angiogenesis. *eLife*, *4*, 2807.

Veninga, H., de Groot, D. M., McCloskey, N., Owens, B. M., Dessing, M. C., Verbeek, J. S., et al. (2011). CD97 antibody depletes granulocytes in mice under conditions of acute inflammation via a Fc receptor-dependent mechanism. *Journal of Leukocyte Biology*, *89*(3), 413–421.

Veninga, H., Hoek, R. M., de Vos, A. F., de Bruin, A. M., An, F.-Q., van der Poll, T., et al. (2011). A novel role for CD55 in granulocyte homeostasis and anti-bacterial host defense. *PLoS One*, *6*, e24431.

Visser, L., de Vos, A. F., Hamann, J., Melief, M.-J., van Meurs, M., van Lier, R. A. W., et al. (2002). Expression of the EGF-TM7 receptor CD97 and its ligand CD55 (DAF) in multiple sclerosis. *Journal of Neuroimmunology*, *132*, 156–163.

Wandel, E., Saalbach, A., Sittig, D., Gebhardt, C., & Aust, G. (2012). Thy-1 (CD90) is an interacting partner for CD97 on activated endothelial cells. *Journal of Immunology*, *188*, 1442–1450.

Wang, W., Jossin, Y., Chai, G., Lien, W.-H., Tissir, F., & Goffinet, A. M. (2016). Feedback regulation of apical progenitor fate by immature neurons through Wnt7-Celsr3-Fzd3 signalling. *Nature Communications*, 7, 10936.

Wang, T., Tian, L., Haino, M., Gao, J.-L., Lake, R., Ward, Y., et al. (2007). Improved antibacterial host defense and altered peripheral granulocyte homeostasis in mice lacking the adhesion class G protein receptor CD97. *Infection and Immunity*, *75*, 1144–1153.

Wang, T., Ward, Y., Tian, L., Lake, R., Guedez, L., Stetler-Stevenson, W. G., et al. (2005). CD97, an adhesion receptor on inflammatory cells, stimulates angiogenesis through binding integrin counterreceptors on endothelial cells. *Blood*, *105*, 2836–2844.

Wang, J.-J., Zhang, L.-L., Zhang, H.-X., Shen, C.-L., Lu, S.-Y., Kuang, Y., et al. (2013). Gpr97 is essential for the follicular versus marginal zone B-lymphocyte fate decision. *Cell Death & Disease*, *4*, e853.

Warschkau, H., & Kiderlen, A. F. (1999). A monoclonal antibody directed against the murine macrophage surface molecule F4/80 modulates natural immune response to Listeria monocytogenes. *Journal of Immunology*, *163*, 3409–3416.

Weng, J., Luo, J., Cheng, X., Jin, C., Zhou, X., Qu, J., et al. (2008). Deletion of G protein-coupled receptor 48 leads to ocular anterior segment dysgenesis (ASD) through down-regulation of Pitx2. *Proceedings of the National Academy of Sciences of the United States of America*, *105*, 6081–6086.

Wilde, C., Fischer, L., Lede, V., Kirchberger, J., Rothemund, S., Schöneberg, T., et al. (2016). The constitutive activity of the adhesion GPCR GPR114/ADGRG5 is mediated by its tethered agonist. *FASEB Journal, 30*, 666–673.

Wobus, M., Bornhäuser, M., Jacobi, A., Kräter, M., Otto, O., Ortlepp, C., et al. (2015). Association of the EGF-TM7 receptor CD97 expression with FLT3-ITD in acute myeloid leukemia. *Oncotarget, 6*, 38804–38815.

Yang, M. Y., Hilton, M. B., Seaman, S., Haines, D. C., Nagashima, K., Burke, C. M., et al. (2013). Essential regulation of lung surfactant homeostasis by the orphan G protein-coupled receptor GPR116. *Cell Reports, 3*, 1457–1464.

Yona, S., Lin, H.-H., Dri, P., Davies, J. Q., Hayhoe, R. P. G., Lewis, S. M., et al. (2008). Ligation of the adhesion-GPCR EMR2 regulates human neutrophil function. *FASEB Journal, 22*, 741–751.

Yona, S., Lin, H.-H., Siu, W. O., Gordon, S., & Stacey, M. (2008). Adhesion-GPCRs: Emerging roles for novel receptors. *Trends in Biochemical Sciences, 33*, 491–500.

Zhou, Y., & Nathans, J. (2014). Gpr124 controls CNS angiogenesis and blood-brain barrier integrity by promoting ligand-specific canonical wnt signaling. *Developmental Cell, 31*, 248–256.

CHAPTER SIX

# GPCR Signaling in *C. elegans* and Its Implications in Immune Response

**Anjali Gupta, Varsha Singh[1]**
Department of Molecular Reproduction, Development and Genetics, Indian Institute of Science, Bangalore, India
[1]Corresponding author: e-mail address: varsha@mrdg.iisc.ernet.in

## Contents

1. Introduction 204
2. Regulation of Innate Immune Response by the Nervous System 206
   2.1 Organization of the Nervous System in *C. elegans* 206
   2.2 NPR-1 GPCR Signaling Promotes Immune Response to *P. aeruginosa* Infection 207
   2.3 OCTR-1 GPCR Suppresses Immune Response to *P. aeruginosa* Infection 209
   2.4 Beta-Arrestin Signaling During Bacterial Infection 210
   2.5 G-Protein and Insulin Signaling During *P. aeruginosa* Infection 211
   2.6 Dopamine Signaling During *P. aeruginosa* Infection 212
   2.7 G-Protein and Serotonin Signaling During *M. nematophilum* Infection 212
3. GPCR Signaling at the Site of Infection Regulates Innate Immune Responses 213
   3.1 FSHR-1 GPCR Signaling in the Intestine During *P. aeruginosa* Infection 213
   3.2 DCAR-1 GPCR Signaling in the Skin During *D. coniospora* Infection 215
4. Behavioral Response of *C. elegans* to Pathogens 217
   4.1 Aversion Response of *C. elegans* on *P. aeruginosa* Lawn 217
   4.2 Aversion Response of *C. elegans* on *Serratia marcescens* Lawn 220
5. Concluding Remarks 221
Acknowledgments 223
References 223

## Abstract

The ability to sense environmental cues is central to the survival of living organisms. G-protein-coupled receptors (GPCRs) are, by far, the most diverse class of sensory receptors and play an important role in surveillance. As *Caenorhabditis elegans* lives in soil and feeds on bacteria, it must have strategies to differentiate between nutritious vs pathogenic bacteria. In *C. elegans*, lacking professional immune cells, GPCRs play a very important role in defense responses, for survival against pathogens. Here, we review a rich body of research to show that *C. elegans* uses GPCRs in different tissues for immune surveillance, immune homeostasis, as well as behavioral responses. Nematode sensory

*Advances in Immunology*, Volume 136
ISSN 0065-2776
http://dx.doi.org/10.1016/bs.ai.2017.05.002

neurons and GPCRs can sense both pathogen-associated molecules as well as damage-associated molecular patterns during infection. Both fight and flight responses, activated upon exposure to pathogens, are driven by GPCRs and trimeric G proteins in the sensory neurons.

## 1. INTRODUCTION

*Caenorhabditis elegans* is a soil dwelling, free living nematode. It derives nutrition by feeding upon microorganisms in the soil and decaying vegetation. The laboratory food of *C. elegans* is monoxenic culture of *Escherichia coli* which is efficiently disrupted by a muscular grinder in the pharynx of the worm, such that very few intact bacteria reach the intestinal lumen. However, some bacteria may pass this defense mechanism and proliferate in the lumen of the intestine. These may be pathogenic bacteria or beneficial ones. Recent gut microbiome study of *C. elegans* isolated from different geographical locations indicate that the gut contains a number of bacteria and fungi (Dirksen et al., 2016). Pathogenic bacteria prefer to colonize different tissues of the nematode (Fig. 1). Most microbes colonize the intestine, or the skin tissue, made up of cuticle and an underlying layer of cells called the hypodermis. Certain bacteria such as *Yersinia pestis* can form biofilm over the mouth of the *C. elegans*, while *Microbacterium nematophilum* can adhere to the cuticle in the posterior part of the intestine and cause a defensive epidermal swelling response called deformed anal region or DAR phenotype (Hodgkin, Kuwabara, & Corneliussen, 2000; Joshua et al., 2003). Most other pathogenic bacteria *Pseudomonas aeruginosa, Salmonella enterica, Staphylococcus aureus, Enterococcus faecalis* proliferate in the intestine of the nematode (Garsin et al., 2001; Labrousse, Chauvet, Couillault, Kurz, & Ewbank, 2000;

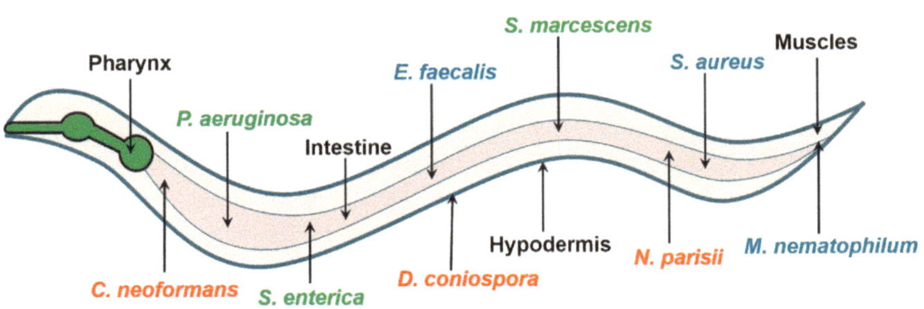


**Fig. 1** *C. elegans* pathogens colonize different tissues. Gram-negative bacteria, *green*; Gram-positive bacteria, *blue*; fungi, *red*.

Tan, Mahajan-Miklos, & Ausubel, 1999). Fungi in the hyphae form can enter the nematode by penetrating the cuticle and hypodermis layer in case of *Drechmeria coniospora* or proliferate in the intestine in yeast form in case of *Cryptococcus neoformans* (Jansson, 1994; Mylonakis, Ausubel, Perfect, Heitman, & Calderwood, 2002). Microsporidia *Nematocida parisii* are fungus-like microbes and cause intracellular invasion of intestinal cells (Troemel, Félix, Whiteman, Barrière, & Ausubel, 2008). Most of the literature reviewed in this chapter is in the context of infection of *C. elegans* by *P. aeruginosa*. *P. aeruginosa* is a rod-shaped, Gram-negative bacterium commonly found in environment. It is a potent human opportunistic pathogen in case of cystic fibrosis, burn wounds, or diabetic foot ulcers. It can infect a wide range of host species including plants, nematodes, flies, fish, and mammals. Most of the virulence factors employed in the host remain conserved, although there may be host-specific virulence mechanisms. *P. aeruginosa* kills *C. elegans* via a number of mechanisms. Fast killing of *C. elegans* is mediated by rapid secretion of toxins such as phenazines and HCN secreted by *Pseudomonas*, while slow killing is mediated by live bacteria that colonizes the worm's gut causing a chronic infection leading to the nematode's death over a period of days (Tan et al., 1999).

Several conserved signaling mechanisms are triggered when *C. elegans* comes in contact with natural or clinically relevant pathogens provided as the food source (Sun, Aballay, & Singh, 2016). The most important of these is p38 mitogen-activated protein (MAP) kinase or PMK-1 signaling pathways which is essential for immune response to all microbes except viruses. p38 MAP kinase is activated at the site of infection, intestine in case of bacteria, or hypodermis in case of hyphae of *D. coniospora*. p38 MAP kinase can turn on overlapping as well as pathogen-specific immune effectors which include antimicrobial peptides (AMP), neuropeptide-like proteins (NLPs), and lectins (Kim et al., 2002; Pujol, Cypowyj, et al., 2008). p38 MAP kinase pathway does not seem to play a role in the clearance of viruses. Natural viral pathogens of *C. elegans* are restricted by the host RNA interference machinery. Insulin signaling receptor DAF-2 and FOXO transcription factor DAF-16 regulate expression of ROS detoxifying enzymes, lysozymes, saposins and also play important role in immune response to bacterial and fungal pathogens, although their site of action is not completely defined (Garsin et al., 2003). ERK MAP kinase is important for hypersensitivity response of *C. elegans* rectal epithelium to *M. nematophilum* (Nicholas & Hodgkin, 2004). *C. elegans wnt* signaling pathway becomes important in response to Gram-positive bacteria *S. aureus* where it turns on production of specific immune effectors expressed in the intestine (Irazoqui, Ng, Xavier, & Ausubel, 2008). Majority of the immune effectors have been found in the pharynx, intestine, or hypodermis and belong to lysozyme, saposins, AMP, NLP, or lectin families.

# 2. REGULATION OF INNATE IMMUNE RESPONSE BY THE NERVOUS SYSTEM

## 2.1 Organization of the Nervous System in *C. elegans*

Innate immune response needs to be activated rapidly in response to pathogen-associated molecular pattern (PAMP) or damage-associated molecular pattern (DAMP) molecules. Innate immunity is also a highly regulated process since too much can lead to inflammation, while too little can led to impaired microbial clearance. Recent reports indicate that the nervous system may be the master regulator of immune homeostasis in *C. elegans* as well in vertebrates. Multicellular animals may have additional cues coming from the site of infection enabling them to exert control on inflammation in a reflexive manner (Chavan & Tracey, 2014). *C. elegans* has a simple nervous system of 302 neurons in a soma of 959 cells. The connectome of *C. elegans* nervous system is defined by thousands of synapses and gap junction allowing one to investigate neural circuits involved in regulation of innate immunity or behavior. Most of the environmental cues are sensed by amphid sensory organ in the head and a phasmid sensory organ in the tail (Fig. 2). The amphid sensory organ is composed of 12 pairs of sensory neurons. Of these eight pairs are chemosensory in nature and can directly access any water soluble cue present in the environment (Inglis, Ou, Leroux, & Scholey, 2007). Three other pairs are olfactory neurons (designated AWA, AWB, AWC) and sense a large range of volatile molecules including secondary metabolites of bacteria (Colbert & Bargmann, 1995). AFD has finger-like cilia and important for thermosensation (Mori & Ohshima, 1995). Amphid neurons are important for food sensing. Another set of sensory neurons, AQR, PQR, URX, have their ciliated endings, enriched in receptors, exposed to the pseudocoelom or false body cavity. Some of these are important for oxygen avoidance and aggregation behavior of the worms (Coates & de Bono, 2002; De Bono & Bargmann, 1998). Pseudocoelom is accessible to all tissues and believed to not only provide hydrostatic structural support to the body but also may serve as vasculature for exchange of information between tissues in an endocrine

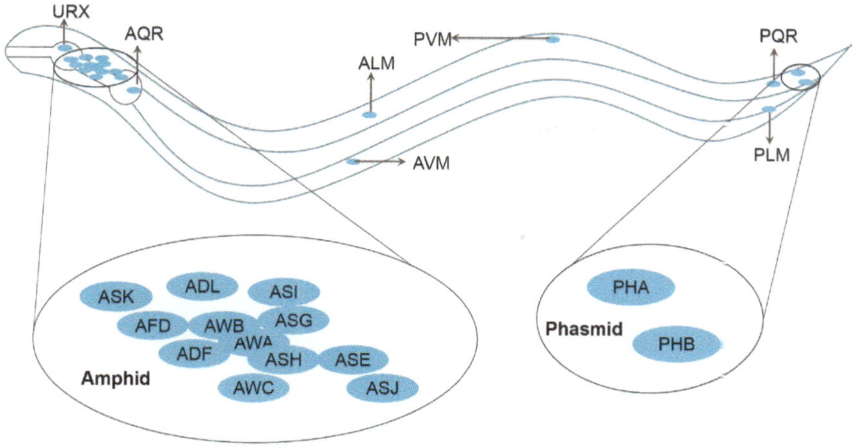


**Fig. 2** Sensory neurons of *C. elegans*. Amphid organ in the head is composed of 12 pairs of chemosensory neurons. The phasmid organ is composed of two pairs of chemosensory neurons known to sense volatile and water-soluble chemicals; AQR, PQR, and URX are known to sense soluble gases like $CO_2$ and $O_2$; ALM (L + R), AVM, PVM, and PLM (L + R) are known to sense mechanical cues.

manner. Ablation of different neurons or perturbation of trimeric G-protein signaling can regulate many physiological processes including adult life span (Alcedo & Kenyon, 2004; Apfeld & Kenyon, 1999; Lans & Jansen, 2007). *C. elegans* neurons secrete numerous peptides of the transforming growth factor-β (TGF-β) family, the insulin family, and neuropeptide family, many of these have antimicrobial properties (Pujol, Zugasti, et al., 2008; Yusuke et al., 2002). These factors can reach other tissues via pseudocoelom and exert their effects on metabolism or immune response. This may lead to noncell autonomous control of immune response by the nervous system.

## 2.2 NPR-1 GPCR Signaling Promotes Immune Response to *P. aeruginosa* Infection

In the first study of its kind, Styer et al. screened 40 mutations affecting *C. elegans* G-protein-coupled receptors (GPCRs) for altered immune responses to *P. aeruginosa* (Styer et al., 2008). While one GPCR positively promoted innate immune responses, three GPCRs suppressed innate immunity in survival assays. Loss of NPR-1, related to mammalian neuropeptide receptor Y, led to susceptibility to infection by *P. aeruginosa, S. enterica* and infection by Gram-positive bacterium *E. faecalis*. This GPCR is expressed in about 20 neurons and is involved in a circuit that responds to environmental oxygen and food. *npr-1* mutant animals prefer low oxygen and show aggregation behavior on *E. coli* lawn under normoxic conditions (21% oxygen) called hyperoxia response or bordering phenotype. This behavior is absent in reference *C. elegans* strain called N2. When N2 and *npr-1* mutant animals were exposed to *P. aeruginosa* in chambers containing 8% oxygen, the *npr-1* mutants were still more susceptible than N2, suggesting a specific role in immune response unrelated to aggregation and bordering. Three neurons, AQR, PQR, and URX, drive the hyperoxia response or aggregation under normoxia. Rescue of NPR-1 function in these neurons can rescue *P. aeruginosa* susceptibility phenotype, suggesting that the oxygen sensing circuit can also regulate the immune response. The transcriptome analysis of N2 and *npr-1* mutant animals shows that many intestinal genes under the control of p38 MAP kinase are downregulated in *npr-1* mutant animals. Although p38 MAP kinase is not transcriptionally regulated by NPR-1 GPCR, the phosphorylation-dependent activation of p38 MAP kinase requires NPR-1 and AQR/PQR/URX neurons. NPR-1 signaling in these neurons also requires guanylyl cyclase GCY-35 and cyclic nucleotide-gated channels TAX-2 and TAX-4 (Fig. 3). This was the first

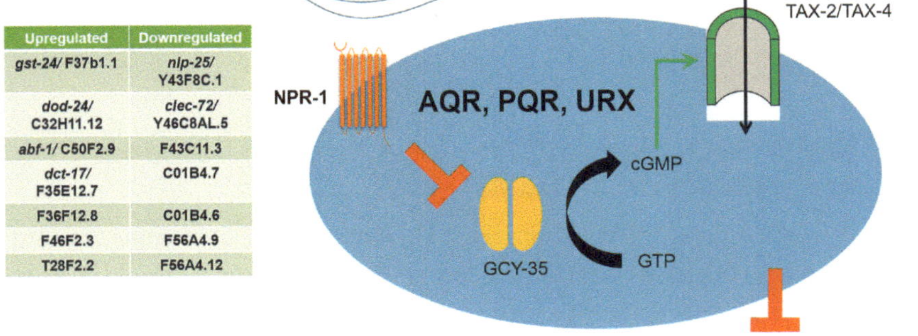


| Upregulated | Downregulated |
|---|---|
| *gst-24/* F37b1.1 | *nlp-25/* Y43F8C.1 |
| *dod-24/* C32H11.12 | *clec-72/* Y46C8AL.5 |
| *abf-1/* C50F2.9 | F43C11.3 |
| *dct-17/* F35E12.7 | C01B4.7 |
| F36F12.8 | C01B4.6 |
| F46F2.3 | F56A4.9 |
| T28F2.2 | F56A4.12 |

**Fig. 3** NPR-1 acts in the AQR, PQR, and URX neurons to promote immune response. NPR-1 suppresses GCY-35 guanylyl cyclase signaling in AQR, PQR, and URX neurons which in turn inhibit immune response in *C. elegans*. NPR-1 regulates activation of p38 MAP kinase and immune effector genes during *P. aeruginosa* infection.

example of a *C. elegans* GPCR controlling immune response. NPR-1 also regulates aversion response of *C. elegans* to *P. aeruginosa* (discussed in Section 4.1.2).

## 2.3 OCTR-1 GPCR Suppresses Immune Response to *P. aeruginosa* Infection

Styer et al. also identified octopamine receptor OCTR-1 as a suppressor of immune response to *P. aeruginosa*. Loss of function of OCTR-1 leads to enhanced resistance to Gram-negative bacteria *P. aeruginosa* as well as *S. enterica*. It is expressed in Amphid sensory neurons ASH, ASI, interneuron AIY, and dopaminergic neurons ADE/CEP. Loss of OCTR-1 function enhances resistance to infection without reducing the bacterial load in the intestine of *C. elegans*. OCTR-1 role in innate immunity depends upon a number of genes from p38 MAP kinase pathway, DAF-2/insulin signaling pathway, and CED-1 cell death pathway. *octr-1* mutant animals have reduced level of phosphorylated p38 MAP kinase which regulates a number of antimicrobial effectors (Fig. 4). OCTR-1 also suppresses a family of pqn (prion-like glutamine[Q] asparagine[*N*] rich domain bearing protein) that are controlled by CED-1. Activity of CED-1 in *octr-1* mutant animals leads to upregulation of pqn/abu genes and resistance to *P. aeruginosa*. This is also the first report of the noncanonical ER stress response pathway including abu/pqn in innate immunity (Sun, Singh, Kajino-Sakamoto, & Aballay, 2011). It is also possible that OCTR-1 GPCR signaling regulated canonical UPR in the endoplasmic reticulum. One of the three canonical UPR pathways is regulated by inositol-requiring protein (IRE) endoribonuclease. IRE removes an intron from the mRNA for UPR transcription factor named X-box binding protein-1 (XBP-1) making it active. During the infection of adult *C. elegans* by *P. aeruginosa*, OCTR-1 suppresses XBP-1 activity. As a result, XBP-1 processing is enhanced in loss of function of OCTR-1 which is also reflected in expression of XBP-1-dependent Hsp70 family ER chaperone HSP-4 (Sun, Liu, & Aballay, 2012). Proteome analysis of animals lacking OCTR-1 by mass spectrometry analysis indicates that this GPCR suppresses many immunity-related proteins, lectin-like molecules, chaperones, and ribosomal proteins (RPs) during infection. RPS-1 and translation factor EIF3, both dysregulated by OCTR-1, support canonical UPR and expression of HSP-4 leading to resistance to *P. aeruginosa* (Liu, Sellegounder, & Sun, 2016). Thus, OCTR-1 appears to suppress immune response in *C. elegans* by inhibiting UPR pathways as well as translation in *C. elegans*.

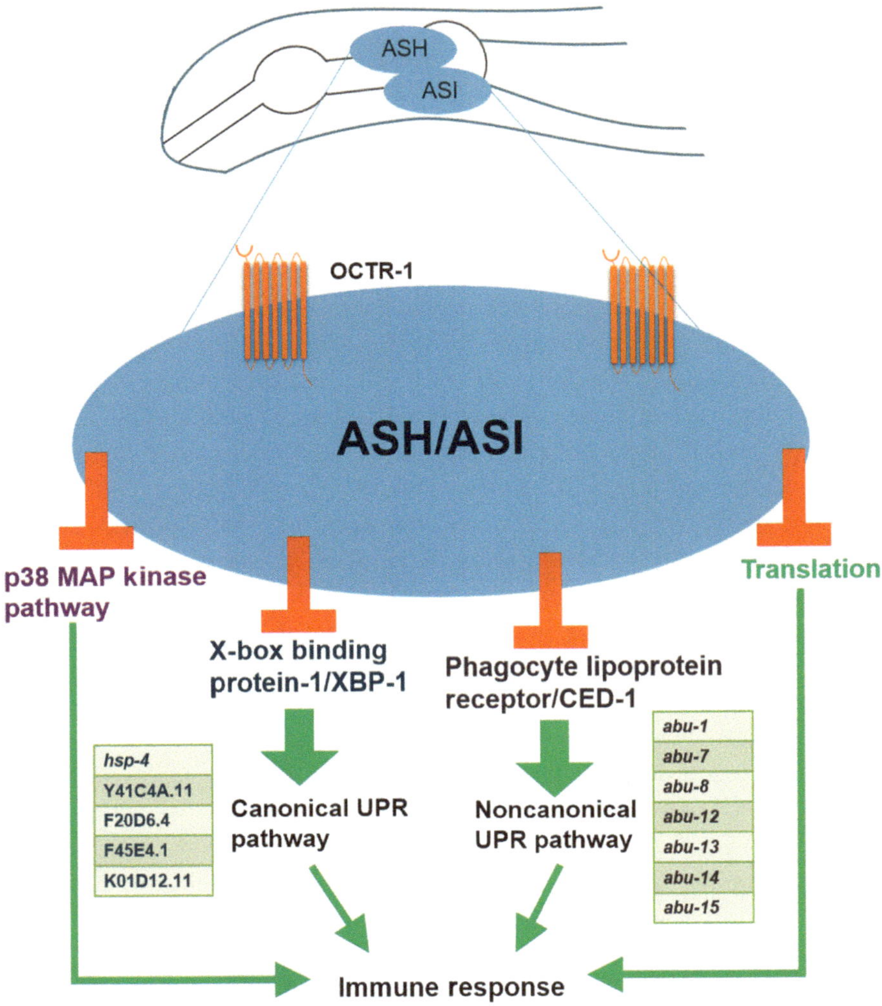


**Fig. 4** OCTR-1 acts in ASH/ASI neurons to suppress immune response. OCTR-1 inhibits p38 MAP kinase activation, DAF-16, and CED-1/lipoprotein receptor. CED-1 controls a noncanonical UPR response composed of pqn/abu genes. OCTR-1 also suppresses canonical UPR pathway controlled by X-box binding protein (XBP-1) and translation machinery. Immune effectors regulated by these pathways are indicated.

## 2.4 Beta-Arrestin Signaling During Bacterial Infection

Beta-arrestin is a adaptor protein for GPCRs, required for receptor desensitization as well as signaling. While human genome has four arrestin proteins, *C. elegans* has only one called ARR-1. *C. elegans* ARR-1 is primarily expressed in the nervous system of *C. elegans* but may also be expressed at lower levels in other tissues. ARR-1 rescue in AQR/PQR/URX neuron and in ASH/ASI neurons shows that immune stimulatory activity of NPR-1 and immune suppressive activity of OCTR-1 is dependent on ARR-1, as expected. ARR-1 activity in thermosensory neuron AFD and chemosensory neurons ADF also contributes to innate immune responses to *P. aeruginosa*. ARR-1-dependent signaling in seven sensory neurons was required for upregulation of noncanonical ER stress response machinery composed of abu/pqn genes (Singh & Aballay, 2012). Arrestin signaling and rescue has similar survival phenotype on *P. aeruginosa* and *S. enterica*, suggesting that neurons identified here may sense bacterial molecules common to Gram-negative bacteria. This study shows a broad role of GPCR signaling in sensory neurons to maintain immune homeostasis in *C. elegans*. Complementation of arr-1 gene in different tissues can be used as an experimental strategy to understand the role of GPCR signaling in individual tissues or cells during infection with Gram-positive bacteria, fungi, and microsporidia.

## 2.5 G-Protein and Insulin Signaling During *P. aeruginosa* Infection

Heterotrimeric G proteins composed of α, β, and γ subunits transduce signals from the membrane through a cycle of guanine nucleotide exchange and hydrolysis. A ligand-bound GPCR acts as a guanine nucleotide exchange factor. *C. elegans* genome has 21 Gα, 2 Gβ, and 2 Gγ encoding genes (Jansen, Thijssen, Werner, Hazendonk, & Plasterk, 1999) expressed in the neurons, muscles, and other tissues of the body. Some of the Gα proteins have expression restricted to a handful of sensory neurons. A typical signaling event in neurons starts with sensing of suitable ligands by receptors on the neuron following which there is a signaling cascade that causes calcium influx and release of neurotransmitter, or neuropeptide secretion via dense core vesicles or DCVs. Studies with defect in neurosecretion enhance the resistance of *C. elegans* to *P. aeruginosa* (Kawli & Tan, 2008). Specifically lack of dense core vesicles secretion due to mutation in calcium activator protein UNC-31 promotes resistance. Dense core vesicles contain peptides, hormones, and serotonin. Neuropeptides are expressed as pre-pro-proteins before processing by protein convertase EGL-3 and carboxypeptidase EGL-21. Loss of EGL-3 and EGL-21 also promotes resistance to infection, suggesting that DCV-mediated neuropeptide secretion control resistance to infection. The DCV mutants have higher expression of AMP such as saposin SPP-1 and ABF-1 which correlated with lower pathogen counts in the intestine of *C. elegans*. In contrast, hypersecretion phenotype due to mutation in Goα subunit encoding gene *goa-1* and diacylglycerol kinase encoding gene *dgk-1* causes susceptibility to *P. aeruginosa* infection. *C. elegans* codes for hundreds of neuropeptide including insulin-like molecules which rely on DCV-mediated secretion. Many insulin family neuropeptides such as INS-7 are regulated by DAF-2 and DAF-16 pathway. Hypersecretion mutant *dgk-1* had increased INS-7 which is further increased during *P. aeruginosa* infection (Kawli & Tan, 2008). This study shows that Gα-dependent neuronal secretion of neuropeptide INS-7 promotes resistance to infection. This is likely dependent on sensing of PAMP or DAMP by many neurons, and may involve more than one receptor.

## 2.6 Dopamine Signaling During *P. aeruginosa* Infection

Dopamine can suppress inflammation in mammals and can also suppress p38 MAP kinase pathway in *C. elegans* (Cai, Cao, & Aballay, 2014; Chavan & Tracey, 2014). A dopamine antagonist chlorpromazine enhances resistance of *C. elegans* to *P. aeruginosa*-mediated killing. Conversely treatment of worms with exogenous dopamine enhances susceptibility to *P. aeruginosa* infection. Dopamine antagonist enhances activation of p38 MAPK and upregulation of p38 MAP kinase-dependent immune effectors *lys-2* and F35E12.5. Of the four dopamine receptors in *C. elegans* genome, only DOP-4 regulates immune responses. Mutation in *dop-4* gene leads to resistance to *P. aeruginosa* infection, p38 MAP kinase activation, and enhanced production of ROS detoxifying enzymes GCS-1, GST-4, GST-7, and GST-10. DOP-4 activity in sensory neuron CEP exposed to the pseudocoelom is important for immune regulation. Ablation of dopaminergic CEP neurons led to enhanced resistance to as well as upregulation of immune effectors *lys-2*, F35E12.5, K08D8.5, and M02F4.7. DOP-4 signaling in amphid neuron ASG appears to play a role in immune regulation during *P. aeruginosa* infection (Cao & Aballay, 2016). Thus, dopamine signaling in *C. elegans* nervous system regulates immune homeostasis at least during *P. aeruginosa* infection.

## 2.7 G-Protein and Serotonin Signaling During *M. nematophilum* Infection

*M. nematophilum* is *C. elegans* natural pathogen found associated with nematodes in the soil. It causes DAR phenotype in worms which is a part of immune defense mechanism localized to the rectal epithelium of the worm.

In this host–pathogen model, serotonin and G-protein signaling happen in two different tissues. Exogenous serotonin inhibits DAR phenotype. Loss-of-function mutation in the rate-limiting enzyme of serotonin biosynthesis, tryptophan hydroxylase or TPH-1, shows enhanced DAR phenotype. Worms have three major serotonergic neurons, ADF, HSN, and NSM. Restoring TPH-1 activity in ADF neurons rescues DAR phenotype, suggesting that ADF neuron may sense some cue from *M. nematophilum*. ADF turns on the DAR inducing circuit in *C. elegans* by releasing serotonin. In *C. elegans* the effects of serotonin on neuronal activity at the neuromuscular junction are mediated by the G-protein GOA-1. EGL-10 is a regulator of G-protein signaling protein and a specific inhibitor of GOA-1(Gao) activity. Mutations in *EGL-10* encoding gene suppress the DAR phenotype (Anderson, Laurenson-Schafer, Partridge, Hodgkin, & McMullan, 2013). This study indicates that serotonin signaling from the ADF neurons inhibits immune response to *M. nematophilum* by altering GPCR and Gαo signaling in the rectal epithelium.

## 3. GPCR SIGNALING AT THE SITE OF INFECTION REGULATES INNATE IMMUNE RESPONSES

### 3.1 FSHR-1 GPCR Signaling in the Intestine During *P. aeruginosa* Infection

FSHR-1 is a GPCR, orthologous to human follicle-stimulating hormone receptor, thyrotropin receptor, and luteinizing hormone receptor. It was initially identified as a predicted neuropeptide receptor which has a role in functioning of cholinergic synapses in nervous system (Sieburth et al., 2005). Later, it was shown to regulate germline differentiation and proliferation in *C. elegans* (Cho, Rogers, & Fay, 2007). Interestingly, FSHR-1 is also a leucine rich repeat (LRR) domain containing GPCR expressed in the nervous system as well as the intestine. LRR domain is found in many pattern recognition receptors in plants. A deletion mutation in *fshr-1* gene causes susceptibility of *C. elegans* to Gram-negative bacteria *P. aeruginosa* and Gram-positive bacteria *S. aureus* and *E. faecalis* (Powell, Kim & Ausubel, 2009). Intestine-specific RNA interference inhibition of *fshr-1* also leads to compromised immune response, suggesting intestine to be the site of activity of FSHR-1 during *P. aeruginosa* infection. Complementation of *fshr-1* in the intestine can rescue susceptibility phenotype almost completely, suggesting that intestinal GPCR is likely activated in the intestine during infection. FSHR-1 suppresses basal expression of immune effectors (F49F1.6, C17H12.8) and promotes *P. aeruginosa*-induced expression of another set of effectors (F01D5.5, F56D6.2). FSHR-1 acts in a pathway parallel to p38 MAP kinase pathway or bZIP-2 during *P. aeruginosa infection* (Fig. 5). Since FSHR-1 positively regulates immune responses against both Gram-negative and Gram-positive bacteria, it is unlikely that it senses a PAMP. It is more likely that FSHR-1 is an indirect sensor of infection. FSHR-1 may sense a DAMP released due to damage to host cells during infection. Indeed a transcription factor bZIP-2 in *C. elegans* has been shown to be a sensor for patterns of pathogenesis (Estes, Dunbar, Powell, Ausubel, & Troemel, 2010). bZIP-2 monitors translation suppression in the nematode host by *P. aeruginosa* exotoxin A. It is conceivable that FSHR-1 also senses a pattern of pathogenesis common to damage caused by both Gram-negative and Gram-positive bacteria. FSHR-1 has recently been found to protect *C. elegans* against xenobiotic stress caused by heavy metal cadmium and oxidative stress caused by paraquat (Miller, Grandi, Giannini, Robinson, & Powell, 2015). *C. elegans* lacking FSHR-1 were deficient in production of ROS detoxifying enzymes, gamma-glutamine cysteine synthetase or GCS-1. This suggests that FSHR-1 may sense ROS or a DAMP released due to oxidative damage to the cells. Further experiments are required to know if that may be the case.

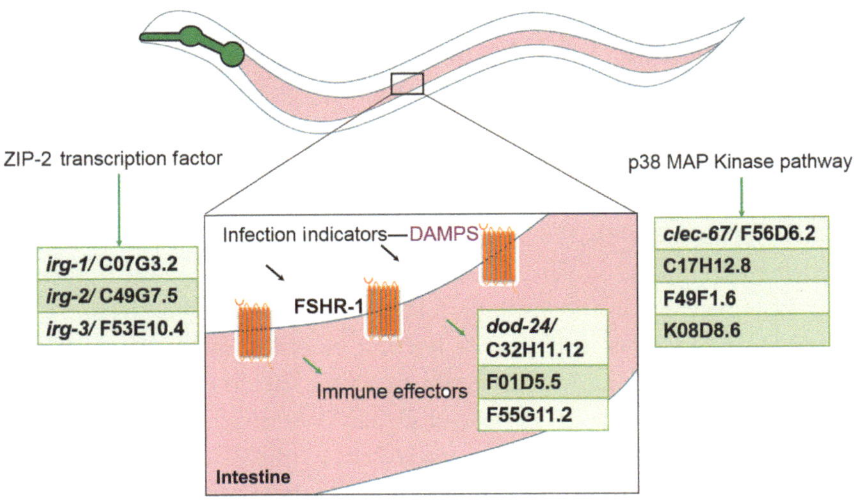


**Fig. 5** FSHR-1 activity in the intestine of *C. elegans* regulates immune responses. A ligand released due to infection or injury leads to activation of FSHR-1. FSHR-1 signaling leads to upregulation of immune effectors. FSHR-1 signaling is parallel to p38 MAP kinase pathway and ZIP-2 transcription factor which control other immune effectors during *P. aeruginosa* infection.

Evidence for G-protein signaling at the site of *P. aeruginosa* also comes from studies on Gqα in *C. elegans*. Gqα/EGL-30 activity is required for the activation of phopsholipase C (PLC). PLC can hydrolyze phosphatidylinositol 4,5 bis phosphate (PIP2) to diacyl glycerol (DAG) and inositol triphosphate. DAG is a second messenger which can activate protein kinase C. In *C. elegans*, Gqα/EGL-30 and PLC/EGL-8 act in the intestine to activate p38 MAP kinase pathway during *P. aeruginosa* infection (Kawli, Wu, & Tan, 2010).

## 3.2 DCAR-1 GPCR Signaling in the Skin During *D. coniospora* Infection

*D. coniospora* is an obligate pathogen often found associated with *C. elegans* in the environment. The infection of *C. elegans* begins with the germination of conidia on the worm cuticle. The hyphae penetrate the cuticle the hypodermis, and rest of the soma of *C. elegans* (Jansson, 1994). This leads to rapid induction of a battery of AMP genes in the hypodermis which requires p38 MAP kinase, STAT-like transcription factor STA-2, protein kinase C TPA-1, and Gα protein GPA-12 (Ziegler et al., 2009). The identification of GPA-12 as regulator of antifungal defense in the skin layer prompted a search for GPCR in the hypodermis. In a genome-wide RNAi screen of 1150 GPCR encoding genes, three GPCRs, DCAR-1, FRPR-11, and SRV-21, were found to regulate induction of NLP-29 AMP (Zugasti et al., 2014). DCAR-1 or dihydrocaffeic acid receptor-mediated regulation of AMP NLP-29 is specific for *D. coniospora* infection but not for osmotic stress which can also turn on NLP-29 expression. DCAR-1 does not affect basal level of hypodermis-specific AMPs but only regulates their induction during infection. This suggests that DCAR-1 activation may be dependent on a PAMP or DAMP linked to *D. coniospora* infection. DCAR-1 does not play a role in longevity on *E. coli* strain OP50, or resistance to *P. aeruginosa* infection of the intestine. This again suggested that DCAR-1 is activated in hypodermis most likely by local damage caused by invading *D. coniospora* hyphae. Interestingly, DCAR-1 is also required for NLP-29 induction in the hypodermis during physical injury caused by mutations in specific cuticle collagen genes. This suggested that DCAR-1 may have an endogenous ligand. As the name suggests, DCAR-1 was isolated as a receptor for dihydrocaffeic acid (DHCA) and other weakly acidic aromatic compounds with a catechol moiety, a carboxyl group on the side chain, and absence of an amino group (Aoki et al., 2011). DCAR-1 is expressed in the nervous system and hypodermis, and the exposure of worms to exogenous DHCA can induce NLP-29 AMP in the hypodermis but not in the intestine. Expression of wild-type copy of DCAR-1 in the hypodermis can rescue NLP-29 expression in the hypodermis, but neuronal expression of DCAR-1 could not. This indicates that DCAR-1 is indeed required at the site of injury in the hypodermis to induce NLP-29. Four other molecules structurally similar to DHCA could also induce AMP expression in the hypodermis. One of these, 4-hydroxyphenyllactic acid (HPLA) is present in the worm lysate during infection and in collagen defective mutants with increased AMP expression. Exogenous HPLA can also induce AMP in *C. elegans* in DCAR-1-dependent manner, suggesting that HPLA might be a natural ligand for GPCR activation. There is no indication that HPLA can be produced by other tissue such as intestine in case of bacterial infection. It is tempting to speculate that DCAR-1 expressed in the nervous system and hypodermis may bind different ligands and mediate behavioral vs physiological immune response, respectively (Fig. 6).

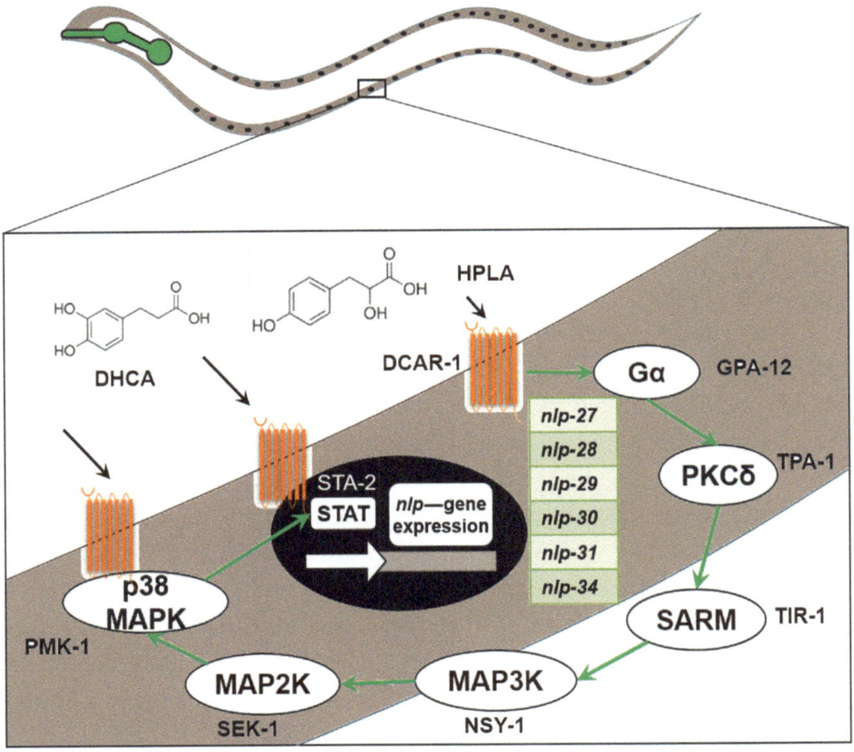


**Fig. 6** DCAR-1 activation in the hypodermis of *C. elegans* regulates immune response to *D. coniospora*. HPLA released during infection can activate DCAR-1 and regulate production of antimicrobial peptides via G-protein signaling, p38 MAP kinase pathway, and STAT transcription factor.

# 4. BEHAVIORAL RESPONSE OF *C. elegans* TO PATHOGENS

## 4.1 Aversion Response of *C. elegans* on *P. aeruginosa* Lawn

### *4.1.1 Amphid Neurons and Serotonin Signaling Regulate* C. elegans *Preference of Bacteria*

The elaborate nervous system of *C. elegans* is geared toward finding food, finding mate, and staying away from noxious stimuli. The nematode has to continually make choices or preferences which are reflected in strong chemotaxis behavior to many bacterial secondary metabolites. Olfactory neurons AWA and AWC drive chemotaxis to attractive odors such as diacetyl and benzaldehyde, while AWB neuron drives chemotaxis away from repulsive odors such as 2-nonanone and 1-octanol. Chemotaxis to the volatiles is dependent on a trimeric G alpha protein ODR-3 expressed in olfactory neurons AWA, AWB, and AWC (Fig. 2). These experiments also imply broader role of *C. elegans* GPCR and trimeric G proteins in the sensing of bacteria in the environment. This also raises the possibility that other bacterial molecules can also induce negative chemotaxis or avoidance behavior, akin to flight response seen in higher animals. The first report of *C. elegans* ability to prefer one bacterium over other came in 2005 (Zhang, Lu, & Bargmann, 2005). When nematodes are allowed to choose between *E. coli* and *P. aeruginosa*, they initially preferred *P. aeruginosa*. However, trained animals, previously exposed to *P. aeruginosa*, prefer *E. coli* over *P. aeruginosa*. This indicates that nematodes can alter their preferences on the basis of olfactory and chemical cues from bacteria. The ability of trained animals to avoid the pathogenic bacteria and move toward the nonpathogenic one also suggests that *C. elegans* can associate an odor with a possible threat of pathogenic infection, perhaps a form of associative memory. Mutants defective in producing catecholamines such as serotonin show significantly reduced learning in the choice assay despite training on the pathogen. Inability to synthesize serotonin causes learning defects, suggesting that olfactory learning requires serotonin. Of the two serotonergic neurons (ADF and NSM) examined, amphid chemosensory neuron ADF is required for *P. aeruginosa* avoidance while NSM for attractive food odors. Although a *Pseudomonas* sensing GPCR is not identified in ADF neuron, it is plausible to think that G-protein signaling may be involved. G-protein signaling likely results in increase in intracellular calcium concentration and release of neuropeptides or neurotransmitters. In a more recent study, olfactory Gα protein ODR-3 in olfactory neurons AWB and AWC has been shown to play important role in olfactory preference between *E. coli* and *P. aeruginosa* (Harris et al., 2014). Amphid chemosensory neuron ADL also promotes preference behavior in *C. elegans*. This is dependent on GPCR SRH-220 and neuropeptide FLP-4 (Yu, Zhi, Guan, Wang, & Wang, 2016). GPCRs involved in AWA- or AWB-mediated preference remain to be found.

### 4.1.2 NPR-1 GPCR Regulates Aversion Response

The immune response of *C. elegans* to *P. aeruginosa* encompasses aversion response and physiological response. *C. elegans* allowed to exhibit lawn-leaving aversion behavior are more resistant than *C. elegans* not allowed to leave the bacterial lawn. This suggests that avoidance or flight response is an important component of defense response against *P. aeruginosa*. *C. elegans* can also choose nonpathogenic *E. coli* over *P. aeruginosa*. The ability to make this choice is a learned behavior and relies on neurotransmitter serotonin as described in the preceding section. NPR-1 GPCR can control both innate immune response and aversion response. Several studies show that NPR-1 controls hyperoxia response and bordering phenotype under normoxia. Bordering phenotypes seen in wild-type nematodes were exposed to mucoidal strains *of E. coli, P. aeruginosa*, or nitrogen fixing Gram-negative bacterium *Sinorhizobium meliloti*. This behavior is dependent on oxygen sensing circuit of *NPR-1* and GCY-35 (Reddy, Hunter, Bhatla, Newman, & Kim, 2011). Inducers of aversion response in *C. elegans* appear to be exopolysaccharide matrix components in different bacterial species. In *Pseudomonas* strains, exopolysaccharide matrix component called alginate can induce bordering behavior in wild-type *C. elegans* (Reddy et al., 2011). This suggested that alginate can suppress oxygen sensing circuit of NPR-1 GPCR and soluble guanylyl cyclase GCY-35.

### 4.1.3 Aversion Response Is Driven by Mechanosensory Neuron OLL

*C. elegans* aversion response relies upon bacterial exopolysaccharide, alginate, and secondary metabolites. *C. elegans* proteins unrelated to GPCRs can also affect aversion response. A mutation affecting an E3 ubiquitin ligase called HECW-1, expressed in several neurons, regulates aversion response of *C. elegans* to *P. aeruginosa*. Mutation in *hecw-1 gene* alone shows increased aversion than N2, while *npr-1* alone shows reduced aversion. HECW-1 acts in OLL neurons to produce the aversion phenotype, and NPR-1 is active in command interneuron RMG. OLL neurons are mechanosensory neurons that respond to light touch on the tip of nematode's nose (Chang, Paek, & Kim, 2011). In this case they likely sense texture of the lawn. Although not explored in the study, the exopolysaccharide produced in the lawn may correlate with the texture of the lawn and may influence sensing by mechanosensory neurons.

### 4.1.4 P. aeruginosa *Secondary Metabolites Drive* C. elegans *Aversion Response*

Although aversion response of *C. elegans* to *P. aeruginosa* lawn has been reported by many research groups, a specific *P. aeruginosa* molecule has not been shown to be sensed by *C. elegans*, until 2014. Meisel et al. showed that TGF-β ligand DAF-7 regulates aversion response (Meisel, Panda, Mahanti, Schroeder, & Kim, 2014) (Fig. 7). DAF-7 is normally secreted

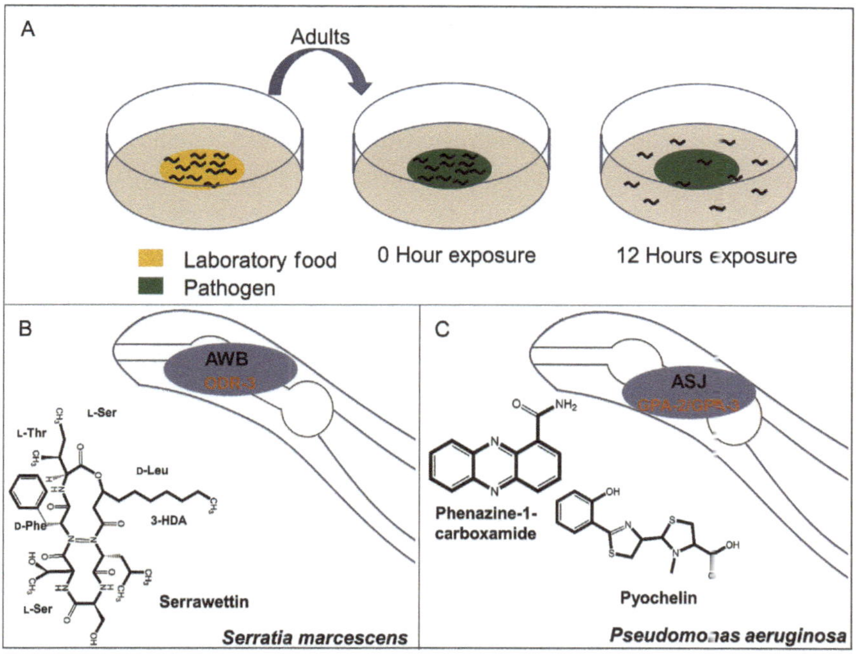


**Fig. 7** (A) Schematic of aversion response in *C. elegans*. Adult nematodes are transferred to small lawn of pathogen, and lawn occupancy is calculated after 12 h. (B) AWB neuron of *C. elegans* can sense serrawettin produced by pathogenic *Serratia marcescens*. (C) ASJ chemosensory neuron can sense pyochelin and phenazine-1-carboxamide produced by pathogenic *Pseudomonas aeruginosa*.

by ASI neurons but induced also in ASJ neuron when exposed to *P. aeruginosa*. At least two bacterial secondary metabolites, pyochelin and pyocyanin, can upregulate DAF-7 in ASJ neuron. The expression of DAF-7 in ASJ within minutes of exposure of nematodes to *P. aeruginosa* indicates a direct sensing of bacterial components by *C. elegans*. DAF-7 expression in ASJ expression requires receptor guanylyl cyclase DAF-11 as well as cGMP-gated ion channels TAX-2/TAX-4, suggesting the involvement of G-protein signaling in regulating *daf-7* expression in ASJ neurons. Lack of Gα protein GPA-2 and GPA-3 also reduces DAF-7 expression in ASJ neurons, suggesting possible involvement of GPCR signaling in sensing of *P. aeruginosa* metabolites. So far there are no reports on GPCR expressed in ASJ neuron.

To summarize the aversion response of *C. elegans* to *P. aeruginosa*, both chemical cues, endogenous or pathogen derived, and mechanical cues are also responsible for the lawn avoidance response.

## 4.2 Aversion Response of *C. elegans* on *Serratia marcescens* Lawn

### *4.2.1 Aversion Response of* C. elegans *to Serrawettin Surfactant From* S. marcescens

*S. marcescens* is a rod-shaped, Gram-negative bacterium which is involved in hospital acquired infections. It is an opportunistic pathogen that affects individuals with compromised immunity. *Serratia* is also pathogenic to *C. elegans* causing death within 4–5 days of feeding and exposure. Similar to *P. aeruginosa*, *S. marcescens* also causes an avoidance response in *C. elegans*. Typically, only 30% of worms remain on *S. marcescens* lawn after 12h. In a screen to identify bacterial component responsible for inducing avoidance response in the host, a surfactant-deficient mutants was found. The surfactant-deficient strain JESM267 was defective in producing surfactant serrawettin, but it was as virulent as wild-type *Serratia* strain Db10 indicating that avoidance inducing principle could be decoupled from bacterial virulence. In JESM267, a mutation in nonribosomal peptide synthetase (NRPS) leads to loss of serrawettin W2, an extracellular cyclodepsipeptide biosurfactant. Two cyclic nucleotide-gated channels TAX-2/TAX-4 are important for aversion response to S. *marcescens*. TAX-2 and TAX-4 are expressed in eight pairs of Amphid neurons—AWB, AWC, AFD, ASE, ASG, ASI, ASJ, and ASK (Fig. 2). Mutation affecting AWB neurons controls lawn-leaving behavior of the nematodes. In addition, Gα-protein/ ODR-3 expressed in AWA, AWB, AWC is indispensable for aversion response. There are three possibilities that could explain the avoidance response of worms toward serrawettin produced by *S. marcescens*: (i) Serrawettin could be directly sensed as a chemorepellant, (ii) Serawettin could be sensed mechanically as a biosurfactant that affects surface tension, or (iii) Serrawettin might regulate production of other downstream repellants. Worms, exposed to purified serrawettin mixed with *E. coli* culture, tend to avoid the lawn ruling out the third possibility. Other biosurfactants such as serrawettin W1, serrawettin W3, and surfactin do not seem to initiate the avoidance response ruling out through the second possibility. Also, when a drop of purified serrawettin is placed on nematode's head, they tend to avoid it by turning backward/reversal. Thus, serrawettin is likely sensed as a chemorepellent by the worms (Pradel et al., 2007). Thus, G-protein signaling in AWB neurons drives aversion response to *S. marcescens*, aiding in survival of the nematode.

#### 4.2.2 Aversion Response Is Driven by $CO_2$ Sensing Neurons

*C. elegans* can sense changes in ambient oxygen and carbon dioxide concentration using BAG and AQR, PQR, and URX neurons whose cilia end in pseudocoelom, the vasculature where gases can be diffuse across from outside. Worms prefer 7%–11% $O_2$ and have a threshold of about $\approx$0.5% $CO_2$ above which worms tend to show avoidance behavior. BAG neurons can sense increase in ambient $CO_2$ via guanylyl cyclase GCY-9. Avoidance to higher $CO_2$ is mediated by TOL-1/Toll-like receptor pathway (Brandt & Ringstad, 2015). TOL-1 is also implicated in the aversion response to *S. marcescens* (Pujol et al., 2001). A possible mechanism of aversion may be linked to $CO_2$ concentration in *S. marcescens* lawn. Nematodes, when exposed to heat-killed *Serratia* lawn or attenuated *Serratia* (Db1140) lawn, do not exhibit avoidance behavior. This indicates that $CO_2$ production by living *Serratia* induces avoidance behavior in *C. elegans*. Since most aerobic bacteria produce similar amounts of $CO_2$, *S. Marcescens* may provide another cue to *C. elegans*. How nematodes integrate $CO_2$ signal from BAG neurons and *Serratia*-specific chemical/mechanical cues from other chemosensory/olfactory/mechanosensory neurons to drive the avoidance behavior is not known.

## 5. CONCLUDING REMARKS

Survival of *C. elegans* against invading pathogen relies on a behavioral component of aversion and a physiological component of innate immune response. Both responses appear to rely heavily on neuronal signaling, primarily in the sensory neurons of *C. elegans*. The sensory perception of pathogen via their secondary metabolite or surfactant may initiate an avoidance response in *C. elegans* to *P. aeruginosa* or *S. marcescens*. Both these are Gram-negative bacteria, commonly found in soil, the natural habitat of *C. elegans*, suggesting evolution of this behavior against frequently encountered pathogens. In case of *P. aeruginosa*, which is better investigated, a number of components can induce aversion response. This includes secondary metabolites pyochelin, phenazine-1-carboximide, and exopolysaccharide alginate. The aversion response relies on serotonin, GPCR involved in oxygen sensing circuit as well as G-protein signaling in chemosensory neuron ASJ. Aversion response also relies on mechanosensory response of *C. elegans*. Extracellular matrix components may play a role in mechanosensation and may turn on aversion response via independent neural circuits. An interesting question is how the worm integrates information from these circuits to engage in lawn-leaving behavior. A comprehensive study of calcium signaling in individual neurons in the background of sensory mutations may pave the way to better understanding of *C. elegans–P. aeruginosa* interactions.

Two GPCRs in *C. elegans* respond to infection at the site of injury and pathogen replication in *C. elegans*. DCAR-1 in the hypodermis can be turned on by the release of an endogenous ligand HPLA. HPLA is released as result of cuticle injury, or by invasion of *D. coniospora* hyphae. DCAR-1 activity in this context turns on a battery of AMP in the hypodermis. FSHR-1 activity in the intestine is essential during injury by *P. aeruginosa* infection, *E. faecalis* infection, heavy metal stress, and paraquat treatment. Oxidative stress created in these scenarios may yield an endogenous molecule or DAMP which may serve as a ligand for FSHR-1, ROS remains a tantalizing target. DCAR-1 and FSHR-1 provide cell autonomous control of immune regulation at the site of infection and injury. Future might reveal other pathogen or tissue-specific regulators of immune response.

Noncell autonomous control of immune homeostasis is regulated by trimeric G proteins, beta-arrestin, and GPCRs in the nervous system of *C. elegans*. These regulate conserved p38 MAP kinase pathway, unfolded protein response machinery of the endoplasmic reticulum, and translation machinery in the intestine. The ability of OCTR-1 and NPR-1 to regulate immune effector production in nonneuronal tissues raises interesting questions on how the nervous system communicates with the intestine which is enervated. The communication between the nervous system and the intestine may occur via an endocrine mechanism where the neurons release neuropeptide in the pseudocoelomic space to be detected by tissues with appropriate receptors. Research along these lines is likely to yield insights into mechanisms of GPCR-mediated immune regulation in multicellular organisms.

## ACKNOWLEDGMENTS

V.S. is supported by Intermediate Fellowship from the Wellcome Trust DBT India Alliance. A.G. is recipient of PhD fellowship under the graduate program in Bio Systems Science and Engineering, Indian Institute of Science.

## REFERENCES

Alcedo, J., & Kenyon, C. (2004). Regulation of C. elegans longevity by specific gustatory and olfactory neurons. *Neuron, 41*(1), 45–55.

Anderson, A., Laurenson-Schafer, H., Partridge, F. A., Hodgkin, J., & McMullan, R. (2013). Serotonergic chemosensory neurons modify the C. elegans immune response by regulating G-protein signaling in epithelial cells. *PLoS Pathogens, 9*(12), e1003787.

Aoki, R., Yagami, T., Sasakura, H., Ogura, K.-i., Kajihara, Y., Ibi, M., et al. (2011). A seven-transmembrane receptor that mediates avoidance response to dihydrocaffeic acid, a water-soluble repellent in Caenorhabditis elegans. *Journal of Neuroscience, 31*(46), 16603–16610.

Apfeld, J., & Kenyon, C. (1999). Regulation of lifespan by sensory perception in Caenorhabditis elegans. *Nature, 402*(6763), 804–809.

Brandt, J. P., & Ringstad, N. (2015). Toll-like receptor signaling promotes development and function of sensory neurons required for a C. elegans pathogen-avoidance behavior. *Current Biology, 25*(17), 2228–2237.

Cai, Y., Cao, X., & Aballay, A. (2014). Whole-animal chemical screen identifies colistin as a new immunomodulator that targets conserved pathways. *MBio, 5*(4), e01235–01214.

Cao, X., & Aballay, A. (2016). Neural inhibition of dopaminergic signaling enhances immunity in a cell-non-autonomous manner. *Current Biology, 26*(17), 2329–2334.

Chang, H. C., Paek, J., & Kim, D. H. (2011). Natural polymorphisms in C. elegans HECW-1 E3 ligase affect pathogen avoidance behaviour. *Nature, 480*(7378), 525–529.

Chavan, S. S., & Tracey, K. J. (2014). Regulating innate immunity with dopamine and electroacupuncture. *Nature Medicine, 20*(3), 239–241.

Cho, S., Rogers, K. W., & Fay, D. S. (2007). The C. elegans glycopeptide hormone receptor ortholog, FSHR-1, regulates germline differentiation and survival. *Current Biology, 17*(3), 203–212.

Coates, J. C., & de Bono, M. (2002). Antagonistic pathways in neurons exposed to body fluid regulate social feeding in Caenorhabditis elegans. *Nature, 419*(6910), 925–929.

Colbert, H. A., & Bargmann, C. I. (1995). Odorant-specific adaptation pathways generate olfactory plasticity in C. elegans. *Neuron, 14*(4), 803–812.

De Bono, M., & Bargmann, C. I. (1998). Natural variation in a neuropeptide Y receptor homolog modifies social behavior and food response in C. elegans. *Cell, 94*(5), 679–689.

Dirksen, P., Marsh, S. A., Braker, I., Heitland, N., Wagner, S., Nakad, R., et al. (2016). The native microbiome of the nematode Caenorhabditis elegans: Gateway to a new host-microbiome model. *BMC Biology, 14*(1), 38.

Estes, K. A., Dunbar, T. L., Powell, J. R., Ausubel, F. M., & Troemel, E. R. (2010). bZIP transcription factor zip-2 mediates an early response to Pseudomonas aeruginosa infection in Caenorhabditis elegans. *Proceedings of the National Academy of Sciences of the United States of America, 107*(5), 2153–2158.

Garsin, D. A., Sifri, C. D., Mylonakis, E., Qin, X., Singh, K. V., Murray, B. E., et al. (2001). A simple model host for identifying Gram-positive virulence factors. *Proceedings of the National Academy of Sciences of the United States of America*, *98*(19), 10892–10897.

Garsin, D. A., Villanueva, J. M., Begun, J., Kim, D. H., Sifri, C. D., Calderwood, S. B., et al. (2003). Long-lived C. elegans daf-2 mutants are resistant to bacterial pathogens. *Science*, *300*(5627), 1921.

Harris, G., Shen, Y., Ha, H., Donato, A., Wallis, S., Zhang, X., et al. (2014). Dissecting the signaling mechanisms underlying recognition and preference of food odors. *Journal of Neuroscience*, *34*(28), 9389–9403.

Hodgkin, J., Kuwabara, P. E., & Corneliussen, B. (2000). A novel bacterial pathogen, Microbacterium nematophilum, induces morphological change in the nematode C. elegans. *Current Biology*, *10*(24), 1615–1618.

Inglis, P. N., Ou, G., Leroux, M. R., & Scholey, J. M. (2007). The sensory cilia of Caenorhabditis elegans. *WormBook*, *8*(1), 22.

Irazoqui, J. E., Ng, A., Xavier, R. J., & Ausubel, F. M. (2008). Role for β-catenin and HOX transcription factors in Caenorhabditis elegans and mammalian host epithelial-pathogen interactions. *Proceedings of the National Academy of Sciences of the United States of America*, *105*(45), 17469–17474.

Jansen, G., Thijssen, K. L., Werner, P., Hazendonk, E., & Plasterk, R. H. (1999). The complete family of genes encoding G proteins of Caenorhabditis elegans. *Nature Genetics*, *21*(4), 414–419.

Jansson, H.-B. (1994). Adhesion of conidia of Drechmeria coniospora to Caenorhabditis elegans wild type and mutants. *Journal of Nematology*, *26*(4), 430.

Joshua, G., Karlyshev, A., Smith, M., Isherwood, K., Titball, R., & Wren, B. (2003). A Caenorhabditis elegans model of Yersinia infection: Biofilm formation on a biotic surface. *Microbiology*, *149*(11), 3221–3229.

Kawli, T., & Tan, M.-W. (2008). Neuroendocrine signals modulate the innate immunity of Caenorhabditis elegans through insulin signaling. *Nature Immunology*, *9*(12), 1415–1424.

Kawli, T., Wu, C., & Tan, M.-W. (2010). Systemic and cell intrinsic roles of Gqα signaling in the regulation of innate immunity, oxidative stress, and longevity in Caenorhabditis elegans. *Proceedings of the National Academy of Sciences of the United States of America*, *107*(31), 13788–13793.

Kim, D. H., Feinbaum, R., Alloing, G., Emerson, F. E., Garsin, D. A., Inoue, H., et al. (2002). A conserved p38 MAP kinase pathway in Caenorhabditis elegans innate immunity. *Science*, *297*(5581), 623–626.

Labrousse, A., Chauvet, S., Couillault, C., Kurz, C. L., & Ewbank, J. J. (2000). Caenorhabditis elegans is a model host for Salmonella typhimurium. *Current Biology*, *10*(23), 1543–1545.

Lans, H., & Jansen, G. (2007). Multiple sensory G proteins in the olfactory, gustatory and nociceptive neurons modulate longevity in Caenorhabditis elegans. *Developmental Biology*, *303*(2), 474–482.

Liu, Y., Sellegounder, D., & Sun, J. (2016). Neuronal GPCR OCTR-1 regulates innate immunity by controlling protein synthesis in Caenorhabditis elegans. *Scientific Reports*, *6*, 36832.

Meisel, J. D., Panda, O., Mahanti, P., Schroeder, F. C., & Kim, D. H. (2014). Chemosensation of bacterial secondary metabolites modulates neuroendocrine signaling and behavior of C. elegans. *Cell*, *159*(2), 267–280.

Miller, E. V., Grandi, L. N., Giannini, J. A., Robinson, J. D., & Powell, J. R. (2015). The conserved G-protein coupled receptor FSHR-1 regulates protective host responses to infection and oxidative stress. *PloS One*, *10*(9), e0137403.

Mori, I., & Ohshima, Y. (1995). Neural regulation of thermotaxis in Caenorhabditis elegans. *Nature*, *376*(6538), 344.

Mylonakis, E., Ausubel, F. M., Perfect, J. R., Heitman, J., & Calderwood, S. E. (2002). Killing of Caenorhabditis elegans by *Cryptococcus neoformans* as a model of yeast pathogenesis. *Proceedings of the National Academy of Sciences of the United States of America, 99*(24), 15675–15680.

Nicholas, H. R., & Hodgkin, J. (2004). The ERK MAP kinase cascade mediates tail swelling and a protective response to rectal infection in C. elegans. *Current Biology, 14*(14), 1256–1261.

Powell, J. R., Kim, D. H., & Ausubel, F. M. (2009). The G protein-coupled receptor FSHR-1 is required for the Caenorhabditis elegans innate immune response. *Proceedings of the National Academy of Sciences of the United States of America, 106*(8), 2782–2787.

Pradel, E., Zhang, Y., Pujol, N., Matsuyama, T., Bargmann, C. I., & Ewbank, J. J. (2007). Detection and avoidance of a natural product from the pathogenic bacterium Serratia marcescens by Caenorhabditis elegans. *Proceedings of the National Academy of Sciences of the United States of America, 104*(7), 2295–2300.

Pujol, N., Cypowyj, S., Ziegler, K., Millet, A., Astrain, A., Goncharov, A., et al. (2008a). Distinct innate immune responses to infection and wounding in the C. elegans epidermis. *Current Biology, 18*(7), 481–489.

Pujol, N., Link, E. M., Liu, L. X., Kurz, C. L., Alloing, G., Tan, M.-W. et al. (2001). A reverse genetic analysis of components of the toll signaling pathway in Caenorhabditis elegans. *Current Biology, 11*(11), 809–821.

Pujol, N., Zugasti, O., Wong, D., Couillault, C., Kurz, C. L., Schulenburg, H., et al. (2008b). Anti-fungal innate immunity in C. elegans is enhanced by evolutionary diversification of antimicrobial peptides. *PLoS Pathogens, 4*(7), e1000105.

Reddy, K. C., Hunter, R. C., Bhatla, N., Newman, D. K., & Kim, D. H. (2011). Caenorhabditis elegans NPR-1-mediated behaviors are suppressed in the presence of mucoid bacteria. *Proceedings of the National Academy of Sciences of the United States of America, 108*(31), 12887–12892.

Sieburth, D., Ch'ng, Q., Dybbs, M., Tavazoie, M., Kennedy, S., Wang, D., et al. (2005). Systematic analysis of genes required for synapse structure and function. *Nature, 436*(7050), 510–517.

Singh, V., & Aballay, A. (2012). Endoplasmic reticulum stress pathway required for immune homeostasis is neurally controlled by arrestin-1. *Journal of Biological Chemistry, 287*(40), 33191–33197.

Styer, K. L., Singh, V., Macosko, E., Steele, S. E., Bargmann, C. I., & Aballay, A. (2008). Innate immunity in Caenorhabditis elegans is regulated by neurons expressing NPR-1/GPCR. *Science, 322*(5900), 460–464.

Sun, J., Aballay, A., & Singh, V. (2016). Cellular responses to infections in Caenorhabditis elegans. In *Encyclopedia of cell biology: Vol. 2* (pp. 845–852). Waltham, MA: Academic Press. http://dx.doi.org/10.1016/b978-0-12-394447-4.20074-6.

Sun, J., Liu, Y., & Aballay, A. (2012). Organismal regulation of xbp-1-mediated unfolded protein response during development and immune activation. *EMBO Reports, 13*(9), 855–860.

Sun, J., Singh, V., Kajino-Sakamoto, R., & Aballay, A. (2011). Neuronal GPCR controls innate immunity by regulating noncanonical unfolded protein response genes. *Science, 332*(6030), 729–732.

Tan, M.-W., Mahajan-Miklos, S., & Ausubel, F. M. (1999). Killing of Caenorhabditis elegans by Pseudomonas aeruginosa used to model mammalian bacterial pathogenesis. *Proceedings of the National Academy of Sciences of the United States of America, 96*(2), 715–720.

Troemel, E. R., Félix, M.-A., Whiteman, N. K., Barrière, A., & Ausubel, F. M. (2008). Microsporidia are natural intracellular parasites of the nematode Caenorhabditis elegans. *PLoS Biology, 6*(12), e309.

Yu, Y., Zhi, L., Guan, X., Wang, D., & Wang, D. (2016). FLP-4 neuropeptide and its receptor in a neuronal circuit regulate preference choice through functions of ASH-2 trithorax complex in Caenorhabditis elegans. *Scientific Reports*, *6*, 21485.

Yusuke, K., Aizawa, T., Hoshino, H., Kawano, K., Nitta, K., & Zhang, H. (2002). Abf-1 and abf-2, ASABF-type antimicrobial peptide genes in Caenorhabditis elegans. *Biochemical Journal*, *361*(2), 221–230.

Zhang, Y., Lu, H., & Bargmann, C. I. (2005). Pathogenic bacteria induce aversive olfactory learning in Caenorhabditis elegans. *Nature*, *438*(7065), 179–184.

Ziegler, K., Kurz, C. L., Cypowyj, S., Couillault, C., Pophillat, M., Pujol, N., et al. (2009). Antifungal innate immunity in C. elegans: PKCδ links G protein signaling and a conserved p38 MAPK cascade. *Cell Host & Microbe*, *5*(4), 341–352.

Zugasti, O., Bose, N., Squiban, B., Belougne, J., Kurz, C. L., Schroeder, F. C., et al. (2014). Activation of a G protein-coupled receptor by its endogenous ligand triggers the innate immune response of Caenorhabditis elegans. *Nature Immunology*, *15*(9), 833–838.

CHAPTER SEVEN

# G Protein-Coupled Receptor Kinases in the Inflammatory Response and Signaling

**Michael D. Steury, Laura R. McCabe, Narayanan Parameswaran[1]**
Michigan State University, East Lansing, MI, United States
[1]Corresponding author: e-mail address: narap@msu.edu

## Contents

1. Introduction 228
   1.1 GRK Structure and Families 230
   1.2 GRK Regulation 232
   1.3 Animal Models of GRK Deficiency 234
2. GRKs in Inflammatory Signaling and Disease 238
   2.1 GRK1 Subfamily 238
   2.2 GRK2 Subfamily 239
   2.3 GRK4 Subfamily 251
3. Conclusion 262
References 262

## Abstract

G protein-coupled receptor kinases (GRKs) are serine/threonine kinases that regulate a large and diverse class of G protein-coupled receptors (GPCRs). Through GRK phosphorylation and β-arrestin recruitment, GPCRs are desensitized and their signal terminated. Recent work on these kinases has expanded their role from canonical GPCR regulation to include noncanonical regulation of non-GPCR and nonreceptor substrates through phosphorylation as well as via scaffolding functions. Owing to these and other regulatory roles, GRKs have been shown to play a critical role in the outcome of a variety of physiological and pathophysiological processes including chemotaxis, signaling, migration, inflammatory gene expression, etc. This diverse set of functions for these proteins makes them popular targets for therapeutics. Role for these kinases in inflammation and inflammatory disease is an evolving area of research currently pursued in many laboratories. In this review, we describe the current state of knowledge on various GRKs pertaining to their role in inflammation and inflammatory diseases.

*Advances in Immunology*, Volume 136
ISSN 0065-2776
http://dx.doi.org/10.1016/bs.ai.2017.05.003

## 1. INTRODUCTION

Cells have remarkable ability to adapt and respond to changes in their environment through recognition of extracellular signals from both the host as well as invading pathogens. These signals come in a variety of forms, not only from the host environment (e.g., hormones, cytokines, and chemokines) but also from microbial substances such as lipopolysaccharides and the breakdown of products from diet. Cells have developed sophisticated mechanisms for detecting and responding to these signals, largely through development of both extracellular and intracellular receptors. Despite the abundance of various receptor classes, G protein-coupled receptors (GPCRs) represent one of the largest proportion of receptors. This family of receptors is encoded by approximately 950 genes and transmits responses to a wide variety of stimuli including odorants, hormones, neurotransmitters, light, extracellular calcium, and many other stimuli regulating a wide variety of biological processes (Takeda, Kadowaki, Haga, Takaesu, & Mitaku, 2002). These receptors are characterized by their seven transmembrane domains that, when active, undergo a conformational change allowing them to activate nearby heterotrimeric GTP-binding proteins (G-proteins). These G-proteins bind to nucleotides guanosine triphosphate (GTP) or guanosine diphosphate (GDP) and form a complex composed of α-subunits (Gα encoded by 16 genes) and βγ dimers (Gβ encoded by 5 genes and Gγ encoded by 12 genes). When active, GPCRs can serve as guanyl nucleotide exchange factors for these G-proteins, catalyzing the release of the bound GDP, and exchanging it for GTP, which in turn causes the dissociation of the heterotrimeric G protein subunits Gα and Gβγ (Cabrera-Vera et al., 2003). These dissociated G-proteins then interact with other effector proteins and secondary messengers to facilitate a large variety of biological outcomes. This activity is terminated when the intrinsic GTPase activity of the Gα subunit causes GTP hydrolysis and the formation Gα-GDP, leading to its reassociation with Gβγ, and the reformation of the heterotrimeric G-protein complex. For a comprehensive review of G-protein activation and regulation, please see other excellent reviews (Hewavitharana & Wedegaertner, 2012; Oldham & Hamm, 2008; Preininger & Hamm, 2004; Ritter & Hall, 2009).

Aside from the intrinsic GTPase activity, G protein-mediated signaling is most often terminated by a conserved two-step mechanism: receptor phosphorylation by G protein-coupled receptor kinases (GRKs) followed by arrestin recruitment and binding to the activated receptor, ultimately leading to internalization and termination of the GPCR signal. Four genes encode arrestins; arrestin-1 and arrestin-4 that are restricted to the retina, whereas arrestin-2 (β-arrestin-1) and arrestin-3 (β-arrestin-2) are ubiquitously expressed throughout the body (Lefkowitz & Shenoy, 2005). This process of receptor desensitization is responsible for decreasing the magnitude of stimulation that GPCRs exhibit in response to prolonged exposure to agonist binding. This process is initiated by various GRKs that phosphorylate the intracellular loops and/or the C-terminal domain of activated GPCRs. This phosphorylation directly leads to the recruitment of multifunctional adaptor proteins, arrestins. Interestingly, some receptors are regulated by a single GRK whereas other receptors can be regulated by multiple GRKs with varying affinities. Recently, it has been suggested that there are distinct phosphorylation patterns termed "barcodes" that determine β-arrestin functionality and contribute to the diverse responses witnessed in GPCR activity after phosphorylation (Nobles et al., 2011). This phosphorylation and subsequent binding of arrestins prevent further activation of G-proteins resulting in the cessation of G-protein signaling despite continued presence of the agonist. Furthermore, GRK–arrestin interaction leads to internalization of the activated GPCR from the cellular surface through formation of clathrin-coated pits. During this internalization, signaling can still occur through the binding of additional signaling proteins to the still attached arrestin proteins where they may serve as scaffolding proteins. This converts the GPCR activation from heterotrimeric G-protein signaling pathway to G-protein-independent pathway (Lefkowitz, 2013). Recently, it was shown that the interaction between β-arrestins and GPCRs form distinct conformational complexes and that these complexes are specialized to carry out different cellular functions depending on the relationship of their binding (Cahill et al., 2017). Moreover, it has been shown that G-protein signaling can be sustained within the endosome through the formation of "megaplexes" indicating a possible new signaling role for these receptors during internalization (Thomsen et al., 2016). These alternative G-protein-independent pathways are addressed in other reviews (Gurevich & Gurevich, 2004; Lefkowitz & Shenoy, 2005; Luttrell, 2005). Because of the wide array of biological stimuli and functionality, regulation and control of these receptors, and their subsequent signaling is a potential target for pharmaceutical intervention. In fact, the GPCR family of receptors represents the largest family of drug targets for a wide variety of conditions including cardiovascular, metabolic, neurodegenerative, infectious, and oncologic diseases (Howard et al., 2001; Thompson, Burnham, & Cole, 2005).

As studies on the regulation of GPCRs continue to unfold, new roles for GRKs have been emerging in addition to the part they play in GPCR phosphorylation. It is now apparent that GRKs (and arrestins) regulate GPCR-independent functions in cell signaling and biology. This non-canonical role for GRKs has been shown to have critical regulatory roles in the context of inflammation and inflammatory diseases. In this review, we will discuss these known and emerging roles of GRKs in both GPCR-dependent and -independent functions in the context of inflammatory processes and pathologies.

## 1.1 GRK Structure and Families

The association between G proteins and receptor-mediated signal transduction developed through a progression of breakthroughs. Initially, it was discovered that GTP was necessary for signal transduction upon receptor activation. When the proteins were characterized for their interaction with GTP to initiate the signaling cascade they were termed "G-proteins" (Rodbell, Krans, Pohl, & Birnbaumer, 1971; Ross & Gilman, 1977). Our understanding of the regulation of GPCRs then progressed through the identification of rhodopsin phosphorylation and that this phosphorylation contributed to the rapid desensitization of rhodopsin signaling. This desensitization was attributed to the "opsin kinase" now known as GRK1 (Bownds, Dawes, Miller, & Stahlman, 1972). It was later in the mid-1980s through the cloning of the β-adrenergic receptor kinase (now named GRK2) that it was demonstrated that GRK1 was, in fact, a member of a family of related kinases that have a common mechanism that can recognize and regulate the active state of a variety of GPCRs (Benovic, Deblasi, Stone, Caron, & Lefkowitz, 1989). During this period, another protein was discovered to have a regulatory impact and could dampen G protein-mediated retinal signaling by also associating with rhodopsin. This protein (termed S antigen because of its role in allergic uveitis) was then renamed "arrestin" because of its "arresting" function in retinal signaling (Wacker, Donoso, Kalsow, Yankeelov, & Organisciak, 1977; Wilden, Hall, & Kühn, 1986; Zuckerman & Cheasty, 1986). It was through further characterization of the relationship between GRKs and arrestins that the canonical two-step inactivation process of desensitization of GPCRs was established. As studies progressed, the family of GRKs expanded to include seven serine/threonine kinases including GRK3 (Benovic et al., 1991), GRK4 (Ambrose et al., 1992), GRK5 (Kunapuli & Benovic, 1993; Premont et al., 1996), GRK6 (Benovic & Gomez, 1993), and GRK7 (Hisatomi et al., 1998; Weiss et al., 1998). These kinases are now grouped based on their sequence similarities and are broken down into three subfamilies: GRK1, composed of GRK1 (rhodopsin kinase) and GRK7 (cone opsin kinase); GRK2, including GRK2 and GRK3; and GRK4, made up of GRK4, GRK5, and GRK6.

Evolutionarily, GRKs are present in most vertebrate and invertebrate species as well as identified in nonmetazoan species, with the kinase domain being the most evolutionarily conserved among the different GRKs (Mushegian, Gurevich, & Gurevich, 2012). All seven GRK members are ~500–700 amino acids long, multidomain proteins containing a family specific N-terminal domain, followed by a regulator of G-protein signaling homology domain (RH domain), an AGC protein kinase domain, and a C-terminal domain (Siderovski, Hessel, Chung, Mak, & Tyers, 1996). The C-terminal is variable (both in length and in structure) among the different subfamilies and contains structural elements that are partly responsible for the diversity of GRK function, regulation, and interaction. Specifically, members of the GRK1 family have C-terminal prenylation sites while GRK2 subfamily members have a pleckstrin homology (PH) domain in their C-termini, that can interact with Gβγ subunits (Touhara, Inglese, Pitcher, Shaw, & Lefkowitz, 1994). The GRK4 family members have palmitoylation sites with the exception of GRK5 that contains a positively charged lipid-binding amphipathic helix in its C-terminus (Premont et al., 1999). GRK5 is also slightly different than its family members due to the presence of a nuclear localization signal that allows it to accumulate in the nucleus where it can act as a histone deacetylase kinase (although it is predominantly cytoplasmic) (Martini et al., 2008). Structurally, GRK1 and GRK7 genes (in humans) contain 4 exons, GRK2 and GRK3 contain 21 exons, and members of GRK4 subfamily have 16 exons. Tissue localization of these kinases varies significantly between subfamilies. GRK1 and GRK7 are termed the "visual kinases" and are localized in the eyes, whereas the nonvisual kinases are distributed nearly ubiquitously throughout the body albeit at different expression levels. This distribution holds true for all nonvisual GRKs with the exception of GRK4 that is restricted to the testis, the proximal tubule of the kidney, and the cerebellum (Inglese, Freedman, Koch, & Lefkowitz, 1993). Although it was originally believed that all GRKs can phosphorylate GPCRs equally, recent studies have shown that the GRK2/3 and the

GRK5/6 subfamilies differ in their ability to phosphorylate inactive GPCRs; GRK5/6 can phosphorylate GPCRs in both the active and the inactive conformations (Li et al., 2015). Since GPCR superfamily represents the largest class of receptors and there are only four GRKs ubiquitously expressed throughout the body, each GRK must be able to phosphorylate and regulate a wide variety of receptors, thus increasing the importance as well as the complexity of their role in signal transduction.

## 1.2 GRK Regulation

GRKs play a vital role in the regulation and behavior of many biological processes through a variety of regulatory mechanisms. It is perhaps due to this variety of functions and interactions that they are so heavily conserved evolutionarily. One of the first proteins that were shown to interact with and activate these GRKs was the Gβγ-subunits (DebBurman, Ptasienski, Benovic, & Hosey, 1996). In addition to G-protein activation of GRKs, lipids also have the ability to regulate GRKs. Furthermore, the other proteins such as calmodulin (CaM), caveolin-1, ERK, and actin can also modulate GRK activity through direct binding (Carman, Lisanti, & Benovic, 1999; Chuang, Paolucci, & De Blasi, 1996). Interestingly, these methods of regulation are not conserved among the various subfamilies of GRKs. For example, GRK2 family members are activated by phospholipids (phosphatidylinositol bisphosphate) through its interaction with the PH domain in the C-terminus, whereas GRK4-like members are activated by phosphatidylinositol bisphosphate through its binding to the N-terminal polybasic region (Pitcher et al., 1996).

Phosphorylation has differing regulatory effects (either activation or deactivation) on GRKs depending on exactly which kinase interacts with a given GRK or whether GRK undergoes autophosphorylation. To illustrate this contrasting regulation in the case of the visual GRKs, phosphorylation by protein kinase-A or for GRK5 by protein kinase-C causes a decrease in kinase activity (Horner, Osawa, Schaller, & Weiss, 2005; Pronin & Benovic, 1997). In contrast, phosphorylation of GRK2 by protein kinase-C enhances its activity (Chuang, LeVine, & De Blasi, 1995). Interestingly, where GRK2 and GRK5 can be phosphorylated by multiple kinases, there is very little evidence of phosphorylation of GRK3, GRK4, or GRK6. In addition to phosphorylation, other posttranslational modifications such as *S*-nitrosylation of GRK2 by nitric oxide synthase has been shown to inhibit its activity (Whalen et al., 2007). These contrasting effects highlight that there are GRK-specific regulation and suggests that there are multiple pathways involved in fine tuning each GRK's response and ability to regulate GPCRs, non-GPCRs, and nonreceptor biological functions. GRKs similar to other protein families that interact with GPCRs engage an interhelical cavity that opens on the cytoplasmic side of the GPCR after activation (Farrens, Altenbach, Yang, Hubbell, & Khorana, 1996) and it is probable that GRKs interact in a similar fashion to discern active vs inactive receptors (Homan & Tesmer, 2014). In recent studies, the crystal structures for various GRKs including GRK1, GRK2, and GRK6 have been characterized and it was observed that the conserved 18–20 amino acids on the N-terminal become helical and form an α-helix stabilizing the closed (active) conformation, allowing it to interact with both lobes of the kinase domain (Boguth, Singh, Huang, & Tesmer, 2010). This suggests then, that GPCR binding helps to promote GRK activity by helping to align catalytic residues of the kinase domain indicating that GPCRs can serve as both the substrate and the allosteric activators (Chen, Dion, Kim, & Benovic, 1993; Palczewski, Buczyłko, Kaplan, Polans, & Crabb, 1991). Furthermore, the mutation of this N-terminus α-helix abolishes GPCR phosphorylation suggesting that this helix formation is indeed the primary step in the activation of GRKs by GPCRs (Pao, Barker, & Benovic, 2009). This mechanism sets apart GRKs from other kinases that phosphorylate accessible residues (Ser, Thr, and Tyr) that are in a specific sequence context. Instead, GRK phosphorylation of Ser and Thr residues are not strongly sequence context-specific and this context independence may help to explain the versatility and flexibility of seven GRKs to regulate hundreds of different GPCRs (Vishnivetskiy et al., 1999). The allosteric binding of GRKs to GPCRs essentially ensures that GRKs, when bound, will phosphorylate nearby proteins with accessible loops, i.e., the intracellular loops and tails of the receptor molecules that they are currently bound to.

In the early years of their discovery, this phosphorylation of GPCRs was believed to be the only function of GRKs. With the identification of nonreceptor substrates, however, it has been shown that these nonreceptor substrates can be phosphorylated more efficiently than peptides derived from cognate GPCRs (Gurevich, Tesmer, Mushegian, & Gurevich, 2012). This suggests that there must be alternative mechanisms of activation other than binding to GPCRs but these mechanisms remain to be identified. One hypothesis on non-GPCR activation of GRKs stems from the idea that any physiochemical environment that resembles the binding pocket found on GPCRs could favor GRK activation.

## 1.3 Animal Models of GRK Deficiency

Studies using knockout and transgenic mice as well as using viral-mediated overexpression models have enabled researchers to identify various pathophysiological roles for GRKs. However, mice lacking these GRKs oftentimes appear to function normally without the addition of a stressor. As indicated before, GRKs also have dual functionality in various tissues by both suppressing G-protein signaling as well as promoting signaling independent of G-proteins. Because of this loss of regulation by any given GRK, knockouts can have differing effects that can either (a) allow enhanced or unregulated receptor signaling because of loss of desensitization or (b) alter the signaling cascade by preventing the switch from G protein signaling to non-G protein pathways or (*c*) alter the functionality of biological processes in a receptor-independent fashion (Premont & Gainetdinov, 2007). It is also possible that certain phenotypes are not exhibited due to compensation and overlapping roles of other GRKs present in the animal. For animals that do have abnormal functionality, the most notable phenotype was in the GRK2 homozygous knockout animal that was embryonic lethal because of defective cardiac development (Jaber et al., 1996). Interestingly, recent work has shown that this lethality, in fact, stems from a role of GRK2 in embryogenesis rather than a specific role in cardiac development; mice with cardiac myocyte-specific GRK2 ablation develop normally (Matkovich et al., 2006). Intrigued by the extreme phenotype, GRK2 heterozygous mice were heavily studied and GRK2 was shown to be important for heart development, lymphocyte chemotaxis, experimental autoimmune encephalomyelitis, sepsis, atherosclerosis and others, making GRK2 one of the most widely researched of all the GRKs. Studies focusing on the other GRK knockout and overexpression animals have shown an important role in several distinct phenotypes including GRK3 in olfaction (Peppel et al., 1997); GRK5 in cholinergic responses and inflammation; and GRK6 in chemotaxis, behavioral responses, locomotor-stimulating effect of cocaine, and others (Gainetdinov et al., 2003). Interestingly, recent studies crossing heterozygous GRK5 and heterozygous GRK6 knockout mice in order to generate GRK5/GRK6 double knockout mice were found to be embryonic lethal, though the mechanism is unknown (Burkhalter, Fralish, Premont, Caron, & Philipp, 2013). A detailed summary of all the phenotypes is provided in Table 1.

**Table 1** Summary of Phenotypes of GRK Knockout or Transgenic Mice

| GRK | Genotype | Phenotype | References |
|---|---|---|---|
| *GRK1* | GRK1 KO | Photon responses are increased and long lasting, loss of phosphorylation of rhodopsin | Chen et al. (1999) |
| | GRK1 overexpression | Increased damage to photoreceptors after intense light | Whitcomb et al. (2010) |
| *GRK2* | Homozygous KO (whole body) | Homozygous KO is embryonic lethal | Jaber et al. (1996) |
| | Heterozygous KO (whole body) | Susceptible to β-arrestin stimulation; altered progression of experimental autoimmune encephalomyelitis through increased infiltration of CNS by lymphocytes and macrophages; increase in lymphocyte chemotaxis toward CCL4 | Matkovich et al. (2006), Vroon, Heijnen, Lombardi, et al. (2004), and Vroon et al. (2005) |
| | Cardiac-specific KO | Enhanced ionotropic sensitivity to isoproterenol; lusitropic, tachyphylaxis | Matkovich et al. (2006) |
| | Cardiac-specific OE | Decreased β-arrestin signaling; attenuated ventricular contractility | Koch et al. (1995) |
| | Cardiac-specific β-ARKct | Enhanced contractility | Koch et al. (1995) |
| | Cardiac-specific β-ARKnt | Cardiac hypertrophy, enhanced β-AR density, and signaling | Keys et al. (2003) |
| | GRK2-C340S | Loss of nitrosylation-mediated cardioprotection from postischemic injury | Huang et al. (2013) |

*Continued*

**Table 1** Summary of Phenotypes of GRK Knockout or Transgenic Mice—cont'd

| GRK | Genotype | Phenotype | References |
|---|---|---|---|
| | Myeloid-specific KO | Increased Inflammation in endotoxemia and polymicrobial sepsis; decreased atherosclerotic lesions in LDL-myeloid dual KO mice | Otten et al. (2013), Parvataneni et al. (2011), and Patial, Saini, et al. (2011) |
| | Vascular smooth muscle OE | β-AR signal attenuation; attenuation of isoproterenol induced vasodilation; elevation of resting mean blood pressure; and hypertrophy | Eckhart, Ozaki, Tevaearai, Rockman, and Koch (2002) |
| | Adrenal-specific KO | Attenuates heart failure progression and improves function postmyocardial infarction | Lymperopoulos et al. (2010) |
| *GRK3* | GRK3 KO | Enhanced sensitivity to olfactory stimuli; altered M2 muscarinic airway regulation; and loss of neuropathic pain-induced opioid tolerance | Lodowski, Pitcher, Capel, Lefkowitz, and Tesmer (2003), Peppel et al. (1997), and Walker, Peppel, Lefkowitz, Caron, and Fisher (1999) |
| | Cardiac-specific OE | Normal β-AR signaling and hemodynamic function; decreased MAPK activity to thrombin | Iaccarino, Rockman, Shotwell, Tomhave, and Koch (1998) |
| | Cardiac-specific GRK3ct | Hypertension; α-1 receptor hypersensitivity | Vinge et al. (2008) |
| *GRK4* | GRK4 KO | No differences detected in fertility or sperm function | Virlon et al. (1998) |
| | GRK4-γ A142V | Hypertension, altered dopamine signaling | Felder et al. (2002) and Wang et al. (2007) |

**Table 1** Summary of Phenotypes of GRK Knockout or Transgenic Mice—cont'd

| GRK | Genotype | Phenotype | References |
|---|---|---|---|
| *GRK5* | GRK5 KO | Muscarinic M2 supersensitivity and impaired desensitization; decreased LPS-induced neutrophil infiltration in lung; decreased cytokine chemokine levels; enhanced hypothermia, hypoactivity, tremor, and salivation by oxotremorine; decreased NFκB activation in thioglycollate-induced peritoneal macrophages and cardiomyocytes; increased NFκB in endothelial cells; increased apoptotic response to genotoxic damage; and decreased thymocyte apoptosis during sepsis | Chen et al. (2010), Gainetdinov et al. (1999), Islam, Bae, Gao, and Koch (2013), Packiriswamy et al. (2013), Patial, Saini, et al. (2011), Rockman et al. (1996), and Sorriento et al. (2008) |
| | Cardiac-specific OE | Increased β-adrenergic receptor desensitization attenuating contractility and heart rate | Rockman et al. (1996) |
| | Cardiac-specific KO | Cardioprotection posttransaortic constriction | Gainetdinov et al. (1999) and Gold, Gao, Shang, Premont, and Koch (2012) |
| | Vascular smooth muscle OE | VSM-specific OE of GRK5 increases blood pressure through β1AR and Ang II receptors | Keys et al. (2003) |
| *GRK6* | GRK6 KO | Altered desensitization of D2-like dopamine receptors and supersensitivity to psychostimulants; deficient lymphocyte chemotaxis; and increased acute inflammation and neutrophil chemotaxis | Fong et al. (2002), Gainetdinov et al. (2003), Kavelaars et al. (2003), and Vroon, Heijnen, Raatgever, et al. (2004) |
| *GRK5/6* | Double KO | Embryonic lethal | Burkhalter et al. (2013) |

Research on the various GRKs ranging from the biochemical and molecular machinations and functionality to the whole body knockout phenotypes demonstrates a unique and essential role for this family of proteins in physiology. However, the role that these GRKs play in specific inflammatory responses is only beginning to be revealed. This review will focus mainly on our recent understanding of the roles that GRKs play in the immune system, specifically their influence on inflammatory signaling pathways and in inflammatory diseases.

# 2. GRKS IN INFLAMMATORY SIGNALING AND DISEASE

## 2.1 GRK1 Subfamily

### *2.1.1 Biochemistry*

The GRK1 subfamily is comprised of the retinal GRKs, GRK1 (rhodopsin kinase) and GRK7 (cone opsin kinase). These kinases are expressed in the retina of vertebrates but the expression pattern within the retinal cells is species dependent, where both GRKs are not always present. Highlighting these differences is the fact that all rod cells in vertebrates express GRK1 but cone cells express either GRK1, GRK7, or both, depending on the species (Osawa & Weiss, 2012). GRK1 and GRK7 share their major domains and sequence homology but much of the current work on this family has been done on GRK1; first, because GRK1 was discovered decades before GRK7 and second, because of the demonstration of the X-ray structure of GRK1. Functionally, after activation of rhodopsin by light, GRK1 facilitates desensitization by phosphorylating rhodopsin on Ser/Thr residues in the C-terminal domain, which facilitates arrestin recruitment and binding. In order to accomplish this, GRK1 and GRK7 must be properly localized to the membrane, and these GRKs are posttranslationally modified to ensure their proper localization within the cell. GRK1 differs from GRK7 in this regard in that GRK1 is posttranslationally modified via farnesylation and blocking this markedly limits its ability to phosphorylate active rhodopsin (Inglese, Koch, Caron, & Lefkowitz, 1992). In contrast, GRK7 is predicted to have a geranylgeranyl modification site on its C-terminus rather than the N-terminus like GRK1. These modifications are believed to serve as hydrophobic anchors that secure GRK1 and GRK7 to the cellular membranes (Inglese et al., 1992) and allow for these kinases to "probe" for activated receptors in a two-dimensional rather than three-dimensional search which is essential for rapid signal termination.

### *2.1.2 Role of GRK1 and GRK7 in Inflammation*

Owing to the limited tissue expression of GRK1 and GRK7, little is known about the role they play in inflammatory conditions of the retina. They do play a critical role in photoreceptors cells and GRK1 null-mice show shorter outer segments and undergo apoptosis if exposed to constant dim light (Chen et al., 1999). Overactivation of these GRKs can be detrimental to photoreceptor health as well and hyperphosphorylation of rhodopsin caused by GRK1 mutations can cause a retinal degenerative disease known as retinal pigmentosa (Shi et al., 2005). GRK1 levels are regulated in rat diabetic models in Brown Norway and Sprague–Dawley after 12 weeks but were unaffected in the Long Evans models (Kim, Kim, et al., 2005). What these altered levels mean in terms of GRK function and if they play a role in diseases such as diabetic retinopathy or other inflammatory conditions remain unknown. Despite their limited role in inflammation, it is important to think of how other therapeutics can influence these GRKs as a side effect of other treatments. For example, HSP90 inhibitors are being developed for oncology, retinal pigmentosa, inherited retinal degeneration, and other neurodegenerative diseases. But HSP90 is critical for the sustained production of GRK1. Treatments that include inhibition of HSP90 for sustained periods of time run this risk of adversely affecting visual function similar to Oguchi's disease (stationary night blindness), a disease that results from genetic mutations and low expression of GRK1 (Chen et al., 1999).

## 2.2 GRK2 Subfamily

### *2.2.1 Biochemistry*

The GRK2 subfamily is comprised of two GRKs, GRK2 and GRK3. These proteins are found ubiquitously throughout the body but show some variability in their localization and expression in different tissues and cell types. One of the fundamental differences that are unique to this subfamily is that their ability to phosphorylate GPCRs is restricted to GPCRs that are actively bound to a ligand (Li et al., 2015). Furthermore, the RGS homology domain (RH domain) in the N-terminus region of GRK2 family members are able to bind to G$\alpha_q$ and G$\alpha$11, inhibiting their downstream signaling capabilities (Carman, Parent, et al., 1999). Interestingly, neither GRK2 nor GRK3 is able to interact with G$\alpha$s, G$\alpha$i, G$\alpha$o, or G$\alpha$12/13, indicating that this functional interaction is specific to those two G-proteins (Carman, Parent, et al., 1999; Day, Carman, Sterne-Marr, Benovic, & Wedegaertner, 2003; Sallese, Mariggiò, D'Urbano, Iacovelli, & De Blasi, 2000). Their RH domain consists of nine α-helices that are analogous to other RGS domains and two additional α-helices derived from a region between the kinase and PH domain. This allows for the RH domain to interact with the PH domain indicating a possible role for regulation of the kinase activity of this subfamily. In addition to the unique binding capabilities of this subfamilies' RH domain, GRK2, and GRK3 contain a PH domain in the C-terminus region that is unique to only these two GRKs. The PH domain consists of seven β-strands and one C-terminal α-helix. This PH domain has the functional capability to bind to phosphatidylinositol 4,5-bisphosphate ($PIP_2$) as well as the Gβγ subunit of the heterotrimeric G protein. This binding to the Gβγ subunit is particularly important for cytoplasmic localization of this subfamily and the subsequent phosphorylation of activated GPCRs. It is believed that through the binding of GRK2 and GRK3 to Gβγ these GRKs can be shuttled to the activated GPCRs and is a potential mechanism for how GRK2 and GRK3 recognize active receptors (Benovic et al., 1989). Another consequence of interaction of Gβγ to GRK2 and GRK3 is that this high-affinity binding can sequester and interfere with downstream targets of Gβγ in a similar manner to $G\alpha_q$. With the discovery of GRK2's crystal structure in complex with Gβγ, we have been able to gain new insights into the regulation of GRK2 (and potentially GRK3). This structure places the three distinct domains at the vertices of a triangle and gives insight on how these proteins with multiple modular domains can act as an integrated signaling molecule (Lodowski et al., 2003). Interestingly, GRK2 and GRK3 are the only isoforms that interact with the free Gβγ subunit and translocate to the membrane in this fashion. Lastly, presence of the PH domain in combination with the cytoplasmic localization of these GRKs may provide a mechanism for this subfamily to influence a large number of nonreceptor substrates both in the presence and absence of GPCR ligand activity.

### 2.2.2 Role of GRK2 in Inflammation

GRK2 is ubiquitously expressed and primarily located in the cytosol. While most studies have looked at GRK2's ability to phosphorylate β-adrenergic and angiotensin II type 1 receptors, GRK2 has the capability to interact with a variety of GPCRs (Carman, Parent, et al., 1999). GRK2 has also been shown to phosphorylate other non-GPCR receptors including the platelet-derived growth factor receptor-β, epithelial Na+channel, and the downstream regulatory element antagonist modulator (DREAM) (Gurevich et al., 2012). It has also been suggested that GRK2 and GRK3 are more efficient at clathrin-mediated endocytosis than GRK5 and GRK6 (Ren et al., 2005). Outside of its role in phosphorylating and regulating GPCR functions, GRK2 has the ability to phosphorylate and/or interact with a large variety of cellular proteins in both tissue- and context-specific fashion. Because of its ability to regulate cellular processes in both canonical and noncanonical fashions, GRK2 has significant influence on a multitude of vital physiological functions including its role in inflammation and inflammatory diseases (Packiriswamy & Parameswaran, 2015).

Among the inflammatory pathways, research from our lab and others has shown that GRKs are important regulators of NFκB signaling pathway. NFκB is a critical player in the regulation of expression of many inflammatory genes and therefore is a target for therapeutic manipulation in several inflammatory diseases. Under unstimulated conditions, NFκB transcription factors (p65 (RelA), p50, RelB, cRel, and p52) are sequestered in the cytoplasm by members of the IκB family of proteins (IκBα, IκBβ, IκBε, p105 (NFκB1), and p100 (NFκB2)). Upon stimulation, IκB is phosphorylated by the IκB kinase complex (IKK) and undergoes ubiquitination and degradation. This releases the sequestered NFκB transcription factors which then translocate to the nucleus and modulate gene transcription (reviewed in Jobin & Sartor, 2000). GRK2 and GRK5 most notably regulate this signaling pathway to influence the outcome of inflammation in immune and nonimmune cell types. GRK2 has been shown to directly interact with proteins involved in this signaling pathway, IκBα, and p105. GRK2 is able to phosphorylate IκBα albeit at very low stoichiometry and able to modulate this pathway in response to TNFα in a macrophage cell line and in HEK293 cells (Patial, Luo, Porter, Benovic, & Parameswaran, 2009). This relationship between GRK2 and NFκB signaling was recently shown in neonatal rat cardiac fibroblasts in response to arginine vasopressin (AVP). AVP increased mRNA and protein levels of IL-6 in these cells and pharmacological inhibition of GRK2 ablated this increase and NFκB activation, linking GRK2 and inflammation in cardiac stress (Xu et al., 2017). This biochemical regulation was also recently demonstrated in colon cancer epithelial cell line and this effect of GRK2 on IκBα was linked to its role in positively inducing some inflammatory genes in these cells (manuscript in revision). Interestingly, GRK2 has also been shown to interact with p105 and modulate ERK pathway downstream of p105 in primary macrophages. In this case, GRK2 was shown to be a negative regulator of this pathway and downstream inflammatory genes (Patial, Saini, et al., 2011). Interestingly,

Toll-like receptor ligands induce GRK2 expression in primary macrophages and neutrophils (Alves-Filho et al., 2009; Loniewski, Shi, Pestka, & Parameswaran, 2008). Additionally, GRK2 has also been shown to localize in the mitochondria after LPS stimulation and regulate the production of reactive oxygen species (Sorriento et al., 2013). Furthermore, immune cells from septic patients exhibit increased levels of GRK2 indicating a possible clinical link between GRK2 levels and the associated signaling pathways and inflammatory diseases (Arraes et al., 2006). This was shown to be the case at least in an animal model of sepsis where IL-33-mediated decrease in GRK2 expression was beneficial for neutrophil migration and therefore better clearance of bacteria and improvement in sepsis. Thus, GRK2 has been suggested as a possible target in sepsis pathogenesis (Alves-Filho et al., 2010).

Another common set of pathways activated in response to inflammatory stimuli are the mitogen-activated protein kinase (MAPKs) pathways. These kinases are a family of serine/threonine protein kinases that respond and react to extracellular signals. These reactions typically mediate a variety of fundamental biological processes ending in the translocation of transcription factors from the cytoplasm to the nucleus, in order to alter gene expression. These MAPK signaling cascades are generally broken down into three major groups: extracellular signal-regulated protein kinases 1 and 2 (ERK 1/2), p38 MAPK, and Jun N-terminus kinase (JNK). These individual kinases are activated upstream by additional MAPKs, specifically: ERK is activated by MAPK kinase 1 (MKK1) and MKK2; p38 MAPK by MKK3, MKK4, and MKK6; and JNK by MKK4 and MKK7.

Like the NFκB pathway, GRK2 most notably regulates this family of signaling proteins in response to both GPCR ligands (i.e., CXCL8) and non-GPCR ligands (e.g., TNFα). Because of the plethora of studies looking at the role of GRK2 in MAPK signaling in response to various GPCR ligands (Evron, Daigle, & Caron, 2012; Penela, Ribas, Aymerich, & Mayor, 2009); here, we focus primarily on the role of GRK2 in MAPK signaling in the context of inflammation. When these ligands activate the ERK1/2 pathway they can induce a wide variety of inflammatory mediators such as TNFα, interleukin 1 (IL-1), IL-8, and prostaglandin E2. The effect of GRK2 on this pathway has been studied in a variety of contexts beginning with its role in macrophages where GRK2 was shown to negatively regulate LPS-induced ERK signaling (Patial, Saini, et al., 2011). GRK2 has also been shown to bind to Raf1 and RhoA that can serve as a scaffolding protein for the ERK pathway enhancing its activation in response to EGF (Robinson & Pitcher, 2013). In a similar study, GRK2 inhibition in cardiomyocytes increased ERK activity through interaction with the RAF MAPK axis (Fu, Koller, Abd Alla, & Quitterer, 2013). Another study focusing on the connection between the scaffolding ability of β-arrestin2 and GRKs found that in HEK cells overexpression of GRK2 completely mitigated the β-arrestin2-mediated increase in ERK1/2 activation (Kim, Ahn, et al., 2005). These effects held true with a variety of receptors including the β2-adrenergic receptor, cannabinoid receptor-2 (Franklin & Carrasco, 2013), angiotensin 1A receptor (Sorriento et al., 2008) and the insulin-like growth factor-1 receptor (Zheng et al., 2012). GRK2 also has the ability to regulate ERK signaling through direct association with MEK (MKK1) and increased levels of GRK2 inhibit ERK activation following chemokine induction (Jimenez-Sainz et al., 2006). Consistent with that, our lab also has recent data in colon epithelial cells that ERK1/2 activity is enhanced in GRK2 knockdown cells not through MKK1 regulation but rather through altering the levels of ROS produced in those cells (manuscript in revision). Interestingly, GRK2 is able to regulate ERK in a variety of fashions; but ERK can also directly interact with and phosphorylate GRK2. Agonist-induced GRK2 phosphorylation by ERK1/2 reduces GRK2's activation by Gβγ and its ability to phosphorylate receptors and marks it for degradation (Elorza, Sarnago, & Mayor, 2000; Pitcher et al., 1999).

The p38 MAPK pathway shares many similarities with the other MAPK signaling responses including inflammation, cell growth, differentiation, and death. Similar to ERK, the p38 MAPK pathway is stimulated by a wide array of stimuli including LPS and other TLR activators such as enterotoxin B and herpes simplex virus (Nick, Avdi, Gerwins, Johnson, & Worthen, 1996; Zachos, Clements, & Conner, 1999). In the context of inflammation and the inflammatory response, p38 plays a critical role in the production of inflammatory mediators to initiate leukocyte recruitment and activation. It accomplishes this task through the regulation of inflammatory genes including TNFα, IL-1β, IL-6, IL-8, cyclooxygenase-2 (COX2), and collagenase 1 and 3 (Lee et al., 1994) all of which are diminished in monocytes/macrophages, neutrophils, and T lymphocytes in the presence of p38 inhibitor SB203580 (Marriott, Clarke, & Dalgleish, 2001). Previous studies have shown that GRK2 and p38 appear to oppose each other where GRK2 can inhibit p38 function by directly phosphorylating it whereas p38 inhibits GRK2-mediated GPCR desensitization (Liu et al., 2012; Peregrin et al., 2006). It regulates GRK2 by directly phosphorylating GRK2 at Ser670 inhibiting its ability to translocate to the outer membrane and interact with activated GPCRs for desensitization. This specifically has been observed in response to MCP-1 and the prevention of the CC chemokine receptor 2 (CCR2) (Liu et al., 2013). GRK2 is able to phosphorylate p38 at its Thr123 residue, which interferes with the ability of p38 to bind to MKK6, therefore, preventing its activation (Peregrin et al., 2006). Furthermore, altering the levels of GRK2 influences the activation of p38 and its functionality including differentiation and cytokine production. This alteration holds true in murine $GRK2^{+/-}$ macrophages and microglial cells; both of these cells exhibit increased p38 activation and TNFα production in response to LPS (Nijboer et al., 2010). In contrast to this, complete GRK2 silencing in mast cells decreases cytokine production (IL-6 and IL-13) in a p38-dependent manner during antigen-induced degranulation (Subramanian, Gupta, Parameswaran, & Ali, 2014). Lastly, GRK2 through its regulation of p38 has been linked to cytokine-induced pain by reducing the neuronal responsiveness to cytokines such as IL-1β and TNFα and thereby reducing cytokine-induced hyperalgesia (Eijkelkamp et al., 2012; Willemen et al., 2010).

Similar to ERK and p38 pathways, JNK can also be activated by mitogens as well as a large variety of stimuli. The role of GRKs in JNK signaling in the immune system and inflammation is not very well characterized. GRK2 has a large and diverse regulatory capacity in the context of NFκB and MAPK signaling pathways that is dependent on expression level, cell type, stimulus, and many other factors. As we continue to study this role both in vitro and in vivo and our understanding of this regulation increases, it remains a likely possibility that novel therapeutic targets and treatments for inflammatory diseases can develop from these interactions.

### *2.2.3 GRK2 Regulation of Immune Processes*

An important component of an effective immune system is the ability for the immune cells to arrive at the site of inflammation. This is achieved through a process called chemotaxis, where cells at the site of inflammation or disease produce chemokines that act on the chemokine receptors (mostly GPCRs) on immune cells to initiate a chemotactic response. The amplitude of that response is dependent on the amount of chemokines produced and their gradient as well as the expression levels of the receptors that are responding to them. This process is also dependent on a series of different steps in the chemotactic process including receptor sensing, cell polarization, membrane protrusion, and adhesion/deadhesion cycles all of which can be cell type and stimulus specific (Insall, 2013). Given that most of the chemokine receptors are GPCRs, these receptors undergo desensitization with constant stimulus and it is not surprising that the GRKs regulating that process play a critical role in chemotaxis and in the immune response. Interestingly, GRK2, GRK3, GRK5, and GRK6 are all expressed at high levels in the immune cells suggesting that alteration or regulation of their expression levels or functionality will drastically influence immune cell chemotaxis and the pathological progression of disease.

GRK2 has been heavily studied in its role in chemotaxis; but looking at the regulation of cellular migration by this kinase is complex. Studies have shown that this regulation is both cell type and stimulus dependent, with GRK2 acting most often in a canonical manner negatively regulating GPCR signaling (Arnon et al., 2011; Liu et al., 2013; Zachos et al., 1999). Overexpression of GRK2 by transfection in cell lines increases phosphorylation and/or desensitization of different chemokine receptors such as CCR2b (Aragay et al., 1998), CCR5 (Olbrich, Proudfoot, & Oppermann, 1999), and CXCR1 (Raghuwanshi et al., 2012). Increased circulating neutrophils as well as increased macrophage localization to inflammatory sites have been observed in mice that are deficient for GRK2 in the myeloid cells or in bone marrow cells of low-density lipoprotein receptor knockout chimeric mice in an atherosclerosis model (Otten et al., 2013). Consistently, both T lymphocytes and splenocytes isolated from GRK2 heterozygous mice had increased migration due to increased activation of ERK and PI3K/Akt pathways in response to CCL5 and CXCL12 compared to WT controls (Vroon, Heijnen, Lombardi, et al., 2004). This, however, was not the case in the sepsis model in the myeloid-specific GRK2 knockout suggesting disease-specific regulation (Parvataneni et al., 2011). Interestingly, a recent study has shown that acute mental stress significantly increased GRK2 levels in peripheral blood mononuclear cells at both 30 and 60 min and had a positive correlation with levels of epinephrine that also increased after the stressor. This study suggests a link between stress and intracellular inflammatory signaling and chemotaxis (Blake Crabb et al., 2016). GRK2 levels are also reduced in cultured human T lymphocytes when exposed to oxidative stress or when cocultured with activated neutrophils (Lombardi et al., 2002). Additionally, GRK2 has been examined in the context of various human diseases and neutrophils from patients with different disease conditions including malaria (Leoratti et al., 2012) and sepsis (Arraes et al., 2006) express increased GRK2 levels that is associated with decreased CXCR2 expression and reduced response to IL-8. When this phenomenon was studied in mice, exposure to IL-33 reversed this overexpression of GRK2 and restored CXCR2 expression leading to increased chemotaxis as well as increased bacterial clearance and survival (Alves-Filho et al., 2010).

In contrast to the above studies, GRK2 can also positively regulate chemotaxis in certain cell types. It's hypothesized that this bimodal role for GRK2 may depend on the cell type as well as the polarization state of the cells. In polarized cells, such as epithelial cells, GRK2 has been shown to positively regulate cellular migration independently of its catalytic activity thus supporting protein–protein interaction as a potential mechanism (Penela et al., 2008). Further supporting this idea, membrane-targeted kinase mutants strongly enhance cell motility through interactions between GRK2 and GIT1 (GRK2-interacting factor) at the leading edge of polarized/migrating cells in wound healing assays (Penela et al., 2008). This association between GRK2 and GIT1 is critical for proper ERK1/2 activation and efficient cellular migration. Recent work in our lab has shown in colon epithelial cells GRK2 knockdown enhances epithelial migration in wound healing assays by critically regulating the activation of ERK1/2 in response to TNFα (manuscript in revision). These studies highlight the importance of ERK1/2 activation in polarized/migrating epithelial cells and the ability for GRK2 to influence migration and chemotaxis in a variety of manner. GRK2 has also been shown to directly phosphorylate histone deacetylase 6, a cytoplasmic histone deacetylase responsible for the deacetylation of tubulin and other substrates critical for cellular migration (Lafarga, Mayor, & Penela, 2012). Additionally, GRK2 can interact with and phosphorylate ERM (ezrin–radixin–moesin) proteins ezrin and radixin, both of which contribute to F-actin polymerization-dependent motility (Cant & Pitcher, 2005; Kahsai, Zhu, & Fenteany, 2010). These novel interactions and nonreceptor substrates for GRK2 contribute to our understanding of GRK2s role in noncanonical regulation of GRK2 in cell motility.

### 2.2.4 GRK2 in Disease

In vitro experiments can give valuable insights into the mechanisms of GRK2 behavior but understanding the role of GRK2 in the context of human disease oftentimes is beyond the scope of an isolated cell line in culture. In mouse and human studies, GRK2 has been implicated in a variety of pathogenesis in multiple tissues and organ systems. GRK2 is a major player in several neurodegenerative diseases such as Alzheimer's disease (AD) (Suo & Li, 2010), multiple sclerosis (MS) (Vroon et al., 2005), and Parkinson's disease (PD) (Bychkov, Gurevich, Joyce, Benovic, & Gurevich, 2008). In AD, GRK2 was shown to serve as a biomarker for early hypoperfusion-induced brain damage, which is associated with mitochondrial damage seen in patients with AD (Obrenovich, Palacios, Gasimov, Leszek, & Aliev, 2009). GRK2 levels are also increased during the early stages of damage in aged human and in AD patients (observed postmortem) (Obrenovich et al., 2009). This increased expression may lead to more interactions with α- and β-synucleins both of which are substrates for GRK2 and have been linked to both PD and AD (Surguchov, 2008). Furthermore, GRK2 was shown to regulate metabotropic glutamate receptor function and expression, whose role has been implicated in AD and MS pathogenesis (Degos et al., 2013; Sorensen & Conn, 2003). Also, studies have shown that downregulation of GRK2 following chronic inflammation sensitizes human and rodent neurons to excitotoxic neurodegeneration via overaction of group I mGluRs (Degos et al., 2013).

GRK2 has also been shown to play a prominent role in inflammatory hyperalgesia. GRK2 heterozygous mice suffer from chronic hyperalgesia due to continued microglial activation via p38-dependent TNFα production (Eijkelkamp, Heijnen, et al., 2010; Willemen et al., 2010) and prolonged activation of prostaglandin E2-mediated pathway. This prolonged activation is mediated through interactions with EPAC1 (exchange protein directly activated cAMP) and activation of protein kinase Cε- and ERK-dependent pathways (Eijkelkamp, Wang, et al., 2010; Wang et al., 2013). Through this same regulation of p38, GRK2 exacerbates brain damage in hypoxic–ischemic injury (Nijboer et al., 2010). The importance of GRK2 in this system is highlighted in that even a transient decrease in GRK2 levels (by intrathecal injection of *GRK2* antisense oligodeoxynucleotides) is sufficient to produce long-lasting neuroplastic changes in nociceptor function that can lead to chronic pain (Ferrari et al., 2012).

Owing to the cardiac developmental defects in the GRK2 homozygous knockout mice, GRK2 has been heavily studied in the context of cardiovascular diseases. GPCR signaling via the β-adrenergic receptor predominates during heart failure in an attempt to improve myocardial contractility and cardiac output. However, increased levels of GRK2 during heart failure cause an increase in β-adrenergic receptor desensitization reducing myocardial contractility and cardiac output worsening heart failure (Iaccarino et al.,

2005). This reduced contractility triggers an influx of catecholamines, which leads to a cycle of activation and persistent desensitization of β-adrenergic receptors. Overexpression of GRK2 in vivo mimics this regulation and decreases myocardial contractility and cardiac output in response to adrenergic stimulation. These effects suggest impaired adrenergic receptor signaling (Brinks et al., 2010; Chen, Bittner, Akhter, Koch, & Davis, 2001). Conversely, inhibition of GRK2 results in increased contractility and enhanced survival in heart failure models (Raake et al., 2008; Vinge et al., 2008). Furthermore, competitive inhibition through βARKct (C-terminal peptide of GRK2) prevented cardiomyopathy and improved heart failure (Harding, Jones, Lefkowitz, Koch, & Rockman, 2001; Rengo et al., 2009; Shah et al., 2001; Sorriento et al., 2010). These benefits seen in the cardiovascular diseases have inspired researchers to try and develop GRK2 inhibitors as a therapeutic option in heart failure and has been met with success at least in preclinical models (Sorriento, Ciccarelli, Cipolletta, Trimarco, & Iaccarino, 2016).

Changes in GRK2 expression has also been implicated in the pathogenesis of other diseases. Overexpression of GRK2 in vascular smooth muscle cells has been shown to lead to the development of hypertension (Eckhart et al., 2002; Keys, Zhou, Harris, Druckman, & Eckhart, 2005) and this increase in GRK2 expression is also found in human patients with hypertension (Gros et al., 1999). Also, GRK2 has been implicated in the development of atherosclerosis. Low-density lipoprotein receptor knockout mice with partial GRK2 deficiency in hematopoietic cells developed fewer atherosclerotic plaques (Otten et al., 2013). This partial GRK2 deficiency led to increased infiltration of macrophages to inflammatory sites resulting in plaques with smaller necrotic cores. Finally, GRK2 has a significant role in the pathogenesis of human and animal models of sepsis (Alves-Filho et al., 2010; Arraes et al., 2006; Packiriswamy et al., 2013; Packiriswamy, Parvataneni, & Parameswaran, 2012; Parvataneni et al., 2011), polymicrobial sepsis (Alves-Filho et al., 2010; Packiriswamy et al., 2013; Parvataneni et al., 2011), endotoxemia (Packiriswamy et al., 2012; Patial, Saini, et al., 2011), and in the regulation of septic shock. Sepsis and a systemic inflammatory response syndrome are leading causes of mortality in intensive care units (Melamed & Sorvillo, 2009) and GPCRs play a major role in the pathophysiological events through regulation of the cardiovascular, immune, and coagulatory responses. Given the role of GRK2 in the pathogenesis of sepsis, inhibition of

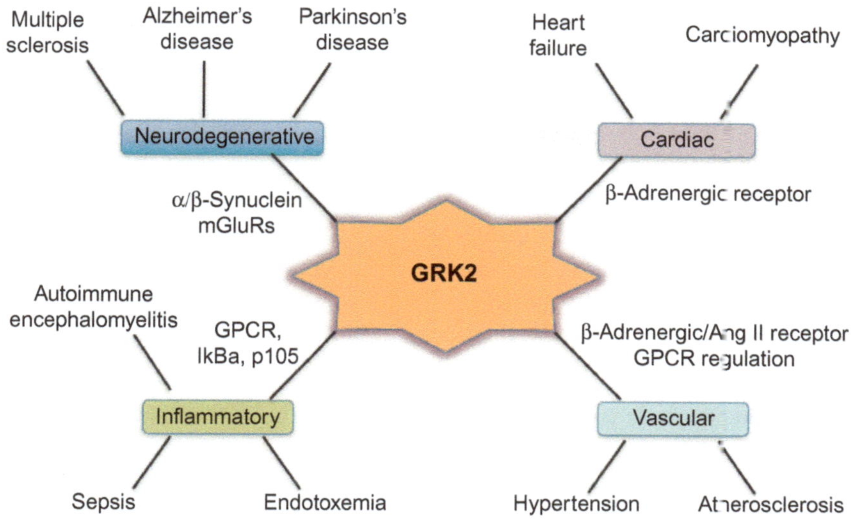


**Fig. 1** GRK2 and its potential role in various disease processes. Some interacting partners/substrates are shown. Please see text for details.

GRK2 expression, and/or activity may be a useful strategy for targeting human sepsis (Alves-Filho et al., 2010; Parvataneni et al., 2011). However, given that GRK2 is ubiquitously expressed and its role varies in different tissues and cell types, inhibition of GRK2 may be beneficial in some diseases whereas it may be deleterious in others. Thus, while GRK2 inhibitors are being developed, it is important to keep in perspective the various roles of GRK2 in different cell signaling pathways, the multitude of substrates/interacting partners and their potential involvement in the various diseases (Fig. 1).

### *2.2.5 GRK3*

GRK3, like GRK2, is ubiquitously expressed throughout the body but its expression levels are considered to be lower than those of GRK2 in all areas except for the central nervous system and the brain (Erdtmann-Vourliotis, Mayer, Ammon, Riechert, & Höllt, 2001). Despite these similarities in tissue expression, GRK3 has an independent role from GRK2 highlighted by its regulation in olfaction and neuronal functions. In the central nervous system, GRK3 is primarily located within the dorsal root ganglions and the olfaction neurons regulating the desensitization of odorant receptors and α2-adrenergic receptors (Walker et al., 1999).

### 2.2.6 GRK3 in Inflammation

Role of GRK3 in inflammatory signaling has been less established than some of the other GRKs. However, GRK3 has been shown to play a role in ERK1/2 signaling where overexpression of GRK3 abolished β-arrestin2-mediated activation of ERK1/2. This was seen in response to activation of a variety receptors including the β2-adrenergic receptor, cannabinoid receptor-2 (Franklin & Carrasco, 2013), angiotensin 1A receptor (Sorriento et al., 2008), and the insulin-like growth factor-1 receptor (Zheng et al., 2012). In support of this, bone marrow-derived leukocytes from GRK3 knockout mice exhibit enhanced ERK1/2 expression in response to CXCL12 (Tarrant et al., 2013). GRK3 can also interact with and modulate these pathways and so further study in this area is necessary to fully understand the regulatory capabilities of this GRK on these pathways.

### 2.2.7 GRK3 Regulation of Immune Processes

As indicated before, GRK2, GRK3, GRK5, and GRK6 are all expressed at high levels in the immune cells suggesting that alteration or regulation of their expression levels or functionality could influence immune cell chemotaxis and therefore progression of disease. Compared to the other GRKs relatively little work has been done on the role of GRK3 in these cells; however, recent work has shown that GRK3 is able to critically regulate immune behavior. GRK3 was identified as a negative regulator of CXCL12 (CXC chemokine ligand 12)/CXCR4 signaling (a defective pathway in human WHIM syndrome) and deficiency of GRK3 results in impaired CXCL12-mediated desensitization, enhanced response to CXCR signaling through ERK1/2 activation, altered granulocyte chemotaxis, and mild myelokathexis (a form of chronic leukopenia) (Tarrant et al., 2013). Work done on human skin cells and leukocytes indicate that overexpression of GRK3 is able to restore CXCL12-induced internalization and desensitization of CXCR4 and normalize chemotaxis (Balabanian et al., 2008). GRK3 is also important in other granulocyte-dependent disease models by mediating the retention of cells in the bone marrow resulting in fewer circulating granulocytes in the blood as well as in the joints during inflammatory arthritis. In conjunction with regulation of CXCR4, GRK3 has also been reported to play an important role in oncology and influence the tumorigenicity, molecular subtype, and metastatic potential of triple negative breast cancer, glioblastoma, ovarian tumors, medulloblastoma, and malignant granulosa cells via dysregulation of GPCR signaling (Billard et al., 2016;

King et al., 2003; Woerner et al., 2012). In a cultured leukemia cell line, GRK3-mediated receptor phosphorylation of the chemokine receptor CC3A and association with β-arrestins were essential for expression of the chemokine CCL2 (DeFea, 2013). GRK3 has also been shown to play a critical role in opioid receptor signaling in that Mu-opioid receptor and kappa-opioid desensitization is significantly slower in the GRK3 knockout mice (Stacey, Lin, Gordon, & McKnight, 2000).

### *2.2.8 GRK3 in Disease*

Unlike GRK2, homozygous knockout mice of GRK3 are viable thus demonstrating different physiological roles for the members of the GRK2 family. GRK2 and GRK5 are well known for their role in cardiovascular diseases but the effect of GRK3 in this tissue is less understood. Transgenic mice expressing cardiac-specific GRK3 inhibitor exhibited enhanced ERK1/2 activation in response to $\alpha_1$-adrenergic receptor stimulation. Additionally, these transgenic mice showed elevated systolic blood pressure and hypertension due to hyperkinetic myocardium and hyperresponsiveness to $\alpha_1$-adrenergic receptor stimulation (Vinge et al., 2008). GRK3 has also been shown to be overexpressed in human prostate cancer and found to promote tumor growth and metastasis by enhancing angiogenesis (Li et al., 2014). Finally, based on its chromosomal location and sequence, GRK3 is believed to be a potential biomarker for bipolar disorder susceptibility through potential alterations in dopamine receptor desensitization (Niculescu et al., 2000). In mice, GRK3 deficiency plays no role in basal locomotor activity but interestingly, knocking out GRK3 significantly reduces locomotor and climbing responses to either cocaine or apomorphine (Gainetdinov, Premont, Bohn, Lefkowitz, & Caron, 2004) potentially through the desensitization of D3 dopamine autoreceptors (Kim, Gainetdinov, Laporte, Caron, & Barak, 2005).

## 2.3 GRK4 Subfamily

The GRK4 subfamily is comprised of GRK4, GRK5, and GRK6 and is located ubiquitously throughout the body with the exception of GRK4, which is located primarily in the testes, kidney, and cerebellum (Inglese et al., 1993). Structurally all members share the AGC kinase domain and a common N-terminus domain that expresses a $PIP_2$-binding site that is similar to the GRK2 subfamily. The C-terminal domain aids in membrane localization and is regulated through palmitoylation sites on GRK4 and

GRK6, but GRK5 is regulated through positively charged lipid-binding amphipathic helices (Cabrera-Vera et al., 2003). This regulation establishes an equilibrium between these GRKs and the membrane allowing for phosphorylation of both activated and inactivated GPCRs (Virlon et al., 1998). Interestingly, GRK4 has low sequence conservation with GRK5 and GRK6 in the C-terminal region and forms a different crystal structure. These differences between GRK4 and the rest of the subfamily have been shown to influence its ability to translocate to the plasma membrane (Xu, Jiang, Shen, Fischer, & Wedegaertner, 2014). It has been hypothesized that these changes in structure and membrane targeting capabilities may be a result of the tissue-specific expression of GRK4 and may account for some of the functional variability among this subfamily.

### *2.3.1 GRK4*

#### 2.3.1.1 Biochemistry, role in inflammation and disease

GRK4 can be regulated and alternatively spliced into four different splice variants that are all capable of independent functions including GCPR regulation. As indicated before, GRK4 has not been examined in detail in immune cells and its role in inflammatory disease is largely unknown. It is known that GRK4α constitutively phosphorylates the dopamine receptor in the proximal tubule reducing its responsiveness hinting at an essential role in the regulation of hypertension. In fact, constitutive phosphorylation of the dopamine receptor is problematic in people with hypertension and polymorphisms in GRK4 can cause dysregulation of dopamine-stimulated salt and fluid excretion in the kidneys (Felder et al., 2002). Similarly, GRK4γ is able to phosphorylate the dopamine receptor as well but only upon agonist activation. These differences between splice variants indicate areas of similarities but also variability among the splice variants leading to questions about the regulation of GRK4 translation that have yet to be answered (Villar et al., 2009). Role of GRK4 in the other tissues where it is expressed is less clear and requires further study.

Expression of GRK4 in the cerebellum seems to be limited to the Purkinje cells and may be involved in the regulation of metabotropic glutamate 1 receptors. The regulation of these receptors suggests a role for GRK4 in motor coordination as well as in learning, but further work remains to be done in this area (Iacovelli et al., 2003). In a recent study, it was shown that GRK4 levels are upregulated in *Fmr*1-null mice, suggesting that RNA-binding protein (FMRP) negatively regulates the expression of GRK4. Fragile X syndrome is caused by the silencing of the *Fmr1* gene and lack of FMRP. Since FMRP negatively regulates GRK4, it is hypothesized that this silencing results in increased GRK4 levels in Fragile X syndrome. Increased GRK4 could contribute to cerebellum-dependent phenotypes due to deregulated desensitization of GABA-B receptors (Maurin et al., 2015). Finally, due to its localization in the cerebellum and testes, studies in GRK4 knockout mice were done investigating its role in these tissues. Interestingly, GRK4 knockout animals showed no differences in basal levels or in response to cocaine for locomotor activity or motor coordination (Gainetdinov et al., 2004). Furthermore, despite the high levels of GRK4 in the testes, there were no apparent changes in fertility or sperm function and no obvious phenotypes were detected in these animals (Virlon et al., 1998). Thus, the role of GRK4 in these tissues remains to be elucidated and more work needs to be done to fully understand the role of this kinase in inflammatory processes.

### *2.3.2 GRK5*

#### 2.3.2.1 Biochemistry

GRK5 is unique among its family members due to the presence of positively charged lipid-binding amphipathic helices on its C-terminus. There is also evidence supporting an interaction with the membrane through this C-terminus involving a region of the RH domain (Xu et al., 2014). These motifs, in combination with the polybasic N-terminus regions, predominantly localize GRK5 to the plasma membrane. This presence at the membrane allows for significant activation-independent phosphorylation of several GPCRs (Li et al., 2015). GRK5 has recently been crystalized in two unique structures, both of which crystalize GRK5 as a monomer with its C-termini forming consistent structures that pack closely to the RH domain. Disrupting this interface between the C-terminus and the RH domain significantly decreased the catalytic activity on GPCRs in vitro as well as in cells. This same study concluded that individual GRK5 subunits were insufficient for persistent membrane association (Xu et al., 2014). Additionally, GRK5 also has a nuclear localization motif and can accumulate in the nucleus where it can interact with proteins including class II histone deacetylase and mediate gene transcription (Martini et al., 2008). This localization is in competition with membrane localization sequences on the C- and N-termini. These termini are in fact key regulatory sites. For example, phosphorylation by PKC on the C-terminus is inhibitory to its function. CaM is inhibitory to most GRKs but is especially preferential to the GRK4 subfamily. $Ca^{2+}$·CaM can bind to both the

N- and C-termini phospholipid-binding motifs and disrupt membrane association of GRK5. This then allows the nuclear localization motif to direct GRK5 to the nucleus (Gold et al., 2013). As mentioned, GRK5 is ubiquitously expressed but is especially expressed at high levels in the heart, lungs, placenta, and kidney with lower expression in the brain (except for the limbic system). Interestingly, expression levels of GRK5 have been shown to increase twofold during neuronal differentiation (Inglese et al., 1993).

#### 2.3.2.2 GRK5 Role in Inflammation

Similar to GRK2, GRK5 plays a major role in NFκB signaling and is able to directly interact with and phosphorylate NFκB p105 (one of the IκB members). This was shown to inhibit Toll-like receptor-4-induced IκB kinase β-mediated phosphorylation of p105, thus negatively regulating lipopolysaccharide-induced ERK activation in macrophages (Parameswaran et al., 2006). Furthermore, GRK5 was subsequently discovered to interact with an additional member of the IκB family, IκBα, to facilitate the nuclear accumulation of IκBα by masking its nuclear export signal. This nuclear accumulation of IκBα led to decreased NFκB activation in endothelial cells (Sorriento et al., 2008). Both of these interactions were shown to be dependent on the RH domain of GRK5 and independent of its kinase domain. Further research has also shown noncanonical regulation of NFκB by GRK5. GRK5 was demonstrated to phosphorylate Ser32/36—the same sites phosphorylated by IκB kinase α and β. This was further supported by decreased levels of cytokines and chemokines in GRK5 knockout mice compared to wild type in the endotoxemia model (Patial et al., 2009; Patial, Shahi, et al., 2011). Further evidence showed that IκBα phosphorylation and p65 nuclear translocation were significantly reduced in LPS-treated GRK5-deficient peritoneal macrophages. As these studies progressed, there have been contrasting results in different models of the GRK5 knockout mouse. Wherein mice generated by the Lefkowitz group (Wu et al., 2012) found that endothelial GRK5 stabilized IκBα similar to earlier studies by Sorriento et al. (2008), studies done in macrophages using different GRK5 knockout mice did not show any role for GRK5 in IκBα phosphorylation or p65 translocation. More recent studies using mice from the Lefkowitz group, however, have shown that GRK5 positively regulates the NFκB pathway in cardiomyocytes (Islam et al., 2013). It is unclear the exact reasoning behind these discrepancies in GRK5 behavior, but it has also been shown that GRK5 itself can be regulated by the NFκB pathway, generating a potential positive feedback loop between GRK5 and NFκB (Islam & Koch, 2012). However, in primary peritoneal macrophages, GRK5 was downregulated after TLR4 activation, suggesting that both regulation of GRK5 expression and its influence on the NFκB pathway are distinct in different cell types and conditions. Interestingly, however, GRK5 was shown to be an evolutionarily conserved kinase in NFκB regulation in human cell lines as well as different species including *Drosophila* and zebrafish (Valanne et al., 2010), thus demonstrating an important role for this kinase in this pathway.

There are some studies examining the relationship between GRK5 and MAPK signaling pathways. GRK5 has been shown to play a role in support of β-arrestin2-mediated ERK activation following stimulation of V2 vasopressin and angiotensin II receptors despite its minor role in desensitization (Kim, Ahn, et al., 2005; Ren et al., 2005). This was later determined to be independent of G-protein coupling (Shenoy et al., 2006). Conversely, in macrophages, through its regulation of p105, GRK5 negatively regulates LPS-stimulated ERK activation (Parameswaran et al., 2006).

JNK can be activated by mitogens as well as a large variety of stimuli including environmental stresses (heat shock, ionizing radiation, and oxidants), genotoxins (topoisomerase inhibitors and alkylating agents), ischemic–reperfusion injury, mechanical shear stress, vasoactive peptides, proinflammatory cytokines, and pathogen-associated molecular pattern molecules/danger-associated molecular patterns (Dabrowski, Grady, Logsdon, & Williams, 1996; Gómez del Arco, Martínez-Martínez, Calvo, Armesilla, & Redondo, 1996; Knight & Buxton, 1996; Komuro et al., 1996; Liu, Gorospe, Yang, & Holbrook, 1995). JNK activation induces the translocation of transcription factors AP-1, c-Jun, ATF-2, and ELK-1 to the nucleus to regulate inflammatory gene transcription (Liu et al., 1995). JNK activation of AP-1 also plays a critical role in the synthesis of TNFα and for the proliferation and differentiation of lymphocytes, giving it a vital role in the regulation of the immune system (Gómez del Arco et al., 1996; Jung et al., 1995). In mice overexpressing GRK5 in cardiac myocytes, constitutively active mutants of $\alpha_{1B}$AR exhibited attenuation of JNK activation in comparison to the controls. GRK5 also had an influence of $\alpha_{1B}$AR signaling, but the complexity of this interaction and regulation remains to be fully elucidated (Eckhart et al., 2000). Finally, there is little known about the relationship between GRK5 and p38,

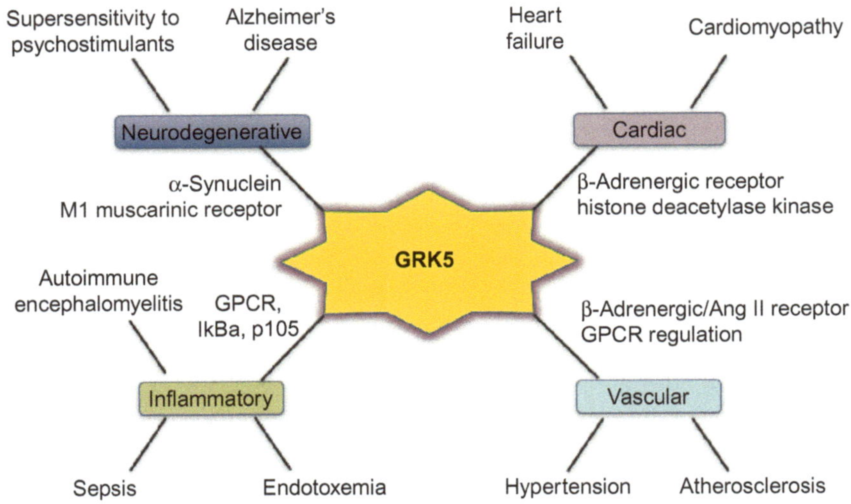


**Fig. 2** GRK5 and its potential role in various disease processes. Some interacting partners/substrates are shown. Please see text for details.

highlighting an area of study that still requires further work to understand the role for this kinase in the MAPK processes.

It is clear that GRK5 plays a critical role in the regulation of NFκB and MAPK signaling in inflammation and immune signaling. With further understanding of this regulation, it may become apparent if these signaling pathways are targetable regulatory sites for therapeutics in inflammatory diseases (Fig. 2).

#### 2.3.2.3 GRK5 Regulation of Immune Processes

GRK5 can regulate GPCRs necessary for chemotaxis in both canonical and noncanonical fashion. Unlike GRK2, GRK5 has been shown to regulate these processes in much fewer cell types, primarily in monocytes. GRK5 modulates chemotaxis in these cells by regulating CCR2, a GPCR for monocyte chemoattractant protein-1 (Wu et al., 2012). GRK5 can also influence these cells in a noncanonical fashion through regulation of colony-stimulating factor-1 receptor, a receptor tyrosine kinase (Wu et al., 2012). By regulating CCR2 desensitization and inhibition of monocyte migration, GRK5 has been shown to attenuate atherosclerosis (Wu et al., 2012). Additionally, GRK5 can regulate CXCR4 desensitization via phosphorylation of HIP (HSP70-interacting protein) in an in vitro model (Barker & Benovic, 2011). Work in our lab, however, has shown that

GRK5 has little to no role in immune chemotaxis in a polymicrobial septic peritonitis model (Packiriswamy et al., 2013), and similar findings were reported in an arthritic model (Tarrant et al., 2008). This, however, was again model specific since GRK5 did regulate neutrophil infiltration in *Escherichia coli* pneumonia model (Packiriswamy, Steury, McCabe, Fitzgerald, & Parameswaran, 2016). GRK5 has also been shown to influence migration and invasion of cancer cells. In prostate cancer, GRK5 has been shown to regulate cell migration and invasion through interactions with moesin (ERM) proteins. It accomplishes this primarily through phosphorylation of moesin on Thr-66 residue regulating its cellular distribution (Chakraborty et al., 2014). Additionally, GRK5 may be involved in neutrophil granulocyte exocytosis. Signal transduction pathways regulating this function are poorly defined although several proteins, including Src, are known to participate in this response. Proteomic studies first identified a potential Src tyrosine kinase motif on GRK5, and phosphorylation of GRK5 by Src was then confirmed in an in vitro kinase assay. Furthermore, immunoprecipitation of phosphorylated GRK5 with Src in intact cells was decreased in the presence of a Src inhibitor, possibly implicating it in the regulation of granulocyte exocytosis (Luerman et al., 2011).

#### 2.3.2.4 GRK5 in Disease

GRK5 also plays a prominent role in a multitude of pathologies (Fig. 2). GRK5 canonically regulates muscarinic receptors but shows a preference for M2 and M4 receptors (Liu et al., 2009). These receptors are present in the presynaptic cells and negatively regulate acetylcholine release in hippocampal memory circuits. Abnormal increases in presynaptic cholinergic activity cause a decrease in acetylcholine release, therefore decreasing postsynaptic muscarinic M1 activity. M1 signaling has been shown to inhibit β-amyloidogenic APP processing and decreased β-amyloid accumulation (Sadot et al., 1996). This connection might be the reason that an AD-like pathology is observed in GRK5-deficient mice. Further connecting GRK5 to AD is its ability to phosphorylate α-synuclein and tubulin, possibly influencing their polymerization and neuronal functions as seen in AD (Carman, Som, Kim, & Benovic, 1998).

GRK5 has many similarities to GRK2 in its role in cardiovascular diseases. Expression and activity of GRK5 are increased during heart failure inducing β-adrenergic receptor desensitization, reducing myocardial contractility, and worsening heart failure (Chen et al., 2001). Overexpression of GRK5 in these cells decreases myocardial contraction and cardiac output in response to adrenergic stimulation (Brinks et al., 2010; Chen et al., 2001). Conversely, its inhibition results in increased cardiac contractility and enhanced survival in heart failure patients (Raake et al., 2008; Vinge et al., 2008). Inhibition of GRK5 with adGRK5-NT (N-terminal peptide of GRK5) prevented cardiomyopathy and improved heart failure (Harding et al., 2001; Rengo et al., 2009; Shah et al., 2001; Sorriento et al., 2010). Again similar to GRK2, overexpression of GRK5 in the vascular smooth muscle cells led to the development of hypertension through regulation of β1-adrenergic receptor and angiotensin II receptors (Eckhart et al., 2002; Keys et al., 2005). In addition to these canonical GPCR-dependent roles, GRK5, through its nuclear localization signal, has been shown to accumulate in the nucleus of cardiomyocytes, function as a histone deacetylase kinase, and promote cardiac hypertrophy (Martini et al., 2008). GRK5 has also been implicated in the development of atherosclerosis and its deficiency in apolipoprotein E knockout mice leads to development of more aortic atherosclerosis than their wild-type controls (Wu et al., 2012). Investigating the role of GRK5 in endotoxemia and polymicrobial sepsis models, our lab has shown that GRK5 deficiency leads to diminished cytokine levels (Packiriswamy et al., 2013, 2012; Patial, Shahi, et al., 2011) likely through decreased NFκB activation in tissues and macrophages (as outlined above). In addition to these reduced cytokine levels, GRK5-deficient mice also showed decreased thymocyte apoptosis and immune suppression (Packiriswamy et al., 2013). This immune suppression was attributed to reduced plasma corticosterone levels in these mice, suggesting regulation of the pituitary–adrenal axis by GRK5. With its diverse role in inflammatory signaling as well as its role in multiple diseases, furthering our understanding of this kinase is critical for our ability to effectively design therapeutics and treatments targeting GRK5 in these various pathologies (Fig. 2).

### 2.3.3 *GRK6*

#### 2.3.3.1 Biochemistry

GRK6 is present ubiquitously throughout the body but is one of the most prominent GRKs in the striatum and other dopaminergic brain areas, particularly in the GABAergic and cholinergic areas. In these tissues, GRK6 localizes primarily in the plasma membrane. GRK6α, one of the four splice variants, however can translocate to the nucleus, although its function there is unclear (Premont & Gainetdinov, 2007). Subcellular localization of GRK6 is regulated through both its C-terminus which can be palmitoylated and can help anchor this GRK to the plasma membrane (Loudon & Benovic, 1997). Interestingly, in the brain GRK6 is primarily localized to synaptic membranes (Ahmed, Bychkov, Gurevich, Benovic, & Gurevich, 2008), suggesting that there must be neuronal-specific mechanisms for altering the localization of GRK6; however, the specific protein partners that govern this remain unknown. For its role in GPCR regulation and desensitization, GRK6 has been shown to phosphorylate GPCRs in both activated and inactivated states with some notable exceptions such as the $D_1$ dopamine receptor (Li et al., 2015). Structurally, GRK6 is currently the only member of the family that has been successfully crystallized in a conformation that is expected to be an active conformation and is one of the only two structures that were able to detect the N-terminal region (along with GRK1). It was shown in these structures that the extreme N-terminal amino acids become helical and are able to interact with both lobes of the kinase domain, helping to align the various catalytic residues contributed by each lobe (Boguth et al., 2010). Mutational studies examining the role of these helical N-terminal motifs supported the model where the exposed hydrophobic residues at the apex of the helix are critical for GPCR phosphorylation (Boguth et al., 2010).

#### 2.3.3.2 GRK6 in Inflammation

GRK6 has been shown to be involved in NFκB signaling in a manner that it can phosphorylate IκBα at Ser32/36, similar to the role of GRK5. This phosphorylation of IκBα by GRK6 regulates inflammation in peritoneal macrophages when stimulated by TNFα (Ohba, Nakaya, Watari, Nagasaka, & Kurose, 2015). GRK6 also plays a role in the regulation of MAPKs by interacting with p38 pathway. GRK6 is able to interact with and downregulate p38 activation. This downregulation has been connected to cytokine-induced pain by reducing the neuronal responsiveness to cytokines such as IL-1β and TNFα and thereby reducing cytokine-induced hyperalgesia (Eijkelkamp et al., 2012; Willemen et al., 2010). Furthermore, GRK6 also plays a role in the dysregulated signaling of ERK and p38 after L-DOPA therapy for PD. This therapy results in a side effect termed L-DOPA-induced dyskinesia that causes decreased expression of GRKs, resulting in defective desensitization of dopamine receptors. Overexpressing GRK6 ameliorated supersensitive ERK and p38 MAP kinase responses to L-DOPA challenges, normalizing the responses of these signaling pathways to control animals (Ahmed et al., 2015). Investigating the role of GRK6 in c-kit-positive cardiac stem cells, it was shown that knockdown of GRK6 significantly decreases stem cell factor (SCF)-induced p38 and ERK through alterations in the phosphorylation state of CXCR4-serine 339. This alteration in MAPK signaling influenced the ability of these cells to regulate c-kit ligand SCF-induced migration during myocardial infarction (Zuo et al., 2016). GRK6 has been shown to play a critical role in β-arrestin2-mediated ERK activation following stimulation of V2 vasopressin and angiotensin II receptors independent of G-protein activation (Kim, Ahn, et al., 2005; Ren et al., 2005; Shenoy et al., 2006). Alterations in this β-arrestin-mediated ERK activation can be seen in pathological conditions and receptor mutations. For example, in the immunodeficiency syndrome, WHIM mutations in the CXCR4 receptor cause impaired ability of GRK6 to phosphorylate these receptors, leading to a lack of β-arrestin-2 recruitment, and subsequent ERK phosphorylation (McCormick, Segarra, Gasperini, Gulino, & Tosato, 2009). GRK6 has a clear role in the regulation of p38 and ERK signaling; however, its role in JNK signaling remains unclear and further work in this area is needed to fully appreciate GRK6's role in MAPK signaling.

Taken together, these signaling data demonstrate that GRK6 plays a role in NFκB and MAPK signaling, although the specific regulation of this pathway remains less understood. More work in this area will help elucidate the relationship between GRK6 and inflammatory pathways in the immune system.

#### 2.3.3.3 GRK6 Regulation of Immune Processes

GRK6 is also expressed at high levels in immune cells and GRK6 has been shown to regulate CXCR2 (Raghuwanshi et al., 2012), CXCR4 (Vroon, Heijnen, Raatgever, et al., 2004), and LTB4 (Kavelaars et al., 2003) influencing immune trafficking. This regulation and desensitization influences neutrophil and lymphocyte recruitment to sites of pathogenesis in several different disease models including inflammatory arthritis and DSS-induced colitis (Eijkelkamp et al., 2007; Fong et al., 2002; Tarrant et al., 2008). In epithelial cells, GRK6 has been identified as a critical mediator in integrin-mediated cell adhesion and migration of tumor cells (Chen et al., 2009). GRK6 deficiency promotes CXCR2-mediated progression and metastasis in a lung carcinoma model (Raghuwanshi et al., 2013). These effects are primarily through canonical regulation of GPCRs. GRK6 can also influence the hematopoietic process and mice lacking GRK6 have been demonstrated to exhibit lymphocytopenia, a loss of hematopoietic stem cells (HSCs), and multiple progenitor populations. This deficiency leads to impaired lymphoid differentiation due to impairment of HSC self-renewal. A possible mechanism for these effects comes from transcriptome and proteomic analysis, suggesting that GRK6 alters these functions through its involvement in reactive oxygen species signaling and/or through its interaction with DNA-PKcs (Le et al., 2016).

#### 2.3.3.4 GRK6 in Diseases

Recent studies have implicated GRK6 in regulating the clearance of apoptotic cells through interaction with GIT1 to activate Rac1, promoting apoptotic engulfment (Nakaya et al., 2013). GRK6 was critical in the removal of apoptotic B cells by splenic white pulp macrophages and in removing senescent red blood cells by splenic red pulp macrophages. In a different study in GRK6-deficient mice, F4/80+ macrophages were shown to have this same effect on senescent red blood cell clearance possibly because of increased iron stores in the splenic red pulp. In response to this uptake of apoptotic cells, macrophages in general induce production and release of immunosuppressive cytokines such as IL-10, transforming growth factor-β, prostaglandin E2, and platelet-activating factor (Fadok et al., 1998) while suppressing the production of proinflammatory cytokines IL1-β, IL-6, IL-12, and TNFα (Peng et al., 2007). Because of these connections, apoptotic cells have a significant impact on the functionality of these phagocytes that can, in turn, modulate inflammatory disease pathogenesis. GRK6 has been shown to modulate these processes in macrophages by phosphorylation of radixin and moesin, both of which are essential for cytoskeleton reorganization and immune cell chemotaxis and phagocytosis. Indeed, GRK6-deficient mice were shown to develop autoimmune disease possibly because of alterations in the response to immune cell phagocytosis. Additionally, GRK6 has been implicated in other diseases and has been shown to modulate arthritis and colitis by regulating immune cell chemotaxis and infiltration (Eijkelkamp et al., 2007; Tarrant et al., 2008). Specifically, in DSS-induced colitis model, GRK6-deficient mice produced more keratinocyte chemokine (the murine equivalent of IL-8), caused increased infiltration of immune cells, and enhanced severity of disease manifestations. Like the colitis model, GRK6-deficient mice exhibited enhanced weight loss and severity of disease in the arthritis model. Finally, studies focusing on the role of GRK6 in the brain have shown that D2 dopamine receptors are direct targets for GRK6. GRK6 knockout mice, unlike other GRK mouse models, show significantly enhanced responsiveness to psychostimulants and other dopamine agonists (Gainetdinov et al., 2003). Due to this enhanced response, GRK6 has been linked to Parkinson's disorder and potentially could influence other diseases related to dopamine supersensitivity including addiction, schizophrenia, and Tourette's syndrome though these have not been thoroughly investigated.

## 3. CONCLUSION

GPCRs comprise the largest class of receptors in the human genome and represent one of the most popular targets for modern therapeutics. With only four universally present GRKs and three others in specific subsets of tissues, understanding their numerous physiological roles in maintaining homeostasis is critical in our continued and future use in targeting these kinases for therapeutics. This task has been greatly aided by the discovery of crystal structures for various GRKs and the pathophysiological role of GRKs in both their canonical GPCR-dependent function and their noncanonical functions in numerous diseases. Despite their limited expression, it is clear that these isoforms are not entirely interchangeable and these GRKs have very defined roles in certain tissues. The use of transgenic and knockout animals has greatly aided our progress in these studies, but with new information, we see overlapping functionalities and compensatory mechanisms from other GRKs and other associated proteins adding to the complexity of these studies. As new functions and substrates are unveiled for GRKs, it is critical to take into consideration the substrate diversity and the role they will play when targeted for therapeutic purposes.

## REFERENCES

Ahmed, M. R., Bychkov, E., Gurevich, V. V., Benovic, J. L., & Gurevich, E. V. (2008). Altered expression and subcellular distribution of GRK subtypes in the dopamine-depleted rat basal ganglia is not normalized by l-DOPA treatment. *Journal of Neurochemistry, 104*(6), 1622–1636. https://doi.org/10.1111/j.1471-4159.2007.05104.x.

Ahmed, M. R., Bychkov, E., Kook, S., Zurkovsky, L., Dalby, K. N., & Gurevich, E. V. (2015). Overexpression of GRK6 rescues L-DOPA-induced signaling abnormalities in the dopamine-depleted striatum of hemiparkinsonian rats. *Experimental Neurology, 266*(4), 42–54. https://doi.org/10.1016/j.expneurol.2015.02.008.

Alves-Filho, J. C., Freitas, A., Souto, F. O., Spiller, F., Paula-Neto, H., Silva, J. S., et al. (2009). Regulation of chemokine receptor by Toll-like receptor 2 is critical to neutrophil migration and resistance to polymicrobial sepsis. *Proceedings of the National Academy of Sciences of the United States of America, 106*(10), 4018–4023. https://doi.org/10.1073/pnas.0900196106.

Alves-Filho, J. C., Sônego, F., Souto, F. O., Freitas, A., Verri, W. A., Auxiliadora-Martins, M., et al. (2010). Interleukin-33 attenuates sepsis by enhancing neutrophil influx to the site of infection. *Nature Medicine*, *16*(6), 708–712. https://doi.org/10.1038/nm.2156.

Ambrose, C., James, M., Barnes, G., Lin, C., Bates, G., Altherr, M., et al. (1992). A novel G protein-coupled receptor kinase gene cloned from 4p16.3. *Human Molecular Genetics*, *1*(9), 697–703.

Aragay, A. M., Mellado, M., Frade, J. M., Martin, A. M., Jimenez-Sainz, M. C., Martinez-A, C., et al. (1998). Monocyte chemoattractant protein-1-induced CCR2B receptor desensitization mediated by the G protein-coupled receptor kinase 2. *Proceedings of the National Academy of Sciences of the United States of America*, *95*(6), 2985–2990.

Arnon, T. I., Xu, Y., Lo, C., Pham, T., An, J., Coughlin, S., et al. (2011). GRK2-dependent S1PR1 desensitization is required for lymphocytes to overcome their attraction to blood. *Science (New York, NY)*, *333*(6051), 1898–1903. https://doi.org/10.1126/science.1208248.

Arraes, S. M. A., Freitas, M. S., da Silva, S. V., de Paula Neto, H. A., Alves-Filho, J. C., Auxiliadora Martins, M., et al. (2006). Impaired neutrophil chemotaxis in sepsis associates with GRK expression and inhibition of actin assembly and tyrosine phosphorylation. *Blood*, *108*(9), 2906–2913. https://doi.org/10.1182/blood-2006-05-024638.

Balabanian, K., Levoye, A., Klemm, L., Lagane, B., Hermine, O., Harriague, J., et al. (2008). Leukocyte analysis from WHIM syndrome patients reveals a pivotal role for GRK3 in CXCR4 signaling. *The Journal of Clinical Investigation*, *118*(3), 1074–1084. https://doi.org/10.1172/JCI33187.

Barker, B. L., & Benovic, J. L. (2011). G protein-coupled receptor kinase 5 phosphorylation of hip regulates internalization of the chemokine receptor CXCR4. *Biochemistry*, *50*(32), 6933–6941. https://doi.org/10.1021/bi2005202.

Benovic, J. L., Deblasi, A., Stone, W. C., Caron, M. G., & Lefkowitz, R. J. (1989). Beta-adrenergic receptor kinase: Primary structure delineates a multigene family. *Science*, *246*(4927), 235–240.

Benovic, J. L., & Gomez, J. (1993). Molecular cloning and expression of GRK6. A new member of the G protein-coupled receptor kinase family. *The Journal of Biological Chemistry*, *268*(26), 19521–19527.

Benovic, J. L., Onorato, J. J., Arriza, J. L., Stone, W. C., Lohse, M., Jenkins, N. A., et al. (1991). Cloning, expression, and chromosomal localization of beta-adrenergic receptor kinase 2. A new member of the receptor kinase family. *The Journal of Biological Chemistry*, *266*(23), 14939–14946.

Billard, M. J., Fitzhugh, D. J., Parker, J. S., Brozowski, J. M., McGinnis, M. W., Timoshchenko, R. G., et al. (2016). G protein coupled receptor kinase 3 regulates breast cancer migration, invasion, and metastasis. *PLoS ONE*, *11*(4), e0152856. https://doi.org/10.1371/journal.pone.0152856.

Blake Crabb, E., Franco, R. L., Bowen, M. K., Huang, C., Caslin, H. L., & Acevedo, E. O. (2016). G protein-coupled receptor kinase-2 in peripheral blood mononuclear cells following acute mental stress. *Life Sciences*, *145*, 184–189. https://doi.org/10.1016/j.lfs.2015.12.035.

Boguth, C. A., Singh, P., Huang, C., & Tesmer, J. J. G. (2010). Molecular basis for activation of G protein-coupled receptor kinases. *The EMBO Journal*, *29*(19), 3249–3259. https://doi.org/10.1038/emboj.2010.206.

Bownds, D., Dawes, J., Miller, J., & Stahlman, M. (1972). Phosphorylation of frog photoreceptor membranes induced by light. *Nature: New Biology*, *237*(73), 125–127.

Brinks, H., Boucher, M., Gao, E., Chuprun, J. K., Pesant, S., Raake, P. W. et al. (2010). Level of G protein-coupled receptor kinase-2 determines myocardial ischemia/reperfusion injury via pro- and anti-apoptotic mechanisms. *Circulation Research*, *107*(9), 1140–1149. https://doi.org/10.1161/CIRCRESAHA.110.221010.

Burkhalter, M. D., Fralish, G. B., Premont, R. T., Caron, M. G., & Philipp, M. (2013). Grk5l controls heart development by limiting mTOR signaling during symmetry breaking. *Cell Reports, 4*(4), 625–632. https://doi.org/10.1016/j.celrep.2013.07.036.

Bychkov, E. R., Gurevich, V. V., Joyce, J. N., Benovic, J. L., & Gurevich, E. V. (2008). Arrestins and two receptor kinases are upregulated in Parkinson's disease with dementia. *Neurobiology of Aging, 29*(3), 379–396. https://doi.org/10.1016/j.neurobiolaging.2006.10.012.

Cabrera-Vera, T. M., Vanhauwe, J., Thomas, T. O., Medkova, M., Preininger, A., Mazzoni, M. R., et al. (2003). Insights into G protein structure, function, and regulation. *Endocrine Reviews, 24*(6), 765–781. https://doi.org/10.1210/er.2000-0026.

Cahill, T. J., Thomsen, A. R. B., Tarrasch, J. T., Plouffe, B., Nguyen, A. H., Yang, F., et al. (2017). Distinct conformations of GPCR-β-arrestin complexes mediate desensitization, signaling, and endocytosis. *Proceedings of the National Academy of Sciences of the United States of America, 114*(10), 2562–2567. https://doi.org/10.1073/pnas.1701529114.

Cant, S. H., & Pitcher, J. A. (2005). G protein-coupled receptor kinase 2-mediated phosphorylation of ezrin is required for G protein-coupled receptor-dependent reorganization of the actin cytoskeleton. *Molecular Biology of the Cell, 16*(7), 3088–3099. https://doi.org/10.1091/mbc.E04-10-0877.

Carman, C. V., Lisanti, M. P., & Benovic, J. L. (1999a). Regulation of G protein-coupled receptor kinases by caveolin. *The Journal of Biological Chemistry, 274*(13), 8858–8864.

Carman, C. V., Parent, J. L., Day, P. W., Pronin, A. N., Sternweis, P. M., Wedegaertner, P. B., et al. (1999b). Selective regulation of Galpha(q/11) by an RGS domain in the G protein-coupled receptor kinase, GRK2. *The Journal of Biological Chemistry, 274*(48), 34483–34492. https://doi.org/10.1074/jbc.274.48.34483.

Carman, C. V., Som, T., Kim, C. M., & Benovic, J. L. (1998). Binding and phosphorylation of tubulin by G protein-coupled receptor kinases. *The Journal of Biological Chemistry, 273*(32), 20308–20316.

Chakraborty, P. K., Zhang, Y., Coomes, A. S., Kim, W.-J., Stupay, R., Lynch, L. D., et al. (2014). G protein-coupled receptor kinase GRK5 phosphorylates moesin and regulates metastasis in prostate cancer. *Cancer Research, 74*(13), 3489–3500. https://doi.org/10.1158/0008-5472.CAN-13-2708.

Chen, E. P., Bittner, H. B., Akhter, S. A., Koch, W. J., & Davis, R. D. (2001). Myocardial function in hearts with transgenic overexpression of the G protein-coupled receptor kinase 5. *The Annals of Thoracic Surgery, 71*(4), 1320–1324.

Chen, C. K., Burns, M. E., Spencer, M., Niemi, G. A., Chen, J., Hurley, J. B., et al. (1999). Abnormal photoresponses and light-induced apoptosis in rods lacking rhodopsin kinase. *Proceedings of the National Academy of Sciences of the United States of America, 96*(7), 3718–3722.

Chen, C. Y., Dion, S. B., Kim, C. M., & Benovic, J. L. (1993). Beta-adrenergic receptor kinase. Agonist-dependent receptor binding promotes kinase activation. *The Journal of Biological Chemistry, 268*(11), 7825–7831.

Chen, Y., Lu, B., Yang, Q., Fearns, C., Yates, J. R., & Lee, J. (2009). Combined integrin phosphoproteomic analyses and small interfering RNA-based functional screening identify key regulators for cancer cell adhesion and migration. *Cancer Research, 69*(8), 3713–3720. https://doi.org/10.1158/0008-5472.CAN-08-2515.

Chen, X., Zhu, H., Yuan, M., Fu, J., Zhou, Y., & Ma, L. (2010). G-protein-coupled receptor kinase 5 phosphorylates p53 and inhibits DNA damage-induced apoptosis. *The Journal of Biological Chemistry, 285*(17), 12823–12830. https://doi.org/10.1074/jbc.M109.094243.

Chuang, T. T., LeVine, H., & De Blasi, A. (1995). Phosphorylation and activation of beta-adrenergic receptor kinase by protein kinase C. *The Journal of Biological Chemistry, 270*(31), 18660–18665.

Chuang, T. T., Paolucci, L., & De Blasi, A. (1996). Inhibition of G protein-coupled receptor kinase subtypes by Ca2+/calmodulin. *The Journal of Biological Chemistry, 271*(45), 28691–28696.

Dabrowski, A., Grady, T., Logsdon, C. D., & Williams, J. A. (1996). Jun kinases are rapidly activated by cholecystokinin in rat pancreas both in vitro and in vivo. *The Journal of Biological Chemistry, 271*(10), 5686–5690.

Day, P. W., Carman, C. V., Sterne-Marr, R., Benovic, J. L., & Wedegaertner, P. B. (2003). Differential interaction of GRK2 with members of the G alpha q family. *Biochemistry, 42*(30), 9176–9184. https://doi.org/10.1021/bi034442+.

DebBurman, S. K., Ptasienski, J., Benovic, J. L., & Hosey, M. M. (1996). G protein-coupled receptor kinase GRK2 is a phospholipid-dependent enzyme that can be conditionally activated by G protein betagamma subunits. *The Journal of Biological Chemistry, 271*(37), 22552–22562.

DeFea, K. A. (2013). Arrestins in actin reorganization and cell migration. *Progress in Molecular Biology and Translational Science, 118*, 205–222. https://doi.org/10.1016/B978-0-12-394440-5.00008-5.

Degos, V., Peineau, S., Nijboer, C., Kaindl, A. M., Sigaut, S., Favrais, G., et al. (2013). G protein-coupled receptor kinase 2 and group I metabotropic glutamate receptors mediate inflammation-induced sensitization to excitotoxic neurodegeneration. *Annals of Neurology, 73*(5), 667–678. https://doi.org/10.1002/ana.23868.

Eckhart, A. D., Duncan, S. J., Penn, R. B., Benovic, J. L., Lefkowitz, R. J., & Koch, W. J. (2000). Hybrid transgenic mice reveal in vivo specificity of G protein-coupled receptor kinases in the heart. *Circulation Research, 86*(1), 43–50.

Eckhart, A. D., Ozaki, T., Tevaearai, H., Rockman, H. A., & Koch, W. J. (2002). Vascular-targeted overexpression of G protein-coupled receptor kinase-2 in transgenic mice attenuates beta-adrenergic receptor signaling and increases resting blood pressure. *Molecular Pharmacology, 61*(4), 749–758.

Eijkelkamp, N., Heijnen, C. J., Carbajal, A. G., Willemen, H. L. D. M., Wang, H., Minett, M. S., et al. (2012). G protein-coupled receptor kinase 6 acts as a critical regulator of cytokine-induced hyperalgesia by promoting phosphatidylinositol 3-kinase and inhibiting p38 signaling. *Molecular Medicine (Cambridge, MA), 18*, 556–564. https://doi.org/10.2119/molmed.2011.00398.

Eijkelkamp, N., Heijnen, C. J., Lucas, A., Premont, R. T., Elsenbruch, S., Schedlowski, M., et al. (2007). G protein-coupled receptor kinase 6 controls chronicity and severity of dextran sodium sulphate-induced colitis in mice. *Gut, 56*(6), 847–854. https://doi.org/10.1136/gut.2006.107094.

Eijkelkamp, N., Heijnen, C. J., Willemen, H. L. D. M., Deumens, R., Joosten, E. A. J., Kleibeuker, W., et al. (2010a). GRK2: A novel cell-specific regulator of severity and duration of inflammatory pain. *The Journal of Neuroscience: The Official Journal of the Society for Neuroscience, 30*(6), 2138–2149. https://doi.org/10.1523/JNEUROSCI.5752-09.2010.

Eijkelkamp, N., Wang, H., Garza-Carbajal, A., Willemen, H. L. D. M., Zwartkruis, F. J., Wood, J. N., et al. (2010b). Low nociceptor GRK2 prolongs prostaglandin E2 hyperalgesia via biased cAMP signaling to Epac/Rap1, protein kinase Cepsilon, and MEK/ERK. *The Journal of Neuroscience: The Official Journal of the Society for Neuroscience, 30*(38), 12806–12815. https://doi.org/10.1523/JNEUROSCI.3142-10.2010.

Elorza, A., Sarnago, S., & Mayor, F. (2000). Agonist-dependent modulation of G protein-coupled receptor kinase 2 by mitogen-activated protein kinases. *Molecular Pharmacology, 57*(4), 778–783.

Erdtmann-Vourliotis, M., Mayer, P., Ammon, S., Riechert, U., & Höllt, V. (2001). Distribution of G-protein-coupled receptor kinase (GRK) isoforms 2, 3, 5 and 6 mRNA in the rat brain. *Brain Research Molecular Brain Research, 95*(1–2), 129–137.

Evron, T., Daigle, T. L., & Caron, M. G. (2012). GRK2: Multiple roles beyond G protein-coupled receptor desensitization. *Trends in Pharmacological Sciences*, *33*(3), 154–164. https://doi.org/10.1016/j.tips.2011.12.003.

Fadok, V. A., Bratton, D. L., Konowal, A., Freed, P. W., Westcott, J. Y., & Henson, P. M. (1998). Macrophages that have ingested apoptotic cells in vitro inhibit proinflammatory cytokine production through autocrine/paracrine mechanisms involving TGF-beta, PGE2, and PAF. *The Journal of Clinical Investigation*, *101*(4), 890–898. https://doi.org/10.1172/JCI1112.

Farrens, D. L., Altenbach, C., Yang, K., Hubbell, W. L., & Khorana, H. G. (1996). Requirement of rigid-body motion of transmembrane helices for light activation of rhodopsin. *Science (New York, NY)*, *274*(5288), 768–770.

Felder, R. A., Sanada, H., Xu, J., Yu, P.-Y., Wang, Z., Watanabe, H., et al. (2002). G protein-coupled receptor kinase 4 gene variants in human essential hypertension. *Proceedings of the National Academy of Sciences of the United States of America*, *99*(6), 3872–3877. https://doi.org/10.1073/pnas.062694599.

Ferrari, L. F., Bogen, O., Alessandri-Haber, N., Levine, E., Gear, R. W., & Levine, J. D. (2012). Transient decrease in nociceptor GRK2 expression produces long-term enhancement in inflammatory pain. *Neuroscience*, *222*(222), 392–403. https://doi.org/10.1016/j.neuroscience.2012.07.004.

Fong, A. M., Premont, R. T., Richardson, R. M., Yu, Y. A., Lefkowitz, R. J., & Patel, D. D. (2002). Defective lymphocyte chemotaxis in beta-arrestin2- and GRK6-deficient mice. *Proceedings of the National Academy of Sciences of the United States of America*, *99*(11), 7478–7483. https://doi.org/10.1073/pnas.112198299.

Franklin, J. M., & Carrasco, G. A. (2013). G-protein receptor kinase 5 regulates the cannabinoid receptor 2-induced up-regulation of serotonin 2A receptors. *The Journal of Biological Chemistry*, *288*(22), 15712–15724. https://doi.org/10.1074/jbc.M113.454843.

Fu, X., Koller, S., Abd Alla, J., & Quitterer, U. (2013). Inhibition of G-protein-coupled receptor kinase 2 (GRK2) triggers the growth-promoting mitogen-activated protein kinase (MAPK) pathway. *The Journal of Biological Chemistry*, *288*(11), 7738–7755. https://doi.org/10.1074/jbc.M112.428078.

Gainetdinov, R. R., Bohn, L. M., Sotnikova, T. D., Cyr, M., Laakso, A., Macrae, A. D., et al. (2003). Dopaminergic supersensitivity in G protein-coupled receptor kinase 6-deficient mice. *Neuron*, *38*(2), 291–303.

Gainetdinov, R. R., Bohn, L. M., Walker, J. K., Laporte, S. A., Macrae, A. D., Caron, M. G., et al. (1999). Muscarinic supersensitivity and impaired receptor desensitization in G protein-coupled receptor kinase 5-deficient mice. *Neuron*, *24*(4), 1029–1036. https://doi.org/10.1016/S0896-6273(00)81048-X.

Gainetdinov, R. R., Premont, R. T., Bohn, L. M., Lefkowitz, R. J., & Caron, M. G. (2004). Desensitization of G protein-coupled receptors and neuronal functions. *Annual Review of Neuroscience*, *27*, 107–144. https://doi.org/10.1146/annurev.neuro.27.070203.144206.

Gold, J. I., Gao, E., Shang, X., Premont, R. T., & Koch, W. J. (2012). Determining the absolute requirement of G protein-coupled receptor kinase 5 for pathological cardiac hypertrophy: Short communication. *Circulation Research*, *111*(8), 1048–1053. https://doi.org/10.1161/CIRCRESAHA.112.273367.

Gold, J. I., Martini, J. S., Hullmann, J., Gao, E., Chuprun, J. K., Lee, L., et al. (2013). Nuclear translocation of cardiac G protein-coupled receptor kinase 5 downstream of select Gq-activating hypertrophic ligands is a calmodulin-dependent process. *PLoS ONE*, *8*(3). https://doi.org/10.1371/journal.pone.0057324.

Gómez del Arco, P., Martínez-Martínez, S., Calvo, V., Armesilla, A. L., & Redondo, J. M. (1996). JNK (c-Jun NH2-terminal kinase) is a target for antioxidants in T lymphocytes. *The Journal of Biological Chemistry*, *271*(42), 26335–26340.

Gros, R., Tan, C. M., Chorazyczewski, J., Kelvin, D. J., Benovic, J. L., & Feldman, R. D. (1999). G-protein-coupled receptor kinase expression in hypertension. *Clinical Pharmacology and Therapeutics*, *65*(5), 545–551. https://doi.org/10.1016/S0009-9236(99)70074-3.

Gurevich, V. V., & Gurevich, E. V. (2004). The molecular acrobatics of arrestin activation. *Trends in Pharmacological Sciences*, *25*(2), 105–111. https://doi.org/10.1016/j.tips.2003.12.008.

Gurevich, E. V., Tesmer, J. J. G., Mushegian, A., & Gurevich, V. V. (2012). G protein-coupled receptor kinases: More than just kinases and not only for GPCRs. *Pharmacology and Therapeutics*, *133*, 40–69. https://doi.org/10.1016/j.pharmthera.2011.08.001.

Harding, V. B., Jones, L. R., Lefkowitz, R. J., Koch, W. J., & Rockman, H. A. (2001). Cardiac beta ARK1 inhibition prolongs survival and augments beta blocker therapy in a mouse model of severe heart failure. *Proceedings of the National Academy of Sciences of the United States of America*, *98*(10), 5809–5814. https://doi.org/10.1073/pnas.091102398.

Hewavitharana, T., & Wedegaertner, P. B. (2012). Non-canonical signaling and localizations of heterotrimeric G proteins. *Cellular Signalling*, *24*(1), 25–34. https://doi.org/10.1016/j.cellsig.2011.08.014.

Hisatomi, O., Matsuda, S., Satoh, T., Kotaka, S., Imanishi, Y., & Tokunaga, F. (1998). A novel subtype of G-protein-coupled receptor kinase, GRK7, in teleost cone photoreceptors. *FEBS Letters*, *424*(3), 159–164. https://doi.org/10.1016/S0014-5793(98)00162-8.

Homan, K. T., & Tesmer, J. J. G. (2014). Structural insights into G protein-coupled receptor kinase function. *Current Opinion in Cell Biology*, *27*, 25–31. https://doi.org/10.1016/j.ceb.2013.10.009.

Horner, T. J., Osawa, S., Schaller, M. D., & Weiss, E. R. (2005). Phosphorylation of GRK1 and GRK7 by cAMP-dependent protein kinase attenuates their enzymatic activities. *The Journal of Biological Chemistry*, *280*(31), 28241–28250. https://doi.org/10.1074/jbc.M505117200.

Howard, A. D., McAllister, G., Feighner, S. D., Liu, Q., Nargund, R. P., Van der Ploeg, L. H., et al. (2001). Orphan G-protein-coupled receptors and natural ligand discovery. *Trends in Pharmacological Sciences*, *22*(3), 132–140.

Huang, Z. M., Gao, E., Fonseca, F. V., Hayashi, H., Shang, X., Hoffman, N. E., et al. (2013). Convergence of G protein-coupled receptor and S-nitrosylation signaling determines the outcome to cardiac ischemic injury. *Science Signaling*, *6*(299), ra95. https://doi.org/10.1126/scisignal.2004225.

Iaccarino, G., Barbato, E., Cipolletta, E., De Amicis, V., Margulies, K. B., Leosco, D., et al. (2005). Elevated myocardial and lymphocyte GRK2 expression and activity in human heart failure. *European Heart Journal*, *26*(17), 1752–1758. https://doi.org/10.1093/eurheartj/ehi429.

Iaccarino, G., Rockman, H. A., Shotwell, K. F., Tomhave, E. D., & Koch, W. J. (1998). Myocardial overexpression of GRK3 in transgenic mice: Evidence for in vivo selectivity of GRKs. *The American Journal of Physiology*, *275*(4 Pt 2), H1298–H1306.

Iacovelli, L., Salvatore, L., Capobianco, L., Picascia, A., Barletta, E., Storto, M., et al. (2003). Role of G protein-coupled receptor kinase 4 and beta-arrestin 1 in agonist-stimulated metabotropic glutamate receptor 1 internalization and activation of mitogen-activated protein kinases. *The Journal of Biological Chemistry*, *278*(14), 12433–12442. https://doi.org/10.1074/jbc.M203992200.

Inglese, J., Freedman, N. J., Koch, W. J., & Lefkowitz, R. J. (1993). Structure and mechanism of the G protein-coupled receptor kinases. *The Journal of Biological Chemistry*, *268*(32), 23735–23738.

Inglese, J., Koch, W. J., Caron, M. G., & Lefkowitz, R. J. (1992). Isoprenylation in regulation of signal transduction by G-protein-coupled receptor kinases. *Nature, 359*(6391), 147–150. https://doi.org/10.1038/359147a0.

Insall, R. (2013). The interaction between pseudopods and extracellular signalling during chemotaxis and directed migration. *Current Opinion in Cell Biology, 25*(5), 526–531. https://doi.org/10.1016/j.ceb.2013.04.009.

Islam, K. N., Bae, J.-W., Gao, E., & Koch, W. J. (2013). Regulation of nuclear factor κB (NF-κB) in the nucleus of cardiomyocytes by G protein-coupled receptor kinase 5 (GRK5). *The Journal of Biological Chemistry, 288*(50), 35683–35689. https://doi.org/10.1074/jbc.M113.529347.

Islam, K. N., & Koch, W. J. (2012). Involvement of nuclear factor κB (NF-κB) signaling pathway in regulation of cardiac G protein-coupled receptor kinase 5 (GRK5) expression. *The Journal of Biological Chemistry, 287*(16), 12771–12778. https://doi.org/10.1074/jbc.M111.324566.

Jaber, M., Koch, W. J., Rockman, H., Smith, B., Bond, R. a., Sulik, K. K., et al. (1996). Essential role of beta-adrenergic receptor kinase 1 in cardiac development and function. *Proceedings of the National Academy of Sciences of the United States of America, 93*(23), 12974–12979.

Jimenez-Sainz, C., Murga, C., Kavelaars, A., Jurado-Pueyo, M., Krakstad, B. F., Heijnen, C. J., et al. (2006). G protein-coupled receptor kinase 2 negatively regulates chemokine signaling at a level downstream from G protein subunits. *Molecular Biology of the Cell, 17*, 25–31. https://doi.org/10.1091/mbc.E05.

Jobin, C., & Sartor, R. B. (2000). The I kappa B/NF-kappa B system: A key determinant of mucosalinflammation and protection. *American Journal of Physiology Cell Physiology, 278*(3), C451–C462.

Jung, S., Yaron, A., Alkalay, I., Hatzubai, A., Avraham, A., & Ben-Neriah, Y. (1995). Costimulation requirement for AP-1 and NF-kappa B transcription factor activation in T cells. *Annals of the New York Academy of Sciences, 766*, 245–252.

Kahsai, A. W., Zhu, S., & Fenteany, G. (2010). G protein-coupled receptor kinase 2 activates radixin, regulating membrane protrusion and motility in epithelial cells. *Biochimica et Biophysica Acta, 1803*(2), 300–310. https://doi.org/10.1016/j.bbamcr.2009.11.002.

Kavelaars, A., Vroon, A., Raatgever, R. P., Fong, A. M., Premont, R. T., Patel, D. D., et al. (2003). Increased acute inflammation, leukotriene B4-induced chemotaxis, and signaling in mice deficient for G protein-coupled receptor kinase 6. *Journal of Immunology (Baltimore, MD: 1950), 171*(11), 6128–6134. https://doi.org/10.4049/jimmunol.171.11.6128.

Keys, J. R., Greene, E. A., Cooper, C. J., Naga Prasad, S. V., Rockman, H. A., & Koch, W. J. (2003). Cardiac hypertrophy and altered beta-adrenergic signaling in transgenic mice that express the amino terminus of beta-ARK1. *American Journal of Physiology. Heart and Circulatory Physiology, 285*(5), H2201–H2211. https://doi.org/10.1152/ajpheart.00112.2003.

Keys, J. R., Zhou, R., Harris, D. M., Druckman, C. A., & Eckhart, A. D. (2005). Vascular smooth muscle overexpression of G protein-coupled receptor kinase 5 elevates blood pressure, which segregates with sex and is dependent on Gi-mediated signaling. *Circulation, 112*(8), 1145–1153. https://doi.org/10.1161/CIRCULATIONAHA.104.531657.

Kim, J., Ahn, S., Ren, X.-R., Whalen, E. J., Reiter, E., Wei, H., et al. (2005a). Functional antagonism of different G protein-coupled receptor kinases for beta-arrestin-mediated angiotensin II receptor signaling. *Proceedings of the National Academy of Sciences of the United States of America, 102*(5), 1442–1447. https://doi.org/10.1073/pnas.0409532102.

Kim, K.-M., Gainetdinov, R. R., Laporte, S. A., Caron, M. G., & Barak, L. S. (2005b). G protein-coupled receptor kinase regulates dopamine D3 receptor signaling by modulating the stability of a receptor-filamin-beta-arrestin complex A case of autoreceptor regulation. *The Journal of Biological Chemistry, 280*(13), 12774–12780. https://doi.org/10.1074/jbc.M408901200.

Kim, Y. H., Kim, Y. S., Noh, H. S., Kang, S. S., Cheon, E. W., Park, S. K., et al. (2005c). Changes in rhodopsin kinase and transducin in the rat retina in early-stage diabetes. *Experimental Eye Research, 80*(6), 753–760. https://doi.org/10.1016/j.exer.2004.09.003.

King, D. W., Steinmetz, R., Wagoner, H. A., Hannon, T. S., Chen, L. Y., Eugster, E. A., et al. (2003). Differential expression of GRK isoforms in nonmalignant and malignant human granulosa cells. *Endocrine, 22*(2), 135–142. https://doi.org/10.1385/ENDO:22:2:135.

Knight, R. J., & Buxton, D. B. (1996). Stimulation of c-Jun kinase and mitogen-activated protein kinase by ischemia and reperfusion in the perfused rat heart. *Biochemical and Biophysical Research Communications, 218*(1), 83–88. https://doi.org/10.1006/bbrc.1996.0016.

Koch, W. J., Rockman, H. A., Samama, P., Hamilton, R. A., Bond, R. A., Milano, C. A., et al. (1995). Cardiac function in mice overexpressing the beta-adrenergic receptor kinase or a beta ARK inhibitor. *Science (New York, NY), 268*(5215), 1350–1353.

Komuro, I., Kudo, S., Yamazaki, T., Zou, Y., Shiojima, I., & Yazaki, Y. (1996). Mechanical stretch activates the stress-activated protein kinases in cardiac myocytes. *FASEB Journal: Official Publication of the Federation of American Societies for Experimental Biology, 10*(5), 631–636.

Kunapuli, P., & Benovic, J. L. (1993). Cloning and expression of GRK5: A member of the G protein-coupled receptor kinase family. *Proceedings of the National Academy of Sciences of the United States of America, 90*(12), 5588–5592.

Lafarga, V., Mayor, F., & Penela, P. (2012). The interplay between G protein-coupled receptor kinase 2 (GRK2) and histone deacetylase 6 (HDAC6) at the crossroads of epithelial cell motility. *Cell Adhesion & Migration, 6*(6), 495–501. https://doi.org/10.4161/cam.21585.

Le, Q., Yao, W., Chen, Y., Yan, B., Liu, C., Yuan, M., et al. (2016). GRK6 regulates ROS response and maintains hematopoietic stem cell self-renewal. *Cell Death & Disease, 7*(11), e2478. https://doi.org/10.1038/cddis.2016.377.

Lee, J. C., Laydon, J. T., McDonnell, P. C., Gallagher, T. F., Kumar, S., Green, D., et al. (1994). A protein kinase involved in the regulation of inflammatory cytokine biosynthesis. *Nature, 372*(6508), 739–746. https://doi.org/10.1038/372739a0.

Lefkowitz, R. J. (2013). A brief history of G-protein coupled receptors (Nobel lecture). *Angewandte Chemie (International Ed. in English), 52*(25), 6366–6378. https://doi.org/10.1002/anie.201301924.

Lefkowitz, R. J., & Shenoy, S. K. (2005). Transduction of receptor signals by beta-arrestins. *Science (New York, NY), 308*(5721), 512–517. https://doi.org/10 1126/science.1109237.

Leoratti, F. M., Trevelin, S. C., Cunha, F. Q., Rocha, B. C., Costa, P. A. C., Gravina, H. D., et al. (2012). Neutrophil paralysis in plasmodium vivax malaria. *PLoS Neglected Tropical Diseases, 6*(6), e1710. https://doi.org/10.1371/journal.pntd.0001710.

Li, W., Ai, N., Wang, S., Bhattacharya, N., Vrbanac, V., Collins, M., et al. (2014). GRK3 is essential for metastatic cells and promotes prostate tumor progression. *Proceedings of the National Academy of Sciences of the United States of America, 111*(4), 1521–1526. https://doi.org/10.1073/pnas.1320638111.

Li, L., Homan, K. T., Vishnivetskiy, S. A., Manglik, A., Tesmer, J. J. G., Gurevich, V. V., et al. (2015). G protein-coupled receptor kinases of the GRK4 protein subfamily phosphorylate inactive G protein-coupled receptors (GPCRs). *The Journal of Biological Chemistry, 290*(17), 10775–10790. https://doi.org/10.1074/jbc.M115.644773.

Liu, Y., Gorospe, M., Yang, C., & Holbrook, N. J. (1995). Role of mitogen-activated protein kinase phosphatase during the cellular response to genotoxic stress. Inhibition of c-Jun N-terminal kinase activity and AP-1-dependent gene activation. *The Journal of Biological Chemistry*, *270*(15), 8377–8380.

Liu, Z., Jiang, Y., Li, Y., Wang, J., Fan, L., Scott, M. J., et al. (2013). TLR4 signaling augments monocyte chemotaxis by regulating G protein-coupled receptor kinase 2 translocation. *Journal of Immunology (Baltimore, MD: 1950)*, *191*(2), 857–864. https://doi.org/10.4049/jimmunol.1300790.

Liu, X., Ma, B., Malik, A. B., Tang, H., Yang, T., Sun, B., et al. (2012). Bidirectional regulation of neutrophil migration by mitogen-activated protein kinases. *Nature Immunology*, *13*(5), 457–464. https://doi.org/10.1038/ni.2258.

Liu, J., Rasul, I., Sun, Y., Wu, G., Li, L., Premont, R. T., et al. (2009). GRK5 deficiency leads to reduced hippocampal acetylcholine level via impaired presynaptic M2/M4 autoreceptor desensitization. *The Journal of Biological Chemistry*, *284*(29), 19564–19571. https://doi.org/10.1074/jbc.M109.005959.

Lodowski, D. T., Pitcher, J. A., Capel, W. D., Lefkowitz, R. J., & Tesmer, J. J. G. (2003). Keeping G proteins at bay: A complex between G protein-coupled receptor kinase 2 and Gbetagamma. *Science (New York, NY)*, *300*(5623), 1256–1262. https://doi.org/10.1126/science.1082348.

Lombardi, M. S., Kavelaars, A., Penela, P., Scholtens, E. J., Roccio, M., Schmidt, R. E., et al. (2002). Oxidative stress decreases G protein-coupled receptor kinase 2 in lymphocytes via a calpain-dependent mechanism. *Molecular Pharmacology*, *62*(2), 379–388. https://doi.org/10.1124/mol.62.2.379.

Loniewski, K., Shi, Y., Pestka, J., & Parameswaran, N. (2008). Toll-like receptors differentially regulate GPCR kinases and arrestins in primary macrophages. *Molecular Immunology*, *45*(8), 2312–2322. https://doi.org/10.1016/j.molimm.2007.11.012.

Loudon, R. P., & Benovic, J. L. (1997). Altered activity of palmitoylation-deficient and isoprenylated forms of the G protein-coupled receptor kinase GRK6. *Journal of Biological Chemistry*, *272*(43), 27422–27427. https://doi.org/10.1074/jbc.272.43.27422.

Luerman, G. C., Powell, D. W., Uriarte, S. M., Cummins, T. D., Merchant, M. L., Ward, R. a., et al. (2011). Identification of phosphoproteins associated with human neutrophil granules following chemotactic peptide stimulation. *Molecular & Cellular Proteomics: MCP*, *10*(3), M110.001552. https://doi.org/10.1074/mcp.M110.001552.

Luttrell, L. M. (2005). Composition and function of g protein-coupled receptor signalsomes controlling mitogen-activated protein kinase activity. *Journal of Molecular Neuroscience: MN*, *26*(2–3), 253–264. https://doi.org/10.1385/JMN:26:2-3:253.

Lymperopoulos, A., Rengo, G., Gao, E., Ebert, S. N., Dorn, G. W., & Koch, W. J. (2010). Reduction of sympathetic activity via adrenal-targeted GRK2 gene deletion attenuates heart failure progression and improves cardiac function after myocardial infarction. *The Journal of Biological Chemistry*, *285*(21), 16378–16386. https://doi.org/10.1074/jbc.M109.077859.

Marriott, J. B., Clarke, I. A., & Dalgleish, A. G. (2001). Inhibition of p38 MAP kinase during cellular activation results in IFN-gamma-dependent augmentation of IL-12 production by human monocytes/macrophages. *Clinical and Experimental Immunology*, *125*(1), 64–70.

Martini, J. S., Raake, P., Vinge, L. E., DeGeorge, B. R., DeGeorge, B., Chuprun, J. K., et al. (2008). Uncovering G protein-coupled receptor kinase-5 as a histone deacetylase kinase in the nucleus of cardiomyocytes. *Proceedings of the National Academy of Sciences of the United States of America*, *105*(34), 12457–12462. https://doi.org/10.1073/pnas.0803153105.

Matkovich, S. J., Diwan, A., Klanke, J. L., Hammer, D. J., Marreez, Y., Odley, A. M., et al. (2006). Cardiac-specific ablation of G-protein receptor kinase 2 redefines its roles in heart development and beta-adrenergic signaling. *Circulation Research*, *99*(9), 996–1003. https://doi.org/10.1161/01.RES.0000247932.71270.2c.

Maurin, T., Melko, M., Abekhoukh, S., Khalfallah, O., Davidovic, L., Jarjat, M., et al. (2015). The FMRP/GRK4 mRNA interaction uncovers a new mode of binding of the Fragile X mental retardation protein in cerebellum. *Nucleic Acids Research*, *43*(17), 8540–8550. https://doi.org/10.1093/nar/gkv801.

McCormick, P. J., Segarra, M., Gasperini, P., Gulino, A. V., & Tosato, G. (2009). Impaired recruitment of Grk6 and beta-Arrestin 2 causes delayed internalization and desensitization of a WHIM syndrome-associated CXCR4 mutant receptor. *PLoS ONE*, *4*(12), e8102. https://doi.org/10.1371/journal.pone.0008102.

Melamed, A., & Sorvillo, F. J. (2009). The burden of sepsis-associated mortality in the United States from 1999 to 2005: An analysis of multiple-cause-of-death data. *Critical Care (London, England)*, *13*(1), R28. https://doi.org/10.1186/cc7733.

Mushegian, A., Gurevich, V. V., & Gurevich, E. V. (2012). The origin and evolution of G protein-coupled receptor kinases. *PLoS ONE*, 7(3), e33806. https://doi.org/10.1371/journal.pone.0033806.

Nakaya, M., Tajima, M., Kosako, H., Nakaya, T., Hashimoto, A., Watari, K., et al. (2013). GRK6 deficiency in mice causes autoimmune disease due to impaired apoptotic cell clearance. *Nature Communications*, *4*, 1532. https://doi.org/10.1038/ncomms2540.

Nick, J. A., Avdi, N. J., Gerwins, P., Johnson, G. L., & Worthen, G. S. (1996). Activation of a p38 mitogen-activated protein kinase in human neutrophils by lipopolysaccharide. *Journal of Immunology (Baltimore, MD: 1950)*, *156*(12), 4867–4875.

Niculescu, A. B., Segal, D. S., Kuczenski, R., Barrett, T., Hauger, R. L., & Kelsoe, J. R. (2000). Identifying a series of candidate genes for mania and psychosis: A convergent functional genomics approach. *Physiological Genomics*, *4*(1), 83–91. https://doi.org/4/1/83 [pii].

Nijboer, C. H., Heijnen, C. J., Willemen, H. L. D. M., Groenendaal, F., Dorn, G. W., van Bel, F., et al. (2010). Cell-specific roles of GRK2 in onset and severity of hypoxic-ischemic brain damage in neonatal mice. *Brain, Behavior, and Immunity*, *24*(3), 420–426. https://doi.org/10.1016/j.bbi.2009.11.009.

Nobles, K. N., Xiao, K., Ahn, S., Shukla, A. K., Lam, C. M., Rajagopal, S., et al. (2011). Distinct phosphorylation sites on the β(2)-adrenergic receptor establish a barcode that encodes differential functions of β-arrestin. *Science Signaling*, *4*(185), ra51. https://doi.org/10.1126/scisignal.2001707.

Obrenovich, M. E., Palacios, H. H., Gasimov, E., Leszek, J., & Aliev, G. (2009). The GRK2 overexpression is a primary hallmark of mitochondrial lesions during early Alzheimer disease. *Cardiovascular Psychiatry and Neurology*, *2009*, 327360. https://doi.org/10.1155/2009/327360.

Ohba, Y., Nakaya, M., Watari, K., Nagasaka, A., & Kurose, H. (2015). GRK6 phosphorylates IκBα at Ser(32)/Ser(36) and enhances TNF-α-induced inflammation. *Biochemical and Biophysical Research Communications*, *461*(2), 307–313. https://doi.org/10.1016/j.bbrc.2015.04.027.

Olbrich, H., Proudfoot, A. E., & Oppermann, M. (1999). Chemokine-induced phosphorylation of CC chemokine receptor 5 (CCR5). *Journal of Leukocyte Biology*, *65*(3), 281–285.

Oldham, W. M., & Hamm, H. E. (2008). Heterotrimeric G protein activation by G-protein-coupled receptors. *Nature Reviews. Molecular Cell Biology*, *9*(1), 60–71. https://doi.org/10.1038/nrm2299.

Osawa, S., & Weiss, E. R. (2012). A tale of two kinases in rods and cones. *Advances in Experimental Medicine and Biology*, *723*, 821–827. https://doi.org/10.1007/978-1-4614-0631-0_105.

Otten, J. J. T., de Jager, S. C. A., Kavelaars, A., Seijkens, T., Bot, I., Wijnands, E., et al. (2013). Hematopoietic G-protein-coupled receptor kinase 2 deficiency decreases atherosclerotic lesion formation in LDL receptor-knockout mice. *FASEB Journal: Official Publication of the Federation of American Societies for Experimental Biology*, *27*(1), 265–276. https://doi.org/10.1096/fj.12-205351.

Packiriswamy, N., Lee, T., Raghavendra, P. B., Durairaj, H., Wang, H., & Parameswaran, N. (2013). G-protein-coupled receptor kinase-5 mediates inflammation but does not regulate cellular infiltration or bacterial load in a polymicrobial sepsis model in mice. *Journal of Innate Immunity*, *5*(4), 401–413. https://doi.org/10.1159/000347002.

Packiriswamy, N., & Parameswaran, N. (2015). G-protein-coupled receptor kinases in inflammation and disease. *Genes and Immunity*, *16*, 367–377. https://doi.org/10.1038/gene.2015.26.

Packiriswamy, N., Parvataneni, S., & Parameswaran, N. (2012). Overlapping and distinct roles of GRK5 in TLR2-, and TLR3-induced inflammatory response in vivo. *Cellular Immunology*, *272*(2), 107–111. https://doi.org/10.1016/j.cellimm.2011.10.019.

Packiriswamy, N., Steury, M., McCabe, I. C., Fitzgerald, S. D., & Parameswaran, N. (2016). Bacterial dose-dependent role of G protein-coupled receptor kinase 5 in Escherichia coli-induced pneumonia. *Infection and Immunity*, *84*(5), 1633–1641. https://doi.org/10.1128/IAI.00051-16.

Palczewski, K., Buczyłko, J., Kaplan, M. W., Polans, A. S., & Crabb, J. W. (1991). Mechanism of rhodopsin kinase activation. *The Journal of Biological Chemistry*, *266*(20), 12949–12955.

Pao, C. S., Barker, B. L., & Benovic, J. L. (2009). Role of the amino terminus of G protein-coupled receptor kinase 2 in receptor phosphorylation. *Biochemistry*, *48*(30), 7325–7333. https://doi.org/10.1021/bi900408g.

Parameswaran, N., Pao, C. S., Leonhard, K. S., Kang, D. S., Kratz, M., Ley, S. C., et al. (2006). Arrestin-2 and G protein-coupled receptor kinase 5 interact with NFkB1 p105 and negatively regulate lipopolysaccharide-stimulated ERK1/2 activation in macrophages. *Journal of Biological Chemistry*, *281*(45), 34159–34170. https://doi.org/10.1074/jbc.M605376200.

Parvataneni, S., Gonipeta, B., Packiriswamy, N., Lee, T., Durairaj, H., & Parameswaran, N. (2011). Role of myeloid-specific G-protein coupled receptor kinase-2 in sepsis. *International Journal of Clinical and Experimental Medicine*, *4*(4), 320–330.

Patial, S., Luo, J., Porter, K. J., Benovic, J. L., & Parameswaran, N. (2009). G-protein-coupled-receptor kinases mediate TNFα-induced NFκB signalling via direct interaction with and phosphorylation of IκBα. *The Biochemical Journal*, *425*(1), 169–178. https://doi.org/10.1042/BJ20090908.

Patial, S., Saini, Y., Parvataneni, S., Appledorn, D. M., Dorn, G. W., Lapres, J. J., et al. (2011a). Myeloid-specific GPCR kinase-2 negatively regulates NFkB1p105-ERK pathway and limits endotoxemic shock in mice. *Journal of Cell Physiology*, *226*(3), 627–637. https://doi.org/10.1016/j.jsbmb.2011.07.002.Identification.

Patial, S., Shahi, S., Saini, Y., Lee, T., Packiriswamy, N., Appledorn, D. M., et al. (2011b). G-protein coupled receptor kinase 5 mediates lipopolysaccharide-induced NFκB activation in primary macrophages and modulates inflammation in vivo in mice. *Journal of Cellular Physiology*, *226*(5), 1323–1333. https://doi.org/10.1002/jcp.22460.

Penela, P., Ribas, C., Aymerich, I., Eijkelkamp, N., Barreiro, O., Heijnen, C. J., et al. (2008). G protein-coupled receptor kinase 2 positively regulates epithelial cell migration. *The EMBO Journal*, *27*(8), 1206–1218. https://doi.org/10.1038/emboj.2008.55.

Penela, P., Ribas, C., Aymerich, I., & Mayor, F. (2009). New roles of G protein-coupled receptor kinase 2 (GRK2) in cell migration. *Cell Adhesion & Migration*, *3*(1), 19–23. https://doi.org/10.4161/cam.3.1.7149.

Peng, Y., Martin, D. A., Kenkel, J., Zhang, K., Ogden, C. A., & Elkon, K. B. (2007). Innate and adaptive immune response to apoptotic cells. *Journal of Autoimmunity*, *29*(4), 303–309. https://doi.org/10.1016/j.jaut.2007.07.017.

Peppel, K., Boekhoff, I., McDonald, P., Breer, H., Caron, M. G., & Lefkowitz, R. J. (1997). G protein-coupled receptor kinase 3 (GRK3) gene disruption leads to loss of odorant receptor desensitization. *The Journal of Biological Chemistry*, *272*(41), 25425–25428.

Peregrin, S., Jurado-Pueyo, M., Campos, P. M., Sanz-Moreno, V., Ruiz-Gomez, A., Crespo, P., et al. (2006). Phosphorylation of p38 by GRK2 at the docking groove unveils a novel mechanism for inactivating p38MAPK. *Current Biology: CB*, *16*(20), 2042–2047. https://doi.org/10.1016/j.cub.2006.08.083.

Pitcher, J. A., Fredericks, Z. L., Stone, W. C., Premont, R. T., Stoffel, R. H., Koch, W. J., et al. (1996). Phosphatidylinositol 4,5-bisphosphate (PIP2)-enhanced G protein-coupled receptor kinase (GRK) activity. Location, structure, and regulation of the PIP2 binding site distinguishes the GRK subfamilies. *The Journal of Biological Chemistry*, *271*(40), 24907–24913.

Pitcher, J. A., Tesmer, J. J., Freeman, J. L., Capel, W. D., Stone, W. C., & Lefkowitz, R. J. (1999). Feedback inhibition of G protein-coupled receptor kinase 2 (GRK2) activity by extracellular signal-regulated kinases. *The Journal of Biological Chemistry*, *274*(49), 34531–34534.

Preininger, A. M., & Hamm, H. E. (2004). G protein signaling: Insights from new structures. *Science's STKE: Signal Transduction Knowledge Environment*, *2004*(218), re3. https://doi.org/10.1126/stke.2182004re3.

Premont, R. T., & Gainetdinov, R. R. (2007). Physiological roles of G protein-coupled receptor kinases and arrestins. *Annual Review of Physiology*, *69*, 511–534. https://doi.org/10.1146/annurev.physiol.69.022405.154731.

Premont, R. T., Macrae, A. D., Aparicio, S. A., Kendall, H. E., Welch, J. E., & Lefkowitz, R. J. (1999). The GRK4 subfamily of G protein-coupled receptor kinases. Alternative splicing, gene organization, and sequence conservation. *The Journal of Biological Chemistry*, *274*(41), 29381–29389. https://doi.org/10.1074/jbc.274.41.29381.

Premont, R. T., Macrae, A. D., Stoffel, R. H., Chung, N., Pitcher, J. A., Ambrose, C., et al. (1996). Characterization of the G protein-coupled receptor kinase GRK4. Identification of four splice variants. *The Journal of Biological Chemistry*, *271*(11), 6403–6410.

Pronin, A. N., & Benovic, J. L. (1997). Regulation of the G protein-coupled receptor kinase GRK5 by protein kinase C. *The Journal of Biological Chemistry*, *272*(6), 3806–3812.

Raake, P. W., Vinge, L. E., Gao, E., Boucher, M., Rengo, G., Chen, X., et al. (2008). G protein-coupled receptor kinase 2 ablation in cardiac myocytes before or after myocardial infarction prevents heart failure. *Circulation Research*, *103*(4), 413–422. https://doi.org/10.1161/CIRCRESAHA.107.168336.

Raghuwanshi, S. K., Smith, N., Rivers, E. J., Thomas, A. J., Sutton, N., Hu, Y., et al. (2013). G protein-coupled receptor kinase 6 deficiency promotes angiogenesis, tumor progression, and metastasis. *Journal of Immunology (Baltimore, MD: 1950)*, *190*(10), 5329–5336. https://doi.org/10.4049/jimmunol.1202058.

Raghuwanshi, S. K., Su, Y., Singh, V., Haynes, K., Richmond, A., & Richardson, R. M. (2012). The chemokine receptors CXCR1 and CXCR2 couple to distinct G protein-coupled receptor kinases to mediate and regulate leukocyte functions. *Journal of Immunology (Baltimore, MD: 1950)*, *189*(6), 2824–2832. https://doi.org/10.4049/jimmunol.1201114.

Ren, X., Reiter, E., Ahn, S., Kim, J., Chen, W., & Lefkowitz, R. J. (2005). Different G protein-coupled receptor kinases govern G protein and beta-arrestin-mediated signaling of V2 vasopressin receptor. *Proceedings of the National Academy of Sciences of the United States of America*, *102*(5), 1448–1453. https://doi.org/10.1073/pnas.0409534102.

Rengo, G., Lymperopoulos, A., Zincarelli, C., Donniacuo, M., Soltys, S., Rabinowitz, J. E., et al. (2009). Myocardial adeno-associated virus serotype 6-betaARKct gene therapy improves cardiac function and normalizes the neurohormonal axis in chronic heart failure. *Circulation*, *119*(1), 89–98. https://doi.org/10.1161/CIRCULATIONAHA.108.803999.

Ritter, S. L., & Hall, R. A. (2009). Fine-tuning of GPCR activity by receptor-interacting proteins. *Nature Reviews. Molecular Cell Biology*, *10*(12), 819–830. https://doi.org/10.1038/nrm2803.

Robinson, J. D., & Pitcher, J. A. (2013). G protein-coupled receptor kinase 2 (GRK2) is a rho-activated scaffold protein for the ERK MAP kinase cascade. *Cellular Signalling*, *25*(12), 2831–3839. https://doi.org/10.1016/j.cellsig.2013.08.031.

Rockman, H. A., Choi, D. J., Rahman, N. U., Akhter, S. A., Lefkowitz, R. J., & Koch, W. J. (1996). Receptor-specific in vivo desensitization by the G protein-coupled receptor kinase-5 in transgenic mice. *Proceedings of the National Academy of Sciences of the United States of America*, *93*(18), 9954–9959. https://doi.org/10.1073/pnas.93.18.9954.

Rodbell, M., Krans, H. M., Pohl, S. L., & Birnbaumer, L. (1971). The glucagon-sensitive adenyl cyclase system in plasma membranes of rat liver. IV. Effects of guanylnucleotides on binding of 125I-glucagon. *The Journal of Biological Chemistry*, *246*(6), 1872–1876.

Ross, E. M., & Gilman, A. G. (1977). Resolution of some components of adenylate cyclase necessary for catalytic activity. *The Journal of Biological Chemistry*, *252*(20), 6966–6969.

Sadot, E., Gurwitz, D., Barg, J., Behar, L., Ginzburg, I., & Fisher, A. (1996). Activation of m1 muscarinic acetylcholine receptor regulates tau phosphorylation in transfected PC12 cells. *Journal of Neurochemistry*, *66*(2), 877–880.

Sallese, M., Mariggiò, S., D'Urbano, E., Iacovelli, L., & De Blasi, A. (2000). Selective regulation of Gq signaling by G protein-coupled receptor kinase 2: Direct interaction of kinase N terminus with activated galphaq. *Molecular Pharmacology*, *57*(4), 826–831.

Shah, A. S., White, D. C., Emani, S., Kypson, A. P., Lilly, R. E., Wilson, K., et al. (2001). In vivo ventricular gene delivery of a beta-adrenergic receptor kinase inhibitor to the failing heart reverses cardiac dysfunction. *Circulation*, *103*(9), 1311–1316.

Shenoy, S. K., Drake, M. T., Nelson, C. D., Houtz, D. A., Xiao, K., Madabushi, S., et al. (2006). Beta-arrestin-dependent, G protein-independent ERK1/2 activation by the beta2 adrenergic receptor. *The Journal of Biological Chemistry*, *281*(2), 1261–1273. https://doi.org/10.1074/jbc.M506576200.

Shi, G. W., Chen, J., Concepcion, F., Motamedchaboki, K., Marjoram, P., Langen, R., et al. (2005). Light causes phosphorylation of nonactivated visual pigments in intact mouse rod photoreceptor cells. *The Journal of Biological Chemistry*, *280*(50), 41184–41191. https://doi.org/10.1074/jbc.M506935200.

Siderovski, D. P., Hessel, A., Chung, S., Mak, T. W., & Tyers, M. (1996). A new family of regulators of G-protein-coupled receptors? *Current Biology: CB*, *6*(2), 211–212. https://doi.org/10.1016/S0960-9822(02)00454-2.

Sorensen, S. D., & Conn, P. J. (2003). G protein-coupled receptor kinases regulate metabotropic glutamate receptor 5 function and expression. *Neuropharmacology*, *44*(6), 699–706. https://doi.org/10.1016/S0028-3908(03)00053-4.

Sorriento, D., Ciccarelli, M., Cipolletta, E., Trimarco, B., & Iaccarino, G. (2016). "Freeze, don't move": How to arrest a suspect in heart failure—A review on available GRK2 inhibitors. *Frontiers in Cardiovascular Medicine*, *3*(December), 48. https://doi.org/10.3389/fcvm.2016.00048.

Sorriento, D., Ciccarelli, M., Santulli, G., Campanile, A., Altobelli, G. G., Cimini, V., et al. (2008). The G-protein-coupled receptor kinase 5 inhibits NFkappaB transcriptional activity by inducing nuclear accumulation of IkappaB alpha. *Proceedings of the National Academy of Sciences of the United States of America*, *105*(46), 17818–17823. https://doi.org/10.1073/pnas.0804446105.

Sorriento, D., Fusco, A., Ciccarelli, M., Rungi, A., Anastasio, A., Carillo, A., et al. (2013). Mitochondrial G protein coupled receptor kinase 2 regulates proinflammatory responses in macrophages. *FEBS Letters*, *587*(21), 3487–3494. https://doi.org/10.1016/j.febslet.2013.09.002.

Sorriento, D., Santulli, G., Fusco, A., Anastasio, A., Trimarco, B., & Iaccarino, G. (2010). Intracardiac injection of AdGRK5-NT reduces left ventricular hypertrophy by inhibiting NF-kappaB-dependent hypertrophic gene expression. *Hypertension (Dallas, TX: 1979)*, *56*(4), 696–704. https://doi.org/10.1161/HYPERTENSIONAHA.110.155960.

Stacey, M., Lin, H. H., Gordon, S., & McKnight, A. J. (2000). LNB-TM7, a group of seven-transmembrane proteins related to family-B G-protein-coupled receptors. *Trends in Biochemical Sciences*, *25*(6), 284–289. https://doi.org/10.1016/S0968-0004(00)01583-8.

Subramanian, H., Gupta, K., Parameswaran, N., & Ali, H. (2014). Regulation of Fc∈RI signaling in mast cells by G protein-coupled receptor kinase 2 and its RH domain. *The Journal of Biological Chemistry*, *289*(30), 20917–20927. https://doi.org/10.1074/jbc.M113.523969.

Suo, W. Z., & Li, L. (2010). Dysfunction of G protein-coupled receptor kinases in Alzheimer's disease. *The Scientific World Journal*, *10*, 1667–1678. https://doi.org/10.1100/tsw.2010.154.

Surguchov, A. (2008). Molecular and cellular biology of synucleins. *International Review of Cell and Molecular Biology*, *270*, 225–317. https://doi.org/10.1016/S1937-6448(08)01406-8.

Takeda, S., Kadowaki, S., Haga, T., Takaesu, H., & Mitaku, S. (2002). Identification of G protein-coupled receptor genes from the human genome sequence. *FEBS Letters*, *520*(1–3), 97–101. https://doi.org/10.1016/S0014-5793(02)02775-8.

Tarrant, T. K., Billard, M. J., Timoshchenko, R. G., McGinnis, M. W., Serafin, D. S., Foreman, O., et al. (2013). G protein-coupled receptor kinase-3-deficient mice exhibit WHIM syndrome features and attenuated inflammatory responses. *Journal of Leukocyte Biology*, *94*(6), 1243–1251. https://doi.org/10.1189/jlb.0213097.

Tarrant, T. K., Rampersad, R. R., Esserman, D., Rothlein, L. R., Liu, P., Premont, R. T., et al. (2008). Granulocyte chemotaxis and disease expression are differentially regulated by GRK subtype in an acute inflammatory arthritis model (K/BxN). *Clinical Immunology (Orlando, FL)*, *129*(1), 115–122. https://doi.org/10.1016/j.clim.2008.06.008.

Thompson, M. D., Burnham, W. M., & Cole, D. E. C. (2005). The G protein-coupled receptors: Pharmacogenetics and disease. *Critical Reviews in Clinical Laboratory Sciences*, *42*(4), 311–392. https://doi.org/10.1080/10408360591001895.

Thomsen, A. R. B., Plouffe, B., Cahill, T. J., Shukla, A. K., Tarrasch, J. T., Dosey, A. M., et al. (2016). GPCR-G protein-β-arrestin super-complex mediates sustained G protein signaling. *Cell*, *166*(4), 907–919. https://doi.org/10.1016/j.cell.2016.07.004.

Touhara, K., Inglese, J., Pitcher, J. A., Shaw, G., & Lefkowitz, R. J. (1994). Binding of G protein beta gamma-subunits to pleckstrin homology domains. *The Journal of Biological Chemistry*, *269*(14), 10217–10220.

Valanne, S., Myllymäki, H., Kallio, J., Schmid, M. R., Kleino, A., Murumägi, A., et al. (2010). Genome-wide RNA interference in drosophila cells identifies G protein-coupled receptor kinase 2 as a conserved regulator of NF-kappaB signaling. *Journal of Immunology (Baltimore, MD: 1950)*, *184*(11), 6188–6198. https://doi.org/10.4049/jimmunol.1000261.

Villar, V. A. M., Jones, J. E., Armando, I., Palmes-Saloma, C., Yu, P., Pascua, A. M., et al. (2009). G protein-coupled receptor kinase 4 (GRK4) regulates the phosphorylation and function of the dopamine D3 receptor. *The Journal of Biological Chemistry*, *284*(32), 21425–21434. https://doi.org/10.1074/jbc.M109.003665.

Vinge, L. E., von Lueder, T. G., Aasum, E., Qvigstad, E., Gravning, J. A., How, O., et al. (2008). Cardiac-restricted expression of the carboxyl-terminal fragment of GRK3 uncovers distinct functions of GRK3 in regulation of cardiac contractility and growth: GRK3 controls cardiac alpha1-adrenergic receptor responsiveness. *The Journal of Biological Chemistry*, *283*(16), 10601–10610. https://doi.org/10.1074/jbc.M708912200.

Virlon, B., Firsov, D., Cheval, L., Reiter, E., Troispoux, C., Guillou, F., et al. (1998). Rat G protein-coupled receptor kinase GRK4: Identification, functional expression, and differential tissue distribution of two splice variants. *Endocrinology*, *139*(6), 2784–2795. https://doi.org/10.1210/endo.139.6.6078.

Vishnivetskiy, S. A., Paz, C. L., Schubert, C., Hirsch, J. A., Sigler, P. B., & Gurevich, V. V. (1999). How does arrestin respond to the phosphorylated state of rhodopsin? *The Journal of Biological Chemistry*, *274*(17), 11451–11454.

Vroon, A., Heijnen, C. J., Lombardi, M. S., Cobelens, P. M., Mayor, F., Caron, M. G., et al. (2004a). Reduced GRK2 level in T cells potentiates chemotaxis and signaling in response to CCL4. *Journal of Leukocyte Biology*, *75*(5), 901–909. https://doi.org/10.1189/jlb.0403136.

Vroon, A., Heijnen, C. J., Raatgever, R., Touw, I. P., Ploemacher, R. E., Premont, R. T., et al. (2004b). GRK6 deficiency is associated with enhanced CXCR4-mediated neutrophil chemotaxis in vitro and impaired responsiveness to G-CSF in vivo. *Journal of Leukocyte Biology*, *75*(4), 698–704. https://doi.org/10.1189/jlb.0703320.

Vroon, A., Kavelaars, A., Limmroth, V., Lombardi, M. S., Goebel, M. U., Van Dam, A.-M., et al. (2005). G protein-coupled receptor kinase 2 in multiple sclerosis and experimental autoimmune encephalomyelitis. *Journal of Immunology*, *174*(7), 4400–4406. https://doi.org/10.4049/jimmunol.174.7.4400.

Wacker, W. B., Donoso, L. A., Kalsow, C. M., Yankeelov, J. A., & Organisciak, D. T. (1977). Experimental allergic uveitis. Isolation, characterization, and localization of a soluble uveitopathogenic antigen from bovine retina. *Journal of Immunology (Baltimore, MD: 1950)*, *119*(6), 1949–1958.

Walker, J. K., Peppel, K., Lefkowitz, R. J., Caron, M. G., & Fisher, J. T. (1999). Altered airway and cardiac responses in mice lacking G protein-coupled receptor kinase 3. *The American Journal of Physiology*, *276*(4 Pt 2), R1214–R1221.

Wang, Z., Armando, I., Asico, L. D., Escano, C., Wang, X., Lu, Q., et al. (2007). The elevated blood pressure of human GRK4gamma A142V transgenic mice is not associated with increased ROS production. *American Journal of Physiology Heart and Circulatory Physiology*, *292*(5), H2083–H2092. https://doi.org/10.1152/ajpheart.00944.2006.

Wang, H., Heijnen, C. J., van Velthoven, C. T. J., Willemen, H. L. D. M., Ishikawa, Y., Zhang, X., et al. (2013). Balancing GRK2 and EPAC1 levels prevents and relieves chronic pain. *The Journal of Clinical Investigation*, *123*(12), 5023–5034. https://doi.org/10.1172/JCI66241.

Weiss, E. R., Raman, D., Shirakawa, S., Ducceschi, M. H., Bertram, P. T., Wong, F., et al. (1998). The cloning of GRK7, a candidate cone opsin kinase, from cone- and rod-dominant mammalian retinas. *Molecular Vision*, *4*(27), 27.

Whalen, E. J., Foster, M. W., Matsumoto, A., Ozawa, K., Violin, J. D., Que, L. G., et al. (2007). Regulation of beta-adrenergic receptor signaling by S-nitrosylation of G-protein-coupled receptor kinase 2. *Cell*, *129*(3), 511–522. https://doi.org/10.1016/j.cell.2007.02.046.

Whitcomb, T., Sakurai, K., Brown, B. M., Young, J. E., Sheflin, L., Dlugos, C., et al. (2010). Effect of g protein-coupled receptor kinase 1 (Grk1) overexpression on rod photoreceptor cell viability. *Investigative Ophthalmology & Visual Science*, *51*(3), 1728–1737. https://doi.org/10.1167/iovs.09-4499.

Wilden, U., Hall, S. W., & Kühn, H. (1986). Phosphodiesterase activation by photoexcited rhodopsin is quenched when rhodopsin is phosphorylated and binds the intrinsic 48-kDa protein of rod outer segments. *Proceedings of the National Academy of Sciences of the United States of America*, *83*(5), 1174–1178.

Willemen, H. L. D. M., Eijkelkamp, N., Wang, H., Dantzer, R., Dorn, G. W., Kelley, K. W., et al. (2010). Microglial/macrophage GRK2 determines duration of peripheral IL-1beta-induced hyperalgesia: Contribution of spinal cord CX3CR1, p38 and IL-1 signaling. *Pain*, *150*(3), 550–560. https://doi.org/10.1016/j.pain.2010.06.015.

Woerner, B. M., Luo, J., Brown, K. R., Jackson, E., Dahiya, S. M., Mischel, P , et al. (2012). Suppression of G-protein-coupled receptor kinase 3 expression is a feature of classical GBM that is required for maximal growth. *Molecular Cancer Research: MCR*, *10*(1), 156–166. https://doi.org/10.1158/1541-7786.MCR-11-0411.

Wu, J.-H., Zhang, L., Fanaroff, A. C., Cai, X., Sharma, K. C., Brian, L., et al. (2012). G protein-coupled receptor kinase-5 attenuates atherosclerosis by regulating receptor tyrosine kinases and 7-transmembrane receptors. *Arteriosclerosis, Thrombosis, and Vascular Biology*, *32*(2), 308–316. https://doi.org/10.1161/ATVBAHA.111.239608.

Xu, H., Jiang, X., Shen, K., Fischer, C. C., & Wedegaertner, P. B. (2014). The regulator of G protein signaling (RGS) domain of G protein-coupled receptor kinase 5 (GRK5) regulates plasma membrane localization and function. *Molecular Biology of the Cell*, *25*(13), 2105–2115. https://doi.org/10.1091/mbc.E13-09-0547.

Xu, F., Sun, S., Wang, X., Ni, E., Zhao, L., & Zhu, W. (2017). GRK2 mediates arginine vasopressin-induced IL-6 production via NF-kB signaling in neonatal rat cardiac fibroblast. *Molecular Pharmacology*. Epub ahead of print. https://doi.org/10.1124/mol.116.107698.

Zachos, G., Clements, B., & Conner, J. (1999). Herpes simplex virus type 1 infection stimulates p38/c-Jun N-terminal mitogen-activated protein kinase pathways and activates transcription factor AP-1. *The Journal of Biological Chemistry*, *274*(8), 5097–5103.

Zheng, H., Worrall, C., Shen, H., Issad, T., Seregard, S., Girnita, A., et al. (2012). Selective recruitment of G protein-coupled receptor kinases (GRKs) controls signaling of the insulin-like growth factor 1 receptor. *Proceedings of the National Academy of Sciences of the United States of America*, *109*(18), 7055–7060. https://doi.org/10.1073/pnas.1118359109.

Zuckerman, R., & Cheasty, J. E. (1986). A 48 kDa protein arrests cGMP phosphodiesterase activation in retinal rod disk membranes. *FEBS Letters*, *207*(1), 35–41.

Zuo, K., Kuang, D., Wang, Y., Xia, Y., Tong, W., Wang, X., et al. (2016). SCF/c-kit transactivates CXCR4-serine 339 phosphorylation through G protein-coupled receptor kinase 6 and regulates cardiac stem cell migration. *Scientific Reports*, *6*(May), 26812. https://doi.org/10.1038/srep26812.

CHAPTER EIGHT

# Canonical and Noncanonical Signaling Roles of β-Arrestins in Inflammation and Immunity

**Mohammad M. Ahmadzai, David Broadbent, Christopher Occhiuto, Canchai Yang, Rupali Das, Hariharan Subramanian**[1]
Michigan State University, East Lansing, MI, United States
[1]Corresponding author: e-mail address: subram46@psl.msu.edu

## Contents

1. Introduction 280
   1.1 Overview 280
2. Properties of β-Arrestins 281
   2.1 Regulation of β-Arrestin Expression and Subcellular Compartmentalization 281
   2.2 Posttranslational Modifications Modulate β-Arrestin Activity 283
3. Canonical Roles of β-Arrestins in the GPCR Life-Cycle 284
   3.1 Structural and Biochemical Properties of GPCRs 284
   3.2 β-Arrestins Orchestrate Desensitization of GPCRs 284
   3.3 β-Arrestins Orchestrate Internalization of GPCRs 286
4. β-Arrestins Modulate Immune Cell Activity via Their Canonical Interactions 288
   4.1 Inflammation and GPCR Activation 289
   4.2 β-Arrestin-Mediated Desensitization and Internalization of CXCR1 and CXCR2 Regulate Neutrophil Function 291
   4.3 β-Arrestins Regulate CXCR4 and CCR5 in T-Lymphocytes and Macrophages 292
   4.4 β-Arrestins Regulate Plasma Membrane PAR2 Levels In Vitro 293
   4.5 β-Arrestins Regulate Complement Receptor-Mediated Degranulation of Mast Cells 294
5. β-Arrestins Modulate Immune Cell Functions via Their Noncanonical Interactions 295
   5.1 β-Arrestin-Bound GPCRs Can Mobilize Signal Transduction Cascades 295
   5.2 β-Arrestins Modulate TLR4-Mediated NFκB Activation 296
   5.3 β-Arrestins Modulate ERK1/2 and JNK3 MAPK Signaling Pathways 297
   5.4 β-Arrestins Modulate Protein Turnover via Interactions With Mdm2 299
6. β-Arrestins Are Implicated in the Pathogenesis of Several Immunological Disorders 300
   6.1 β-Arrestins in $T_H$1-Driven Inflammation 301
   6.2 β-Arrestins in $T_H$2-Driven Inflammation 303
7. Future Perspectives and Conclusions 304
Acknowledgments 306
References 306

*Advances in Immunology*, Volume 136
ISSN 0065-2776
http://dx.doi.org/10.1016/bs.ai.2017.05.004

## Abstract

β-Arrestins are a highly conserved family of cytosolic adaptor proteins that contribute to many immune functions by orchestrating the desensitization and internalization of cell-surface G protein-coupled receptors (GPCRs) via well-studied canonical interactions. In cells of the innate and adaptive immune system, β-arrestins also subserve a parallel but less understood role in which they propagate, rather than terminate, intracellular signal transduction cascades. Because β-arrestins are promiscuous in their binding, they are capable of interacting with several different GPCRs and downstream effectors; in doing so, they vastly expand the repertoire of cellular responses evoked by agonist binding and the scope of responses that may contribute to inflammation during infectious and sterile insults. In this chapter, we attempt to provide an overview of the canonical and noncanonical roles of β-arrestins in inflammatory diseases.

# 1. INTRODUCTION

## 1.1 Overview

Mammalian cells employ a host of strategies in order to fine-tune the effects of receptor activation on cell function. One important mechanism by which this occurs involves the desensitization and downregulation of plasma membrane receptors by the arrestin family of proteins. By altering both the sensitivity and the sheer abundance of receptors on the cell surface, arrestins mitigate the deleterious effects of receptor overstimulation and thereby maintain homeostasis (Freedman & Lefkowitz, 1996; Kang, Tian, & Benovic, 2014; Moore, Milano, & Benovic, 2007; Pierce & Lefkowitz, 2001; Shenoy & Lefkowitz, 2011).

To date, three distinct subfamilies of arrestins have been described in humans: visual-arrestin, β-arrestin, and α-arrestin (Sharma & Parameswaran, 2015). The visual-arrestins were the first arrestins to be discovered and were named for their pivotal role in visual phototransduction in rod and cone cells of the retina (Craft & Whitmore, 1995; Freedman & Lefkowitz, 1996; Song, Raman, Gurevich, Vishnivetskiy, & Gurevich, 2006). Subsequently, investigations into agonist–receptor interactions unveiled an instrumental role for β-arrestins in the downregulation of G protein-coupled receptors (GPCRs) that fundamentally altered our understanding of GPCR biology (Laporte et al., 1999; Lohse, Benovic, Codina, Caron, & Lefkowitz, 1990; Rajagopal & Shenoy, 2017; Shenoy & Lefkowitz, 2011). The α-arrestins, though relatively new additions to the arrestin family, are structurally similar to their visual- and β-arrestin counterparts, but regulate lipolysis and the pathogenesis of obesity

(Kang et al., 2014). While each arrestin mediates some degree of receptor downregulation, it is important to note that this is largely context-specific: the expression patterns of the visual- and the α-arrestins are relatively restricted, whereas β-arrestins are ubiquitously expressed (Gurevich & Gurevich, 2006; Lefkowitz, Rajagopal, & Whalen, 2006; Shenoy, McDonald, Kohout, & Lefkowitz, 2001). Indeed, β-arrestins engage the broadest array of signaling pathways in humans and rodents, suggesting that they are physiologically indispensible and may participate in the pathogenesis of numerous diseases (Moore et al., 2007; Pierce & Lefkowitz, 2001).

The involvement of β-arrestins in GPCR desensitization and internalization is a canonical feature of these proteins that is now widely accepted (Freedman & Lefkowitz, 1996; Lefkowitz, Ahn, Lefkowitz, & Shenoy, 2007; Pierce & Lefkowitz, 2001). Over the last decade, however, a growing body of evidence has shown that β-arrestins per se initiate intracellular signaling cascades that modulate a host of cellular processes, including: gene transcription, oncogenesis, apoptosis, and cell migration (Lefkowitz et al., 2006; Pierce & Lefkowitz, 2001). Indeed, the finding that this noncanonical signaling transpires in a GPCR-independent manner is striking and raises many questions, particularly with regards to its physiologic relevance and therapeutic potential. To these ends, we seek here to provide a global overview of the canonical and noncanonical signaling roles of β-arrestins, and to discuss their putative roles in the classical inflammatory response and select immune pathologies.

## 2. PROPERTIES OF β-ARRESTINS

### 2.1 Regulation of β-Arrestin Expression and Subcellular Compartmentalization

β-Arrestins are cytosolic adaptor proteins that exist as two distinct isoforms, referred to as β-arrestin1 and β-arrestin2 (Kang et al., 2014; Pierce & Lefkowitz, 2001; Shenoy & Lefkowitz, 2011). In rats and humans, each β-arrestin isoform is encoded by separate genes and is subjected to varying degrees of alternative splicing, which vastly diversifies their structural and functional properties (Jiang, Xie, Liang, & Noble, 2013; Kang et al., 2014; Komori, Cain, Roch, Miller, & Matsumoto, 1998; Parruti et al., 1993). The parent gene encoding β-arrestin1, *ARRB1*, is localized to 11q13.4 (Parruti et al., 1993). The parent gene encoding β-arrestin2 is *ARRB2* and is localized to 17p13.2 (Zody et al., 2006). In humans, full-length β-arrestin1 is 418 amino acids long and approximately 47 kDa, whereas full-length β-arrestin2 is 409 amino acids long and possesses a molecular mass of 46 kDa (Gurevich & Gurevich, 2006; Sharma & Parameswaran, 2015). β-Arrestins are the most highly conserved members of the arrestin family, with 39%–50% sequence homology between vertebrates and invertebrates and 44%–84% sequence homology within vertebrate species (Craft & Whitmore, 1995).

Overall, β-arrestin1 and β-arrestin2 demonstrate 78% amino acid sequence similarity, with most differences arising in their C-termini (DeWire, Ahn, Lefkowitz, & Shenoy, 2007). Key structural features shared between the two β-arrestin isoforms include an N-terminal nuclear localization signal that facilitates nuclear entry, in addition to C-terminal clusters of serine and threonine residues that can be modified in order to regulate β-arrestin activity (see Fig. 1 and Section 2.2, later) (Lefkowitz et al., 2006; Wang, Wu, et al., 2003).

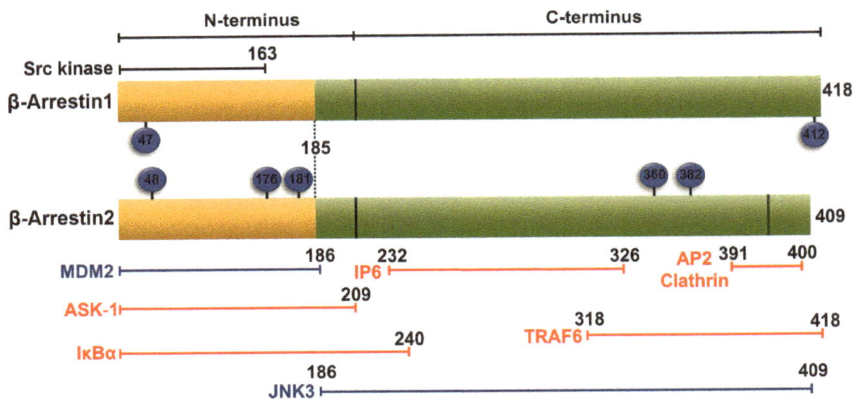


**Fig. 1** Structural properties and binding sites of β-arrestin1 and β-arrestin2. Both β-arrestin isoforms are capable of binding to the same soluble proteins (*highlighted in red*). The similarity of their binding sites arises from the ~78% sequence similarity between these two isoforms. Both β-arrestin1 and β-arrestin2 can bind similar proteins that are critical to receptor internalization and to the initiation of signal transduction cascades. Importantly, clathrin binds to a similar stretch of amino acids located in the C-terminus of both β-arrestin isoforms, demonstrating that both proteins can effectively partake in the downregulation and internalization of receptors. The nuclear export signal shown in the C-terminus of β-arrestin2 (*shaded in dark green*) is a unique feature of this isoform. Because β-arrestin1 lacks this isoform, β-arrestin2 localizes to the cytoplasm where it engages a broad number of cytoplasmic targets. Heterodimerization can occur, however, allowing for β-arrestin1 to be readily transported to the cytoplasm along with β-arrestin2. *Blue circles* indicate phosphorylation target sites (Milano, Kim, Stefano, Benovic, & Brenner, 2006; Scott et al., 2002; Wang, Wu, et al., 2003).

Nevertheless, there are functionally relevant differences in amino acid sequence that contribute to the unique localization patterns of the β-arrestins. Specifically, β-arrestin2 possesses a C-terminal nuclear export signal, whereas this feature is absent in β-arrestin1 (Scott et al., 2002; Wang, Wu, et al., 2003). As a result, β-arrestin2 is better poised to undertake specific cellular processes within the cytoplasm, while β-arrestin1 may interact with myriad nuclear targets (Gurevich & Gurevich, 2006; Kang et al., 2005; Lefkowitz et al., 2006; Moore et al., 2007).

Despite the absence of a nuclear export signal in β-arrestin1, however, it is noteworthy that both β-arrestin isoforms are readily detected within the cytoplasm of various cells. Indeed, β-arrestins can undergo homo- and heterooligomerization such that β-arrestin1 may be transported to the cytoplasm when complexed with β-arrestin2 (Kang et al., 2014). X-ray crystallography studies conducted over the last decade have strongly supported this finding and revealed that β-arrestin oligomerization depends on binding of inositol hexakisphosphate ($IP_6$) to $IP_6$-binding sites found in the N- and C-termini of both β-arrestin isoforms (Milano et al., 2006). In this regard, inherent amino acid sequence differences and intermolecular interactions can drastically alter the subcellular localization patterns of β-arrestin1 and β-arrestin2.

## 2.2 Posttranslational Modifications Modulate β-Arrestin Activity

Amino acid sequence differences also permit posttranslational modifications that fine-tune β-arrestin activity in an isoform-specific manner. To these ends, β-arrestin1 possesses a C-terminal Ser412 residue that remains phosphorylated under agonist-free conditions (Kang et al., 2014; Lin et al., 1997). On the other hand, β-arrestin2 features several additional modifiable sites, including: Pro176, Pro179, Pro181, Ser360, and Thr382 (Gurevich & Gurevich, 2006; Kang et al., 2014; Yan et al., 2011). Importantly, casein kinase II mediates the phosphorylation of Thr382 in β-arrestin2, whereas prolyl hydroxylase 2 catalyzes the hydroxylation of Pro176, Pro176, and Pro181 (Kim, Barak, Caron, & Benovic, 2002; Lin et al., 2002; Yan et al., 2011).

In the case of β-arrestin1, phosphorylation sterically hinders the protein's C-terminus from interacting with endocytic machinery involved in receptor downregulation (Kang et al., 2014). In stark contrast, phosphorylation of β-arrestin2 is inconsequential to its role in receptor downregulation; rather, the overall activity level of β-arrestin2 depends upon the hydroxylation status of its Pro176, Pro179, and Pro181 residues (Kim et al., 2002; Lin et al., 1997).

Clearly, the signaling pathways and cellular machinery involved in the posttranslational modification of β-arrestins are critical determinants of GPCR trafficking and signaling per se. As β-arrestins engage with GPCRs, however, they form scaffolds that unleash a great number of intracellular signaling events that are critical to cell function. To these ends, any meaningful discussion regarding the canonical signaling roles of β-arrestins must therefore begin with an overview of the structural and biochemical properties of these receptors.

# 3. CANONICAL ROLES OF β-ARRESTINS IN THE GPCR LIFE-CYCLE

## 3.1 Structural and Biochemical Properties of GPCRs

GPCRs constitute a ubiquitous superfamily of plasmalemmal proteins that transmit signals across biological membranes (Pierce & Lefkowitz, 2001). GPCRs are implicated in myriad physiological and disease processes and are consequently the target of a growing number of drug therapies (Imming, Sinning, & Meyer, 2006). Structurally, GPCRs are highly conserved throughout all genera of life and prototypically possess seven-membrane-spanning alpha-helical domains that are arranged in a serpentine pattern (de Mendoza, Sebé-Pedrós, & Ruiz-Trillo, 2014). For this reason, GPCRs are also referred to as seven-transmembrane receptors. The extracellular N-terminal region of the GPCR typically contains the orthosteric agonist binding site, while the intracellular C-terminal region interacts with targets pertinent to signal transduction (Cherezov et al., 2007). In the agonist-bound state, the C-terminus of the receptor binds and activates effector guanine nucleotide-binding proteins (G-proteins). Activated G-proteins subsequently engage an array of membrane-bound or cytosolic targets in order to stimulate or inhibit the production of intracellular second messengers (Kobilka, 2007) (see Fig. 2).

## 3.2 β-Arrestins Orchestrate Desensitization of GPCRs

In the agonist-bound state, the affinity with which GPCRs engage their affiliated G-proteins dictates whether or not a cellular response will ensue (Kenakin, 2002). β-Arrestins disrupt the GPCR signaling axis by inhibiting the interactions between G-proteins and the C-terminus of the receptor. In order to accomplish this, β-arrestins collaborate with the G-protein receptor

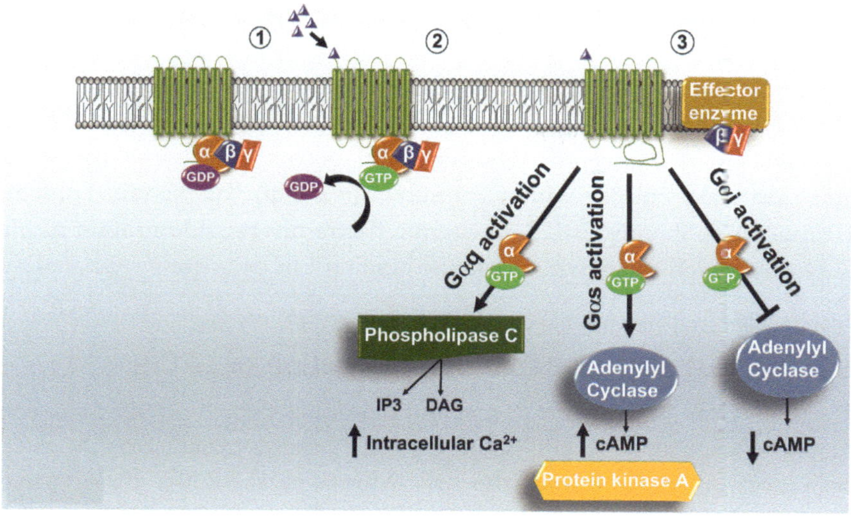


**Fig. 2** GPCR activation can mobilize one of several signal transduction cascades. Agonist binding to a GPCR (1) induces a conformational change in the structure of the receptor that activates guanine nucleotide-binding proteins (G-proteins) via the exchange of GDP for GTP by guanine nucleotide exchange factor (GEF) proteins (2). The resultant dissociation of the heterotrimeric G-protein complex liberates the active $G_{\alpha}$ subunit (3), which proceeds to bind and activate one of several highly conserved downstream effectors. Specifically, $G_{\alpha s}$ and $G_{\alpha i}$ stimulate and inhibit the catalytic conversion of ATP to cAMP by adenylate cyclase at the cell membrane, respectively, and thereby modulate the activity of cAMP-dependent protein kinase A (PKA). Activation of $G_{\alpha q}$ stimulates phospholipase C-mediated cleavage of phosphoinositol-4,5-bisphosphate ($PIP_2$) into inositol-1,4,5-trisphosphate ($IP_3$) and diacylglyceride. $IP_3$ diffuses into the cytosol, where it binds and activates $IP_3$-gated calcium channels expressed on the surface of the endoplasmic reticulum. The ensuing release of $Ca^{2+}$ from the intracellular stores further propagates this signal transduction cascade. Ultimately, the three signal transduction cascades illustrated here only provide a mere snapshot of the very many players that have been shown to mediate classical GPCR functions. For a more detailed overview of other pathways, readers are referred to other excellent reviews on this subject (Kobilka, 2007; Rosenbaum, Rasmussen, & Kobilka, 2009).

kinase (GRK) family of proteins, which comprise several well-characterized isoforms, namely: GRK1, GRK2, GRK3, GRK4, GRK5, and GRK6 (Benovic et al., 1987; DeFea, 2007; Freedman & Lefkowitz, 1996; Rajagopal & Shenoy, 2017). In the prototypical example of the β2-adrenergic receptor, receptor-stimulation activates GRK2, which phosphorylates serine and threonine residues located in the C-terminus of the receptor (Rajagopal & Shenoy, 2017). Receptor-phosphorylation enables β-arrestin2 to bind to the receptor and thereby sterically hinder receptor–G-protein interactions.

In this manner, agonist binding fails to generate any further intracellular signal, and the receptor is said to be desensitized (Rajagopal & Shenoy, 2017). Typically, receptor desensitization occurs on a rapid time-scale and can take place within minutes (<2 min) of agonist binding (Rajagopal & Shenoy, 2017). As a result, β-arrestins provide a robust mechanism by which the effects of agonist binding can be acutely regulated.

## 3.3 β-Arrestins Orchestrate Internalization of GPCRs

In conjunction to their role in receptor desensitization, β-arrestins also regulate the abundance of GPCRs on the cell-surface by facilitating the internalization of receptors into the cell (Moore et al., 2007; Rajagopal & Shenoy, 2017). As highly versatile adaptor proteins, β-arrestins provide the physical scaffold needed for binding of clathrin and adaptor protein 2 to the cell membrane (see Fig. 3). Clathrin recruitment, in turn, heralds formation of coated-pits that are subsequently engulfed as endocytic vesicles (Goodman et al., 1996; Han, Gurevich, Vishnivetskiy, Sigler, & Schubert, 2001; Laporte, Oakley, Holt, Barak, & Caron, 2000). Once internalized, the receptor may be targeted to the lysosome for degradation or recycled back to the cell membrane (Rajagopal & Shenoy, 2017).

Both β-arrestin isoforms specifically interact with clathrin via C-terminal sequences that engage the LIEL/P-binding domain of the clathrin protein heavy chain motif (Goodman et al., 1996). Interestingly, Goodman, Krupnick, Gurevich, Benovic, and Keen (1997) have shown that recruitment of the endocytic machinery by β-arrestin1 requires dephosphorylation of Ser412, whereas the phosphorylation status β-arrestin2 does not affect this interaction. In particular, mutation in the Ser412 residue of β-arrestin1 completely abolishes clathrin recruitment and attenuates receptor-internalization, suggesting that the activity of β-arrestin1 is primarily governed by its phosphorylation state, whereas β-arrestin2 is constitutively active and/or regulated by other mechanisms, including proline hydroxylation of C-terminal Pro176, Pro179, and Pro181 residues (see Section 2.2) (Goodman et al., 1997; Gurevich & Gurevich, 2006). Ultimately, clathrin-recruitment and β-arrestin-mediated receptor downregulation are immensely context-specific events: phosphorylation of Ser412 in β-arrestin1 abrogates receptor-internalization but has no effect on the ability of β-arrestin1 to bind to and desensitize its cognate receptor.

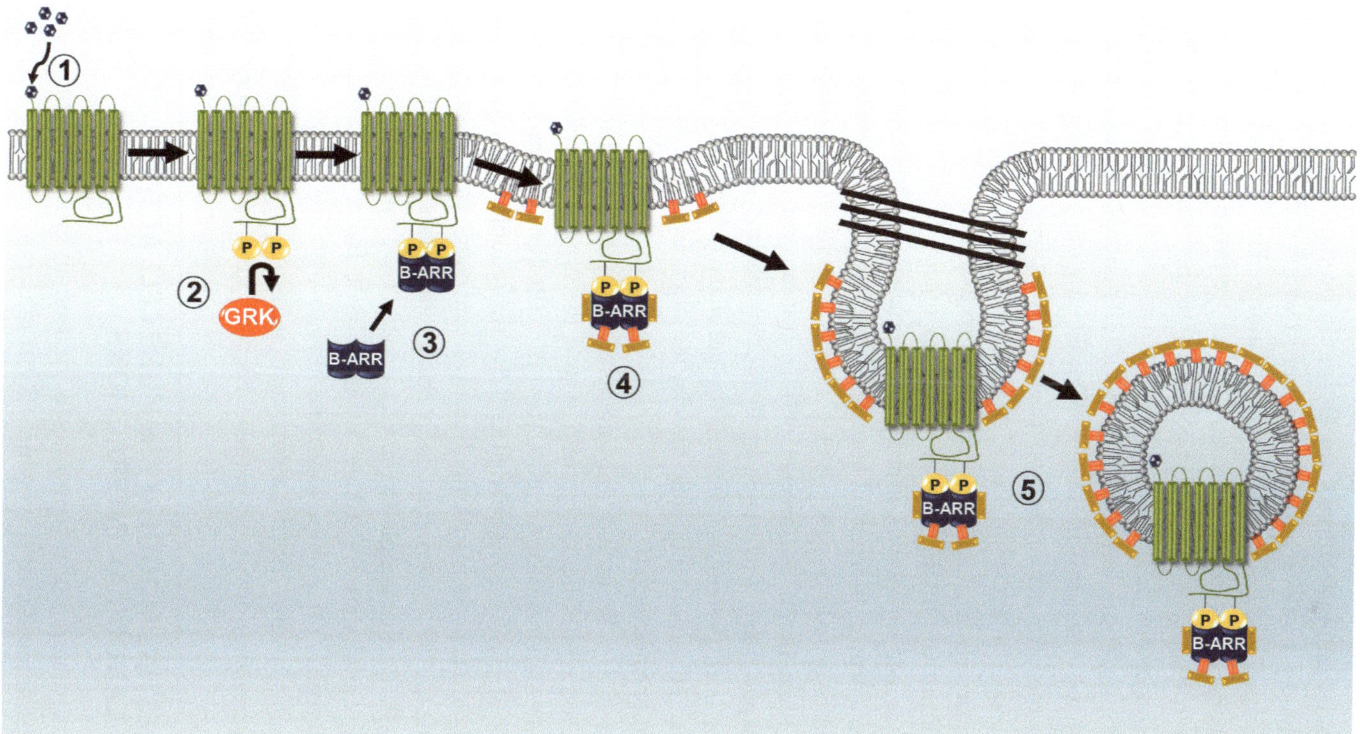


**Fig. 3** β-Arrestins orchestrate GPCR internalization in a clathrin-dependent manner. Following agonist binding (1), a number of soluble GRKs are activated, which catalyze the phosphorylation of the agonist-bound receptor (2). The phosphorylated C-terminus of the GPCR then serves to recruit β-arrestins (B-ARR) to the cell membrane (3 and 4), which in turn bind and polymerize clathrin protein into cages that stabilize the receptor complex during the subsequent engulfment of the cell membrane. The engulfed receptor, now sequestered within an endocytic vesicle, may be shuttled further into the cell for degradation within lysosomes, or it may be recycled back to the cell membrane (5). While the arcane mechanisms governing the fate of the receptor and its degradation vs recycling have yet to be fully characterized, it is increasingly clear that β-arrestins fulfill a crucial role in each of these processes (DeWire et al., 2007; Rajagopal & Shenoy, 2017).

The finding that β-arrestins almost universally interact with GPCRs is a well established and defining feature of the β-arrestin protein family. It is important to emphasize, however, that the manner in which GPCRs interact with β-arrestins is highly variable (Oakley, Laporte, Holt, Caron, & Barak, 2000; Williams et al., 2008). Indeed, GPCRs are functionally subdivided into Class A and Class B GPCRs according to the nature of their interactions with β-arrestins (Oakley et al., 2000).

In general, the interactions between Class A GPCRs and β-arrestins are short-lived such that β-arrestin is no longer colocalized within endosomal vesicles following receptor internalization (Kang et al., 2014; Sharma & Parameswaran, 2015). In addition, Class A GPCRs appear to bind to β-arrestin2 with relatively higher affinity compared to β-arrestin1. Because of the fleeting nature of this interaction, Class A receptors are rapidly recycled back to the cell membrane after being internalized ($<30$ min), as dissociation of the receptor–β-arrestin2 complex likely diverts the receptor away from degradative pathways (Kang et al., 2014; Moore et al., 2007; Oakley, Laporte, Holt, Barak, & Caron, 1999). In contrast, the interaction of Class B GPCRs and β-arrestins are markedly more stable and are of comparable affinity for both β-arrestin isoforms, the corollary being that Class B GPCRs are retained within endocytic vesicles for extended time periods ($>1$ h) (Rajagopal & Shenoy, 2017).

Notably, Class A GPCRs include the adrenergic, μ-opioid, and the D1 dopamine receptors, while the angiotensin II and the type I neurokinin (NK1R) receptors are prototypical Class B GPCRs (Sharma & Parameswaran, 2015). The finding that all GPCRs that couple to the $G_{\alpha/i}$ pathway are Class A receptors is striking, however, and suggests that structural features of the GPCR C-terminus likely contribute to this feature. On the other hand, GPCRs that couple to the $G_{q/11}$ or the $G_{\alpha/s}$ pathways have been categorized in both classes (Oakley, Laporte, Holt, Barak, & Caron, 2001). While there is no clear physiologic reason as to why this functional dichotomy exists, it is possible that the downstream signaling cascades mobilized by GPCR activation dictate how these receptors bind to β-arrestins.

## 4. β-ARRESTINS MODULATE IMMUNE CELL ACTIVITY VIA THEIR CANONICAL INTERACTIONS

The current paradigm of receptor desensitization and internalization illustrates an elegant mechanism by which β-arrestins can fine-tune the responsiveness of cells to various stimuli. In light of their role in regulating the GPCR life-cycle, it is conceivable that β-arrestins drastically alter the physiological properties of cells and drive the pathogenesis of human disease. Therefore, it is not surprising that β-arrestins have been the subject of intense scrutiny in the context of inflammation and immunological disease (Jiang et al., 2013; Sharma & Parameswaran, 2015).

## 4.1 Inflammation and GPCR Activation

The immune system comprises a dynamic network of cellular and non-cellular elements that control an array of life-sustaining functions. These include defense against infectious pathogens; repair of damaged host tissues; clearance of cellular debris; apoptosis and proliferation; and, coagulation and hemodynamic stability. It is noteworthy, however, that most chronic diseases and acute illnesses prominently feature an inflammatory component that defines or perpetuates the underlying pathology (Okin & Medzhitov, 2012). Atherosclerotic heart disease, for example, is a leading cause of morbidity and mortality in the United States that results from the sub-endothelial deposition of cholesterol and the formation of lipid-laden plaques, which are liable to rupture and occlude blood flow to the myocardium (Libby et al., 2009). Ultimately, the formation of these so-called atheromas constitutes a hallmark pathophysiological feature of the disease that is, to a large extent, mediated by macrophages that infiltrate the lesion (Libby et al., 2009).

Over the last decade, many studies that sought to characterize the immunological roles of β-arrestins focused on a handful of GPCRs that are well known for their indispensible roles in inflammation. These included chemokine receptors, the proteinase-activated receptor 2 (PAR2), and complement receptors (Bennett, Fox, & Signoret, 2011; DeFea, Zalevsky, et al., 2000; Jiang et al., 2013; Vibhuti, Gupta, Subramanian, Guo, & Ali, 2011).

Broadly speaking, the inflammatory response begins with the release of proinflammatory and chemotactic cytokines that mobilize leukocytes to foci of tissue damage and/or infection (Medzhitov, 2010). The resultant tissue neutrophilia constitutes a hallmark histological feature of inflammation and is dependent upon the actions of interleukin-8 (IL-8) (Middleton, Patterson, Gardner, Schmutz, & Ashton, 2002; Su & Richmond, 2015). IL-8 binds to the chemokine receptors, CXCR1 and CXCR2, which are abundantly expressed on the surface of neutrophils in addition to several other cells involved in the acute phase of inflammation (Middleton et al., 2002).

Activation of CXCR1 and -2 induces neutrophil-homing to the tissues as part of a multistep process referred to as extravasation (Middleton et al., 2002). While CXCR1 is more selective for IL-8 than CXCR2, stimulation of either receptor evokes neutrophil (1) adhesion to the postcapillary endothelium; (2) diapedesis into the tissue compartment; and (3) directional migration along the chemokine gradient toward the inflammatory focus (Su & Richmond, 2015) (see Fig. 4).

In addition to orchestrating changes in cytoskeletal architecture and actin dynamics, CXCR1 and CXCR2 activation stimulates the protein kinase A (PKA) and the mitogen-activated protein kinase (MAPK) pathways, which promote release of cytotoxic granules and translocation of nuclear factor kappa B (NFκB) to the nucleus (Borish & Steinke, 2003; Murdoch & Finn, 2000; Richardson, Ali, Pridgen, Haribabu, & Snyderman, 1998). Activation of the NFκB pathway, in turn, upregulates expression of other proinflammatory mediators and can thereby accelerate the incipient inflammatory response (Lawrence, 2009).

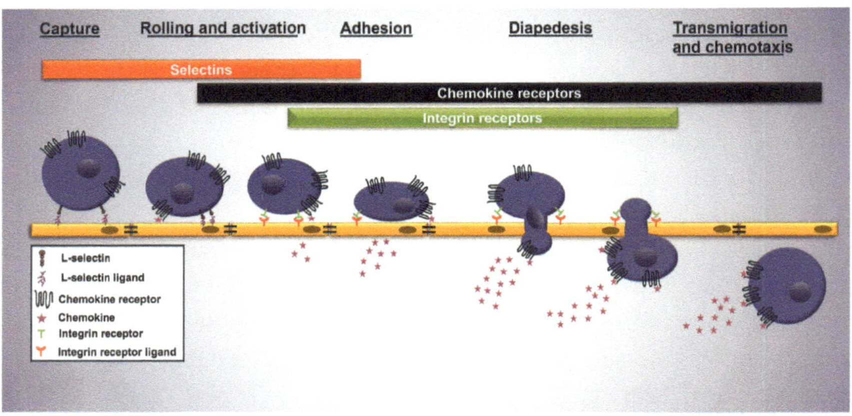


**Fig. 4** Leukocytes extravasate into the tissue compartment during the inflammatory response. Chemokine-receptor activation induces a host of cellular responses that facilitate extravasation of leukocytes to foci of tissue damage and/or infection. This begins with the capture and rolling phases, which involve the formation of low-affinity, transient interactions between L-selectins and their ligands expressed on the surface of blood-borne leukocytes and vascular endothelial cells, respectively. Subsequently, chemokines immobilized on the endothelial surface bind and activate their cognate receptors, which then initiate an inside-out signaling cascade that results in activation of chemokine receptors, which leads to integrin receptor activation and the firm adhesion of leukocytes on the surface of postcapillary endothelial cells. Adhered cells then form cytoskeletal projections that protrude between vascular endothelial cells, through a process referred to as diapedesis, which permit circulating immune cells to adhere to capillary endothelial cells and transmigrate into the tissue compartment.

## 4.2 β-Arrestin-Mediated Desensitization and Internalization of CXCR1 and CXCR2 Regulate Neutrophil Function

Aberrant activation of neutrophils can incur tremendous collateral damage to host tissues (Jonsson, Mancardi, Albanesi, & Bruhns, 2013; Navegantes et al., 2017). As a result, the mobilization and activation of neutrophils are partly controlled by regulation of CXCR1 and/or CXCR2 expression levels (Bennett et al., 2011). Similar to their interactions with prototypical GPCRs, β-arrestins bind to phosphorylated serine residues located in the C-termini of CXCR1 and CXCR2, and thereby uncouple the receptor from downstream G-proteins (Barlic et al., 1999; Su et al., 2005). Additionally, β-arrestins target CXCR1 and -2 to clathrin-coated pits for eventual internalization and sequestration within the cell (Barlic et al., 1999; Gilliland, Salanga, Kawamura, Trejo, & Handel, 2013; Lohse & Hoffmann, 2014). Early studies utilizing radiolabeling techniques showed that neutrophil exposure to high concentrations of IL-8 promotes internalization of cell-surface CXCR1 and CXCR2 (Barlic et al., 2000) Elsewhere, these findings were confirmed in mouse models of β-arrestin1 or β-arrestin2 knockout mice (Basher et al., 2008; Su et al., 2005). As expected, in vitro migration assays showed that neutrophils isolated from β-arrestin-deficient mice displayed greater migratory responses consequent to IL-8-stimulation (Su et al., 2005). Furthermore, greater numbers of neutrophils were isolated from tissues of β-arrestin2-deficient mice in both dorsal air-pouch and excision wounding models of inflammation (Su et al., 2005).

It is noteworthy, however, that the interactions between β-arrestins and CXCR2 modulate several additional aspects of neutrophil function beyond just chemotaxis (Zhao, Wimmer, Trieu, DiScipio, & Schraufstatter, 2004). Specifically, β-arrestin1- and β-arrestin2-null mice demonstrated exaggerated activation of the respiratory burst pathway. Taken together, it appears that β-arrestin2 dampens agonist-induced neutrophil activation and chemotaxis by negatively modulating activation of CXCR2 (Basher et al., 2008; Su et al., 2005). In contrast, both β-arrestin1 and β-arrestin2 are readily capable of evoking these same responses via their interactions with CXCR1, demonstrating that the functional outcome evoked by receptor activation varies according to the β-arrestin isoform that is recruited (Barlic et al., 1999).

It is important to note that the early landmark studies investigating the roles of β-arrestin in the internalization of CXCR1 and -2 signaling were conducted using HEK293 and other artificial cell systems that were transfected to overexpress the chemokine receptors, in the presence and absence of GRK2, GRK3, GRK5, GRK6, and both β-arrestin isoforms (Barlic et al., 1999; Basher et al., 2008; Fan, Yang, Wang, Qian, & Richmond, 2001). Strikingly, coexpression of β-arrestins and GRK2 was necessary for internalization and translocation of CXCR1 to clathrin-coated pits (Barlic et al., 1999). Nevertheless, CXCR1 and CXCR2 drastically differ in their propensity to undergo internalization and recycling, despite demonstrating ~80% sequence homology in both mice and humans (Holmes, Lee, Kuang, Rice, & Wood, 1991). Between the two, β-arrestin-mediated CXCR1 internalization is readily followed by recycling to the cell membrane, whereas CXCR2 recycling is not β-arrestin-dependent (Gilliland et al., 2013; Raghuwanshi et al., 2012; Richardson, Marjoram, Barak, & Snyderman, 2003). Ultimately, differences in the fates of CXCR1 and CXCR2 following β-arrestin-mediated internalization are likely amenable to sequence differences in the C-termini of the receptors, which feature varying numbers of serine and threonine residues that may undergo phosphorylation (Richardson et al., 2003).

## 4.3 β-Arrestins Regulate CXCR4 and CCR5 in T-Lymphocytes and Macrophages

As the effects of β-arrestins on neutrophil recruitment and activation became clearer, subsequent groups sought to elucidate the role of β-arrestins in the regulation of other immune cell types. In the setting of primary infection by human immunodeficiency virus (HIV)-1, viral entry into CD4+ T-lymphocytes (T-cells) and macrophages requires the expression of CXCR4 and CCR5, respectively (Bennett et al., 2011). In the setting of the normal immune response, however, both CXCR4 and CCR5 participate in the recruitment of T-cells and macrophages, such that loss of these receptors can lead to an immunocompromised state (Contento et al., 2008; Zou, Kottmann, Kuroda, Taniuchi, & Littman, 1998).

Early studies using transfected artificial cell lines successfully demonstrated that β-arrestins directly mediate desensitization and internalization of CXCR4 and CCR5 (Aramori et al., 1997; Bhandari, Trejo, Benovic, & Marchese, 2007; Cheng et al., 2000; Haribabu et al., 1997; Huttenrauch, Nitzki, Lin, Höning, & Oppermann, 2002; Hüttenrauch, Pollok-Kopp, & Oppermann, 2005). In rat basophilic leukemia and HEK293 cell lines, β-arrestin2 mediated the internalization of truncated CXCR4, which lacked the C-terminal phosphorylation residues to which β-arrestins normally bind (Cheng et al., 2000; Haribabu et al., 1997). In stark contrast to previous studies, these findings raised the possibility that GPCRs possess redundant binding domains through which β-arrestins can bypass desensitization and evoke internalization (Cheng et al., 2000). In this case, the second and/or third intracellular loops of CXCR4 were thought to undergo agonist-dependent phosphorylation and to recruit β-arrestins to the cell membrane (Cheng et al., 2000).

Hüttenrauch et al. (2005) used transfected HEK293 cells to similarly show that β-arrestin2 orchestrates desensitization and internalization of CCR5 in a GRK2-dependent manner. Since CCR5 was shown to undergo homodimerization as well as heterodimerization with the complement fragment receptor 5a (C5aR), truncated CCR5 could still be internalized despite the absence of C-terminal phosphorylation residues in CCR5 (Hüttenrauch et al., 2005). In this case, however, receptor-internalization was dependent upon β-arrestin1 or β-arrestin2 recruitment to the phosphorylated C-terminus of the associated CCR5. A more striking finding, however, was that internalization of CCR5 occurred following activation of C5aR by its endogenous agonist, C5a, raising the possibility that any event leading to β-arrestin recruitment per se could alter the global expression levels of other receptors on the cell surface (Hüttenrauch et al., 2005). Although these findings were described in an idealized cell-system, the consequences of such heterologous receptor-internalization on normal physiology are undoubtedly profound.

## 4.4 β-Arrestins Regulate Plasma Membrane PAR2 Levels In Vitro

The type 2 proteinase-activated receptor (PAR2) is abundantly expressed on the surface of neurons and epithelial cells of the gastrointestinal tract (Cenac et al., 2002; Vergnolle, Wallace, Bunnett, & Hollenberg, 2001). In contrast to other GPCRs, PAR2 exhibits a unique activation profile in that the endogenous ligand is generated via proteolytic cleavage of the extracellular N-terminal domain of the receptor (Schmidlin & Bunnett, 2001). In the setting of the allergic response, mast cell-derived tryptases promote the cleavage and activation of PAR2, which precipitates clinical manifestation of allergic enteropathy (Cenac et al., 2002). Given the fundamentally irreversible nature of PAR2 activation, the mechanisms that terminate PAR2 signaling are of paramount interest.

In contrast to the events that unfold subsequent to the internalization of other receptor families, internalized PAR2 must undergo degradation with high fidelity in order to mitigate the risk of reactivation at the cell membrane. Using a rat kidney cell line transfected with GFP-tagged β-arrestin1, Déry, Thoma, Wong, Grady, and Bunnett (1999) elegantly showed that β-arrestin1 translocates to the plasmalemma following trypsin-mediated activation of PAR2 (Déry et al., 1999). Within minutes of agonist binding, β-arrestin1 and the PAR2 receptor undergo clathrin-mediated endocytosis. While β-arrestin1 is subsequently released from endocytic vesicles, it is noteworthy that PAR2 is almost exclusively targeted to the *endo*-lysosomal pathway for degradation (Déry et al., 1999).

Moreover, Nichols et al. (2012) demonstrated that the interactions between β-arrestin2 and PAR2 contribute to both pro- and antiinflammatory signaling pathways in an OVA-sensitized mouse model of airway disease. Specifically, β-arrestin2$^{-/-}$ mice demonstrated a drastic reduction in tissue leukocyte numbers and a concurrent decrease in the production of proinflammatory cytokines, including IL-4, IL-6, IL-13, and TNFα (DeFea, Zalevsky, et al., 2000; Nichols et al., 2012). With regards to its antiinflammatory role, epithelial PAR2 stimulates the release of prostaglandin $E_2$, which promotes bronchodilation. Consequently, the fact that the anticonstrictor effects of PAR2 were unaffected in β-arrestin2-deficient mice suggests that its putative antiinflammatory role is independent of β-arrestin activation (Nichols et al., 2012). Taken together, it is nevertheless clear that both β-arrestin isoforms can dramatically alter the PAR2 signaling program, thereby contributing to the onset of allergic inflammation.

## 4.5 β-Arrestins Regulate Complement Receptor-Mediated Degranulation of Mast Cells

Complement receptors are GPCRs that are expressed on several immune cell types and are activated by complement protein fragments that are generated in the plasma during an acute inflammatory response (Dunkelberger & Song, 2010). While complement proteins subserve a vital role as opsonins by coating the surface of pathogens and thereby directing pathogen engulfment by phagocytes, they are also chemoattractants that guide leukocyte migration to infectious foci.

Notably, complement receptors are abundantly expressed on the surface of mast cells, which are tissue-resident effectors of inflammation that release histamine, bradykinin, tryptases, and other inflammatory mediators consequent to activation (Gilfillan & Tkaczyk, 2006). In conjunction with eosinophils and basophils, mast cells are mobilized during infections by complex, multicellular organisms like helminth worms (Mackay, Rosen, & Kay, 2001). Importantly, mast cells are also central culprits of anaphylactic and allergic responses, although the pathways that converge on mast cell activation under both physiological and pathological circumstances are usually antibody-mediated (Mackay et al., 2001). To these ends, several recent studies have revealed complex and divergent roles for β-arrestins in C3a receptor (C3aR)-mediated mast cell activation (Gupta, Subramanian, Klos, & Ali, 2012; Vibhuti et al., 2011).

In particular, Vibhuti et al. (2011) demonstrated that the lentivirus-mediated knock-down of β-arrestin2 in human mast cell lines abolishes desensitization and internalization of C3aR and thereby augments NFκB activity and the consequent production of chemokines. Loss of β-arrestin1, however, had no such effect on C3aR activity, although degranulation was drastically impaired in cells that were deficient in β-arrestin1 (Vibhuti et al., 2011). Indeed, subsequent studies confirmed these findings and further implicated β-arrestin2 as a negative regulator of the NFκB pathway (Gupta et al., 2012). Specifically, site-directed mutagenesis techniques were used to identify the C3aR phosphorylation sites that are critical in recruiting β-arrestin2 to the receptor consequent to agonist binding (see Section 5.3; Gupta et al., 2012; Vibhuti et al., 2011).

Even though a multitude of diverse insults can lead to inflammation, the overall process appears to unfold rather seamlessly and harmoniously. Through their canonical actions, β-arrestins act at a subcellular level in order to bolster the functional versatility of the immune response. Typically, this involves regulation of the biochemical parameters that mediate activation of lymphoid and myeloid cells through short- and long-term changes in cell-surface receptor density. In this regard, the staggering complexity of the inflammatory response is predicated on redundancy and overlap; indeed, even the mechanisms through which β-arrestins engage their putative targets differ according to the receptor and/or cell-type in question. To these ends, the interactions described between the β-arrestins and the chemokine-, complement-, and proteinase-receptors provide a mere snapshot of just how extensively the canonical roles of β-arrestin modulate immune cell physiology and cell function.

# 5. β-ARRESTINS MODULATE IMMUNE CELL FUNCTIONS VIA THEIR NONCANONICAL INTERACTIONS

## 5.1 β-Arrestin-Bound GPCRs Can Mobilize Signal Transduction Cascades

Chemokine receptor activation promotes numerous changes in leukocyte function, including increased cellular adhesiveness, rearrangements in cyto-skeletal architecture, and the release of cytotoxic granules by inflammatory effector cells (Murdoch & Finn, 2000). Interestingly, chemokine receptor activation does not evoke these functional outcomes in isolation. For example, binding of the chemokine, SDF-1α, to its cognate receptor, CXCR4, elicits both chemotaxis and maturation of T-lymphocytes in vitro (Bleul, Fuhlbrigge, Casasnovas, Aiuti, & Springer, 1996; Suzuki, Rahman, & Mitsuya, 2001). While chemotaxis results from a rapid series of biochemical events, cellular maturation entails slow changes in gene transcription and protein synthesis.

It is now evident that GPCRs mobilize multiple, parallel signaling pathways following agonist binding via β-arrestin-dependent and -independent mechanisms (DeWire et al., 2007). In this regard, the very proteins that orchestrate receptor downregulation events also generate their own signal transduction cascades. Although the exact mechanisms by which this process unfolds has been obscure, it appears that the promiscuous manner in which β-arrestins engage their cellular targets is a recurring theme that is central to this process.

A wealth of evidence has uncovered that, in addition to clathrin and adaptor protein 2 (AP2), β-arrestins bind to cytosolic and membrane-bound proteins that are prolific regulators of gene transcription, protein turnover, chemotaxis, and apoptosis (DeWire et al., 2007). The promiscuous interactions between β-arrestins and their intracellular targets allow this to occur, typically via the formation of diverse signaling scaffolds that possess the ability to transactivate members of other signaling pathways (Lefkowitz & Shenoy, 2005; Paradis et al., 2015). Some of the best-studied cellular targets that are recruited by β-arrestins to agonist-bound GPCRs include the NFκB pathway and its inhibitor, IκB; the MAPK proteins, extracellular signal-regulated kinase 1/2 (ERK 1/2); the Janus kinase proteins (JNK1, 2, and 3); the E3 ubiquitin ligase; and the Mouse double minute 2 homolog (Mdm2).

## 5.2 β-Arrestins Modulate TLR4-Mediated NFκB Activation

While chemokines provide the chemical signal necessary for the recruitment of toxin-laden granulocytes to foci of infection, their extravasation is prefaced by the activation of the toll-like family of receptors (TLRs) expressed on the surface of tissue host- and antigen-presenting cells (Jiang et al., 2013; Middleton et al., 2002). During infections, gram-negative bacteria release lipopolysaccharide (LPS) from their outer cell membrane (Lu, Yeh, & Ohashi, 2008). LPS is a prototypical TLR4 (CD14) agonist that promotes translocation of NFκB to the nucleus (Lu et al., 2008). As a transcription factor, NFκB regulates the expression of genes encoding acute phase proteins and proinflammatory cytokines, including IL-6, IL-8, TNFα in addition to several others (Jiang et al., 2013; Lu et al., 2008; Sharma & Parameswaran, 2015). IκB is a cytosolic protein that binds to NFκB in order to mitigate its unwarranted activation. Ultimately, TLR4-mediated inflammation begins with degradation of IκB and culminates in the liberation of NFκB and the expression of proinflammatory cytokines (Kawai & Akira, 2007).

β-Arrestins modulate the interaction between IκB and NFκB in several immune cell types. LPS-mediated activation of TLR4, for example, promotes binding of β-arrestin2 to IκB in the Raw264.7 macrophage cell line (Kizaki et al., 2008). By binding to IκB, β-arrestin2 stabilizes IκB and thereby potentiates its inhibition of the NFκB signaling pathway. This effect, however, is dependent upon stimulation of the β2-adrenergic receptor, suggesting that sympathetic stimuli can further modify how β-arrestins interface with the NFκB pathway and the inflammatory response (Kizaki et al., 2008). By this same token, however, it is clear that the expression levels of β-arrestins directly determine the fidelity of the NFκB pathways. It is consequently unsurprising that LPS-mediated TLR4-activation precipitates decreased expression and increased turnover of β-arrestin1, although isoform-involvement varies according to the cell model and various other parameters (Kizaki et al., 2008).

Indeed, the functional roles of β-arrestins in the context of NFκB signaling appear highly dependent upon the cell type, the receptor system, and the experimental conditions. In mouse embryonic fibroblasts (MEFs), for instance, both β-arrestin isoforms attenuate NFκB activity, such that knockdown of β-arrestin1 or β-arrestin2 using siRNA abolishes expression of IL-8 (Fan et al., 2007). Elsewhere, knockout of β-arrestin2 in bone marrow-derived macrophages leads to a drastic reduction in IL-6 and TNFα production in vitro following LPS-stimulation, suggesting that β-arrestin2 promotes NFκB-mediated expression of these cytokines (Fan et al., 2007; Gao et al., 2004). While each β-arrestin isoform affects NFκB in some way, it appears that the nuclear export signal located in β-arrestin2 is a critical structural element enabling much of its interaction with IκB and/or NFκB (Scott et al., 2002; Wang, Wu, et al., 2003).

## 5.3 β-Arrestins Modulate ERK1/2 and JNK3 MAPK Signaling Pathways

The MAPK pathway regulates several critical cellular processes, including proliferation and survival of cells (Pierce & Lefkowitz, 2001). Activation of the MAPK pathway entails the sequential recruitment and phosphorylation of a series of cytosolic kinases that eventually mobilize various transcription factors to the nucleus, where they induce expression of cell-cycle genes (Luttrell et al., 2001). Classically, MAPK signaling begins at multimolecular scaffolds that allow the various members of the kinase cascade to interact with each other. Furthermore, there are several widely expressed and redundant members within the MAPK family, each with the ability to cross-activate other MAPK proteins (Luttrell et al., 2001).

ERK1/2 and c-Jun N-terminal kinase Type 3 (JNK3) are distinct end-point effectors that are activated via MAPK signaling (Fan et al., 2007; McDonald et al., 2000; Paradis et al., 2015; Zhao et al., 2004). Broadly speaking, ERK1/2 promotes expression of genes pertinent to cellular proliferation and development, whereas JNK3 is typically associated with the expression of genes involved in inflammation and apoptosis (Pierce & Lefkowitz, 2001). To these ends, β-arrestins subserve divergent signaling roles in MAPK signaling in that their recruitment to agonist-bound receptors can either suppress or promote ERK1/2 activity. For example, both β-arrestin1 and β-arrestin2 can attenuate activation of ERK1/2 consequent to C3aR stimulation in human mast cells in vitro (Vibhuti et al., 2011). In the case of PAR2, however, β-arrestin mediates the internalization of the receptor into endocytic vesicles, followed by the recruitment of Raf-1 and ERK1/2 to the PAR2–β-arrestin complex. Importantly, this association with the β-arrestin–receptor complex serves to divert active ERK1/2 away from its cellular targets, thereby leading to decreased expression of growth- and proliferation-related gene products (DeFea, Zalevsky, et al., 2000).

Similarly, McDonald et al. (2000) showed that β-arrestin2, but not β-arrestin1, can colocalizes with JNK3 in transfected COS-7 following stimulation of the Type 1A angiotensin II receptor ($AT_1R$). Interestingly, the interaction between JNK3 and β-arrestin2 is thought to provide a mechanism by which JNK3 can be sequestered in the cytoplasm and thereby regulate its involvement in gene transcription events (McDonald et al., 2000).

In stark contrast, stimulation of the neurokinin 1 receptor (NK1R) by its endogenous agonist, substance P, leads to formation of a multimolecular scaffold that comprises NK1R, β-arrestin, and Src, the latter being yet another member of the MAPK family (DeFea, Vaughn, et al., 2000). This signaling complex subsequently serves to phosphorylate ERK1/2, the corollary being that activated pERK1/2 undergoes translocation to the nucleus, where it promotes expression of proliferation-related genes (DeFea, Vaughn, et al., 2000). While β-arrestin thus subserves a negative role in ERK1/2 activation following PAR2 and C3aR stimulation, it enhances

ERK1/2 signaling following NK1R activation. Taken together, these results suggest that the interaction between β-arrestins and the ERK1/2 signaling pathway varies from cell to cell, and that its consequence on cell proliferation or apoptosis may vary accordingly in the setting of inflammation.

## 5.4 β-Arrestins Modulate Protein Turnover via Interactions With Mdm2

The E3 ubiquitin ligase, Mdm2, is a primarily cytosolic protein that catalyzes the addition of ubiquitin onto proteins (Shenoy et al., 2001; Wang, Gao, et al., 2003). Ubiquitinated proteins are shuttled to the proteasome complex, which degrade ubiquitin-tagged proteins in an ATP-dependent manner. The activity of E3 ubiquitin ligases controls the half-life of various proteins and is critical to ensuring the timely destruction of transcription-factors that orchestrate global cellular processes, including cell division, apoptosis, and differentiation (Song et al., 2006). Importantly, Mdm2 mediates the ubiquitination and degradation of the critical cell-cycle regulator, p53, and thereby controls cellular proliferation/death (Wang, Gao, et al., 2003). It is noteworthy that even though p53 protects against cancer by inducing apoptosis of cells that have sustained irreparable DNA damage, the clonal expansion of $CD4^+$ T-lymphocytes during host-infection requires acute downregulation of p53 (Watanabe, Moon, Vacchio, Hathcock, & Hodes, 2014).

β-Arrestin2 regulates progression of the cell-cycle and cell death via indirect interactions with Mdm2 (Wang, Gao, et al., 2003). Following stimulation of the $\beta_2$-adrenergic or the $D_2$-dopamine receptor, for example, the ligated-receptor sequentially recruits β-arrestin2 and Mdm2 to the plasma membrane and/or endocytosed receptor complex (Shenoy et al., 2001). Binding of β-arrestin2 to Mdm2 effectively sequesters the ubiquitin ligase and prevents it from interacting with p53 (Shenoy et al., 2001). In this manner, excessive levels of β-arrestin2 can trap Mdm2 and prevent turnover of p53: the resultant excess in p53 consequently promotes apoptosis and cell death. In contrast, deficiencies in β-arrestin2 function may conceivably overburden the cell with Mdm2-mediated p53 degradation, rendering cells incapable of invoking apoptosis of (pre-) cancerous cells or halting the proliferation of immune cells. Of the two β-arrestin isoforms, β-arrestin2 possesses a nuclear export signal that permits its localization within the cytosol. It is therefore likely that β-arrestin2 is better poised to modulate the activity of Mdm2 compared to β-arrestin1, which lacks this nuclear export signal (Wang, Wu, et al., 2003).

## 6. β-ARRESTINS ARE IMPLICATED IN THE PATHOGENESIS OF SEVERAL IMMUNOLOGICAL DISORDERS

While it is striking that β-arrestins communicate with several signaling pathways not typically associated with GPCRs, these findings complicate our understanding of how β-arrestins contribute to health and/or disease. Because of the divergent roles subserved by β-arrestins, it has proven difficult to extrapolate how they might contribute to several well-studied immune-pathologies. Loss of β-arrestin1 function favors decreased migration in lymphocytes, for example, while that same β-arrestin deficiency promotes the opposite phenotype in neutrophils (Fong et al., 2002).

It is nevertheless undeniable that the noncanonical interactions of β-arrestins modulate inflammation in several ways. In the following sections, we consequently seek to provide a focused overview of the noncanonical roles of β-arrestins in autoimmune and infectious inflammation. In doing so, we hope to develop a meaningful framework of how β-arrestins contribute to states of health and disease. For a more exhaustive review of the noncanonical signaling roles of β-arrestins, we refer the reader to two excellent reviews by Sharma and Parameswaran (2015) and Jiang et al. (2013).

The inflammatory responses that underpin autoimmunity constitute misdirection of host machinery that normally eliminates infectious pathogens (Medzhitov, 2010; Navegantes et al., 2017). In fact, patterns of cytokine-production and immune cell-recruitment to inflammatory foci are often similar between certain infectious and autoimmune disease states. Of note, hypersensitivity reactions that feature a cell-mediated or an immune complex-generating component involve excessive secretion of interferon (IFN)-γ, IL-2, and TNFα by Type 1 CD4+ T-lymphocytes ($T_H1$) (Mosmann & Coffman, 1989; Romagnani, 1996). In $T_H1$-shifted states of disease and autoimmunity, tissue infiltrates characteristically feature macrophages and neutrophils as the predominant immune cell type. In addition, $T_H1$ inflammatory states promote a bias in antibody class switching to IgA and IgG (Moldaver, Larché, & Rudulier, 2016). In the setting of viral or intracellular infections, a $T_H1$-shifted phenotype bolsters host immunity in a controlled manner, and thereby facilitates pathogen destruction.

When hosts are infected by complex organisms (e.g., helminthic parasites), the resultant immune response must be catered toward the elimination

**Table 1** Overview of the Roles of β-Arrestins in Other $T_H1$-Shifted Autoimmune Disease States

| Disease State | Putative Pathophysiological Mechanism | Change in β-Arrestin Expression | Effect | References |
|---|---|---|---|---|
| Multiple sclerosis | Progressive, cell-mediated destruction of CNS myelin sheath | β-arr1$^{-/-}$ | Protective | Shi et al. (2007) |
| | | β-arr1$^{trans}$ | Susceptible | |
| | | β-arr2$^{-/-}$ | Susceptible | |
| Colitis | Remitting–relapsing neutrophilic intestinal inflammation | β-arr1$^{-/-}$ | Protective | Lee et al. (2013) |
| | | β-arr1$^{-/-}$ | Protective | |

*trans*, transgenic.

of extracellular pathogens that cannot be readily engulfed. Such infections bias the immune response toward the release of IL-4, IL-5, and IL-10 by Type 2 CD4+ T-helper ($T_H2$) lymphocytes. This so-called $T_H2$-shifted phenotype diametrically opposes $T_H1$ inflammation and is vital to eliminating multicellular infections by recruiting downstream effectors of inflammation, like mast cells and eosinophils (Mackay et al., 2001). Nevertheless, over activation of these very same protective mechanisms contributes to the pathogenesis of allergic asthma, in which hypersecretion of IL-4 stimulates production of IgE antibodies that sensitize mast cells against innocuous environmental allergens, and hypersecretion of IL-5 recruits eosinophils to the asthmatic airways (Mackay et al., 2001; Moldaver et al., 2016).

Ultimately, classifying inflammation according to the phenotype of the involved T-cell population provides a convenient framework in which to describe the noncanonical signaling roles of β-arrestins (see Table 1). In order to disentangle how β-arrestins might contribute to the pathogenesis of autoimmune pathologies, it will therefore be useful to similarly explore their role in host-defense against infectious insults since any role subserved by β-arrestins in one circumstance would be recapitulated in the other (Moldaver et al., 2016).

## 6.1 β-Arrestins in $T_H1$-Driven Inflammation

### *6.1.1 Rheumatoid Arthritis and Viral Infections*

Rheumatoid arthritis (RA) is a chronic autoimmune arthropathy that features progressive degeneration of joints, which results in pain and substantial decrease in quality of life (Choy, 2012; Robinson, 1966). Although RA primarily affects large joints, it can also involve soft tissues and tendinous insertions. In advanced presentations, for instance, RA can result in restricted diaphragm movement and atlanto-axial instability of the cervical spine (Robinson, 1966). Overall, the disease has a marked predilection for women of childbearing age. Although the pathophysiological basis of RA is unclear, ~80% of patients with RA inherit the human leukocyte antigen DRB1 isoform (Choy, 2012; Kerlan-Candon et al., 2001).

RA classically features a $T_H1$-shifted phenotype, in which hypersecretion of IFNγ, IL-1, and TNFα promotes neutrophil and macrophage infiltration of the joint space and the production of host-specific IgG autoantibodies that accelerate joint destruction (Choy, 2012; Romagnani, 1996). In mouse models of RA, joint injection with collagen in the presence of adjuvant stimulates expression of β-arrestin1 and β-arrestin2, while inflammatory cytokine production evoked by hyaluronan injection is largely dependent upon the expression levels of β-arrestin1 and β-arrestin2 (Li et al., 2011). In the case of β-arrestin1 overexpression, injection with hyaluronan appears to augment cytokine secretion, while overexpression of β-arrestin2 appeared to attenuate cytokine secretion consequent to hyaluronan injection (Li et al., 2011). Interestingly, hyaluronan-mediated TLR4 activation and the downstream production of proinflammatory cytokines were inhibited by β-arrestin2 (Campo et al., 2015), suggesting that β-arrestin2 might subserve a more direct role in the pathogenesis of RA than β-arrestin1. While studies have overall reported conflicting results, it appears that β-arrestin2 protects against the onset of RA, whereas β-arrestin1 has less of an effect on this process.

Because of its nuclear export signal, β-arrestin2 primarily resides within the cytoplasm and is therefore more likely to engage with signaling pathways involved in viral infections. All viruses interact with the cytoplasmic milieu to some extent. A subset of viruses, however, relies on the nuclear compartment for successful replication. Studies regarding the role of β-arrestins in the setting of the antiviral immune response suggest that both β-arrestin1 and β-arrestin2 attenuate the antiviral response (Mo et al., 2008; Seregin et al., 2010). In MEFs and HeLa cells, for example, β-arrestin1 attenuates the IFNγ-mediated antiviral response and knockout of β-arrestin1 was associated with decreased viral loads and robust elimination of the offending pathogen (Mo et al., 2008). Similarly, antiviral activity of natural killer cells was inversely related to β-arrestin2 activity: mice that were overexpressing β-arrestin2 exhibited poorer clearance of murine cytomegalovirus, whereas knockouts of β-arrestin2 demonstrated improved antiviral responses (Yu et al., 2008).

The inflammation that results from a viral infection classically involves a $T_H1$-shifted phenotype that is well suited to annihilate intracellular pathogens (Mosmann & Coffman, 1989). In the setting of infection, β-arrestins hamper the inflammatory axis and virus removal; in the setting of autoimmune disease, β-arrestins protect against RA. At least in the case of β-arrestin2, it appears that β-arrestins negatively modulate $T_H1$-associated cytokine responses and are therefore critical to both physiologic and pathological processes.

## 6.2 β-Arrestins in $T_H2$-Driven Inflammation

### *6.2.1 Allergic Asthma and Helminthic Infection*

Allergic asthma is an obstructive airway disease characterized by episodes of reversible bronchoconstriction and symptoms of wheezy breathlessness (Moldaver et al., 2016). Allergic asthma is the most common subtype of the disease, resulting from host-sensitization against airborne allergens and the production of allergen-specific IgE antibodies that bind to the type I high affinity IgE receptor ($Fc_εRI$) on the surface of mast cells (Gilfillan & Tkaczyk, 2006; Moldaver et al., 2016). Subsequent reexposure to the allergen cross-links IgE-bound $Fc_εRI$, leading to mast cell activation and several downstream effects, including the release of histamine, production of lipid mediators, secretion of tryptase, and contraction of airway smooth muscle cells (Gilfillan & Tkaczyk, 2006).

In a mouse model of allergic asthma, knockout of β-arrestins led to decreased airway inflammation and $T_H2$-proliferation when expression of β-arrestin2 was ablated (Walker et al., 2003). By virtue of its nuclear export signal, β-arrestin2 is thought to readily engage with a number of transcriptional regulators that drive the production of proinflammatory cytokines (Gurevich & Gurevich, 2006; Scott et al., 2002). Strikingly, when β-arrestin2 expression was abolished using silencing RNA after the disease had already been established, there was no effect on allergen-induced airway contraction suggesting that β-arrestin2 is more critical for the establishment of the disease rather than its perpetuation (Chen et al., 2015; Hollingsworth et al., 2010; Nials & Uddin, 2008). Consistent with these findings, ablation of β-arrestin2 expression in airway epithelial cells abrogated allergen-induced airway contraction in animals that were sensitized to airborne allergen (Hollingsworth et al., 2010). These findings lie in contrast to the role of β-arrestin1, which does not contribute to the pathogenesis of allergic asthma (Nichols et al., 2012). Specifically, knockout of β-arrestin1 had no effect on the infiltration of eosinophils or the degree of methacholine-evoked airway contractions. Furthermore, loss of β-arrestin1 function also did not alter either the onset or the persistence of asthmatic symptoms in an OVA-allergen mouse model of asthma (Nichols et al., 2012).

Strikingly, the overall immunological response to helminth infection recapitulates many of the defining elements of allergic asthma (Hepworth, Maurer, & Hartmann, 2012; Mackay et al., 2001; Romagnani, 1996). Unlike asthmatics, however, parasitized individuals generate IgE antibodies against antigenic determinants located on the surface of the foreign pathogen rather than against otherwise innocuous environmental allergens (Mackay et al., 2001). It is consequently surprising that, to the best of our knowledge, there has been no investigation to date into the putative role of β-arrestins in parasitic/multicellular infections. In particular, a deeper understanding of the expression, regulation, and signaling patterns of β-arrestins in the setting of helminthic disease might clarify their role(s) in allergic asthma and thereby provide a more definitive answer as to how β-arrestin levels might possibly contribute to the pathogenesis of allergic asthma.

## 7. FUTURE PERSPECTIVES AND CONCLUSIONS

β-Arrestins modulate the desensitization, internalization, and degradation of a vast number of cellular receptors involved in the activation of immune cells. In this regard, the canonical roles of β-arrestins can directly affect a vast number of cell surface receptors that are involved in chemotaxis, chemokine secretion, and inflammatory mediator release. Because of this, the classical roles of β-arrestins are critical determinants of the immune response. Remarkably, shortcomings in β-arrestin function are associated with divergent immune effects; for example, loss of β-arrestin2 can precipitate increased tissue infiltration of neutrophils but also obliterate migration of $T_H2$ lymphocytes (Contento et al., 2008; Su et al., 2005; Zou et al., 1998). Such seemingly contradictory effects are unsurprising, however, given that inflammation is a multilayered and extremely interconnected phenomenon.

As the noncanonical roles of β-arrestins came to light, subsequent studies focused on how these proteins interact with distinct intracellular processes, including protein turnover, cell division, migration, and apoptosis. While the notion that β-arrestins collaborate with several MAPK pathway proteins adds a novel layer of functional complexity, it also complicates how we approach therapies of inflammatory and immune diseases. Strikingly, the possibility that a "biased agonist" could co-opt this β-arrestin-mediated interaction in order to invoke an entirely novel cellular response is tantalizing (see Fig. 5).

Indeed, the long-standing "one agonist-one response" paradigm will be under increasing scrutiny as newer compounds are discovered that are capable of invoking only the noncanonical signal transduction cascades of GPCRs by selectively recruiting β-arrestins to the cell membrane (Violin & Lefkowitz, 2007). Nevertheless, current progress has been drastically hampered by the absence of adequate screening methods that can identify compounds with high affinity for GPCRs and that are capable of activating noncanonical signaling responses that are rapid and readily detectable.

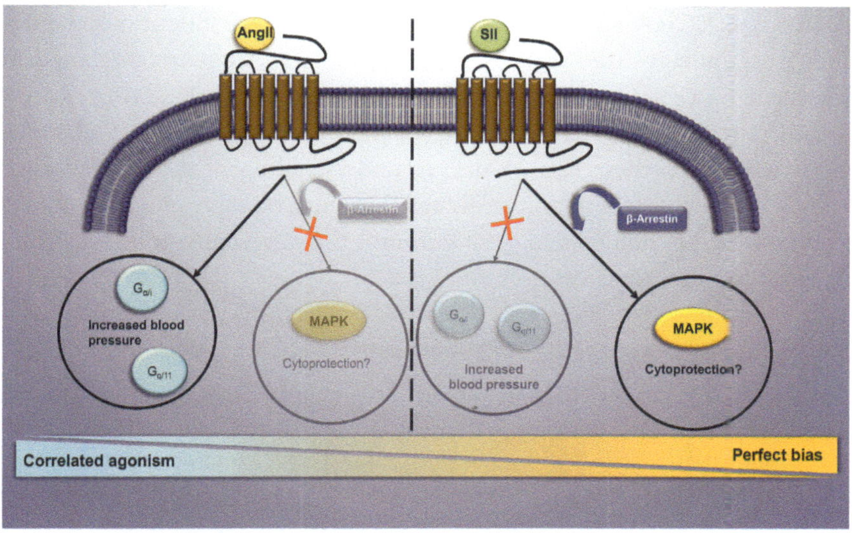


**Fig. 5** β-Arrestins are preferentially recruited to GPCRs in response to specific agonists. Activation of the angiotensin II type 1A receptor ($AT_{1A}R$) provokes mobilization of classical $G_{\alpha/s}$ and $G_{q/11}$ G-proteins, which promote increased intracellular levels of cAMP and $Ca^{2+}$, respectively. A modified synthetic analogue (SII), on the other hand, engages the $AT_{1A}R$ but only promotes activation of a MAPK pathway. In contrast to the endogenous agonist, angiotensin II, this modified peptide exhibits perfect bias toward a β-arrestin-mediated signal transduction cascade. As angiotensin II-mediated overstimulation of $AT_{1A}R$ is thought to directly contribute to the pathogenesis of hypertension, the prospect of developing novel therapeutic agents that are biased agonists would be groundbreaking (Bologna, Teoh, Bayoumi, Tang, & Kim, 2017).

The immune system employs a staggering array of cellular and humoral elements in order to swiftly eliminate pathogens. Consequently, inflammatory effectors and mediators are only mobilized on an as-needed basis and in a time-sensitive manner. Given that inflammatory cells wield tremendous destructive potential, however, myriad control mechanisms exist in order to control how inflammation develops, propagates, and terminates. Evidently, aberrancies in the regulatory signaling programs that control inflammation can be devastating, leading to immunodeficiency on one hand, but autoimmune disease on the other.

The canonical and noncanonical signaling roles of β-arrestins are thus indispensible to multiple aspects of immune cell function. In keeping with their well-characterized interactions with $\beta_2$-adrenergic receptors, β-arrestins modulate leukocyte and lymphocyte function by altering the sensitivity and the expression of critical cell surface receptors. In the context of immune function, many early studies show that β-arrestins orchestrate the desensitization and internalization of the chemokine receptors, CXCR1, CXCR2, CXCR4, and CCR5 and therefore affect the function of neutrophils, macrophages, and T-cells. Despite the fact that β-arrestins mediate divergent processes in immune cells, it is as yet too early and premature to draw conclusions as to the role of β-arrestins in the setting of inflammatory diseases. To these ends, a deeper understanding of the noncanonical roles of β-arrestins will undoubtedly revolutionize how we conceive of normal immune physiology and greatly inform the development of drug therapies for immunologic diseases.

## ACKNOWLEDGMENTS

We thank the H.S. and R.D. laboratory members for exciting discussions and critical review of the book chapter. Research work in the laboratory of H.S. and R.D. is supported by the NIH Grants R00HL121073 (H.S.) and K22CA188149 (R.D.). We apologize to any authors whose work might have been omitted unintentionally due to limited space and/or the specific focus of the chapter.

## REFERENCES

Aramori, I., Zhang, J., Ferguson, S. S. G., Bieniasz, P. D., Cullen, B. R., & Caron, M. G. (1997). Molecular mechanism of desensitization of the chemokine receptor CCR-5: Receptor signaling and internalization are dissociable from its role as an HIV-1 co-receptor. *The EMBO Journal*, *16*, 4606–4616. http://dx.doi.org/10.1093/emboj/16.15.4606.

Barlic, J., Andrews, J. D., Kelvin, A. A., Bosinger, S. E., DeVries, M. E., Xu, L., et al. (2000). Regulation of tyrosine kinase activation and granule release through beta-arrestin by CXCRI. *Nature Immunology*, *1*, 227–233.

Barlic, J., Khandaker, M. H., Mahon, E., Andrews, J., DeVries, M. E., Mitchell, G. B., et al. (1999). Beta-arrestins regulate interleukin-8-induced CXCR1 internalization. *The Journal of Biological Chemistry, 274*, 16287–16294.

Basher, F., Fan, H., Zingarelli, B., Borg, K. T., Luttrell, L. M., Tempel, G. E. et al. (2008). Beta-arrestin 2: A negative regulator of inflammatory responses in Polymorphonuclear leukocytes. *International Journal of Clinical and Experimental Medicine, 1*, 32–41. e-Century Publishing Corporation.

Bennett, L. D., Fox, J. M., & Signoret, N. (2011). Mechanisms regulating chemokine receptor activity. *Immunology, 134*, 246–256.

Benovic, J. L., Kühn, H., Weyand, I., Codina, J., Caron, M. G., & Lefkowitz, R. J. (1987). Functional desensitization of the isolated beta-adrenergic receptor by the beta-adrenergic receptor kinase: Potential role of an analog of the retinal protein arrestin (48-kDa protein). *Proceedings of the National Academy of Sciences of the United States of America, 84*, 8879–8882.

Bhandari, D., Trejo, J., Benovic, J. L., & Marchese, A. (2007). Arrestin-2 interacts with the ubiquitin-protein isopeptide ligase atrophin-interacting protein 4 and mediates endosomal sorting of the chemokine receptor CXCR4. *The Journal of Biological Chemistry, 282*, 36971–36979. http://dx.doi.org/10.1074/jbc.M705085200.

Bleul, C. C., Fuhlbrigge, R. C., Casasnovas, J. M., Aiuti, A., & Springer, T. A. (1996). A highly efficacious lymphocyte chemoattractant, stromal cell-derived factor 1 (SDF-1). *The Journal of Experimental Medicine, 184*, 1101–1109.

Bologna, Z., Teoh, J.-P., Bayoumi, A. S., Tang, Y., & Kim, I.-M. (2017). Biased G protein-coupled receptor signaling: New player in modulating physiology and pathology. *Biomolecules & Therapeutics, 25*, 12–25.

Borish, L. C., & Steinke, J. W. (2003). 2. Cytokines and chemokines. *The Journal of Allergy and Clinical Immunology, 111*, S460–S475.

Campo, G. M., Avenoso, A., D'Ascola, A., Scuruchi, M., Calatroni, A., & Campo, S. (2015). Beta-arrestin-2 negatively modulates inflammation response in mouse chondrocytes induced by 4-mer hyaluronan oligosaccharide. *Molecular and Cellular Biochemistry, 399*, 201–208.

Cenac, N., Coelho, A.-M., Nguyen, C., Compton, S., Andrade-Gordon, P., MacNaughton, W. K., et al. (2002). Induction of intestinal inflammation in mouse by activation of proteinase-activated receptor-2. *The American Journal of Pathology, 161*, 1903–1915. American Society for Investigative Pathology.

Chen, M., Hegde, A., Choi, Y. H., Theriot, B. S., Premont, R. T., Chen, W., et al. (2015). Genetic deletion of β-arrestin-2 and the mitigation of established airway hyperresponsiveness in a murine asthma model. *American Journal of Respiratory Cell and Molecular Biology, 53*, 346–354.

Cheng, Z. J., Zhao, J., Sun, Y., Hu, W., Wu, Y. L., Cen, B., et al. (2000). Beta-arrestin differentially regulates the chemokine receptor CXCR4-mediated signaling and receptor internalization, and this implicates multiple interaction sites between beta-arrestin and CXCR4. *The Journal of Biological Chemistry, 275*, 2479–2485.

Cherezov, V., Rosenbaum, D. M., Hanson, M. A., Rasmussen, S. G. F., Thian, F. S., Kobilka, T. S., et al. (2007). High-resolution crystal structure of an engineered human 2-adrenergic G protein-coupled receptor. *Science, 318*, 1258–1265.

Choy, E. (2012). Understanding the dynamics: Pathways involved in the pathogenesis of rheumatoid arthritis. *Rheumatology (Oxford), 51*(Suppl. 5), v3–v11.

Contento, R. L., Molon, B., Boularan, C., Pozzan, T., Manes, S., Marullo, S., et al. (2008). CXCR4-CCR5: A couple modulating T cell functions. *Proceedings of the National Academy of Sciences, 105*, 10101–10106.

Craft, C. M., & Whitmore, D. H. (1995). The arrestin superfamily: Cone arrestins are a fourth family. *FEBS Letters, 362*, 247–255.

de Mendoza, A., Sebé-Pedrós, A., & Ruiz-Trillo, I. (2014). The evolution of the GPCR signaling system in eukaryotes: Modularity, conservation, and the transition to metazoan multicellularity. *Genome Biology and Evolution*, *6*, 606–619.

DeFea, K. A. (2007). Stop that cell! Beta-arrestin-dependent chemotaxis: A tale of localized actin assembly and receptor desensitization. *Annual Review of Physiology*, *69*, 535–560.

DeFea, K. A., Vaughn, Z. D., O'Bryan, E. M., Nishijima, D., Déry, O., & Bunnett, N. W. (2000a). The proliferative and antiapoptotic effects of substance P are facilitated by formation of a beta-arrestin-dependent scaffolding complex. *Proceedings of the National Academy of Sciences of the United States of America*, *97*, 11086–11091.

DeFea, K. A., Zalevsky, J., Thoma, M. S., Déry, O., Mullins, R. D., & Bunnett, N. W. (2000b). Beta-arrestin-dependent endocytosis of proteinase-activated receptor 2 is required for intracellular targeting of activated ERK1/2. *The Journal of Cell Biology*, *148*, 1267–1281.

Déry, O., Thoma, M. S., Wong, H., Grady, E. F., & Bunnett, N. W. (1999). Trafficking of proteinase-activated receptor-2 and beta-arrestin-1 tagged with green fluorescent protein. Beta-arrestin-dependent endocytosis of a proteinase receptor. *The Journal of Biological Chemistry*, *274*, 18524–18535.

DeWire, S. M., Ahn, S., Lefkowitz, R. J., & Shenoy, S. K. (2007). Beta-arrestins and cell signaling. *Annual Review of Physiology*, *69*, 483–510.

Dunkelberger, J. R., & Song, W.-C. (2010). Complement and its role in innate and adaptive immune responses. *Cell Research*, *20*, 34–50.

Fan, H., Luttrell, L. M., Tempel, G. E., Senn, J. J., Halushka, P. V., & Cook, J. A. (2007). Beta-arrestins 1 and 2 differentially regulate LPS-induced signaling and proinflammatory gene expression. *Molecular Immunology*, *44*, 3092–3099.

Fan, G.-H., Yang, W., Wang, X.-J., Qian, Q., & Richmond, A. (2001). Identification of a motif in the carboxyl terminus of CXCR2 that is involved in adaptin 2 binding and receptor internalization. *Biochemistry*, *40*, 791–800.

Fong, A. M., Premont, R. T., Richardson, R. M., Yu, Y.-R. A., Lefkowitz, R. J., & Patel, D. D. (2002). Defective lymphocyte chemotaxis in beta-arrestin2- and GRK6-deficient mice. *Proceedings of the National Academy of Sciences of the United States of America*, *99*, 7478–7483.

Freedman, N. J., & Lefkowitz, R. J. (1996). Desensitization of G protein-coupled receptors. *Recent Progress in Hormone Research*, *51*, 319–351.

Gao, H., Sun, Y., Wu, Y., Luan, B., Wang, Y., Qu, B., et al. (2004). Identification of beta-arrestin2 as a G protein-coupled receptor-stimulated regulator of NF-kappaB pathways. *Molecular Cell*, *14*, 303–317.

Gilfillan, A. M., & Tkaczyk, C. (2006). Integrated signalling pathways for mast-cell activation. *Nature Reviews Immunology*, *6*, 218–230.

Gilliland, C. T., Salanga, C. L., Kawamura, T., Trejo, J., & Handel, T. M. (2013). The chemokine receptor CCR1 is constitutively active, which leads to G protein-independent, β-arrestin-mediated internalization. *The Journal of Biological Chemistry*, *288*, 32194–32210. http://dx.doi.org/10.1074/jbc.M113.503797.

Goodman, O. B., Krupnick, J. G., Gurevich, V. V., Benovic, J. L., & Keen, J. H. (1997). Arrestin/clathrin interaction. Localization of the arrestin binding locus to the clathrin terminal domain. *The Journal of Biological Chemistry*, *272*, 15017–15022.

Goodman, O. B., Krupnick, J. G., Santini, F., Gurevich, V. V., Penn, R. B., Gagnon, A. W., et al. (1996). Beta-arrestin acts as a clathrin adaptor in endocytosis of the beta2-adrenergic receptor. *Nature*, *383*, 447–450.

Gupta, K., Subramanian, H., Klos, A., & Ali, H. (2012). Phosphorylation of C3a receptor at multiple sites mediates desensitization, β-arrestin-2 recruitment and inhibition of NF-κB activity in mast cells. *PLoS One*, 7, e46369.

Gurevich, V. V., & Gurevich, E. V. (2006). The structural basis of arrestin-mediated regulation of G-protein-coupled receptors. *Pharmacology & Therapeutics, 110*, 465–502.
Han, M., Gurevich, V. V., Vishnivetskiy, S. A., Sigler, P. B., & Schubert, C. (2001). Crystal structure of beta-arrestin at 1.9 A: Possible mechanism of receptor binding and membrane translocation. *Structure, 9*, 869–880.
Haribabu, B., Richardson, R. M., Fisher, I., Sozzani, S., Peiper, S. C., Horuk, R., et al. (1997). Regulation of human chemokine receptors CXCR4. Role of phosphorylation in desensitization and internalization. *The Journal of Biological Chemistry, 272*, 28726–28731.
Hepworth, M. R., Maurer, M., & Hartmann, S. (2012). Regulation of type 2 immunity to helminths by mast cells. *Gut Microbes, 3*, 476–481. Taylor & Francis.
Hollingsworth, J. W., Theriot, B. S., Li, Z., Lawson, B. L., Sunday, M., Schwartz, D. A., et al. (2010). Both hematopoietic-derived and non-hematopoietic-derived {beta}-arrestin-2 regulates murine allergic airway disease. *American Journal of Respiratory Cell and Molecular Biology, 43*, 269–275.
Holmes, W. E., Lee, J., Kuang, W. J., Rice, G. C., & Wood, W. I. (1991). Structure and functional expression of a human interleukin-8 receptor. *Science, 253*, 1278–1280.
Huttenrauch, F., Nitzki, A., Lin, F.-T., Höning, S., & Oppermann, M. (2002). Beta-arrestin binding to CC chemokine receptor 5 requires multiple C-terminal receptor phosphorylation sites and involves a conserved Asp-Arg-Tyr sequence motif. *The Journal of Biological Chemistry, 277*, 30769–30777.
Hüttenrauch, F., Pollok-Kopp, B., & Oppermann, M. (2005). G protein-coupled receptor kinases promote phosphorylation and β-arrestin-mediated internalization of CCR5 homo- and hetero-oligomers. *The Journal of Biological Chemistry, 280*, 37503–37515. http://dx.doi.org/10.1074/jbc.M500535200.
Imming, P., Sinning, C., & Meyer, A. (2006). Drugs, their targets and the nature and number of drug targets. *Nature Reviews Drug Discovery, 5*, 821–834. Nature Publishing Group.
Jiang, D., Xie, T., Liang, J., & Noble, P. W. (2013). Beta-arrestins in the immune system. *Progress in Molecular Biology and Translational Science, 118*, 359–393. http://dx.doi.org/10.1016/B978-0-12-394440-5.00014-0.
Jonsson, F., Mancardi, D. A., Albanesi, M., & Bruhns, P. (2013). Neutrophils in local and systemic antibody-dependent inflammatory and anaphylactic reactions. *Journal of Leukocyte Biology, 94*, 643–656.
Kang, J., Shi, Y., Xiang, B., Qu, B., Su, W., Zhu, M., et al. (2005). A nuclear function of beta-arrestin1 in GPCR signaling: Regulation of histone acetylation and gene transcription. *Cell, 123*, 833–847.
Kang, D. S., Tian, X., & Benovic, J. L. (2014). Role of beta-arrestins and arrestin domain-containing proteins in G protein-coupled receptor trafficking. *Current Opinion in Cell Biology, 27*, 63–71.
Kawai, T., & Akira, S. (2007). Signaling to NF-κB by toll-like receptors. *Trends in Molecular Medicine, 13*, 460–469.
Kenakin, T. (2002). Efficacy at G-protein-coupled receptors. *Nature Reviews. Drug Discovery, 1*, 103–110. Nature Publishing Group.
Kerlan-Candon, S., Combe, B., Vincent, R., Clot, J., Pinet, V., & Eliaou, J. F. (2001). HLA-DRB1 gene transcripts in rheumatoid arthritis. *Clinical and Experimental Immunology, 124*, 142–149. Wiley-Blackwell.
Kim, Y.-M., Barak, L. S., Caron, M. G., & Benovic, J. L. (2002). Regulation of arrestin-3 phosphorylation by casein kinase II. *The Journal of Biological Chemistry, 277*, 16837–16846.
Kizaki, T., Izawa, T., Sakurai, T., Haga, S., Taniguchi, N., Tajiri, H., et al. (2008). β2-Adrenergic receptor regulates toll-like receptor-4-induced nuclear factor-κB activation

through β-arrestin 2. *Immunology, 124*, 348–356. http://dx.doi.org/10.1111/j.1365-2567.2007.02781.x.
Kobilka, B. K. (2007). G protein coupled receptor structure and activation. *Biochimica et Biophysica Acta, 1768*, 794–807. NIH Public Access.
Komori, N., Cain, S. D., Roch, J. M., Miller, K. E., & Matsumoto, H. (1998). Differential expression of alternative splice variants of beta-arrestin-1 and -2 in rat central nervous system and peripheral tissues. *The European Journal of Neuroscience, 10*, 2607–2616.
Laporte, S. A., Oakley, R. H., Holt, J. A., Barak, L. S., & Caron, M. G. (2000). The interaction of beta-arrestin with the AP-2 adaptor is required for the clustering of beta 2-adrenergic receptor into clathrin-coated pits. *The Journal of Biological Chemistry, 275*, 23120–23126.
Laporte, S. A., Oakley, R. H., Zhang, J., Holt, J. A., Ferguson, S. S., Caron, M. G., et al. (1999). The beta2-adrenergic receptor/beta-arrestin complex recruits the clathrin adaptor AP-2 during endocytosis. *Proceedings of the National Academy of Sciences of the United States of America, 96*, 3712–3717.
Lawrence, T. (2009). The nuclear factor NF-kappaB pathway in inflammation. *Cold Spring Harbor Perspectives in Biology, 1*, a001651. Cold Spring Harbor Laboratory Press.
Lee, T., Lee, E., Irwin, R., Lucas, P. C., McCabe, L. R., & Parameswaran, N. (2013). β-Arrestin-1 deficiency protects mice from experimental colitis. *The American Journal of Pathology, 182*, 1114–1123.
Lefkowitz, R., Ahn, S., Lefkowitz, R. J., & Shenoy, S. K. (2007). Introduction to special section on β-arrestins. *Annual Review of Physiology, 69*. http://dx.doi.org/10.1146/annurev.ph.69.013107.100021
Lefkowitz, R. J., Rajagopal, K., & Whalen, E. J. (2006). New roles for β-arrestins in cell signaling: Not just for seven-transmembrane receptors. *Molecular Cell, 24*, 643–652.
Lefkowitz, R. J., & Shenoy, S. K. (2005). Transduction of receptor signals by beta-arrestins. *Science, 308*, 512–517.
Li, P., Cook, J. A., Gilkeson, G. S., Luttrell, L. M., Wang, L., Borg, K. T., et al. (2011). Increased expression of beta-arrestin 1 and 2 in murine models of rheumatoid arthritis: Isoform specific regulation of inflammation. *Molecular Immunology, 49*, 64–74.
Libby, P., Ridker, P. M., Hansson, G. K., & Leducq Transatlantic Network on Atherothrombosis. (2009). Inflammation in atherosclerosis: From pathophysiology to practice. *Journal of the American College of Cardiology, 54*, 2129–2138. NIH Public Access.
Lin, F.-T., Chen, W., Shenoy, S., Cong, M., Exum, S. T., & Lefkowitz, R. J. (2002). Phosphorylation of beta-arrestin2 regulates its function in internalization of beta(2)-adrenergic receptors. *Biochemistry, 41*, 10692–10699.
Lin, F.-T., Krueger, K. M., Kendall, H. E., Daaka, Y., Fredericks, Z. L., Pitcher, J. A., et al. (1997). Clathrin-mediated endocytosis of the -adrenergic receptor is regulated by phosphorylation/dephosphorylation of -arrestin1. *The Journal of Biological Chemistry, 272*, 31051–31057. American Society for Biochemistry and Molecular Biology.
Lohse, M. J., Benovic, J. L., Codina, J., Caron, M. G., & Lefkowitz, R. J. (1990). Beta-arrestin: A protein that regulates beta-adrenergic receptor function. *Science, 248*, 1547–1550.
Lohse, M. J., & Hoffmann, C. (2014). Arrestin interactions with G protein-coupled receptors. *219*, 15–56.
Lu, Y.-C., Yeh, W.-C., & Ohashi, P. S. (2008). LPS/TLR4 signal transduction pathway. *Cytokine, 42*, 145–151.
Luttrell, L. M., Roudabush, F. L., Choy, E. W., Miller, W. E., Field, M. E., Pierce, K. L., et al. (2001). Activation and targeting of extracellular signal-regulated kinases by -arrestin scaffolds. *Proceedings of the National Academy of Sciences, 98*, 2449–2454.
Mackay, I. R., Rosen, F. S., & Kay, A. B. (2001). Allergy and allergic diseases. *The New England Journal of Medicine, 344*, 30–37.

McDonald, P. H., Chow, C. W., Miller, W. E., Laporte, S. A., Field, M. E., Lin, F. T., et al. (2000). Beta-arrestin 2: A receptor-regulated MAPK scaffold for the activation of JNK3. *Science, 290*, 1574–1577.

Medzhitov, R. (2010). Inflammation 2010: New adventures of an old flame. *Cell, 140*, 771–776.

Middleton, J., Patterson, A. M., Gardner, L., Schmutz, C., & Ashton, B. A. (2002). Leukocyte extravasation: Chemokine transport and presentation by the endothelium. *Blood, 100*, 3853–3860.

Milano, S. K., Kim, Y.-M., Stefano, F. P., Benovic, J. L., & Brenner, C. (2006). Nonvisual arrestin oligomerization and cellular localization are regulated by inositol hexakisphosphate binding. *The Journal of Biological Chemistry, 281*, 9812–9823.

Mo, W., Zhang, L., Yang, G., Zhai, J., Hu, Z., Chen, Y., et al. (2008). Nuclear β-Arrestin1 functions as a scaffold for the dephosphorylation of STAT1 and moderates the antiviral activity of IFN-γ. *Molecular Cell, 31*, 695–707.

Moldaver, D. M., Larché, M., & Rudulier, C. D. (2016). An update on lymphocyte subtypes in asthma and airway disease. *Chest*. http://dx.doi.org/10.1016/j.chest.2016.10.038

Moore, C. A. C., Milano, S. K., & Benovic, J. L. (2007). Regulation of receptor trafficking by GRKs and arrestins. *Annual Review of Physiology, 69*, 451–482.

Mosmann, T. R., & Coffman, R. L. (1989). TH1 and TH2 cells: Different patterns of lymphokine secretion lead to different functional properties. *Annual Review of Immunology, 7*, 145–173.

Murdoch, C., & Finn, A. (2000). Chemokine receptors and their role in inflammation and infectious diseases. *Blood, 95*, 3032–3043.

Navegantes, K. C., de Souza, G. R., Pereira, P. A. T., Czaikoski, P. G., Azevedo, C. H. M., & Monteiro, M. C. (2017). Immune modulation of some autoimmune diseases: The critical role of macrophages and neutrophils in the innate and adaptive immunity. *Journal of Translational Medicine, 15*, 36. BioMed Central.

Nials, A. T., & Uddin, S. (2008). Mouse models of allergic asthma: Acute and chronic allergen challenge. *Disease Models & Mechanisms, 1*, 213–220.

Nichols, H. L., Saffeddine, M., Theriot, B. S., Hegde, A., Polley, D., El-Mays, T., et al. (2012). β-Arrestin-2 mediates the proinflammatory effects of proteinase-activated receptor-2 in the airway. *Proceedings of the National Academy of Sciences of the United States of America, 109*, 16660–16665. National Academy of Sciences.

Oakley, R. H., Laporte, S. A., Holt, J. A., Barak, L. S., & Caron, M. G. (1999). Association of beta-arrestin with G protein-coupled receptors during clathrin-mediated endocytosis dictates the profile of receptor resensitization. *The Journal of Biological Chemistry, 274*, 32248–32257.

Oakley, R. H., Laporte, S. A., Holt, J. A., Barak, L. S., & Caron, M. G. (2001). Molecular determinants underlying the formation of stable intracellular G protein-coupled receptor-beta-arrestin complexes after receptor endocytosis*. *The Journal of Biological Chemistry, 276*, 19452–19460.

Oakley, R. H., Laporte, S. A., Holt, J. A., Caron, M. G., & Barak, L. S. (2000). Differential affinities of visual arrestin, beta arrestin1, and beta arrestin2 for G protein-coupled receptors delineate two major classes of receptors. *The Journal of Biological Chemistry, 275*, 17201–17210.

Okin, D., & Medzhitov, R. (2012). Evolution of inflammatory diseases. *Current Biology, 22*, R733–R740.

Paradis, J. S., Ly, S., Blondel-Tepaz, É., Galan, J. A., Beautrait, A., Scott, M. G. H., et al. (2015). Receptor sequestration in response to β-arrestin-2 phosphorylation by ERK1/2 governs steady-state levels of GPCR cell-surface expression. *Proceedings of the National Academy of Sciences of the United States of America, 112*, E5160–E5168. National Academy of Sciences.

Parruti, G., Peracchia, F., Sallese, M., Ambrosini, G., Masini, M., Rotilio, D., et al. (1993). Molecular analysis of human beta-arrestin-1: Cloning, tissue distribution, and regulation of expression. Identification of two isoforms generated by alternative splicing. *The Journal of Biological Chemistry*, *268*, 9753–9761.

Pierce, K. L., & Lefkowitz, R. J. (2001). Classical and new roles of β-arrestins in the regulation of G-protein-coupled receptors. *Nature Reviews. Neuroscience*, *2*, 727–733.

Raghuwanshi, S. K., Su, Y., Singh, V., Haynes, K., Richmond, A., & Richardson, R. M. (2012). The chemokine receptors CXCR1 and CXCR2 couple to distinct G protein-coupled receptor kinases to mediate and regulate leukocyte functions. *Journal of Immunology*, *189*, 2824–2832. NIH Public Access.

Rajagopal, S., & Shenoy, S. K. (2017). GPCR desensitization: Acute and prolonged phases. *Cellular Signalling*. 1–8, Elsevier Inc. http://dx.doi.org/10.1016/j.cellsig.2017.01.024. in press.

Richardson, R. M., Ali, H., Pridgen, B. C., Haribabu, B., & Snyderman, R. (1998). Multiple signaling pathways of human interleukin-8 receptor A. Independent regulation by phosphorylation. *The Journal of Biological Chemistry*, *273*, 10690–10695.

Richardson, R. M., Marjoram, R. J., Barak, L. S., & Snyderman, R. (2003). Role of the cytoplasmic tails of CXCR1 and CXCR2 in mediating leukocyte migration, activation, and regulation. *Journal of Immunology*, *170*, 2904–2911.

Robinson, H. S. (1966). Rheumatoid arthritis—Atlanto-axial subluxation and its clinical presentation. *Canadian Medical Association Journal*, *94*, 470–477. Canadian Medical Association.

Romagnani, S. (1996). Th1 and Th2 in human diseases. *Clinical Immunology and Immunopathology*, *80*, 225–235.

Rosenbaum, D. M., Rasmussen, S. G. F., & Kobilka, B. K. (2009). The structure and function of G-protein-coupled receptors. *Nature*, *459*, 356–363. Nature Publishing Group.

Schmidlin, F., & Bunnett, N. W. (2001). Protease-activated receptors: How proteases signal to cells. *Current Opinion in Pharmacology*, *1*, 575–582.

Scott, M. G. H., Le Rouzic, E., Périanin, A., Pierotti, V., Enslen, H., Benichou, S., et al. (2002). Differential nucleocytoplasmic shuttling of beta-arrestins. Characterization of a leucine-rich nuclear export signal in beta-arrestin2. *The Journal of Biological Chemistry*, *277*, 37693–37701.

Seregin, S. S., Appledorn, D. M., Patial, S., Bujold, M., Nance, W., Godbehere, S., et al. (2010). Beta-arrestins modulate adenovirus-vector-induced innate immune responses: Differential regulation by beta-arrestin-1 and beta-arrestin-2. *Virus Research*, *147*, 123–134.

Sharma, D., & Parameswaran, N. (2015). Multifaceted role of β-arrestins in inflammation and disease. *Genes and Immunity*, *16*, 499–513.

Shenoy, S. K., & Lefkowitz, R. J. (2011). β-Arrestin-mediated receptor trafficking and signal transduction. *Trends in Pharmacological Sciences*, *32*, 521–533.

Shenoy, S. K., McDonald, P. H., Kohout, T. A., & Lefkowitz, R. J. (2001). Regulation of receptor fate by ubiquitination of activated beta 2-adrenergic receptor and beta-arrestin. *Science*, *294*, 1307–1313.

Shi, Y., Feng, Y., Kang, J., Liu, C., Li, Z., Li, D., et al. (2007). Critical regulation of CD4+ T cell survival and autoimmunity by beta-arrestin 1. *Nature Immunology*, *8*, 817–824.

Song, X., Raman, D., Gurevich, E. V., Vishnivetskiy, S. A., & Gurevich, V. V. (2006). Visual and both non-visual arrestins in their "inactive" conformation bind JNK3 and Mdm2 and relocalize them from the nucleus to the cytoplasm. *The Journal of Biological Chemistry*, *281*, 21491–21499.

Su, Y., Raghuwanshi, S. K., Yu, Y., Nanney, L. B., Richardson, R. M., & Richmond, A. (2005). Altered CXCR2 signaling in beta-arrestin-2-deficient mouse models. *Journal of Immunology*, *175*, 5396–5402.

Su, Y., & Richmond, A. (2015). Chemokine regulation of neutrophil infiltration of skin wounds. *Advances in Wound Care, 4*, 631–640. Mary Ann Liebert, Inc.

Suzuki, Y., Rahman, M., & Mitsuya, H. (2001). Diverse transcriptional response of CD4+ T cells to stromal cell-derived factor SDF-1: Cell survival promotion and priming effects of SDF-1 on CD4+ T cells. *Journal of Immunology, 167*, 3064–3073.

Vergnolle, N., Wallace, J. L., Bunnett, N. W., & Hollenberg, M. D. (2001). Protease-activated receptors in inflammation, neuronal signaling and pain. *Trends in Pharmacological Sciences, 22*, 146–152.

Vibhuti, A., Gupta, K., Subramanian, H., Guo, Q., & Ali, H. (2011). Distinct and shared roles of β-arrestin-1 and β-arrestin-2 on the regulation of C3a receptor signaling in human mast cells. *PLoS One, 6*, e19585.

Violin, J. D., & Lefkowitz, R. J. (2007). β-Arrestin-biased ligands at seven-transmembrane receptors. *Trends in Pharmacological Sciences, 28*, 416–422.

Walker, J. K. L., Fong, A. M., Lawson, B. L., Savov, J. D., Patel, D. D., Schwartz, D. A., et al. (2003). β-Arrestin-2 regulates the development of allergic asthma. *The Journal of Clinical Investigation, 112*, 566–574.

Wang, P., Gao, H., Ni, Y., Wang, B., Wu, Y., Ji, L., et al. (2003a). Beta-arrestin 2 functions as a G-protein-coupled receptor-activated regulator of oncoprotein Mdm2. *The Journal of Biological Chemistry, 278*, 6363–6370.

Wang, P., Wu, Y., Ge, X., Ma, L., & Pei, G. (2003b). Subcellular localization of beta-arrestins is determined by their intact N domain and the nuclear export signal at the C terminus. *The Journal of Biological Chemistry, 278*, 11648–11653.

Watanabe, M., Moon, K. D., Vacchio, M. S., Hathcock, K. S., & Hodes, R. J. (2014). Downmodulation of tumor suppressor p53 by T cell receptor signaling is critical for antigen-specific CD4(+) T cell responses. *Immunity, 40*, 681–691.

Williams, T. J., Pease McDonagh, J. E., Petit, S. J., Saran, N., Clark, P. C., Messer, A., et al. (2008). The chemokine receptor CXCR3 is degraded following internalization and is replenished at the cell surface by de novo synthesis. *Journal of Immunology, 180*, 6713–6724.

Yan, B., Huo, Z., Liu, Y., Lin, X., Li, J., Peng, L., et al. (2011). Prolyl hydroxylase 2: A novel regulator of β2-adrenoceptor internalization. *Journal of Cellular and Molecular Medicine, 15*, 2712–2722.

Yu, M.-C., Su, L.-L., Zou, L., Liu, Y., Wu, N., Kong, L., et al. (2008). An essential function for β-arrestin 2 in the inhibitory signaling of natural killer cells. *Nature Immunology, 9*, 898–907. Nature Publishing Group.

Zhao, M., Wimmer, A., Trieu, K., DiScipio, R. G., & Schraufstatter, I. U. (2004). Arrestin regulates MAPK activation and prevents NADPH oxidase-dependent death of cells expressing CXCR2. *The Journal of Biological Chemistry, 279*, 49259–49267.

Zody, M. C., Garber, M., Adams, D. J., Sharpe, T., Harrow, J., Lupski, J. R., et al. (2006). DNA sequence of human chromosome 17 and analysis of rearrangement in the human lineage. *Nature, 440*, 1045–1049.

Zou, Y.-R., Kottmann, A. H., Kuroda, M., Taniuchi, I., & Littman, D. R. (1998). Function of the chemokine receptor CXCR4 in haematopoiesis and in cerebellar development. *Nature, 393*, 595–599.

CHAPTER NINE

# Emerging Roles of Regulators of G Protein Signaling (RGS) Proteins in the Immune System

**Kirk M. Druey**[1]
Molecular Signal Transduction Section, Laboratory of Allergic Diseases, NIAID/NIH, Bethesda, MD, United States
[1]Corresponding author: e-mail address: kdruey@niaid.nih.gov

## Contents

1. Introduction 316
2. The RGS Protein Family 320
   2.1 Discovery of RGS Proteins 320
   2.2 RGS Protein Structure 320
3. Mechanisms by Which RGS Proteins Control GPCR Signaling 322
   3.1 Structural and Biochemical Basis for RGS GAP Activity 322
   3.2 RGS Regulation of GPCR Signal Transduction 323
4. RGS Proteins and Cell Migration Signaling 325
   4.1 Signaling Pathways Mediating Leukocyte Chemotaxis 325
   4.2 How RGS Proteins Might Affect Chemotaxis 326
5. RGS Proteins in Innate Immune Cells 328
   5.1 Granulocytes 328
   5.2 Monocytes/Macrophages 329
   5.3 Mast Cells 331
   5.4 Natural Killer Cells 332
   5.5 Dendritic Cells 332
6. RGS Functions in Adaptive Immune Responses 333
   6.1 B Lymphocytes 333
   6.2 Plasma Cells 335
   6.3 T Lymphocytes 335
7. Noncanonical Functions of RGS Proteins in the Immune System 339
   7.1 IgE-Mediated Anaphylaxis 339
   7.2 Transcriptional Regulation in B Lymphocytes 340
   7.3 Phagocytosis 340
   7.4 Macrophage Survival and Proinflammatory Functions 341
8. RGS Association With Diseases of the Immune System 341
   8.1 RGS1 and Celiac Disease 341
   8.2 RGS1 and Autoimmune Diabetes 342
   8.3 RGS1 and IgA Nephropathy 342

*Advances in Immunology*, Volume 136
ISSN 0065-2776
http://dx.doi.org/10.1016/bs.ai.2017.05.001

8.4 RGS1 and RGS10 and Multiple Sclerosis 343
8.5 RGS1 and RGS13 and B Cell Malignancies 343
9. Conclusions 343
Acknowledgment 344
References 344

## Abstract

The regulators of G protein signaling (RGS) proteins are a large, evolutionarily conserved group of intracellular proteins expressed in every cell type and tissue throughout the body including the immune system. Through their signature GTPase-activating protein (GAP) activity on heterotrimeric G proteins and interactions with signaling complexes and membrane constituents (e.g., lipids), RGS proteins determine the intensity and duration of G protein-coupled receptor-induced responses. They may also have a function in generating intracellular signaling gradients necessary for the directional migration of leukocytes to inflamed tissues containing local accumulations of chemoattractants. Although physiological functions of most RGS proteins in leukocytes and lymphoid organs are largely unknown, it appears thus far that deficiency of individual RGS proteins in mice does not affect homeostatic immune responses in the absence of immunogenic challenge and/or microbial infection. Although aberrant expression of some RGS proteins has been linked to dysregulated immunity and/or neoplasia in humans, there are no human diseases attributed to specific RGS dysfunction. Here, we highlight mostly published work describing expression and functions of the core group of RGS proteins that were among the first discovered, in both innate and adaptive immune processes, with particular emphasis on cell trafficking.

## 1. INTRODUCTION

A key property of the immune system is cell motility. Cells mediating either innate or adaptive host defenses originate in the bone marrow (macrophages, granulocytes, B lymphocytes) or thymus (T lymphocytes). Yet unlike the mostly static cells composing tissues and solid organs, leukocytes emigrate from their tissue of origin to be shared by the body through continual circulation between the bloodstream, mucosal surfaces, and secondary lymphoid organs such as spleen and lymph nodes (LNs). Perturbations to protective barriers such as skin and GI tract due to injury or mucosal exposure to inhaled or ingested pathogens elicit active recruitment of immune cells to these sites to facilitate microbial destruction and wound repair. Neutrophils, which are often the body's first line of defense against invading microbes, rapidly accumulate in "swarms" at sites of tissue infection; swarming is a highly coordinated process involving directed cell migration (chemotaxis) in response to inflammatory mediators as well as feed forward amplification of cell recruitment due to paracrine actions of chemoattractants secreted from neutrophils themselves (Kienle & Lammermann, 2016).

Immune cells are drawn to sites of inflammation by gradients of a wide array of chemoattractants—humoral mediators including peptides and lipids (e.g., sphingosine 1-phosphate, S1P; platelet-activating factor, PAF)—and chemokines, small peptides secreted by resident tissue cells and leukocytes. The majority of chemoattractant substances act on G protein-coupled receptors (GPCRs), which signal through heterotrimeric G protein complexes (Gαβγ). Ligand binding induces conformational changes in the receptor that trigger the exchange of GTP for GDP on the α subunit, which releases Gα-GTP from the Gβγ dimer (Hepler & Gilman, 1992) (Fig. 1). Gβγ and Gα-GTP each activate a distinct set of effectors that ultimately

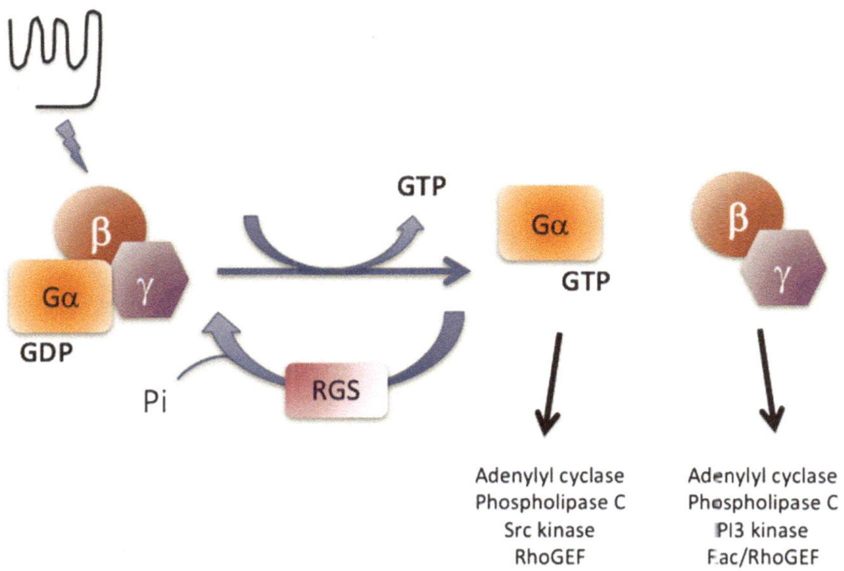


**Fig. 1** RGS proteins and the G protein nucleotide cycle. Seven transmembrane receptors induce exchange of GTP for GDP on the alpha subunit and dissociation of the heterotrimeric complex. Gα subunits activate or inhibit several downstream effectors including adenylyl cyclase, phospholipase C, protein kinases, and guanine nucleotide exchange factors (GEFs). βγ activates its own set of effectors including PI3 kinase and GEFs for Rac and Rho family GTPases. The cycle is terminated by the intrinsic GTPase activity of Gα, which releases inorganic phosphate (Pi). RGS proteins bind Gα in a transition state between the GTP- and GDP-bound forms, which lowers the free energy required for GTP hydrolysis.

result in cell polarization and migration toward the apex of the chemokine gradient. Four classes of heterotrimeric G protein complexes have been identified on the basis of distinct Gα subunits: Gαi/o, whose principal activity is to inhibit adenylyl cyclase (AC) and reduce generation of the second messenger cyclic AMP; Gαq, which activates phospholipase C and causes increased cytosolic $Ca^{2+}$; Gα12/13, which activates monomeric GTPases of the Rho family, leading to cytoskeleton reorganization; Gαs, which activates AC and increases cAMP, has not been implicated in chemotactic signaling (Fig. 2).

Although studies have demonstrated a primary role for Gβγ and its downstream effectors in chemotaxis, most chemoattractant receptors couple to Gαi, suggesting that it also has nonredundant function(s) in chemoattractant signaling. Pertussis toxin (PTX), which ADP-ribosylates Gαi/o subunits and prevents GPCR coupling, inhibits leukocyte chemotaxis to chemokines in solution, and in vivo administration of PTX to mice has profound effects on leukocyte trafficking (Chaffin & Perlmutter, 1991; Cyster & Goodnow, 1995; Spangrude, Sacchi, Hill, Van Epps, & Daynes, 1985). Recent studies have suggested a unique and direct role for Gαi and cAMP

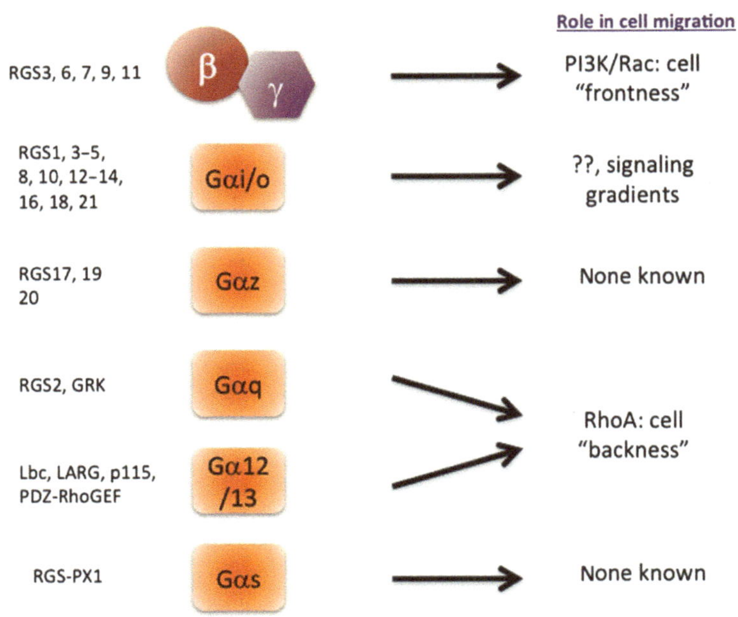


**Fig. 2** Heterotrimeric G proteins, RGS proteins, and their functions in cell migration. RGS proteins and their G protein targets are shown on the *left*. Potential roles for individual G proteins and their effectors in chemotaxis and migration signaling are listed on the *right*.

in the establishment of intracellular signaling gradients that promote polarization in the presence of shallow chemoattractant gradients, which is discussed later. An alternative chemotaxis pathway mediated by Gαq may be involved in chemotaxis of some, but not all, leukocytes. Gαq-deficient neutrophils and dendritic cells (DCs) had impaired chemotaxis in response to *N*-formyl methionyl leucyl phenylalanine and multiple chemokines including CCL2, CCL21, and CXCL12, whereas Gαq deficiency did not affect chemotaxis of T lymphocytes to CXCL12 (Shi et al., 2007). Gα13 subunits were shown to be required for chemotactic responses of a T lymphocyte cell line (Jurkat) to the chemokine CXCL12 through a RhoA-dependent pathway (Tan, Martin, & Gutkind, 2006), and Gα13 has also been implicated in egress of B cells from LN germinal centers, the anatomical sites of their maturation into antibody-secreting cells, via the receptors S1P3 and P2RY8 (Muppidi, Lu, & Cyster, 2015). The overall importance of Gα12/13 for chemoattractant signaling in other leukocytes remains unclear.

While rapid initiation of chemotaxis signal transduction pathways is crucial for the mobilization of leukocytes to inflammatory sites, several coordinate mechanisms of adaptation or desensitization are necessary to regulate the retention or egress of immune cells as infection or injury resolves. Similarly, under homeostatic conditions, leukocyte trafficking to and from lymphoid organs is highly regulated by specific chemoattractant/receptor interactions, which determine cell positioning therein and facilitate intercellular communications and production of mediators including cytokines and reactive oxygen species required for proper immune responses. For example, B lymphocytes negotiate S1P1 gradients to enter LNs from circulation, whereas CXCL13/CXCR5 interactions are important for their migration to B-cell zones within LN follicles (Cyster & Schwab, 2012). As with most GPCRs, desensitization of signaling downstream of chemoattractant receptors occurs through receptor downregulation. Upon ligand binding, the scaffolding protein β-arrestin is recruited to the receptor, which activates a distinct set of signaling pathways, some of which are involved in chemotaxis (Min & Defea, 2011). In addition, β-arrestin leads to recruitment of GPCR kinases (GRKs), which phosphorylate receptors, leading to internalization and recycling to the cell surface (Liggett, 2011). The physiological importance of this mechanism for leukocyte trafficking is supported by the observation that lymphocytes cycle between exit from LNs and spleen back into circulation toward a gradient of S1P followed by rapid return into lymphoid tissue against the gradient, which is associated with GRK2-mediated desensitization and downregulation of surface S1PR1 (Arnon et al., 2011).

Another mechanism by which cells modulate both the amplitude and duration of GPCR signaling is mediated by the regulators of G protein signaling (RGS) proteins. This is a large family of proteins whose biochemical mechanism of action—acting as GTPase-activating proteins (GAPs) for heterotrimeric G protein α subunits—has been extensively characterized in vitro, yet whose physiological functions in the immune system are by and large poorly understood (Sjogren, 2017). In this review, we discuss RGS structure and biochemical properties as well as emerging evidence for the functions of individual RGS proteins in the immune system.

# 2. THE RGS PROTEIN FAMILY

## 2.1 Discovery of RGS Proteins

The lifetime of activated (GTP-bound) Gα subunits determines the intensity and interval of GPCR signaling, yet the rate of GTP hydrolysis by purified Gα proteins in solution-based assays is far too slow to account for the much faster deactivation of GPCR signaling observed in vivo. For instance, purified transduction, the Gαi orthologue mediating retinal phototransduction through the GPCR rhodopsin, hydrolyzes GTP with a half-life of ~15 s, whereas retinal responses to light exposure decay much faster on the subsecond scale (Kimple, Bosch, Giguere, & Siderovski, 2011). The discovery of the founding member of the RGS family, the yeast protein Sst2, helped to reconcile this discrepancy. Yeast *Saccharomyces cerevisiae* respond to the mating pheromone α-factor, which acts on the GPCR Ste2 to induce growth arrest via activation of the Gα subunit GPA1. Mutants defective in Sst2 were found to be supersensitive to pheromone, suggesting that Sst2 mediates a desensitization mechanism by inhibiting this GPCR pathway (Venkatapurapu et al., 2015). Subsequently, work in diverse systems including yeast, mammalian cells, and *C. elegans* uncovered the presence of a large family of proteins orthologous to Sst2 that could not only complement the absence of Sst2 in yeast but also negatively regulate signaling by GPCRs in mammalian cells (Stewart & Fisher, 2015; Woodard, Jardin, Berna-Erro, Salido, & Rosado, 2015).

## 2.2 RGS Protein Structure

More than 30 RGS proteins have been identified in mammals. RGSs are universally defined by the presence of the RGS domain or RGS "box." This domain mediates physical interaction with Gα subunits and GAP activity

(Popov, Yu, Kozasa, & Wilkie, 1997). RGS or RGS-homology domains have also been identified in a number of proteins, including G protein-receptor kinases (GRKs), Rho GEFs (e.g., LARG, p115RhoGEF, Lbc), sorting nexins (PX1), and even GPCRs (AtRGS1 in *Arabidopsis thaliana*) (Chen & Jones, 2004; Siderovski & Willard, 2005; Sjogren, Blazer, & Neubig, 2010). Because prominent functions in the immune system have not been ascribed to many of these RGS-like proteins, we focus here on the core group of RGS proteins among the first to be identified, which in most cases act as GAPs for Gi and Gq (Fig. 2). These canonical RGS proteins can be further subdivided into four subfamilies based on the degree of homology of their RGS domains and by the presence of additional domains or motifs mediating additional protein–protein interactions (Fig. 3). The B/R4 subfamily (RGS1–5, 8, 13, 16, 21), whose structure consists of little more than the RGS domain (except for RGS3 which has an N-terminal PSD-95/Dlg/ZO-1 (PDZ) domain), has been most extensively characterized in the immune system (Xie, Chan, & Druey, 2016). The A/RZ subgroup contains an amino terminal cysteine string, which is the site of palmitoylation, and interacts with Gi/o and Gz classes of G proteins. This group has been primarily studied in nervous system (Nunn, Mao, Chidiac, & Albert, 2006).

The C/R7 RGS subfamily contains a G-gamma like (GGL) domain, which mediates obligate heterodimerization with Gβ5. These proteins are

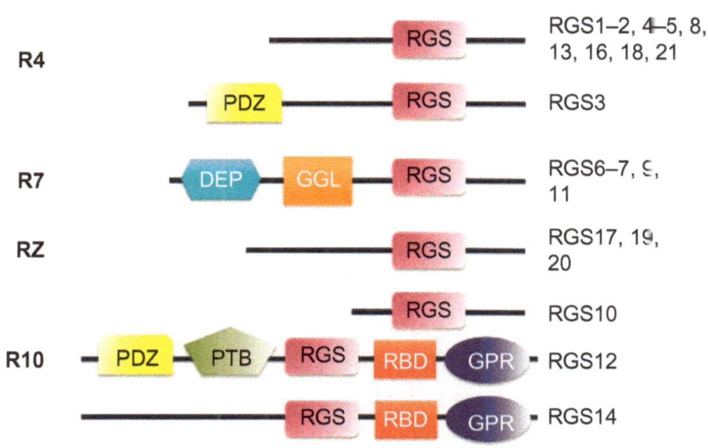


**Fig. 3** The core RGS subfamilies. *DEP*, disheveled/*Egl10*/pleckstrin homology; *GGL*, Gγ-like; *GPR*, G protein regulatory; *RGS*, regulator of G protein signaling; *PDZ*, PSD-95/Dlg/ZO-1; *PTB*, protein tyrosine binding; *RBD*, Ras/Rap binding.

also highly expressed in neurons, where they may act as a composite signaling unit at cell membranes through interactions with scaffolding proteins (R7 binding protein, R7 anchoring protein) and GPCRs (GPR158 and GPR179) (Anderson, Posokhova, & Martemyanov, 2009). DEP domain-dependent homodimers of RGS7 were recently identified in brain, although the physiological significance of this interaction is uncertain (Tayou et al., 2016). Although RGS10 is a small protein consisting of little more than the RGS motif, the other members of the D/R10 subfamily (RGS12 and 14) contain additional protein–protein interaction modules outside of the RGS domain. RGS12 contains PDZ and protein tyrosine binding (PTB) domains, and both RGS12 and RGS14 harbor two tandem Rap/Ras interacting domains and a G protein regulatory (GPR) motif, which mediates guanine nucleotide exchange (GEF) activity on Gαi (Evans, Dudek, & Hepler, 2015; Gerber, Squires, & Hepler, 2016; Siderovski & Willard, 2005). The functions of this subclass of proteins in the immune system are largely unknown. The GPR motif of RGS14 binds GDP-bound Gαi1/3 and may be involved in the recruitment of RGS14 to the plasma membrane (PM) (Brown et al., 2015). However, the binding of RGS14 to inactive Gα subunits does not appear to preclude its GAP activity on neighboring Gα-GTP (Brown, Lambert, & Hepler, 2016).

# 3. MECHANISMS BY WHICH RGS PROTEINS CONTROL GPCR SIGNALING

## 3.1 Structural and Biochemical Basis for RGS GAP Activity

Early biochemical and structural studies revealed that the principal property of the core group of RGS proteins was GAP activity. In vitro "single-turnover" assays in which recombinant Gα proteins preloaded with $P^{32}$-GTP in the absence of divalent cations ($Mg^{2+}$) required for GTP hydrolysis revealed that addition of nanomolar amounts of RGS proteins and $Mg^{2+}$ accelerated the hydrolysis of $P^{32}$-GTP 10–100-fold compared to Gα alone (Hepler, Berman, Gilman, & Kozasa, 1997; Watson, Linder, Druey, Kehrl, & Blumer, 1996). Lipid vesicles reconstituted with GPCR, Gα, and RGS proteins confirmed RGS GAP activity in the context of a receptor and heterotrimeric G protein complex; these studies suggested some level of selectivity depending on the receptor and G protein present (Zeng et al., 1998). Protein–protein binding studies and X-ray crystallographic analysis revealed that RGS proteins lack intrinsic enzymatic activity but instead lower the free energy needed for GTP hydrolysis by stabilizing a transition state of Gα lying between the GTP and GDP-bound states (Popov et al., 1997; Srinivasa, Watson, Overton, & Blumer, 1998). This intermediary complex is mimicked by the presence of aluminum fluoride, GDP, and $Mg^{2+}$, which creates a stable complex of Gα–GDP–RGS. The crystal structure of the Gα–RGS complex defined key interactions between key residues in the RGS domain and so-called switch residues in Gα that undergo a conformational change upon GTP binding (Srinivasa et al., 1998).

## 3.2 RGS Regulation of GPCR Signal Transduction

Studies of ion-gated channel kinetics suggested both positive and negative regulatory effects of some RGS proteins on GPCR–G protein nucleotide cycling. RGS8 enhanced both the onset and decay of currents mediated by Gβγ activation of inward-rectifying $K^+$ channels (GIRKs) in heterologous expression systems without changing the amplitude of the peak response (Doupnik, Davidson, Lester, & Kofuji, 1997; Saitoh, Kubo, Miyatani, Asano, & Nakata, 1997). Because RGS proteins have no bearing on the rate of GDP–GTP exchange by Gα (with notable exceptions, as discussed later), a non-GAP function of RGS proteins was impugned, perhaps through scaffolding of accessory proteins or effects on receptor–G protein coupling. Alternatively, a model of "kinetic" or "spatial" scaffolding was hypothesized, in which the availability of G proteins available for coupling to activated GPCRs clustered at the cell membrane becomes rate limiting for the overall response (Zhong et al., 2003). In theory the presence of an RGS protein could alleviate the "roadblock": by increasing the rate of GTP hydrolysis, there would be a continuous supply of Gα-GDP available for nucleotide cycling.

Recent studies have partly clarified these conundrums. Lambert et al. examined the effects of a Gα containing mutations in the switch II region (G202A in human Gαi1, G203A in human Gαo) previously found to increase greatly the intrinsic rate of GTP hydrolysis in the absence of an RGS protein. The structure of these mutants closely resembles the Gα·GDP·$AlF_4^-$ transition state complex preferred for RGS binding (Lambert et al., 2010). Heterologous expression of this supraphysiological Gα GTPase (which also contained a point mutation conferring complete RGS insensitivity, discussed later) recapitulated patterns observed by overexpression of RGS8: increased signal onset and decay following agonist stimulation, yet overall decreased agonist sensitivity. These results suggested that RGS GAP activity was sufficient to explain its effects on GPCR signaling kinetics. Subsequently, Katanaev and colleagues discovered an RGS protein in *Drosophila*, Dhit, which, in addition to possessing GAP activity, acts as a GDP dissociation inhibitor (GDI) for Gαo. GDI activity was mediated by a conserved Arg residue in the RGS domain that is shared by homologous members of the RZ family of mammalian RGSs (RGS17, 19, 20) but not other RGS subfamilies. RGS19 had GDI activity on both Gαi3 and Gαo, whereas the R12 *Drosophila* homologue Loco had no GDI activity (Lin et al., 2014). Collectively, these studies have provided support for hypothesis that RGS proteins influence GPCR signaling primarily through their effects on heterotrimeric G protein nucleotide cycling. Nonetheless, as RGS proteins have also been shown to physically interact directly with GPCRs, any number of G-protein accessory proteins (spinophilin, effectors such as AC and PLCβ) (Abramow-Newerly, Roy, Nunn, & Chidiac, 2006; Ma et al., 2015), and receptor-independent G protein activators (AGSs) (Zhao & Chidiac, 2015), the full effects of RGS GAP and non-GAP functions on GPCR-mediated responses are probably complex and remain to be fully characterized in native systems.

Most immune cell types express several RGS proteins, and genetic manipulation of individual RGS family members suggests that they serve nonredundant roles in regulating immune cell functions, particularly leukocyte chemotaxis and trafficking in vivo. Studies over the last 20 years have uncovered critical residues in both Gα and RGS that are critical for efficiency and possibly selectivity. Recent work has revealed several residues in the RGS domain in RGS2 that confer selectivity toward Gαq relative to Gαi (Nance et al., 2013). Mutation of a Gly residue in switch I region of the Gα homologue Gpa1 to Ser (G184S) in yeast demonstrated that this residue was dispensible for Gα functions including receptor, GTP, and Gβγ binding, yet this mutant was completely resistant to RGS binding and GAP activity (Lan et al., 1998). Knockin alleles of Gαi2 (G184S) have been established in mice to enable interrogation of RGS activity in multiple organ systems (Neubig, 2015). Most studies of mammalian cells and tissues containing hypomorphic alleles of individual RGS proteins have demonstrated that the overall intensity of effector activation (e.g., $Ca^{2+}$ flux, protein phosphorylation, cAMP generation) in response to GPCR ligands is increased in the absence of individual RGS proteins, suggesting that they negatively regulate GPCR signaling (Bansal, Druey, & Xie, 2007; Miao et al., 2016).

# 4. RGS PROTEINS AND CELL MIGRATION SIGNALING

## 4.1 Signaling Pathways Mediating Leukocyte Chemotaxis

Circulating leukocytes respond to chemoattractant gradients in tissues by polarizing in the direction of the gradient, forming protrusions (lamellipodia) at the "front" of the cell, followed by integrin-mediated adhesion to the endothelium and eventual diapedesis into interstitial space due to retraction at the cell "back." These events propel leukocytes into inflamed tissues (Xu & Jin, 2015). The respective roles for Gαi and Gβγ signaling in these processes remain incompletely defined (Fig. 2). Smrcka and colleagues used probes that selectively stimulate either G protein subunit in neutrophils (Surve et al., 2016). A small molecule activator of Gβγ (12155, which does not affect Gαi–nucleotide cycling) elicited directed cell migration (chemotaxis) in solution-based Transwell assays (Surve, Lehmann, & Smrcka, 2014), yet neutrophils adhered to glass surfaces coated with ICAM (to mimic integrin-mediated adhesion to the endothelium) were completely unable to polarize and migrate following 12155 stimulation. In contrast, neutrophil stimulation with Mas7, which bypasses GPCRs to promote Gi-GTP binding, or expression of a constitutively active Gi mutant (Q204L), led to polarization in the absence of chemoattractant. These results suggested parallel and complementary roles for Gi and Gβγ in directional motility. The effects of G protein signaling on polarization and adhesion were dependent on cAMP and PKA stimulation: Gβγ activates AC type 9 through a noncanonical pathway, which generates cAMP and activates PKA, inhibiting cell polarity. By inhibiting AC and simultaneously delimiting intracellular cAMP generation, Gi-GTP may counteract the effects of PKA signaling and thereby promote cell polarization.

The roles of Gβγ in cell polarity and chemotaxis are in general better defined. Free Gβγ activates PI3Kγ, which in turn generates phosphatidylinositol 3,4,5-trisphosphate (PIP3), which recruits GEFs for small GTPases including Rac and Cdc42 including (PIP3)-dependent Rac exchanger protein (pREX), DOCKs (dedicator of cytokinesis), and ELMO (Wang, Xu, Pan, & Jin, 2016; Zhao et al., 2007). In a feed forward process, PI3K-mediated PIP3 accumulation at the cell front amplifies Rac- and Cdc42-GTP formation and actin polymerization at the leading edge. Activation of Rho family GTPases at the cell back results in myosin light chain kinase phosphorylation by the RhoA coiled-coil kinase (ROCK). Increased myosin light chain phosphorylation in turn enhances myosin bundling with actin filaments and formation of a trailing "uropod" that promotes cell retraction (Xu et al., 2003). The mechanisms by which cells generate differential intracellular signaling gradients and the full spectrum of heterotrimeric G proteins involved in chemotaxis remain incompletely defined.

## 4.2 How RGS Proteins Might Affect Chemotaxis

Given the prominent role of Gβγ in chemoattractant signaling, it has been hypothesized that RGS proteins may only indirectly affect chemotaxis by regulating the availability of Gβγ subunits freed from Gα-GTP via its acceleration of GTP hydrolysis. In general, cells deficient in individual RGS family members exhibit augmented chemotactic responses to a variety of chemoattractants (Bansal, DiVietro, et al., 2008; Han, Huang, Kim, & Kehrl, 2006; Shankar et al., 2012). However, since leukocytes have the capacity to polarize in response to uniform chemoattractant stimuli or in the presence of shallow chemokine gradients, the relative importance of RGS activity for generating differential intracellular signaling gradients remains uncertain. In a recent study, Masuda and colleague visualized RGS–G protein interactions using fluorescence resonance energy transfer (FRET) probes containing the RGS domain of RGS10 fused to Gαi in chemotaxing HL60 cells (neutrophil-like) (Masuda et al., 2017). Increased FRET signals would theoretically reflect Gαi activation since RGS10 only binds Gαi-GTP. The RGS10-Gαi probe was expressed uniformly along the PM at baseline, and uniform application of chemoattractant leukotriene B4 (LTB4, a strong chemoattractant for neutrophils) led to broadly distributed FRET signals. In contrast, cells exposed to a shallow gradient of LTB4 demonstrated increased FRET only at the leading edge. These results suggest that cells sense the direction of a chemoattractant gradient through focal Gαi activation at the apex of the stimulus. These studies did not directly address the role of RGS10 in this process or other mechanisms that might spatially restrict Gαi activation.

Because chemoattractant GPCRs are relatively uniformly distributed on the cell surface, cells may have the ability to generate foci of concentrated RGS activity that might to confine G protein activation to certain regions. Recently, O'Neill and Gautam used optogenetic techniques to determine the importance of intracellular G protein signaling gradients for motility and directional sensing (O'Neill & Gautam, 2014). In cells treated uniformly with a chemokine stimulus, discrete light signals were given to induce recruitment of RGS4 fused to light-activated membrane targeting domains

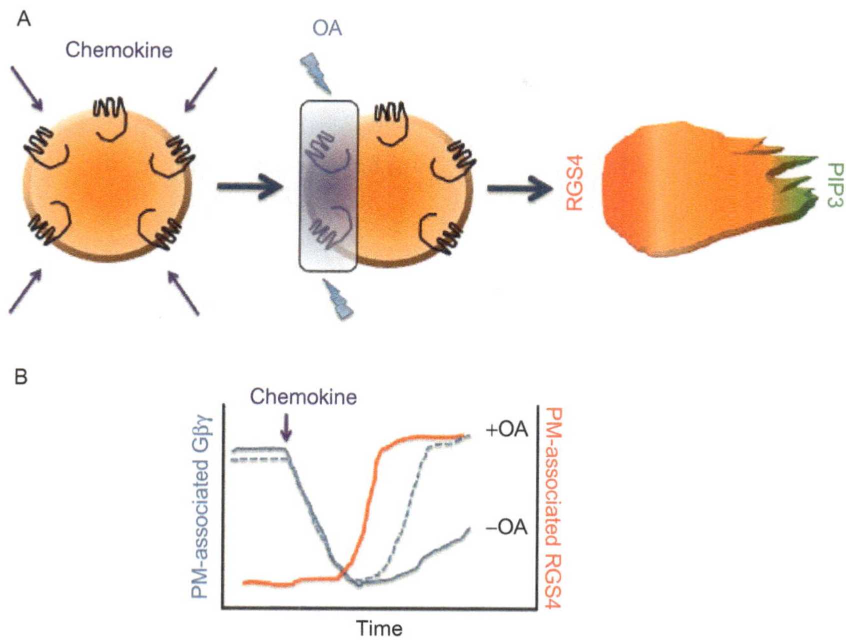


**Fig. 4** Intracellular signaling gradients induced by RGS activity may drive cell polarization and motility. (A) Localized light stimulation of cells expressing a photoactivatable RGS4 accumulates RGS4 protein at the cell "back" and generation of phosphatidylinositol 3,4,5-trisphosphate (PIP3) at the cell "front" due to a gradient of Gαi activity induced by chemokine. Cells polarize and form lamellipodia even in the absence of a chemoattractant gradient. (B) Chemokine stimulation releases Gβγ from the plasma membrane over time. In optically activated (OA) cells, increased RGS4 GAP activity accelerates return of βγ to the PM due to increased formation of Gαi-GDP and heterotrimers. *Adapted from O'Neill, P.R., & Gautam, N. (2014). Subcellular optogenetic inhibition of G proteins generates signaling gradients and cell migration.* Molecular Biology of the Cell, 25, *2305–2314.*

to a specific region in order to achieve localized GAP activity (Fig. 4A). Cellular localization of RGS4 and Gβγ was tracked through fusion with GFP or its derivatives. In unstimulated cells Gβγ was localized primarily at the PM but diffused to intracellular membranes upon cell stimulation with chemokine (Fig. 4B). In optically activated cells expressing photoactivatable RGS4, Gβγ returned to the PM much faster than in cells lacking RGS4, which reflects Gα GTPase activity and reassembly of the Gα–GDP–βγ complex. Chemokine-stimulated HeLa cells expressing these constructs exhibited a rapid loss of PM-associated Gβγ reporter over time, whereas optical activation (OA) of a specific membrane region resulted in rapid accumulation of RGS4 in the light-stimulated area and increased Gβγ retention at the PM.

Gβγ was not similarly concentrated in cells lacking RGS4 or not treated with blue light, indicating that Gβγ translocation indeed reflected enhanced deactivation of Gi-GTP mediated by RGS4 GAP activity.

Further mechanistic studies revealed that RGS4 GAP activity may contribute to cell polarization by restricting PIP3 accumulation to the leading edge, which can be tracked by the translocation of the membrane anchoring pleckstrin homology domain of Akt (GFP-PH-Akt) (Ko, Guo, Su, Gao, & Gelman, 2014). RAW 264.7 cells (macrophage cell line) expressing photoactivatable RGS4 at one side of the cell and then stimulated with light concentrated PIP3 at the opposite side in response to a uniform chemokine stimulus, followed by polarization and formation of lamellipodia in proximity to PIP3 accumulation (Fig. 4A). Cells that had adapted to the initial stimulus could also repolarize in the presence of chemokine with a second light pulse bringing RGS4 to the back of the cell. These findings indicate that experimentally induced, localized inhibition of Gαi (and indirectly Gβγ) activity at one end of the cell is sufficient for cell polarization in the presence of a uniform chemoattractant stimulus.

The near complete unresponsiveness of B cells and lymphocytes from mice expressing RGS-resistant Gαi2 (G184S mutation) to several chemokines may be consistent with these findings (Cho et al., 2012; Hwang et al., 2015) in that these cells are unable to generate intracellular gradients of Gαi2 activation. Since RGS4 and many other RGS proteins are expressed diffusely throughout the cytosol or at the PM (Huang & Fisher, 2009) constitutively, further live tracking studies of fluorescent and/or endogenous RGS proteins in cells exposed to either global or spatially focused chemoattractants are needed to determine how they localize to specific cell regions.

# 5. RGS PROTEINS IN INNATE IMMUNE CELLS

## 5.1 Granulocytes

Neutrophils from G184S knockin (KI mice) exhibited aberrant trafficking in vivo. In a sterile peritonitis model induced by intraperitoneal injection of thioglycollate, KI neutrophils were retained in bone marrow and recruited poorly to the peritoneum compared to WT controls. Although KI cells display increased baseline motility compared to controls, chemotactic responses to chemokines involved in these responses (CXCL12, CXCL2) were nearly absent (Cho et al., 2012). Although peak signaling intensity (intracellular $Ca^{2+}$ flux) in response to chemokine was relatively intact, the kinetics of response decay were prolonged, and KI neutrophils responded poorly to a repeat chemokine challenge. As a result, the antimicrobial function of KI neutrophils was severely impaired following intravenous infection with *Staphylococcus aureus*. Although several *Rgs* transcripts are detectable in mouse neutrophils including *Rgs2, 14, 18*, and *19* (Cho et al., 2012), the functions of individual RGS proteins in these cells are virtually unknown. Our preliminary studies indicate that loss of RGS5 impairs in vivo trafficking patterns of neutrophils in mice. Neutrophil development in bone marrow of *Rgs5*$^{-/-}$ mice is normal, yet they have increased numbers of circulating neutrophils in peripheral blood and increased neutrophil recruitment to the peritoneum in response to thioglycollate challenge. Bone marrow-derived neutrophils from RGS5-deficient mice gene exhibit increased chemotaxis in solution-based Transwell assays, which may underlie their aberrant trafficking patterns in vivo (our unpublished data).

The expression and functions of RGS proteins in other granulocyte populations including basophils and eosinophils have not been reported in detail. However, a role for RGS2 in myeloid-derived suppressor cells (MDSCs) was suggested by a recent study. MDSCs are stromal cells highly associated with malignancy; they promote tumor growth by inhibiting the host inflammatory response and by promoting angiogenesis. RGS2 was found to be highly upregulated in tumor-derived MDSCs compared to control MDSCs, especially under hypoxic conditions. *Rgs2*$^{-/-}$ mice were more resistant to tumor growth, which was associated with decreased tumor vascularization. Transfer of RGS2-deficient MDSCs into tumor-bearing mice significantly reduced their tumor-promoting activity relative to WT MDSCs, possibly due to impaired production of angiogenic cytokines such as MCP-1 (Boelte et al., 2011).

## 5.2 Monocytes/Macrophages

Macrophages play a central role in innate host immune responses via their capacity to act as professional pathogen scavengers and antigen presenting cells. Microbial-associated molecular patterns, which are pathogen-derived substances including glycans, nucleic acids, and glycoproteins, activate macrophages through Toll-like receptors and other pattern recognition receptors (PRRs) (Mahla, Reddy, Prasad, & Kumar, 2013). Macrophages express multiple RGS proteins including RGS1, 2, 5, 10, 12, 16, and 19. RGS1 and RGS2 were reciprocally regulated by stimulation of murine bone marrow-derived macrophages with several TLR ligands (Lee et al., 2010;

Riekenberg et al., 2009). Published work has demonstrated impaired TLR signaling in macrophages from Giα2(G184S) KI mice or in RAW cells overexpressing Gαi2. Either manipulation reduced cytokine production induced by lipopolysaccharide (LPS), which acts on TLR4 (Li et al., 2012). These findings suggest that Gαi2 and RGS proteins regulate innate inflammatory responses of macrophages.

A number of recent studies have demonstrated that RGS proteins modulate macrophage trafficking during inflammation. Macrophages are an essential component of inflammatory atherosclerotic plaques in blood vessels. Strong *Rgs1* expression was detected in macrophages associated with aortic plaques and aneurysms in atherosclerosis-prone $ApoE^{-/-}$ mice and with carotid artery plaques in humans (Patel et al., 2015). Macrophages from $Rgs1^{-/-}$ mice exhibited hyperresponsive and sustained chemotaxis in response to CCL2, CCL3, CCL5, CXCL12 compared to WT, and $Rgs1^{-/-}ApoE^{-/-}$ mice were significantly less susceptible to plaque formation. An independent assessment of monocyte/macrophage trafficking in vivo using zymosan-induced sterile peritonitis revealed that macrophages from $Rgs1^{-/-}ApoE^{-/-}$ were initially recruited more robustly to the peritoneum during the early phases of inflammation compared to controls. However, at later time points during resolution of inflammation WT macrophages were retained in the peritoneum longer than RGS1-deficient macrophages following zymosan injection (Patel et al., 2015). Thus, RGS1 might regulate macrophage-dependent vascular inflammation by increasing retention of macrophages in atherosclerotic plaques.

RGS5 is also expressed in macrophages, but in contrast to RGS1 it may negatively regulate atherosclerosis in mice because of its pleiotropic expression in several plaque-associated cells including endothelium, vascular smooth muscle, and macrophages. $Rgs5^{-/-}$ mice were more prone to develop atherosclerosis due to increased accumulation of lipids, apoptotic macrophages, and endothelial cells in plaques (Cheng et al., 2015). A separate study in which hypercholesterolemic LDL receptor-deficient mice were treated with angiotensin II (AngII), which accelerates the development of atherosclerosis, demonstrated that RGS5 expression is downregulated in peritoneal macrophages by AngII. Stimulation with agonists acting on antiinflammatory peroxisome proliferator-activated receptor (PPAR) family of nuclear receptors prevented the decrease in *Rgs5* expression induced by AngII, which attenuated AngII-induced recruitment of macrophages to atherosclerotic plaques and AngII signaling (Erk phosphorylation). Thus, PPAR agonists might reduce atherosclerotic plaque formation by inhibiting proinflammatory responses of macrophages through increased *Rgs5* expression (Takata et al., 2008).

RGS10 is strongly expressed in microglia, which are proinflammatory macrophages resident in the brain. Microglia may contribute to neuronal cell death in neurodegenerative central nervous system (CNS) disorders linked to chronic systemic inflammation including Parkinson's disease (PD). *Rgs10*$^{-/-}$ mice had increased numbers of microglia in CNS at baseline and were more susceptible to inflammation-induced neurodegeneration resembling PD compared to WT mice (Lee et al., 2008). Inflammatory stressors such as TNFα and LPS downregulated RGS10 in primary mouse microglia and in the microglial cell line BV2. RNAi-mediated knockdown (KD) of RGS10 in BV2 microglia cells led to increased proinflammatory cytokine secretion in response to LPS and enhanced their cytotoxicity on dopaminergic neurons. LPS-induced NFκB activation and proinflammatory cytokine production was increased in microglia from RGS10-deficient mice compared to WT, while LPS-evoked chemotaxis or phagocytic functions were not affected (Lee, Chung, McAlpine, & Tansey, 2011). These findings suggest that RGS10 limits neurotoxicity in PD and possibly other neuroinflammatory conditions by regulating microglial cytokine production induced by TLR stimulation.

The R4 RGS member RGS16 regulates proinflammatory cytokine production by monocytes. Overexpression of RGS16 in THP-1 cells, a prototypic human monocyte cell line, increased cytokine production by the TLR2 agonist Pam3CysSK4, whereas gene silencing had the opposite effect (Suurvali et al., 2015). *RGS16* is also detectable in human monocytes and is upregulated by infection and LPS, among other stimuli. No studies have addressed directly whether RGS10, 16, or other RGS proteins regulate proinflammatory TLR and/or NFκB signaling through their GAP activity or through other mechanisms.

## 5.3 Mast Cells

Mast cells (MCs) have functions in both innate and adaptive immunity. Long-lived tissue MCs are located in close proximity to nerves and blood vessels and produce a number of cytokines in response to PRR stimulation by pathogens. In addition, cross-linking of IgE-receptors on MCs by antigen-specific IgE induces their adaptive immune functions, which include degranulation and release of preformed and newly synthesized inflammatory mediators. RGS13 expression has been described in both human and mouse MCs. Migration of circulating MC progenitors from blood to tissues is partly controlled by specific chemoattractants including metabolites of arachidonic acid (PGE2 and LTB4) and chemokines and their receptors (CXCR2, CCL2) (Dahlin & Hallgren, 2015). HMC-1 and LAD2 cells (human MC lines) expressed transcripts for several RGSs including *RGS1, 2, 5, 10,* and *13*; RGS13 protein was upregulated by chemoattractant (C5a, CXCL12) treatment of HMC-1 cells. RGS13 KD enhanced chemoattractant-induced chemotaxis of HMC-1 cells and cytokine (IL-8) secretion in part due to exaggerated signaling ($Ca^{2+}$ mobilization, Akt phosphorylation) compared to controls, whereas RGS13 overexpression inhibited such responses (Bansal, DiVietro, et al., 2008). RGS13 shRNA also enhanced degranulation of LAD2 cells stimulated with S1P. Thus, RGS13 may control GPCR-mediated immune responses of human MCs including recruitment to tissues. The roles of other RGS proteins expressed in MCs have not been reported.

## 5.4 Natural Killer Cells

Natural killer (NK) cells induce cytotoxicity of cancer cells and pathogen-infected cells through both inhibitory (Ly49D and activating (Ly49A) receptors. Rat NK cells expressed mRNAs for *Rgs1, 2, 5, 8, 16,* and *18*; RGS2 was upregulated by IL-2. Ly49D inhibitory receptor ligation increased *Rgs2* abundance, whereas Ly49A ligation upregulated *Rgs2, 16,* and *18* expression (Kveberg, Ryan, Rolstad, & Inngjerdingen, 2005). Although these findings suggest that these RGS proteins have a function in NK cell functions, their physiological role in NK-mediated immunity has not been formally studied.

## 5.5 Dendritic Cells

DCs are professional APCs that specialize in priming T cell activation. There are both monocyte- and myeloid-derived DCs that originate in the bone marrow and mature in tissues in response to PRR and cytokine stimulation. Human monocyte-derived DCs expressed *RGS16*, which was dramatically upregulated by LPS and IL-10, strong signals for DC activation (Perrier et al., 2004). A separate study found that human monocyte-derived DCs express mRNAs for *RGS2, 3, 10, 13, 14, 18,* and *19*, and stimulation of these cells or murine bone marrow-derived DCs with TLR3 or TLR4 agonists increased expression of *RGS1, 16,* and *20* while downregulating *RGS14* and *18*. Overexpression of RGS1, RGS13, or RGS18 inhibited human monocyte-derived DC chemotaxis to CXCL12 and CCL21 (Shi, Harrison, Han, Moratz, & Kehrl, 2004). BM-derived DCs from *Rgs1*$^{-/-}$ mice chemotaxed more to these chemokines compared to WT. These findings suggest that these RGS proteins may regulate the migration of immature DCs into tissues. It is unclear whether RGS proteins affect APC functions of DCs. A recent study found that DCs from *Rgs10*$^{-/-}$ mice presented antigen and induced T cell proliferation comparably to DCs from WT mice (Lee et al., 2016).

# 6. RGS FUNCTIONS IN ADAPTIVE IMMUNE RESPONSES

## 6.1 B Lymphocytes

Immature B cells enter LNs and spleen from circulation, where the traverse lymphoid follicles to interact with follicular T helper (Tfh) cells and follicular dendritic cells (FDCs), allowing interrogation of foreign antigens and evolution into antibody-secreting plasma cells or memory B cells. Several studies have implicated prominent roles for G proteins and RGS proteins (in particular, Gαi2, and RGS1 and 13, respectively) in the homing of B cells to lymphoid tissue, navigation of follicles, and exit into the bloodstream. Although B lymphocytes express both Gαi2 and Gαi3, B cells deficient in Gαi2 exhibit strongly impaired chemotaxis to chemokines, homing to LNs, and motility with them, suggesting a nonredundant role for Gai2 in B cell migration (Han et al., 2006).

Within lymphoid follicles, germinal centers (GCs) are prominent microanatomical structures that form in response to antigen exposure. GCs compartmentalize appropriate B-Tfh and B-FDC interactions in part based on internal chemokine gradients; such cell positioning is required for stimulation of B cells with maturation-promoting cytokines and other factors. A chemokine gradient mediates migration of follicular B cells within lymphoid follicles to form GCs. In the GC dark zone, B cells sample antigens, which induce expansion and somatic hypermutation of their immunoglobulin (Ig) genes to generate Ag-specific B cells. As they mature, these cells migrate to the light zone to interact with FDCs and Tfh cells and undergo antibody class switching and affinity maturation to become either memory B cells or Ab-secreting plasma cells. CXCL12 and CXCL13 gradients promote B cell positioning within GCs, and perturbations in the chemokine milieu, chemokine receptor expression, or chemokine responsiveness of B cells may interrupt the humoral immune process to generate immunodeficiency or autoimmune responses (Bannard & Cyster, 2017).

Within the B cell lineage, expression of RGS1, 2, 3, 10, 13, 14, and 19 has been described in mice and humans (Moratz, Harrison, & Kehrl, 2004). GC B lymphocytes express high quantities of RGS1 and RGS13, and these RGS family members have been the best studied in B cells overall (Shi, Harrison, Wilson, Moratz, & Kehrl, 2002). *Rgs1*$^{-/-}$ mice had larger follicles in spleen with GCs formed spontaneously in the absence of immunization. Direct visualization of B lymphocyte trafficking in mice injected with B cells from *Rgs1*$^{-/-}$ mice demonstrated that they homed to LN microvasculature more than WT cells and also exhibited faster rates of exit from lymphoid tissue (Han et al., 2005). These B cells had increased chemotaxis in Transwell assays in vitro and more robust and sustained signaling responses to chemokines ($Ca^{2+}$ flux) (Hwang, Park, Harrision, Huang, & Kehrl, 2010). Despite the abnormal architecture of lymphoid organs and aberrant B cell trafficking, immunoglobulin levels were relatively comparable to WT mice, and they were competent to mount high-titer antibody responses following immunization with TNP-KLH (Moratz, Hayman, Gu, & Kehrl, 2004). In mice deficient in Gαi2 there were fewer follicles containing GCs in LNs, and the loss of *Rgs1* alleles could partially compensate for these defects. Thus, the stoichiometry of RGS1 and Gαi2 in spleen appears to be crucial for the lymphoid follicle development, although the relative importance of each for antibody responses has not been fully defined (Hwang et al., 2010).

RGS13 is prominently expressed in hematopoietic cells including B and T lymphocytes and MCs, although it is also highly expressed in specialized neuroendocrine epithelial cells in the GI mucosa (Johnson & Druey, 2002; Shi et al., 2002; our unpublished data). RGS13 KD in human B cell lines resulted in greater peak $Ca^{2+}$ responses to several chemoattractants including CXCL12, CXCL13, and platelet-activating factor (PAF) and impaired their ability to desensitize to a repeat chemokine stimulus (Han et al., 2006). As a result, RGS13-deficient B cells polarized to a greater extent than control cells and remained polarized for longer periods of time following chemokine exposure. Recent detailed studies of *Rgs13*$^{-/-}$ mice in which GFP is knocked into the *Rgs13* locus (RGS13-GFP) revealed that RGS13 expression increased substantially in GC B cells following immunization and declined during GC B cell differentiation into memory B cells or plasma cells (Hwang, Hwang, Park, Harrison, & Kehrl, 2013). Surprisingly, GFP$^{+}$ GC B cells lacking RG13 expression from LN or spleen showed equivalent or reduced chemotaxis to multiple chemokines including CXCL12 and CXCL13 compared to WT cells. In contrast, similar cell populations from mixed BM chimeras exhibited enhanced chemotaxis, suggesting a B cell-extrinsic defect leading to poor chemokine responses in *Rgs13*-GFP mice. These mice also had enlarged GC structures with more GC B cells in spleen and LNs following immunization than WT mice, yet mounted antibody responses to TNP-KLH immunization that were similar to WT and did not display evidence of autoimmunity. However, in a separate study of autoimmune prone mice (BXD2) mice, loss of *Rgs13* attenuated manifestations of autoimmunity including serum autoantibodies and renal dysfunction due to immune complex deposition in the kidney. Reduced size of spontaneously formed GCs containing fewer GC B cells and immature ($IgM^+$) plasmablasts in spleen and LNs appeared to underlie this phenotype (Wang et al., 2013). Collectively, these studies suggest that under physiological conditions, RGS13 functions to slow the transit of B cells through GCs, thereby prolonging Ag and T cell interactions that promote maturation and high-affinity antibody responses.

## 6.2 Plasma Cells

High RGS1 expression has been observed in antibody-secreting plasma cells. *Rgs1*$^{-/-}$ mice also had enhanced GC responses to immunization with ovalbumin and aberrant plasma cell trafficking (Moratz, Harrison, et al., 2004). *Rgs1*$^{-/-}$ mice had increased numbers of OVA-specific plasma cells in circulation compared to WT, and they persisted for longer periods of time. In a separate study, LPS was found to upregulate RGS1 expression in a human PC line (RPMI 8226), which inhibited CXCL12-induced chemotaxis and Akt activation; this phenotype was prevented by RGS1 KD (Pak et al., 2015). High expression of RGS1 in myeloma (plasma cell malignancy) cells within the bone marrow was detected by immunohistochemistry, and RGS1 protein abundance was significantly associated with poor overall survival in one study of myeloma patients (Roh et al., 2017).

## 6.3 T Lymphocytes

Early studies showed that T cells pretreated with PTX and injected intravenously into mice effectively failed to enter into LNs from circulation (Warnock, Askari, Butcher, & von Andrian, 1998). Splenic T cells from *Gnai2*$^{-/-}$ mice were also unable to effectively home to LNs following adoptive transfer, consistent with the importance of Giα2 for T cell trafficking in vivo (Hwang, Park, & Kehrl, 2007). Consistent with these defects, loss of Gαi2 resulted in markedly impaired chemotaxis and $Ca^{2+}$ mobilization in response to CXCL12 and CCL19. After maturing in the thymus, naïve CD4 and CD8 T cells can further differentiate into various effector subsets upon encounter with Ag, each of which are associated with expression of a prototypical set of chemokine receptors. Th1 (CCR5, CXCL3), Th2 (CCR4, 8), Th17 (CCR6), and Tfh (CXCR5) receptors may mediate their recruitment to inflamed tissues and lymphoid organs (Valverde-Villegas, Matte, de Medeiros, & Chies, 2015; Yu, Cavero, Lu, & Li, 2015; Zhang, Zhang, Gu, & Sun, 2014). Similarly, various RGS proteins expressed in each of these subsets may help each T cell subset to navigate the inflammatory microenvironment.

RGS1 has been detected in T cells in humans and mice, with particularly high expression (10-fold+) in CD4 and CD8 cells isolated from the GI tract compared to circulating T cells (Gibbons et al., 2011). The distribution of thymocyte subsets (Moratz, Harrison, et al., 2004) or peripheral CD4 and CD8 T cells numbers in spleen, LN, and gut of *Rgs1*$^{-/-}$ mice did not differ from WT at baseline, nor did splenic T cells from *Rgs1*$^{-/-}$ mice exhibit chemotactic defects in response to CCL19, which promotes egress of cells from circulation into LNs, compared to WT cells, suggesting that RGS1 does not have a prominent role in trafficking of circulating T cells to homeostatic chemokines.

In a recent study, overexpression of RGS1 in peripheral blood T cells reduced chemotaxis to both CXCL12 and CCL19 but did not reduce overall motility. By contrast, T cells obtained from the GI tract of *Rgs1*$^{-/-}$ mice displayed had increased chemotaxis to CXCL12 and CCL19 but not to gut homing-associated chemokines CCL25 and CCL20 (ligands of CCR9 and CCR6, respectively). Intraperitoneal injection of CD4+ T cells into *Rag2*$^{-/-}$ (T and B cell-deficient) mice induces gut inflammation resembling inflammatory bowel disease (IBD) in humans. Injection of CD4 cells from *Rgs1*$^{-/-}$ mice resulted in a less severe phenotype than did transfer of WT cells (Gibbons et al., 2011). Expression of *RGS1* was also found to be significantly higher in gut-derived T cells from patients with IBD than those from healthy controls. Collectively, these findings suggest that RGS1 regulates retention of T cells in gut-associated lymphoid tissue and that excess *RGS1* expression may contribute to immunopathology in IBD.

RGS2 is expressed in human and mouse CD4 and CD8 cells and, like RGS1, is upregulated by TCR stimulation and in T cells associated with gut (Gibbons et al., 2011; Oliveira-Dos-Santos et al., 2000). RGS2 functions in T cells have not been studied in detail. In mice, LN T cells from *Rgs2*$^{-/-}$ mice displayed reduced proliferation and IL-2 production in response to PMA/ionomycin or TCR stimulation with anti-CD3/CD28, whereas B cell proliferation to mitogens was intact. Although chemotaxis of splenic T cells from *Rgs2*$^{-/-}$ mice to CXCL12, CCL3, and CCL5 was reported to be normal, cutaneous inflammation (footpad swelling) induced by injection of lymphocytic choriomeningitis virus was significantly reduced in *Rgs2*$^{-/-}$ mice compared to heterozygotes, suggesting impaired adaptive immunity in the absence of RGS2. Further mechanistic studies of T cell migration patterns at baseline and during inflammatory pathology will need to be done to determine the functions of RGS2 in adaptive immune responses.

Unlike other members of the R4 subfamily that consist of little more than the RGS domain, RGS3 has several splice variants including a larger protein containing a PDZ motif, which has been shown to bind B-ephrin receptors to modulate their signaling (Kehrl, Srikumar, Harrison, Wilson, & Shi, 2002; Lu, Sun, Klein, & Flanagan, 2001). RGS3 was upregulated in splenic Th1 and Th2 cells differentiated from naïve splenocytes in vitro, and RGS3 KD augmented chemotaxis of T cell lines (EL4) to CXCL12 and CCL19 (Williams et al., 2013). To investigate the role of RGS GAP activity in T lymphocyte functions in vivo, Dulin and colleagues generated mice with an *Rgs3* allele lacking the RGS domain sequence. Similar to mice lacking RGS1, 2, and 16, architecture of the thymus and peripheral lymphoid organs and B and T lymphocyte numbers in the periphery were relatively similar to WT mice at baseline. However, following sensitization and respiratory inoculation with house dust mite allergen, Rgs3ΔRGS mice had exaggerated allergen-induced lung inflammation, with increased numbers of activated T cells and formation of structures resembling lymphoid follicles surrounding blood vessels. Thus, RGS3 may regulate proinflammatory T cell trafficking in response to immune challenge. PDZ–RGS3 interactions with B class ephrins through the PDZ domain mediate fundamental processes in developing neurons including migration and axon guidance through interaction with ephrin receptor tyrosine kinases on neighboring cells (Lu et al., 2001). Ephrin B ligands are expressed on T cells and DCs and have been suggested to act as costimulatory molecules during T cell activation (Mimche et al., 2015). Global and conditional knockouts of the PDZ encoding exons of *PDZ–RGS3* have been generated; the short RGS3 isoform is intact. These mice had defects in cortical neuronal development due to early cell cycle exit and enhanced differentiation of neural progenitors in the developing cortex, which resembles defects observed in *EphB1* gene-deleted mice or mice expressing RGS-resistant Gαi2 (G184S) (Qiu, Wang, Tsark, & Lu, 2010). Collectively, these studies suggest independent functions of the PDZ domain and RGS box of RGS3 in cell migration, and it will be of interest to examine T cell trafficking and intercellular communications in mice deficient in RGS3 in specific leukocyte populations.

RGS10 is well expressed in human and mouse T and B lymphocytes and is upregulated by chemokine stimulation (CXCL12) of T cells (Garcia-Bernal et al., 2011; Hwang et al., 2013). There are two RGS10 isoforms, RGS10S and RGS10L, which differ only by a few amino acids in their amino termini. T cells express both isoforms, while B cells express only RGS10L (Lee & Tansey, 2015). It is unknown whether these isoforms have distinct physiological functions. In contrast to RGS1–2, RGS10 expression was similar or reduced in gut-associated CD4 and CD8 cells relative to peripheral blood counterparts. Overexpression of RGS10 in T cell lines (Jurkat, Molt-4) or peripheral blood-derived human T cells inhibited migration toward CXCL12 gradients, while RGS10 KD enhanced chemokine-induced chemotaxis and integrin-mediated adhesion (Garcia-Bernal et al., 2011; Gibbons et al., 2011). The role of RGS10 in T cell-mediated immunity in vivo was recently examined by Tansey and colleagues in *Rgs10*$^{-/-}$ mice (Lee et al., 2016). Similar to RGS1 or RGS16 deficiency, global loss of RGS10 did not affect baseline numbers of B or T cells in peripheral blood or lymphoid tissue, and splenic CD4 cells actually exhibited reduced chemotaxis to CCL19 and CXCL12. Naïve splenic T cells from *Rgs10*$^{-/-}$ mice had intact TCR-induced proliferation and cytokine production and also differentiated into and Th1 or Th17 cells in vitro under polarizing conditions comparably to WT cells. However, in the setting of experimental autoimmune encephalomyelitis (EAE), a mouse Th1/Th17-dependent model of the human disease multiple sclerosis (MS), recruitment of both Th1 and Th17 cells to CNS induced by immunization with EAE-inducing myelin oligodendrocyte protein ($MOG_{35-55}$) was significantly reduced in *Rgs10*$^{-/-}$ mice compared to WT. These T cells produced fewer cytokines upon reexposure to the immunogen and the mice had significantly less clinical disease. Transfer of in vitro polarized Th1 (but not Th17) cells from *Rgs10*$^{-/-}$ mice into WT recipients attenuated the clinical EAE phenotype compared to mice receiving WT cells. Further detailed studies of RGS10 expression in Th1 and Th17 cells and its regulation of chemotaxis of these subsets should help clarify how RGS10 controls T cell migration in vivo.

RGS16 was initially cloned as an IL-2 response gene in human T cells (Beadling, Druey, Richter, Kehrl, & Smith, 1999), but it is expressed widely including in platelets and DCs (Karim et al., 2016). RGS16 is rapidly upregulated in human T cells polarized under both Th1 and Th2 conditions and in mouse Th1, Th2, and Th17 subsets (Lippert et al., 2003; Shankar et al., 2012). Accordingly, transgenic overexpression or gene deletion of *Rgs16* in mice has little impact on chemotaxis of naïve T cells but clearly enhances responsiveness of Th1 and Th2 cells to their cognate chemokines (CXCL9 and CCL17, respectively). Increased numbers of cytokine-producing Th2 cells and more of their associated cytokines were detected in lungs and LNs of *Rgs16*$^{-/-}$ mice challenged with allergen compared to WT mice, which enhanced inflammation and lung remodeling (Shankar et al., 2012). These findings indicate that RGS16 may modulate allergic airway inflammation by constraining Th2 trafficking.

## 7. NONCANONICAL FUNCTIONS OF RGS PROTEINS IN THE IMMUNE SYSTEM

Given the array of protein interaction domains other than the RGS box present in many RGS proteins and their proximity to signaling complexes at the PM, it is not surprising that several GAP-independent functions have been ascribed to them, particularly in endocrine function and cancer. In addition, regulation of RGS GAP activity occurs through protein–protein interactions, modifications in cellular localization, and posttranslational modification, which have been reviewed recently (Kach, Sethakorn, & Dulin, 2012). Many noncanonical activities have also been extensively reviewed elsewhere (Sethakorn, Yau, & Dulin, 2010), but we highlight a few here with particular relevance to the immune system.

### 7.1 IgE-Mediated Anaphylaxis

Systemic MC degranulation triggered by antigen cross-linking of high-affinity IgE receptors (FcER1) and release of inflammatory mediators such as histamine can lead to life-threatening anaphylactic reactions including hypotensive shock. IgE-mediated MC degranulation is partly dependent on phosphatidylinositol-3-OH kinase (PI3 kinase) activity (Hawkins & Stephens, 2015). In addition to its role in regulating chemokine-induced MC migration, RGS13 was found to reduce IgE-dependent MC degranulation by inhibiting PI3K activity. RGS13 bound the phosphorylated form of the regulatory p85α subunit of Class 1A PI3K and restricted access of the enzyme to signaling complexes assembled by FcERI cross-linking. MCs from *Rgs13*$^{-/-}$ mice exhibited increased IgE-mediated degranulation compared to WT, and the mice were much more susceptible to systemic anaphylaxis induced by Ag (Bansal, Xie, Rao, Nocka, & Druey, 2008). Subsequently, several other R4 RGSs including RGS1, 2, 4, and 5 have been shown to interact with PI3K (Damera et al., 2012; Liang, Bansal, Xie, & Druey, 2009). Given the widespread importance of PI3K activity for the function of many leukocytes including neutrophils and lymphocytes, it will be of interest to explore this interaction in other immune cell types.

## 7.2 Transcriptional Regulation in B Lymphocytes

RGS13 was also shown to regulate the activity of the transcription factor cyclic AMP response element binding protein (CREB) through a direct interaction occurring at a site outside of the RGS13 RGS domain. cAMP or $Ca^{2+}$ signaling elicited translocation of cytosolic RGS13 to the nucleus, and binding of RGS13 to phosphorylated CREB in complex with the coactivator CREB binding protein (CBP/p300) displaced this transcriptional complex from promoters (Xie, Geiger, Johnson, Nyborg, & Druey, 2008). CREB target genes including *c-fosb*, *Obf1*, and *Mta2* are important for GC B cell maturation and antibody responses. These and other CREB target genes were upregulated in GC B cells from *Rgs13*$^{-/-}$ mice and BXD2/*Rgs13*$^{-/-}$ mice compared to controls (Hwang et al., 2013; Wang et al., 2013), although it is unknown whether this was due to the direct regulation of CREB by RGS13.

## 7.3 Phagocytosis

RGS14 was identified as a novel effector of integrin-dependent phagocytosis in macrophages. Complement receptor 3 is a surface integrin (αMβ2) that mediates phagocytic uptake of opsonized particles through the cytoskeletal protein talin, Rap1 GTPase, and $Ca^{2+/}$calmodulin-dependent kinase II. RGS14 was previously identified as a Rap1 effector via its two canonical Ras/Rap binding domains (R1 and R2) (Lim, Dupuy, Critchley, & Caron, 2010). Direct interaction of RGS14 and Rap1 was dependent on a single residue (R333) in the R1 domain. Silencing of RGS14 in J774. A1 macrophages reduced CR3-dependent binding to complement component C3b-coated sheep red blood cells, which is indicative of decreased αMβ2 activation (Lim, Thompson, May, Hotchin, & Caron, 2013). The role of RGS14 in macrophage phagocytosis of microbial particles in vivo in *Rgs14*$^{-/-}$ mice has not yet been reported.

### 7.4 Macrophage Survival and Proinflammatory Functions

Notch signaling plays a broadly important role in immunity by affecting survival, cytokine production, and antigen-presenting capacity of leukocytes including macrophages (Shang, Smith, & Hu, 2016). Notch receptor activation is evoked by LPS treatment of macrophages through receptor cleavage by TNFα-converting enzyme and gamma secretase-dependent mechanisms. This results in translocation of the receptor to the nucleus, where it induces gene transcription by interacting with the transcription factor RBP-Jk/CSL. Notch activation cross-regulates LPS-induced proinflammatory functions of macrophages both positively and negatively. A phosphoproteomics screen identified RGS19 phosphorylation in LPS-treated macrophages, which was inhibited by Notch signaling. KD of RGS19 reduced LPS-mediated Akt phosphorylation and increased cell death (Sangphech, Osborne, & Palaga, 2014). Although a prior study revealed that RGS19 is phosphorylated by casein kinase 2 (CK2) in vitro at Ser24 (Fischer, Elenko, Wan, Thomas, & Farquhar, 2000), the site of RGS19 phosphorylation and kinase activated by LPS/Notch signaling was not determined. The mechanism by which pRGS19 acts as effector of Notch pathway activation to regulate Akt phosphorylation is also unclear. Although Gαi2 may be involved in TLR signaling as noted earlier, some studies have suggested that RGS19 might not interact with Gαi2 (De Vries, Mousli, Wurmser, & Farquhar, 1995). A separate study found that RGS19 stimulated growth of cancer cell lines through the interaction with an accessory protein, GAIP-interacting protein C-terminus (GIPC), through its PDZ motif (Tso et al., 2010).

## 8. RGS ASSOCIATION WITH DISEASES OF THE IMMUNE SYSTEM

### 8.1 RGS1 and Celiac Disease

Celiac disease (CD) is a chronic inflammatory disorder of the small intestine attributed to sensitivity to dietary gluten. It is characterized by increased Th1 cells specific for gluten-derived components (gliadin). Recent meta-analysis of 20 GWAS (more than 14,000 cases and 24,000 controls) found a negative correlation between rs2816316 (C/A) in *RGS1* and CD, with an odds ratio (OR) of 0.77. The minor allele (C) frequency was 14% in the CD group and

17.5% in the control group. This variant maps 8 kb from the 5′ end of *RGS1*. Gut expression of RGS1 in CD has not been evaluated, but as noted earlier, gut T cells have much higher *Rgs1* expression than circulating cells, and T cells from $Rgs1^{-/-}$ mice enhanced susceptibility of mice to chronic intestinal inflammation compared to WT (Guo et al., 2016).

## 8.2 RGS1 and Autoimmune Diabetes

Type 1 autoimmune diabetes mellitus (DM) cosegregates with CD at the population level, suggesting overlapping genetic origins. The CD-associated variant rs2816316 also displayed significant association with type I DM (OR 0.72; $P = 2.58 \times 10^{-11}$) (Caballero-Franco & Kissler, 2016; Smyth et al., 2008). Tfh cells are located in lymphoid GCs, where they promote GC affinity maturation. Tfh frequencies are elevated in patients with type I DM and in a mouse model of DM, suggesting that Tfh dysregulation plays a role in the disease. Tfh cells must downregulate CCR7 signaling and negotiate CXCL13 gradients to reach B cell follicles (Haynes et al., 2007; Linterman et al., 2011). RGS1 is upregulated in Tfh compared naïve T cells. The potential role of RGS1 in Tfh function and DM was studied using an RGS1 KD mouse strain on the nonobese diabetic (NOD) background (Caballero-Franco & Kissler, 2016). There were fewer Tfh cells in lymphoid follicles than in WT, and they displayed increased chemotactic responses to CCL19 and 21 (CCR7 ligands) but not CXL13 (CXCR5 ligand). The RGS1 KD NOD mice had larger and more numerous GCs following immunization, but GC structures contained significantly fewer Tfh cells. These findings support a role for RGS1 in the downregulation of CCR7 responses needed to position Tfh cells within the GC niche. However, RGS1 KD did not alter the overall risk of developing DM relative to WT, making its contribution to the pathogenesis of type I DM and other autoimmune diseases in which Tfh cells have a prominent role still unclear.

## 8.3 RGS1 and IgA Nephropathy

IgA nephropathy is a progressive renal disease characterized by IgA deposits in glomeruli, which can lead to renal failure in 20%–40% of patients. It is the most common form of primary glomerulonephritis and exhibits familial clustering, but susceptibility loci identified thus far only can only account for a small proportion of heritability. Variants previously associated with autoimmunity were genotyped in a recent study of ~1200 patients with IgA nephropathy and ~1200 controls (Zhou et al., 2016). rs12022418, located 9.7 kb upstream of the 5′ end of *RGS1*, was found to be significantly associated with the disease (OR 1.24, *P* value $6.27 \times 10^{-4}$). No proxy SNPs (potentially causal SNPs in linkage disequilibrium with the most highly associated GWAS SNP) could be identified, making the functional significance of this association unclear. However, rs12022418 was found to be an "eSNP" (expression regulatory SNP that could potentially affect *RGS1* expression) in the Genotype-Tissue Expression (GTEx) portal.

### 8.4 RGS1 and RGS10 and Multiple Sclerosis

MS is a chronic, relapsing/remitting inflammatory disease of unknown etiology characterized by demyelination of the CNS. Several studies have identified *RGS1*, *7*, *9*, and *14* as susceptibility loci for MS, and RGS1 and RGS10 expression in PBMCs is higher in patients with MS compared to controls. IFNβ, a standard treatment for MS, rapidly upregulated RGS1 expression in PBMCs from patients with MS and healthy controls (Tran et al., 2010). As discussed earlier, RGS10 deficiency attenuates clinical severity of EAE, which is a mouse model of MS. This phenotype was associated with a marked reduction in CNS infiltration with pathogenic Th1 cells, although it remains unclear whether this is due to altered differentiation, survival, and/or impaired trafficking (Lee et al., 2016).

### 8.5 RGS1 and RGS13 and B Cell Malignancies

RGS1 and RGS13 are particularly well expressed in Burkitt lymphomas, thought to be a malignancy of GC B cells (Piovan et al., 2007). In Burkitt lymphoma cells, hypoxia upregulated RGS1, which may contribute to the poor responsiveness of these cells to CXCL12. RGS1 abundance was also demonstrated to be high in primary sinonasal tract diffuse large B cell lymphoma (DLBCL), a non-Hodgkin lymphoma, which may be due to copy number increases at 1q31, which contains the *RGS1* locus. High RGS1 protein expression by immunohistochemistry was associated with poor overall survival of patients with DLBCL (Carreras et al., 2017).

## 9. CONCLUSIONS

Since the discovery of RGS proteins over 20 years ago, considerable progress has been made in the determination of their structural features, biochemical functions in vitro, and regulation of GPCR-mediated processes at the cellular level. However, the physiological functions of most RGS proteins at the organismal level have not been fully defined, particularly within the immune system. It remains unclear why most leukocytes express several RGS family members that would appear to be redundant based on their shared biochemical activities in vitro. Unique protein–protein interactions, covalent modifications, or differential subcellular localization of individual RGS proteins might specify unique functions in a given cell type. Further studies using optogenetic techniques and high-resolution imaging will be needed to determine if and how RGS proteins may be recruited to specific cellular regions to generate intracellular signaling gradients that promote cell polarization and chemotaxis.

Notably, no systemic immune diseases, including those associated with immunodeficiency or autoimmunity, have been attributed to pathogenic alleles in *RGS* genes in humans. A number of GWAS have identified polymorphic variants in specific *RGS*s (in particular *RGS1*) linked to complex autoimmune, inflammatory, or neoplastic diseases that are probably driven by multigenic defects. We have highlighted several examples of potential candidate driver variants in specific *RGS* genes together with existing evidence from mechanistic studies, if any, supporting their relevance to a specific disease. The explosion of whole exome sequencing and other genomic discovery techniques may yet uncover pathogenic mutations in RGS proteins associated with human immune diseases, whose functional significance can be tested in cellular and animal models.

## ACKNOWLEDGMENT

This work was supported by the Intramural Research Program, NIAID/NIH.

## REFERENCES

Abramow-Newerly, M., Roy, A. A., Nunn, C., & Chidiac, P. (2006). RGS proteins have a signalling complex: Interactions between RGS proteins and GPCRs, effectors, and auxiliary proteins. *Cellular Signalling*, *18*, 579–591.

Anderson, G. R., Posokhova, E., & Martemyanov, K. A. (2009). The R7 RGS protein family: Multi-subunit regulators of neuronal G protein signaling. *Cell Biochemistry and Biophysics*, *54*, 33–46.

Arnon, T. I., Xu, Y., Lo, C., Pham, T., An, J., Coughlin, S., et al. (2011). GRK2-dependent S1PR1 desensitization is required for lymphocytes to overcome their attraction to blood. *Science*, *333*, 1898–1903.

Bannard, O., & Cyster, J. G. (2017). Germinal centers: Programmed for affinity maturation and antibody diversification. *Current Opinion in Immunology*, *45*, 21–30.

Bansal, G., DiVietro, J. A., Kuehn, H. S., Rao, S., Nocka, K. H., Gilfillan, A. M., et al. (2008). RGS13 controls G protein-coupled receptor-evoked responses of human mast cells. *Journal of Immunology*, *181*, 7882–7890.

Bansal, G., Druey, K. M., & Xie, Z. (2007). R4 RGS proteins: Regulation of G-protein signaling and beyond. *Pharmacology & Therapeutics*, *116*, 473–495.

Bansal, G., Xie, Z., Rao, S., Nocka, K. H., & Druey, K. M. (2008). Suppression of immunoglobulin E-mediated allergic responses by regulator of G protein signaling 13. *Nature Immunology*, *9*, 73–80.

Beadling, C., Druey, K. M., Richter, G., Kehrl, J. H., & Smith, K. A. (1999). Regulators of G protein signaling exhibit distinct patterns of gene expression and target G protein specificity in human lymphocytes. *Journal of Immunology*, *162*, 2677–2682.

Boelte, K. C., Gordy, L. E., Joyce, S., Thompson, M. A., Yang, L., & Lin, P. C. (2011). Rgs2 mediates pro-angiogenic function of myeloid derived suppressor cells in the tumor microenvironment via upregulation of MCP-1. *PLoS One*, *6*, e18534.

Brown, N. E., Goswami, D., Branch, M. R., Ramineni, S., Ortlund, E. A., Griffin, P. R., et al. (2015). Integration of G protein alpha (Galpha) signaling by the regulator of G protein signaling 14 (RGS14). *The Journal of Biological Chemistry*, *290*, 9037–9049.

Brown, N. E., Lambert, N. A., & Hepler, J. R. (2016). RGS14 regulates the lifetime of Galpha-GTP signaling but does not prolong Gbetagamma signaling following receptor activation in live cells. *Pharmacology Research & Perspectives*, *4*, e00249.

Caballero-Franco, C., & Kissler, S. (2016). The autoimmunity-associated gene RGS1 affects the frequency of T follicular helper cells. *Genes and Immunity*, *17*, 228–238.

Carreras, J., Kikuti, Y. Y., Bea, S., Miyaoka, M., Hiraiwa, S., Ikoma, H., et al. (2017). Clinicopathological characteristics and genomic profile of primary sinonasal tract diffuse large B cell lymphoma (DLBCL) reveals gain at 1q31 and RGS1 encoding protein; high RGS1 immunohistochemical expression associates with poor overall survival in DLBCL not otherwise specified (NOS). *Histopathology*, *70*(4), 595–621.

Chaffin, K. E., & Perlmutter, R. M. (1991). A pertussis toxin-sensitive process controls thymocyte emigration. *European Journal of Immunology*, *21*, 2565–2573.

Chen, J. G., & Jones, A. M. (2004). AtRGS1 function in *Arabidopsis thaliana*. *Methods in Enzymology*, *389*, 338–350.

Cheng, W. L., Wang, P. X., Wang, T., Zhang, Y., Du, C., Li, H., et al. (2015). Regulator of G-protein signalling 5 protects against atherosclerosis in apolipoprotein E-deficient mice. *British Journal of Pharmacology*, *172*, 5676–5689.

Cho, H., Kamenyeva, O., Yung, S., Gao, J. L., Hwang, I. Y., Park, C., et al. (2012). The loss of RGS protein-Galpha(i2) interactions results in markedly impaired mouse neutrophil trafficking to inflammatory sites. *Molecular and Cellular Biology*, *32*, 4561–4571.

Cyster, J. G., & Goodnow, C. C. (1995). Pertussis toxin inhibits migration of B and T lymphocytes into splenic white pulp cords. *The Journal of Experimental Medicine*, *182*, 581–586.

Cyster, J. G., & Schwab, S. R. (2012). Sphingosine-1-phosphate and lymphocyte egress from lymphoid organs. *Annual Review of Immunology*, *30*, 69–94.

Dahlin, J. S., & Hallgren, J. (2015). Mast cell progenitors: Origin, development and migration to tissues. *Molecular Immunology*, *63*, 9–17.

Damera, G., Druey, K. M., Cooper, P. R., Krymskaya, V. P., Soberman, R. J., Amrani, Y., et al. (2012). An RGS4-mediated phenotypic switch of bronchial smooth muscle cells promotes fixed airway obstruction in asthma. *PLoS One*, 7, e28504.

De Vries, L., Mousli, M., Wurmser, A., & Farquhar, M. G. (1995). GAIP, a protein that specifically interacts with the trimeric G protein G alpha i3, is a member of a protein family with a highly conserved core domain. *Proceedings of the National Academy of Sciences of the United States of America*, *92*, 11916–11920.

Doupnik, C. A., Davidson, N., Lester, H. A., & Kofuji, P. (1997). RGS proteins reconstitute the rapid gating kinetics of gbetagamma-activated inwardly rectifying K+ channels. *Proceedings of the National Academy of Sciences of the United States of America*, *94*, 10461–10466.

Evans, P. R., Dudek, S. M., & Hepler, J. R. (2015). Regulator of G protein signaling 14: A molecular brake on synaptic plasticity linked to learning and memory. *Progress in Molecular Biology and Translational Science*, *133*, 169–206.

Fischer, T., Elenko, E., Wan, L., Thomas, G., & Farquhar, M. G. (2000). Membrane-associated GAIP is a phosphoprotein and can be phosphorylated by clathrin-coated vesicles. *Proceedings of the National Academy of Sciences of the United States of America, 97*, 4040–4045.

Garcia-Bernal, D., Dios-Esponera, A., Sotillo-Mallo, E., Garcia-Verdugo, R., Arellano-Sanchez, N., & Teixido, J. (2011). RGS10 restricts upregulation by chemokines of T cell adhesion mediated by alpha4beta1 and alphaLbeta2 integrins. *Journal of Immunology, 187*, 1264–1272.

Gerber, K. J., Squires, K. E., & Hepler, J. R. (2016). Roles for regulator of G protein signaling proteins in synaptic signaling and plasticity. *Molecular Pharmacology, 89*, 273–286.

Gibbons, D. L., Abeler-Dorner, L., Raine, T., Hwang, I. Y., Jandke, A., Wencker, M., et al. (2011). Cutting edge: Regulator of G protein signaling-1 selectively regulates gut T cell trafficking and colitic potential. *Journal of Immunology, 187*, 2067–2071.

Guo, C. C., Wang, M., Cao, F. D., Huang, W. H., Xiao, D., Ye, X. G., et al. (2016). Meta-analysis on associations of RGS1 and IL12A polymorphisms with celiac disease risk. *International Journal of Molecular Sciences, 17*, 457.

Han, J. I., Huang, N. N., Kim, D. U., & Kehrl, J. H. (2006). RGS1 and RGS13 mRNA silencing in a human B lymphoma line enhances responsiveness to chemoattractants and impairs desensitization. *Journal of Leukocyte Biology, 79*, 1357–1368.

Han, S. B., Moratz, C., Huang, N. N., Kelsall, B., Cho, H., Shi, C. S., et al. (2005). Rgs1 and Gnai2 regulate the entrance of B lymphocytes into lymph nodes and B cell motility within lymph node follicles. *Immunity, 22*, 343–354.

Hawkins, P. T., & Stephens, L. R. (2015). PI3K signalling in inflammation. *Biochimica et Biophysica Acta, 1851*, 882–897.

Haynes, N. M., Allen, C. D., Lesley, R., Ansel, K. M., Killeen, N., & Cyster, J. G. (2007). Role of CXCR5 and CCR7 in follicular Th cell positioning and appearance of a programmed cell death gene-1high germinal center-associated subpopulation. *Journal of Immunology, 179*, 5099–5108.

Hepler, J. R., Berman, D. M., Gilman, A. G., & Kozasa, T. (1997). RGS4 and GAIP are GTPase-activating proteins for Gq alpha and block activation of phospholipase C beta by gamma-thio-GTP-Gq alpha. *Proceedings of the National Academy of Sciences of the United States of America, 94*, 428–432.

Hepler, J. R., & Gilman, A. G. (1992). G proteins. *Trends in Biochemical Sciences, 17*, 383–387.

Huang, J., & Fisher, R. A. (2009). Nuclear trafficking of regulator of G protein signaling proteins and their roles in the nucleus. *Progress in Molecular Biology and Translational Science, 86*, 115–156.

Hwang, I. Y., Hwang, K. S., Park, C., Harrison, K. A., & Kehrl, J. H. (2013). Rgs13 constrains early B cell responses and limits germinal center sizes. *PLoS One, 8*, e60139.

Hwang, I. Y., Park, C., Harrision, K. A., Huang, N. N., & Kehrl, J. H. (2010). Variations in Gnai2 and Rgs1 expression affect chemokine receptor signaling and the organization of secondary lymphoid organs. *Genes and Immunity, 11*, 384–396.

Hwang, I. Y., Park, C., Harrison, K., Boularan, C., Gales, C., & Kehrl, J. H. (2015). An essential role for RGS protein/Galphai2 interactions in B lymphocyte-directed cell migration and trafficking. *Journal of Immunology, 194*, 2128–2139.

Hwang, I. Y., Park, C., & Kehrl, J. H. (2007). Impaired trafficking of Gnai2+/− and Gnai2 −/− T lymphocytes: Implications for T cell movement within lymph nodes. *Journal of Immunology, 179*, 439–448.

Johnson, E. N., & Druey, K. M. (2002). Functional characterization of the G protein regulator RGS13. *The Journal of Biological Chemistry, 277*, 16768–16774.

Kach, J., Sethakorn, N., & Dulin, N. O. (2012). A finer tuning of G-protein signaling through regulated control of RGS proteins. *American Journal of Physiology. Heart and Circulatory Physiology, 303*, H19–H35.

Karim, Z. A., Alshbool, F. Z., Vemana, H. P., Conlon, C., Druey, K. M., & Khasawneh, F. T. (2016). CXCL12 regulates platelet activation via the regulator of G-protein signaling 16. *Biochimica et Biophysica Acta, 1863*, 314–321.

Kehrl, J. H., Srikumar, D., Harrison, K., Wilson, G. L., & Shi, C. S. (2002). Additional 5′ exons in the RGS3 locus generate multiple mRNA transcripts, one of which accounts for the origin of human PDZ-RGS3. *Genomics, 79*, 860–868.

Kienle, K., & Lammermann, T. (2016). Neutrophil swarming: An essential process of the neutrophil tissue response. *Immunological Reviews, 273*, 76–93.

Kimple, A. J., Bosch, D. E., Giguere, P. M., & Siderovski, D. P. (2011) Regulators of G-protein signaling and their Galpha substrates: Promises and challenges in their use as drug discovery targets. *Pharmacological Reviews, 63*, 728–749.

Ko, H. K., Guo, L. W., Su, B., Gao, L., & Gelman, I. H. (2014). Suppression of chemotaxis by SSeCKS via scaffolding of phosphoinositol phosphates and the recruitment of the Cdc42 GEF, Frabin, to the leading edge. *PLoS One, 9*, e111534.

Kveberg, L., Ryan, J. C., Rolstad, B., & Inngjerdingen, M. (2005). Expression of regulator of G protein signalling proteins in natural killer cells, and their modulation by Ly49A and Ly49D. *Immunology, 115*, 358–365.

Lambert, N. A., Johnston, C. A., Cappell, S. D., Kuravi, S., Kimple, A. J., Willard, F. S., et al. (2010). Regulators of G-protein signaling accelerate GPCR signaling kinetics and govern sensitivity solely by accelerating GTPase activity. *Proceedings of the National Academy of Sciences of the United States of America, 107*, 7066–7071.

Lan, K. L., Sarvazyan, N. A., Taussig, R., Mackenzie, R. G., DiBello, P. R., Dohlman, H. G., et al. (1998). A point mutation in Galphao and Galphai1 blocks interaction with regulator of G protein signaling proteins. *The Journal of Biological Chemistry, 273*, 12794–12797.

Lee, J. K., Chung, J., McAlpine, F. E., & Tansey, M. G. (2011). Regulator of G-protein signaling-10 negatively regulates NF-kappaB in microglia and neuroprotects dopaminergic neurons in hemiparkinsonian rats. *The Journal of Neuroscience, 31*, 11879–11888.

Lee, J. K., Kannarkat, G. T., Chung, J., Joon Lee, H., Graham, K. L., & Tansey, M. G. (2016). RGS10 deficiency ameliorates the severity of disease in experimental autoimmune encephalomyelitis. *Journal of Neuroinflammation, 13*, 24.

Lee, J. K., McCoy, M. K., Harms, A. S., Ruhn, K. A., Gold, S. J., & Tansey, M. G. (2008). Regulator of G-protein signaling 10 promotes dopaminergic neuron survival via regulation of the microglial inflammatory response. *The Journal of Neuroscience, 28*, 8517–8528.

Lee, J. K., & Tansey, M. G. (2015). Physiology of RGS10 in neurons and immune cells. *Progress in Molecular Biology and Translational Science, 133*, 153–167.

Lee, H. K., Yeo, S., Kim, J. S., Lee, J. G., Bae, Y. S., Lee, C., et al. (2010). Protein kinase C-eta and phospholipase D2 pathway regulates foam cell formation via regulator of G protein signaling 2. *Molecular Pharmacology, 78*, 478–485.

Li, P., Neubig, R. R., Zingarelli, B., Borg, K., Halushka, P. V., Cook, J. A., et al. (2012). Toll-like receptor-induced inflammatory cytokines are suppressed by gain of function or overexpression of Galpha(i2) protein. *Inflammation, 35*, 1611–1617.

Liang, G., Bansal, G., Xie, Z., & Druey, K. M. (2009). RGS16 inhibits breast cancer cell growth by mitigating phosphatidylinositol 3-kinase signaling. *The Journal of Biological Chemistry, 284*, 21719–21727.

Liggett, S. B. (2011). Phosphorylation barcoding as a mechanism of directing GPCR signaling. *Science Signaling, 4*, pe36.

Lim, J., Dupuy, A. G., Critchley, D. R., & Caron, E. (2010). Rap1 controls activation of the alpha(M)beta(2) integrin in a talin-dependent manner. *Journal of Cellular Biochemistry, 111*, 999–1009.

Lim, J., Thompson, J., May, R. C., Hotchin, N. A., & Caron, E. (2013). Regulator of G-protein signalling-14 (RGS14) regulates the activation of alphaMbeta2 integrin during phagocytosis. *PLoS One*, *8*, e69163.

Lin, C., Koval, A., Tishchenko, S., Gabdulkhakov, A., Tin, U., Solis, G. P., et al. (2014). Double suppression of the Galpha protein activity by RGS proteins. *Molecular Cell*, *53*, 663–671.

Linterman, M. A., Pierson, W., Lee, S. K., Kallies, A., Kawamoto, S., Rayner, T. F., et al. (2011). Foxp3 + follicular regulatory T cells control the germinal center response. *Nature Medicine*, *17*, 975–982.

Lippert, E., Yowe, D. L., Gonzalo, J. A., Justice, J. P., Webster, J. M., Fedyk, E. R., et al. (2003). Role of regulator of G protein signaling 16 in inflammation-induced T lymphocyte migration and activation. *Journal of Immunology*, *171*, 1542–1555.

Lu, Q., Sun, E. E., Klein, R. S., & Flanagan, J. G. (2001). Ephrin-B reverse signaling is mediated by a novel PDZ-RGS protein and selectively inhibits G protein-coupled chemoattraction. *Cell*, *105*, 69–79.

Ma, P., Ou, K., Sinnamon, A. J., Jiang, H., Siderovski, D. P., & Brass, L. F. (2015). Modulating platelet reactivity through control of RGS18 availability. *Blood*, *126*, 2611–2620.

Mahla, R. S., Reddy, M. C., Prasad, D. V., & Kumar, H. (2013). Sweeten PAMPs: Role of sugar complexed PAMPs in innate immunity and vaccine biology. *Frontiers in Immunology*, *4*, 248.

Masuda, K., Kitakami, J. I., Kozasa, T., Kodama, T., Ihara, S., & Hamakubo, T. (2017). Visualization of ligand-induced Gi protein activation in chemotaxing cells. *The FASEB Journal*, *31*(3), 910–919.

Miao, R., Lu, Y., Xing, X., Li, Y., Huang, Z., Zhong, H., et al. (2016). Regulator of G-protein signaling 10 negatively regulates cardiac remodeling by blocking mitogen-activated protein kinase-extracellular signal-regulated protein kinase 1/2 signaling. *Hypertension*, *67*, 86–98.

Mimche, P. N., Brady, L. M., Keeton, S., Fenne, D. S., King, T. P., Quicke, K. M., et al. (2015). Expression of the receptor tyrosine kinase EphB2 on dendritic cells is modulated by Toll-like receptor ligation but is not required for T cell activation. *PLoS One*, *10*, e0138835.

Min, J., & Defea, K. (2011). Beta-arrestin-dependent actin reorganization: Bringing the right players together at the leading edge. *Molecular Pharmacology*, *80*, 760–768.

Moratz, C., Harrison, K., & Kehrl, J. H. (2004). Regulation of chemokine-induced lymphocyte migration by RGS proteins. *Methods in Enzymology*, *389*, 15–32.

Moratz, C., Hayman, J. R., Gu, H., & Kehrl, J. H. (2004). Abnormal B-cell responses to chemokines, disturbed plasma cell localization, and distorted immune tissue architecture in Rgs1 −/− mice. *Molecular and Cellular Biology*, *24*, 5767–5775.

Muppidi, J. R., Lu, E., & Cyster, J. G. (2015). The G protein-coupled receptor P2RY8 and follicular dendritic cells promote germinal center confinement of B cells, whereas S1PR3 can contribute to their dissemination. *The Journal of Experimental Medicine*, *212*, 2213–2222.

Nance, M. R., Kreutz, B., Tesmer, V. M., Sterne-Marr, R., Kozasa, T., & Tesmer, J. J. (2013). Structural and functional analysis of the regulator of G protein signaling 2-galphaq complex. *Structure*, *21*, 438–448.

Neubig, R. R. (2015). RGS-insensitive G proteins as in vivo probes of RGS function. *Progress in Molecular Biology and Translational Science*, *133*, 13–30.

Nunn, C., Mao, H., Chidiac, P., & Albert, P. R. (2006). RGS17/RGSZ2 and the RZ/a family of regulators of G-protein signaling. *Seminars in Cell & Developmental Biology*, *17*, 390–399.

Oliveira-Dos-Santos, A. J., Matsumoto, G., Snow, B. E., Bai, D., Houston, F. P., Whishaw, I. Q., et al. (2000). Regulation of T cell activation, anxiety, and male aggression by RGS2. *Proceedings of the National Academy of Sciences of the United States of America, 97*, 12272–12277.

O'Neill, P. R., & Gautam, N. (2014). Subcellular optogenetic inhibition of G proteins generates signaling gradients and cell migration. *Molecular Biology of the Cell, 25*, 2305–2314.

Pak, H. K., Gil, M., Lee, Y., Lee, H., Lee, A. N., Roh, J., et al. (2015). Regulator of G protein signaling 1 suppresses CXCL12-mediated migration and AKT activation in RPMI 8226 human plasmacytoma cells and plasmablasts. *PLoS One, 10*, e0124793.

Patel, J., McNeill, E., Douglas, G., Hale, A. B., de Bono, J., Lee, R., et al. (2015). RGS1 regulates myeloid cell accumulation in atherosclerosis and aortic aneurysm rupture through altered chemokine signalling. *Nature Communications, 6*, 6614.

Perrier, P., Martinez, F. O., Locati, M., Bianchi, G., Nebuloni, M., Vago, G., et al. (2004). Distinct transcriptional programs activated by interleukin-10 with or without lipopolysaccharide in dendritic cells: Induction of the B cell-activating chemokine, CXC chemokine ligand 13. *Journal of Immunology, 172*, 7031–7042.

Piovan, E., Tosello, V., Indraccolo, S., Masiero, M., Persano, L., Esposito, G., et al. (2007). Differential regulation of hypoxia-induced CXCR4 triggering during B-cell development and lymphomagenesis. *Cancer Research, 67*, 8605–8614.

Popov, S., Yu, K., Kozasa, T., & Wilkie, T. M. (1997). The regulators of G protein signaling (RGS) domains of RGS4, RGS10, and GAIP retain GTPase activating protein activity in vitro. *Proceedings of the National Academy of Sciences of the United States of America, 94*, 7216–7220.

Qiu, R., Wang, J., Tsark, W., & Lu, Q. (2010). Essential role of PDZ-RGS3 in the maintenance of neural progenitor cells. *Stem Cells, 28*, 1602–1610.

Riekenberg, S., Farhat, K., Debarry, J., Heine, H., Jung, G., Wiesmuller, K. H., et al. (2009). Regulators of G-protein signalling are modulated by bacterial lipopeptides and lipopolysaccharide. *The FEBS Journal, 276*, 649–659.

Roh, J., Shin, S. J., Lee, A. N., Yoon, D. H., Suh, C., Park, C. J., et al. (2017) RGS1 expression is associated with poor prognosis in multiple myeloma. *Journal of Clinical Pathology, 70*(3), 202–207.

Saitoh, O., Kubo, Y., Miyatani, Y., Asano, T., & Nakata, H. (1997). RGS8 accelerates G-protein-mediated modulation of K+ currents. *Nature, 390*, 525–529

Sangphech, N., Osborne, B. A., & Palaga, T. (2014). Notch signaling regulates the phosphorylation of Akt and survival of lipopolysaccharide-activated macrophages via regulator of G protein signaling 19 (RGS19). *Immunobiology, 219*, 653–660.

Sethakorn, N., Yau, D. M., & Dulin, N. O. (2010). Non-canonical functions of RGS proteins. *Cellular Signalling, 22*, 1274–1281.

Shang, Y., Smith, S., & Hu, X. (2016). Role of Notch signaling in regulating innate immunity and inflammation in health and disease. *Protein & Cell, 7*, 159–174.

Shankar, S. P., Wilson, M. S., DiVietro, J. A., Mentink-Kane, M. M., Xie, Z., Wynn, T. A., et al. (2012). RGS16 attenuates pulmonary Th2/Th17 inflammatory responses. *Journal of Immunology, 188*, 6347–6356.

Shi, G. X., Harrison, K., Han, S. B., Moratz, C., & Kehrl, J. H. (2004). Toll-like receptor signaling alters the expression of regulator of G protein signaling proteins in dendritic cells: Implications for G protein-coupled receptor signaling. *Journal of Immunology, 172*, 5175–5184.

Shi, G. X., Harrison, K., Wilson, G. L., Moratz, C., & Kehrl, J. H. (2002). RGS13 regulates germinal center B lymphocytes responsiveness to CXC chemokine ligand (CXCL)12 and CXCL13. *Journal of Immunology, 169*, 2507–2515.

Shi, G., Partida-Sanchez, S., Misra, R. S., Tighe, M., Borchers, M. T., Lee, J. J., et al. (2007). Identification of an alternative G{alpha}q-dependent chemokine receptor signal transduction pathway in dendritic cells and granulocytes. *The Journal of Experimental Medicine, 204*, 2705–2718.

Siderovski, D. P., & Willard, F. S. (2005). The GAPs, GEFs, and GDIs of heterotrimeric G-protein alpha subunits. *International Journal of Biological Sciences, 1*, 51–66.

Sjogren, B. (2017). The evolution of RGS proteins as drug targets—20 years in the making. IUPHAR review: "X" *British Journal of Pharmacology, 174*(6), 427–437.

Sjogren, B., Blazer, L. L., & Neubig, R. R. (2010). Regulators of G protein signaling proteins as targets for drug discovery. *Progress in Molecular Biology and Translational Science, 91*, 81–119.

Smyth, D. J., Plagnol, V., Walker, N. M., Cooper, J. D., Downes, K., Yang, J. H., et al. (2008). Shared and distinct genetic variants in type 1 diabetes and celiac disease. *The New England Journal of Medicine, 359*, 2767–2777.

Spangrude, G. J., Sacchi, F., Hill, H. R., Van Epps, D. E., & Daynes, R. A. (1985). Inhibition of lymphocyte and neutrophil chemotaxis by pertussis toxin. *Journal of Immunology, 135*, 4135–4143.

Srinivasa, S. P., Watson, N., Overton, M. C., & Blumer, K. J. (1998). Mechanism of RGS4, a GTPase-activating protein for G protein alpha subunits. *The Journal of Biological Chemistry, 273*, 1529–1533.

Stewart, A., & Fisher, R. A. (2015). Introduction: G protein-coupled receptors and RGS proteins. *Progress in Molecular Biology and Translational Science, 133*, 1–11.

Surve, C. R., Lehmann, D., & Smrcka, A. V. (2014). A chemical biology approach demonstrates G protein betagamma subunits are sufficient to mediate directional neutrophil chemotaxis. *The Journal of Biological Chemistry, 289*, 17791–17801.

Surve, C. R., To, J. Y., Malik, S., Kim, M., & Smrcka, A. V. (2016). Dynamic regulation of neutrophil polarity and migration by the heterotrimeric G protein subunits Galphai-GTP and Gbetagamma. *Science Signaling*, 9, ra22.

Suurvali, J., Pahtma, M., Saar, R., Paalme, V., Nutt, A., Tiivel, T., et al. (2015). RGS16 restricts the pro-inflammatory response of monocytes. *Scandinavian Journal of Immunology, 81*, 23–30.

Takata, Y., Liu, J., Yin, F., Collins, A. R., Lyon, C. J., Lee, C. H., et al. (2008). PPARdelta-mediated antiinflammatory mechanisms inhibit angiotensin II-accelerated atherosclerosis. *Proceedings of the National Academy of Sciences of the United States of America, 105*, 4277–4282.

Tan, W., Martin, D., & Gutkind, J. S. (2006). The Galpha13-Rho signaling axis is required for SDF-1-induced migration through CXCR4. *The Journal of Biological Chemistry, 281*, 39542–39549.

Tayou, J., Wang, Q., Jang, G. F., Pronin, A. N., Orlandi, C., Martemyanov, K. A., et al. (2016). Regulator of G protein signaling 7 (RGS7) can exist in a homo-oligomeric form that is regulated by Galphao and R7-binding protein. *The Journal of Biological Chemistry, 291*, 9133–9147.

Tran, T., Paz, P., Velichko, S., Cifrese, J., Belur, P., Yamaguchi, K. D., et al. (2010). Interferonbeta-1b induces the expression of RGS1 a negative regulator of G-protein signaling. *International Journal of Cell Biology, 2010*, 529376.

Tso, P. H., Wang, Y., Wong, S. Y., Poon, L. S., Chan, A. S., & Wong, Y. H. (2010). RGS19 enhances cell proliferation through its C-terminal PDZ motif. *Cellular Signalling, 22*, 1700–1707.

Valverde-Villegas, J. M., Matte, M. C., de Medeiros, R. M., & Chies, J. A. (2015). New insights about Treg and Th17 cells in HIV infection and disease progression. *Journal of Immunology Research, 2015*, 647916.

Venkatapurapu, S. P., Kelley, J. B., Dixit, G., Pena, M., Errede, B., Dohlman, H. G., et al. (2015). Modulation of receptor dynamics by the regulator of G protein signaling Sst2. *Molecular Biology of the Cell, 26*, 4124–4134.

Wang, J. H., New, J. S., Xie, S., Yang, P., Wu, Q., Li, J., et al. (2013). Extension of the germinal center stage of B cell development promotes autoantibodies in BXD2 mice. *Arthritis and Rheumatism, 65*, 2703–2712.

Wang, Y., Xu, X., Pan, M., & Jin, T. (2016). ELMO1 directly interacts with Gbetagamma subunit to transduce GPCR signaling to Rac1 activation in chemotaxis. *Journal of Cancer, 7*, 973–983.

Warnock, R. A., Askari, S., Butcher, E. C., & von Andrian, U. H. (1998). Molecular mechanisms of lymphocyte homing to peripheral lymph nodes. *The Journal of Experimental Medicine, 187*, 205–216.

Watson, N., Linder, M. E., Druey, K. M., Kehrl, J. H., & Blumer, K. J. (1996). RGS family members: GTPase-activating proteins for heterotrimeric G-protein alpha-subunits. *Nature, 383*, 172–175.

Williams, J. W., Yau, D., Sethakorn, N., Kach, J., Reed, E. B., Moore, T. V., et al. (2013). RGS3 controls T lymphocyte migration in a model of Th2-mediated airway inflammation. *American Journal of Physiology. Lung Cellular and Molecular Physiology, 305*, L693–L701.

Woodard, G. E., Jardin, I., Berna-Erro, A., Salido, G. M., & Rosado, J. A. (2015). Regulators of G-protein-signaling proteins: Negative modulators of G-protein-coupled receptor signaling. *International Review of Cell and Molecular Biology, 317*, 97–183.

Xie, Z., Chan, E. C., & Druey, K. M. (2016). R4 regulator of G protein signaling (RGS) proteins in inflammation and immunity. *The AAPS Journal, 18*, 294–304.

Xie, Z., Geiger, T. R., Johnson, E. N., Nyborg, J. K., & Druey, K. M. (2008). RGS13 acts as a nuclear repressor of CREB. *Molecular Cell, 31*, 660–670.

Xu, X., & Jin, T. (2015). The novel functions of the PLC/PKC/PKD signaling axis in G protein-coupled receptor-mediated chemotaxis of neutrophils. *Journal of Immunology Research, 2015*, 817604.

Xu, J., Wang, F., Van Keymeulen, A., Herzmark, P., Straight, A., Kelly, K., et al. (2003). Divergent signals and cytoskeletal assemblies regulate self-organizing polarity in neutrophils. *Cell, 114*, 201–214.

Yu, M., Cavero, V., Lu, Q., & Li, H. (2015). Follicular helper T cells in rheumatoid arthritis. *Clinical Rheumatology, 34*, 1489–1493.

Zeng, W., Xu, X., Popov, S., Mukhopadhyay, S., Chidiac, P., Swistok, J., et al. (1998). The N-terminal domain of RGS4 confers receptor-selective inhibition of G protein signaling. *The Journal of Biological Chemistry, 273*, 34687–34690.

Zhang, Y., Zhang, Y., Gu, W., & Sun, B. (2014). TH1/TH2 cell differentiation and molecular signals. *Advances in Experimental Medicine and Biology, 841*, 15–44.

Zhao, P., & Chidiac, P. (2015). Regulation of RGS5 GAP activity by GPSM3. *Molecular and Cellular Biochemistry, 405*, 33–40.

Zhao, T., Nalbant, P., Hoshino, M., Dong, X., Wu, D., & Bokoch, G. M. (2007). Signaling requirements for translocation of P-Rex1, a key Rac2 exchange factor involved in chemoattractant-stimulated human neutrophil function. *Journal of Leukocyte Biology, 81*, 1127–1136.

Zhong, H., Wade, S. M., Woolf, P. J., Linderman, J. J., Traynor, J. R., & Neubig, R. R. (2003). A spatial focusing model for G protein signals. Regulator of G protein signaling (RGS) protien-mediated kinetic scaffolding. *The Journal of Biological Chemistry, 278*, 7278–7284.

Zhou, X. J., Nath, S. K., Qi, Y. Y., Sun, C., Hou, P., Zhang, Y. M., et al. (2016). Novel identified associations of RGS1 and RASGRP1 variants in IgA nephropathy. *Scientific Reports, 6*, 35781.

CHAPTER TEN

# Roles of NHERF Family of PDZ-Binding Proteins in Regulating GPCR Functions

**David Broadbent, Mohammad M. Ahmadzai, Ananth K. Kammala, Canchai Yang, Christopher Occhiuto, Rupali Das, Hariharan Subramanian[1]**

Michigan State University, East Lansing, MI, United States

[1]Corresponding author: e-mail address: subram46@psl.msu.edu

## Contents

1. Introduction 354
2. PDZ Proteins 355
   2.1 Structure of PDZ Domains 355
   2.2 PDZ Domain Classification 356
3. NHERF Family of PDZ-Binding Proteins 356
   3.1 NHERF Isoforms 356
   3.2 NHERFs Are Class I PDZ Proteins 357
4. NHERFs Regulate Phosphate Reabsorption 358
   4.1 PTHR Activation Inhibits Npt2a Inorganic Phosphate (Pi) Transport 358
   4.2 Pathogenesis of Pi Dysregulation by PTHR 358
   4.3 NHERF1 Is Required for Apical Localization of Npt2a 359
   4.4 NHERF1/2 Are Essential for PTHR Inhibition of Npt2a 359
   4.5 Phosphorylation of NHERF1 at Ser77 Results in Npt2a Endocytosis and Degradation 360
5. NHERF-Mediated Regulation of Ion Transport and Disease 361
   5.1 Overview 361
   5.2 Cystic Fibrosis and CFTR 362
   5.3 NHERFs Regulate CFTR Localization and Function via Direct Interactions 363
   5.4 NHERF2/3 Dimerization Regulates NHE3 Function 365
   5.5 Several GPCRs Indirectly Regulate CFTR Functions via NHERFs 366
   5.6 β2-AR–NHERF1 Interactions Regulate CFTR Functions 366
   5.7 Purinergic Receptor P2Y1R–NHERF Interactions Regulate CTFR Functions 367
   5.8 NHERF2 Interacts With the Lysophosphatidic Acid Receptor LPA2R and PLC-β3 368
6. NHERFs and Cancer 369
   6.1 Overview 369
   6.2 NHERF1 Regulates β-Catenin/E-Cadherin Epithelial-to-Mesenchymal Transition 370

*Advances in Immunology*, Volume 136
ISSN 0065-2776
http://dx.doi.org/10.1016/bs.ai.2017.05.008

6.3 NHERF1 Is Necessary for PTEN Tumor Suppressor Function 372
6.4 Cellular Transition From Membrane NHERF1 to Cytosolic/Nuclear NHERF1 373
7. NHERFs in Immunity and Inflammation 373
7.1 Overview 373
7.2 NHERF1 and NHERF2 Promote Mast Cell Responses 374
7.3 NHERF1 Inhibits T-Cell Signaling 374
8. Clinical Relevance and Future Perspectives 377
Acknowledgments 377
References 378

## Abstract

Multicellular organisms are equipped with an array of G-protein-coupled receptors (GPCRs) that mediate cell–cell signaling allowing them to adapt to environmental cues and ultimately survive. This is mechanistically possible through complex intracellular GPCR machinery that encompasses a vast network of proteins. Within this network, there is a group called scaffolding proteins that facilitate proper localization of signaling proteins for a quick and robust GPCR response. One protein family within this scaffolding group is the PSD-95/Dlg/ZO-1 (PDZ) family which is important for GPCR localization, internalization, recycling, and downstream signaling. Although the PDZ family of proteins regulate the functions of several receptors, this chapter focuses on a subfamily within the PDZ protein family called the $Na^+/H^+$ exchanger regulatory factors (NHERFs). Here we extensively review the predominantly characterized roles of NHERFs in renal phosphate absorption, intestinal ion regulation, cancer progression, and immune cell functions. Finally, we discuss the future perspectives and possible clinical application of targeting NHERFs in several disorders.

## 1. INTRODUCTION

A hallmark of multicellular organisms is the ability of individual cells to coordinate together for the benefit of the whole organism. One way that cells communicate with each other is through secreting peptide ligands that bind to cell surface receptors on target cells and initiate intracellular responses. Secreting cells control the "signal" strength by regulating the amount of excreted ligand while target cells modulate the sensitivity and abundance of target receptors. G-protein-coupled receptors (GPCRs), in particular, are important ubiquitous receptors involved in many aspects of human physiology. Organisms rely on GPCR signaling for quick physiological responses and the effects can often be seen within minutes of stimulation. Classical GPCR signaling acts through Gα and Gβγ coupled subunits, ultimately resulting in phosphorylation of several signaling proteins and activation of transcription factors that regulate cell migration, proliferation, and growth throughout the human body.

Due to their vast physiological functions, GPCRs are excellent pharmacologic targets. In fact, 50%–60% of drugs target GPCRs and an ongoing area of research focuses on developing new strategies to target them (Lundstrom, 2009). In the past few decades, scaffolding proteins such as the PSD-95/Dlg/ZO-1 (PDZ) protein family have been shown to regulate GPCR function and an astounding 20% of GPCRs are predicted to interact with this protein family (Dunn & Ferguson, 2015; Lee & Zheng, 2010). For this reason, understanding PDZ proteins has become an important step toward targeting GPCRs for the ultimate treatment of patients.

In this chapter, it is not possible to provide an in-depth understanding of all PDZ proteins. Instead, we will review a subfamily within the PDZ protein family called the $Na^+/H^+$ exchanger regulatory factors (NHERFs). These are model PDZ proteins and a deeper understanding of how NHERF proteins modulate GPCR activity may provide valuable mechanistic insight that can be generalized to other members of the PDZ protein family as a whole. Readers interested in other PDZ domain binding proteins are directed to these excellent review articles for more information (Dunn & Ferguson, 2015; Kim & Sheng, 2004; Lee & Zheng, 2010; Sheng & Sala, 2001; Songyang et al., 1997).

We will consider NHERFs regulation of a variety of GPCRs including, but not limited to: parathyroid hormone receptor type 1 (PTHR), β2-adrenergic receptor (β2-AR), P2Y1 purinergic receptors (P2Y1Rs), the lysophosphatidic acid receptors (LPARs), complement anaphylatoxin C3a receptor (C3aR) and highlight the physiological implications of these interactions. We will also review well-characterized NHERF regulation of physiological mechanisms that are independent of GPCRs which are important to consider for pharmacological targeting of NHERF proteins. Taken together, this chapter will focus on NHERF-mediated regulation of ion homeostasis in the kidneys and intestine, cancer progression, inflammation, and immunity.

## 2. PDZ PROTEINS

### 2.1 Structure of PDZ Domains

PDZ proteins contain PDZ domains. PDZ domains are conserved ~90 amino acid stretches consisting of six β-sheets, sandwiched by two α-helices. They are named after the postsynaptic density protein

(PSD-95), discs-large tumor suppressor (Dlg), and tight junction protein-1 (ZO-1) in which they were originally found (Songyang et al., 1997). Since their discovery, ~250 identified PDZ domains in over 100 proteins have now been recognized (Tonikian et al., 2008). The sheer number of these domains suggests that they regulate a diverse number of receptor families, and therefore impact a broad range of functions throughout the human body.

PDZ domains possess a central globular binding groove that lies between the α2-helix and β2-sheet. This region features conserved several amino acid residues collectively referred to as the "carboxylate-binding loop," characterized by the following sequence: R/L-X-X-X-G-L-G-F. As the name suggests, the carboxylate-binding loop allows PDZ domains to interact with the carboxyl tails of other proteins. Additionally, PDZ domains can interact with internal motifs that form β-sheet "finger"-like projections that mimic carboxyl tail chemistry and insert similarly into the binding groove (Harris & Lim, 2001; Hillier, Christopherson, Prehoda, Bredt, & Lim, 1999).

### 2.2 PDZ Domain Classification

Among the canonical PDZ domains, there is significant sequence variability and this is how PDZ proteins have a diverse interactome (Sheng & Sala, 2001). For this reason, it is easier to classify PDZ proteins according to the carboxyl tail-binding motifs with which they interact. Generally, these PDZ-binding motifs can be split into three classes. Class I domains recognize the carboxyl tail motif X-S/T-X-Φ, where X represents any amino acid and Φ indicates any hydrophobic amino acid. Class II and class III domains recognize X-Φ-X- Φ and X-D/E-X-Φ, respectively (Songyang et al., 1997).

## 3. NHERF FAMILY OF PDZ-BINDING PROTEINS

### 3.1 NHERF Isoforms

The NHERF family consists of four isoforms encoded by several genes throughout the human genome. Each NHERF isoform was originally named with relation to how it was discovered. Their names and genomic locations are summarized in Table 1. Presently, the rationale for grouping these proteins into an NHERF family is based on their shared role in regulating the sodium–hydrogen exchanger 3 (NHE3) channel. Additionally, genomic sequence blast alignment identified NHERF1–4 as containing sequence homology. It also placed NHERF1/NHERF2 and

**Table 1** Reference Table for NHERF Family Proteins

| Isoform | Gene Name | Chromosomal Location | Common Names |
|---|---|---|---|
| NHERF1 | *SLC9A3R1* | 17q25.1 | EBP50 |
| NHERF2 | *SLC9A3R2* | 16p13.3 | E3KARP; TKA-1; SIP-1 |
| NHERF3 | *PDZK1* | 1q21.1 | CAP70; CLAMP; Napi-Cap-1; PDZD1 |
| NHERF4 | *PDZD3* | 11q23.3 | IKEPP; PDZK2; Napi-Cap-2 |

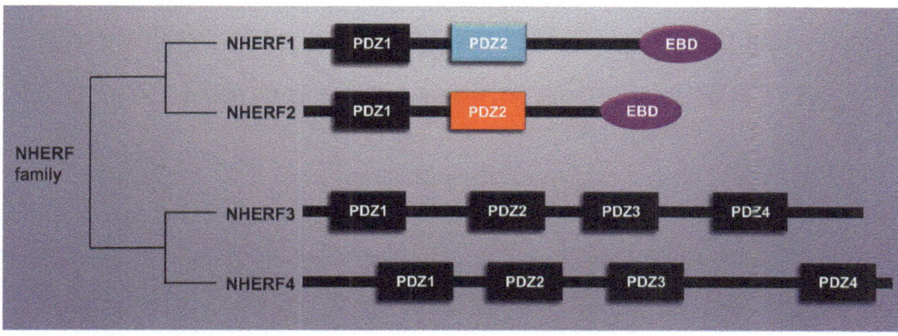


**Fig. 1** Schematic representation of the human NHERF family of PDZ proteins. This figure depicts the quantity and location of PDZ domains contained within the NHERF proteins. The presence of ezrin-binding domains (EBD) is included as well. It also highlights the evolutionary relationship among the NHERF family proteins.

NHERF3/NHERF4 in further subgroups of relatedness (Thelin, Hodson, & Milgram, 2005). NHERF1 and NHERF2 share 44% sequence homology with tandem PDZ domains and an ezrin-binding domain (EBD) (Ardura & Friedman, 2011). In contrast, NHERF3 and NHERF4 both contain four PDZ domains, but only the third PDZ domain (PDZ3) between them is related in structure and function (Thelin et al., 2005). A schematic describing the location, quantity, familial relationships, and relevant domains is provided in Fig. 1.

## 3.2 NHERFs Are Class I PDZ Proteins

NHERF1 and NHERF2 are generally considered class I PDZ proteins. However, additional mutagenesis analyses demonstrate that NHERF1 PDZ1 deviates slightly from canonical class I binding and prefers to bind to an aspartic acid in -4 position, D-S/T-X-L (Hall et al., 1998). While mutagenesis analyses have not been performed on NHERF2, it shares

~70% homology with the NHERF1 PDZ domains and hence it most likely binds to the same motifs as NHERF1. Due to their structural similarities, NHERF1/2 share many of the same interacting partners. Nevertheless, both NHERF1/2 can subserve distinct functions. NHERF3 has been shown to interact with class I PDZ motifs, but most of its PDZ domain interactions remain largely uncharacterized (Walther, Caetano, Dunn, & Ferguson, 2015). NHERF4 has the most divergent PDZ domains in regards to structure and function with exception of PDZ3 which binds to class I receptors (Lee et al., 2012; Thelin et al., 2005).

# 4. NHERFs REGULATE PHOSPHATE REABSORPTION

## 4.1 PTHR Activation Inhibits Npt2a Inorganic Phosphate (Pi) Transport

PTHR is a GPCR that is abundantly expressed in bones and kidneys. Activation of PTHR serves to maintain phosphate (Pi) homeostasis by inhibiting Pi reabsorption by epithelial cells of the proximal tubule in the kidney. In this regard, PTHR facilitates excretion of excess Pi in the urine. This effect was first described over 90 years ago by Greenwald and Gross, who showed that parathyroid gland extract increased urinary Pi excretion (Greenwald & Gross, 1925). The mechanism by which stimulation of the PTHR enhanced Pi excretion was unclear until the type II sodium-phosphate cotransporter (Npt2a) was cloned by Murer and colleagues (Magagnin et al., 1993; Murer, Hernando, Forster, & Biber, 2000). Npt2a is the main sodium phosphate cotransporter in the kidney that reabsorbs Pi (Beck et al., 1998; Villa-Bellosta & Sorribas, 2010). Importantly, the parathyroid hormone (PTH)-PTHR system negatively regulates Npt2a in order to inhibit Pi transport, which results in excess Pi secretion in the urine (Hernando, Forgo, Biber, & Murer, 2000; Werner et al., 1991).

## 4.2 Pathogenesis of Pi Dysregulation by PTHR

Dysregulation of Npt2a results in low serum Pi levels (due to increased Pi excretion), leading to impaired bone growth and development. Accordingly, patients with hypoparathyroidism and PTH knockout mice exhibit mild hyperphosphatemia and hypocalcemia (Bai, Miao, Goltzman, & Karaplis, 2007; Christov & Jüppner, 2013). This effect is also observed when Npt2a channels are highly expressed, as is seen in patients with rare mutations in Npt2a transcription factors, resulting in calcinosis (tumor-like deposits in tissue made up of phosphate and calcium precipitates)

(Benet-Pages, Orlik, Strom, & Lorenz-Depiereux, 2005; Sun et al., 2016). Conversely, hyperparathyroidism can result in bone pain and pathological fractures (Bandeira et al., 2014). These symptoms are due to phosphate wasting which is the excretion of high levels of Pi through urine and low Pi serum levels (Beck et al., 1998).

## 4.3 NHERF1 Is Required for Apical Localization of Npt2a

Importantly, phosphate wasting is present in Npt2a-null mice, NHERF1-null mice, and humans harboring NHERF1 mutations (Beck et al., 1998; Karim et al., 2008; Shenolikar, Voltz, Minkoff, Wade, & Weinman, 2002; Wagner, Rubio-Aliaga, Biber, & Hernando, 2014). These studies demonstrate a vital role for NHERF1 in the regulation of Npt2a function and Pi absorption. Both NHERF1/2 and Npt2a localize to the apical membrane of proximal tubule epithelial cells in the kidneys (Wade et al., 2003; Wade, Welling, Donowitz, Shenolikar, & Weinman, 2001). Npt2a has a class I PDZ-binding motif at its intracellular C-terminal tail. This PDZ motif interacts with PDZ1 of NHERF1 and is important for Npt2a localization to the apical membrane (Courbebaisse et al., 2012). This has been shown in vitro using a low-expressing NHERF1 subclone of opossum kidney cells, OKH cells, that concurrently exhibited enhanced cytosolic expression of Npt2a (Mahon, Cole, Lederer, & Segre, 2003). The cytoplasmic localization of Npt2a was also observed in the renal brush-border membrane of NHERF1-null mice (Shenolikar et al., 2002). Interestingly, although NHERF2 binds to Npt2a, it is unable to compensate for the loss of NHERF1 in NHERF1-null mice (Cunningham et al., 2008; Shenolikar et al., 2002). For this reason, only NHERF1 is considered necessary for localization of Npt2a to the apical membrane.

## 4.4 NHERF1/2 Are Essential for PTHR Inhibition of Npt2a

Most channels and receptors in polarized cells will localize to either the basolateral or apical compartments. PTHR, unlike NHERF1/2 or Npt2a, actually resides at both the apical and basolateral membranes in the proximal tubule (Kaufmann et al., 1994). The apical localization of PTHR is dependent on NHERF1/2 ezrin complexes (Wheeler et al., 2008). The spatial segregation of PTHR in polarized cells, such as the proximal tubule cells, suggests differential regulation. Indeed, activation of PTHR at these membranes results in different downstream signals (Traebert, Völkl, Biber, Murer, & Kaissling, 2000). Basolateral PTHR signals through classical protein kinase A (PKA) activation but also activates protein kinase C (PKC), independent of phospholipase β (PLCβ) signaling (Picotto, Massheimer, & Boland, 1997). On the other hand, apical PTHR signals predominantly through PKC, while inhibiting PKA (Kaufmann et al., 1994; Traebert et al., 2000). It is still unclear how PTHR expressed at two different sites in the same cell exhibits different signaling pathways, but a possible explanation for this would involve its association with NHERFs. A current working theory suggests that NHERF proteins bind PTHR only when coupled to specific G-proteins at the apical surface. Wang and coworkers showed this through a series of coimmunoprecipitation experiments, demonstrating that both NHERF1 and 2 interact with Gαq, but only NHERF2 interacts with Gαi which inhibits Gαs signaling (Wang et al., 2010). The physiological relevance of this differential regulation of PTHR remains unknown, but knowing this mechanism does provide insight into how NHERF1/2 may be regulating other GPCRs apart from PTHR. It is likely that apical or basolateral stimulation of PTHR results in a more robust inhibition of Npt2a.

While it is currently unknown whether PKA or PKC is the main pathway in Npt2a inhibition, sufficient evidence demonstrates that both PKC and PKA activation inhibit Npt2a phosphate transport and is dependent on the presence of NHERF1. Proximal tubule cells from NHERF1-null mice were resistant to PTH inhibition or downstream PKC or PKA inhibition of Npt2a (Cunningham et al., 2006). Furthermore, addition of NHERF1 increased basal Pi absorption and restored PTH inhibition of Npt2a, suggesting that NHERF1 regulates the PTH-Npt2a axis for excretion of Pi (Cunningham et al., 2006).

## 4.5 Phosphorylation of NHERF1 at Ser77 Results in Npt2a Endocytosis and Degradation

Under homeostatic conditions, the PDZ1 motif of NHERF1 is associated with Npt2a, tethering it to the apical membrane. Another cytoskeletal protein, ezrin plays a dual purpose by interacting with actin to maintain Npt2a tethering and with PKA through A-kinase anchoring protein (AKAP) (Dransfield et al., 1997; Gisler et al., 2001; Wang et al., 2012). Thus, NHERF1, Npt2a and ezrin form a ternary complex at the plasma membrane that is associated with PTHR signaling. PTHR activation results in NHERF1 phosphorylation at Thr95. This modification of NHERF1 is thought to stabilize Ser77 which attenuates NHERF1 and Npt2a binding (Weinman et al., 2010). It is still unclear how the Ser77 residue of NHERF1 is phosphorylated but it possibly involves both PKC and PKA. NHERF1 is

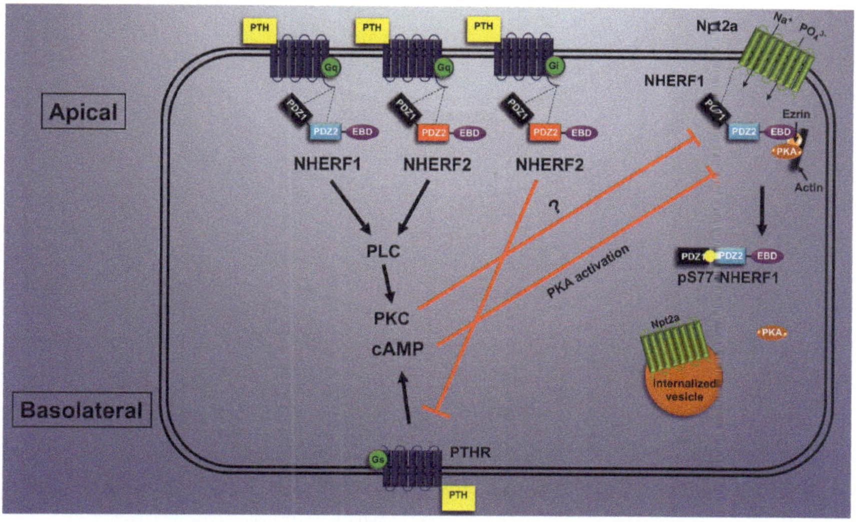


**Fig. 2** NHERF-mediated regulation of Npt2a by PTHR. NHERF1 stabilizes Npt2a at the apical membrane. PTH activation of PTHR at the basolateral membrane results in Gαs signaling, activating PKA through increasing cAMP levels. Phosphorylation of NHERF1 through PKA destabilizes the Npt2a-NHERF1 complex, initiating Npt2a internalization. PTH activation of PTHR at the apical membrane results in Gαq or Gαq and Gαi signaling depending on whether PTHR is bound by NHERF1 or NHERF2, respectively. Gαq signaling activates PKC and through an unknown mechanism inhibits Npt2a by initiating its internalization. Gαi signaling inhibits basolateral PTH-activated PTHR, downregulating PKA. By this mechanism, PTHR fine tunes downregulation of Npt2a.

subsequently phosphorylated at Ser77, initiating the dissociation of the ternary complex; Npt2a is subsequently endocytosed and degraded through a lysosomal pathway (Murer & Biber, 2010; Wang et al., 2012; Weinman et al., 2007). In summary, PTH activation increases PKA and PKC activation, depending on the expression and presence of NHERF proteins. NHERF1 is phosphorylated and dissociates from the ternary complex (Fig. 2). Npt2a is subsequently endocytosed via clathrin-coated pits and is degraded by lysosomes. Npt2a is not recycled back, and must be translated de novo for increased Pi transport (Tröhler, Bonjour, & Fleisch, 1976).

# 5. NHERF-MEDIATED REGULATION OF ION TRANSPORT AND DISEASE

## 5.1 Overview

Improper ion transport across epithelial layers in the lung, kidney, and intestine can lead to various diseases. In this section, we provide a brief background into the role of the cystic fibrosis transmembrane conductance regulator (CFTR) as a central protein regulator that interacts with a variety of receptors and channels to regulate ion absorption and secretion in the lung, kidney, and intestine. We then highlight CFTR regulation via NHERFs and consider other well-defined receptors that also modulate CFTR through NHERF proteins.

## 5.2 Cystic Fibrosis and CFTR

Cystic fibrosis (CF) occurs through mutations in CFTR causing improper excretion of $HCO_3^-/Cl^-$ and enhanced $Na^+$ absorption via the NHE3 channel (Borowitz, 2015). CF patients exhibit thickening of mucus in the lung, resulting in decreased mucus clearance. Mucus buildup harbors an environment for bacterial infections. Repeated infections cause permanent bronchial damage, eventually leading to respiratory failure. Patients that live past the age of 40 experience a myriad of symptoms including meconium ileus, male infertility, liver cirrhosis, pancreatitis, and heart failure (Elborn, 2016).

The pathophysiologic basis of CF is thought to involve poor mucociliary clearance secondary to dysregulation of the airway surface liquid (ASL) (Elborn, 2016). A commonly accepted theory for ineffective mucociliary clearance is defined by the absence of $Cl^-$ excretion with associated $Na^+$ hyperabsoption, resulting in an osmotic imbalance between lung tissue and ASL. Water readily diffuses through lung tissue and ASL volume decreases until the osmotic gradient equilibrates. Therefore, the lack of proper hydration is thought to cause increased mucus viscosity and reduced clearance (Boucher, 2007).

In addition, low ASL pH was found to exacerbate these symptoms by inactivating antimicrobial proteins such as lysozyme and lactoferrin (Pezzulo et al., 2012). This suggested that CF patients also suffered from reduced $HCO_3^-$ excretion. Results from CFTR knockout studies hinted that CFTR would be a likely candidate for $HCO_3^-$ transport and studies have shown that $HCO_3^-$ excretion is heavily dependent on CFTR (Kim, Liao, & Hanrahan, 2014). Interestingly, others have also suggested that $HCO_3^-$ excretion is dependent on a CFTR-dependent $HCO_3^-/Cl^-$ exchange transporter, pendrin (Garnett et al., 2011). Elsewhere, it has been shown that intestinal bicarbonate excretion occurs mainly through CFTR- coupled $HCO_3^-/Cl^-$ exchangers DRA (Slc26a3) and PAT1 (Slc26a6), suggesting that CFTR may often regulate bicarbonate excretion through channel coupling (Field, 1993; Keely & Barrett, 2000). Experiments with CFTR coupling to DRA and PAT1 suggest reciprocal activation, where expression of DRA or PAT1 will stimulate CFTR activity and cAMP activation of CFTR increases DRA and PAT1 activity (Ko et al., 2002, 2004; Lee, Choi, et al., 1999; Lee, Wigley, et al., 1999). These studies highlight the importance of both CFTR and coupled bicarbonate channels in regulating $HCO_3^-$ transport across epithelium.

CFTR also modulates intestinal salt and water absorption through regulation of NHE3. NHE3-null mice and patients with mutations in DRA have poor water absorption and severe diarrhea (Kere, Lohi, & Höglund, 1999; Schultheis et al., 1998). Similar symptoms are exhibited by overactive CFTR disorders caused by *Vibrio cholera* and *Clostridium difficile* infection (Thiagarajah & Verkman, 2003). Conversely, increased water absorption in patients with CFTR loss-of-function mutations suffers from intestinal impaction and constipation (van der Doef et al., 2010). Additionally, CFTR regulates duodenal $HCO_3^-$ secretion. Thus, it is clear that cellular CFTR effects are varied and this is possible through its interaction with adapter proteins such as NHERFs.

## 5.3 NHERFs Regulate CFTR Localization and Function via Direct Interactions

Several studies have shown that NHERF proteins play an integral role in CFTR regulation. Moyer et al. showed that mutations in the C-terminal CFTR PDZ-binding motif resulted in decreased CFTR localization in polarized epithelial cells (Moyer et al., 1999). Another group, using a similar mutant and a pulse-chase method, determined that both WT and the truncated mutant were able to traffic to the membrane, but had a shorter half-life of 13 h instead of 24 (Swiatecka-Urban et al., 2002). More recent studies confirm that this C-terminal motif binds to NHERF1 and this binding is required for membrane localization of CFTR (Sharma et al., 2016). In addition to the PDZ motif in the C-terminal domain, another regulatory region exists in CFTR that, when mutated, increased NHERF1–CFTR binding and apical localization of the receptor (Sharma et al., 2016).

In addition to CFTR localization, NHERF1 has been shown to modulate CFTR activity. Lamprecht et al. provide a clear analysis of CFTR gating. Classical CFTR channel gating consists of three main steps: (1) phosphorylation of regulatory domain, (2) ATP binding to nucleotide domain where it is hydrolyzed providing energy for channel function, and (3) regulatory interactions between CFTR and other proteins (Lamprecht & Seidler, 2006). NHERFs play a role in steps 1 and 3. In step 1, NHERF1/2 facilitate CFTR phosphorylation by acting as scaffolds bringing activated PKA in proximity with the regulatory domain of CFTR (Lamprecht & Seidler, 2006; Naren et al., 2003). The rational for this model is supported by NHERF1/2 interacting indirectly with PKA via ezrin binding. The role of NHERFs in step 3 is more complex, but they essentially promote protein interactions with CFTR. Evidence has shown that CFTR homodimerization increases the open probability of the CFTR channel and that NHERF1/3 promotes CFTR oligomerization (Li, Dai, Jana, Callaway, & Bu, 2005; Raghuram, Hormuth, & Foskett, 2003; Wang, Yue, Derin, Guggino, & Li, 2000). The physiological consequence of NHERF1-mediated CFTR homodimerization was brought into question by another group that showed that only about 2% of NHERF1 bound to CFTR, suggesting that other PDZ-binding proteins might also be involved (Li, Roy, Dandridge, & Naren, 2004). NHERF1/2 have also been shown to stabilize CFTR coupling of DRA and PAT1 (Ko et al., 2002, 2004). These mechanisms are not perfectly understood and it is likely that NHERFs regulate CFTR function through more proteins than we have listed here. It is also likely that NHERF dimerization affects CFTR oligomerization and channel coupling.

Despite these studies demonstrating an important role for NHERF1 in CFTR regulation, the physiological relevance of the NHERF1-CFTR can be questioned. Clinical studies characterizing rare PDZ-motif altering, CFTR mutations led to surprising results. One patient's genotype was F508del/Q1476X, F508del being the most abundant mutation for CFTR, while Q1476X mutation results in a termination codon exactly before the PDZ-binding motif (Audrézet et al., 2002). The other patient had a truncation mutant S1455X, which left off the last 26 amino acids (Mickle et al., 1998). Surprisingly, both patients exhibited normal lung function, yet elevated sweat chloride concentrations. While the elevated sweat chloride test is consistent in CF patients, the severe pathology for CF is in the lung and these patients were asymptomatic. This suggests that while NHERF1 does play a role in CFTR regulation, altered PDZ-binding motifs within CFTR do retain at least partial function in the lungs of these patients.

## 5.4 NHERF2/3 Dimerization Regulates NHE3 Function

NHERF proteins form heterodimers and these associations are required for regulation of ion transport. In vivo studies using NHERF2 or NHERF3-null mice have shown that both NHERF2 and NHERF3 must be present for $Ca^{2+}$-induced inhibition of NHE3 (Cinar et al., 2007; Murtazina et al., 2011; Zachos et al., 2006). The same phenotype was reproduced in caco-2 cells where either NHERF2 or NHERF3 were deleted using shRNAs (Sarker et al., 2011; Zachos et al., 2009). Using various methods including coimmunoprecipitation and fluorescence resonance energy transfer (FRET) microscopy, Yang and coworkers showed that NHERF2/NHERF3 do dimerize and that this interaction is likely to be the strongest binding dimer formed by the NHERF family of proteins (Yang et al., 2014). NHERF2/NHERF3 interaction is dependent on the C-terminal PDZ-binding motif on NHERF3 (D-T-E-M) that interacts with PDZ1 or PDZ2 domain of NHERF2; PDZ1 is preferred over PDZ2. Interestingly, NHERF3 PDZ1, which increases localization of NHE3 to epithelial membrane, did so independently of NHERF2. However, carbachol (a cholinergic agonist)-mediated inhibition of NHE3 was dependent on NHERF2/NHERF3 dimerization. In this complex, NHERF2 is thought to bridge NHERF3 and NHE3 together (Yang et al., 2014). Carbachol inhibition was abrogated when NHERF3 was knocked down, or when its C-terminal PDZ-binding motif was mutated to alanines, but not when PDZ1 of NHERF3 domain was absent (Yang et al., 2014). This suggests that carbachol inhibition of NHE3 is dependent on NHERF2 binding with NHERF3 and that the PDZ1 domain of NHERF3 likely does not play a role in this process.

These findings suggest that NHERF dimerization is important for the regulation of NHE3 activity. The importance of this association is significant: NHERF2, which normally has two PDZ domains available for interaction with other proteins, has five available when complexed with NHERF3 allowing for more complex signaling (Yang et al., 2014). Many experiments characterizing NHERF interactions often only used a single NHERF family member in the study. Most of the in vitro studies characterizing the role of NHERFs used PS120, OK, and caco-2 cells that do not necessarily express all of the NHERF proteins, and hence many physiological functions of the receptors they regulate could have been missed. As such, NHERF dimerization provides a mechanism where these adapter molecules will form macromolecular complexes to regulate the activity of receptors and/or channels.

## 5.5 Several GPCRs Indirectly Regulate CFTR Functions via NHERFs

GPCR activation can regulate CFTR functions as well. In this section, we will discuss how the scaffolding functions of NHERF1 with several GPCRs may regulate CFTR activity. In particular, we will review findings from several groups that have reported that NHERFs can interact with the β2-AR, P2Y1, and LPA2R to possibly regulate CFTR functions.

## 5.6 β2-AR–NHERF1 Interactions Regulate CFTR Functions

In vivo activation of the GPCR β2-AR stimulated PKA-mediated CFTR $Cl^-$ excretion in airway epithelium (Walker et al., 1997). Two groups have also demonstrated that β2-AR complexes with CFTR through NHERF1 (Naren et al., 2003; Taouil et al., 2003). Naren and coworkers demonstrated that deletion of CFTR PDZ-binding motif (where NHERF1 associates with CFTR) in the lung tissue caused physical and functional dissociation of β2-AR from CFTR. Interestingly, the complex exists at basal conditions and PKA-mediated phosphorylation of CFTR inhibited complex formation. Thus, complex formation is inversely proportional to CFTR phosphorylation (Naren et al., 2003). Taken together, these studies suggest a model where complex formation between β2-AR and CFTR contributes a basal $HCO_3^-/Cl^-$ secretion rate and β2-AR activation upregulates $HCO_3^-/Cl^-$ secretion through further phosphorylation of CFTR. An in vivo study further demonstrated a role for the β2-AR–NHERF1–CFTR complex in mice. Specifically, NHERF1-null mice did not secrete duodenal $HCO_3^-$ in response to the β2-AR agonist, clenbuterol, demonstrating a physiologically relevant role of β2-AR–NHERF1–CFTR complex formation and ion regulation in the intestinal epithelia (Singh et al., 2009).

β2-AR–NHERF1–CFTR complex also regulates the activity of NHE3. Several groups have demonstrated that cAMP-mediated inhibition of NHE3 changes if CFTR is coexpressed (Clarke & Harline, 1996; Lamprecht & Seidler, 2006; Yun et al., 1997). When NHE3 is solely expressed in cells, NHERF1/2 bind to the C-terminal PDZ-binding motif of NHE3, scaffolding a NHE3–ezrin–PKA complex (Yun et al., 1997). When PKA is activated, NHE3 is phosphorylated on S552 and S605 and NHE3 inhibited. On the other hand, when NHE3 is coexpressed with CFTR, NHE3 is activated. Mechanistically, it is plausible that agonist-activated β2-AR recruits NHERF1/2 from NHE3 to the CFTR–β2-AR complex. Dissociation of NHERF1/2 and NHE3 likely attenuates phosphorylation of S552/PS605 due to uncoupling with PKA, initiating $Na^+/H^+$ transport. Interestingly, recruitment of NHERF1 to β2-AR is excitatory, while NHERF2 inhibits CFTR channel transport (Singh et al., 2009). The implications of differential regulation of NHE3 by CFTR have significant physiological effects. Such a model allows a signaling switch from an absorbent cell (NHE3 alone) to a secretory cell (CFTR/NHE3) (Lamprecht & Seidler, 2006). These findings are summarized in Figs. 3 and 4.

## 5.7 Purinergic Receptor P2Y1R–NHERF Interactions Regulate CTFR Functions

Purinergic receptors (P2Y) are GPCRs that are activated by phospho nucleotide ligands such as ATP or UTP. P2Y1R specifically recognizes ADP and ATP. P2Y receptors are expressed widely in tissues, including the brain, heart, and kidney. P2Y1R contains a class I NHERF1 PDZ-binding motif D-T-S-L, which is able to bind to both NHERF1/2 (Fam et al , 2005; Hall et al., 1998). P2Y activation by ATP results in Gαq signaling through PLCβ (Ralevic & Burnstock, 1998). In comparison to β2-AR, NHERF1/2 do not interact robustly with P2Y1R (Hall et al., 1998). The PDZ1 domain of NHERF1 and both the PDZ1 and PDZ2 domains of NHERF2 have

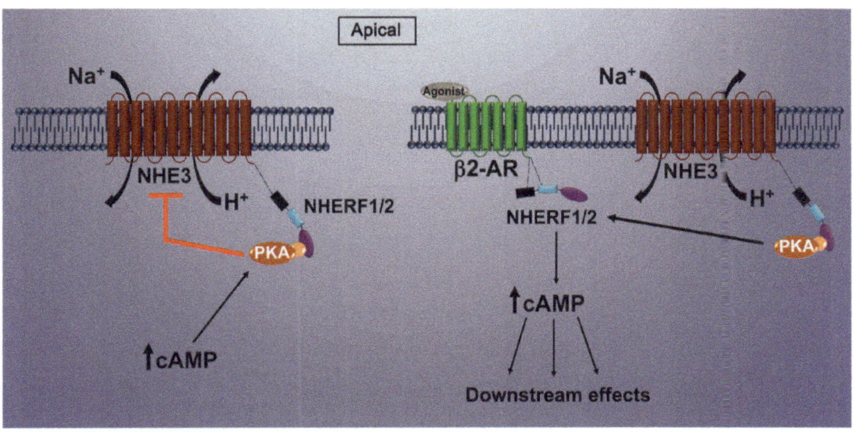


**Fig. 3** β2-AR stimulation attenuates cAMP inhibition of NHE3. *Left*: Classical cAMP inhibition of NHE3. NHERF1/2 interact with the PDZ motif on the cytoplasmic tail of NHE3. Increased local levels of cAMP activate PKA which subsequently inhibits NHE3 through a phosphorylation-dependent mechanism. *Right*: Agonist activation of β2-AR recruits NHERF1/2 and initiates cAMP-dependent signaling. β2-AR recruitment of NHERF1/2 results in an active NHE3 channel despite increased cellular levels of cAMP.

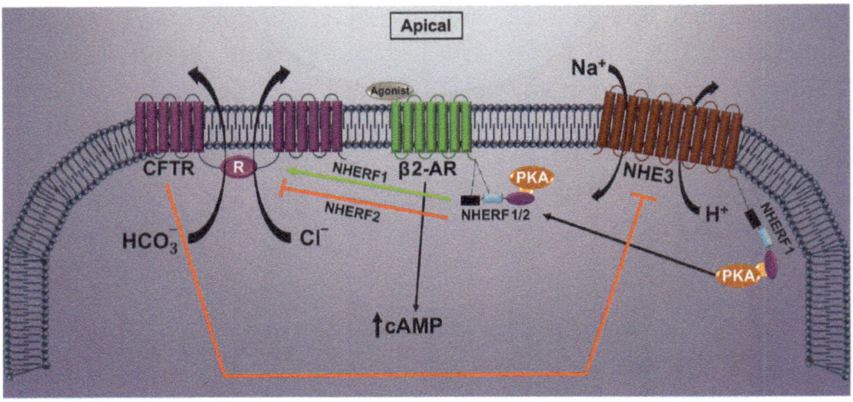


**Fig. 4** Stimulation of β2-AR regulates CFTR functions. β2-AR agonist stimulation recruits NHERF1/2 from NHE3 and increases local levels of cAMP. Recruitment of NHERF1 facilitates PKA-mediated activation of CFTR. CFTR activation increases anion secretion and inhibits NHE3 $Na^+$ absorption. This is essentially a secretory mechanism resulting in a net increase of ions in the lumen, reducing intestinal $H_2O$ absorption. Conversely, NHERF2 recruitment inhibits CFTR through an uncharacterized mechanism. CFTR down-regulation decreases anion secretion while increasing $Na^+$ absorption. In this manner, β2-AR differentially regulates the intestinal epithelia toward a secretory or absorptive phenotype.

been shown to interact with P2Y1R (Fam et al., 2005; Guerra et al., 2004; Hall et al., 1998).

Stimulation of *Xenopus* A6 monolayers with P2Y1R agonist MeSADP (a modified form of ADP) increased intracellular $Ca^{2+}$, cAMP signaling, and $Cl^-$ secretion (Guerra et al., 2004). An interesting result occurred when MeSADP was added to the basolateral compartment and glibenclamide, a cAMP-activated $Cl^-$ channel blocker, was added apically. The result showed abrogation of $Cl^-$ apical efflux from the P2Y1R basolateral signal. Further work identified that CFTR was activated by PKA and agonist induced P2Y1R was unable to activate CFTR in the presence of a NHERF2-PDZ1 deletion mutant (Guerra et al., 2004). Such a mechanism provides a system where apical secretion or absorption can be controlled by basolateral environmental cues in polarized cells.

## 5.8 NHERF2 Interacts With the Lysophosphatidic Acid Receptor LPA2R and PLC-β3

Lysophosphatidic acid (LPA) is found in serum and is released by a variety of mammalian cells including platelets, adipocytes, fibroblasts, and some cancer cells (Pages, Simon, Valet, & Saulnier-Blache, 2001). It binds to and stimulates a group of GPCRs known as LPARs. LPARs are coupled to G-proteins: Gαq and Gαi. As stated earlier, Gαq and Gαi activate PLC-β and inhibit cAMP signaling, respectively. Through a series of coimmunoprecipitation studies, NHERF2 PDZ2 was found to bridge PLC-β3 to LPA2R; other PLC-β and LPA isoforms were not found to bind NHERF2 (Oh et al., 2004). LPA2R interacted with NHERF2 through its PDZ-binding motif D-S-T-L, activating G-protein signaling that led to activation of the downstream serine/threonine protein kinase (AKT) and Ras-Raf-MEK-ERK pathways (Yun et al., 2005).

Another study identified a PDZ-binding motif necessary for NHERF2 binding on PLC-β3, N-T-Q-L (Hwang et al., 2000). Interestingly, deletion of PDZ2 abolished PLC-β3 and LPA2R interaction and their downstream signals (Oh et al., 2004). Oh et al. provided a model where NHERF2 dimerized through PDZ1, leaving two PDZ2 regions available for PLC-β 3 binding. This model does not explain recent results from Holcomb et al. who solved the first crystal structure for LPAR–NHERF2 interactions (Holcomb et al., 2014). Surprisingly, the crystal structure they analyzed was for NHERF2 PDZ1 interaction, instead of NHERF2 PDZ2 association. Coimmunoprecipitation studies, however, suggest that PDZ1 is not necessary for the LPA2R and PLC-β3 interaction (Oh et al., 2004). It is currently unclear whether PDZ1 and PDZ2 domains of NHERF1 bind to the same sites on LPA2R. Other surprising results to consider for the LPA2R–NHERF2 interaction came from another group studying somatostatin receptors (SSTR) who reported a very similar complex between SSTR–NHERF3–PLC-β3. In place of PDZ2 of NHERF2, PDZ1 of NHERF3 was shown to link the GPCR with PLC-β3 (Kim et al., 2012). Though the physiological relevance of the interaction of NHERFs with LPA2R or SSTR is not very clear, it is plausible that like for β2-AR and P2Y1R, the complexes of NHERFs with these GPCRs may regulate CFTR functions.

# 6. NHERFs AND CANCER

## 6.1 Overview

One of the most interesting and currently relevant controversies surrounding NHERF1 is its role in cancer. As the only characterized NHERF to play a role in cancer signaling, NHERF1 has been labeled as both a tumor suppressor and an oncogene, having dual roles in cancer progression. This contradiction was partially resolved by Georgescu et al. who hypothesized that apical membrane interactions of NHERF1 suppress tumor formation while cytosolic interactions facilitate tumor formation (Georgescu, Morales, Molina, & Hayashi, 2008). Since then, it was discovered that NHERF1 also traffics into the nucleus and that this translocation is also associated with a malignant phenotype in breast cancer cells (Du et al., 2016). It is unknown whether NHERF1 is more of a tumor suppressor or a tumor facilitator, but it seems likely that these functions of NHERF1 are context dependent. In ovarian and breast cancer, for example, 50% of tumors are reported to have a loss of NHERF1 heterozygosity, suggesting an important role as a tumor suppressor (Orsetti, Courjal, Cuny, Rodriguez, & Theillet, 1999; Plummer, Adams, Simmons, & Casey, 1997; Presneau et al., 2005). In contrast, others have shown that aggressive breast cancer tumors overexpress NHERF1 (Cardone et al., 2007). In support of the findings of Georgescu et al., Mangia et al. demonstrated that the subcellular location of NHERF1 changed significantly from membrane localization in normal tissues to its cytosolic presence in breast cancer tumors (Mangia et al., 2009). This has also been shown to be the case in colorectal cancer (CRC) (Du et al., 2016; Hayashi, Molina, Hamilton, & Georgescu, 2010). In this section, we consider mechanisms of NHERF1 localization and give a brief overview of their interactions with membrane, cytosolic, and nuclear proteins and their role in cancer progression or suppression.

## 6.2 NHERF1 Regulates β-Catenin/E-Cadherin Epithelial-to-Mesenchymal Transition

An important characteristic of metastatic epithelial derived cancer is its ability to undergo epithelial-to-mesenchymal transition. This is a vital developmental program where cells will initiate antiapoptotic pathways and modulate cell adhesion proteins in order to promote cell motility and proper placement (Hanahan & Weinberg, 2011). Unfortunately, cancer cells essentially "hijack" this program which leads to metastatic cancer and poor patient prognosis (Wu & Zhou, 2008).

One such pathway is the Wnt-β-catenin pathway (Vogelstein & Kinzler, 2004). β-Catenin is a transcription factor that is also involved in adherens junction formation through binding the intracellular domain of E-cadherin. E-cadherin is essential for epithelial cell–cell connections and downregulation of E-cadherin results in increased cell motility (Paul, Bhattacharya, Chatterjee, & Ghosh, 2013; Tian et al., 2011). This is partially due to the decreased physical connections made between epithelial cells and due to increased cytosolic β-catenin signaling which promotes cell motility.

In steady-state conditions, most of the β-catenin in the cytosol is anchored to E-cadherins. Importantly, β-catenin signaling is also kept in check through a variety of proteins and a basal cytosolic destruction complex that primes cytosolic β-catenin for proteasome degradation (Georgescu, Gagea, & Cote, 2016). Cells can reverse this process when a Wnt ligand binds to a Frizzled family receptor (Tian et al., 2011). Around 85% of sporadic CRC deregulate Wnt-β-catenin pathway by various means, all resulting in cytosolic β-catenin stability (Georgescu et al., 2016). This ultimately leads to nuclear translocation of β-catenin where it interacts with T-cell factor (TCF)/lymphoid enhancer factor (LEF) and enhances subsequent transcription of oncogenic growth genes such as c-myc and cyclin D1 (Tian et al., 2011). Both of these proteins promote cell proliferation and migration through multiple mechanisms. This includes direct interactions and through acting as transcription factors, upregulating proteins involved in cancer progression. Reviews for these pathways are cited here (Musgrove, Caldon, Barraclough, Stone, & Sutherland, 2011; Wolfer & Ramaswamy, 2011).

Recent evidence has demonstrated that NHERF1 has a pivotal role in promoting transcription of Wnt signaling genes. At the membrane, NHERF1 interacts with β-catenin and is essential for the E-cadherin/β-catenin complex. By facilitating β-catenin binding of E-cadherin, NHERF1 acts as a tumor suppressor by promoting adherens junction formation and basal reduction of β-catenin-mediated transcription (Kreimann et al., 2007). However, due to mechanisms that are not fully understood, NHERF1 will dissociate from this complex, shifting to the cytosol. β-Catenin is then free to traffic to the nucleus where it interacts with the transcription factor T-cell factor (TCF-1). Thus, the lack of membrane NHERF1 upregulates β-catenin translocation to the nucleus. However, NHERF1 has also been shown to play a role as an oncogene by promoting Wnt signaling gene transcription and enhancing β-catenin activity within the nucleus. Normal tissue downregulates β-catenin-mediated transcription through cleaving TCF-1 at its N-terminal; thus, cleaving the β-catenin binding site, which abolishes the β-catenin–TCF-1 interaction. However, evidence shows that nuclear NHERF1 circumvents this regulation. When cells have undergone an NHERF1 membrane to cytosol transition, NHERF1 translocates into the nucleus where it forms a ternary complex with β-catenin through PDZ2 and the truncated form of TCF-1 through PDZ1 (Lim & Jou, 2016). This complex restores Wnt-mediated gene transcription, promoting proliferation, and cell motility. Interestingly, β-catenin trafficking is independent of NHERF1 (Lin et al., 2012). Also the mechanism regulating TCF truncation is poorly understood, but experiments have shown that this particular truncation is only able to upregulate Wnt transcription targets when NHERF1 is present in the nucleus. Thus, NHERF1 acts as both an oncogene and a tumor suppressor through Wnt-β-catenin signaling (Fig. 5).

## 6.3 NHERF1 Is Necessary for PTEN Tumor Suppressor Function

Phosphatase and tensin homolog (PTEN) is a well-documented tumor suppressor that is often inactivated in many cancers (Li et al., 1997). Its primary role is to suppress AKT signaling through dephosphorylating the upstream regulator phosphatidylinositol 3,4,5-trisphosphate (PIP3) to phosphatidylinositol 4,5-bisphosphate (PIP2) (Maehama & Dixon, 1998). This is in direct opposition to phosphatidylinositol-3-OH kinase (PI3K), which is activated in response to a tyrosine kinase receptor called platelet-derived growth factor receptor (PDGFR).

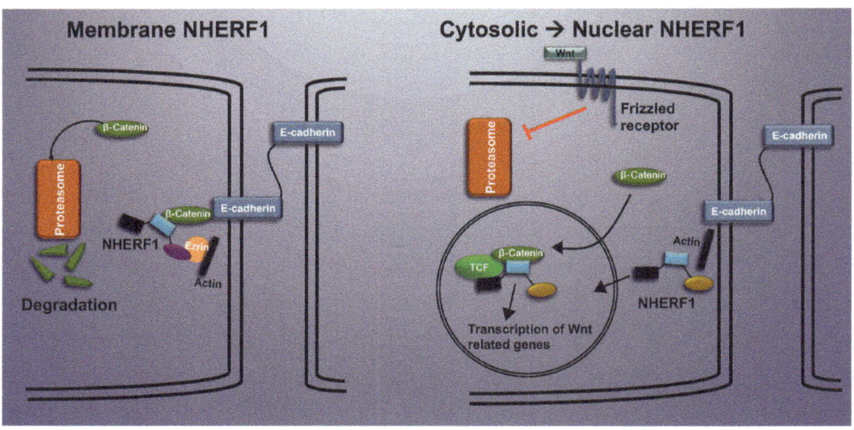


**Fig. 5** Role of NHERF1 in E-cadherin stabilization and Wnt signaling. *Left*: NHERF1 localizes β-catenin at the membrane and promotes E-cadherin stability through NHERF1–β-catenin–E-cadherin complex. At basal conditions, a complex of proteins promotes proteasome degradation of cytosolic β-catenin, eliminating free β-catenin and its role as a transcription factor in the nucleus. *Right*: Through mechanisms not entirely understood, NHERF1-β–catenin–E-cadherin dissociates and NHERF1 and β-catenin traffic independently into the nucleus where they interact to form a complex of transcription factors. NHERF1 mediates binding between a truncated form of T-cell factor (TCF), lacking a β-catenin-binding domain, and β-catenin. This nuclear transcription factor complex promotes cancer progression by transcribing genes downstream of Wnt important for proliferation and survival. This process is amplified by Wnt signaling through a frizzled receptor which inhibits β-catenin degradation, thus enhancing the trafficking of β-catenin into the nucleus.

PDGFR activity correlates with chemotaxis and proliferative diseases such as cancer or atherosclerosis (Heldin & Westermark, 1999). PTEN counteracts these effects when recruited to PDGFR by NHERF1/2. Mechanistically, NHERF1/2 bridge a ternary complex between PDGFR and PTEN through PDZ-binding motifs at their C-terminal end (Pan, Weinman, & Dai, 2008; Takahashi, Morales, Kreimann, & Georgescu, 2006). This effect has been verified in many cancers. NHERF1-null mice have increased AKT signaling in mammary glands compared to wild type controls (Pan et al., 2008). In glioblastoma, PTEN was unable to suppress AKT signaling unless cytosolic NHERF1 was relocated to the membrane (Molina, Morales, Hayashi, Aldape, & Georgescu, 2010). In summary, cells that undergo a NHERF1 membrane-to-cytosolic transition essentially have an inactivated PTEN phenotype. In this manner, membrane localized NHERF1 is a significant tumor suppressor.

### 6.4 Cellular Transition From Membrane NHERF1 to Cytosolic/ Nuclear NHERF1

From the reviewed NHERF1 mechanisms in cancer, understanding how NHERF1 translocation to the cytosol/nucleus occurs is vital for possible treatment options. Even so, the mechanism behind NHERF1 subcellular localization remains poorly understood. Several studies provide insight into this mechanism, but a clear trend has yet to emerge. One study achieved cytosolic/nuclear localization simply by using CRC cell lines at low confluency (Lin et al., 2012). They demonstrated that nonfunctional PDZ1 localized to the membrane. Contradictory results were found by Du et al. who demonstrated that truncation or mutation of PDZ1 in MCF7 breast cancer cells and HEK 293 cells resulted in nuclear NHERF1 localization (Du et al., 2016). A more recent finding also revealed a PDZ2 mutation that resulted in increased nuclear NHERF1 expression (Yang et al., 2017). These last two studies involved point mutations that were identified in cancer cells and as such may have physiological relevance.

## 7. NHERFs IN IMMUNITY AND INFLAMMATION

### 7.1 Overview

The majority of research on NHERF1 is focused on ion transport and cancer progression. However, there are a few research groups that have shown that NHERF1 also plays a role in modulating immune cell functions. The roles of NHERF3 and NHERF4 are, for the most part, uncharacterized, but it is very likely that they may play a role in regulating immune functions and inflammation as well.

## 7.2 NHERF1 and NHERF2 Promote Mast Cell Responses

Mast cells are tissue resident immune cells that cause allergic reactions in humans. The IgE antibody that is produced by activated B cells in response to an allergen or antigen binds to the high-affinity FcεRI receptor expressed on mast cells. When challenged with the same allergen or antigen, the IgE occupied FcεRI receptor on mast cells becomes cross-linked, which leads to their activation. Once activated, mast cells rapidly release preformed mediators such as histamine and several proteases by a process termed as "degranulation" that accelerate the inflammatory/allergic response. Specifically, it has been reported that tryptase, a protease released via mast cell degranulation, can convert the complement component C3 to C3a, which is an anaphylatoxin. C3a can further bind to C3aR on mast cells amplifying the degranulation response in a positive feedback loop (Fig. 6). This process is central to allergic responses and can result in life threatening anaphylaxis (Fukuoka et al., 2008; Schaefer et al., 2013; Ali, 2010). Current medication for IgE-mediated anaphylaxis does not treat its cause, but only reduces symptoms. Therefore, identifying novel strategies that down-regulate this process would be highly significant and NHERF1/2 have been reported to be a potential target.

Depletion of NHERF1 or NHERF2 from mast cells resulted in decreased mast cell degranulation. It also decreased NF-κB transcription and C—C motif chemokine ligand 4 (CCL4) chemokine production (Subramanian, Gupta, & Ali, 2012). C3aR is one of the few GPCRs in mast cells that possess a class I PDZ domain. Classical NHERF1 interactions would suggest that NHERF1 interacts with the carboxyl tail of C3aR, but surprisingly NHERF1 did not associate with the receptor. It is still not clear how NHERF1/2 interactions modulate mast cell function (Subramanian et al., 2012).

## 7.3 NHERF1 Inhibits T-Cell Signaling

T-cells were originally named after the thymus gland in which they mature and are characterized by a specialized T-cell receptor (TCR). T-cell function is critically important for the adaptive and innate immune system and is required for clearance of pathogens or cancerous tissue. Abnormal regulation of T-cells can result in a variety of autoimmune diseases or a compromised immune system. To keep the balance between overactive or

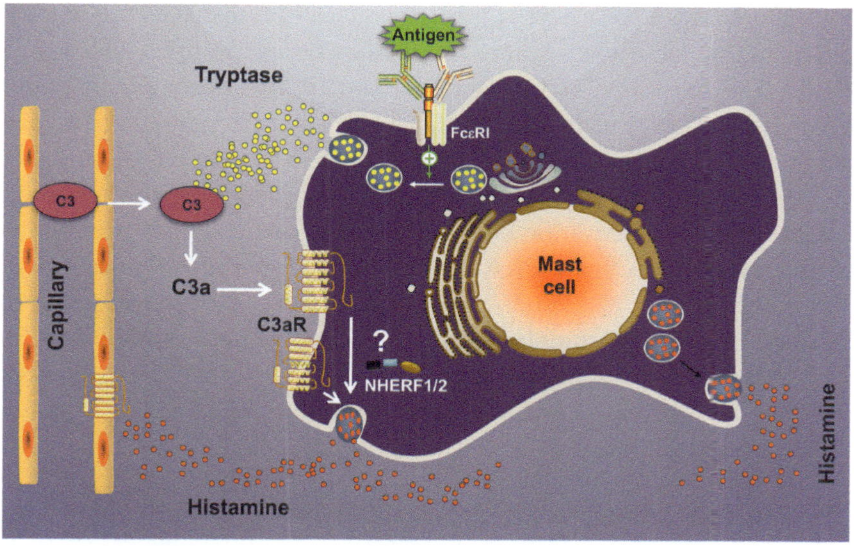


**Fig. 6** NHERF1/2 promote mast cell degranulation. Antigen cross-linking of FcεRI releases tryptase filled granules into the extracellular matrix. The complement component (C3) produced by epithelial cells in the blood vessels is cleaved to become C3a (by tryptase) which initiates a positive feedback loop by binding to its receptor (C3aR) and upregulating degranulation and histamine release (figure adapted from Ali, 2010). Despite the presence of a class I PDZ motif in the C3aR, NHERF1 and NHERF2 do not associate with the receptor, but the downstream signals are upregulated by them. The mechanism through which NHERF1/2 promote mast cell degranulation is yet to be elucidated.

unreactive T-cells, organisms have adapted highly regulated mechanisms of activation and downregulation. These are equally important for maintaining homeostasis within any organism.

T-cell activation occurs when foreign pathogens are presented by antigen-presenting cells (APCs) at the TCR. Activated T-cells will proliferate and begin to release cytokines that promote pathogen clearance. This process is mediated by key proximal signals just downstream of the TCR. Many of these key signaling proteins are recruited to lipid rafts where the immune synapse (IS)—area of interaction for the APC and TCR—will form. Proximal signals that occur after TCR ligation include lymphocyte-specific protein tyrosine kinase (Lck) which phosphorylates intracellular regions of the TCR and recruits zeta-chain-associated protein kinase 70 (Zap-70). Zap-70 then interacts with scaffolding molecules such a linker of activated T-cells (Lat) and SH2 domain containing leukocyte protein of 76 kDa (SLP-76) which facilitate the majority of downstream signals (Mosenden & Tasken, 2011).

Importantly, T-cells have regulatory mechanisms that keep the immune system in check and in vitro biochemical studies have identified NHERF1 as likely playing a role in TCR downregulation. To date, the physiological relevance of NHERF1 in downregulating TCR has yet to be validated in vivo, but the identified mechanism is as follows: NHERF1 negatively regulates TCR activation by facilitating C-terminal Src kinase (Csk) activation. Csk reduces TCR activation by inhibiting Lck. NHERF1 has been shown to activate Csk through a PKA-ezrin-NHERF1 complex (Jarnaess et al., 2008; Ruppelt et al., 2007; Stokka, Mosenden, Ruppelt, Lygren, & Tasken, 2010). This complex brings PKA into close proximity to Csk, which can then be activated when local levels of cAMP are increased (Fig. 7). This increase in cAMP levels can be mediated through classical GPCRs coupled with Gs.

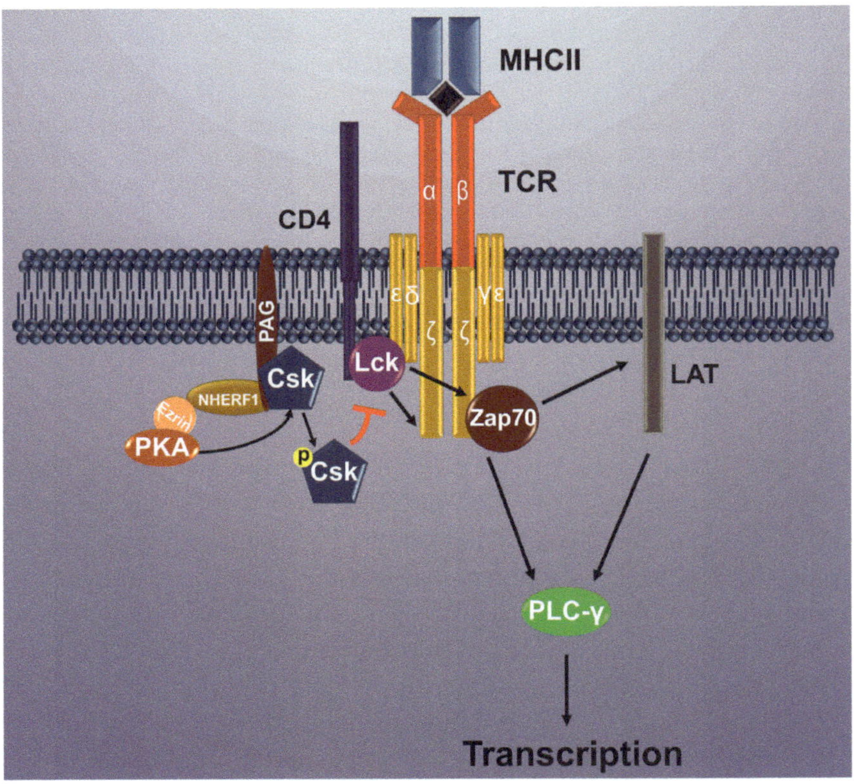


**Fig. 7** NHERF1 negatively regulates T-cell signaling via Csk. Depiction of immune synapse along with key downstream T-cell signaling proteins including LAT, Lck, Zap-70, and PLC-γ. NHERF1 traffics PKA into close proximity to the TCR. Increases in local cAMP as a result of TCR stimulation activate PKA which phosphorylates Csk (p-Csk). p-Csk inhibits Lck, reducing key TCR-induced downstream phosphorylation events.

## 8. CLINICAL RELEVANCE AND FUTURE PERSPECTIVES

In summary, the NHERF family of proteins plays a significant role in renal phosphate absorption, ion absorption/secretion in the kidneys and intestine, cancer progression, and regulating immune cell functions. Therefore, pharmacologically targeting NHERFs is an attractive strategy for alleviating several diseases. For, e.g., modulating NHERF1/2 functions would be relevant in treating secretory diarrhea which is cause by increased $Cl^-$ excretion (an active CFTR) and decreased $Na^+$ uptake (an inhibited NHE3) in the intestine (Li & Naren, 2010). As previously discussed, NHERF1 promotes CFTR activity and NHERF2 is central to the machinery that inhibits NHE3 (Sarker et al., 2011; Seidler et al., 2009). Pharmacologically targeting both would hypothetically alleviate symptoms of secretory diarrhea. It could also reduce anaphylactic shock by reducing mast cell degranulation or promote T-cell function. However, targeting NHERF proteins has obvious challenges.

The NHERF family is ubiquitously expressed throughout the body. While NHERF1/2 knockout mice have served as valuable tools to study NHERF functions, these mice exhibit developmental defects. NHERF1 depletion is especially detrimental, causing early death in female mice. A more effective approach may be to develop a conditional knockout mouse, which would allow differentiation between developmental defects versus functional defects. The functional defects would allow us to predict possible side effects of a drug targeting these ubiquitous proteins. One possible strategy for treatment would be localized treatment, such as targeting only epithelial cells in the gut. This approach might also work for tissue localized diseases such as cancer, but due to the dual role of NHERF1 in cancer, targeting NHERF1 pharmaceutically is complex; an ideal strategy would be to use a drug that disrupts its cytosolic interactions while maintaining its membrane localization pattern. Despite these challenges, NHERF proteins remain promising drug targets for treating various diseases and as our understanding of NHERF interactions continues to expand, so do the possible uses for NHERF inhibitors.

## ACKNOWLEDGMENTS

We thank the H.S. and R.D. laboratory members for exciting discussions and critical review of the book chapter. Research work in the laboratory of H.S. and R.D. is supported by the NIH grants R00HL121073 (H.S.) and K22CA188149 (R.D.). We apologize to any authors whose work might have been omitted unintentionally due to limited space and/or the specific focus of the chapter.

## REFERENCES

Ali, H. (2010). Regulation of human mast cell and basophil function by anaphylatoxins C3a and C5a. *Immunology Letters, 128*, 36–45.

Ardura, J. A., & Friedman, P. A. (2011). Regulation of G protein-coupled receptor function by Na+/H+ exchange regulatory factors. *Pharmacological Reviews, 63*, 882–900.

Audrézet, M. P., Chen, J. M., Le Maréchal, C., Ruszniewski, P., Robaszkiewicz, M., Raguénès, O., et al. (2002). Determination of the relative contribution of three genes-the cystic fibrosis transmembrane conductance regulator gene, the cationic trypsinogen gene, and the pancreatic secretory trypsin inhibitor gene-to the etiology of idiopathic chronic pancreatitis. *European Journal of Human Genetics, 10*, 100–106.

Bai, X., Miao, D., Goltzman, D., & Karaplis, A. C. (2007). Early lethality in Hyp mice with targeted deletion of PTH gene. *Endocrinology, 148*, 4974–4983.

Bandeira, F., Cusano, N. E., Silva, B. C., Cassibba, S., Almeida, C. B., Machado, V. C., et al. (2014). Bone disease in primary hyperparathyroidism. *Arquivos Brasileiros de Endocrinologia e Metabologia, 58*, 553–561.

Beck, L., Karaplis, A. C., Amizuka, N., Hewson, A. S., Ozawa, H., & Tenenhouse, H. S. (1998). Targeted inactivation of Npt2 in mice leads to severe renal phosphate wasting, hypercalciuria, and skeletal abnormalities. *Proceedings of the National Academy of Sciences of the United States of America, 95*, 5372–5377.

Benet-Pages, A., Orlik, P., Strom, T. M., & Lorenz-Depiereux, B. (2005). An FGF23 missense mutation causes familial tumoral calcinosis with hyperphosphatemia. *Human Molecular Genetics, 14*, 385–390.

Borowitz, D. (2015). CFTR, bicarbonate, and the pathophysiology of cystic fibrosis. *Pediatric Pulmonology, 50*, S24–S30.

Boucher, R. C. (2007). Evidence for airway surface dehydration as the initiating event in CF airway disease. *Journal of Internal Medicine, 261*, 5–16.

Cardone, R. A., Bellizzi, A., Busco, G., Weinman, E. J., Dell'Aquila, M. E., Casavola, V., et al. (2007). The NHERF1 PDZ2 domain regulates PKA-RhoA-p38-mediated NHE1 activation and invasion in breast tumor cells. *Molecular Biology of the Cell, 18*, 1768–1780.

Christov, M., & Jüppner, H. (2013). Insights from genetic disorders of phosphate homeostasis. *Seminars in Nephrology, 33*, 143–157.

Cinar, A., Chen, M., Riederer, B., Bachmann, O., Wiemann, M., Manns, M., et al. (2007). NHE3 inhibition by cAMP and Ca2+ is abolished in PDZ-domain protein PDZK1-deficient murine enterocytes. *Journal of Physiology (London), 581*, 1235–1246.

Clarke, L. L., & Harline, M. C. (1996). CFTR is required for cAMP inhibition of intestinal Na+ absorption in a cystic fibrosis mouse model. *The American Journal of Physiology, 270*, G259–G267.

Courbebaisse, M., Leroy, C., Bakouh, N., Salaün, C., Beck, L., Grandchamp, B., et al. (2012). A new human NHERF1 mutation decreases renal phosphate transporter NPT2a expression by a PTH-independent mechanism. *PLoS One, 7*, e34764.

Cunningham, R., Esmaili, A., Brown, E., Biswas, R. S., Murtazina, R., Donowitz, M., et al. (2008). Urine electrolyte, mineral, and protein excretion in NHERF-2 and NHERF-1 null mice. *American Journal of Physiology. Renal Physiology, 294*, F1001–F1007.

Cunningham, R., Steplock, D., Xiaofei, E., Biswas, R. S., Wang, F., Shenolikar, S., et al. (2006). Adenoviral expression of NHERF-1 in NHERF-1 null mouse renal proximal tubule cells restores Npt2a regulation by low phosphate media and parathyroid hormone. *American Journal of Physiology. Renal Physiology, 291*, F896–F901.

Dransfield, D. T., Bradford, A. J., Smith, J., Martin, M., Roy, C., Mangeat, P. H., et al. (1997). Ezrin is a cyclic AMP-dependent protein kinase anchoring protein. *The EMBO Journal, 16*, 35–43.

Du, G., Gu, Y., Hao, C., Yuan, Z., He, J., Jiang, W. G., et al. (2016). The cellular distribution of Na+/H+ exchanger regulatory factor 1 is determined by the PDZ-I domain and regulates the malignant progression of breast cancer. *Oncotarget*, *7*, 29440–29453.

Dunn, H. A., & Ferguson, S. S. (2015). PDZ protein regulation of G protein-coupled receptor trafficking and signaling pathways. *Molecular Pharmacology*, *88*, 624–639.

Elborn, J. S. (2016). Cystic fibrosis. *The Lancet*, *388*, 2519–2531.

Fam, S. R., Paquet, M., Castleberry, A. M., Oller, H., Lee, C. J., Traynelis, S. F., et al. (2005). P2Y1 receptor signaling is controlled by interaction with the PDZ scaffold NHERF-2. *Proceedings of the National Academy of Sciences of the United States of America*, *102*, 8042–8047.

Field, M. (1993). Intestinal electrolyte secretion. History of a paradigm. *Archives of Surgery*, *128*, 273–278.

Fukuoka, Y., Xia, H.-Z., Sanchez-Munoz, L. B., Dellinger, A. L., Escribano, L., & Schwartz, L. B. (2008). Generation of anaphylatoxins by human beta-tryptase from C3, C4, and C5. *Journal of Immunology*, *180*, 6307–6316.

Garnett, J. P., Hickman, E., Burrows, R., Hegyi, P., Tiszlavicz, L., Cuthbert, A. W., et al. (2011). Novel role for pendrin in orchestrating bicarbonate secretion in cystic fibrosis transmembrane conductance regulator (CFTR)-expressing airway serous cells. *Journal of Biological Chemistry*, *286*, 41069–41082.

Georgescu, M. M., Gagea, M., & Cote, G. (2016). NHERF1/EBP50 suppresses Wnt-β-catenin pathway-driven intestinal neoplasia. *Neoplasia*, *18*, 512–523.

Georgescu, M. M., Morales, F. C., Molina, J. R., & Hayashi, Y. (2008). Roles of NHERF1/EBP50 in cancer. *Current Molecular Medicine*, *8*, 459–468.

Gisler, S. M., Stagljar, I., Traebert, M., Bacic, D., Biber, J., & Murer, H. (2001). Interaction of the type IIa Na/Pi cotransporter with PDZ proteins. *The Journal of Biological Chemistry*, *276*, 9206–9213.

Greenwald, I., & Gross, J. (1925). The effect of the administration of a potent parathyroid extract upon the excretion of nitrogen, phosphorus, calcium, and magnesium, with some remarks on the solubility of calcium phosphate in serum and on the pathogenesis of tetany. *Journal of Biological Chemistry*, *66*, 217–227.

Guerra, L., Favia, M., Fanelli, T., Calamita, G., Svetlo, M., Bagorda, A., et al. (2004). Stimulation of Xenopus P2Y1 receptor activates CFTR in A6 cells. *Pflügers Archiv*, *449*, 66–75.

Hall, R. A., Ostedgaard, L. S., Premont, R. T., Blitzer, J. T., Rahman, N., Welsh, M. J., et al. (1998). A C-terminal motif found in the beta2-adrenergic receptor, P2Y1 receptor and cystic fibrosis transmembrane conductance regulator determines binding to the Na+/H+ exchanger regulatory factor family of PDZ proteins. *Proceedings of the National Academy of Sciences of the United States of America*, *95*, 8496–8501.

Hanahan, D., & Weinberg, R. A. (2011). Hallmarks of cancer: The next generation. *Cell*, *144*, 646–674.

Harris, B. Z., & Lim, W. A. (2001). Mechanism and role of PDZ domains in signaling complex assembly. *Journal of Cell Science*, *114*, 3219–3231.

Hayashi, Y., Molina, J. R., Hamilton, S. R., & Georgescu, M.-M. (2010). NHERF1/EBP50 is a new marker in colorectal cancer. *Neoplasia*, *12*, 1013–1022.

Heldin, C. H., & Westermark, B. (1999). Mechanism of action and in vivo role of platelet-derived growth factor. *Physiological Reviews*, *79*, 1283–1316.

Hernando, N., Forgo, J., Biber, J., & Murer, H. (2000). PTH-induced downregulation of the type IIa Na/P-i-cotransporter is independent of known endocytic motifs. *Journal of the American Society of Nephrology*, *11*, 1961–1968.

Hillier, B. J., Christopherson, K. S., Prehoda, K. E., Bredt, D. S., & Lim, W. A. (1999). Unexpected modes of PDZ domain scaffolding revealed by structure of nNOS-syntrophin complex. *Science*, *284*, 812–815.

Holcomb, J., Jiang, Y., Lu, G., Trescott, L., Brunzelle, J., Sirinupong, N., et al. (2014). Structural insights into PDZ-mediated interaction of NHERF2 and LPA(2), a cellular event implicated in CFTR channel regulation. *Biochemical and Biophysical Research Communications, 446*, 399–403.

Hwang, J. I., Heo, K., Shin, K. J., Kim, E., Yun, C. H. C., Ryu, S. H., et al. (2000). Regulation of phospholipase C-beta 3 activity by Na+/H+ exchanger regulatory factor 2. *Journal of Biological Chemistry, 275*, 16632–16637.

Jarnaess, E., Ruppelt, A., Stokka, A. J., Lygren, B., Scott, J. D., & Tasken, K. (2008). Dual specificity A-kinase anchoring proteins (AKAPs) contain an additional binding region that enhances targeting of protein kinase A type I. *Journal of Biological Chemistry, 283*, 33708–33718.

Karim, Z., Gérard, B., Bakouh, N., Alili, R., Leroy, C., Beck, L., et al. (2008). NHERF1 mutations and responsiveness of renal parathyroid hormone. *The New England Journal of Medicine, 359*, 1128–1135.

Kaufmann, M., Muff, R., Stieger, B., Biber, J., Murer, H., & Fischer, J. A. (1994). Apical and basolateral parathyroid hormone receptors in rat renal cortical membranes. *Endocrinology, 134*, 1173–1178.

Keely, S. J., & Barrett, K. E. (2000). Regulation of chloride secretion. Novel pathways and messengers. *Annals of the New York Academy of Sciences, 915*, 67–76.

Kere, J., Lohi, H., & Höglund, P. (1999). Genetic disorders of membrane transport III. Congenital chloride diarrhea. *The American Journal of Physiology, 276*, G7–G13.

Kim, J. K., Kwon, O., Kim, J., Kim, E. K., Park, H. K., Lee, J. E., et al. (2012). PDZ domain-containing 1 (PDZK1) protein regulates phospholipase C-β3 (PLC-β3)-specific activation of somatostatin by forming a ternary complex with PLC-β3 and somatostatin receptors. *The Journal of Biological Chemistry, 287*, 21012–21024.

Kim, D., Liao, J., & Hanrahan, J. W. (2014). The buffer capacity of airway epithelial secretions. *Frontiers in Physiology, 5*, 188.

Kim, E. J., & Sheng, M. (2004). PDZ domain proteins of synapses. *Nature Reviews Neuroscience, 5*, 771–781.

Ko, S. B., Shcheynikov, N., Choi, J. Y., Luo, X., Ishibashi, K., Thomas, P. J., et al. (2002). A molecular mechanism for aberrant CFTR-dependent HCO(3)(−) transport in cystic fibrosis. *The EMBO Journal, 21*, 5662–5672.

Ko, S. B., Zeng, W., Dorwart, M. R., Luo, X., Kim, K. H., Millen, L., et al. (2004). Gating of CFTR by the STAS domain of SLC26 transporters. *Nature Cell Biology, 6*, 343–350.

Kreimann, E. L., Morales, F. C., de Orbeta-Cruz, J., Takahashi, Y., Adams, H., Liu, T. J., et al. (2007). Cortical stabilization of beta-catenin contributes to NHERF1/EBP50 tumor suppressor function. *Oncogene, 26*, 5290–5299.

Lamprecht, G., & Seidler, U. (2006). The emerging role of PDZ adapter proteins for regulation of intestinal ion transport. *American Journal of Physiology. Gastrointestinal and Liver Physiology, 291*, G766–G777.

Lee, M. G., Choi, J. Y., Luo, X., Strickland, E., Thomas, P. J., & Muallem, S. (1999a). Cystic fibrosis transmembrane conductance regulator regulates luminal $Cl^-/HCO_3^-$ exchange in mouse submandibular and pancreatic ducts. *The Journal of Biological Chemistry, 274*, 14670–14677.

Lee, J. H., Nam, J. H., Park, J., Kang, D. W., Kim, J. Y., Lee, M. G., et al. (2012). Regulation of SLC26A3 activity by NHERF4 PDZ-mediated interaction. *Cellular Signalling, 24*, 1821–1830.

Lee, M. G., Wigley, W. C., Zeng, W., Noel, L. E., Marino, C. R., Thomas, P. J., et al. (1999b). Regulation of $Cl^-/HCO_3^-$ exchange by cystic fibrosis transmembrane conductance regulator expressed in NIH 3T3 and HEK 293 cells. *The Journal of Biological Chemistry, 274*, 3414–3421.

Lee, H. J., & Zheng, J. J. (2010). PDZ domains and their binding partners: Structure, specificity, and modification. *Cell Communication and Signaling: CCS, 8*, 8.

Li, J., Dai, Z., Jana, D., Callaway, D. J., & Bu, Z. (2005). Ezrin controls the macromolecular complexes formed between an adapter protein $Na^+/H^+$ exchanger regulatory factor and the cystic fibrosis transmembrane conductance regulator. *The Journal of Biological Chemistry, 280*, 37634–37643.

Li, C. Y., & Naren, A. P. (2010). CFTR chloride channel in the apical compartments: Spatiotemporal coupling to its interacting partners. *Integrative Biology, 2*, 161–177.

Li, C., Roy, K., Dandridge, K., & Naren, A. P. (2004). Molecular assembly of cystic fibrosis transmembrane conductance regulator in plasma membrane. *The Journal of Biological Chemistry, 279*, 24673–24684.

Li, J., Yen, C., Liaw, D., Podsypanina, K., Bose, S., Wang, S. I., et al. (1997). PTEN, a putative protein tyrosine phosphatase gene mutated in human brain, breast, and prostate cancer. *Science, 275*, 1943–1947.

Lim, H., & Jou, T.-S. (2016). Ras-activated RSK1 phosphorylates EBP50 to regulate its nuclear localization and promote cell proliferation. *Oncotarget*, 7, 10283–10296.

Lin, Y. Y., Hsu, Y. H., Huang, H. Y., Shann, Y. J., Huang, C. Y., Wei, S. C., et al. (2012). Aberrant nuclear localization of EBP50 promotes colorectal carcinogenesis in xenotransplanted mice by modulating TCF-1 and β-catenin interactions. *The Journal of Clinical Investigation, 122*, 1881–1894.

Lundstrom, K. (2009). An overview on GPCRs and drug discovery: Structure-based drug design and structural biology on GPCRs. *Methods in Molecular Biology, 552*, 51–66.

Maehama, T., & Dixon, J. E. (1998). The tumor suppressor, PTEN/MMAC1, dephosphorylates the lipid second messenger, phosphatidylinositol 3,4,5-trisphosphate. *Journal of Biological Chemistry, 273*, 13375–13378.

Magagnin, S., Werner, A., Markovich, D., Sorribas, V., Stange, G., Biber, J. et al. (1993). Expression cloning of human and rat renal cortex Na/Pi cotransport. *Proceedings of the National Academy of Sciences of the United States of America, 90*, 5979–5983

Mahon, M. J., Cole, J. A., Lederer, E. D., & Segre, G. V. (2003). $Na^+/H^+$ exchanger-regulatory factor 1 mediates inhibition of phosphate transport by parathyroid hormone and second messengers by acting at multiple sites in opossum kidney cells. *Molecular Endocrinology, 17*, 2355–2364.

Mangia, A., Chiriatti, A., Bellizzi, A., Malfettone, A., Stea, B., Zito, F. A., et al. (2009). Biological role of NHERF1 protein expression in breast cancer. *Histopathology, 55*, 600–608.

Mickle, J. E., Macek, M., Fulmer-Smentek, S. B., Egan, M. M., Schwiebert, E., Guggino, W., et al. (1998). A mutation in the cystic fibrosis transmembrane conductance regulator gene associated with elevated sweat chloride concentrations in the absence of cystic fibrosis. *Human Molecular Genetics*, 7, 729–735.

Molina, J. R., Morales, F. C., Hayashi, Y., Aldape, K. D., & Georgescu, M.-M. (2010). Loss of PTEN binding adapter protein NHERF1 from plasma membrane in glioblastoma contributes to PTEN inactivation. *Cancer Research, 70*, 6697–6703.

Mosenden, R., & Tasken, K. (2011). Cyclic AMP-mediated immune regulation—Overview of mechanisms of action in T cells. *Cellular Signalling, 23*, 1009–1016.

Moyer, B. D., Denton, J., Karlson, K. H., Reynolds, D., Wang, S., Mickle, J. E., et al. (1999). A PDZ-interacting domain in CFTR is an apical membrane polarization signal. *The Journal of Clinical Investigation, 104*, 1353–1361.

Murer, H., & Biber, J. (2010). Phosphate transport in the kidney. *Journal of Nephrology, 23*(Suppl. 16), S145–S151.

Murer, H., Hernando, N., Forster, I., & Biber, J. (2000). Proximal tubular phosphate reabsorption: Molecular mechanisms. *Physiological Reviews, 80*, 1373–1409.

Murtazina, R., Kovbasnjuk, O., Chen, T.-E., Zachos, N. C., Chen, Y., Kocinsky, H. S., et al. (2011). NHERF2 is necessary for basal activity, second messenger inhibition, and LPA stimulation of NHE3 in mouse distal ileum. *American Journal of Physiology-Cell Physiology, 301*, C126–C136.

Musgrove, E. A., Caldon, C. E., Barraclough, J., Stone, A., & Sutherland, R. L. (2011). Cyclin D as a therapeutic target in cancer. *Nature Reviews. Cancer, 11*, 558–572.

Naren, A. P., Cobb, B., Li, C., Roy, K., Nelson, D., Heda, G. D., et al. (2003). A macromolecular complex of beta 2 adrenergic receptor, CFTR, and ezrin/radixin/moesin-binding phosphoprotein 50 is regulated by PKA. *Proceedings of the National Academy of Sciences of the United States of America, 100*, 342–346.

Oh, Y. S., Jo, N. W., Choi, J. W., Kim, H. S., Seo, S. W., Kang, K. O., et al. (2004). NHERF2 specifically interacts with LPA(2) receptor and defines the specificity and efficiency of receptor-mediated phospholipase C-beta 3 activation. *Molecular and Cellular Biology, 24*, 5069–5079.

Orsetti, B., Courjal, F., Cuny, M., Rodriguez, C., & Theillet, C. (1999). 17q21-q25 aberrations in breast cancer: Combined allelotyping and CGH analysis reveals 5 regions of allelic imbalance among which two correspond to DNA amplification. *Oncogene, 18*, 6262–6270.

Pages, C., Simon, M. F., Valet, P., & Saulnier-Blache, J. S. (2001). Lysophosphatidic acid synthesis and release. *Prostaglandins & Other Lipid Mediators, 64*, 1–10.

Pan, Y., Weinman, E. J., & Dai, J. L. (2008). Na(+)/H(+) exchanger regulatory factor 1 inhibits platelet-derived growth factor signaling in breast cancer cells. *Breast Cancer Research, 10*, R5.

Paul, I., Bhattacharya, S., Chatterjee, A., & Ghosh, M. K. (2013). Current understanding on EGFR and Wnt/β-catenin signaling in glioma and their possible crosstalk. *Genes & Cancer, 4*, 427–446.

Pezzulo, A. A., Tang, X. X., Hoegger, M. J., Abou Alaiwa, M. H., Ramachandran, S., Moninger, T. O., et al. (2012). Reduced airway surface pH impairs bacterial killing in the porcine cystic fibrosis lung. *Nature, 487*, 109–113.

Picotto, G., Massheimer, V., & Boland, R. (1997). Parathyroid hormone stimulates calcium influx and the cAMP messenger system in rat enterocytes. *The American Journal of Physiology, 273*, C1349–C1353.

Plummer, S. J., Adams, L., Simmons, J. A., & Casey, G. (1997). Localization of a growth suppressor activity in MCF7 breast cancer cells to chromosome 17q24-q25. *Oncogene, 14*, 2339–2345.

Presneau, N., Dewar, K., Forgetta, V., Provencher, D., Mes-Masson, A. M., & Tonin, P. N. (2005). Loss of heterozygosity and transcriptome analyses of a 1.2 Mb candidate ovarian cancer tumor suppressor locus region at 17q25.1-q25.2. *Molecular Carcinogenesis, 43*, 141–154.

Raghuram, V., Hormuth, H., & Foskett, J. K. (2003). A kinase-regulated mechanism controls CFTR channel gating by disrupting bivalent PDZ domain interactions. *Proceedings of the National Academy of Sciences of the United States of America, 100*, 9620–9625.

Ralevic, V., & Burnstock, G. (1998). Receptors for purines and pyrimidines. *Pharmacological Reviews, 50*, 413–492.

Ruppelt, A., Mosenden, R., Groenholm, M., Aandahl, E. M., Tobin, D., Carlson, C. R., et al. (2007). Inhibition of T cell activation by cyclic adenosine 5′-monophosphate requires lipid raft targeting of protein kinase a type I by the A-kinase anchoring protein Ezrin. *Journal of Immunology, 179*, 5159–5168.

Sarker, R., Valkhoff, V. E., Zachos, N. C., Lin, R., Cha, B., Chen, T. E., et al. (2011). NHERF1 and NHERF2 are necessary for multiple but usually separate aspects of basal and acute regulation of NHE3 activity. *American Journal of Physiology-Cell Physiology, 300*, C771–C782.

Schaefer, B., Piliponsky, A. M., Oka, T., Song, C. H., Gerard, N. P., Gerard, C., et al. (2013). Mast cell anaphylatoxin receptor expression can enhance IgE-dependent skin inflammation in mice. *Journal of Allergy and Clinical Immunology, 131*, 541.

Schultheis, P. J., Clarke, L. L., Meneton, P., Miller, M. L., Soleimani, M., Gawenis, L. R., et al. (1998). Renal and intestinal absorptive defects in mice lacking the NHE3 $Na^+/H^+$ exchanger. *Nature Genetics, 19*, 282–285.

Seidler, U., Singh, A. K., Cinar, A., Chen, M. M., Hillesheim, J., Hogema, B., et al. (2009). The role of the NHERF family of PDZ scaffolding proteins in the regulation of salt and water transport lessons learned from knockout mice. *Molecular Structure and Function of the Tight Junction: From Basic Mechanisms to Clinical Manifestations, 1165*, 249–260.

Sharma, N., LaRusch, J., Sosnay, P. R., Gottschalk, L. B., Lopez, A. P., Pellicore, M. J., et al. (2016). A sequence upstream of canonical PDZ-binding motif within CFTR COOH-terminus enhances NHERF1 interaction. *American Journal of Physiology. Lung Cellular and Molecular Physiology, 311*, L1170–L1182.

Sheng, M., & Sala, C. (2001). PDZ domains and the organization of supramolecular complexes. *Annual Review of Neuroscience, 24*, 1–29.

Shenolikar, S., Voltz, J. W., Minkoff, C. M., Wade, J. B., & Weinman, E. J. (2002). Targeted disruption of the mouse NHERF-1 gene promotes internalization of proximal tubule sodium-phosphate cotransporter type IIa and renal phosphate wasting. *Proceedings of the National Academy of Sciences of the United States of America, 99*, 11470–11475.

Singh, A. K., Riederer, B., Krabbenhöft, A., Rausch, B., Bonhagen, J., Lehmann, U., et al. (2009). Differential roles of NHERF1, NHERF2, and PDZK1 in regulating CFTR-mediated intestinal anion secretion in mice. *The Journal of Clinical Investigation, 119*, 540–550.

Songyang, Z., Fanning, A. S., Fu, C., Xu, J., Marfatia, S. M., Chishti, A. H., et al. (1997). Recognition of unique carboxyl-terminal motifs by distinct PDZ domains. *Science, 275*, 73–77.

Stokka, A. J., Mosenden, R., Ruppelt, A., Lygren, B., & Tasken, K. (2010). The adaptor protein EBP50 is important for localization of the protein kinase A-Ezrin complex in T-cells and the immunomodulating effect of cAMP. *Biochemical Journal, 425*, 381–388.

Subramanian, H., Gupta, K., & Ali, H. (2012). Roles for NHERF1 and NHERF2 on the regulation of C3a receptor signaling in human mast cells. *Plos One*, 7, e51355.

Sun, L., Zhao, L., Du, L., Zhang, P., Zhang, M., Li, M., et al. (2016). Identification of two novel mutations in the GALNT3 gene in a Chinese family with hyperphosphatemic familial tumoral calcinosis. *Bone Research, 4*, 16038.

Swiatecka-Urban, A., Duhaime, M., Coutermarsh, B., Karlson, K. H., Collawn, J., Milewski, M., et al. (2002). PDZ domain interaction controls the endocytic recycling of the cystic fibrosis transmembrane conductance regulator. *The Journal of Biological Chemistry*, 277, 40099–40105.

Takahashi, Y., Morales, F. C., Kreimann, E. L., & Georgescu, M. M. (2006). PTEN tumor suppressor associates with NHERF proteins to attenuate PDGF receptor signaling. *EMBO Journal, 25*, 910–920.

Taouil, K., Hinnrasky, J., Hologne, C., Corlieu, P., Klossek, J. M., & Puchelle, E. (2003). Stimulation of beta 2-adrenergic receptor increases cystic fibrosis transmembrane conductance regulator expression in human airway epithelial cells through a cAMP/protein kinase A-independent pathway. *The Journal of Biological Chemistry, 278*, 17320–17327.

Thelin, W. R., Hodson, C. A., & Milgram, S. L. (2005). Beyond the brush border: NHERF4 blazes new NHERF turf. *The Journal of Physiology, 567*, 13–19.

Thiagarajah, J. R., & Verkman, A. S. (2003). CFTR pharmacology and its role in intestinal fluid secretion. *Current Opinion in Pharmacology, 3*, 594–599.

Tian, X., Liu, Z., Niu, B., Zhang, J., Tan, T. K., Lee, S. R., et al. (2011). E-cadherin/beta-catenin complex and the epithelial barrier. *Journal of Biomedicine & Biotechnology, 2011*, 567305.

Tonikian, R., Zhang, Y., Sazinsky, S. L., Currell, B., Yeh, J. H., Reva, B., et al. (2008). A specificity map for the PDZ domain family. *PLoS Biology, 6*, e239.

Traebert, M., Völkl, H., Biber, J., Murer, H., & Kaissling, B. (2000). Luminal and contraluminal action of 1-34 and 3-34 PTH peptides on renal type IIa Na-P(i) cotransporter. *American Journal of Physiology. Renal Physiology*, *278*, F792–F798.

Tröhler, U., Bonjour, J. P., & Fleisch, H. (1976). Inorganic phosphate homeostasis. Renal adaptation to the dietary intake in intact and thyroparathyroidectomized rats. *The Journal of Clinical Investigation*, *57*, 264–273.

van der Doef, H. P., Kokke, F. T., Beek, F. J., Woestenenk, J. W., Froeling, S. P., & Houwen, R. H. (2010). Constipation in pediatric cystic fibrosis patients: An underestimated medical condition. *Journal of Cystic Fibrosis*, *9*, 59–63.

Villa-Bellosta, R., & Sorribas, V. (2010). Compensatory regulation of the sodium/phosphate cotransporters NaPi-IIc (SCL34A3) and Pit-2 (SLC20A2) during Pi deprivation and acidosis. *Pflügers Archiv*, *459*, 499–508.

Vogelstein, B., & Kinzler, K. W. (2004). Cancer genes and the pathways they control. *Nature Medicine*, *10*, 789–799.

Wade, J. B., Liu, J., Coleman, R. A., Cunningham, R., Steplock, D. A., Lee-Kwon, W., et al. (2003). Localization and interaction of NHERF isoforms in the renal proximal tubule of the mouse. *American Journal of Physiology. Cell Physiology*, *285*, C1494–C1503.

Wade, J. B., Welling, P. A., Donowitz, M., Shenolikar, S., & Weinman, E. J. (2001). Differential renal distribution of NHERF isoforms and their colocalization with NHE3, ezrin, and ROMK. *American Journal of Physiology. Cell Physiology*, *280*, C192–C198.

Wagner, C. A., Rubio-Aliaga, I., Biber, J., & Hernando, N. (2014). Genetic diseases of renal phosphate handling. *Nephrology, Dialysis, Transplantation*, *29*(Suppl. 4), iv45–iv54.

Walker, L. C., Venglarik, C. J., Aubin, G., Weatherly, M. R., McCarty, N. A., Lesnick, B., et al. (1997). Relationship between airway ion transport and a mild pulmonary disease mutation in CFTR. *American Journal of Respiratory and Critical Care Medicine*, *155*, 1684–1689.

Walther, C., Caetano, F. A., Dunn, H. A., & Ferguson, S. S. (2015). PDZK1/NHERF3 differentially regulates corticotropin-releasing factor receptor 1 and serotonin 2A receptor signaling and endocytosis. *Cellular Signalling*, *27*, 519–531.

Wang, B., Ardura, J. A., Romero, G., Yang, Y., Hall, R. A., & Friedman, P. A. (2010). Na/H exchanger regulatory factors control parathyroid hormone receptor signaling by facilitating differential activation of G(alpha) protein subunits. *The Journal of Biological Chemistry*, *285*, 26976–26986.

Wang, B., Means, C. K., Yang, Y., Mamonova, T., Bisello, A., Altschuler, D. L., et al. (2012). Ezrin-anchored protein kinase A coordinates phosphorylation-dependent disassembly of a NHERF1 ternary complex to regulate hormone-sensitive phosphate transport. *The Journal of Biological Chemistry*, *287*, 24148–24163.

Wang, S., Yue, H., Derin, R. B., Guggino, W. B., & Li, M. (2000). Accessory protein facilitated CFTR-CFTR interaction, a molecular mechanism to potentiate the chloride channel activity. *Cell*, *103*, 169–179.

Weinman, E. J., Biswas, R. S., Peng, G., Peng, Q., Shen, L., Turner, C. L., et al. (2007). Parathyroid hormone inhibits renal phosphate transport by phosphorylation of serine 77 of sodium-hydrogen exchanger regulatory factor-1. *The Journal of Clinical Investigation*, *117*, 3412–3420.

Weinman, E. J., Steplock, D., Zhang, Y., Biswas, R., Bloch, R. J., & Shenolikar, S. (2010). Cooperativity between the phosphorylation of Thr95 and Ser77 of NHERF-1 in the hormonal regulation of renal phosphate transport. *The Journal of Biological Chemistry*, *285*, 25134–25138.

Werner, A., Moore, M. L., Mantei, N., Biber, J., Semenza, G., & Murer, H. (1991). Cloning and expression of CDNA for a NA/PI cotransport system of kidney cortex. *Proceedings of the National Academy of Sciences of the United States of America*, *88*, 9608–9612.

Wheeler, D., Garrido, J. L., Bisello, A., Kim, Y. K., Friedman, P. A., & Romero, G. (2008). Regulation of parathyroid hormone type 1 receptor dynamics, traffic, and signaling by the $Na^+/H^+$ exchanger regulatory factor-1 in rat osteosarcoma ROS 17/2.8 cells. *Molecular Endocrinology*, *22*, 1163–1170.

Wolfer, A., & Ramaswamy, S. (2011). MYC and metastasis. *Cancer Research*, *71*, 2034–2037.

Wu, Y., & Zhou, B. P. (2008). New insights of epithelial-mesenchymal transition in cancer metastasis. *Acta Biochimica et Biophysica Sinica*, *40*, 643–650.

Yang, X., Du, G., Yu, Z., Si, Y., Martin, T. A., He, J., et al. (2017). A novel NHERF1 mutation in human breast cancer and effects on malignant progression. *Anticancer Research*, *37*, 67–73.

Yang, J., Singh, V., Chen, T. E., Sarker, R., Xiong, L., Cha, B., et al. (2014). NHERF2/NHERF3 protein heterodimerization and macrocomplex formation are required for the inhibition of NHE3 activity by carbachol. *Journal of Biological Chemistry*, *289*, 20039–20053.

Yun, C. H., Oh, S., Zizak, M., Steplock, D., Tsao, S., Tse, C. M., et al. (1997). cAMP-mediated inhibition of the epithelial brush border Na+/H+ exchanger, NHE3, requires an associated regulatory protein. *Proceedings of the National Academy of Sciences of the United States of America*, *94*, 3010–3015.

Yun, C. C., Sun, H., Wang, D., Rusovici, R., Castleberry, A., Hall, R. A. et al. (2005). LPA2 receptor mediates mitogenic signals in human colon cancer cells. *American Journal of Physiology. Cell Physiology*, *289*, C2–11.

Zachos, N. C., Li, X., Kovbasnjuk, O., Hogema, B., Sarker, R., Lee, L. J., et al. (2009). NHERF3 (PDZK1) contributes to basal and calcium inhibition of NHE3 activity in Caco-2BBe cells. *Journal of Biological Chemistry*, *284*, 23708–23718.

Zachos, N. C., Li, X., Kovbasnjuk, O., Sarkar, R., Hogema, B., de Jonge, H. et al. (2006). Elevated intracellular calcium (Ca2+ i) inhibits NFIE3 activity by a PDZK1 (NHERF3) dependent process in which NHE3 and PDZK1 bind basally and dissociate with elevated Ca2+ i. *Gastroenterology*, *130*, A51.

FSC
www.fsc.org
MIX
Papier | Fördert
gute Waldnutzung
FSC® C083411

Kolibri 360 GmbH
Ferdinand-Jühlke-Straße 7
99095 Erfurt, Deutschland
produktsicherheit@kolibri360.de

# Game Theory for Applied Econometricians

Over the last 30 years the practice and use of game theory has changed dramatically, yet textbooks continue to present game theory with algebraic formalism and toy models. This book, on the other hand, illustrates game theory concepts using real-world data and analyses problems with real policy implications. The focus is on applying current learning to real world problems by providing an introduction to game theory and econometric analysis based on game theoretic principles using the computer language R.

The book covers the standard topics of an introductory game theory course including dominant strategies, Nash equilibrium and Bayes Nash equilibrium. It layers on top of this an approach to statistics and econometrics called Structural Modeling. In this approach, key parameter estimates rely upon game theoretic analysis. The real-world examples used to illustrate these concepts vary in scope and include an analysis of bargaining between hospitals and insurers, equilibrium entry of retail tire stores, bid rigging in timber auctions and contracts in 19th century whaling.

This book is aimed at the general reader with the equivalent of a bachelor's degree in economics, statistics or some more technical field. The book could be used as a text for an upper level undergraduate course or a lower level graduate course in economics or business.

**Christopher P. Adams** was born and raised in Melbourne Australia and is a lifelong Carlton supporter. Chris received his PhD in economics from the University of Wisconsin - Madison. He has taught at the University of Vermont, Dartmouth College, University Maryland and Johns Hopkins. He has 17 years experience working in merger regulation and antitrust for the US Federal Trade Commission. He is currently a Principal Analyst at the US Congressional Budget Office. Chris' work and research focus on econometrics, empirical industrial organization, pharmaceutical innovation and auctions. His work has been published in various academic journals including *The Econometrics Journal, Health Affairs, Health Economics, Marketing Science* and *Economics Letters*. Chris is a colon cancer survivor and a research patient advocate with ECOG-ACRIN. Most importantly, he is a father to CJ and husband to Deena.

**CHAPMAN & HALL/CRC Series on Statistics in Business and Economics**

**Recently Published Titles**

**Empirical Research in Accounting**
Tools and Methods
*Ian D. Gow and Tongqing Ding*

**Game Theory for Applied Econometricians**
Data Analytics with R
*Christopher P. Adams*

For more information about this series, please visit: https://www.routledge.com/Chapman-and-HallCRC-Series-on-Statistics-in-Business-and-Economics/book-series/CHSBE

# Game Theory for Applied Econometricians

## Data Analytics with R

Christopher P. Adams

CRC Press
Taylor & Francis Group
Boca Raton London New York

CRC Press is an imprint of the
Taylor & Francis Group, an **informa** business
A CHAPMAN & HALL BOOK

Designed cover image: Christopher P. Adams

First edition published 2025
by CRC Press
2385 NW Executive Center Drive, Suite 320, Boca Raton FL 33431

and by CRC Press
4 Park Square, Milton Park, Abingdon, Oxon, OX14 4RN

*CRC Press is an imprint of Taylor & Francis Group, LLC*



ISBN: 978-1-032-39833-4 (hbk)
ISBN: 978-1-032-39835-8 (pbk)
ISBN: 978-1-003-35160-3 (ebk)

DOI: 10.1201/b23262

Typeset in CMR10 font
by KnowledgeWorks Global Ltd.

*Publisher's note:* This book has been prepared from camera-ready copy provided by the authors.

# Contents

Introduction xiii

I Static Games of Complete Information 1

1 Games 3
1.1 Introduction . . . 3
1.2 Some Strategic Situations . . . 3
1.2.1 Aliens Attack! . . . 4
1.2.2 Can AI Cope with Traffic? . . . 4
1.3 End of Skyjacking . . . 5
1.3.1 More Security? . . . 5
1.3.2 Hijacking Data using **R** . . . 5
1.3.3 United Arlines Flight 93 . . . 6
1.4 A Game . . . 7
1.4.1 Definitions . . . 7
1.4.2 Examples . . . 8
1.5 Prisoner's Dilemma . . . 10
1.5.1 The Game . . . 10
1.5.2 What is the Best Outcome? . . . 11
1.5.3 What should You Do? . . . 11
1.5.4 Does This Make Sense? . . . 12
1.6 Empirical Analysis: *Friend or Foe* using **R** . . . 12
1.6.1 Trust Box . . . 12
1.6.2 Data Analysis in **R** . . . 13
1.6.3 Analysis in **R** . . . 14
1.7 Discussion and Further Reading . . . 17

2 Nash Equilibrium 18
2.1 Introduction . . . 18
2.2 What is a Nash Equilibrium? . . . 19
2.2.1 Dominance . . . 19
2.2.2 Prisoner's Dilemma . . . 20
2.2.3 Coordination Game . . . 21
2.3 Nash Equilibrium . . . 22
2.3.1 Definition . . . 22

2.3.2 Algorithm for Finding Nash Equilibrium . . . . . . . . . . 22
2.3.3 Prisoner's Dilemma Game . . . . . . . . . . . . . . . . . 23
2.3.4 Coordination Game . . . . . . . . . . . . . . . . . . . . 23
2.4 Entry Games . . . . . . . . . . . . . . . . . . . . . . . . . 24
2.4.1 Bresnahan and Reiss . . . . . . . . . . . . . . . . . . . 24
2.4.2 Two Firm Entry Game . . . . . . . . . . . . . . . . . . 25
2.4.3 Many Firm Entry Game . . . . . . . . . . . . . . . . . . 27
2.4.4 Nash Equilibrium . . . . . . . . . . . . . . . . . . . . . 28
2.4.5 Fixed Cost of Entry . . . . . . . . . . . . . . . . . . . . 28
2.4.6 Equilibrium Number of Firms . . . . . . . . . . . . . . . 28
2.4.7 Simulation of Entry Game in **R** . . . . . . . . . . . . . . 29
2.5 Empirical Analysis: Tire Markets using **R** . . . . . . . . . . . 30
2.5.1 Data . . . . . . . . . . . . . . . . . . . . . . . . . . . 31
2.5.2 Structural Model . . . . . . . . . . . . . . . . . . . . . 32
2.5.3 Entry Estimator . . . . . . . . . . . . . . . . . . . . . 33
2.5.4 Estimation in **R** . . . . . . . . . . . . . . . . . . . . . 33
2.5.5 Structural Estimation . . . . . . . . . . . . . . . . . . . 35
2.5.6 Policy Simulation in **R** . . . . . . . . . . . . . . . . . . 36
2.6 Discussion and Further Reading . . . . . . . . . . . . . . . . 38

**3 Oligopoly** **39**
3.1 Introduction . . . . . . . . . . . . . . . . . . . . . . . . . 39
3.2 Cournot's Model . . . . . . . . . . . . . . . . . . . . . . . 40
3.2.1 Two Firm Model . . . . . . . . . . . . . . . . . . . . . . 40
3.2.2 Best Response . . . . . . . . . . . . . . . . . . . . . . . 40
3.2.3 Nash Equilibrium . . . . . . . . . . . . . . . . . . . . . 41
3.2.4 Cournot Model with N Firms . . . . . . . . . . . . . . . 42
3.2.5 Cournot Model in **R** . . . . . . . . . . . . . . . . . . . 43
3.2.6 Solve for the Nash Equilibrium with **R** . . . . . . . . . . 43
3.2.7 Simulation of Cournot Model in **R** . . . . . . . . . . . . 44
3.3 Bertrand's Model . . . . . . . . . . . . . . . . . . . . . . . 45
3.3.1 Two-Firm Game . . . . . . . . . . . . . . . . . . . . . . 45
3.3.2 Nash Equilibrium . . . . . . . . . . . . . . . . . . . . . 46
3.4 Hotelling's Model . . . . . . . . . . . . . . . . . . . . . . . 47
3.4.1 Hotelling's Line . . . . . . . . . . . . . . . . . . . . . . 47
3.4.2 Differentiated Goods Game . . . . . . . . . . . . . . . . 48
3.4.3 Best Response . . . . . . . . . . . . . . . . . . . . . . . 48
3.4.4 Nash Equilibrium . . . . . . . . . . . . . . . . . . . . . 49
3.5 Empirical Analysis: Hamburger Competition with **R** . . . . . . 49
3.5.1 Data . . . . . . . . . . . . . . . . . . . . . . . . . . . 49
3.5.2 Estimation (Part 1) . . . . . . . . . . . . . . . . . . . . 51
3.5.3 Empirical Equilibrium . . . . . . . . . . . . . . . . . . . 51
3.5.4 Estimator . . . . . . . . . . . . . . . . . . . . . . . . . 53
3.5.5 Estimator in **R** . . . . . . . . . . . . . . . . . . . . . . 54
3.5.6 Distances . . . . . . . . . . . . . . . . . . . . . . . . . 55

3.5.7 Ownership . . . . . 56
3.5.8 Ownership of Outlets . . . . . 57
3.5.9 Estimation (Part 2) . . . . . 58
3.5.10 Goodness of Fit . . . . . 59
3.5.11 A Merger of McDonald's Outlets . . . . . 59
3.6 Discussion and Further Reading . . . . . 61

**4 Empirical Entry Games** **63**
4.1 Introduction . . . . . 63
4.2 Empirical Model . . . . . 63
4.2.1 Single Firm Entry Model . . . . . 64
4.2.2 Multiple Firm-Independent Entry Model . . . . . 65
4.2.3 Entry Game . . . . . 67
4.3 Empirical Analysis: Bookstore Entry with **R** . . . . . 67
4.3.1 Single Firm Entry Model in **R** . . . . . 68
4.3.2 Multiple Firm-Independent Entry in **R** . . . . . 70
4.3.3 Entry Game in **R** . . . . . 71
4.4 Empirical Analysis: Bookstore Entry using **R** . . . . . 72
4.4.1 Data . . . . . 73
4.4.2 Estimation of Single Firm Entry . . . . . 73
4.4.3 Estimation of Two Firm Entry Model . . . . . 75
4.4.4 Estimation of the Entry Game Model . . . . . 76
4.4.5 Model Fit . . . . . 76
4.5 Policy Analysis using **R** . . . . . 77
4.6 Discussion and Further Reading . . . . . 79

**5 Mixed Strategies** **80**
5.1 Introduction . . . . . 80
5.2 Zero-Sum Games . . . . . 80
5.2.1 Rock-Paper-Scissors . . . . . 80
5.2.2 Nash Equilibrium . . . . . 81
5.2.3 Mixed Strategies . . . . . 82
5.2.4 Penalty Kicks . . . . . 82
5.2.5 Algorithim for Mixed Strategy Nash Equilibrium . . . . . 83
5.3 Rock-Paper-Scissors . . . . . 84
5.3.1 Mixed Strategy Nash Equilibrium . . . . . 85
5.3.2 More Complicated Version . . . . . 85
5.3.3 Using **R** to Determine when Player 1 is Indifferent . . . . . 86
5.3.4 Solving for MSNE using **R** . . . . . 87
5.4 Empirical Analysis: Penalty Kicks using **R** . . . . . 89
5.4.1 Penalty Kick Game . . . . . 89
5.4.2 Data . . . . . 90
5.4.3 GMM Estimator . . . . . 91
5.4.4 GMM Estimator in **R** . . . . . 92
5.4.5 Difference By Footedness . . . . . 94

5.5 Multiple Equilibria . . . . . . . . . . . . . . . . . . . . . . . . . 95
5.5.1 Coordination Game . . . . . . . . . . . . . . . . . . . . 96
5.5.2 Bookstore Entry . . . . . . . . . . . . . . . . . . . . . 96
5.5.3 MSNE Estimator in **R** . . . . . . . . . . . . . . . . . . 97
5.6 Empirical Analysis: Bookstore Entry with MSNE using **R** . . . 98
5.7 Discussion and Further Reading . . . . . . . . . . . . . . . . . 99

# II Dynamic Games of Complete Information 101

## 6 Dynamic Games 103

6.1 Introduction . . . . . . . . . . . . . . . . . . . . . . . . . . . 103
6.2 Extensive Form . . . . . . . . . . . . . . . . . . . . . . . . . 104
6.2.1 Coordination Game . . . . . . . . . . . . . . . . . . . . 104
6.2.2 Strategies Revisited . . . . . . . . . . . . . . . . . . . . 105
6.2.3 Nash Equilibrium of the Coordination Game . . . . . . 105
6.2.4 Game Tree . . . . . . . . . . . . . . . . . . . . . . . . 106
6.3 Subgame Perfection . . . . . . . . . . . . . . . . . . . . . . . 107
6.3.1 Definition . . . . . . . . . . . . . . . . . . . . . . . . . 108
6.3.2 Coordination Game . . . . . . . . . . . . . . . . . . . . 108
6.3.3 Empirical Entry Game . . . . . . . . . . . . . . . . . . 109
6.3.4 Equilibrium in Entry Game . . . . . . . . . . . . . . . 109
6.3.5 Empirical Implications . . . . . . . . . . . . . . . . . . 110
6.4 Empirical Analysis: Bookstore Entry with Subgame Perfection in **R** . . . . . . . . . . . . . . . . . . . . . . . . . . . . . . . 110
6.4.1 SPNE Estimator . . . . . . . . . . . . . . . . . . . . . 110
6.4.2 SPNE Estimates . . . . . . . . . . . . . . . . . . . . . 111
6.5 Discussion and Further Reading . . . . . . . . . . . . . . . . . 111

## 7 Repeated Games 113

7.1 Introduction . . . . . . . . . . . . . . . . . . . . . . . . . . . 113
7.2 Repeated Prisoner's Dilemma . . . . . . . . . . . . . . . . . . 114
7.2.1 Prisoner's Dilemma . . . . . . . . . . . . . . . . . . . . 114
7.2.2 Normal Form Representation . . . . . . . . . . . . . . 115
7.2.3 Finitely Repeated Game . . . . . . . . . . . . . . . . . 115
7.2.4 Infinitely Repeated Prisoner's Dilemma . . . . . . . . . 116
7.2.5 Cooperation as a Nash Equilibrium . . . . . . . . . . . 117
7.3 Bertrand Competition . . . . . . . . . . . . . . . . . . . . . . 118
7.3.1 Two Firm Model . . . . . . . . . . . . . . . . . . . . . 118
7.3.2 $J$ Firm Model . . . . . . . . . . . . . . . . . . . . . . . 118
7.3.3 Nash Equilibrium in **R** . . . . . . . . . . . . . . . . . . 119
7.4 Empirical Analysis: Retail Gasoline Pricing using **R** . . . . . . 120
7.4.1 Perth Gas Price Data . . . . . . . . . . . . . . . . . . . 120
7.4.2 Estimating Parameters . . . . . . . . . . . . . . . . . . 120
7.4.3 Parameter Estimates . . . . . . . . . . . . . . . . . . . 122

7.5 Repeated Oligopoly . . . 124
7.5.1 Collusive Equilibrium . . . 124
7.5.2 Identifying Collusion . . . 125
7.5.3 Choosing Super Markups . . . 126
7.5.4 Estimating Collusive Prices in **R** . . . 126
7.6 Empirical Analysis: Collusion in Perth Gas Stations using **R** . . 127
7.6.1 Super Markups . . . 128
7.6.2 Analyzing Mergers with Collusion . . . 128
7.7 Discussion and Further Reading . . . 130

**8 Bargaining** **133**
8.1 Introduction . . . 133
8.2 Ultimatum Game . . . 134
8.2.1 The Game . . . 134
8.2.2 Game Tree . . . 135
8.2.3 Normal Form . . . 135
8.2.4 Nash Equilibrium . . . 136
8.2.5 Subgame Perfection . . . 138
8.3 Empirical Analysis: Ultimatum Game in India using **R** . . . 138
8.3.1 Data . . . 138
8.3.2 Equilibrium Play or Fairness? . . . 139
8.3.3 Do Stakes Matter? . . . 140
8.4 Two Period Alternating Offers Game . . . 143
8.4.1 The Game . . . 143
8.4.2 Game Tree . . . 144
8.4.3 Subgame Perfection . . . 144
8.5 Infinite Alternating Offers Game . . . 145
8.5.1 The Game . . . 145
8.5.2 Game Tree . . . 146
8.5.3 Subgame Perfection . . . 147
8.5.4 Game Ends in Period 1 . . . 148
8.5.5 Infinite Alternating Offers Game in **R** . . . 148
8.6 Nash Bargaining Model . . . 150
8.6.1 The Model . . . 150
8.6.2 Nash Solution using **R** . . . 151
8.7 Modeling Hospital Competition and Pricing . . . 151
8.7.1 Bargaining Model . . . 152
8.7.2 Demand for Hospitals . . . 153
8.7.3 Willingness-to-Pay . . . 153
8.8 Empirical Analysis: Hospital Competition with **R** . . . 154
8.8.1 Data . . . 155
8.8.2 Pricing and WTP . . . 156
8.8.3 Mergers and Willingness To Pay . . . 158
8.9 Discussion and Further Reading . . . 159

# III Static Games of Incomplete Information 161

**9 Bayes Nash Equilibrium 163**
9.1 Introduction . . . 163
9.2 A Bayesian Game . . . 164
9.2.1 A Grading Game . . . 164
9.2.2 Grading Game Simulation using **R** . . . 164
9.2.3 Definitions . . . 166
9.2.4 The Game . . . 167
9.2.5 Course Grading Game . . . 167
9.2.6 Equilibrium . . . 168
9.2.7 Equilibrium of Course Grade Game . . . 169
9.3 Empirical Entry Game . . . 169
9.3.1 The Game . . . 169
9.3.2 Equilibrium . . . 170
9.3.3 Estimating Entry Games . . . 171
9.3.4 Estimating Entry Games in **R** . . . 172
9.4 Empirical Analysis: Mega Bookstore Entry (Again) using **R** . . 173
9.4.1 Estimates . . . 173
9.4.2 Policy . . . 173
9.5 Discussion and Further Reading . . . 175

**10 Auctions 176**
10.1 Introduction . . . 176
10.2 Sealed Bid Auctions . . . 177
10.2.1 Sealed Bid Model . . . 178
10.2.2 Bayes Nash Equilibrium . . . 178
10.2.3 Sealed Bid Simulation in **R** . . . 179
10.2.4 Sealed Bid Estimator . . . 180
10.2.5 Sealed Bid Estimator in **R** . . . 181
10.3 English Auctions . . . 183
10.3.1 Second Price Auction Game . . . 184
10.3.2 Bayes Nash Equilibrium . . . 184
10.3.3 Order Statistics . . . 185
10.3.4 Identifying the Value Distribution . . . 186
10.3.5 English Auction Estimator . . . 186
10.3.6 English Auction Estimator in **R** . . . 187
10.4 Empirical Analysis: Testing the Rationality of Loggers using **R** 188
10.4.1 Timber Data . . . 188
10.4.2 Sealed Bid Auctions . . . 189
10.4.3 Comparing English Auctions to Sealed Bid Auctions . . 190
10.5 Empirical Analysis: Testing for Collusion using **R** . . . 192
10.5.1 A Test of Collusion . . . 192
10.5.2 "Large" English Auctions . . . 193
10.5.3 Large English Auction Estimator . . . 193

10.5.4 Large English Auction Estimator in **R** . . . . . . . . . . . 194
10.5.5 Evidence of Collusion . . . . . . . . . . . . . . . . . . . . 194
10.6 Discussion and Further Reading . . . . . . . . . . . . . . . . . 196

**11 Auctions with Affiliated Valuations 197**
11.1 Introduction . . . . . . . . . . . . . . . . . . . . . . . . . . 197
11.2 Auctions with Common Values . . . . . . . . . . . . . . . . . 197
11.2.1 Simple Model . . . . . . . . . . . . . . . . . . . . . . 198
11.2.2 Winner's Curse . . . . . . . . . . . . . . . . . . . . . 198
11.2.3 Bayes Nash Equilibrium . . . . . . . . . . . . . . . . 199
11.2.4 Common Value Auction in **R** . . . . . . . . . . . . . . . 200
11.2.5 Simulation of Common Value Auction using **R** . . . . . 201
11.2.6 Estimator for Common Values Auctions . . . . . . . . . 203
11.2.7 Common Values Estimator in **R** . . . . . . . . . . . . . 204
11.3 Empirical Analysis: Signal Distribution from OCS Auctions using **R** . . . . . . . . . . . . . . . . . . . . . . . . . . . . 205
11.3.1 Data . . . . . . . . . . . . . . . . . . . . . . . . . . . 206
11.3.2 Estimating the Signal Distribution . . . . . . . . . . . . 207
11.4 Auctions with Coalitions . . . . . . . . . . . . . . . . . . . . 209
11.4.1 The Benefit of Coalitions . . . . . . . . . . . . . . . . 210
11.4.2 Estimating Coalitions in **R** . . . . . . . . . . . . . . . 211
11.4.3 Policy Simulation . . . . . . . . . . . . . . . . . . . . 212
11.5 Discussion and Further Reading . . . . . . . . . . . . . . . . . 215

**IV Dynamic Games of Incomplete Information 217**

**12 Moral Hazard 219**
12.1 Introduction . . . . . . . . . . . . . . . . . . . . . . . . . . 219
12.2 Principal-Agent Game . . . . . . . . . . . . . . . . . . . . . 220
12.2.1 Simple No Contract Game . . . . . . . . . . . . . . . . 220
12.2.2 Bayes Nash Equilibrium . . . . . . . . . . . . . . . . 221
12.2.3 Simple Contract Game . . . . . . . . . . . . . . . . . 222
12.2.4 More Complicated Contract Game . . . . . . . . . . . 223
12.2.5 Bayes Nash Equilibrium . . . . . . . . . . . . . . . . 224
12.2.6 Parameterized Model . . . . . . . . . . . . . . . . . . 224
12.2.7 Simulation with **R** . . . . . . . . . . . . . . . . . . . 225
12.3 Empirical Analysis: Whaling Corporations in the 19th Century using **R** . . . . . . . . . . . . . . . . . . . . . . . . . . . . 226
12.3.1 Whaling Data . . . . . . . . . . . . . . . . . . . . . . 227
12.3.2 Regressions . . . . . . . . . . . . . . . . . . . . . . . 227
12.3.3 Calibrating the Model . . . . . . . . . . . . . . . . . . 230
12.4 Discussion and Further Reading . . . . . . . . . . . . . . . . . 232

**13 Adverse Selection 233**
13.1 Introduction . . . . . . . . . . . . . . . . . . . . . . . . . . 233

13.2 Akerlof's Lemons . . . . . . . . . . . . . . . . . . . . . . 233
13.2.1 The Lemons Market . . . . . . . . . . . . . . . . . . 234
13.2.2 Lemons Game . . . . . . . . . . . . . . . . . . . . . 234
13.2.3 Bayes Nash Equilibrium . . . . . . . . . . . . . . . . 235
13.3 Insurance . . . . . . . . . . . . . . . . . . . . . . . . . 235
13.3.1 Model . . . . . . . . . . . . . . . . . . . . . . . . . 236
13.3.2 Game with Unknown Types . . . . . . . . . . . . . . 236
13.3.3 Bayes Nash Equilibrium . . . . . . . . . . . . . . . . 237
13.3.4 Simulation using **R** . . . . . . . . . . . . . . . . . 237
13.3.5 Mandatory Insurance . . . . . . . . . . . . . . . . . 239
13.4 Empirical Analysis: Health Insurance using **R** . . . . . . . . . . 240
13.4.1 Willingness to Pay . . . . . . . . . . . . . . . . . . . 240
13.4.2 Premiums . . . . . . . . . . . . . . . . . . . . . . . 240
13.4.3 Tax Policy . . . . . . . . . . . . . . . . . . . . . . . 243
13.5 Discussion and Further Reading . . . . . . . . . . . . . . . 245

**Bibliography** **247**

**Subject Index** **253**

**Author Index** **257**

# Introduction

## The Research Assistant

You are back in DC! Yes. It is as hot and humid as it was for your internship, maybe even hotter and humiditier. You managed to secure a job as an "RA" at the US Federal Trade Commission. You are in the Bureau of Economics in AT1. You have an awesome place in Navy Yard, not far from the Nats stadium, the Wharf and the home for DC United. A mate from college is your room mate and they started at the Fed last week. Your career is about to begin!

## First Day

What the heck is he talking about? A colleague, was it John or Jeff, maybe Dave, has been talking 100 mph for last few minutes and you are just catching snippets. "We need to estimate WTP." "Can you believe they are using Elzinga-Hogarty." "I hope Alabama joins the suit." "Bill Town is great." You have been nodding but you quickly lost the thread of the conversation.

You decided to go with a suit, which was definitely a mistake. Even though you were hardly outside, you are sweating buckets and no one else seems to be in a suit, not even the managers. The managers are lovely and seemed very excited that you were starting. They liked that you had coding experience in **R** and had done some cool work in your internship on minimum wages. Apparently, you are going to jump straight into a case.

John/Jeff/maybe-Dave is explaining the case to you. It seems to have something to do with hospitals in Alabama. Your screen is full of data, there are columns with things you recognize like zip codes, age, and then things like DRG which you have no idea about. Apparently, this data is from a payer, although you are not sure what that is. The rows are claims. There is a column with the price, but they seem wrong. The numbers are enormous, 150008, 25020, 83251. Is this dollars? You were in hospital a few weeks back getting some stitches after an incident playing kick ball. You are pretty sure it was $25.00

Your job is to determine the relationship between price and WTP. You just wish you knew what WTP stood for.

## Second Day

Yesterday you were able to get **R** and `RStudio` set up and get access to the payer data. You were able to calculate WTP for the hospitals and you showed Dave the results.

OK. Now we need to calculate WTP post-merger. You responded that you didn't think you had the data, you were pretty sure that there was no indicator for merger in the data. Dave laughed out loud. Then he saw the expression on your face, caught himself, and stated matter-of-factly, no we have to simulate the merger.

Simulate the merger? How would we know what would happen to prices after the hospitals merge? How do we know how prices are determined now? From what you could gather so far there were three hospitals in the city and five insurance companies serving beneficiaries in the area. The prices in the data were determined by bargaining between these hospitals and the insurance companies. You search your memory back to microeconomics, you remember the class on monopoly pricing, when the seller had market power to determine price. Do these hospitals have market power? You remember something in the text book about monopsony, when the buyer has market power to determine price. Do the insurers have market power? Do both the hospitals and the insurers have market power?

# The Book

This is an empirical game theory book. Traditionally, game theory is presented as a theoretical subject. Generally, applications are discussed but there are no explicit empirical applications. Yet, since the 1990s, game theory has been at the heart of empirical analysis of competition and markets in the economics sub-field of industrial organization. In economics, this layering of theory on to empirical analysis is called **structural econometrics**. This book is focused on the game theory, not necessarily the econometrics.

## What Does it Cover?

The book aims to provide an introduction game theory, a mathematical approach to understanding economic relationships. The goal is for the reader to understand what a game is and how the mathematics works. The reader will be able to create a game that explains behavior of economic actors observed in her data. She will be able to ask *how* questions. She will be able to see how parameters of the game relate to characteristics of the data. She may be able to use the game to ask *what if* questions. What would the economic actors observed in the data do if the world was different? What if the government introduced a new policy? What if technology changed. What if the hospitals merged?

The book covers standard game theory concepts such as **normal form games**, **exentensive form games**, **Nash equilibrium**, **mixed strategy Nash equilibrium**, and **subgame perfection**. It also covers standard empirical methods such as **linear regression**, **two-way fixed effects**, **logit models**, and **maximum likelihood estimation**. But it covers a number of concepts that are important to the intersection of game theory and data analysis such as **generalized method of moments** and **two-step estimators**.

## What is the Approach?

The book teaches game theory through code, in particular, it will use the scripting language **R**. It is not primarily aimed at teaching **R**. Rather, it is primarily aimed at teaching game theory. This idea of using computer programming as a tool of instruction goes back to at least Seymour Papert and MIT's AI lab in the 1970s.[1] Papert helped develop a programming language called `Logo`. The goal of `Logo` was to teach geometry by programming how a **turtle** moves around the screen. You may have used one of the offspring of `Logo`, such as `Scratch` or `Lego Mindstorms`.

The book uses Papert's ideas to teach game theory. You will learn the math of the game or estimation method and then how to program that game or estimation method. The book makes particular use of the computer's ability to simulate data. This allows us to experiment with more complicated and realistic games than is possible with pen and paper.

The book is written in `RStudio` using `Sweave`. `Sweave` allows LaTeX to be integrated into **R**. LaTeX is a free type-setting language that is designed for writing math. Much of the code that is used in the book is actually presented in the book. Sometimes it is more practical to create a data set outside the of book. In those cases, the data and the code that created the data are available here `https://github.com/christopherpadams/EmpiricalGameTheory`. In a couple of other cases, the preferred code does not produce nice output for the book, so it is left out.

## What is How Analysis?

Often we want to understand patterns we see in the data. How are prices related to the number of firms in the market? Why do firms enter some markets and not others?

For those questions, we can use game theory to guide our thinking and our analysis. A game theoretic model of firm entry can help us understand why some markets had both Borders and Barnes & Noble, other markets had just one, and many more have none.

[1] `https://el.media.mit.edu/logo-foundation/`

## What is What If Analysis?

Game theory is also important in *what if* analysis. In those analysis, we want to understand how behavior will change when faced with a situation not observed in the data. What would have happened if Borders and Barnes & Noble would have merged in the mid-1990s? Would there have been more or fewer mega bookstores? Should the federal government allow bidders to collude in oil drilling auctions? We can estimate the parameters of game theoretic model using observed data and then make changes to the model to simulate what would happen in the case that is not observed in the data. We can then run the model and predict the outcome. We can simulate a merger between Borders and Barnes & Noble in the mid-1990s or what happens to bids on oil drilling leases when bidders are allowed to collude.

## What About the Real World?

The book presents interesting and important questions. The book presents an analysis of competition between various types of firms and asks what happens if mergers are allowed or not. It considers price regulation policies for retail gasoline and whether these regulations lead to higher prices for consumers. It looks at how the US federal government runs auctions for timber and oil drilling leases. Hopefully, the book points you to new questions and new data to answer existing questions.

The book does not recommend policies. The government economist, Alice Rivlin, argued that it is extremely important to provide policy makers with objective analysis. In a memo to staff of the Congressional Budget Office (CBO), she said the following.[2]

> We are not to be advocates. As private citizens, we are entitled to our own views on the issues of the day, but as members of CBO, we are not to make recommendations, or characterize, even by implication, particular policy questions as good or bad, wise or unwise.

Economists in government, the private sector and the academy, work on important policy questions. Economists are most effective when they do not advocate for policy positions, but present objective analysis of the economics and the data. This book presents an objective analysis of interesting policy questions but doesn't state whether the policy positions are good or bad, wise or unwise.

[2] https://www.cbo.gov/sites/default/files/Public_Policy_Issues_Memo_Rivlin_1976.pdf

## The Outline

The book is laid out the same way Robert Gibbons laid out his classic text, *Game Theory for Applied Economists.* There are four parts. **Static games of complete information**, dynamic games of complete information, static games of incomplete information, and dynamic games of incomplete information.

### Static Games of Complete Information

This part presents the simplest version of a game.

Chapter 1 introduces the basic mathematical concepts of game theory. It analyzes the most famous game in game theory, the **prisoner's dilemma**. The game is used in a TV game show and the chapter uses data on actual behavior in the game where the players of the game make choices worth thousands of dollars. Do real people on a TV game show play the game as the mathematics predicts?

Chapter 2 introduces two important equilibrium concepts, **dominant strategy equilibrium** and **Nash equilibrium**. The chapter uses Nash equilibrium to understand how the number of tire retailers varies from city to city.

Chapter 3 studies **oligopoly**, markets with a small number of competitors, and three models of how these markets work, **Cournot**, **Bertrand**, and **Hotelling**. The most general model allows competing firms to be similar but not the same. This model is used to understand pricing competition between McDonald's outlets in late 1990s Santa Clara county.

Chapter 4 considers the implications of multiple Nash equilibria in a game where two firms are choosing whether or not to enter the same market. This game is used to analyze the entry decisions by the mega bookstores in the 1990s, Borders and Barnes & Noble. The chapter analyzes problems where the game does not always make a single prediction.

Chapter 5 analyzes **mixed strategies** in the context of both **coordination games** and **zero-sum games**. It uses mixed strategies to model entry by the mega bookstores, Borders, and Barnes & Noble. Zero-sum games were the first types of games analyzed using game theory. Many parlor games like chess, droughts, and poker are zero-sum games. The chapter uses mixed strategies and zero-sum games to understand the choices made by soccer players when kicking and defending penalty kicks in the English Premier League.

### Dynamic Games of Complete Information

These games are substantially more complicated than the games presented in the first part of the book. In response, we need to make a number of simplifying assumptions that allow us to use the richness of the dynamics without being overwhelmed by the complexities.

Chapter 6 introduces **subgame perfection** and uses the concept to analyze the entry dynamics of the mega bookstores, Borders and Barnes & Noble.

Chapter 7 presents three different models of **bargaining**. It asks whether the simplest, the **ultimatum game**, makes predictions that are consistent with actual behavior of actual people when making decisions involving large sums of money. The answer is no, not really. A more complicated game makes more reasonable predictions. Luckily the **Rubenstein alternating offers game** makes predictions similar to a much simpler analytical tool, the **Nash bargaining solution**. This tool is used to analyze mergers between hospitals in Palm Beach County, the home county for the publisher of this book.

Chapter 8 returns to **oligopoly markets** but allows more complicated interactions between the firms. The chapter analyzes the pricing behavior of gasoline retailers in Perth Australia. The chapter presents a model to explain the weird saw tooth pattern in retail gas prices. It considers the extent of the prediction error when a merger model assumes firms choose prices more independently than they actually do.

## Static Games of Incomplete Information

In the first two parts of the book, the players of the game are assumed to know everything. In the second two parts of the book, that assumption is relaxed.

Chapter 9 revisits analysis of entry by mega bookstores, Borders and Barnes & Noble. The difference is that each firm observes its own entry costs but not the entry costs of the other firm. In this model equilibrium, the firms do not know their competitors unobserved costs of entry. The firms know their own costs of entering a new market, but not their competitors costs of entering that same market. While the game is more complicated than the game presented in Chapter 4, it is sometimes simpler to use in data analysis.

Chapter 10 and Chapter 11 analyze auctions. Chapter 10 uses two of the main auction types, **sealed bid auctions** and **English auctions** to analyze bidding behavior and collusion in US Forestry auctions conducted in the 1970s.

Chapter 11 considers auctions where the bidders don't know exactly how much to value the item they are bidding on. The classic example is oil drilling auctions on the US Outer Continental Shelf (OCS). The chapter asks whether the government should allow firms to collude in those auctions.

## Dynamic Games of Incomplete Information

The fourth part of the book analyzes the problems of moral hazard and adverse selection.

Chapter 12 considers the **principal-agent problem** and uses it to analyze the use of corporate financing in 19th-Century whaling in New England. What sort of contracts were used by firms and families to finance whaling expeditions, where the whaling ships were literally sailing around the world?

Chapter 13 brings game theory to health insurance markets. The chapter presents a model of the used car market, which suggests that information problems will cause the market to fail. Similarly, information problems in health insurance markets suggest that government interventions such as large tax subsidies for people insured through their work are necessary in order to have people insured.

## Notation

As you have seen above, the book uses particular fonts and symbols for various important things. It uses the symbol **R** to refer to the scripting language. It uses `typewriter font` to represent code in **R**. Initial mentions of an important term are in **bold face font**.

When discussing actual data, it uses $x_i$ to refer to the observed characteristic for some individual $i$. It uses $\mathbf{x}$ to denote a vector of the $x_i$'s. For matrices, it uses $\mathbf{X}$ for a matrix and $\mathbf{X}'$ for the matrix transpose. A row of that matrix is $\mathbf{X}_i$ or $\mathbf{X}_i'$ to highlight that it is a row vector. Lastly, for parameters of interest it uses Greek letters. For example, $\beta$ generally refers to a vector of parameters, although in some cases it is a single parameter of interest, while $\hat{\beta}$ refers to the estimate of the parameter.

# Hello R World

To use this book you need to download **R** and `RStudio` on your computer. Both are free.

## Download R and RStudio

First, download the appropriate version of `RStudio` here: `https://www.rstudio.com/products/rstudio/download/#download`. Then you can download the appropriate version of **R** here: `https://cran.rstudio.com/`.

Once you have the two programs downloaded and installed open up `RStudio`. To open up a *script* go to "File > New File > R Script." You should have 4 windows, a script window, a console window, a global environment window, and a window with help, plots, and other things.

## Using the Console

Go to the console window and click on the `>`. Then type `print("Hello R World")` and hit enter. Remember to use the quotes. In general, **R** functions have the same basic syntax, `functionname` with parentheses, and some input inside the parentheses. Inputs in quotes are treated as text while inputs without quotes are treated as variables.

```
> print("Hello R World")
  [1] "Hello R World"
```

Try something a little more complicated.

```
> a = "Chris"  # or write your own name
> print(paste("Welcome",a,"to R World"))
  [1] "Welcome Chris to R World"
```

Here we are creating a variable called `a`. The `#` is used in **R** to "comment out" lines in codes. **R** does not read the line following the hash.

In **R** we can place one function inside another function. The function `paste` is used to join text and variables together. The function `paste()` defaults to placing a space between the inputs. When placing one function inside another make sure to keep track of all of the parentheses. A common error is to have more or less closing parentheses than opening parentheses.

```
> paste(
  "Welcome",
   a,
   "to R World"
) |>
  print(
  )
  [1] "Welcome Chris to R World"
```

**R** can also accept code that looks like above. Using spaces and new lines helps a human reader understand what the code is saying. The symbol `|>` says take the result from above and use it in the following function. For some reason, there are two different symbols used `%>%` or `|>`. You can select which type under global options. Here `|>` is used.

## Thanks

I'm grateful to my wife, Deena Ackerman, for allowing me time to work on this project. Thank you to numerous friends and colleagues for providing feedback and suggestions on the book. I am particularly grateful to Devesh Raval and Emek Basker who gave extensive feedback on early drafts. Thanks also to a number of researchers who have been willing to provide me with the data used in the book including Raph Thomadsen, Eric Hilt and David Byrne. All errors are my own.

## Discussion and Further Reading

The book is laid out the same way as Gibbons (1992). If you are looking for a more detailed or technical description of the various types of games then see Fudenberg and Tirole (1991). A lot of the applications presented in this book are from the sub-field of economics called industrial organization. The classic theory text for that field is Tirole (1988). Recently more empirical-oriented industrial organization books have come out including Aguirregabiria (2021) and Hortaçsu and Joo (2023). Paarsch and Hong (2006) have a similar orientation with a focus on auctions. If you are interesting in some of the economic experiments presented in the book, then Camerer (2003) is a good but somewhat dated overview of experimental game theory.

The book uses the coding language **R** to illustrate models and empirical problems. It generally uses the `tidyverse` and `data.table` flavors. The best introduction to the language is the books by Hadley Wickham and coauthors, in particular *R for Data Science* which is here `https://r4ds.had.co.nz/`.

# Part I

# Static Games of Complete Information

# 1

# *Games*

## 1.1 Introduction

Game theory provides answers to life's many questions. Which side of the street should you walk on when there is no sidewalk? Will we ever talk to aliens? When should you turn across traffic at an intersection? Which mergers should be allowed? Should government intervene to stop bank runs? Who rides shotgun? During the COVID-19 emergency how many potential vaccines should the US government promise to purchase? Should NFL teams punt on 4th down more than they do?[1]

When there is no side walk, you should walk on side facing the oncoming traffic. The reason is that you can see the car coming toward you, but more importantly, you can see that the car coming toward you, can see you. You know that they know you are there. And, importantly, they know that you know, that you are there. Of course, you know that they know that you know that you are there. And they even know that. In game theory, we call this infinite circle of knowledge, **common knowledge**, and it is a fundamental assumption of most games. Players in the games analyzed in this book know everything, they even know that.

The chapter provides an introduction to games and game theory. It informally introduces a couple of strategic situations. It asks why there is no hijackings of airplanes any more. It presents data showing a significant drop off in hijackings early in the 2000s. The chapter presents the most famous game in game theory, the **prisoner's dilemma**. The chapter ends with an empirical analysis of how actual people play the prisoner's dilemma game. It uses data from the TV game show, *Friend or Foe*.

## 1.2 Some Strategic Situations

The section introduces two strategic situations: communicating with aliens and turning into traffic.

[1]The answers are: The opposite side of which cars drive. No. It depends where you live. How close competitors are the firms? Yes. Winner of Rock-Paper-Scissors. Many more than they actually did. No.

DOI: 10.1201/9781003351603-1

### 1.2.1 Aliens Attack!

Many years ago you could sign up to have your computer used for something called the Search for Extraterrestrial Intelligence (SETI). The idea is that the program would use all the spare computing power around the world to work through a huge data set of radio signals from the world's radio telescopes. It was like Bitcoin mining, but for aliens, this search would look for evidence that the signal was generated by an intelligent life form from another planet.

To determine the reasonableness of this exercise, consider a game. This game has two players, an Earthling and an Alien. Each player has two choices. They can listen for a signal from the other player or they could send a signal to the other player. Because of physics we will assume that for both players listening for the signal is substantially cheaper than sending the signal. Consider the Earthling's choices. If they listen for a signal, it is pretty cheap and if they hear a signal then jackpot. We learn that we are no longer alone in the Universe! If the Earthling sends a signal, then it is expensive and there is a possibility that the Alien may hear the signal and they will learn that we exist. For the Alien, the outcomes are similar for their two choices.

What is an Earthling to do? If they listen, it is cheap and there is some possibility of finding out something amazing. If they send a signal, it is expensive, and while the Alien may learn something, the Earthling may not. The Earthling is probably better off listening for the signal. The Alien is also better off just listing for the signal. Given that both the Earthling and the Alien are listening, they are never going to hear anything and the search for extraterrestrial intelligence may not in fact be very intelligent.

### 1.2.2 Can AI Cope with Traffic?

It is not clear that we can or should have humans and AI-driven cars interact on the road. Consider the problem of turning across traffic at a traffic light. Most Americans, when confronted with the question of when to turn across traffic, will generally wait for the green arrow. While the through traffic often speeds up on the yellow. In Australia in the 1980s, there were few lights with green arrows and so drivers were pretty aggressive at turning across traffic. Australians tend to slow and stop at a yellow light, lest you T-bone the person aggressively turning across traffic. In Vermont, you may be surprised to learn that it is customary to allow drivers to turn across traffic on the green light. It is a bit of a shock to hit the gas only to look up and see a car turning across you. As in Australia, Vermont drivers learn to go easy when the light goes green in order to allow the turning traffic through. There are three different equilibrium outcomes in three different places. Would an AI-driven car know these rules and know how they change from place to place?

## 1.3 End of Skyjacking

Airplane hijacking used to be a thing. The high point was the late 1960s and early 1970s with over 50 a year, but even as late as 2000, there were over 20 hijackings a year. By the Twenty teens, there were some years with zero. To see what happened? See Figure 1.1

### 1.3.1 More Security?

Was there an increase in security? Yes. Security increased in the 1970s and 1980s. Obviously, there was a huge increase in security in the 2000s after the terrorist attacks on September 11, 2001. But that is not the reason, at least not the reason for the difference in the number hijackings before and after September 11, 2001. What changed was the game between passengers and hijackers.

When hijackers take over a plane, they are way outnumbered by the passengers. Hijacking relies on an implicit understanding between hijackers and passengers. Passengers agree to sit quietly in order to be safely released. Hijackers agree to release passengers so that they can get the plane or the ransom or whatever it is that hijackers want. You might think that hijackers have weapons while passengers don't, but neither may be true. The September 11 hijackers used box cutters, not exactly an AR-15.

### 1.3.2 Hijacking Data using R

We can see what happened by importing and plotting data on airline hijackings.[2] Here we use `read.csv()` which creates a `data.frame` object called `df`. This object is basically the stock standard data set we use in Stata or any other statistical programming language. If you go over to the Environment window and click on `df`, you will see an excel-like sheet with variable names as the column names and years on the rows. To call out variables in the data set `df`, we use the symbol $ and then the name of the variable, for example `df$Year`.

```
> # dir = "ENTER LOCATION OF DATA FILES"
> file = paste0(dir, "Airline_Hijackings.csv")
> df = read.csv(file)
```

```
> plot(df$Year, df$Hijackings, type = "l")
> abline(v = 2001)
```

[2]Data from the Aviation Safety Network, https://aviation-safety.net/statistics/period/stats.php?cat=A1.

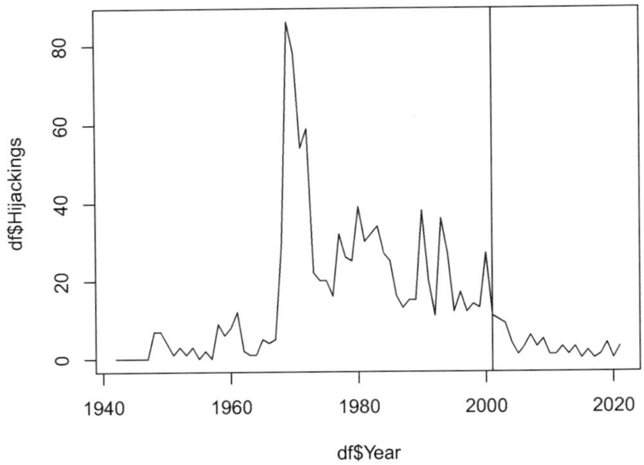


**FIGURE 1.1**
Annual count of airline hijackings around the world from 1940 to 2021. The plot shows the increase up to the 1970s and then a slow decline into the 90s, then a dramatic dropoff after 2001.

Figure 1.1 presents the annual count of airline hijackings around the world. Hijacking really spiked in the late 60s early 70s. It then steadily declined to 2001. The figure shows the dramatic change in the number of hijackings before and after 2001.

Figure 1.1 is made with the base **R** plotting system. The function `plot()` creates a scatter plot with years on the x-axis and the number of hijacking in the y-axis. In order to get the line, use the option `type = "l"`. In addition, a vertical line is added using the function `abline()` setting `v = 2001`.

### 1.3.3 United Arlines Flight 93

Figure 1.1 suggests that something changed in the early 2000s. We actually know exactly when and where the game between hijackers and passengers changed. It changed on United Airlines Flight 93 over central Pennsylvania on the morning of September 11, 2001. There had been three flights hijacked earlier in the morning, the hijackers took over the plane and the passengers sat quietly. The passengers and crew assumed that they would be safe as long as they obeyed the instructions. The hijackers were playing a different game. They didn't want to take the plane to somewhere or get the ransom money.

They wanted to fly the plane into New York's World Trade Center or the Pentagon in Arlington, VA. Passengers and crew on Flight 93 learned that the game had changed through text messages and phone calls. They knew that sitting quietly was not going to get them home to their families safely. Passengers and crew overwhelmed the hijackers and brought the plane down in a field killing all on board.

The game had changed. Hijackers can no longer rely on the compliance of crew and passengers. We see this even with people being rowdy on planes. The passengers and crew overwhelm them and the rowdy passenger ends up tied down to their seat with duct tape.

## 1.4 A Game

A game is a formal mathematical object. This formalism is super important. New students to game theory will often substitute their own intuition for the formal mathematics. This is a big error. Keep your intuition in check. Intuition is useful for understanding the result you get or suggesting that you may have made an error in your calculation. To find the solution to a game, use math.

The section presents the formal mathematical objects used in game theory.

### 1.4.1 Definitions

A basic game has three formal parts.

- A set of **players**
- A set of **strategies**
- A set of **payoffs**

**Definition 1.** *A* player *is a strategic actor in the game. They have choices and make their choices based on what other players in the game do.*

What makes game theory different from other mathematical models of economic activity is the assumption that players in the game are strategic actors. They account for the reaction of other players in the game when making their choice. In Econ 101, we assumed that the players of the game, the consumers or businesses, are not strategic. These player may optimize utility or profits, but they take the actions of the other players as fixed. Players in a game account for the strategic behavior of other players in the game.

**Definition 2.** *A* strategy *is a complete plan of actions for every possible circumstance faced by the player in the game.*

When we start in this book, the strategies will be very simple. They will just be actions. Each player will have a limited set of choices and their strategies correspond to those choices. In the second part of the book things get a whole lot more complicated. A **strategy** is a complete plan. Think about how complicated this object can be. Think about the game chess. There are millions and millions of different configurations. It is estimated that there are $10^{44}$ possible legal positions in a game of chess.[3] That is 1 followed by 44 zeros. A strategy for chess must state what the player will do in each possible case!

**Definition 3.** Payoffs *are the outcomes of interest to the players of the game, these may be winning or losing, monetary values or utilities.*

The third object of a game is the payoffs. The **payoffs** to players depend directly upon the combination of actions taken by players in the game. It is this dependence that makes it a strategic situation.

Often in games the absolute magnitude of the payoffs is less important than the relative magnitude. Students new to game theory may be put off by the weird precision that seems to exist in the games. In reality, the numbers are just stand ins, what really matters is the relative ordering of the outcomes.

### 1.4.2 Examples

To get a bit more of a feel for what a game is, consider the examples mentioned above.

First up is the SETI game. There are three formal parts of the game, players, strategies and payoffs. Assume that when the player chooses to signal they also listen for a signal.

- Players: Earthling, Alien
- Strategies:
  - Earthling: Listen, Signal
  - Alien: Listen, Signal
- Payoffs:
  - Earthling: Listen, Alien: Listen
    - Earthling: Nada
    - Alien: Nada
  - Earthling: Listen, Alien: Signal
    - Earthling: Earthling learns there are Aliens and it changes the course of human history!
    - Alien: Nada plus a large cost for sending the signal.

[3] `https://github.com/tromp/ChessPositionRanking` accessed 4/14/23.

  - Earthling: Signal, Alien: Listen
    * Earthling: Earthling pays a large cost for nothing
    * Alien: Aliens learn there are Earthlings changing the whole course of Alien history!
  - Earthling: Signal, Alien: Signal
    * Earthling: Earthling pays a large cost but learn their are Aliens!
    * Alien: Aliens pay a large cost but learn there are Earthlings!

  Earthlings and Aliens pay a large cost but both learn each other exists.

We can do the same thing and formally present the hijacking game. Hijacking game:

- Players: Hijacker, Passenger
- Strategies:
  - Hijacker: Kill passengers, Don't kill passengers
  - Passenger: Fight hijacker, Don't Fight hijacker
- Payoffs:
  - Hijacker: Kill, Passenger: Fight
    * Hijacker: Dead
    * Passenger: Dead
  - Hijacker: Kill, Passenger: Don't Fight
    * Hijacker: Dead or arrested
    * Passenger: Dead
  - Hijacker: Don't Kill, Passenger: Fight
    * Hijacker: Dead or injured
    * Passenger: Dead or injured
  - Hijacker: Don't Kill, Passenger: Don't Fight
    * Hijacker: Get money or where they are going
    * Passenger: Go home to their families

Prior to Flight 93, everyone thought that game was such that the Don't Kill, Don't Fight outcome was the most likely to occur. In the first three flights of September 11, 2001, the hijackers took advantage of the fact that crew and passengers thought that by choosing Don't Fight, they would have a reasonable possibility of going home to their families. During Flight 93, passengers and crew realized that hijackers were playing the Kill strategy and adjusted their behavior accordingly. Today passengers and crew will almost certainly play Fight. Given that strategy, Hijackers are better off not hijacking in the first place.

## 1.5 Prisoner's Dilemma

It is probably the most famous game in game theory. If anyone has heard all about game theory, then they have heard of the **prisoner's dilemma** game. The classic version of the tale is that there are two suspects to a crime. The suspects are taken into two different cells. The detectives visit each suspect in turn. They offer the same deal to both. The deal is that if the suspect rolls over and cops to the crime implicating the other suspect then they get to walk. While if they stay quiet and their mate rolls over they go away for a long time. Each suspect is presented with this deal and made aware that their mate also got the deal.

The suspects also know that if they stay mum, they will probably get charged with a less serious crime. The police don't have anything to tie them to the more serious crime except each other. Finally, if they both squeal, then they both get an intermediate sentence.

The section presents a formal analysis of the prisoner's dilemma game.

### 1.5.1 The Game

Table 1.1 presents the **normal form** representation of the prisoner's dilemma game. Each player has two choices, stay mum or squeal, with Suspect 1's choice on the rows. The payoffs for each outcome are in the cells, with Suspect 1 listed first. The payoffs are in number of years of prison, with bigger numbers being worse, and thus negative.

**TABLE 1.1**
Normal form representation prisoner's dilemma game, with two players $S_1$ and $S_2$. For $S_1$ their choices are the rows and their payoffs are listed first in each cell.

| $S_1, S_2$ | Mum | Squeal |
|---|---|---|
| Mum | $-1, -1$ | $-5, 0$ |
| Squeal | $0, -5$ | $-3, -3$ |

We can also write the game out like we did above.

- Players: Suspect 1, Suspect 2
- Strategies:
  - Suspect 1: Mum or Squeal
  - Suspect 2: Mum or Squeal

- Payoffs:
  - {Mum, Mum}: {−1, −1}
  - {Mum, Squeal}: {−5, 0}
  - {Squeal, Mum}: {0, −5}
  - {Squeal, Squeal}: {−3, −3}

There are four outcomes, both choose Mum, both choose Squeal, Suspect 1 chooses Mum and Suspect 2 chooses Squeal, and Suspect 1 chooses Squeal and Suspect 2 chooses Mum.

### 1.5.2 What is the Best Outcome?

The outcome with the highest payoff for both players is {Mum, Mum}. NOTE, the outcome is not payoffs, but strategies.

This outcome gives a total of −2 points, while other outcomes give both players less in total (−5 and −6). This outcome is **Pareto optimal**. The only way to make one player better off is by making the other player worse off. The only outcome that is not Pareto optimal is {Squeal, Squeal}. Moving from {Squeal, Squeal} to {Mum, Mum} makes both players better off.

### 1.5.3 What should You Do?

Assume you are Suspect 1. What should you do?

- Assume Suspect 2 chooses Mum. The payoffs from your choices are as follows:
  - Mum: −1
  - Squeal: 0
- Assume Suspect 2 chooses Squeal. The payoffs from your choices are as follows:
  - Mum: −5
  - Squeal: −3

If your mate is going to stay mum, then you get a big payoff from squealing. You get to walk out a free person. So in that case you are better off squealing. What if your mate squeals? That rat bastard! Well, if you stay mum then you go down for a long time. If you also squeal you at least reduce your sentence. So in that case you should squeal. In fact, in both cases you should squeal.

We say that squealing is a **dominant strategy** because for every possible strategy of the other player, you are always better off squealing.

### 1.5.4 Does This Make Sense?

The reason that the **prisoner's dilemma** game is so famous is that it suggests an outcome that some find unintuitive. It suggests that both players would squeal when both players would be better off if they both agreed to stay mum. By squealing they both end up spending three years in prison, while if they had stayed mum they would have both only spent one year in prison. Why would they end up at the outcome that makes them worse off?

## 1.6 Empirical Analysis: *Friend or Foe* using R

What happens when real people play games like the prisoner's dilemma presented above? One way economists test this is by running experiments. These are often done on college students for small payoffs, say money or a Starbucks gift card. What about the real world?

In the early 2000s, a game show aired on the Game Show Network called *Friend or Foe*. The show had players pair up and compete against other teams to answer trivia questions. The teams earned cash by answering the questions correctly. However, before taking their winnings home, the two players in the winning team had to play a game. Not a game show game, well sure a game show game, but a game theory game as well. A prisoner's dilemma game to be precise, the game was called *Trust Box*.

The cool thing is that someone watched these games and created a data set with the payoffs, the outcomes, the strategies chosen and some demographic information on the players. Note also that amounts here are in thousands of dollars. So real people playing for real money in a completely unreal situation.

The section presents the game and uses **R** to analyze data collected on how actual contestants played the game.

### 1.6.1 Trust Box

**TABLE 1.2**
Normal form representation *Trust Box* game, with two players $P_1$ and $P_2$. For $P_1$ their choices are the rows and their payoffs are listed first in each cell. The team's winnings is $x$. It is amount which will vary based on how well the team did in the trivia part of the show.

| $P_1, P_2$ | Friend | Foe |
|---|---|---|
| Friend | $0.5x, 0.5x$ | $0, x$ |
| Foe | $x, 0$ | $0, 0$ |

Table 1.2 presents the normal form representation of the *Trust Box* game.

Each player is in a box with a button to press, *Friend or Foe*. If both players press Friend then they share the proceeds evenly, $0.5x$ each, where $x$ is the team's winnings from the trivia part of the game. If the first player chooses Foe and the second player chooses Friend, then the first takes home all the winnings and the second gets nothing, $x$ and 0. If both players choose Foe, then they take home nothing, 0.

While the strategies have different names, this game has the same ranking of payoffs as the prisoner's dilemma game presented in the previous section. Remember it is the ranking of the payoffs that matters.

The choice Foe is a **weakly dominant strategy** for Player 1. Let Player 2 play Friend, for Player 1 Foe is better as $x > 0.5x$. Let Player 2 play Foe, for Player 1 Foe and Friend give the same payoff (0). If a strategy gives the highest payoff against some strategies and the equal highest against all other strategies, we say it is **weakly dominant**. Player 1 is better off choosing Foe. Player 2 is also better off choosing Foe.

### 1.6.2 Data Analysis in R

We can bring in the data from the game show that is available from the `Ecdat` package.[4] You can see that using the function `data()` you create an object called `FriendFoe` which is the data set. Each row is a different game and the two players are differentiated by whether or not there is a `1` after the variable name. These are the data described in Kalist (2004).

We can ask how actual people play the prisoner's dilemma game and look at which cells of the game described in Table 1.2 the players end up in. The syntax uses the **pipe operator** (`|>`). The result of the code on one line is inserted into the function on the next line. The syntax creates an object called `game_mat` which is eventually going to end up as a $2 \times 2$ matrix. The columns and rows of `game_mat` are named with the names of the action in the game. The table is printed using the `xtable` package.

```
> library(Ecdat)
> library(tidyverse)
> data("FriendFoe")
> game_mat = FriendFoe |>
+   summarize(
+     mean(play == "friend" & play1 == "friend"),
+ # this creates a vector of 0 and 1s depending whether
+ # the statement inside the parenthesis is true
+ # note the use of "==" to ask whether something is true
+     mean(play == "foe" & play1 == "friend"),
+     mean(play == "friend" & play1 == "foe"),
+     mean(play == "foe" & play1 == "foe")
+     ) |>
```

[4]To use this package you need to install it using the menu or `install.packages()`.

```
+       matrix(, nrow = 2)
> colnames(game_mat) = rownames(game_mat) = c("Friend","Foe")
```

```
> library(xtable)
> print(xtable(game_mat, digits = 2), floating = FALSE)
```

**TABLE 1.3**
Fraction of games that finish in each cell. Almost a third of the time the players end up at {Foe, Foe} and take home nothing.

| | Friend | Foe |
|---|---|---|
| Friend | 0.23 | 0.26 |
| Foe | 0.19 | 0.32 |

Table 1.3 presents the results of the game. It shows which cells everyone ends up in. Was this what you were expecting? The largest group of people end up playing Foe and Foe and getting nothing. Why is this? Do you think the game show designed it this way? Was the game show hoping that even more people would choose this option? What change to the game could you make to get more people to choose Foe?

### 1.6.3 Analysis in R

Table 1.3 raises lots of questions. Does the outcome of the game depend on the stakes of the game, that is, how much prize money the two players have won? Does it depend on the gender of the players or how similar the two players are or how much experience they had together.

The next bit of code creates a similarity index based on whether the two contestants are the same gender, the same race and similar in age. This uses the function `mutate()`.

```
> FriendFoe = FriendFoe |>
+   mutate(
+     similar = (sex == sex1) +
+       (max(abs(age - age1)) - abs(age - age1))/10 +
+          (white == white1)
+   )
```

For some of the analysis, it is helpful to reshape the data. The following code takes the columns associated with the first player and stacks them on top of the columns associated with the second player. The index `c(1:4, 12)` keeps the variables of interest for first player listed, while `c(8:11,13)` does the equivalent for the second player listed. The first data set is called `df1` and the second is `df2`. The code then labels the columns of `df2` using the column names of `df1`. It then joins the two data set together using the function `rbind()`, where the `r` is for row and calls the new data set `df`.

```
> df1 = FriendFoe[ , c(1:4, 12)]
> df2 = FriendFoe[ , c(8:11, 13)]
> colnames(df2) = colnames(df1)
> df = rbind(df1,
+            df2)
```

The data then create some new variables or recreate some previous variables lost from the original data. The function `rep()` repeats the vector or value in the first space by the number in the second. The variable `game` creates an id for each game in the data.

```
> df$game = rep(1:227, 2)
> df$round = rep(FriendFoe$round, 2)
> df$season = rep(FriendFoe$season, 2)
> df$cash = rep(FriendFoe$cash, 2)
> df$similar = rep(FriendFoe$similar, 2)
```

Now we can do some analysis on this data. What is the relationship between how much a player earns during the trivia part of the game, their demographics and how far they get in the game? Also, does their strategy choice in the prisoner's dilemma depend on their demographics, or the amount of cash on the table? For the first part we estimate the relationship with standard **linear regression** using the `lm()` function.`lm1` is an object created in **R** that can be used by other code to create regression result tables. For the second part, we use a **logit model** and a **probit model**. These are standard methods for analyzing problems where the outcome has two possibilities. Here `play` could be either `"friend"` or `"foe"`. In **R** we use the `glm()` function for generalized linear model. To specify a logit model, you need to state `family = binomial(link = "logit")`.

```
> lm1 = lm(cash ~ sex + white + age + round + season,
+          data = df)
> glm1 = glm(play == "friend" ~ sex + white + age +
+              similar + cash,
+            family = binomial(link = "logit"),
+            data = df)
> glm2 = glm(play == "friend" ~ sex + white + age +
+              similar + cash,
+            family = binomial(link = "probit"),
+            data = df)
```

We can display the results using the `stargazer` package.[5]

[5]Here the package is transforming the **R** output into LateX code.

**TABLE 1.4**
Linear regression estimates for the relationship between the amount earned in the trivia part of the game as well as logit and probit estimates of the probability of playing friend in the *Trust Box* part of the game.

```
> require(stargazer)
> stargazer(list(lm1, glm1, glm2),
+           keep.stat = c("n", "rsq"),
+           float = FALSE)
```

| | *Dependent variable:* | | |
|---|---|---|---|
| | cash | play == "friend" | |
| | *OLS* | *logistic* | *probit* |
| | (1) | (2) | (3) |
| sexmale | 0.102 | −0.105 | −0.067 |
| | (0.175) | (0.193) | (0.120) |
| whiteyes | 0.309 | 0.441* | 0.275* |
| | (0.234) | (0.265) | (0.163) |
| age | −0.005 | 0.041*** | 0.025*** |
| | (0.011) | (0.013) | (0.008) |
| round2 | 2.242*** | | |
| | (0.215) | | |
| round3 | 5.090*** | | |
| | (0.206) | | |
| season2 | −0.553*** | | |
| | (0.183) | | |
| similar | | 0.077 | 0.049 |
| | | (0.121) | (0.075) |
| cash | | 0.007 | 0.005 |
| | | (0.034) | (0.021) |
| Constant | 1.246*** | −2.028*** | −1.270*** |
| | (0.449) | (0.770) | (0.476) |
| Observations | 454 | 454 | 454 |
| $R^2$ | 0.591 | | |
| *Note:* | *p<0.1; **p<0.05; ***p<0.01 | | |

Table 1.4 presents the regression results. The first column of results is for the **linear regression**. It shows the relationship between the size of the prize, demographic characteristics, and characteristics of the team's performance. The last two columns present the logit and probit regressions respectively. These show the relationship between a player's demographic characteristics and playing the strategy friend in the *Trust Box* game.

Table 1.4 shows that the amount of cash earned increases significantly in the latter rounds. It also seems that the budget decreased in season 2. In addition, the table shows that the strategies of the players do seem to change with the player demographics. Older players seem to be much more willing to cooperate. The stakes, although in thousands of dollars, seem less important. Similarity of demographics does not seem important either. That said, according to List (2006), there is a selection round where players are matched into teams and that selection may account for the results.

## 1.7 Discussion and Further Reading

Games are all around us. Strategic behavior is very common. This book aims to give you the tools to understand how the real world works. It aims to make you comfortable understanding the strategic behavior generating the data you are analyzing.

The *Trust Box* analysis is based on List (2006) and Kalist (2004). The American economist, John List, has made his career pioneering the running of economic experiments in the real world, **field experiments**. One objective of these experiments is to test the predictions of the game theory models (Camerer, 2003). The book uses data from another field experiment in Chapter 8.

# 2

# *Nash Equilibrium*

## 2.1 Introduction

The modern version of game theory began with the PhD dissertation of the American mathematician, John Nash. You may recognize the name from the hit movie, *A Beautiful Mind* with Australian Russell Crowe playing West Virginian Nash. Nash showed that for a large set of games, there is at least one outcome that is stable. Nash was interested in which outcome of the game is likely to occur. Stability seems like a prerequisite for any outcome that we are likely to observe. Stability says that if an outcome happens to occur then the outcome of the game is unlikely to change. In contrast, instability says that if an outcome happens to occur then the outcome of the game is likely to change. We are unlikely to observe an unstable outcome. The fact that Nash's solution concept is both a prerequisite for an outcome that is likely to occur and exists for a large set of games, makes it a super valuable idea. Today we call his proposed outcome a **Nash equilibrium**

Nash worked in the sub field of game theory called non-cooperative game theory. As he himself points out in his Nobel essay, this style of game theory was not in style.[1] It differed substantially from the ideas of Johnny von Neumann who was a senior mathematician at Princeton where Nash did his dissertation. Hungarian-American von Neumann helped develop game theory and was one of the first to suggest its use in economics. His collaboration with German-American economist, Oskar Morgenstern, produced the first game theory text, *Theory of Games and Economic Behavior* (1944). Luckily for us, Nash was stubborn and his efforts helped turn non-cooperative game theory into what today we just call game theory.

In this chapter, we will introduce this workhorse equilibrium concept as well as the concept of a **dominant strategy equilibrium**. The chapter illustrates these concepts by looking at the market structure in retail tire stores. How many stores are there going to be in the market? How does the number of stores relate to the price of tires in the market?

[1] `https://www.nobelprize.org/prizes/economic-sciences/1994/ceremony-speech/`. Accessed on 2/23/23.

DOI: 10.1201/9781003351603-2

## 2.2 What is a Nash Equilibrium?

What is an **equilibrium**? We are looking for an outcome that is a reasonable prediction of the game. We are not looking for the best outcome or the worst outcome, but the outcome with the most empirical content. The one that we are likely to observe in the wild. This outcome is not chosen in some ad hoc manner but based on a set of assumptions. A set of rules.

**Definition 4.** *An* equilibrium concept *is a set of rules for determining which outcome(s) of the game will occur.*

There are various assumptions we may want to hold in order to say that a particular outcome is a reasonable prediction of the game. The section presents the idea of **dominance** and how we may come up with a prediction of the game based only on assuming that the players of the game make choices that are rational.

### 2.2.1 Dominance

Before we get to **Nash equilibrium**, consider an alternative equilibrium concept, called **dominant strategy equilibrium**. Chapter 1 introduced the concept of a **dominant strategy** and **weakly dominant strategy**.

**Definition 5.** *A strategy is* dominant *if it gives the player the highest payoff irrespective of the strategies of the other players. A strategy is* weakly dominant *if for some strategies of the other players, the payoff is equal highest.*

A dominant strategy is always the best choice. It does not matter what any other player of the game does. Given that it is always the best choice, then presumably it is the one that a rational player is most likely to choose.

**Definition 6.** *A (weakly)* dominant strategy equilibrium *is an outcome of the game where all players play strategies that are (weakly) dominant strategies.*

If we assume that players are rational, and there exists an outcome which is a **dominant strategy equilibrium**, then we would expect that outcome to be the one we observe. The issue is that such outcomes may not exist in the game we are studying.

In the discussion above there is a subtle but important assumption. If a player has two choices, $A$ and $B$, where $A$ is part of a proposed equilibrium but $B$ gives the player the exact same payoff, then $A$ is still part of the equilibrium. For $A$ not to be part of an equilibrium, there must be a $B$ that makes the player strictly better off.

### 2.2.2 Prisoner's Dilemma

Consider a version of the prisoner's dilemma game presented in Chapter 1. In this version, the actions have generic names, but the payoffs have the same ordering as they did in the version presented in the previous chapter. Remember it is the ordering that matters, just like in Formula 1.

**TABLE 2.1**
Normal form representation of a prisoner's dilemma game with players $P_1$ and $P_2$ and actions BLACK and RED. $P_1$'s actions are on the rows. The payoffs are in the cells with $P_1$'s in the first position.

| $P_1, P_2$ | BLACK | RED |
|---|---|---|
| BLACK | 3, 3 | 0, 5 |
| RED | 5, 0 | 2, 2 |

Table 2.1 presents the generic prisoner's dilemma game with two players $P_1$ and $P_2$ and two actions BLACK and RED. The payoffs are written out in **normal form**, which is a matrix-like representation of the game.

The outcome that has both players playing **dominant strategies** is {RED, RED}.

If you are player 1, you see that you are better off choosing RED irrespective of what player 2 does. In the table, player 1's payoff is the first element in the cell. Player 1's strategies are the rows. If player 2 chooses BLACK, we see that RED has a payoff of 5 which is greater than 3. If player 2 chooses RED, we that RED has a payoff of 2 which is greater than 0. RED is the **dominant strategy** for player 1. Similarly for player 2.

The reason the prisoner's dilemma is so famous is that it predicts an outcome which is bad for the players of the game. Even though two rational players will end up at the dominant strategy equilibrium, both players would be better off choosing {BLACK, BLACK} which has a payoff of {3, 3}. This outcome has a payoff that is strictly better for both players than the predicted outcome which has a payoff of {2, 2}. Our equilibrium concept predicts the outcome that we believe is most likely to occur not necessarily the outcome that is the best for the players.

The prediction of this game drives real world policy. The US Department of Justice (DOJ) has an explicit policy of giving leniency to firms that are the first to provide evidence of collusion in a market.[2] The policy explicitly aims to create a prisoner's dilemma among colluding firms.

Does dominant strategy equilibrium predict outcomes in real games? Chapter 1 analyzes data from a real TV program where players are playing a prisoner's dilemma game for real money. In *Friend or Foe*, if both players choose Foe then they each leave the show without any of their winnings. While

[2] `https://www.justice.gov/atr/leniency-program` accessed on November 12 2023.

if they both choose Friend then they would share their winnings. In the data, we see that {Foe, Foe} has the highest probability of occurring.[3]

### 2.2.3 Coordination Game

Now consider a slightly different game. We call the game presented in normal form in Table 2.2, a coordination game. We will soon see why.

**TABLE 2.2**
Normal form representation of a **coordination game** with players $P_1$ and $P_2$ and actions RED and BLACK. $P_1$'s action choices are on the rows. Payoffs are in brackets with $P_1$ listed first.

| $P_1, P_2$ | BLACK | RED |
|---|---|---|
| BLACK | 2, 5 | 0, 0 |
| RED | 0, 0 | 5, 2 |

Assume you are player 1. Can you work out your best strategy? We can go through the best choice for each choice made by player 2.

- Assume $P_2$ chooses BLACK. $P_1$'s payoffs are:
  - BLACK: 2
  - RED: 0
- Assume $P_2$ chooses RED. $P_1$'s payoffs are:
  - BLACK: 0
  - RED: 5

Is there a dominant strategy to this game? No. The best choice for you depends on the choice of player 2. If player 2 plays BLACK, then you should also play BLACK because $2 > 0$. If player 2 plays RED, then you should also play RED because $5 > 0$. See a **coordination game**! Both players should coordinate on the color.

What outcome do you predict will happen in this game? While both players want to coordinate they do not agree on which outcome to coordinate. Player 1 prefers that they coordinate on RED and Player 2 prefers that they coordinate on BLACK.

[3]The second part of the book considers changes to the prisoner's dilemma game that lead to predicted outcomes where the players have higher payoffs.

## 2.3 Nash Equilibrium

The section discusses the definition of a **Nash equilibrium** and illustrates how to find one or more Nash equilibria in games that have been presented earlier in the chapter, the prisoner's dilemma and the coordination game.

### 2.3.1 Definition

**Definition 7.** *A* Nash equilibrium *is a set of strategies such that each player's strategy has the highest payoff given the strategies of the other players.*

Like with the dominant strategy equilibrium, we assume that each player chooses the strategy that gives the highest payoff. Our base assumption is that the players of the game are rational. The difference here is that choice is dependent on the choice of the other players.

Stability is baked into Nash's concept. Each player is choosing the best option given the choices of all the other players. If all players are choosing the best option, then there is no need for any player to change. The outcome is stable.

### 2.3.2 Algorithm for Finding Nash Equilibrium

While Nash equilibrium has a number of nice properties, finding it is not one of them. We will use the following cumbersome algorithm. One issue new students have with game theory is that it seems like the teacher just picked some outcome at random and voila, it is an equilibrium! Of course, the teacher did not just pick at random. They already knew the answer. In reality, you do not know which one to pick and so unfortunately you just have to try them all. There are no short cuts.

- Step 1: Choose a candidate outcome (set of strategies).
- Step 2: Hold Player 1's strategy fixed. Is Player 2's strategy optimal?
  - Yes: Go to Step 3.
  - No: Not a Nash equilibrium.
- Step 3: Hold Player 2's strategy fixed. Is Player 1's strategy optimal?
  - Yes: Nash equilibrium.
  - No: Not a Nash equilibrium.

If the game has more than two players, then the algorithm can be expanded to go through each player in turn.

### 2.3.3 Prisoner's Dilemma Game

Let's test out our algorithm on a game we already know, the prisoner's dilemma. For this game, there exists a more efficient algorithm, but we are illustrating how this more general algorithm works. Remember the first step is just picking an outcome. No magic.

- Step 1: $\{BLACK, BLACK\}$
- Step 2: $P_1$ plays BLACK. Is BLACK optimal for $P_2$?
  - BLACK: 3, RED: 5
  - No: Not a Nash equilibrium.

Let's pick another one.

- Step 1: $\{RED, RED\}$
- Step 2: $P_1$ plays RED. Is RED $P_2$'s optimal strategy?
  - BLACK: 0, RED: 2
  - Yes. Go to Step 3.
- Step 3: $P_2$ plays RED. Is RED $P_1$'s optimal strategy?
  - BLACK: 0, RED: 2
  - Yes!
  - $\{RED, RED\}$ is a Nash equilibrium.

OK. There is some magic. You should try the algorithm on the other two outcomes.

### 2.3.4 Coordination Game

Let's try something a little more complicated, the coordination game presented in Table 2.2.

- Step 1: $\{BLACK, BLACK\}$
- Step 2: $P_1$ plays BLACK. Is BLACK optimal for $P_2$?
  - BLACK: 5, RED: 0
  - Yes. Go to Step 3.
- Step 3: $P_2$ plays BLACK. Is BLACK optimal for $P_1$?
  - BLACK: 2, RED: 0
  - Yes.
  - $\{BLACK, BLACK\}$ is a Nash equilibrium.

That was easy/lucky. Are there any other Nash equilibria of this game?

- Step 1: $\{RED, RED\}$
- Step 2: $P_1$ plays RED. Is RED $P_2$'s optimal strategy?
  - BLACK: 0, RED: 2
  - Yes. Go to Step 3.
- Step 3: $P_2$ plays RED. Is RED $P_1$'s optimal strategy?
  - BLACK: 0, RED: 5
  - Yes.
  - $\{RED, RED\}$ is a Nash equilibrium.

There are two Nash equilibria. Actually, there are even more, but we will come back to that in Chapter 5.

One of the valuable things about a Nash equilibrium is that it always exists.[4] Unfortunately, the cost is that there may be more than one Nash equilibrium in any game. Is one of the Nash equilibria of the coordination game more reasonable than the other? What is the most reasonable prediction of the game? This question of whether there exists more reasonable equilibria is called refinement. In this book, we will consider a number of refinements of Nash equilibrium. Chapter 4 analyzes what happens when the game we want to take to the data has multiple equilibria.

## 2.4 Entry Games

The section introduces empirical entry games. It presents the framework for how economists generally think about determinants of the number of firms in the market. You will often hear policy makers complain that prices are high because there is a lack of competition. There are too few firms in the market. Rarely, do policy makers step back and ask why there are too few firms in the market. The section presents a game in which a small number of firms choose to enter the market.

### 2.4.1 Bresnahan and Reiss

In a series of papers published in the early 1990s, the American economists, Tim Bresnahan and Peter Reiss, analyze the empirical implications of a simple entry game. Bresnahan and Reiss (1991b) analyze average prices for retail

[4] A Nash equilibrium exists for any finite game, that is any game where the number of players and the number of strategies is finite.

tire stores in various U.S. towns. The data shows that there is basically no relationship between observed average prices for tires and the number of tire retailers in the market.[5] Why do you think that is?

The problem is **endogeneity**.[6] The raw observations do not account for the fact that these tire retailers choose whether or not to be in the market based on various factors including the cost of selling tires. We are more likely to see more firms in larger markets, but some of these firms are going to have higher costs of selling tires and these higher costs will drive up prices. We have two countervailing forces, more competition driving down prices and less efficient firms driving up prices.

This is not just some academic issue. It was a central concern in the US Federal Trade Commission's (FTC) case against the merger of Staples and Office Depot (Ashenfelter et al., 2006). In the late 1990s, the two office supply super stores wanted to merge. As evidence against the merger, the FTC showed that prices where lower in cities with more stores. Is that a causal statement? Did the increase in the number of stores cause prices to fall? Moreover, would the opposite happen. If the merger took place and the number of independent firms in the market fell, would prices go up?

### 2.4.2 Two Firm Entry Game

Consider a relatively simple version of an entry game where there are just two firms. If only one firm enters then that firm earns monopoly profits and pays a fixed cost of entry, represented by the number 2. If that firm doesn't enter, nothing happens, which is represented by 0. If both firms enter there is competition but the firms also have to pay the fixed costs of entry. This payoff is represented by $-1$, which is bad. So a firm is willing to enter the market, but only if they are a monopolist.

More formally we have two players, the firm's strategies are entered or don't enter and the payoffs depend on whether the other firm enters as well.

- Players: Firm 1, Firm 2
- Strategies:
  - Firm 1: Enter, Don't Enter
  - Firm 2: Enter, Don't Enter
- Payoffs:
  - {Enter, Enter}: $\{-1, -1\}$
  - {Enter, Don't}: $\{2, 0\}$

---

[5] Surprisingly the authors state that their data shows that "entry lowers margins". It is unclear how they come to this conclusion.

[6] In econometrics we use the term **endogeneity** to mean that different cases observed in the data may not be determined at random. Importantly, case assignment may be directly related to the observed outcome.

- {Don't, Enter}: {0, 2}
- {Don't, Don't Enter}: {0, 0}

We can also represent this game in a normal form payoff matrix.

**TABLE 2.3**
A normal form representation of a two firm entry game. Firm 1's strategies are the rows and the Firm 2's strategies are the columns. The payoffs are in the cells, with Firm 1's payoff first.

| Firm 1, Firm 2 | Enter | Don't |
|---|---|---|
| Enter | $-1$, $-1$ | 2, 0 |
| Don't | 0, 2 | 0, 0 |

Table 2.3 presents the normal representation of the game. What is the Nash equilibrium of the game? It seems unlikely that both firms will enter and we see quickly that is not an equilibrium. If both firms enter, then Firm 1 would have been better off not entering as 0 is greater than $-1$. See the first column and the payoffs of the first element which are for Firm 1.

How about the case where both firms don't enter? Again we see that Firm 1 is better off entering. Look at the second column and see that 2 is greater than 0.

Let's check the other outcomes more systematically using our algorithm.

- Step 1: {Enter, Don't}
- Step 2: Firm 1 plays Enter. Is Don't optimal for Firm 2?
  - Enter: $-1$, Don't: 0
  - Yes. Go to Step 3.
- Step 3. Firm 2 plays Don't. Is Enter optimal for Firm 1?
  - Enter: 2, Don't: 0
  - Yes. It is a Nash equilibrium!

Is that the only Nash equilibrium of the game?

No. There is another Nash equilibrium where Firm 2 enters but Firm 1 does not.[7] This entry game is a coordination-type game.

Game theory makes an interesting prediction. It predicts that the market will be a monopoly, but it doesn't predict which firm will be the monopoly. This turns out to have some implications for the empirical analysis of bookstores analyzed in Chapter 4.

[7]Again there are other equilibria but we will get to them. Be patient!

### 2.4.3 Many Firm Entry Game

We can make the game more general by having up to $\bar{N}$ firms choosing whether or not to enter.

- Players: $\bar{N} > 0$ firms
- Strategies: Enter, Don't Enter
- Payoffs:
  - Enter: $\frac{(a-c)^2}{b(N+1)^2} - F$, where $N \leq \bar{N}$, where $N$ is the number of firms that choose Enter.
  - Don't Enter: 0

In this case we have a small number of firms, $\bar{N}$. Each firm has a entry cost $F$. Entry costs may include finding a retail space, developing the space to sell tires, contracting with wholesalers and manufacturers, etc. Once firms enter (or not), the market mechanism determines the profits each firm will make.

The profit function is on the complicated side. Notice that the profits the firms make in the market are determined by the number of firms that enter ($N$). The more firms that enter, the lower the profits. Competition drives down prices and profits. Moreover, profits could be so low that they are below the fixed cost of entry. In that case, the firms are better off not entering.

To see where the complicated profit function comes from, assume price is determined by the following function.[8]

$$p = \frac{a}{N+1} + \frac{Nc}{N+1} \tag{2.1}$$

where $c$ is the **marginal cost of production**, $a$ is a demand parameter, and $N$ is the number of firms that enter the market. As the number of firms enter the price falls and it converges to marginal cost as $N$ gets large. The first part is close to zero when N becomes big. The second part is close to c because the fraction of $N$ over $N+1$ is close to 1 when $N$ is big. The **marginal cost of production** refers to the incremental costs of selling a tire, this may include hourly wages and the wholesale cost of the tire itself.

If a firm enters the market their profits are determined by quantity that they sell ($q$) multiplied by their profit margin ($p-c$). Demand in the market is assumed to be determined by the following linear function $Q(p) = a - bp$. The parameter $a$ determines the level of demand and the parameter $b$ determines the sensitivity of demand to price. Demand falls more dramatically for a particular price increase if $b$ is larger. Each firm in the market is identical and so they just split demand evenly, $q = \frac{Q}{N}$.

$$q \times (p-c) = \frac{(a-c)^2}{b(N+1)^2} \tag{2.2}$$

Multiplying demand by margin gives the profit function.

[8]This is the solution to a Cournot game with $N$ symmetric firms which is a game discussed in Chapter 3.

### 2.4.4 Nash Equilibrium

What is the outcome of this game? A Nash equilibrium requires that each firm is playing its optimal strategy given the actions of all the other firms.

Each firm will enter if and only if the following inequality holds.

$$\frac{(a-c)^2}{b(N+1)^2} > F \tag{2.3}$$

This means that if there are $N$ firms in the market in equilibrium. The inequality above must hold. If it didn't some of the firms would not enter the market. Also, it can't be profitable for another firm to enter. What if one more firm enters? In that case, profits for each firm must fall below the fixed cost of entry.

$$\frac{(a-c)^2}{b(N+2)^2} < F \tag{2.4}$$

If there are $N < \bar{N}$ firms in the market, then it must be profitable for all of those firms to enter, but not profitable for any more firms to enter.

### 2.4.5 Fixed Cost of Entry

Let's add one more complication to the game. Let the fixed costs be a function of the number of firms that enter, $F = \theta(N+1)^K$, where $K$ and $\theta$ are some parameters of costs. This assumption states that when more firms are in the market their costs of entry are higher. It may be that land or facility space becomes more expensive when there are more firms looking to use the land or space to sell tires. This is the idea that different firms have different fixed costs of entry. In markets with a small number of firms, those that enter will have low fixed cost of entering. In markets with a large number of firms in equilibrium, the firms will have higher fixed cost of entering.

### 2.4.6 Equilibrium Number of Firms

The number of firms in the market is an integer, a counting number, 1, 2, 85, etc. Unfortunately, while these numbers are easy they are annoying to use for solving equations. To make solving the equation easier we will make the unrealistic assumption that the number of firms in the market is a real number.[9] Real numbers are useful because they have the property that there exists a solution to our equilibrium equation.

In equilibrium the following equality holds.

$$\frac{(a-c)^2}{b(N+1)^2} = \theta(N+1)^K \tag{2.5}$$

[9] Real numbers include integers and rational numbers (fractions) but also more exotic numbers like $\pi$ or $\sqrt(2)$.

If $N$ firms enter then for each firm they get the same profits from entering or not entering. The Nash equilibrium does not predict which $N$ firms enter just that there will be $N$ firms that enter. Solving, the equilibrium number of firms is $N = \left(\frac{(a-c)^2}{b\theta}\right)^{\frac{1}{K+2}} - 1$. Below we will use logs and it will look nicer.

We see that the equilibrium number of firms is increasing in the profitability of the market and decreasing in the customer's sensitivity to price.

### 2.4.7 Simulation of Entry Game in R

To understand how this game works, we can simulate entry into thousands of markets. In the simulation, there are 1,000 markets with a maximum of 10 firms. Demand and cost parameters vary from market to market as do entry costs.

The code uses the function `runif()`. This generates a set of random numbers that are uniformly distributed between 0 and 1. The probability of drawing any particular number between 0 and 1 is the same. To be able to replicate the results exactly it uses `set.seed()`.[10] The different values used are just made up.

```
> set.seed(123456789)
> M = 1000
> N_bar = 10
> a = 4 + 1*runif(M)
> b = 0 + 0.6*runif(M)
> c = 1 + 2*runif(M)
> theta = 0 + 0.05*runif(M)
> K = 0.9
```

We can put the solution to Equation (2.5) in code form. Given the equilibrium entry, we can then plug the numbers back into the pricing equation to determine equilibrium prices in each of the markets. This is an example of where **R** shines. We can write out something that looks pretty similar to the math, but is hiding a lot more complexity. These two lines actually determine the equilibrium number of firms and equilibrium prices for all 1,000 markets.

```
> N = (((a - c)^2)/(b*theta))^(1/(K+2)) - 1
> p = a/(N+1) + (N*c)/(N+1)
```

We can compare equilibrium prices to the theoretical relationship between prices and the number of firms. To determine the equilibrium prices we calculate the average price at each equilibrium level of the number of firms. To do this, we use the package `data.table`.[11] This package is very useful for doing calculations

[10]Computers don't actually generate random numbers. The numbers come from a complicated non-linear function. They look random, but if you know the previous number in the sequence and the function used, then you can exactly determine the next number in the sequence.

[11]The syntax used by `data.table` is different from base **R** and also from `tidyverse()`.

on lots of different subsets of the data. Here we want to do the calculation for each market with the same number of equilibrium entrants. Round the number of firms using `round()` so that the number of firms is an integer. The resulting data set `dt`, for data table, has two variables "N" and "p". The next line takes the average price for each market with $N$ firms and creates a new data set `dt1`.

```
> library(data.table)
> dt = data.table(N = round(N),
+                 p = p)
> dt1 = dt[, .(p = mean(p)), by = N]
```

To calculate the theoretical equivalent, we take the average for the parameter values and calculate the price if the number of firms in the market was determined exogenously rather than in equilibrium. The last line uses the price formula from Equation (2.1).

```
> a_m = mean(a)
> c_m = mean(c)
> N_m = 1:N_bar
> p_m = a_m/(N_m + 1) + (N_m*c_m)/(N_m + 1)
```

Figure 2.1 presents the relationship between the number of firms in the market and the price. This is only an example, but it does show that the theoretical relationship between price and the number of firms can differ from the empirical relationship. The theoretical relationship shows that prices fall with competition. In the simulated data, the relationship between the number of firms and the price in the market is less negative, at least for a small number of firms. The reason is that the empirical relationship is an equilibrium where the firms are taking account of what the price will be when deciding to enter the market.

If we just naively take the number of firms in a market and regress that on price, we are not finding the true relationship between price and competition. We are estimating a relationship mediated by equilibrium decisions of firms. Once we acknowledge this, we have two choices. Throw up our hands and give up or think seriously about these equilibrium decisions and how the data is generated.

## 2.5 Empirical Analysis: Tire Markets using R

Bresnahan and Reiss wrote a series of papers where they thought carefully about the empirical implications of entry games. In one of those papers, the authors get data on the number of firms in geographically distinct towns over a wide variety of industries including retail tire stores.

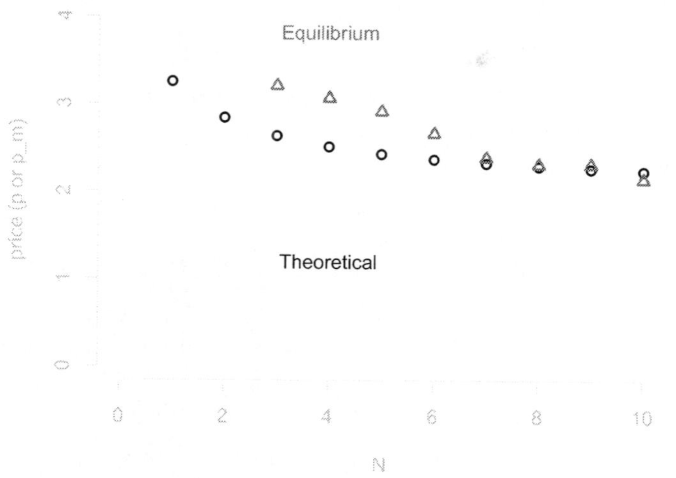


**FIGURE 2.1**
Scatter plot of prices from simulated data gives the number of firms ($N$) in the market. The equilibrium prices (`dt1$p`) (red triangles) fall less quickly than the theoretical prices (`p_m`) (black circles).

The section discusses the data on retail tire markets and how to take the model presented earlier in the chapter to the data. It estimates the parameters of the game and uses the game to simulate changes to policy such as reducing the costs associated with setting up a new firm.

### 2.5.1 Data

Those data provide information on the number of tire stores in each town, the population of each town, the number of commuters into the town, various economic indicators such as house prices and land prices and various demographic indicators such as age and family income.[12]

The code below also loads the data and plots the data. The data are in a csv file and is read in using `read.csv()`. The variable `dir` needs to be defined. This is set equal to the path where the data resides. The data is then plotted using `ggplot()` from the package `ggplot2` (or `tidyverse`).

[12]The data used here comes from Jeremy Bejara and his Github repository from a 2019 structural industrial organization course, `https://github.com/jmbejara/comp-econ-sp19/blob/master/lectures/5-14_Structural_IO_with_MLE/bresnahan-reiss-1991-discussion.ipynb`

```
> file = paste0(dir, "BresnahanAndReiss1991_DATA.csv")
> read.csv(file) |>
+   ggplot(mapping = aes(TPOP,
+                        TIRE)) +
+   geom_point() +
+   labs(x = "Total Population (000s log scale)",
+        y = "Number of Tire Stores") +
+   scale_x_log10()
```

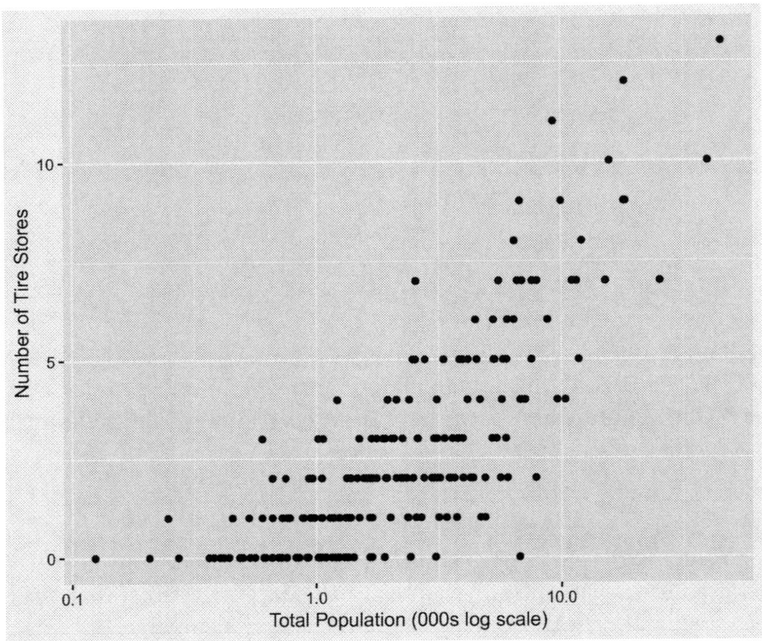


**FIGURE 2.2**
Scatter plot of the empirical relationship between the number of tire stores and the population of the town. It shows the general positive relationship with larger towns having more retail tire stores.

Our interest is figuring out what determines the number of stores in a market. Figure 2.2 shows that as the market gets larger, the number of firms increases.

### 2.5.2 Structural Model

We will follow Bresnahan and Reiss (1991b) and build a structural model. The basic idea is that the game theory provides the empirical relationship which we can then match to the data. From that matching, we can back out the parameters of the game theoretic model. Once we have the model parameters we can run policy simulations. At least that is the theory.

We can write Equation (2.5) in logs. We take logs of both sides of the equation and then rearrange to put the log of the number of firms on the left-hand side. This makes the equation look more like a linear regression equation.

$$
\begin{aligned}
&(a-c)^2 = b(N+1)^2\theta(N+1)^K \\
&or \\
&(2+K)\log(N+1) = 2\log(a-c) - \log(b) - \log(\theta) \\
&or \\
&\log(N+1) = \tfrac{2}{2+K}\log(a-c) - \tfrac{1}{2+K}\log(b) - \tfrac{1}{2+K}\log(\theta)
\end{aligned}
\tag{2.6}
$$

This transformation is useful for enabling us to use a standard **linear regression** estimator.

The relationship from the structural model states that the number of firms is increasing in average profits of the firm $(a-c)$ and decreasing in the substitution to other products $(b)$ and the entry costs $(\theta)$. The rate of increase is determined by the parameter $K$.

### 2.5.3 Entry Estimator

Equation (2.6) states that in equilibrium the number of firms (log of the number of firms) is determined by factors determining demand size such as total population and per-capita income, factors determining costs such as wages, factors determining the slope of the demand function such as closeness of substitutes and finally factors determining the cost of entry such as property rental costs.

Unfortunately, it is not clear how our data maps into our theoretical parameters. We do have good measures for overall demand, like total population, but we don't have information on wages. We have some information about land prices which may affect rental prices and entry costs. We have information on the number of commuters which may be a measure of both market size and substitution out of the market.

### 2.5.4 Estimation in R

To do the estimation we can run a linear regression with the log of the number of retail tire stores against characteristics of the town such as the population, number of commuters, income, and land values. The code below reads in the data and then runs the regression. The `data.frame` called `data` removes any observations with missing values using `na.omit()`.[13] The code then runs two regressions. The first regression includes only the population of the town.

[13]**R** uses `NA` to represent missing values.

**TABLE 2.4**
OLS estimates of the relationship between market characteristics and the number of retail tire stores.

| | *Dependent variable:* | |
|---|---|---|
| | log_stores | |
| | (1) | (2) |
| population | 0.08*** | 0.07*** |
| | (0.01) | (0.01) |
| cummuters | | 0.04 |
| | | (0.07) |
| income | | 0.13*** |
| | | (0.04) |
| land_value | | 0.07 |
| | | (0.19) |
| Constant | 0.74*** | −0.03 |
| | (0.05) | (0.22) |
| Observations | 202 | 202 |
| $R^2$ | 0.36 | 0.40 |
| *Note:* | *p<0.1; **p<0.05; ***p<0.01 | |

```
> data = read.csv(file) |>
+   mutate(
+     log_stores = log(TIRE + 1),
+     population = TPOP,
+     cummuters = OCTY,
+     income = PINC,
+     land_value = LANDV
+   ) |>
+   na.omit()
> lm1 = lm(log_stores ~ population, data)
> lm2 = lm(log_stores ~ population + cummuters +
+              income + land_value, data)
```

Table 2.4 shows the empirical relationship between number of stores and various economic factors. It shows that there is strong positive relationship between market size and the number of firms. The empirical relationship between other factors is not obvious, at least in this data.

### 2.5.5 Structural Estimation

The estimates presented in Table 2.4 provide the empirical relationship between the number of firms in the market and observed characteristics of the market. We are interested in mapping those estimates into our game theoretic model. Assume that the observed relationship captured by the second column of Table 2.4 is generated by the game described above (Equation (2.6)). Specifically, it represents a set of Nash equilibria of that game.

We will make the following assumptions.

- $a = a_0 \exp(population)^{a_1} \exp(commuters)^{a_2} \epsilon_a$
- $c = \epsilon_c$
- $b = b_0 \exp(income)^{b_1} \epsilon_b$
- $\theta = \theta_0 \exp(\text{land\_value})^{\theta_1} \epsilon_\theta$

That is the demand level is a function of population and commuters and some unobserved characteristic ($\epsilon_a$). You will see in a sec why we wrote this down in such a weird way.[14] The marginal cost is unobserved. We will be forced to assume that $\epsilon_c = 0$ for what we do below.[15] The slope of demand is a function of income and an unobserved characteristic ($\epsilon_b$). Lastly, the cost of entry is a function of land values and an unobserved characteristic ($\epsilon_\theta$).

Given these assumptions and the regression results in Table 2.4 we have the following relationships.[16] The coefficient estimates are on the changes in the values of the observed characteristics.

$$
\begin{aligned}
0.07 &= \tfrac{2}{2+K} a_1 \\
0.04 &= \tfrac{2}{2+K} a_2 \\
0.13 &= -\tfrac{1}{2+K} b_1 \\
0.07 &= -\tfrac{1}{2+K} \theta_1
\end{aligned}
\tag{2.7}
$$

We can't uniquely determine the parameter values from these equations. In order to move forward we will assume that $a_1 = 1$. Given this assumption $K = 0.86$, $a_2 = 0.06$, $b_1 = -0.37$ and $\theta_1 = -0.20$. The parameter estimates are determined relative to the coefficient on population.

We have one more empirical relationship up our sleeve. The constant in the regression is -0.03. Given our assumptions, we have that $0.70 \log(a_0) - 0.35 \log(b_0) - 0.35 \log(\theta_0) = -0.03$. Again, we don't have enough information to pin down all the parameters. If we assume that $a_0 = b_0 = 1$, then $\theta_0 = 1.09$.

[14] In economics we generally refer to this specification as a Cobb-Douglas function.

[15] Given that we don't observe marginal costs our policy estimates will be OK as long as we don't try to model big changes.

[16] We are just taking the coefficient values from Model (2) and not worring about how well those coefficients are estimated.

### 2.5.6 Policy Simulation in R

What happens if the market size increases or if entry costs fall?

In this case to determine the equilibrium number of tire stores and equilibrium prices we create two functions using `function()`. These functions take in values for log of total population, log of commuters, log of income, and log of land values. They are based on Equations (2.6) and (2.1) respectively. The `p()` includes the equilibrium number of firms through the function `N()`.

```
> N = function() {
+    exp((2/(2 + K))*(log(a_0) + a_1*population + a_2*commuters) -
+       (1/(2 + K))*(log(b_0) + b_1*income) -
+          (1/(2 + K))*(log(theta_0) + theta_1*land_value) - 1)
+ }
> p = function() {
+    exp((log(a_0) + a_1*population + a_2*commuters))/
+       (N()+1)
+ }
```

We can use the estimates of the parameters of the game theoretic model to understand the likely impact of various policy changes.

```
> K = 0.86
> a_0 = 1
> a_1 = 1
> a_2 = 0.06
> b_0 = 1
> b_1 = -0.37
> theta_0 = 1.09
> theta_1 = -0.20
```

These are our baseline estimates given the observed data and our baseline estimates for the number of firms and the price in the market, `N0` and `p0` respectively.[17]

```
> population = mean(data$TPOP)
> commuters = mean(data$OCTY)
> income = mean(data$PINC)
> land_value = mean(data$LANDV)
> N0 = N()
> p0 = p()
```

Consider a policy that increases commuters by 10%. This may be a policy that makes it easier to drive to the town or increases railway capacity or

[17]Defining variables outside any function means that they are **global** variables that can be accessed by any function. In this case those variables can be accessed by `N()` and `p()`.

reduces parking costs downtown or requires government workers to return to the office. Note the weird way that the change comes in, this is because the measures are in logs.

```
> commuters = mean(data$OCTY) + log(1.1)
> N()/N0
  [1] 1.004007
> p()/p0
  [1] 1.002058
```

The policy has a very small effect, increasing the number of firms in the market by 0.4% and prices by about 0.2%.

Alternatively, consider a policy that increases the population by 10%. This might be something that increases the connectedness of the town to near by towns or allows more housing to be built.

```
> population = mean(data$TPOP) + log(1.1)
> commuters = mean(data$OCTY)
> N()/N0
  [1] 1.068922
> p()/p0
  [1] 1.034692
```

This policy increases the number of firms by 7% and prices by about 3.5%. Both of these policies increase the number of firms in the market but also increase the retail price of tires. Demand increases, which increases prices, but the impact is mediated by the equilibrium change in the number of firms and competition.

Lastly, consider a policy that decreases entry costs by 10%. This may be a policy that makes it easier for a retail tire store to enter the market, like reducing the permits required to set up the store or construct the store.

```
> population = mean(data$TPOP)
> theta_0 = 0.9*theta_0
> N()/N0
  [1] 1.037526
> p()/p0
  [1] 0.9667752
```

This policy increases the number of firms by 4% and reduces prices by 3%.

We may naively expect a policy that reduces entry costs to have a significant impact on both entry and prices. Both effects are mitigated in equilibrium. As it becomes cheaper to enter the market, firms understand that entering may not be that profitable because prices will fall. Similarly, because entry response is modest the price response is modest.

## 2.6 Discussion and Further Reading

In a series of papers, Tim Bresnahan and Peter Reiss showed how game theory could be used to improve empirical analysis of market structures. This chapter uses data from Bresnahan and Reiss (1991b) to illustrate a structural model of firm entry in the retail tire market. In the original paper, the authors examine a number of markets and include information on prices for the retail tire market.

This chapter introduces the idea of a Nash equilibrium. In the entry game, firms will enter as long as it is profitable given all the other firms that will also enter. Because firms can strategically respond to the decisions of other firms, policies that we may naively believe to significantly lower prices, may not.

The chapter introduces the idea of using a structural model to interpret the data. It assumes that the observed relationships in the data are a Nash equilibrium of a particular entry game. While a lot of questionable assumptions are required given the data available, the analysis helps us to understand why various policies may have pretty modest impact on the number of firms in the market and the prices of tires.

In the FTC's case against Staple's acquisition of Office Depot, the FTC presented evidence that there was a positive relationship between prices and the number of retail office supply stores (Ashenfelter et al., 2006). A concern with the analysis is that both the prices and the number of stores may be driven by other factors such as land values (Manuszak and Moul, 2008). The game theory can help us understand potential pitfalls in the empirical analysis and which econometric methods may provide solutions. Chapter 4 revisits firm entry games by analyzing the decisions of Barnes & Noble and Borders to enter various markets in the United States.

# 3

# *Oligopoly*

## 3.1 Introduction

When most people, even most economists, discuss competition they have in mind the model presented in Econ 101. There are many firms all adding a small amount to the market. No firm has control over the price that they receive. Each firm can leave or enter the market at will, so any large profits get bid away as more firms enter. In this world, prices are equal to marginal cost and the amount of goods is such that consumers and firms cannot be made better off (without one or the other being made worse off). This is not what industrial organization economists mean when they discuss competition. Economists that work in antitrust and competition policy, distinguish between firms who work together to determine price and firms that work independently of each other.

This chapter presents the standard model of how firms set prices and output when they are doing so independently of each other. Chapter 8 will return to oligopoly and allow less independence in price setting. The chapter presents three standard models of competition, Cournot, Bertrand, and Hotelling.[1] The chapter uses **R** to simulate the Nash equilibrium of a Cournot game with three firms.

The Hotelling model is used to understand pricing of hamburgers in Santa Clara County California in the late 1990s. In particular, the chapter analyzes data on the price of the McDonald's Big Mac and the Burger King Whopper at the outlets throughout Santa Clara County. The model allows us to estimate how prices are affected by competition between firms that are differentiated by location. If one franchisee purchased all the McDonald's outlets in Santa Clara County, what would happen to the price of both the Big Mac and the Whopper in the county?

[1]Confusingly the Hotelling model is often called a Bertrand model. While this chapter presents two examples of a price setting game, many people assign the Bertrand name to any static price setting game where the products are not homogeneous.

DOI: 10.1201/9781003351603-3

## 3.2 Cournot's Model

In the 1850s a French applied mathematician made a far fetched claim. Augustine Cournot suggested having competing firms does not necessarily mean that prices will be equal to marginal cost. Tabarnak! How could this be? It is almost by definition that prices equal marginal cost in economics. Some fifty years later, another Frenchman, François Bertrand, argued that Cournot was full of merde. Even with just two firms, prices would equal marginal cost.

The section presents a two-firm version of the Cournot model and then generalizes that to a $N$-firm version. It uses **R** to numerically simulate the Nash equilibrium in a three-firm version of the model.

### 3.2.1 Two Firm Model

Formally we have the following game where we have two firms choosing quantity $(q_i)$ and contemplating the impact of their collective choices on profits.

- Players: Firm 1 and Firm 2.
- Strategies:
  - Firm 1: $q_1 \geq 0$
  - Firm 2: $q_2 \geq 0$
- Payoffs:
  - Firm 1: $p(q_1, q_2) \times q_1 - c(q_1)$
  - Firm 2: $p(q_1, q_2) \times q_2 - c(q_2)$

The first part of Firm 1's payoff is revenue. The price $p(q_1, q_2)$ is a function of each firm's output in the market. The price is multiplied by the output produced by Firm 1 which is $q_1$. The second part is costs which is a function of the amount of output produced. To keep things simple let $p(q_1, q_2) = a - b \times (q_1 + q_2)$ and $c_j(q_j) = c \times q_j$. That is, we have linear demand and constant marginal cost. Assume also that $a - c > 0$. This will become important later.

Cournot's game assumes that both firms make identical goods. This may be a good model of a wheat market or an electricity market. In such markets each firm decides how much to produce and supplies to a centralized exchange that takes in demand and determines the price everyone gets. In the game, each firm does not know how much the other firm produces.

### 3.2.2 Best Response

One issue that people new to game theory find very confusing is the idea of a **best response function**. In the presentation of the game above, it is clearly stated that each firm does not know the other firm's choice when choosing its action. Then five minutes later we claim that firms have a function such that its action is a best response to the other firm's action. How can they both not know what the other firm is doing and have a best response to it? Makes no sense.

Both statements can be true because they refer to different concepts. One is a description of the game and the other is a description of the algorithm used to find the Nash equilibrium of the game. The games discussed in this part of the book assume the players choose actions once and simultaneously. In this sense, each firm does not know how much the other firm is producing. The best response function is an analytical tool used to find the Nash equilibrium. The Nash equilibrium is where each firm chooses the optimal output given the output chosen by the other firm. That is the Nash equilibrium is where each firm's output choice is a best response to the other firm's output choice.

The best response function is the solution to the following optimization problem.

$$\max_{q_1} \quad (a - bq_1 - bq_2)q_1 - cq_1 \tag{3.1}$$

The solution to this optimization problem is given by the first order condition.

$$\begin{aligned} & a - bq_1 - bq_2 - c - bq_1 = 0 \\ & or \\ & q_1 = \tfrac{a-c-bq_2}{2b} \end{aligned} \tag{3.2}$$

The best response function states that Firm 1's quantity is increasing in the difference $a - c$, decreasing in the quantity of Firm 2 ($q_2$) and decreasing in the willingness of customers to substitute out of the market ($b$). In this model, the actions of the two players are strategic substitutes, in words, when one firm increases its output, the other firm responds by decreasing its output.

### 3.2.3 Nash Equilibrium

The Nash equilibrium is where each firm is playing the best response to the other firm. That is, where the first order condition for Firm 1 (Equation (3.2)) and the equivalent condition for Firm 2 both hold. To solve for the equilibrium we need to solve for two unknowns from a system of two linear equations. Simply counting the number of unknowns and the number of independent linear equations we know that the solution exists and is unique.

In this case, we can use a nice trick to solve it. Because both firms are identical then it must be that in equilibrium they produce the exact same amount. That is $q_1 = q_2 = q$. Substituting this into Equation (3.2) allows us to find the solution.

$$\begin{aligned} q &= \tfrac{a-c-bq}{2b} \\ 2bq &= a - c - bq \\ 3bq &= a - c \\ q &= \tfrac{a-c}{3b} \end{aligned} \tag{3.3}$$

This is each firm's output in equilibrium. Equation (3.3) states that output will fall when marginal cost ($c$) increases and when substitutability ($b$) increases. In order for equilibrium output to be positive it must be the case that $a - c > 0$, which is an assumption made above.

If we substitute this back into demand we can determine price. We have two firms so we need to substitute back both equilibrium quantities.

$$\begin{aligned} p &= a - 2b\tfrac{a-c}{3b} \\ &= \tfrac{3a-2a+2c}{3} \\ &= \tfrac{a+2c}{3} \end{aligned} \tag{3.4}$$

Equation (3.4) states that price will increase with marginal cost, but are they higher than marginal cost?

$$\begin{aligned} &p > c \\ &\Leftrightarrow \tfrac{a+2c}{3} > c \\ &\Leftrightarrow a + 2c > 3c \\ &\Leftrightarrow a - c > 0 \end{aligned} \tag{3.5}$$

As long as there is positive output in the market, prices will be greater than marginal cost. Mayhem. Cats and dogs living together!

### 3.2.4 Cournot Model with N Firms

What happens as the number of firms increase? Do we get back to perfect competition?

In this case, Firm 1's best response becomes.

$$q_1 = \frac{a - c - b\sum_{j=2}^{N} q_j}{2b} \tag{3.6}$$

Again to solve for equilibrium with symmetric firms we can do the trick of setting all the output levels to be the same.

$$\begin{aligned} &q = \tfrac{a-c-b(N-1)q}{2b} \\ &2bq = a - c - b(N-1)q \\ &(N+1)bq = a - c \\ &q = \tfrac{a-c}{(N+1)b} \end{aligned} \tag{3.7}$$

So in equilibrium, output is decreasing proportionately with the number of firms in the market.

Substituting equilibrium quantities back into demand we can determine prices. Remember we have to multiply by $N$.

$$\begin{aligned} p &= a - bN\tfrac{a-c}{(N+1)b} \\ &= \tfrac{(N+1)a-Na+Nc}{N+1} \\ &= \tfrac{a+Nc}{N+1} \\ &= \tfrac{a}{N+1} + \tfrac{N}{N+1}c \end{aligned} \tag{3.8}$$

Look familiar? It is the equation used to determine price in Chapter 2.

We can see that as $N$ gets large, $p$ converges to $c$. That is Cournot's model does give perfect competition but only for a large number of firms in the market.

### 3.2.5 Cournot Model in R

In order to better understand how the model works it helps to program it up in **R**. The function `price_cournot()` is used to determine the price given market output `q`. The function `br_cournot()` determine the firm's output given the output of the other two firms. This function is based on Equation (3.6). The notation `q[-i]` is used to refer to all the elements of `q` except the `i`th element.

```
> price_cournot = function(q) a - b*sum(q)

> br_cournot = function(q, i) {
+   if (price_cournot(q) > 0) {
+     return(max(c(0, (a - c[i] - b*sum(q[-i]))/(2*b))))
+   } else {
+     return(0)
+   }
+ }
```

The best response function checks to make sure prices and quantities are positive. It is based on Equation (3.6).

### 3.2.6 Solve for the Nash Equilibrium with R

The algorithm below looks for an equilibrium where the best response's of each firm lead to the same quantity. The algorithm chooses a starting level of output and then calculates the best response for each firm to get a new level of output. It then checks whether the new level of output is the same as the old level of output. If it is, then we have a Nash equilibrium. Remember a Nash equilibrium is where each firm is choosing a level of output that is a best response to all the other firms. If the new level of output is different from the old level of output, the algorithm sets the old level of output to the output just calculated and finds the best response to that amount. The algorithm stops when the new and old amounts become equal or close to equal.

In the code we have a default maximum number of iterations (`maxit = 100`) and a default level of convergence (`epsilon = 1e-5`).[2] The code uses `epsilon` to refer to a small number. It uses a `while()` loop to run until one of the conditions fails to hold.

The initial value for the output is determined by Equation (3.7) assuming all firms in the game have the same costs. The algorithm then calculates the

[2]`1e-5` is a way to write small numbers, in this case 0.00001.

best response for each firm and checks whether the new output is the same as the old output. The algorithm sets the old output to the new output and calculates the best response to that output. The algorithm continues until the difference between the new and old output is less than `epsilon` or the number of iterations exceeds `maxit`. The algorithm returns the output in equilibrium and whether the algorithm converged.

```
> ne_cournot = function(maxit=100, epsilon=1e-5,
+                             trace=FALSE, converged=TRUE) {
+   diff = 10000 # some big number
+   iter = 1
+   N = length(c)
+   q_old = rep((a - mean(c))/((N+1)*b), N)
+   # initial values for q_old
+   while(diff > epsilon & iter < maxit) {
+     q_new = rep(0, N)
+     for(i in 1:N) {
+       q_new[i] = br_cournot(q_old, i)
+     }
+     diff = sum(abs(q_new - q_old))
+     iter = iter + 1
+     if(trace) {
+       print(diff)
+       print(iter)
+     }
+     q_old = q_new
+   }
+   if(iter == maxit) {
+     converged = FALSE
+   }
+   return(list(q_star = q_new, converged = converged))
+ }
```

This algorithm is not super sophisticated and it takes advantage of uniqueness of the result. The algorithm does allow us to simulate more interesting models than the simple symmetric-firm model presented above.

### 3.2.7 Simulation of Cournot Model in R

The following simulation allows the costs to vary between firms and shows how variation in marginal costs leads to differences in market share for firms.

```
> set.seed(123456789)
> N = 3
> a = 0.5
> b = 0.2
```

```
> c = a*runif(N) # so costs vary between firms.
> q_star = ne_cournot()$q_star
> p_star = price_cournot(q_star)
```

The results are as follows. The marginal costs (`c`) are determined randomly. The output (`q_star`) and the price (`p_star`) are determined by the equilibrium algorithm.

```
> c
 [1] 0.3465879 0.3364405 0.3269508
> q_star
 [1] 0.1545377 0.2052715 0.2527168
> p_star
 [1] 0.3774948
```

In this example Firm 1 has the highest costs and the lowest quantity, while Firm 3 has the lowest cost and the highest quantity.

This simulation illustrates a standard result of the Cournot model, the less efficient (higher marginal cost) firms have lower market share and the more efficient (lower marginal cost) firms have higher market share.

It also makes it clear that less efficient firms could be in the market. There is nothing about the basic Cournot game that forces them to leave the market.

## 3.3 Bertrand's Model

Bertrand was having none of it. Perfect competition was not some edge case. In Cournot's game the firms choose quantity, but what happens if the firms choose price?

The section presents a two-firm version of Bertrand's game. Think about two ice-cream stands that are next to each other. Each stand displays the price of a cone of ice-cream cone. If one ice-cream stand charges more than the other, then everyone will buy from the cheaper stand. If both stands charge the same price, then they split the market.

### 3.3.1 Two-Firm Game

This time the two firms choose a price and payoffs are determined by the quantity which is a function of both prices.

- Players: Firm 1 and Firm 2.
- Strategies:

  - Firm 1: $p_1 \geq 0$
  - Firm 2: $p_2 \geq 0$

- Payoffs:
  - Firm 1: $p_1 \times q_1(p_1, p_2) - c_1 \times q_1(p_1, p_2)$
  - Firm 2: $p_2 \times q_2(p_2, p_1) - c_2 \times q_2(p_2, p_1)$

Again the payoffs for each firm are revenue less costs. This time, the firm chooses the price to charge and the market mechanism determines how much quantity the firm will sell.

$$q_1 = \begin{cases} 1 & \text{if } q_1 < q_2 \\ 0.5 & \text{if } q_1 = q_2 \\ 0 & \text{if } q_1 > q_2 \end{cases} \tag{3.9}$$

Let's assume that the market size is 1. Equation (3.9) says that if you have the lowest price, then everyone buys from you and if you have the highest price then no one buys from you. If both firms have the same price, they split the market.

This is a market where it is very easy for customers to substitute. One slight reduction in price can cause the whole market to shift.

A real-world example might be a supply contract request for quotes. Which ever firm offers the lowest price gets the full supply contract.

### 3.3.2 Nash Equilibrium

Consider a simple case where both firms have the same marginal cost, $c_1 = c_2 = c$. The unique Nash equilibrium is $p_1 = p_2 = c$.

To confirm that it is an equilibrium, assume that $p_1 = c$. What happens if $p_2 > c$? In this case, Firm 2's output is 0 and profits are 0. What if $p_2 = c$. In this case, Firm 2's output is 0.5, but Firm 2's profits are 0. Lastly, if $p_2 < c$, then Firm 2's output is 1, but Firm 2's profits are negative. So Firm 2 is indifferent between choosing $p_2 = c$ or $p_2 > c$. Given that Firm 2 cannot do better than the Nash equilibrium, this confirms $p_2 = c$ is optimal for Firm 2. We can make the exact same argument for Firm 1.

It is not only an equilibrium but a unique equilibrium. A candidate equilibrium is, $p_1 = p_2 > c$. Again, keep Firm 1's price at $p_1 > c$. If Firm 2 charges $p_2 > p_1$, then Firm 2's profits are 0. If Firm 2 charges $p_2 = p_1$ then Firm 2's profits are positive, $0.5(p_1 - c)$. If Firm 2 charges $p_2 = p_1 - \epsilon$ where $\epsilon$ is some small number, then Firm 2's profits are $(p_1 - c - \epsilon)$. These profits are a lot higher. Sure price is slightly lower, but Firm 2 went from selling to half the market to selling to the whole market. In this game there is a huge incentive to slightly undercut your rival in this market. Because of this incentive, there is no other Nash equilibrium.

Bertrand proved his point. With just two firms, price equals marginal cost, an important characteristic of perfect competition.

What happens in Bertrand's model if the two firms have different marginal costs ($c_j$)?

## 3.4 Hotelling's Model

To go from perfect competition being some edge case to it being a constant of the model all we needed to do was assume firms choose price instead of quantity! Nonsense. If you look more closely, the models proposed by the two Frenchmen are very different from each other. In particular, the demand in the Bertrand model is very particular.

Early in the Twentieth Century, the American statistician and economist, Harold Hotelling, suggested a compromise. He suggested a model where firms choose price but where demand was not nearly so particular as Bertrand assumes.

The section presents Hotelling's original model. It then presents a model of differentiated goods with linear demand and determines the Nash equilibrium in prices for that model.

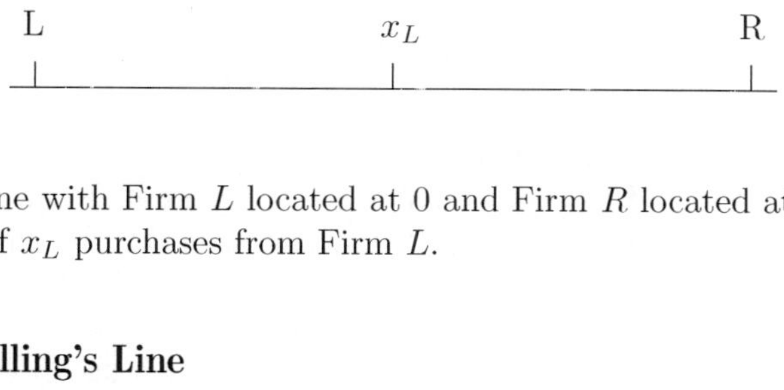


**FIGURE 3.1**
Hotelling's line with Firm $L$ located at 0 and Firm $R$ located at 1. Everyone "living" left of $x_L$ purchases from Firm $L$.

### 3.4.1 Hotelling's Line

Figure 3.1 represents Hotelling's game. There are two firms $L$ and $R$. Customers for the two firms "live" along the line. Consider two frozen custard (ice cream) places located at each end of a beach board walk. Your beach chairs may be located closer to one frozen custard place than the other. Customers prefer to go to the closer firm if the products and prices are otherwise the same. Hotelling's key insight is that while firms often compete by selling similar products, these products may not be identical. Moreover, some people may prefer one product to the other. Some people actually prefer Pepsi to Coke. The location on the line represents how much the customer prefers $L$ to $R$.

The line and the distance between the customer and the firm represent how willing they are to purchase from a particular firm. Importantly, it represents how much the two firm's products are substitutes for each other. In the hamburger example presented below, we use the actual distance between stores to measure substitutability but more generally Hotelling's line is a metaphor for how similar or different two products are from each other. The closer the two firms are, the easier it is to substitute between them and the lower the price is likely to be.

### 3.4.2 Differentiated Goods Game

Now consider a game where there are $N$ firms that all sell a product that is similar but not the same as each other. For example, a set of hamburger restaurants in Santa Clara. While the Big Mac may be the same, the location of each McDonald's outlet is different.

- Players: $N$ firms where each firm has a location $\{x_i, y_i\} \in \Re^2$.
- Strategies: $p_i \geq 0$ for all $i \in \{1, ..., N\}$
- Payoffs: $(p_i - c_i) \times q_i(p_i, p_{-i})$

Each of the $N$ firms chooses a price $p_i$. The profits are the price less marginal cost ($c_i$) multiplied by the quantity sold $q_i$. This quantity is determined by both the price the firm chargers $p_i$, by all the prices all the other firms charge $p_{-i}$ and by the distances between the firms.[3]

To make things simpler and more concrete, assume that there are just two firms $i$ and $j$, Firm $i$'s sales are affected by $p_i$ and $p_j$ in the following way.

$$q_i(p_i, p_j) = \alpha + \beta p_i + \frac{\gamma p_j}{d_{ij}} + \epsilon_i \tag{3.10}$$

It is a linear demand model similar to the model presented earlier in the chapter. The quantity sold by Firm $i$ is a function of the price Firm $i$ charges, the price Firm $j$ charges, and some unobserved term $\epsilon_i$. Assuming, $\beta < 0$, then the higher the price the lower the demand for $i$'s product. Demand is also a function of the competitor's price $p_j$. The extent of this is determined by $\gamma$, which is positive. The demand for $i$ is higher when the competitor charges a higher price. The extent of the competitor's price matters depends on the distance between the outlets. The larger that distance ($d_{ij}$), the less the two firms compete with each other for customers. Again, $d_{ij}$ could represent physical distance or just a measure of the difference between the two products.[4]

### 3.4.3 Best Response

Given all this, what will be the price in the market? We assume that the price is determined by the Nash equilibrium. The Nash equilibrium is the price such that Firm $i$ is unwilling to change their price given the price charged by Firm $j$, and Firm $j$ is unwilling to change their price given Firm $i$'s price.

[3] $-i$ means not $i$.

[4] We will Euclidean distance, $d_{ij} = \sqrt{((x_i - x_j)^2 + (y_i - y_j)^2)}$. This is the as-the-crow-flies distance.

Firm $i$'s problem is as follows.

$$max_{p_i} \quad (p_i - c_i)(\alpha + \beta p_i + \tfrac{\gamma p_j}{d_{ij}}) \tag{3.11}$$

The solution to the optimization problem is the solution to the first order condition.

$$(\alpha + \beta p_i + \frac{\gamma p_j}{d_{ij}}) + \beta(p_i - c_i) = 0 \tag{3.12}$$

Firm $j$'s problem is similar. The two equations from the first order conditions are the best response functions for each firm.

### 3.4.4 Nash Equilibrium

Given the two equations derived from the first-order conditions, we can write down a system of equations. Each firm's best response function is as follows.

$$\begin{aligned} p_i &= \tfrac{c_i}{2} - \tfrac{\alpha}{2\beta} - \tfrac{\gamma p_j}{2\beta d_{ij}} \\ p_j &= \tfrac{c_j}{2} - \tfrac{\alpha}{2\beta} - \tfrac{\gamma p_i}{2\beta d_{ij}} \end{aligned} \tag{3.13}$$

The prices are higher when costs are higher. Prices are strategic complements. While there is a negative sign in front of $\gamma$, we said above that $\beta$ is negative. When $p_j$ increases, then Firm $i$'s best response is to *increase* $p_i$. Again, how much the two firm's interact depends on $\gamma$ and the distance $d_{ij}$.

## 3.5 Empirical Analysis: Hamburger Competition with **R**

When he was doing his PhD at Stanford, Wash U Marketing professor, Raph Thomadsen, decided to study competition for hamburgers in Santa Clara County. Raph was interested in competition between the two big chains, McDonald's and Burger King as well as competition within the chain brands.

To study competition he got information on each outlet from the health department. Most importantly he got the outlet's location. He then physically visited each outlet determining the price of the hamburgers offered and whether the outlet had other features like a drive through or a playland.

This section models the prices for hamburgers in the late 90s in Santa Clara County using the model presented earlier.

### 3.5.1 Data

The data is provided by Raph Thomadsen and used in his RAND paper (Thomadsen, 2005).[5] For each outlet there is information about the brand,

[5] There is a slight descrepancy between the coordinates in the code and the location of the outlets using Google maps.

ownership structure, features of the outlet, age of the outlet, and price of the sandwich.

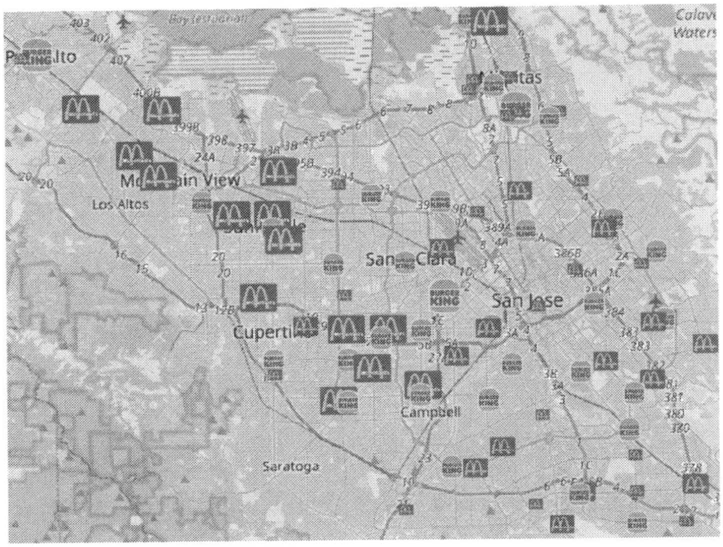


**FIGURE 3.2**
McDonald's and Burger King outlets in Santa Clara County. The size of the logo indicates whether the price of the sandwich is in the top third of prices, the middle third, or the bottom third.

Figure 3.2 is created using the `leaflet` package in **R** with the coordinates from the code and importing the icons. The figure gives some idea how competition works in Santa Clara county. There is a logo at each brand's location. The size of the logo represents the price charged for the signature sandwich. The higher the price, the bigger the icon. The outlets up near the Stanford campus have fairly high prices. Often a Burger King is paired with a McDonald's, but neither has really low prices. Prices on the west and south side of San Jose seem lower than on the east side and north east side.

You may be surprised by how much prices for a Big Mac vary across one city. You may also be surprised that the competition of the most interest is competition between McDonald's outlets. How can that be? Doesn't the McDonald's corporate headquarters set prices for the Big Mac? No. It depends. Some outlets in this data are owned by the McDonald's corporation and for those, yes, corporate headquarters would have a lot of say over price. But most of the outlets are owned by franchisees. Under California law, corporate headquarters is restricted in what it can require of them. Headquarters cannot determine the price charged by the outlet for its sandwiches.

### 3.5.2 Estimation (Part 1)

It is traditional when estimating the parameters of a game theory model, to first present estimates from a standard model like a linear regression model. The game theory helps us think about what to put into that regression model but also makes clear that we should be careful in interpreting the results. The raw prices are adjusted to make the regression results look nicer.

The data used in the rest of the analysis is restricted to outlets that are not corporate owned. The issue is that their pricing strategies may look very different from the franchise locations.

```
> file = paste0(dir,"outlets_gt_ch3.csv")
> data = fread(file) |>
+   mutate(
+     LPrice1 = log(100*(Price - 2.49)),
+     BK = BK == 1
+   )
> lm1 = lm(LPrice1 ~ BK, data)
> lm2 = lm(LPrice1 ~ BK + Playland +
+              DriveThru + Mall, data)
> lm3 = lm(LPrice1 ~ BK + Playland +
+              DriveThru + Mall + Race + Male, data)
```

Table 3.1 presents some regression results of log of sandwich price on characteristics of the outlet. This set up seems to match relatively closely to the table presented in Thomadsen (2005). Prices are lower for The Whopper (on average) and are lower at outlets with various amenities and in Malls as well as in various locations based on demographic characteristics. The coefficients on most of the characteristics are not statistically significantly different from zero. This is probably due to the small sample size. One coefficient that is statistically significantly different from zero is the dummy on being a Burger King outlet. The Whopper is cheaper than the Big Mac.

### 3.5.3 Empirical Equilibrium

As we did in the previous chapter, we will assume that the prices observed in the data are determined by the Nash equilibrium of the game described above. The implication is that for the whole set of prices we see from all the outlets, the whole set of best responses must hold. In the final data set we have prices and information for 79 outlets in Santa Clara County.

Our estimation problem is to find the $\alpha$, $\beta$, and $\gamma$ such that our 79 best response equations hold given the 79 prices we observe and the distance between each location. Actually, we are going to make our empirical model slightly more complicated by adding parameters for observed characteristics of the outlet.

$$2\beta p_i - \beta c_i + \mathbf{X}_i'\alpha + \frac{\gamma p_j}{d_{ij}} + \epsilon_i = 0 \qquad (3.14)$$

**TABLE 3.1**
OLS estimates of the equilibrium relationship between price and observed characteristics of the outlet and their location.

| | *Dependent variable:* | | |
|---|---|---|---|
| | LPrice1 | | |
| | (1) | (2) | (3) |
| BK | $-0.21^{***}$ | $-0.23^{***}$ | $-0.22^{***}$ |
| | (0.05) | (0.05) | (0.05) |
| Playland | | $-0.08$ | $-0.07$ |
| | | (0.06) | (0.06) |
| DriveThru | | $-0.02$ | $-0.04$ |
| | | (0.06) | (0.06) |
| Mall | | $-0.13$ | $-0.12$ |
| | | (0.10) | (0.10) |
| Race | | | 0.50 |
| | | | (0.64) |
| Male | | | 0.16 |
| | | | (0.19) |
| Constant | $4.54^{***}$ | $4.59^{***}$ | $4.50^{***}$ |
| | (0.03) | (0.06) | (0.11) |
| Observations | 79 | 79 | 79 |
| $R^2$ | 0.19 | 0.22 | 0.23 |
| *Note:* | $^{*}$p<0.1; $^{**}$p<0.05; $^{***}$p<0.01 | | |

where $\mathbf{X}_i$ is a vector of observed characteristics for outlet $i$, e.g. $(1,\ brand_i,\ drivethru_i)$, where $brand_i$ indicates the brand of the outlet and $drivethru_i$ is 1 if the outlet has a drive through and 0 otherwise.[6] The parameter $\alpha$ is now a vector of parameters $(\alpha_0, \alpha_1, \alpha_2)$.[7]

Writing this down for all the outlets we have 79 equations that need to hold $(J = 79)$. How much they affect each other depends on the parameter $\gamma$ and the distance between each outlet, $d_{ij}$. The other thing to notice is the assumption that the demand parameters $(\alpha,\ \beta,\ \gamma)$ are the same for every outlet. What differs between the outlets is the observed characteristics $\mathbf{X}_i$, their marginal cost $c_i$ and the distance to the other outlets $d_{ij}$ and unobserved characteristics

[6] The matrix notation $\mathbf{X}$ to represent a block of data. In this case, each row is an outlet and each column is a characteristic of the outlet.

[7] We are using matrix notation where two vectors of three elements $x'y$ means $(x_1y_1 + x_2y_2 + x_3y_3)$. As you can see the matrix notation is a lot more compact.

of the outlet $\epsilon_i$.

$$\begin{array}{l} 2\beta p_1 - \beta c_1 + \mathbf{X}_1'\alpha + \gamma \sum_{j=2}^{J} \frac{p_j}{d_{1j}} + \epsilon_1 = 0 \\ 2\beta p_2 - \beta c_2 + \mathbf{X}_2'\alpha + \gamma (\frac{p_1}{d_{12}} + \sum_{j=3}^{J} \frac{p_j}{d_{2j}}) + \epsilon_2 = 0 \\ \dots \\ 2\beta p_J - \beta c_J + \mathbf{X}_J'\alpha + \gamma \sum_{j=1}^{J-1} \frac{p_j}{d_{Jj}}) + \epsilon_J = 0 \end{array} \tag{3.15}$$

This is quite a mess. It is a lot more compact to write this out using matrix notation.

To see what is going on look at the case where there are just two firms ($J = 2$). Also let's assume that marginal costs are zero and there are no observed characteristics or unobserved characteristics of the outlets.

$$\begin{array}{l} 2\beta p_1 + \gamma \frac{p_2}{d_{12}} = 0 \\ 2\beta p_2 + \gamma \frac{p_1}{d_{21}} = 0 \end{array} \tag{3.16}$$

We can write this out with matrices.

$$2\beta \begin{bmatrix} p_1 \\ p_2 \end{bmatrix} + \gamma \begin{bmatrix} 0 & \frac{1}{d_{12}} \\ \frac{1}{d_{21}} & 0 \end{bmatrix} \begin{bmatrix} p_1 \\ p_2 \end{bmatrix} = \begin{bmatrix} 0 \\ 0 \end{bmatrix} \tag{3.17}$$

Remembering the matrix multiplication rules, we can write this out to get the equations above. In full matrix notation, we have the following.

$$2\beta p + \gamma \mathbf{D} p = 0 \tag{3.18}$$

where $\mathbf{D}$ is a matrix representing the distances between the stores with zeros on the diagonal and the inverse distance between the stores in each cell, and $p$ is a vector of prices for the sandwiches.

The full empirical equilibrium can be written in matrix notation.

$$2\beta p - \beta c + \mathbf{X}\alpha + \gamma \mathbf{D} p + \epsilon = 0 \tag{3.19}$$

where $p$ is the vector of prices for the sandwich, $c$ is a vector of marginal costs for each outlet, $\mathbf{X}$ is the matrix of observed characteristics for each outlet, $\alpha$ is a vector of representing how customers value those characteristics and $\mathbf{D}$ is a full matrix with the distances between all of the outlets in Santa Clara County. You have to admit, it looks a lot nicer.

### 3.5.4 Estimator

Our estimation problem is to find the parameters of the model such that equilibrium prices in the model most closely matche the observed prices. To do this we will use a **method of moments** estimator. It sounds pretty fancy but it is just least squares. The idea is that because there are some unobserved characteristics that are determining the price, represented by $\epsilon$ in Equation (3.19) the equation will not precisely hold.

However we will assume that the Equation (3.19) holds on average for each outlet. We are assuming that the unobserved term is zero on average in equilibrium. Assumption 1 makes this idea formal.

**Assumption 1.** $\mathbb{E}(\epsilon_i | \mathbf{X_i}, p, c_i, \mathbf{D}_i) = 0$ *for all* $i \in \{1, ..., N\}$, *where* $\mathbf{D}_i$ *is the vector of distances from outlet* $i$ *to every other outlet.*

Now we don't actually know the set of average prices for the outlets in the market. Rather we only observed a set prices for each outlet once. So we are going to assume that the analog of Assumption 1 holds in the data we observe. Rather than requiring that this equation is exactly zero, we will look for the parameters that make the average of $\epsilon_i$ squared as small as possible. We are going to minimize the sum of squares, or least squares.

$$\min_{\{\alpha,\beta,\gamma\}} \quad \tfrac{1}{N}\textstyle\sum_{i=1}^{N}(2\beta p_i - \beta c_i + \mathbf{X}_i\alpha + \gamma \mathbf{D}_i p)^2 \tag{3.20}$$

In the actual estimation we will add one more complication. We don't observe $c_i$, the marginal cost at the outlets. We don't think that is a big deal because within a brand, the marginal costs will be very similar across the outlets in the data. One requirement from corporate headquarters is that each outlet use a supplier of similar quality to corporate's preferred supplier. They can't require a particular supplier be used but they can require certain standards be maintained. It is reasonable to think the cost of ingredients into the sandwich is pretty similar across all the outlets within a brand. In addition, all the outlets face similar labor markets and would pay similar wages. All this means that we assume $c_i = brand_i + \epsilon_c$.

### 3.5.5 Estimator in R

Now we need to translate all the math above into code so that we run the estimator on the data. The function `outlet_price_f()` maps Equation (3.20) into **R** code. It looks pretty similar but it is not quite the same. There is an extra matrix `Omega`. We will worry about this matrix in the next section. For the moment it is going to be the **identity matrix**, that is a matrix with all 1's on the diagonal and zeros every where else. Given this, the function is identical to Equation (3.20).[8] To do matrix multiplication in **R** the operation is `A%*%B` where `A` and `B` are two matrices where the number of columns of `A` is equal to the number of rows of `B`. While the operation `A*B` means that each cell of `A` is multiplied by each cell of `B`.[9]

```
> outlet_price_f = function(price, D, Omega, cost, X,
+                                 alpha, beta, gamma) {
+     epsilon = beta*price +
```

[8]Do some matrix algebra with the identity matrix to convince yourself.

[9]`A` and `B` should have the same dimensions, but it is not strictly necessary but does lead to weird results if it does not hold.

```
+         gamma*D%*%price +
+         beta*(price - cost) +
+         gamma*(Omega*D)%*%(price - cost) +
+         X%*%alpha
+       sos = mean(epsilon^2)
+       return(sos)
+ }
```

The first function tries to look as close to the math as possible. The second function is a translation function. It translates from what works best as a function to optimize to what looks nicest. This translation function (`outlet_price_f_int()`) makes it clear to **R** that `X` is a matrix.[10]. It then translates a vector `par` which is used by the optimization function `optim()`, into our parameters, `alpha`, `beta`, and `gamma`. We force `beta` to be negative and `gamma` to be positive. We do this using the `exp()` function, which takes a number and makes it greater than 0.[11] As a rule optimization algorithms are not good at subtlety. Better to let it choose what ever parameter values it likes and then translate its choice into what ever restrictions you want to place on the parameter.

```
> outlet_price_f_int = function(par, price, D, Omega, BK, X) {
+   X = as.matrix(X)
+   J = dim(X)[2]
+   alpha = par[1:J]
+   beta = -exp(par[J+1])
+   gamma = exp(par[J+2])
+   cost = cbind(1,BK)%*%par[c(J+3,J+4)]
+   return(outlet_price_f(price, D, Omega, cost,
+                         X, alpha, beta, gamma))
+ }
```

### 3.5.6 Distances

In order to use the distance between outlets in our estimation we need to calculate distance between outlets. The code uses the `dist()` function to calculate the Euclidean distance between two points ("as the crow flies distance"). It is used in the loop to calculate the distance between all the outlets in the data set. The code runs a loop in **R** using `for()`. Whenever running a loop in **R** it is good practice to create an empty object, here it is a matrix `dist_mat`, that gets filled in during the loop.[12]

[10]**R** sometimes gets confused about what is a matrix and what isn't, so you need to repeat yourself a bit.

[11]It is the exponential function.

[12]This little trick speeds up the **R** substantially.

```
> dist = function(lat0, lon0, lat1, lon1) {
+   return(sqrt((lat0 - lat1)^2 + (lon0 - lon1)^2))
+ }
> J = dim(data)[1]
> dist_mat = matrix(NA, J, J)
> for(i in 1:J) {
+   for(j in 1:J) {
+     dist_mat[i,j] = dist(data$lat[i], data$lon[i],
+                          data$lat[j], data$lon[j])
+   }
+ }
```

Now we can create our matrix `D` by finding the inverse of the distance for each outlet combination and setting the values on the diagonal to 0.[13] Lastly we need to set the stores that have the same location to have a distance that is very close.[14]

```
> D = 1/dist_mat
> diag(D) = 0
> D[is.infinite(D)] = 1/0.006
```

### 3.5.7 Ownership

Almost there. There is one more thing to discuss before estimating prices and that is ownership. In the pricing model, we made the simplifying assumption that each outlet is priced independently. That is not true in the data. In the data, there are people that own multiple franchises. In this case, the owner is going to price their sandwiches accounting for the fact that they own other outlets.

Assume that one person owns two outlets and must choose the optimal price for each. Remember the demand function for each outlet depends on the prices of both outlets.

$$\max_{\{p_1,p_2\}} \quad q_1(p_1,p_2)(p_1 - c_1) + q_2(p_2,p_1)(p_2 - c_2) \tag{3.21}$$

For this case the first order conditions are as follows.

$$\begin{aligned} p_1: &\quad q_1(p_1,p_2) + \frac{dq_1}{dp_1}(p_1 - c_1) + \frac{dq_2}{dp_1}(p_2 - c_2) = 0 \\ p_2: &\quad q_2(p_2,p_1) + \frac{dq_1}{dp_2}(p_1 - c_1) + \frac{dq_2}{dp_1}(p_2 - c_2) = 0 \end{aligned} \tag{3.22}$$

So this time if they raise the price of the sandwich in outlet 1, they lose sales ($\frac{dq_1}{dp_1}$) but they gain back profits in proportion of the people that switch to

[13] In **R** the operation `1/D` calculates the inverse at each cell of the matrix, not the matrix inverse. To calculate the matrix inverse use `solve()`.

[14] 0.006 is the minimum distance of stores that don't have the same location.

the other outlet ($\frac{dq_2}{dp_1}$). Because the cost to the outlet of increasing price is mitigated by the recapturing of customers, prices will be higher.

In our problem, we have the following.

$$\begin{aligned} \frac{dq_1}{dp_1} &= \beta \\ \frac{dq_1}{dp_2} &= \frac{\gamma}{d_{12}} \end{aligned} \tag{3.23}$$

The omega notation allows us to keep track of the various ownership possibilities. Each outlet is assumed to own itself ($\omega_{11} = \omega_{22} = 1$), but there is a possibility that both outlets are owned by the same person ($\omega_{12} = \omega_{21} = 1$) or by different people ($\omega_{12} = \omega_{21} = 0$).

$$\begin{aligned} \beta p_1 + \gamma \frac{p_2}{d_{12}} + \omega_{11}\beta(p_1 - c_1) + \omega_{12}\gamma \frac{p_2 - c_2}{d_{12}} &= 0 \\ \beta p_2 + \gamma \frac{p_1}{d_{21}} + \omega_{22}\beta(p_2 - c_2) + \omega_{21}\gamma \frac{p_1 - c_1}{d_{21}} &= 0 \end{aligned} \tag{3.24}$$

We can write it in the following form.

$$2\beta p - \beta c + \gamma \mathbf{D} p + \gamma \mathbf{\Omega} \cdot \mathbf{D}(p - c) = 0 \tag{3.25}$$

where $\mathbf{\Omega} \cdot \mathbf{D}$ means cell by cell multiplication and $\mathbf{\Omega}$ is the ownership matrix.

$$\mathbf{\Omega} = \begin{bmatrix} \omega_{11} & \omega_{12} \\ \omega_{21} & \omega_{22} \end{bmatrix} \tag{3.26}$$

Remember that the diagonals of the matrix $\mathbf{D}$ are zeros.

### 3.5.8 Ownership of Outlets

In the data, there is an index for the different owners. If the ownership variable is 0 the outlet is independent and individually owned. If the variable is 1, then it is corporate owned (and has been dropped). If the variable is greater than 1 then the number is used to identify which outlets have the same owner. For example, outlets with ownership set to 6 have the same owner. The code below uses `which()` to find the column where the ownership code is the same. Each outlet is assumed to be owned by itself so the diagonal of the matrix is 1s.

```
> Omega = matrix(0, J, J)
> for(j in 1:J) {
+   if(data$Ownership[j] > 1) {
+     Omega[j,which(data$Ownership == data$Ownership[j])] = 1
+   }
+ }
> diag(Omega) = 1
```

### 3.5.9 Estimation (Part 2)

The final piece of the puzzle is to use the optim() function. This is a basic optimization algorithm used in **R**. It is used because it was made freely available by John C. Nash, no relation to John F. Nash, of Nash equilibrium fame. To use this function we give it a set of starting values (init). Here these are set to zero for alpha and very small numbers for beta and gamma. When the first function is called, it hits these values with exp() and $\exp(\log(a)) = a$. That is, log is the inverse of exp. The diag() function creates a matrix with values on the diagonal and zeros every where else. The function dim() finds the dimensions of a matrix, where the first element is the number of rows. We give optim() the initial values of the parameters, the function to optimize, then any extra information that the function will need. In this case the values for D, price, Omega, BK, and X. Lastly we can control the maximum number of iterations it will use before stopping. The default is small, so you may want to make this big using maxit. The parameter trace gives you a print out of what the function is doing when it is set to 1.

```
> init = c(rep(0, 5), log(0.001), log(0.001), c(0,0))
> a1 = optim(par = init,
+             fn = outlet_price_f_int,
+             D = D,
+             price = exp(data$LPrice1),
+             Omega = Omega,
+             BK = data$BK,
+             X = cbind(data$Playland,
+                       data$Mall,
+                       data$DriveThru,
+                       data$Race,
+                       data$BK),
+             control=list(trace = 0,maxit = 10000000))


> # alpha
> a1$par[1:5]
  [1] -0.005985475 -0.017518129  0.005481921 -0.006670482
      -0.005309910
> # beta
> -exp(a1$par[6])
  [1] -2.308111e-05
> # gamma
> exp(a1$par[7])
  [1] 2.583274e-08
> # cost
> a1$par[8:9]
  [1] 4.717501 8.238029
```

Standard errors have not been calculated for these estimates. How would you do that?

### 3.5.10 Goodness of Fit

It is standard in structural econometrics to present some type of goodness of fit analysis. Here we can use the parameter values to generate the predicted price. We can then compare the prediction to the observed prices.

To do the analysis we need to create a new function. This function finds the new set of prices given the parameter values we found above.

```
> outlet_price_f_int2 = function(par, D, Omega,
+                                 cost, X, alpha, beta, gamma) {
+   X = as.matrix(X)
+   price = exp(par)
+   return(outlet_price_f(price, D, Omega,
+                          cost, X, alpha, beta, gamma))
+ }
```

```
> init = data$LPrice1
> b1 = optim(par = init,
+            fn = outlet_price_f_int2,
+            D = D,
+            Omega = Omega,
+            cost = cbind(1,data$BK)%*%a1$par[8:9],
+            X = cbind(data$Playland,
+                      data$Mall,
+                      data$DriveThru,
+                      data$Race,
+                      data$BK),
+            alpha = a1$par[1:5],
+            beta = -exp(a1$par[6]),
+            gamma = exp(a1$par[7]))
```

Figure 3.3 presents the goodness of fit of the simulation using the estimated parameters. In this exercise, we use the estimated parameters and then simulate the prices that satisfy the Nash equilibrium condition. The fitted curve is a little higher than the actual prices.

### 3.5.11 A Merger of McDonald's Outlets

It is not clear if the FTC has ever analyzed the impact of concentration among McDonald's franchisee owners. The economic theory is not different from analyzing the impact of a merger between hospitals or supermarkets.

Our experiment is for the independent McDonald's outlets to be purchased by the same person.

```
> data$Ownership2 = ifelse(data$BK == 2,
+                          10,
+                          data$Ownership)
> J = length(data$Ownership2)
> Omega2 = matrix(0, J, J)
> for(j in 1:J) {
+   if(data$Ownership2[j] > 1) {
+     Omega2[j, which(data$Ownership2 == data$Ownership2[j])] = 1
+   }
+ }
> diag(Omega2) = 1
```

We will assume the merger changes the ownership of the outlets but not the existence of the outlets. In our math and our code, this change is captured using `Omega` matrix. The code above finds all the McDonald's outlets that are independently owned and sets them to all have the same ownership.

The new firm's pricing decision for any outlet accounts for how that outlet's price affects demand at their other outlets. When one McDonald's increases the price for their Big Mac, some customers will switch to another outlet. After the merger many of these customers switch to outlets that are owned by the same firm. The merger reduces the loss in profits when prices are increased. The merger will lead to higher prices for Big Macs.

```
> init = b1$par
> c1 = optim(par = init,
+            fn = outlet_price_f_int2,
+            D = D,
+            Omega = Omega2,
+            cost = cbind(1,data$BK == 1)%*%a1$par[c(8,9)],
+            X = cbind(data$Playland,
+                      data$Mall,
+                      data$DriveThru,
+                      data$Race,
+                      data$BK),
+            alpha = a1$par[1:5],
+            beta = -exp(a1$par[6]),
+            gamma = exp(a1$par[7]))
```

The code below generates the `ggplot()` of the density of prices for the actual prices, the predicted prices from the estimated model and the predicted prices from the model of the merger of independent McDonald's outlets.

```
> ggplot_dens_outlets = data.frame(
+   Price = exp(data$LPrice1),
+   sim = exp(b1$par),
+   merger = exp(c1$par)
+ ) |>
+   ggplot(aes(x = Price)) +
+   geom_density(aes(y = ..scaled..), alpha = 0.5) +
+   geom_density(aes(x = sim, y = ..scaled..),
+                alpha = 0.5,
+                linetype = "dashed") +
+   geom_density(aes(x = merger, y = ..scaled..),
+                alpha = 0.5,
+                linetype = "dotted") +
+   geom_text(aes(x = 50, y = 1, label = "Actual")) +
+   geom_text(aes(x = 75, y = 0.3, label = "Sim")) +
+   geom_text(aes(x = 140, y = 0.4, label = "Merger Sim")) +
+   labs(x = "Price (normalized)",
+        y = "",
+        title = "") +
+   ## no numbers on y axis
+   scale_y_continuous(breaks = NULL) +
+   theme_minimal()

> ggplot_dens_outlets
```

Figure 3.3 shows both the goodness of fit of the model and the simulated impact on price of a merger between all the independent McDonald's outlets. The roll-up of independent McDonald's outlets would lead to a substantial increase in the prices of outlets in the market. Not only the McDonald's outlets but also the Burger King outlets that they compete with.

## 3.6 Discussion and Further Reading

Using game theory to analyze oligopoly models of competition actually predates game theory. In fact, all three models presented in this chapter predate Nash's analysis of game theory, even though all rely on the equilibrium concept.

Hotelling's 1929 paper, *Stability in Competition*, provides much of the intuition for the way many industrial organization economists think about competition for retail products (Hotelling, 1929). In the model, it is the "distance" between products that matters for competition. Not so much the exact number of competitors but how close the competitors are to each other in the minds of consumers.

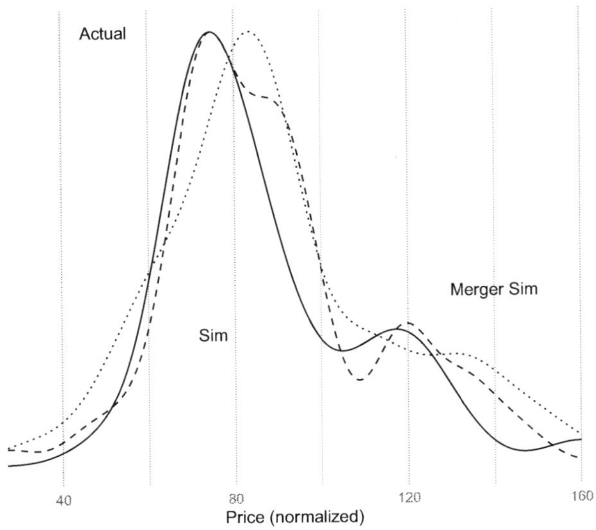


**FIGURE 3.3**
Plot of density of actual prices (solid line), predicted prices (dashed line) from the estimating model and simulated prices from the model of the merger of independent McDonald's outlets (dotted line). The dotted line is shifted up from the actual and simulated price distributions.

Somewhat confusingly we refer to a general model of differentiated price competition as a Bertrand model.[15] The classic paper taking this model to the data is Berry et al. (1995).

The subtlety of the Hotelling model didn't fit well into how US antitrust conducted merger review. The standard merger screen is a measure called the Herfindahl-Hirschman index (HHI) and the change the index caused by the merger. It is calculated by determining which firms are in the market, calculating each firm's share, squaring them and adding them up. While HHI is not a bad approximation of competition in homogeneous goods markets modeled by Cournot, it doesn't make a lot of sense for differentiated goods markets modeled by Hotelling. In the simple Hotelling line, the extent of competition can vary substantially without any change in the HHI.

The 2010 Merger Guidelines from the Department of Justice and the Federal Trade Commission, made an adjustment suggesting that a different screen may be better for differentiated goods mergers. The Upward Price Pressure screen measures how close two firms are by how many customers are diverted for a price increase.[16]

[15] Chapter 7 works through this model.

[16] `https://www.justice.gov/atr/horizontal-merger-guidelines-08192010` accessed on 11/21/23.

# 4

# *Empirical Entry Games*

## 4.1 Introduction

This chapter revisits the entry game presented in Chapter 2. In this game, we have two firms considering whether or not to enter a market. The issue is that it is costly to enter and both firms would prefer to have the market to themselves. In equilibrium, we will tend to have one firm in the market, but the game theory may not predict which one.

The chapter applies an entry game to the question of which markets the mega bookstores, Borders and Barnes & Noble entered in the 1990s. The 90s saw massive changes in book retailing. Changes in distribution technology made it profitable for big box book stores to enter the market with a huge range of titles as well as other products such as music, games, and even coffee. The chains, Borders and Barnes & Noble led the charge by purchasing smaller chains and entering green fields sites. Which markets did these firms enter? What would have happened if they had been allowed to merge?

The chapter introduces an empirical model of entry based on the decision making model of Dan McFadden. The empirical model is extended to allow the firms to make decisions that are dependent on each other, a game. The chapter analyzes the effect of a merger, if that merger would have occurred in the early 90s.

## 4.2 Empirical Model

The section begins with a description of a standard choice model. This model was originally developed by Dan McFadden in the early 1970s to analyze consumer choice problems. McFadden was interested in who would use the new subway system that had been built in San Francisco, the Bay Area Rapid Transit (BART) system. Here we are adapting the model to analyze the choice of which market the firm will enter. The model is generalized to allow unobserved characteristics of the market to be correlated across firms. It is generalized again to allow firms to make choices that are dependent upon each other.

DOI: 10.1201/9781003351603-4

### 4.2.1 Single Firm Entry Model

By the year 2000, Barnes & Noble had bookstores in 283 counties in the United States, but this is out of over 3,000 counties. Placing these large stores in a location is not cheap, particularly if they involve new construction. So which counties dd Barnes & Noble choose to locate?

Assume that the firm's latent profits from locating in a particular county are as follows.

$$\pi_{1i} = \mathbf{X}_i'\beta_1 + \xi_{1i} \tag{4.1}$$

where $\mathbf{X}_i$ are market characteristics such as population size, $\beta_1$ determines how these characteristics are mapped into profits for the firm and $\xi_{1i}$ is unobserved characteristics of the market and can be thought of as representing entry costs for the firm. These unobserved characteristics are unobserved by the econometrician (that's you) but they are observed by the firm itself.

If the data include information on the market such as population size ($pop$) and median income ($income$), then the decision to enter can be written as follows.

$$\pi_{1i} = \beta_{10} + \beta_{11} pop_i + \beta_{12} income_i + \xi_{1i} \tag{4.2}$$

For Firm 1, their profits in market $i$ depend on $pop_i$ and $income_i$ as determined by the parameters $\beta_{11}$ and $\beta_{12}$, respectively.

In matrix notation we have that the vector of parameters is written as follows.

$$\beta_1 = \begin{bmatrix} \beta_{10} \\ \beta_{11} \\ \beta_{12} \end{bmatrix} \tag{4.3}$$

Similarly, the matrix of observed characteristics for $N$ markets as follows.

$$\mathbf{X} = \begin{bmatrix} 1 & pop_1 & income_1 \\ 1 & pop_2 & income_2 \\ \dots & & \\ 1 & pop_N & income_N \end{bmatrix} \tag{4.4}$$

The first column is just 1's. In the matrix algebra, this column is multiplied by the $\beta_{10}$ parameter to give a constant across all the markets. We use the notation $\mathbf{X}_i'$ to emphasize that we are using the $i$th row of the matrix and that is being multiplied by the vector of parameters $\beta_1$.

Barnes & Noble enter the market if and only if the following inequality holds.

$$\begin{aligned} \mathbf{X}_i'\beta_1 + \xi_{1i} &> 0 \text{ or} \\ \xi_{1i} > -\mathbf{X}_i'\beta_1 & \end{aligned} \tag{4.5}$$

So if the unobserved entry costs are low enough, then Barnes & Noble will enter the market.

We expect firm profits to determined by various factors affecting demand for books, costs associated with selling books, and fixed costs associated with the location of the store. If we assume that the unobserved costs of entry are distributed standard normal, $\xi_i \sim \mathcal{N}(0,1)$, then the probability of entry is $\Phi(-\mathbf{X}_i'\beta_1)$, where $\Phi()$ is the cumulative distribution of the standard normal. This is the **probit** model introduced in Chapter 1. We can estimate the parameters $\beta_1$ using the `glm()` procedure we used in Chapter 1

$$\max_{\beta_1} \sum_{i=1}^{N} y_{1i} \log(\Phi(-\mathbf{X}_i'\beta_1)) + (1 - y_{1i}) \log((1 - \Phi(-\mathbf{X}_i'\beta_1)) \qquad (4.6)$$

To estimate the model we can find the $\beta_1$ that maximizes the likelihood of the data, where $y_{1i}$ is the observed entry decision for each county.[1]

### 4.2.2 Multiple Firm-Independent Entry Model

Taking baby steps to our full model, consider a model where we have two firms entering a market but the two firms are making decisions independently of each other. This could be a model of entry of Barnes & Noble and Best Buy. Both are big box stores whose decision to enter a market will be based on similar things, both observed and unobserved by the econometrician. However, with one focused on books and the other focused on electronics, it is unlikely that their decisions to enter a particular market will be dependent on each other.

The two firms will enter market $i$ if and only if the following inequalities hold.

$$\begin{aligned} \mathbf{X}_i'\beta_1 + \xi_{1i} &> 0 \\ \mathbf{X}_i'\beta_2 + \xi_{2i} &> 0 \end{aligned} \qquad (4.7)$$

Again $\mathbf{X}_i$ represents observed characteristics of the market such as population size. These characteristics are mapped into each firm's profit function by the parameter vectors $\beta_1$ and $\beta_2$.

So far this doesn't seem to be different from the model in the previous section. The difference is that we can allow the unobserved characteristics of the market to be correlated across the two firms. Assume that $\{\xi_{1i}, \xi_{2i}\} \sim \mathcal{N}(\mu, \mathbf{\Sigma})$ where $\mu = \{0, 0\}$ and

$$\mathbf{\Sigma} = \begin{bmatrix} 1 & \rho \\ \rho & 1 \end{bmatrix} \qquad (4.8)$$

In words, we allow the unobserved characteristics of the markets to be distributed standard bivariate normal. The parameter $\rho \in [-1, 1]$ represents the correlation across the two firms. This parameter is likely to depend on how similar the two firms are in terms of their customer base and costs of setting up the store. A fast food restaurant and big box book store may have unobserved characteristics that are not vary correlated $\rho = 0$, while two big box stores may have a highly positive correlation (a high $\rho$).

[1]We log the probabilities so that we don't run into problems because the numbers are too small for the computer to represent.

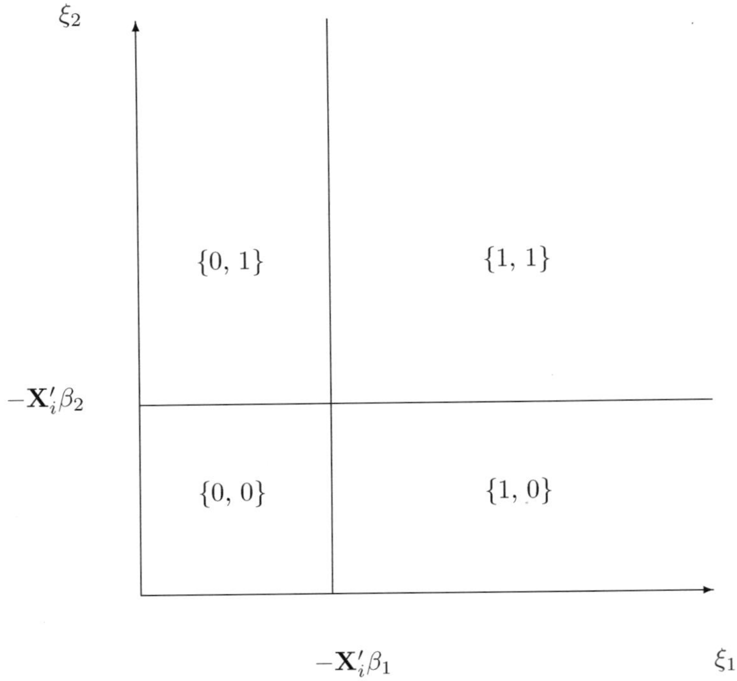


**FIGURE 4.1**
Empirical implications of the two firm independent entry model. Firm 1's entry is denoted in the first position of the brackets. If both firms have unobserved characteristics that are low then neither will enter $\{0, 0\}$, if Firm 1's unobserved characteristic is high then Firm 1 will enter $\{1, 0\}$, if Firm 2's unobserved characteristic is also high the Firm 2 will also enter $\{1, 1\}$.

Let 1 denote entry and 0 denote the choice not to enter. In this model we observe four cases, neither firm enters $\{0,0\}$, Firm 1 enters but Firm 2 does not $\{1,0\}$, Firm 1 doesn't enter but Firm 2 does $\{0,1\}$, and both firms enter $\{1,1\}$.

Figure 4.1 presents the four cases given the distribution of the unobserved characteristics of the markets for the two firms. The pattern of entry across markets will tell us if the unobserved characteristics are correlated across firms. If we see lots of cases where both firms do the same thing, both enter the market or neither enters the market, then that is consistent with positively correlated unobserved characteristics. If we see lots of cases where there is just one firm in the market but which is pretty evenly distributed, then that is consistent with the unobserved characteristics being uncorrelated or negatively correlated across the two firms.

Again we want to find the $\beta_1$, $\beta_2$ and $\rho$ that maximize the likelihood of the observed data which is represented by $y_1$ and $y_2$.

### 4.2.3 Entry Game

After all of that set up, we can write down our empirical entry game. Like in the previous section we have two firms considering whether or not they should enter market $i$. They will enter if the following inequalities hold.

$$\begin{aligned} \mathbf{X}_i'\beta_1 - D_{2i}\alpha_1 + \xi_{1i} > 0 \\ \mathbf{X}_i'\beta_2 - D_{1i}\alpha_2 + \xi_{2i} > 0 \end{aligned} \tag{4.9}$$

where $D_{1i} \in \{0,1\}$ and $D_{2i} \in \{0,1\}$. There are a couple of differences between these inequalities and the inequalities in Equation (4.7). The first is that there are two extra parameters $\alpha_1$ and $\alpha_2$. These represent the reduced profits associated with competing with the other firm. This only occurs if the other firm enters the market. The profits for Firm 1 are lower if $D_{2i} = 1$ by the amount $\alpha_1$, where $D_{2i} = 1$ if and only if the second inequality holds. The profits for Firm 2 are lower if $D_{1i} = 1$ which only occurs if the first inequality holds.

Now you may be able to start seeing the issue. The first inequality depends on the outcome of the second inequality which depends on the outcome of the first inequality! To see what is going on consider a version of the figure presented above.

Figure 4.2 represents the game. If the unobserved benefits for both firms are high (fixed cost of entry is low), then both firms will enter $\{1,1\}$, if they are low, then neither will enter $\{0,0\}$. If they are very high for Firm 1 and very low for Firm 2, then Firm 1 will enter $\{1,0\}$. Similarly the other way $\{0,1\}$. There is also the intermediate outcome where the model does not make a clear prediction on what will happen. It could be that Firm 1 enters while Firm 2 does not, or it could be the other way around. The model clearly predicts that one firm will enter, it just does not predict which.

## 4.3 Empirical Analysis: Bookstore Entry with R

The section presents the code for estimating the empirical models presented above. To show the progression from the original entry model to the entry game, the section presents code for all three models. The first model is just a probit, so you could use the `glm()` function baked into **R**. However including

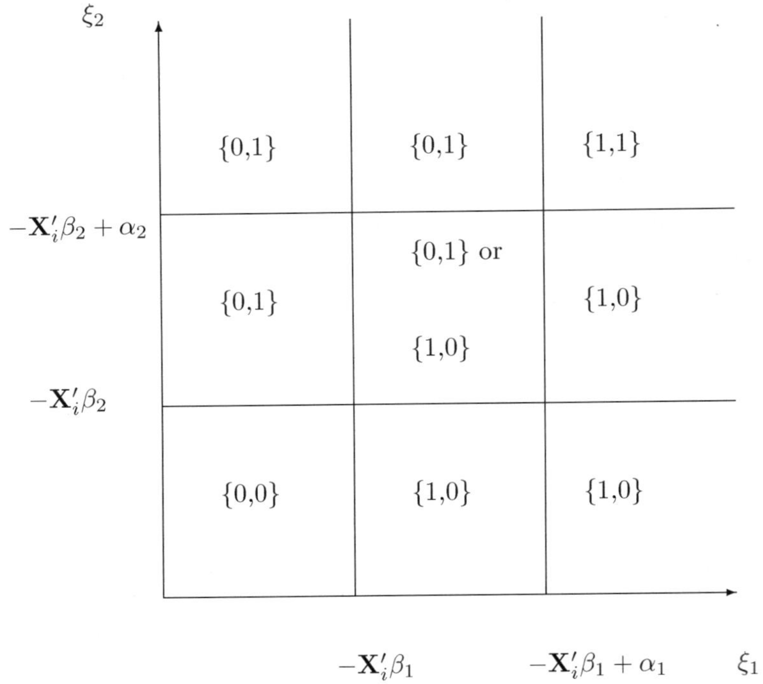


**FIGURE 4.2**
Empirical implications of the entry game. Firm 1's entry is denoted in the first position of the brackets. The region in the middle square has an in determinant outcome. One firm will enter, but it is not clear which.

the code for this case makes it easier to see how the more complicated models work.

### 4.3.1 Single Firm Entry Model in R

The estimation algorithm is **maximum likelihood**. In this method, we find the parameter values that maximize the likelihood that the data we observe was generated by a particular model. We will need a function that calculates the probability that the observed data occurred given a particular parameter value. There are a couple of different ways of doing that in **R**. In this section, we will calculate the probabilities **numerically**. That is the section presents a method for approximating the probability by having the computer do a lot of calculations. This method is not as efficient as using a built-in `C`-based function, but it is easier to see what is going on, particularly as the problem gets more complicated.

```
> set.seed(123456789)
> K = 10000
> Z_1 = rnorm(K)
```

This code generates `K = 10000` pseudo-random draws from a standard normal distribution. The function `rnorm()` is the **R** function for drawing from a standard normal. These are **global** variables, meaning that they are available to any function we write. The function `set.seed()` is used to make sure that the results can be exactly replicated.

```
> f_entry = function(X, beta1) {
+   X1 = as.matrix(cbind(1, X))
+   N = dim(X1)[1]
+   xi = Z_1
+   Xb = X1%*%beta1
+   p = rep(0, N)
+   for(k in 1:K) {
+     pi_k = Xb + xi[k]
+     p = p + (pi_k > 0)
+   }
+   return(p/K)
+ }
> f_loglik = function(X, y, beta) {
+   epsilon = 1e-5
+   p = f_entry(X, beta)
+   return(-mean((y == 1)*log(p + epsilon) +
+                  (y == 0)*log(1 - p + epsilon)))
+ }
```

The function `f_entry()` takes in the matrix of data `X` (without the column of 1's) and the vector of parameters `beta`. The matrix `X` consists of columns of data stating observed characteristics of each market, such as population size and median income.

The probability of entering market $i$ is determined by the probability that profits will be positive given a large number of possible values for the unobserved characteristic `xi`.

The function `f_loglik` calculates the probability that the model is true given the observed data. This code is equivalent to Equation (4.6). This function takes a vector of outcomes, `y`. The vector `y` states whether or not the firm entered each of the markets in the data where 1 denotes entry and 0 denotes that the firm did not enter the market. The function transforms the probability into logs so that we don't run into problems where the numbers we are calculating are smaller than the smallest number the computer can handle. Later when we apply this function to an actual problem, we will benefit from the fact that the optimal value for `beta` is the same for the log-transformed function as it is for the original function. Notice that there is a negative sign in front of the mean() function. This is there because the optimize algorithm used defaults to find the minimum so the minimum of the negative log-likelihood is the maximum log-likelihood. The value epsilon is a small number designed to make sure that the computer doesn't crash if it tries to calculate the log of zero.

### 4.3.2 Multiple Firm-Independent Entry in R

```
> Z_2 = rnorm(K)
```

This time we need a distribution with two dimensions. So the first step is to create another large set of random numbers drawn from a standard normal function.

```
> f_2entry = function(X, beta_1, beta_2, rho) {
+   N = dim(X)[1]
+   xi_1 = Z_1
+   xi_2 = Z_2*sqrt(1 - rho^2) + rho*Z_1
+   Xb_1 = X%*%beta_1
+   Xb_2 = X%*%beta_2
+   p_00 = p_01 = p_11 = rep(0, N)
+   for(k in 1:K) {
+     pi_1k = Xb_1 + xi_1[k]
+     pi_2k = Xb_2 + xi_2[k]
+     p_00 = p_00 + (pi_1k < 0 & pi_2k < 0)
+     p_01 = p_01 + (pi_1k < 0 & pi_2k > 0)
+     p_11 = p_11 + (pi_1k > 0 & pi_2k > 0)
+   }
+   return(list(p_00 = p_00/K,
+               p_01 = p_01/K,
+               p_11 = p_11/K))
+ }
> f_loglik_2 = function(X, y, beta_1, beta_2, rho) {
+   epsilon = 1e-10
+   Lik = f_2entry(X, beta_1, beta_2, rho)
+   return((y[,1] == 0 & y[,2] == 0)*log(Lik$p_00 + epsilon) +
+          (y[,1] == 0 & y[,2] == 1)*log(Lik$p_01 + epsilon) +
+          (y[,1] == 1 & y[,2] == 1)*log(Lik$p_11 + epsilon) +
+          (y[,1] == 1 & y[,2] == 0)*log(1 -
+                                        Lik$p_00 -
+                                        Lik$p_01 -
+                                        Lik$p_11 +
+                                        epsilon))
+ }
> f_loglik_2_int = function(par, X, y) {
```

```
+   X = as.matrix(cbind(1, X))
+   J = dim(X)[2]
+   beta_1 = par[1:J]
+   beta_2 = par[(J+1):(2*J)]
+   rho = -1 + 2*exp(par[2*J+1])/(1 + exp(par[2*J+1]))
+   return(-mean(f_loglik_2(X, y, beta_1, beta_2, rho)))
+ }
```

Just by counting lines of code, we see that things are a lot more complicated when we have two firms whose decisions are independent but correlated. The function `f_2entry()` has parameters for Firm 1 (`beta_1`) and Firm 2 (`beta_2`) and the correlation between the unobserved term (`rho`). The unobserved term `xi_2` is a function of both `Z_1` and `Z_2`. The higher `rho` the more it weights `Z_1`. The function determines the probability of observing three cases neither enter, only Firm 1 enters, and both enter. The fourth case is calculated as the residual because probabilities must add to 1.

The code now includes an additional function `f_loglik_2_int()`. This is an intermediate function designed to be used by the **R**'s optimization algorithm `optim()`. In this case, the parameter `rho` is restricted to be between $-1$ and 1. It is good coding practice to allow the optimization algorithm to choose what ever values it likes, but then transform those values into the restricted set required by the model. To do this the code uses `exp(x)/(1 + exp(x))`, also known as the **softmax** function. This function takes any value of `x` and turns it into a number that lies between 0 and 1.

### 4.3.3 Entry Game in R

While the entry game is quite a bit more complicated than the previous model, the estimator is not that different. There are two additional parameters `alpha_1` and `alpha_2` but everything else is pretty much the same The big difference is that the model cannot distinguish two cases in the data. The model predicts when one firm will enter the market, but it does not predict which firm that will be. Therefore, we need to combine those two cases in order to estimate our model.

```
> f_entry_game = function(X, beta_1, beta_2, alpha_1, alpha_2
+                               , rho) {
+   N = dim(X)[1]
+   xi_1 = Z_1
+   xi_2 = Z_2*sqrt(1 - rho^2) + rho*Z_1
+   Xb_1 = X%*%beta_1
+   Xb_2 = X%*%beta_2
+   p_00 = p_11 = rep(0, N)
+   for(k in 1:K) {
+     pi_1k = Xb_1 + xi_1[k]
+     pi_2k = Xb_2 + xi_2[k]
```

```
+       p_00 = p_00 + (pi_1k < 0 & pi_2k < 0)
+       p_11 = p_11 +
+         (pi_1k - alpha_1 > 0 & pi_2k - alpha_2 > 0)
+     }
+     return(list(p_00 = p_00/K,
+                 p_11 = p_11/K))
+ }
```

The function `f_loglik_game_int()` is pretty similar to the equivalent function in the previous section. The difference is that there are two additional parameters. We will restrict these two parameters to be positive. We will impose the result that an increase in competition lowers profits (or does nothing).

```
> f_loglik_game = function(X, y, beta_1, beta_2, alpha_1,
+                          alpha_2, rho) {
+   epsilon = 1e-10
+   Lik = f_entry_game(X, beta_1, beta_2, alpha_1, alpha_2, rho)
+   return((y[,1] == 0 & y[,2] == 0)*log(Lik$p_00 + epsilon) +
+          (y[,1] == 1 & y[,2] == 1)*log(Lik$p_11 + epsilon) +
+          ((y[,1] == 1 & y[,2] == 0) +
+            (y[,1] == 0 & y[,2] == 1))*log(1 -
+                                           Lik$p_00 -
+                                           Lik$p_11 +
+                                           epsilon))
+ }
> f_loglik_game_int = function(par, X, y) {
+   X = as.matrix(cbind(1, X))
+   J = dim(X)[2]
+   beta_1 = par[1:J]
+   beta_2 = par[(J+1):(2*J)]
+   alpha_1 = exp(par[2*J+1])
+   alpha_2 = exp(par[2*J+2])
+   rho = -1 + 2*exp(par[2*J+3])/(1 + exp(par[2*J+3]))
+   return(-mean(f_loglik_game(X, y, beta_1, beta_2, alpha_1,
+                              alpha_2, rho)))
+ }
```

## 4.4 Empirical Analysis: Bookstore Entry using R

Barnes & Noble dates itself to the 1800s, but in the early 1990s, the firm developed the super bookstore. The big box of bookstores. The objective was to carry a huge range of books, music, games, and even food and coffee. It revolutionized book retail in the United States.

Where did Barnes & Noble choose to enter? What determined those locations?

### 4.4.1 Data

The data used here is from Adams and Basker (2025). The authors combine information from publicly available census data and published directories of retail bookstores.

**TABLE 4.1**
Mean County Characteristics by Presence of Barnes & Noble and Borders.

| | Population | Income | College Share (%) | Bookstores (#) |
|---|---|---|---|---|
| None | 37,857 | 33,513 | 15 | 2 |
| Only Barnes & Noble | 275,435 | 43,210 | 27 | 14 |
| Only Borders | 548,218 | 49,118 | 31 | 40 |
| Both | 272,191 | 46,476 | 26 | 12 |

Notes: Presence of Barnes & Noble and Borders referes to the year 2000. There are 2,792 counties with neither chain, 155 counties with only Barnes & Noble, 36 counties with only Borders, and 128 counties with both chains. Population, income, and college share use data from 2000. Income refers to median county-level household income in 2000. College share is the share of population aged 25 and older with a college degree in 2000. Bookstores is the total number of bookstores in the county in 1990.

Table 4.1 shows the difference between counties with and without Borders and Barnes & Noble stores. Unsurprisingly, the firms entered counties with larger populations, richer counties, counties with higher educated population, and counties with more bookstores in 1990.

### 4.4.2 Estimation of Single Firm Entry

The model presented in Equation (4.5) can be estimated by combining data on the location of Barnes & Noble stores with various economic and demographic information at the county level. In addition to these data, we have information on the number of book stores in the county in 1990, which is generally earlier than the Barnes & Noble super bookstores came into existence. We can similarly model the entry of Borders.

In this analysis, there is no game. Barnes & Noble and Borders are assumed to make their entry decisions optimally, but these decisions are completely independent of each other. To estimate the parameters of the model, we can either use a maximum-likelihood estimator and the functions `f_loglik()` and `f_enter()` or we can use the `glm()` procedure introduced in Chapter 1.

```
> file = paste0(dir, "book_2000.csv")
> dt = fread(file)

> glm1 = glm(enter ~ log_pop_2000 + med_income + college +
+               stores_1990,
+             data = dt,
+             family = binomial(link = "probit"))
```

This function is used to estimate cases where the outcome is binary, enter or did not enter, and where the unobserved characteristic of the market is distributed as a normal distribution with mean of 0 and standard deviation of 1. This is a **probit model**.

```
> init = glm1$coefficients
> data = data.frame(y = dt$enter,
+                   pop = dt$log_pop_2000,
+                   income = dt$med_income,
+                   college = dt$college,
+                   stores_1990 = dt$stores_1990)
> data = na.omit(data)
> res1 = bs(init, f_loglik,
+           y = data$y,
+           X = cbind(data$pop,
+                     data$income,
+                     data$college,
+                     data$stores_1990))
```

The `glm1` results are used as the initial starting value for the maximum likelihood estimator using `f_loglik()`. The code then creates the data to be used by the estimator, including information on which stores Barnes & Noble entered, population size of the county, percentage of college graduates in the county and the number of book stores in the county in 1990. The object `res1` stores the results. This uses a function called `bs()` which creates pseudo samples from the data set and uses the `optim()` function to determine the parameter values that maximize the likelihood using the `f_loglik()` function.[2]

```
> dt$enter2 = ifelse(dt$borders > 0, 1, 0)
> glm2 = glm(enter2 ~ log_pop_2000 +
+              med_income +
+              college + stores_1990, data = dt,
+            family=binomial(link="probit"))
```

The data set is adjusted to remove observations with missing values. The function `na.omit()` is used to do this.[3]

[2]The `bs()` function is available from the github page for the book.
[3]**R** uses `NA` to denote missing values.

```
> init = glm2$coefficients
> data = data.frame(y = dt$enter2,
+                   pop = dt$log_pop_2000,
+                   income = dt$med_income,
+                   college = dt$college,
+                   stores_1990 = dt$stores_1990)
> data = na.omit(data)
> res2 = bs(init, f_loglik,
+           y = data$y,
+           X = cbind(data$pop,
+                     data$income,
+                     data$college,
+                     data$stores_1990))
```

The code above similarly creates a data set for estimating the choice of market to enter for Borders stores. It estimates the probit model using the `bs()` function and stores the results in the object `res2`.

### 4.4.3 Estimation of Two Firm Entry Model

Table 4.2 presents the results from the different entry models discussed above. The first model assumes that the two firms make optimal decisions that are independent of each other. In addition, it assumes that the unobserved characteristics faced by the two firms in each market are independent across the two firms.

The two-firm model is similar to single firm entry model. The entry decisions of the two firms are independent of each other but the unobserved characteristics of the markets may be correlated across firms. We say that the firms are strategically independent but the markets are statistically dependent. This is a **biprobit model**. The code again creates the data set for the analysis removes missing values and uses the `bs()` function calling `f_loglik_2_int` It stores the results in `res3`.

```
> init = c(glm1$coefficients,
+          glm2$coefficients,
+          0)
> data = data.frame(y_1 = dt$enter,
+                   y_2 = dt$enter2,
+                   pop = dt$log_pop_2000,
+                   income = dt$med_income,
+                   college = dt$college,
+                   stores_1990 = dt$stores_1990)
> data = na.omit(data)
> res3 = bs(init, f_loglik_2_int,
+           y = cbind(data$y_1,
+                     data$y_2),
```

```
+             X = cbind(data$pop,
+                       data$income,
+                       data$college,
+                       data$stores_1990))
```

### 4.4.4 Estimation of the Entry Game Model

For the third model, we have two firms entering the markets and again the unobserved characteristics of the two firms are correlated across markets. This time the decision to enter depends on the choice of the other firm. Decisions are now dependent. This model uses the same data as the previous model. It uses the `bs()` function with the `f_loglik_game_int()` likelihood function and saves the results in the object `res4`.

```
> init = c(glm1$coefficients,
+          glm2$coefficients,
+          0,
+          0,
+          0)
> res4 = bs(init, f_loglik_game_int,
+             y = cbind(data$y_1,
+                       data$y_2),
+             X = cbind(data$pop,
+                       data$income,
+                       data$college,
+                       data$stores_1990))
```

The table of results also presents information on the standard errors of the estimates of the parameters. These values are determined using a **bootstrap method**. This method approximates how estimates may vary when a different sample is used.[4]

Table 4.2 presents the estimated parameter values for the three models discussed above. The $\alpha$ parameters state that having two firms lowers profits. The store locations are positively associated with population size, education, and the number of existing bookstores. It is unclear if income has any impact. Finally, the unobserved entry costs are highly correlated across the two firms.

### 4.4.5 Model Fit

We can use the estimated parameters to simulate the model and compare the predicted entry decisions to the actual entry decisions. This provides one test of the model's fit. The code, not shown, simulates the model 1,000 times

[4]The bootstrap method approximates estimates from a new sample by resampling the current data and restimating the model using the new pseudo-sample.

**TABLE 4.2**
Results from estimates of the three models. The first set of columns labeled "Probit" are estimates assuming that the two firms are making entry decisions that are both strategically independent and statistically independent. The second set of columns labeled "BiProbit" assumes the firm entry decisions are strategically independent. It allows unobserved characteristics of the market to be correlated across firms. The two columns labeled "Game" refer to the case where the entry decisions of the two firms are both strategically and stochastically dependent. The columns labeled "SD" refer to the bootstrap standard errors.

| | Probit | SD | BiProbit | SD | Game | SD |
|---|---|---|---|---|---|---|
| const_1 | −15.06 | 0.22 | −15.03 | 0.09 | −15.11 | 0.25 |
| Pop_1 | 1.08 | 0.02 | 1.09 | 0.01 | 1.07 | 0.01 |
| Income_1 | −1.03 | 0.44 | −1.04 | 0.25 | −0.76 | 0.48 |
| College_1 | 5.68 | 0.66 | 5.51 | 0.34 | 5.65 | 0.55 |
| Stores_1990_1 | 0.28 | 0.09 | 0.28 | 0.07 | 0.37 | 0.11 |
| const_2 | −11.57 | 0.26 | −11.54 | 0.11 | −11.37 | 0.20 |
| Pop_2 | 0.66 | 0.03 | 0.66 | 0.01 | 0.65 | 0.02 |
| Income_2 | 1.75 | 0.51 | 1.74 | 0.25 | 1.31 | 0.30 |
| College_2 | 2.49 | 0.70 | 2.59 | 0.33 | 2.70 | 0.55 |
| Stores_1990_2 | 0.52 | 0.08 | 0.49 | 0.06 | 0.79 | 0.11 |
| alpha_1 | | | | | 0.73 | 0.18 |
| alpha_2 | | | | | 0.70 | 0.12 |
| rho | | | −0.08 | 0.10 | 0.47 | 0.10 |

and compares the predicted entry decisions to the actual entry decisions. The results are presented in Table 4.3.

Table 4.3 presents model fit results for the three models. There is not much difference between the three estimators, although the game theory model is slightly better at predicting the one-firm market.

## 4.5 Policy Analysis using R

What would happen if Barnes & Noble and Borders had merged? Here we are not going to worry about the effect on prices but on which counties the merged firm would enter. Assume that the $\alpha$ parameters remain the same post merger.[5] The difference the merger brings is that it allows the firms to coordinate entry. This analysis assumes that the merged firm will keep the two distinct brands.

To determine what would happen in this alternate universe, we can simulate

[5] Given that the $\alpha$ parameters are accounting for both diversion between stores and prices, we would expect them to be smaller but not zero post merger.

**TABLE 4.3**
Comparison of predictions of the three models to the actual outcomes. The higher percentage on the diagonal the better fit of the model. All models are good at predicting the high probability event, which is no entry by either firm. The non-strategic models are best at predicting when there will be two firms in two-firm markets, while the game theory model is best at predicting the one-firm market.

| | None | One Firm | Two Firm |
|---|---|---|---|
| Probit: None | 97.2 | 42.3 | 9.4 |
| Probit: One | 2.5 | 38.8 | 34.9 |
| Probit: Two | 0.3 | 18.9 | 55.6 |
| BiProbit: None | 97.1 | 41.6 | 9.0 |
| BiProbit: One | 2.6 | 40.0 | 35.9 |
| BiProbit: Two | 0.3 | 18.4 | 55.1 |
| Game: None | 97.1 | 41.0 | 9.0 |
| Game: One | 2.6 | 41.7 | 36.6 |
| Game: Two | 0.3 | 17.3 | 54.4 |

thousands of outcomes in the markets for which have data using the parameter values estimated above.

**TABLE 4.4**
Comparison of actual entry to simulated entry in 2000 and simulated entry under a merger.

| | Actual | Sim | Merge |
|---|---|---|---|
| none | 2919 | 2895 | 2895 |
| BN or Borders | 170 | 191 | 238 |
| both | 128 | 93 | 46 |

Table 4.4 presents summary of simulations of a merger between Borders and Barnes & Noble. The simulations tend to predict more counties with stores than we actually see, but fewer counties where both stores are present. The model predicts that the merger will lead to fewer markets with a Borders and Barnes & Noble. Not presented is the variation in the estimates, but it is pretty clear that there will be fewer markets with both stores. As some consumers prefer one or the other, people in those markets are worse off. We can think of that as a quality reduction. In addition, the reduced competition in those markets is likely to lead to higher prices.[6]

The value in the first row of columns 2 and 3 is the same. The merger changes whether or not the market is will have two firms or one, but not whether or not at least one firm will enter. The reason is that the game explicitly assumes that the two firms coordinate on entry. Chapters 5 and 8

[6]See the analysis presented in Chapter 3.

consider games where the firms cannot coordinate. In these cases, the merger may actually lead more firms to enter the market.

## 4.6 Discussion and Further Reading

The analysis is based on Adams and Basker (2025). The authors analyze the entry of mega bookstores in the US using publicly available census data and directories of retail bookstores. The entry game analysis is based on Bresnahan and Reiss (1991a) and Tamer (2003). Chapters 5, 6, and 8 revisit this problem under different modeling assumptions. Chapter 5 revisits the problem assuming that the outcome of the game in the data is a mixed strategy Nash equilibrium.

Game theory provides a better explanation of the entry decisions we observe, however it can also makes the analysis more challenging. Entry problems are a type of coordination game and like other coordination games they tend to have multiple equilibria. This means for a set of parameter values a model makes multiple predictions about what we will see in the data. This makes the models a challenge to estimate. The solution presented here follows Bresnahan and Reiss (1991a) which simply combines all the equilibria into one observed outcome. Tamer (2003) presents a method for estimating a more efficient model. Many papers in empirical industrial organization make alternative assumptions that lead to unique predictions from the game. Chapter 6 discusses this option.

Borders and Barnes & Noble did not propose to merger in the 1990s or 2000s. Did they contemplate it? Were they dissuaded by the FTC's challenge of the merger between Blockbuster and Hollywood Video in 1999 and again in 2005?[7] Borders went into bankruptcy proceedings in 2011 and Barnes & Noble acquired some of its intellectual property.[8]

[7] `https://www.nytimes.com/2005/03/26/business/media/blockbuster-ends-bid-for-rival.html`

[8] `https://www.barnesandnobleinc.com/about-bn/history/`. Accessed on 2/20/23.

# 5

# *Mixed Strategies*

## 5.1 Introduction

Chapter 2 stated that John Nash showed that the Nash equilibrium exists in a large set of games. This is a powerful result and is part of the reason why Nash equilibrium is so important. But Nash's proof relies on **mixed strategies**. Mixed strategies are a more complicated method for determining the outcome of a game. Mixed strategies are vital for the analysis of parlor games like chess or cards, and useful for the analysis of games with multiple equilibria like the **coordination game** presented in earlier chapters.

The chapter introduces mixed strategies using the classic childhood game Rock-Paper-Scissors. It analyzes penalty kicks in soccer's English Premier League. The chapter revisits the entry of big box bookstores analyzed in Chapter 4. This time it assumes players are playing a mixed strategy when there are multiple equilibria.

## 5.2 Zero-Sum Games

**Zero-sum games** are the oldest types of games studied in game theory. They are a natural way to represent parlor games like chess. In those games, there is a winner and a loser. What the winner wins, the loser loses. This section introduces some important zero-sum games and discusses how to find the equilibrium.

### 5.2.1 Rock-Paper-Scissors

Rock-Paper-Scissors is one of the more famous childhood games. It requires two players with at least one hand. The game provides solutions to many previously unsolvable problems - who gets the last slice of cake, who gets to ride shotgun, who gets to mow the lawn.

- Players: Player 1 and Player 2

DOI: 10.1201/9781003351603-5

- Strategies: Rock, Paper, Scissors
- Payoffs: See Table 5.1

Table 5.1 provides a normal form representation for the game. There are only three outcomes in the game, win, lose, and draw. We can represent these outcomes as payoffs, 1, $-1$ and 0 respectively. If both players choose the same shape, it is a draw. Then Rock beats Scissors, Paper beats Rock, and Scissors beats Paper. Notice that the numbers in each cell add to 0, this is what we mean by zero-sum.

**TABLE 5.1**
Normal form representation of Rock-Paper-Scissors with two players $P_1$ and $P_2$, the strategies for $P_1$ are on the rows and $P_2$'s strategies are on the columns. Payoffs are each cell, with $P_1$'s payoff listed first.

| $P_1, P_2$ | ROCK | PAPER | SCISSORS |
|---|---|---|---|
| ROCK | 0, 0 | $-1$, 1 | 1, $-1$ |
| PAPER | 1, $-1$ | 0, 0 | $-1$, 1 |
| SCISSORS | $-1$, 1 | 1, $-1$ | 0, 0 |

### 5.2.2 Nash Equilibrium

Does Rock-Paper-Scissors have a Nash equilibrium? To check whether a particular outcome is a Nash equilibrium we can follow the standard algorithm. In the first step, we posit the Nash equilibrium. Then in the second step, assume all but one player plays the posited strategy, determine the optimal strategy for the remaining player(s). If the posited strategy is not optimal, then it is not a Nash equilibrium.

- Step 1: ROCK, ROCK
- Step 2: $P_1$ plays ROCK. Is ROCK optimal for $P_2$?
  - ROCK: 0, PAPER: 1, SCISSORS: $-1$
  - No: Not a Nash equilibrium.

We see that if we posit that $P_1$ plays Rock, then it is optimal for $P_2$ to play Paper. Remember, Paper beats Rock. So $\{ROCK, ROCK\}$ is not a Nash equilibrium.

Similarly we can use the algorithm to determine if $\{PAPER, ROCK\}$ is a Nash equilibrium.

- Step 1: PAPER, ROCK
- Step 2: $P_1$ plays PAPER. Is ROCK optimal for $P_2$?

– ROCK: −1, PAPER: 0, SCISSORS: 1

– No: Not a Nash equilibrium.

Can you check other outcomes?

Chapter 2 states that there exists a Nash equilibrium for this game. Why can't we find it?

### 5.2.3 Mixed Strategies

**Theorem 1.** *For any finite game, there exists a Nash equilibrium, where that equilibrium may be in* mixed strategies.

Nash found that for any game with a finite number of players and strategies there exists a Nash equilibrium, but only if you allow players to play mixed strategies.

**Definition 8.** *A* mixed strategy *is a strategy that puts weights on each action, such that those weights are positive and sum to one.*

An example of a mixed strategy in Rock-Paper-Scissors is $\{0.2, 0.3, 0.5\}$, where that is 20% of the time the player chooses ROCK, 30% of the time they choose PAPER and 50% of the time they choose SCISSORS. This may not be a good mixed strategy but it is a mixed strategy. A strategy that puts 100 percent of the weight on a single action is a **pure strategy**. A **pure strategy Nash equilibrium** is just the Nash equilibrium discussed in Chapter 2.

### 5.2.4 Penalty Kicks

Let's consider another famous **zero-sum game**. In soccer there are situations where one player can kick the ball at the goal with only the goal keeper to try to stop it. This may occur when the defending team fouls the team with the ball. The result of the foul is that the team with the ball gets a free shot on goal. In soccer, goal keepers are small relative to the goal and the Kicker is close enough that it is probably the case that the Kicker will score if they kick to one side or the other, unless the Goalie moves that direction at the same time as the kick is made. Because the Kicker is so close, the Goalie can't wait to see where the ball is aimed before diving.

Here is a basic version of the game.

- Players: Kicker, Goalie
- Strategies:
  - Kicker: Kick Left, Kick Right
  - Goalie: Dive Left, Dive Right
- Payoffs: See Table 5.2

**TABLE 5.2**
Normal form representation of a penalty kicks game, where the Kicker's actions are in the rows, the payoffs are in the cells and the Kicker's payoff is listed first.

| Kicker, Goalie | LEFT | RIGHT |
|---|---|---|
| LEFT | $-1, 1$ | $1, -1$ |
| RIGHT | $1, -1$ | $-1, 1$ |

Table 5.2 presents the normal form representation of the game. Again, the values in each cell sum to zero. If the Kicker scores a goal, they get 1 and the Goalie gets $-1$. It is set up assuming that if the Kicker kicks left and the Goalie dives left then the goal is saved. The Goalie gets 1 and the Kicker gets $-1$. If the Goalie dives to the left and the kick goes right it is a goal! The Goalie gets $-1$ and the Kicker gets 1.

Below the chapter looks at real data from penalty kicks. In the real game, a score increases the probability that the scoring team wins (and decreases the probability that the Goalie's team wins).

### 5.2.5 Algorithim for Mixed Strategy Nash Equilibrium

To understand the algorithm for determining a mixed strategy, think about what must be true in equilibrium. If we are in a mixed strategy equilibrium then each player must be indifferent between their strategy choices when the other player's strategy is taken as given. The Goalie must be indifferent between diving to the left or diving to the right, given the Kicker's strategy. If that wasn't the case, the Goalie would prefer to dive left (or dive right) which is a **pure strategy**. So for there to exist a mixed strategy equilibrium both players must be exactly indifferent to each pure strategy. They must be sitting on this knife's edge. Sounds painful.

Let $p$ be the probability that Kicker chooses LEFT. Let $q$ be the probability Goalie chooses LEFT.

- Find $q$ such that for Kicker, the payoff for LEFT equals the payoff for RIGHT.

- Find $p$ such that for Goalie, the payoff for LEFT equals the payoff for RIGHT.

If the Goalie's strategy is $q$, the weight on diving LEFT, then the Kicker's payoffs are:

- LEFT: $(q)(-1) + (1 - q)(1) = 1 - 2q$

- RIGHT: $(q)(1) + (1 - q)(-1) = -1 + 2q$

Reading Table 5.2 and the first row, we see that if the first column is chosen, the Kicker gets $-1$ and 1 if second column is chosen (the first element).

The algorithm states that we need to find the $q$ such that the Kicker is indifferent between LEFT and RIGHT. What strategy by the Goalie makes the Kicker indifferent between kicking LEFT or RIGHT? What strategy of the Goalie gives the Kicker the same expected payoff? What $q$ is such that $1 - 2q = -1 + 2q$? To repeat. We need to find the strategy of the Goalie that makes the Kicker indifferent between her two choices.

$$\begin{aligned} &-4q = -2 \\ &4q = 2 \\ &q = \tfrac{2}{4} \\ &q = 0.5 \end{aligned} \tag{5.1}$$

Now let Kicker choose $p$, the weight on kicking LEFT. Goalie's payoffs are:

- LEFT: $(p)(1) + (1 - p)(-1) = -1 + 2p$
- RIGHT: $(p)(-1) + (1 - p)(1) = 1 - 2p$

To see this look at the first column of Table 5.2 and the second element, the Goalie gets 1 if LEFT is chosen and $-1$ if RIGHT is chosen.

To find the equilibrium level for $p$, determine where the Goalie is indifferent between LEFT and RIGHT. This time, we need to find the strategy of the Kicker that makes the Goalie indifferent between his choices.

$$\begin{aligned} &-1 + 2p = 1 - 2p \\ &4p = 2 \\ &p = \tfrac{2}{4} \\ &p = 0.5 \end{aligned} \tag{5.2}$$

Nash equilibrium is $\{p = 0.5, q = 0.5\}$

One of the greatest penalty scorers was the Argentinian, Diego Maradona. According to Maradona, his secret to penalty kicks was to wait and see what the goalie did and then do the opposite. He cheated! Just like you used to do with your younger sibling when playing Rock-Paper-Scissors. Maradona was playing a different game. A dynamic game.

## 5.3 Rock-Paper-Scissors

We saw above that there is no pure strategy Nash equilibrium for the Rock-Paper-Scissors game. This section finds the mixed strategy Nash equilibrium (MSNE). It then explores how the equilibrium changes when the game becomes more complicated.

### 5.3.1 Mixed Strategy Nash Equilibrium

Now we know how to find a mixed strategy Nash equilibrium, consider a standard version of Rock-Paper-Scissors. Can you guess what the equilibrium is?

Let Player 2 choose $\{q_1, q_2\}$. So $q_1$ is the probability Player 2 chooses ROCK, $q_2$ is the probability that Player 2 chooses PAPER and $1 - q_1 - q_2$ is the probability that Player 2 chooses SCISSORS. Find $\{q_1, q_2, 1 - q_1 - q_2\}$ such that Player 1 is indifferent.

Quick algorithm: Guess and confirm.

Guess: $\{q_1 = \frac{1}{3}, q_2 = \frac{1}{3}\}$ Confirm:

- ROCK: $(q_1)(0) + (q_2)(-1) + (1 - q_1 - q_2)(1) = \frac{0}{3} + \frac{-1}{3} + \frac{1}{3} = 0$
- PAPER: $(q_1)(1) + (q_2)(0) + (1 - q_1 - q_2)(-1) = \frac{1}{3} + \frac{0}{3} + \frac{-1}{3} = 0$
- SCISSORS: $(q_1)(-1) + (q_2)(1) + (1 - q_1 - q_2)(0) = \frac{-1}{3} + \frac{1}{3} + \frac{0}{3} = 0$

Confirmed!

Given that the two players are identical we can guess that the Nash equilibrium choices are the same: $\{\{p_1 = \frac{1}{3}, p_2 = \frac{1}{3}\}, \{q_1 = \frac{1}{3}, q_2 = \frac{1}{3}\}\}$. Can you check that this is in fact the Nash equilibrium?

Are there any others? How would you check?

### 5.3.2 More Complicated Version

Now let's make things a bit more interesting. Assume Player 1 gets a high value of playing ROCK when Player 2 plays SCISSORS. And because this is a zero sum game, Player 2 gets a low value from playing SCISSORS when Player 1 plays ROCK. Probably best not to ask too many more questions. Given this preference for ROCK, what do you think will be the Nash equilibrium?

That's a good guess, but it is wrong.

**TABLE 5.3**
Normal form representation of a more complicated Rock-Paper-Scissors. The payoff changed in the top right cell.

| $P_1, P_2$ | ROCK | PAPER | SCISSORS |
|---|---|---|---|
| ROCK | 0, 0 | −1, 1 | 2, −2 |
| PAPER | 1, −1 | 0,0 | −1, 1 |
| SCISSORS | −1, 1 | 1, −1 | 0, 0 |

What are the payoffs for Player 1? Let Player 2's strategy be $\{q_1, q_2, 1 - q_1 - q_2\}$.

- ROCK: $(q_1)(0) + (q_2)(-1) + (1 - q_1 - q_2)(2)$
- PAPER: $(q_1)(1) + (q_2)(0) + (1 - q_1 - q_2)(-1)$

- SCISSORS: $(q_1)(-1) + (q_2)(1) + (1 - q_1 - q_2)(0)$

This lets us work out which mixed strategy of Player 2 that makes Player 1 indifferent between their choices. OK. That is, for what values of $q_1$ and $q_2$ are all three values equal to each other? This looks complicated. Let's use the computer.

### 5.3.3 Using R to Determine when Player 1 is Indifferent

We can write the payoffs for Player 1 has functions of Player 2's mixed strategy choice. We can then plot the functions to see if there is a point where Player 1 is indifferent. We can keep adjusting $q_1$ and $q_2$ until all three lines intersect.

```
> rock = function(q1, q2) {
+   q1*0 + q2*(-1) + (1 - q1 - q2)*(2)
+ }
> paper = function(q1, q2) {
+   q1*1 + q2*0 + (1 - q1 - q2)*(-1)
+ }
> scissors = function(q1, q2) {
+   q1*(-1) + q2*(1) + (1 - q1 - q2)*(0)
+ }
```

The code creates the object ggplot_rps which is a plot of the expected value of playing ROCK, PAPER, and SCISSORS for different 10 different values of $q_1$ and $q_2 = 0.42$. We are cheating a bit by assuming we already know the value of $q_2$.

```
> ggplot_rps = data.frame(
+   q_1 = seq(0, 1, by = 0.1),
+   rock = rock(seq(0, 1, by = 0.1), rep(0.42, 10)),
+   paper = paper(seq(0, 1, by = 0.1), rep(0.42, 10)),
+   scissors = scissors(seq(0, 1, by = 0.1), rep(0.42, 10))
+ ) |>
+   ggplot(aes(x = q_1)) +
+     geom_line(aes(y = rock)) +
+     geom_line(aes(y = paper), linetype = "dotted") +
+     geom_line(aes(y = scissors), linetype = "dashed") +
+     geom_vline(xintercept = 0.33,
+               linetype = "dashed",
+               color = "gray") +
+     scale_x_continuous(breaks = seq(0, 1,
+                                     by = 0.1)) +
+     scale_y_continuous(breaks = seq(-2, 2,
+                                     by = 1)) +
+     labs(title = "Payoff to Player 1",
```

```
+          x = "Probability P2 chooses ROCK",
+          y = "") +
+      geom_text(aes(x = 0.5, y = -1, label = "rock")) +
+      geom_text(aes(x = 0.5, y = 1, label = "paper")) +
+      geom_text(aes(x = 0.8, y = 0, label = "scissors"))

> ggplot_rps
```

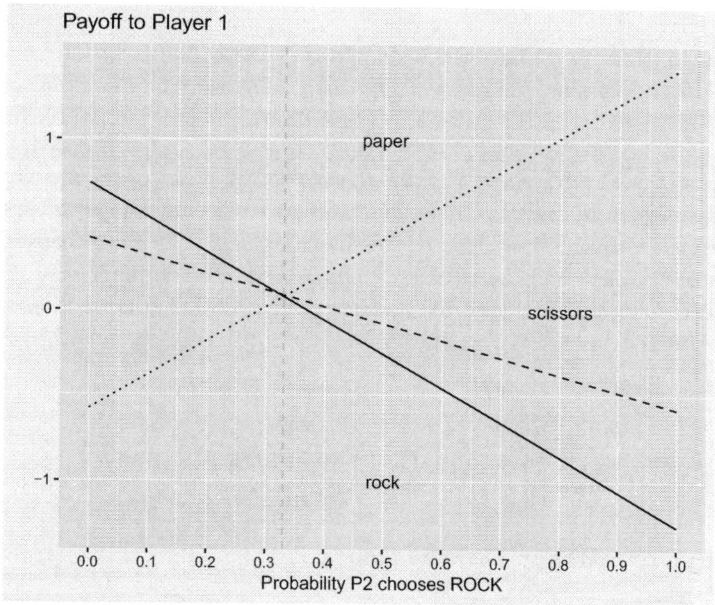


**FIGURE 5.1**
The figure plots the expected value of playing ROCK, PAPER and SCISSORS when $q_2 = 0.42$ for different values of $q_1$. Player 1 is indifferent between all three strategies when $q_1 = 0.33$.

Figure 5.1 shows where Player 1 is indifferent for different values of $q$. Player 2 tends to play PAPER! If Player 2 chooses ROCK about one-third of the time and PAPER 42% of the time, then Player 1 is indifferent.

### 5.3.4 Solving for MSNE using R

What about Player 2? Let Player 1 choose $\{p_1, p_2\}$.

- ROCK: $(p_1)(0) + (p_2)(1) + (1 - p_1 - p_2)(-1)$
- PAPER: $(p_1)(1) + (p_2)(0) + (1 - p_1 - p_2)(-1)$

- SCISSORS: $(p_1)(-2) + (p_2)(1) + (1 - p_1 - p_2)(0)$

For what strategy choice, is Player 2 indifferent between their three choices? For what values of $p_1$ and $p_2$ are these three values equal to each other?

```
> rock1 = function(p1, p2) {
+   p1*0 + p2*(1) + (1 - p1 - p2)*(-1)
+ }
> paper1 = function(p1, p2) {
+   p1*1 + p2*0 + (1 - p1 - p2)*(-1)
+ }
> scissors1 = function(p1, p2) {
+   p1*(-2) + p2*(1) + (1 - p1 - p2)*(0)
+ }
```

For this case, let's use computer muscle to solve the problem. The function `f_rps()` determines the values of $p_1$ and $p_2$ such that Player 2 is indifferent between the three choices. It does this by calculating the Euclidean distance between the choices, or the sum of squared differences. The function `f_rps_int()` is the intermediate function for `optim()`. This function translates the values chosen by the optimization algorithm into probabilities using the soft-max function.

```
> f_rps = function(p1, p2) {
+   (rock1(p1, p2) - paper1(p1, p2))^2 +
+              (rock1(p1, p2) - scissors1(p1, p2))^2
+ }
> f_rps_int = function(par) {
+   p = exp(par)/(1 + sum(exp(par)))
+   return(f_rps(p[1], p[2]))
+ }
> init = c(0,0)
> a_rps = optim(init, f_rps_int)
> exp(a_rps$par)/(1 + sum(exp(a_rps$par)))
  [1] 0.2499962 0.2500044
```

Player 1 plays ROCK with probability of about one quarter and plays PAPER with probability one quarter and SCISSORS half the time.

What the heck is going on? We said that Player 1 prefers ROCK but ends up playing SCISSORS most of the time. The reason is that Player 2 doesn't like it if Player 1 chooses ROCK so they tend to play PAPER and because Player 2 plays PAPER, Player 1 tends to play SCISSORS. Clear? No?

## 5.4 Empirical Analysis: Penalty Kicks using R

In soccer penalty kicks occur for two reasons. The first may occur if a player with the ball is fouled near goal, actually in what is known as the penalty box. This area extends 16 meters either side of the goal and 16 meters forward of the goal. Once fouled in this area, the team with the ball gets a penalty kick. The second is when there is a drawn game and penalty kicks are used to determine the winner. In both cases, the team with the ball gets to kick the ball from 11 meters in front of goal with only the goal keeper between the player and the goal. The Goalie has so little time to react to the kick that we can think of the Goalie and the Kicker choosing their strategies simultaneously.

The section analyzes data from the English Premier League, which is the highest league in English and Welsh soccer.

### 5.4.1 Penalty Kick Game

**TABLE 5.4**
Normal form representation of a penalty kicks game, where the Kicker's actions are in the rows, the payoffs are in the cells and the payoffs are the parameters to be estimated. It is a zero-sum game.

| Kicker, Goalie | LEFT | CENTER | RIGHT |
|---|---|---|---|
| LEFT | $p_{ll}$ | $p_{lc}$ | $p_{lr}$ |
| CENTER | $p_{cl}$ | $p_{cc}$ | $p_{cr}$ |
| RIGHT | $p_{rl}$ | $p_{rc}$ | $p_{rr}$ |

Table 5.4 presents the normal form representation of the game we will take to the data. Each player has three action choices, LEFT, CENTER, and RIGHT. The payoffs are given by the 9 parameters to be estimated. The parameter $p_{ij}$ is the probability that the Kicker scores a goal given that the Kicker chooses $i \in \{l, c, r\}$ and the Goalie chooses $j \in \{l, c, r\}$. The value to the Goalie is just the negative. It is a zero-sum game.

We will allow the two players to use mixed strategies. The vector representing the Kicker's strategy is as follows.

$$q_k = \left[ \begin{array}{ccc} q_{kl} & q_{kc} & q_{kr} \end{array} \right] \tag{5.3}$$

The weights must some to one, $\sum_{i \in \{l,c,r\}} q_{ki} = 1$. Similarly, we can represent the Goalie's strategy.

$$q_g = \left[ \begin{array}{ccc} q_{gl} & q_{gc} & q_{gr} \end{array} \right] \tag{5.4}$$

### 5.4.2 Data

The data used here are from Kaggle and covers penalty kicks in the English Premier League during the 2016/2017 season.[1]

```
> file = paste0(dir, "penalty_data.csv")
> data = fread(file) |>
+   filter(
+     Kick_Direction != ""
+   )
```

The code brings in the data and removes the cases where the kick direction is missing. The code uses `fread()` which is part of `data.table`. The first step is calculating the strategies for both players. The assumption is that every Kicker and Goalie is using the same strategy. Alternatively, you can think of this as the average strategy choice from a distribution of strategies. The code uses `table()` to count the number of each case and then divides by the total number of observations to get the probability of each strategy choice. The vector $q_k$ is then reordered to match the order in Table 5.4. The same is done for the Goalie.

```
> q_k = table(data$Kick_Direction)
> q_k = q_k/sum(q_k)
> q_k = q_k[c(2, 1, 3)]
```

```
> q_g = table(data$Kick_Direction)
> q_g = q_g/sum(q_g)
> q_g = q_g[c(2, 1, 3)]
```

We can also calculate the parameter value from Table 5.4.

```
> action = c("L", "C", "R")
> res_mat = matrix(NA, 3, 3)
> for(i in 1:3) {
+   for(j in 1:3) {
+     dt_ij = data[data$Kick_Direction == action[i] &
+                  data$Keeper_Direction == action[j]]
+     res_mat[i,j] = mean(dt_ij$Scored == "Scored")
+   }
+ }
```

Table 5.5 presents the empirical results of the penalty kick game. Does it look like what you expected? We would expect a lower probability of scoring

[1] https://www.kaggle.com/datasets/mauryashubham/english-premier-league-penalty-dataset-201617 license: CCO: Public Domain, ShubhamMaurya, English Premier League Penalty Dataset, 201617.

**TABLE 5.5**
Strategies and conditional probabilities for penalty kicks in the English Premier League in the 2016/2017 season. The first column is Kicker's strategy and the first row is the Goalie's strategy. The cells the probability that the Kicker scores given the action chosen by the Kicker and the Goalie.

| | q_k | G:LEFT | G:CENTER | G:RIGHT |
|---|---|---|---|---|
| q_g | | 0.46 | 0.17 | 0.38 |
| K:LEFT | 0.46 | 0.65 | 1.00 | 0.88 |
| K:CENTER | 0.17 | 1.00 | 0.00 | 0.89 |
| K:RIGHT | 0.38 | 0.83 | 1.00 | 0.56 |

when the Goalie chooses the same direction as the Kicker. These are on the diagonal and we see that in fact the lower probabilities are on the diagonal. Kicking to the center leads to a really low probability of scoring when the Goalie also chooses not to move. You may be surprised how often the Kicker still scores even when the Goalie chooses the same direction. It is 65 percent when kicking LEFT and 56 percent when kicking RIGHT. It is also interesting that it is not symmetric. The probability of scoring when kicking LEFT is higher than the probability of scoring when kicking RIGHT. Kickers are a little more likely to kick LEFT than RIGHT.

Right footed kickers may have a better chance of scoring when kicking to the LEFT. From a right footed kicker, the ball will naturally curve across and away from the Goalie. We can check to see if there are differences in payoffs and strategies between left and right footed kickers.

### 5.4.3 GMM Estimator

One issue with looking at differences by the footedness of the kicker is that the data is limited. A solution is to bring in more information from the game theory.

We know a few things that must be true in the game. In particular, in equilibrium, the strategy for the Kicker must make the Goalie indifferent between the choices. And the strategy for the Goalie must make the Kicker indifferent between the choices.

The Kicker and the Goalie choose $q_k$ and $q_g$, respectively, such that the following equalities hold. The periods in the subscripts mean that it represents any choice.

$$\begin{aligned} q_k' p_{.l} &= q_k' p_{.c} = q_k' p_{.r} \\ q_g' p_{l.} &= q_g' p_{c.} = q_g' p_{r.} \end{aligned} \tag{5.5}$$

The vector $q_k$ is laid down on its side and we use matrix multiplication rules to multiply it with the conditional probabilities when the Goalie chooses LEFT. These conditional probabilities are the first column of Table 5.4. The elements in $q_k$ are multiplied with the corresponding elements in $p_{.l}$ and the three numbers are summed together to give the probability of a goal when the Kicker plays strategy $q_k$ and the Goalie plays LEFT.

We will guess the matrix of strategies, $p$, then use equilibrium relationship to solve for $q_k$ and $q_g$. In addition, we directly observe $q_k$, $q_g$ and $p$ in the data. Don't we have too many conditions? Yes we do. Won't requiring that there is a mixed strategy Nash equilibrium lead to different estimates? Yes. Yes it will. The data combined with the game theory provide too many conditions and thus too many estimates. The solution is to average over the estimates. We are over identified. The generalized method of moments (GMM) algorithm provides a way to find that average. A method of moments estimator was used in Chapter 3. Here we have multiple **moments**.[2]

### 5.4.4 GMM Estimator in R

The first step is to create functions `f_mixed()` and `f_mixed_int()`. These functions are used to determine the equilibrium strategy of the Kicker and the Goalie given a set of conditional probabilities of scoring. Given a set of conditional probabilities `p`, `optim()` is used to determine the equilibrium strategies `q_k` and `q_g`. This is done by finding the strategies that minimize the sum of squared differences between payoffs of the three options for the other player.

```
> f_mixed = function(q_k, q_g, p) {
+   p = matrix(unlist(p), nrow=3)
+   pi_1 = t(q_k)%*%p[,1]
+   pi_2 = t(q_k)%*%p[,2]
+   pi_3 = t(q_k)%*%p[,3]
+
+   pi_4 = t(q_g)%*%p[1,]
+   pi_5 = t(q_g)%*%p[2,]
+   pi_6 = t(q_g)%*%p[3,]
+   return((pi_1 - pi_2)^2 + (pi_1 - pi_3)^2 +
+              (pi_4 - pi_5)^2 + (pi_4 - pi_6)^2)
+ }
> f_mixed_int = function(par, p) {
+   q_k = exp(par[1:3])/sum(exp(par[1:3]))
+   q_g = exp(par[4:6])/sum(exp(par[4:6]))
+   return(f_mixed(q_k, q_g, p))
+ }
```

The function `f_penalty()` takes in the proposed estimate of the conditional probabilities `p` and the observed actions and outcomes `X`. It then calculates

[2]The term refers to a characteristic of the distribution such as the mean (the first moment) or the variance (related to the second moment). But here we use the term more generally and include characteristics of the equilibrium.

7 **moments**. For instance, it determines the difference between the estimate of the Kicker's probability of kicking LEFT against the observed probability that the Kicker kicked LEFT. It also compares the proposed estimate of the conditional probabilities to the actual conditional probabilities in the data. The function `f_penalty_int()` is an intermediate function called by `optim()`. It turns the parameter values into numbers between 0 and 1. It then calls `f_gmm()`, which is a generic function for calculating the GMM optimization problem.[3] The standard errors are calculated using the bootstrap.

```
> f_penalty = function(p, X) {
+   N = dim(X)[1]
+   init = rep(0, 6)
+   p = matrix(p, nrow = 3)
+   q = optim(par = init, fn = f_mixed_int, p = p)
+   q_k = exp(q$par[1:3])/sum(exp(q$par[1:3]))
+   q_g = exp(q$par[4:6])/sum(exp(q$par[4:6]))
+   g_k = g_g = matrix(0, 3, N)
+   for(i in 1:3) {
+     g_k[i,] = q_k[i] - (X$kicker == action[i])
+     g_g[i,] = q_g[i] - (X$goalie == action[i])
+   }
+   g_s = rep(0, N)
+   for(i in 1:3) {
+     for(j in 1:3) {
+       index_ij = which(
+         X$kicker == action[i] & X$goalie == action[j]
+       )
+       g_s[index_ij] = p[i,j] -
+         (X$score[index_ij] == "TRUE")
+     }
+   }
+   G = rbind(g_k,
+             g_g,
+             g_s)
+   return(G)
+ }
> f_penalty_int = function(par, X) {
+   p = exp(par)/(1 + exp(par))
+   p = matrix(p, nrow = 3)
+   G = f_penalty(p, X)
+   return(f_gmm(G, K = 7))
+ }
```

[3] The `f_gmm()` function is available from the github site for the book.

### 5.4.5 Difference By Footedness

Does the behavior of the Kicker and the Goalie change depending on the which foot the Kicker generally kicks with? In general a right-footed Kicker will have an easier time scoring to the LEFT. The Kicker will generally strike the ball with their foot slightly to the right of center, that will put a counter-clockwise rotation on the ball which will then curve in the air from right to left.

So kicking LEFT for a right-footed Kicker will generally be better, holding the Goalie's strategy constant. Of course the Goalie gets a say in this. The Goalie may want to choose LEFT more often and thus be more likely to stop the ball when the right-footed Kicker chooses LEFT. For left-footed Kickers, the opposite is true.

```
> dt_R = dt[dt$Foot == "R"]
> p = matrix(NA, 3, 3)
> for(i in 1:3) {
+   for(j in 1:3) {
+     index_ij = which(dt_R$Kick_Direction == action[i] &
+                       dt_R$Keeper_Direction == action[j])
+     p[i,j] = mean(dt_R$Scored[index_ij] == "Scored")
+   }
+ }
> p[is.nan(p)] = NA
```

In the code the function `is.nan()` is used replace infinite values with missing values (`NA`). The matrix `p` gives the conditional probability that the Kicker scores given the Kicker is right footed.

```
> X = data.frame(
+   "kicker" = dt_R$Kick_Direction,
+   "goalie" = dt_R$Keeper_Direction,
+   "score" = dt_R$Scored == "Scored"
+ )
> epsilon = 1e-10
> init = log(p + epsilon)
> a = optim(par = init,
+           fn = f_penalty_int,
+           X = X,
+           control = list(trace = 0,
+                          maxit = 100000))
```

Similarly for Left footers.

```
> dt_L = dt[dt$Foot == "L"]
> p = matrix(NA, 3, 3)
> for(i in 1:3) {
+   for(j in 1:3) {
```

```
+      index_ij = which(dt_L$Kick_Direction==action[i] &
+                         dt_L$Keeper_Direction == action[j])
+      p[i,j] = mean(dt_L$Scored[index_ij] == "Scored")
+    }
+ }
> p[is.nan(p)] = 0
```

**TABLE 5.6**
Strategy estimates for Right and Left footed players. The first and fourth columns are the averages from the bootstrap for the two cases respectively. The top three rows are the strategies for the Kicker, while the bottom three rows are the strategies for the Goalie.

| | Right | 0.05 | 0.95 | Left | 0.05 | 0.95 |
|---|---|---|---|---|---|---|
| Kicker - L | 0.53 | 0.44 | 0.61 | 0.47 | 0.28 | 0.69 |
| Kicker - C | 0.19 | 0.13 | 0.26 | 0.01 | 0.00 | 0.06 |
| Kicker - R | 0.28 | 0.20 | 0.35 | 0.52 | 0.31 | 0.67 |
| Goalie - L | 0.54 | 0.44 | 0.62 | 0.19 | 0.03 | 0.36 |
| Goalie - C | 0.07 | 0.02 | 0.12 | 0.01 | 0.00 | 0.11 |
| Goalie - R | 0.39 | 0.31 | 0.48 | 0.80 | 0.64 | 0.95 |

Table 5.6 presents the strategies given the footedness of the Kicker. Right-footed kickers will generally more accurate kicking LEFT and we see that they choose LEFT 53% of the time, while left-footed kickers choose RIGHT 52% of the time. Goalie's respond by playing LEFT to a right-footed kicker 54% of the time and RIGHT to a left-footed kickers 80% of the time!

The analysis suggests that the Goalie's strategy is completely different depending on the footedness of the kicker.

## 5.5 Multiple Equilibria

Another reason why we may see mixed strategies in a game is because there are multiple equilibria. We saw this in the coordination game. The two players would like to coordinate but it is unclear which choice they should coordinate on. In that situation a mixed strategy Nash equilibrium may be a more reasonable prediction of the outcome. Similarly, there was a coordination problem in our analysis of entry of mega bookstores. This section reconsiders the problem assuming that Borders and Barnes & Noble are playing a mixed strategy Nash equilibrium.

### 5.5.1 Coordination Game

Here is the normal form representation presented in Chapter 2.

**TABLE 5.7**
Normal form representation of a coordination game from Chapter 2.

| $P_1, P_2$ | BLACK | RED |
|---|---|---|
| BLACK | 2, 5 | 0, 0 |
| RED | 0, 0 | 5, 2 |

We use the same algorithm as for zero-sum games to find the MSNE. What is your guess on the equilibrium? Remember that we don't have the zero-sum game weirdness here.

Let $q_1$ and $q_2$, respectively, be the probability that Player 1 and Player 2 choose BLACK. What is the $q_2$ that makes Player 1 indifferent between BLACK and RED?

$$\begin{aligned} 2q_2 &= 5(1-q_2) \\ 7q_2 &= 5 \\ q_2 &= \tfrac{5}{7} \end{aligned} \tag{5.6}$$

What is the $q_1$ that makes Player 2 indifferent between BLACK and RED?

$$\begin{aligned} 5q_1 &= 2(1-q_1) \\ 7q_1 &= 2 \\ q_1 &= \tfrac{2}{7} \end{aligned} \tag{5.7}$$

Yes. Player 1 prefers RED and so weights their play to RED, while Player 2 prefers BLACK and weights their play to BLACK.

### 5.5.2 Bookstore Entry

Chapter 4 presents a game describing entry of two mega bookstores, Barnes & Noble and Borders. In the game Barnes & Noble will enter market $i$ if the following inequality holds.

$$\mathbf{X}_i'\beta_1 - D_{2i}\alpha_1 + \xi_{1i} \geq 0 \tag{5.8}$$

where $D_{2i} \in \{0, 1\}$ represents the choice to enter market $i$ of Borders. Under the assumptions of the equilibrium the value of $D_{2i}$ is known by Barnes & Noble when they make their entry decision. Both bookstores also know $\xi_{1i}$, which is the value the econometrician doesn't observe.

In Chapter 4 we showed that there is a case where Barnes & Noble will enter but only if Borders does not, and similarly Borders will enter but only if Barnes & Noble does not. At these values of the $\xi$s, we have a coordination game with multiple equilibria.

We can use the same algorithm to determine the mixed strategy Nash equilibrium. Let $q_1$ be the probability that Barnes & Noble enters and $q_2$ be the probability that Borders enters. What $q_2$ makes Barnes & Noble indifferent between entering and not entering?

$$\begin{aligned} &\mathbf{X}_i'\beta_1 - q_{2i}\alpha_1 + \xi_{1i} = 0 \\ &or \\ &q_{2i} = \frac{\mathbf{X}_i'\beta_1 + \xi_{1i}}{\alpha_1} \end{aligned} \tag{5.9}$$

Similarly we can solve for $q_1$

$$q_{1i} = \frac{\mathbf{X}_i'\beta_2 + \xi_{2i}}{\alpha_1} \tag{5.10}$$

If we assume that Barnes & Noble and Borders are at a mixed strategy Nash equilibrium, then we no longer have a indeterminacy problem with our estimator. For every value of $\xi_1$ and $\xi_2$ we have a known probability over which outcome will occur.

### 5.5.3 MSNE Estimator in R

The estimator is pretty similar to the one used in Chapter 4. There are a couple of differences with `f_entry_mix()`. First, the area where there is indeterminacy about which firm will enter, the mixed strategy Nash equilibrium determines what outcome will occur. Second, because there is no indeterminacy we can ask the probability that Firm 2 enters while Firm 1 does not.

To determine the mixed strategy Nash equilibrium we use Equations (5.10) and (5.9). The mixed strategy only occurs when the profits are such that it is only profitable for one of the two firms to enter. Because it is a mixed strategy, there is a possibility that any of the four outcomes occur. If there is an indeterminant outcome the mixed strategy Nash equilibrium determines the probabilities. In the code these are given by `q_1k` and `q_2k`. The probability of an indeterminant outcome is given by `p_ind`.

```
> f_entry_mix = function(X, beta_1, beta_2, alpha_1, alpha_2,
+                         rho) {
+   N = dim(X)[1]
+   xi_1 = Z_1
+   xi_2 = Z_2*sqrt(1 - rho^2) + rho*Z_1
+   Xb_1 = X%*%beta_1
+   Xb_2 = X%*%beta_2
+   p_00 = p_01 = p_11 = rep(0, N)
+   for(k in 1:K) {
+     pi_1k = Xb_1 + xi_1[k]
+     pi_2k = Xb_2 + xi_2[k]
+     q_1k = max(c(min(c((pi_2k)/alpha_2,1)),0))
+     q_2k = max(c(min(c((pi_1k)/alpha_1,1)),0))
+     p_ind = (pi_1k > 0 &
```

```
+                pi_1k - alpha_1 < 0 &
+                pi_2k > 0 &
+                pi_2k - alpha_2 < 0)
+      p_00 = p_00 +
+        (pi_1k < 0 & pi_2k < 0) +
+        p_ind*(1 - q_1k)*(1 - q_2k)
+      p_01 = p_01 +
+        (pi_1k < 0 & pi_2k > 0) +
+        p_ind*(1 - q_1k)*q_2k
+      p_11 = p_11 +
+        (pi_1k - alpha_1 > 0 & pi_2k - alpha_2 > 0) +
+        p_ind*q_1k*q_2k
+    }
+    return(list(p_00 = p_00/K,
+                p_01 = p_01/K,
+                p_11 = p_11/K))
+ }
```

## 5.6 Empirical Analysis: Bookstore Entry with MSNE using R

The analysis presented in this section uses exactly the same data used in Chapter 4. It also uses much of same empirical machinery.

Table 5.8 shows that the two different assumptions lead to similar results. Assuming that the outcome is a mixed strategy Nash equilibrium leads to an estimate that Borders is less impacted by competition with Barnes & Noble than the other way around. It also estimates less statistical dependence between the two stores.

Table 5.9 simulates the effect of a merger between Borders and Barnes & Noble on the willingness to have both brands in a market. The welfare impact of the merger is ambiguous. While it is the case that we see a reduction in competition. The merged firm is less like to have both brands in the market. In this way the merger leads to fewer brands in a market which reduces quality and the reduced head to head competition leads to higher prices. However, you see that there is a small reduction in the number of markets without a bookstore. The merger has the benefit of allowing the brands to coordinate their entry decision. While we have a game of complete information, the players cannot coordinate their choice in the mixed strategy Nash equilibrium. Because there is a possibility of having too many firms enter a market, firms reduce their willingness to enter some markets. The merger solves the coordination problem, reduces the possibility of having too many firms enter a market, and increases the willingness for the combined firm to enter some markets.

**TABLE 5.8**
Results from estimates of the game theory model from Chapter 4 and the model assuming a mixed strategy Nash equilibrium. The two columns labeled "Pure" refer to the case where the entry decisions of the two firms are both strategically and statistically dependent, but we assume a pure strategy Nash equilibrium. The two columns labeled "Mix" refer to the same model but assuming the outcome is a mixed strategy Nash equilibrium.

| | Pure | SD | Mix | SD |
|---|---|---|---|---|
| const_1 | −15.11 | 0.25 | −14.82 | 0.20 |
| Pop_1 | 1.07 | 0.01 | 1.05 | 0.02 |
| Income_1 | −0.76 | 0.48 | −0.97 | 0.23 |
| College_1 | 5.65 | 0.55 | 5.71 | 0.28 |
| Stores_1990_1 | 0.37 | 0.11 | 0.39 | 0.07 |
| const_2 | −11.37 | 0.20 | −11.48 | 0.20 |
| Pop_2 | 0.65 | 0.02 | 0.63 | 0.02 |
| Income_2 | 1.31 | 0.80 | 1.91 | 0.38 |
| College_2 | 2.70 | 0.55 | 2.82 | 0.39 |
| Stores_1990_2 | 0.79 | 0.11 | 0.77 | 0.10 |
| alpha_1 | 0.73 | 0.18 | 0.70 | 0.21 |
| alpha_2 | 0.70 | 0.12 | 0.56 | 0.15 |
| rho | 0.47 | 0.10 | 0.30 | 0.10 |

**TABLE 5.9**
Comparison of actual entry to simulated entry in the year 2000 and simulated entry under a merger. These results are based on the assumption that Borders and Barnes & Noble are playing a mixed strategy Nash equilibrium

| | Actual | Sim | Merge |
|---|---|---|---|
| none | 2919 | 2895 | 2881 |
| BN | 155 | 135 | 176 |
| Borders | 15 | 54 | 72 |
| both | 128 | 95 | 51 |

## 5.7 Discussion and Further Reading

Mixed strategy Nash equilibria are weird. Many people find them unintuitive and the algorithm for finding them is less than obvious. But for some games, they seem like the correct prediction.

While soccer penalty kicks may not represent the types of games you are interested in, it does provide an example of real people making real decisions with real consequences. Sporting contests provide researchers with access to large amounts of data on relatively simple strategic situations. This makes sports a good laboratory for testing game theory's predictions. Adams (2020) presents an analysis of play choice in NFL games.

The chapter introduces one of the most important empirical methodologies in structural estimation, the **generalized method of moments** (Hansen, 1982). Games often provides more moments than parameters to estimate, GMM provides a way to average over the moments.

The chapter also revisits the entry game analyzed in Chapter 4. The entry game has multiple equilibria. This chapter assumes that the outcome is the result of a mixed strategy Nash equilibrium. It revisits the merger simulation under the mixed strategy Nash equilibrium assumption. As before the merger reduces the number of markets with both brands, but unlike Chapter 4, the merger increases the number of markets with at least one firm.

# Part II

# Dynamic Games of Complete Information

# 6

# *Dynamic Games*

## 6.1 Introduction

The first part of the book considers static games of complete information. These games have very simple strategies. Perhaps with the exception of games with mixed strategies, these games are straightforward to analyze. This part of the book considers dynamic games. These games have strategies that can be substantially more complex than those we have seen. This complexity will force us to make important simplifying assumptions.

A **dynamic game** refers to the information available to the player at the time of their choice of action. If the player has no information about what the other players are doing, then it is a static game. If at least one player is able to observe the actions of the other player prior to making their choice, then we have a dynamic game.

To restate whether or not a game of complete information is static or dynamic has to do with information, not time. We can have static games where players move at different times, but they don't get to observe the other player's action prior to their move. We can have dynamic games where players move more or less at the same time, but where at least one player observes the action of the other player prior to their move.

**Definition 9.** *In a* dynamic game *at least one player observes information about the other player's actions prior to choosing their action.*

This chapter introduces an alternative representation of a game, the **extensive form** representation. Chapter 1 introduced the **normal form** representation of a game. This is the matrix like object with one player on the columns and the other on the rows. The normal form representation is associated with static games while the **extensive form** representation is associated with dynamic games. There is nothing to stop you from using normal form representations for dynamic games or extensive form for static games, but we will see the value of the extensive form representation for dynamic games.

The chapter introduces a new equilibrium concept, **subgame perfection**. **Perfection** refers to equilibrium refinements. We are interested in reducing the number of predictions for the game that we believe are reasonable. The chapter revisits our analysis of entry game and the choices of Borders and

DOI: 10.1201/9781003351603-6

Barnes & Noble about which counties to enter. We can reduce the multiplicity of equilibria in the entry game by assuming that Barnes & Noble moves first and using subgame perfection.

## 6.2 Extensive Form

The section makes an adjustment to the **coordination game** introduced in Chapter 2 to illustrate the **extensive form** representation.

### 6.2.1 Coordination Game

Consider a variation on the coordination game. Assume that Player 1 chooses their action first and Player 2 observes that action before choosing their action.

- Players: Player 1 and Player 2
- Strategies:
  - Player 1: {BLACK, RED}
  - Player 2:
    - If Player 1 plays BLACK, then BLACK. If Player 1 plays RED, then BLACK.
    - If Player 1 plays BLACK, then RED. If Player 1 plays RED, then BLACK.
    - If Player 1 plays BLACK, then BLACK. If Player 1 plays RED, then RED.
    - If Player 1 plays BLACK, then RED. If Player 1 plays RED, then RED.
- Payoffs
  - {BLACK, {{BLACK: BLACK}, {RED: BLACK}}}: $\{2, 5\}$
  - {BLACK, {{BLACK: RED}, {RED: BLACK}}}: $\{0, 0\}$
  - {BLACK, {{BLACK: BLACK}, {RED: RED}}}: $\{2, 5\}$
  - {BLACK, {{BLACK: RED}, {RED: RED}}}: $\{0, 0\}$
  - {RED, {{BLACK: BLACK}, {RED, BLACK}}}: $\{0, 0\}$
  - {RED, {{BLACK: RED}, {RED: BLACK}}}: $\{0, 0\}$
  - {RED, {{BLACK: RED}, {RED: RED}}}: $\{5, 2\}$
  - {RED, {{BLACK: BLACK}, {RED: RED}}}: $\{5, 2\}$

Wow. Things got complicated right quick! Player 1's strategy is simple. This player does not have any information before they choose their action, so their strategies are just their actions. Things are a lot more complicated for Player 2. Player 2 gets to observe Player 1's action prior to making their choice. Therefore Player 2's strategy must account for this information. Player 2 has four possible strategies. This is because there are two possible states, Player 1's action choice, and two choices for each state. Two times two is four.

We will use the notation {BLACK: BLACK} to mean if Player 1 plays BLACK, then Player 2 plays BLACK.

### 6.2.2 Strategies Revisited

**Strategies** are one of the three main components of a game. In Chapter 1, we stated the definition of a strategy.

**Definition 10.** *A* strategy *is a function that maps from the player's information set to the player's actions.*

While in the first part of the book, a strategy was simply an action. Here it is a function. It maps from the information observed to an action. Moreover, it is a function that maps from every possibility to an action. Sure, this is just another way of saying it is a function, but it is really important to remember that it is a complete plan. It states what the player will do in every possible and every conceivable case.

### 6.2.3 Nash Equilibrium of the Coordination Game

Remember the Nash equilibrium algorithm asks us to posit a candidate set of strategies for all the players and then check that each player's strategy is optimal given the posited strategies.

Player 1's strategy is just RED or BLACK, while Player 2's strategy describes what they will do for the two possible cases. The strategy states what Player 2 will do for a situation that never actually occurs. This is what is meant by a complete plan. Player 2's strategy states what they will do under *every* circumstance, not just what happens in the proposed equilibrium.

Candidate: $\{RED, \{\{BLACK : BLACK\}, \{RED : RED\}\}\}$

- Assume Player 1 plays RED
- Player 2's payoffs
  - $\{\{BLACK : BLACK\}, \{RED : RED\}\}$: 2
  - $\{\{BLACK : RED\}, \{RED : RED\}\}$: 2
  - $\{\{BLACK : RED\}, \{RED : BLACK\}\}$: 0
  - $\{\{BLACK : BLACK\}, \{RED : BLACK\}\}$: 0
  - Yes. Player 2's strategy is optimal (not dominated).

- Assume Player 2 plays $\{\{BLACK : BLACK\}, \{RED : RED\}\}$
- Player 1's payoffs
  - RED: 5
  - BLACK: 2
  - Yes. It is a Nash equilibrium.

It is an equilibrium for Player 1 to choose RED and for Player 2 to also choose RED.

Are there any others?
Candidate: $\{BLACK, \{\{BLACK : BLACK\}, \{RED : BLACK\}\}\}$

- Assume Player 1 plays BLACK
- Player 2's payoffs
  - $\{\{BLACK : BLACK\}, \{RED : RED\}\}$: 5
  - $\{\{BLACK : RED\}, \{RED : RED\}\}$: 0
  - $\{\{BLACK : RED\}, \{RED : BLACK\}\}$: 0
  - $\{\{BLACK : BLACK\}, \{RED : BLACK\}\}$: 5
  - Yes. Player 2's strategy is optimal (not dominated).
- Assume Player 2 plays $\{\{BLACK : BLACK\}, \{RED : BLACK\}$
- Player 1's payoffs
  - RED: 0
  - BLACK: 1
  - Yes. It is a Nash equilibrium.

There are at least two Nash equilibrium.

Is $\{BLACK, \{\{BLACK : BLACK\}, \{RED : BLACK\}\}\}$ a likely outcome from the game? The two players coordinate on black, which is fine. What about the idea that Player 2 would choose BLACK when they know Player 1 has chosen RED. Is that reasonable?

### 6.2.4 Game Tree

One way to see the issue with the Nash equilibria is to look at the **game tree**, the extensive form representation presented in Figure 6.1.

A **game tree** is a directed graph, it consists of **nodes** and **edges**. The nodes represent places where a player makes a decision. An edge is the line and arrow from node to a node further along the game tree. The edge represents the flow of information.

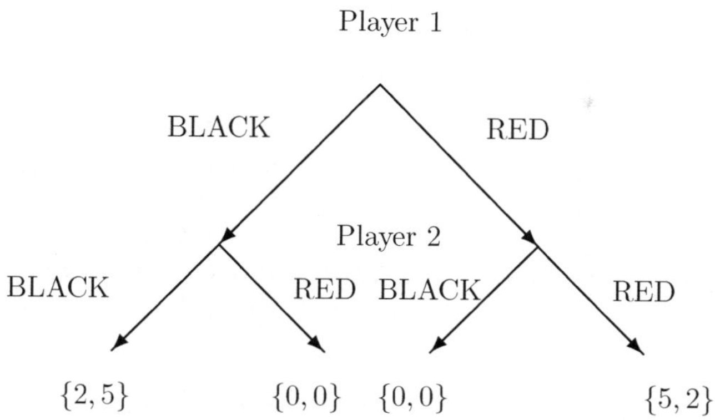


**FIGURE 6.1**
Extensive form representation of the dynamic coordination game tree.

**Definition 11.** *A* node *of a game tree is a place where the player makes a choice.*

Consider the Nash equilibrium $\{BLACK, \{BLACK : BLACK, RED : BLACK\}\}$. We checked that it was in fact a Nash equilibrium above. Look at what happens on the tree. If Player 1 plays RED, then Player 2 has a choice of BLACK or RED. In the equilibrium, they choose BLACK, but their payoff would have been higher if they had chosen RED.

Working down the tree, the equilibrium states that Player 1 chooses RED, so we go down the right branch. Now it is Player 2's turn. They can choose RED or BLACK. If they choose RED they get 2, while if they choose BLACK they get 0. In equilibrium they choose BLACK. Does that make sense? In the next section, we will consider an equilibrium refinement that rules out this case.

## 6.3 Subgame Perfection

The first equilibrium refinement we will consider is **subgame perfection**. An equilibrium refinement is an additional property of the predicted outcome that must be true. **Subgame perfection** requires a set of strategies be a Nash equilibrium but also that the subset of strategies associated with each **subgame** be a Nash equilibrium of the subgame.

The section defines subgame perfection and then works through the implications for the coordination game presented above.

### 6.3.1 Definition

**Definition 12.** *A* subgame *is a game that can be played from any node of the game tree.*

Definition 12 introduces the idea of a **subgame**. When you look at the game tree in Figure 6.1, there are three distinct nodes. For Player 2, there is the node after which Player 1 plays black and the node after which Player 1 plays red. There is also the initial node where Player 1 makes their choice of black or red. At each of these nodes we describe a separate game. This game is a subgame.

**Definition 13.** *A* subgame perfect Nash equilibrium *is a Nash equilibrium where the strategies in each subgame are a Nash equilibrium from that subgame.*

If for a particular Nash equilibrium strategy set we can look at each subgame and associate the strategies with a Nash equilibrium for that subgame, then the strategy set is **subgame perfect**.

### 6.3.2 Coordination Game

In the **coordination game**, there are three subgames. The whole game is a subgame. The other two begin at Player 2's decision node. In these two subgames, there is just one player (Player 2) and their strategies are just the actions {BLACK, RED}.

Is {BLACK, {{BLACK: BLACK}, {RED: BLACK}}} subgame perfect?

Consider the subgame at the node where Player 1 choose RED.

- Players: Player 2
- Strategies: {BLACK, RED}
- Payoffs: BLACK: 0, RED: 2

It is not a Nash equilibrium of the subgame for Player 2 to choose BLACK because they would be better off choosing RED.

Subgame perfection removes outcomes that allow non-credible strategies. Nash equilibrium requires the players to choose strategies that are optimal given the strategies of the other players. Subgame perfection requires players to choose actions that are optimal even if in equilibrium these actions will never be played.

Is it better to be Player 1 or Player 2 in this game? You may think Player 2, as they get to see what Player 1 does and react to it. If we rule out non-credible threats, then Player 1 always gets their way. Player 1 has a **first-mover advantage**. If the first move in the game can commit to a strategy then they have a distinct advantage. They can force the other player to choose actions that the first player prefers.

### 6.3.3 Empirical Entry Game

Consider a different version of entry game analyzed in Chapter 4. Instead of having Borders and Barnes & Noble enter at the same time, assume that Barnes & Noble moves first. Borders observed Barnes & Noble's decision to enter or not and then decides to enter.

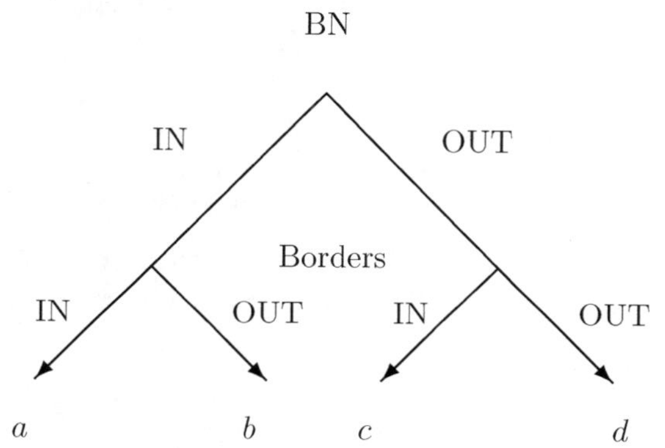


**FIGURE 6.2**
Dynamic entry game tree. $a = \{\mathbf{X}_i'\beta_1 - \alpha_1 + \xi_{1i}, \mathbf{X}_i'\beta_2 - \alpha_2 + \xi_{2i}\}$, $b = \{\mathbf{X}_i'\beta_1 + \xi_{1i}, 0\}$, $c = \{0, \mathbf{X}_i'\beta_2 + \xi_{2i}\}$ and $d = \{0, 0\}$.

Figure 6.2 represents the entry game. Barnes & Noble moves first and chooses whether or not to enter the market and then Borders chooses. The payoffs state that upon entry the firm earns $\mathbf{X}_i'\beta_1$, but if they face competition then those profits fall by $\alpha_1$. The entry costs are captured by $\xi_{1i}$.

### 6.3.4 Equilibrium in Entry Game

To determine the **subgame perfect Nash equilibrium**, we solve the last subgame first. If Barnes & Noble entered the market, Borders will enter if and only if the following inequality holds.

$$\mathbf{X}_i'\beta_2 - \alpha_2 + \xi_{2i} \geq 0 \tag{6.1}$$

Borders will only enter if the duopoly profits more than outweigh the entry costs. While if Barnes & Noble has not entered, Borders will enter if the following inequality holds.

$$\mathbf{X}_i'\beta_2 + \xi_{2i} \geq 0 \tag{6.2}$$

This time, Borders enters if monopoly profits are high enough.

Now we go back and consider Barnes & Noble's choices. If they enter, there are two cases. Case 1 is that Borders enters ($\mathbf{X}_i'\beta_2 - \alpha_2 + \xi_{2i} \geq 0$), they will also enter if the following inequality holds.

$$\mathbf{X}_i'\beta_1 - \alpha_1 + \xi_{1i} \geq 0 \tag{6.3}$$

Case 2 is that Borders does not enter, and so Barnes & Noble will enter if and only if $\mathbf{X}_i'\beta_1 + \xi_{1i} \geq 0$.

If Barnes & Noble chooses not to enter, then there are also two cases. In Case 1, where Borders enters, it is an equilibrium if and only if $\mathbf{X}_i'\beta_1 - \alpha_1 + \xi_{1i} \leq 0$. In Case 2, it is an equilibrium if and only if $\mathbf{X}_i'\beta_1 + \xi_{1i} \leq 0$.

### 6.3.5 Empirical Implications

To summarize our four observed outcomes are a subgame perfect Nash equilibrium if the following inequalities hold.

- Both enter: $\mathbf{X}_i'\beta_1 - \alpha_1 + \xi_{1i} \geq 0$ and $\mathbf{X}_i'\beta_2 - \alpha_2 + \xi_{2i} \geq 0$.
- BN enters only: $\mathbf{X}_i'\beta_1 + \xi_{1i} \geq 0$ and $\mathbf{X}_i'\beta_2 - \alpha_2 + \xi_{2i} \leq 0$
- Borders enters only: $\mathbf{X}_i'\beta_1 - \alpha_1 + \xi_{1i} \leq 0$ and $\mathbf{X}_i'\beta_2 + \xi_{2i} \geq 0$.
- Neither firm enters: $\mathbf{X}_i'\beta_1 + \xi_{1i} \leq 0$ and $\mathbf{X}_i'\beta_2 + \xi_{2i} \leq 0$

How do these compare to the outcomes in Chapter 4? There is no indeterminacy. If you go back and look at Figure 4.2 and the middle square it has $\{0, 1\}$ or $\{1, 0\}$. Under this set up with subgame perfection it becomes $\{1, 0\}$. Only Barnes & Noble will enter. Where before it there were two equilibria (more if you include mixed strategies), here there is always a unique subgame perfect Nash equilibrium. Of course, the price is that we need to make very strong assumptions about how the game is played.

## 6.4 Empirical Analysis: Bookstore Entry with Subgame Perfection in R

The estimator for the subgame perfect Nash equilibrium is basically the same as in Chapter 4. The difference is that `f_entry_spne()` can separately estimate the cases where there is just one firm. Where previously there were multiple equilibria, now Barnes & Noble enters, while Borders does not.

### 6.4.1 SPNE Estimator

The code for the estimator is somewhat longer than the estimator used in Chapter 4. The reason is that we now assume a unique equilibrium and so we can separately estimate all four possible cases.

```
> f_entry_spne = function(X, beta_1, beta_2, alpha_1,
+                            alpha_2, rho) {
+   N = dim(X)[1]
+   xi_1 = Z_1
+   xi_2 = Z_2*sqrt(1 - rho^2) + rho*Z_1
+   Xb_1 = X%*%beta_1
+   Xb_2 = X%*%beta_2
+   p_00 = p_01 = p_11 = rep(0, N)
+   for(k in 1:K) {
+     pi_1k = Xb_1 + xi_1[k]
+     pi_2k = Xb_2 + xi_2[k]
+     p_00 = p_00 + (pi_1k < 0 & pi_2k < 0)
+     p_01 = p_01 + (pi_1k - alpha_1 < 0 & pi_2k > 0)
+     p_11 = p_11 + (pi_1k - alpha_1 > 0 &  pi_2k - alpha_2 > 0)
+   }
+   return(list(p_00 = p_00/K,
+               p_01 = p_01/K,
+               p_11 = p_11/K))
+ }
```

### 6.4.2 SPNE Estimates

Table 6.1 shows that the assumption about simultaneous move gives very similar estimates to the assumption that Barnes & Noble moves first. The big difference is on the estimates of the impact of competition on the two firms. In order to reconcile the observed entry decisions with modeling assumptions, the estimator states Barnes & Noble is not affected much by competition, while Borders is.

Table 6.2 compares the model predictions to the actual data for the model presented in Chapter 4 and for a model presented here where Barnes & Noble moves first and there is a subgame perfect Nash equilibrium. The subgame perfect Nash equilibrium model does a better job of predicting the case where there is only one firm, but a substantially worse job of predicting the two-firm case.

## 6.5 Discussion and Further Reading

Notice that so far in this part of the book, there has been no discussion of time in these so-called dynamic games. This goes back to the point made earlier, dynamics has to do with information not time. Chapters 7 and 8 introduce time.

**TABLE 6.1**
Results from estimates of the game theory model from Chapter 4 and the model assuming Barnes & Noble moves first with a subgame perfect Nash equilibrium. The two columns labeled "Multi" refer to the case where we assume there could between two pure strategy Nash equilibrium. The two columns labeled "SPNE" refer to the model where Barnes & Noble moves first and there is a subgame perfect Nash equilibrium.

| | Multi | SD | SPNE | SD |
|---|---|---|---|---|
| const_1 | −15.11 | 0.25 | −15.12 | 0.06 |
| Pop_1 | 1.07 | 0.01 | 1.04 | 0.02 |
| Income_1 | −0.76 | 0.48 | −1.01 | 0.21 |
| College_1 | 5.65 | 0.55 | 5.61 | 0.19 |
| Stores_1990_1 | 0.37 | 0.11 | 0.15 | 0.06 |
| const_2 | −11.37 | 0.20 | −11.15 | 0.15 |
| Pop_2 | 0.65 | 0.02 | 0.65 | 0.02 |
| Income_2 | 1.31 | 0.80 | 1.93 | 0.17 |
| College_2 | 2.70 | 0.55 | 2.85 | 0.17 |
| Stores_1990_2 | 0.79 | 0.11 | 0.63 | 0.08 |
| alpha_1 | 0.73 | 0.18 | 0.50 | 0.16 |
| alpha_2 | 0.70 | 0.12 | 1.08 | 0.14 |
| rho | 0.47 | 0.10 | 0.39 | 0.06 |

**TABLE 6.2**
Comparison of predictions of the two models. The subgame perfect Nash equilibrium is somewhat better at fitting the case where there is only one firm, but not when there are two firms.

| | None | One Firm | Two Firm |
|---|---|---|---|
| Multi: None | 97.1 | 41.0 | 9.0 |
| Multi: One | 2.6 | 41.7 | 36.6 |
| Multi: Two | 0.3 | 17.3 | 54.4 |
| SPNE: None | 97.5 | 48.8 | 14.0 |
| SPNE: One | 2.4 | 43.2 | 53.3 |
| SPNE: Two | 0.1 | 8.0 | 32.7 |

The empirical analysis in this chapter is based on work presented in Adams and Basker (2025). The authors use information on the location of Barnes & Nobel and Borders collected from bookstore directories and firm websites to analyze the dynamics of the retail bookstore industry.

# 7

# *Repeated Games*

## 7.1 Introduction

Repeated games are the lens through which we are adapting our thinking on competition and competition policy. In the late 1990s, the economics of competition policy changed pretty dramatically. Game theory and structural econometrics provided antitrust authorities with new tools for modeling mergers. Chapters 3 and 4 introduce methods used to analyze the impact of retail mergers. While these models substantially improved our ability to understand competition and predict the outcomes of mergers, something was not quite right. Our approach to collusion and markets with collusive pricing remained rudimentary. Using the new models to analyze collusion was like pushing a square peg into a round hole.

The analysis presented here on collusive pricing is influenced by two papers. First, economists Nathan Miller, Gloria Sheu, and Matthew Weinberg presented compelling evidence that our new models worked poorly when used to analyze the US beer industry. Their paper, "Oligopolistic Price Leadership and Mergers: The United States Beer Industry," was published in the *American Economic Review* in 2021, and suggested a major rethink in the models we need for analyzing competition. Second, Canadian and Australian economists, David Byrne and Nic de Roos, analyzed pricing in the retail gasoline market in Perth Western Australia. Their paper, "Learning to Coordinate: A study in retail gasoline" was published in the *American Economic Review* in 2019. The authors use daily pricing data from a large number of retailers to show evidence of price leadership and coordination.

This chapter shows how the lens of repeated games can be used to explain firm behavior and pricing. Repeated interactions change the strategic relationships quite dramatically. In a single shot game, players don't have to account for the consequences of their actions. In repeated games they do.

Using data from Perth retail gasoline stations, the chapter presents two models of competition and pricing based on repeated interactions. The first model is a standard static pricing model that was introduced in Chapter 3. The second is a collusive oligopoly pricing model. This model shows how pricing is constrained by the incentives of firms to cheat and choose a lower price. The chapter estimates the parameters using data from gas stations in Perth

DOI: 10.1201/9781003351603-7

in a period where margins where low in 2008. It then compares the profit margins predicted by the collusive model to the actual profit margins of the same firms in 2012. While well-intentioned, price transparency regulation by the state government seems to cause Perth motorists to pay more for their petrol (gasoline).

The chapter runs a merger simulation using each of the two models of pricing behavior. It suggests that merger analysis should account for these changes when analyzing the likely effect of the merger. The predicted price increase using a repeated game model may be substantially higher than the predicted price increase using the standard static game presented earlier in the book.

## 7.2 Repeated Prisoner's Dilemma

The prisoner's dilemma is the most famous games in game theory. The game shows that the predicted outcome may not be the outcome that is best for both players. It may not be Pareto efficient. How much of that result is due to the set up of the game? What if players had to account for each other's actions? What if the game repeated?

This section presents the prisoner's dilemma game and shows how the Nash equilibrium change when dynamics are added. In particular, it shows when playing Cooperate is supported as a subgame perfect Nash equilibrium.

### 7.2.1 Prisoner's Dilemma

The game set up is as follows. We will relabel the strategies of the game presented in Chapters 1 and 2.

- Players: Player 1 and Player 2
- Strategies: Cooperate or Defect
- Payoffs:
  - {Cooperate, Cooperate}: {3, 3}
  - {Cooperate, Defect}: {0, 5}
  - {Defect, Cooperate}: {5, 0}
  - {Defect, Defect}: {2, 2}

In this game, the Nash equilibrium is {Defect, Defect}, even though both would be better off with the {Cooperate, Cooperate} outcome. The question is whether making a slight change to the set up of the game changes the predicted outcome.

### 7.2.2 Normal Form Representation

Chapter 1 introduced the normal form representation of a game.

**TABLE 7.1**
Normal form representation prisoner's dilemma game, with two players, Player 1 and Player 2. For Player 1, their choices are the rows and their payoffs are listed first in each cell.

| $P_1, P_2$ | Cooperate | Defect |
|---|---|---|
| Cooperate | 3, 3 | 0, 5 |
| Defect | 5, 0 | 2, 2 |

Table 7.1 presents the **normal form** of the prisoner's dilemma game. We can determine the Nash equilibrium by looking at the second column and seeing whether Player 1 is better off choosing the top row or the bottom row. Player 1 gets nothing from choosing the top row and 2 from the bottom. Similarly for Player 2, we can check the Nash equilibrium by looking at the bottom row and seeing if Player 2 wants the first column or the second column. Player 2 gets 0 from the first column and 2 from the second column. {Defect, Defect} is a Nash equilibrium.

### 7.2.3 Finitely Repeated Game

Consider a version of the game where the static prisoner's dilemma (above) is repeated a finite number of times ($T$ times). This could be 20 periods, for example.

The game is different. The players are the same but the strategies are completely different. A **strategy** is a function that maps from the complete history of the game to an action. In period 1, the complete history is null, so the strategy is just the action choices {Cooperate, Defect}. In period 2, the complete history is what ever the outcome was in period 1. There are 4 possibilities as listed above. From each possibility, the strategy states which action the player will choose. In period 3, things are even more complicated. Now the history includes 16 possibilities. For each of the 4 possible outcomes in period 1, there are 4 possible outcomes in period 2. Again the strategy states what the player will do in period 3 given each of the 16 possible histories. As you can imagine, for a game with twenty periods, there are an awful lot of possibilities.

What is the Nash equilibrium of this game? It is probably better to ask, what isn't. Consider any set of strategies where one player plays Cooperate in the last period. The game in the last period is essentially the same as a one period game. We know from the analysis of the one period game that the best response must be a strategy where they play Defect in the last period. Now consider sets of strategies where the player always plays Defect in the last period but Cooperate in the second to last period. Again, the best response must be a strategy that states the player plays Defect in the second to last period and Defect in the last period. Using this logic, we can show that the Nash equilibria are associated with playing Defect in every period. That is to say, despite adding a lot of complexity to the game, the prediction doesn't change.

### 7.2.4 Infinitely Repeated Prisoner's Dilemma

What if we make one more change to the game? This time, we repeat the game above for an infinite number of periods. Does this seem like a reasonable change? Infinity is quite a long time. For the strategic behavior to change, we don't literally need an infinite number of periods, we just need the players to be unsure when the game is going to end. It is better to think of a finite period game as one where every player knows exactly when the game is going to end. While an infinitely repeated game is one where players don't know exactly when the game will end.

For this case, we need to add another parameter to the payoffs, $r \in [0,1)$. This represents a discount rate and is a number between 0 and 1. The practical reason for adding a discount rate is that with an infinite number of periods we cannot analyze the game. The payoffs are infinite under all possible strategies. As you may remember from calculus, an infinite sum of a non-decreasing sequence is infinite. But with a discount rate, we can create a decreasing sequence that decreases fast enough that our infinite sum sums to a finite number. A player's utility given a particular strategy $s$ is an infinite sum of a discounted sequence of per period payoffs.

$$U(s) = \sum_{t=1}^{\infty} r^t \pi_t(s) \tag{7.1}$$

where $s$ is the strategy. If we assume that the payoff to the players in each period ($\pi_t(s)$) is bounded and there is a discount rate so that the payoff is lower in the future than today ($r < 1$), then the infinite sum, $U(s) < \infty$. Now we can analyze the game because different strategies may have different payoffs.

We can also use a very useful trick. If $\pi_t(s) = \pi(s)$, that is, if the per-period payoff is constant for a given strategy $s$ and $r$ is strictly between 0 and 1, which it is, we have the following simplification.

$$U(s) = \frac{\pi(s)}{1-r} \tag{7.2}$$

So if the discount rate is 0.9, then the total payoff is 10 times the per-period payoff for the strategy $s$.

Where does this discount rate come from? What does it mean? The most obvious way to think about it is as an interest rate. Actually, one minus the interest rate. When we are thinking about dynamic decisions it makes sense for decision makers to refer to the interest rate when determining the value of future decisions. Here, it may be reasonable to think of the discount rate as representing the probability that the game continues into the next period.

### 7.2.5 Cooperation as a Nash Equilibrium

Can having both players cooperate every period be supported as a Nash equilibrium of the infinitely repeated prisoner's dilemma? Yes.

To construct the supporting strategies, we need to allow players to punish defection. There are many strategies that have this feature but the simplest is called the **grim-trigger strategy**. In this strategy, the player plays Cooperate unless the history includes one of the players playing Defect. In that case, the player plays Defect forever. So "grim" is for the fact that the threat is the worst possible kind and "trigger" is for the fact that the strategy involves a simple event that changes behavior.

A strategy maps from the complete history of the game into the choice of Defect or Cooperate in each period. It is a set of functions for each period $t$, $\{s_t\}_{t=1}^{\infty}$. Each function is $s_t : H_t \rightarrow \{Defect, Cooperate\}$, where $H_t$ is the complete history of possible actions that could have happened in the previous $t-1$ periods. In this case, if $H_t$ includes Defect, then $s_t(H_t) = Defect$ while if it doesn't, then $s_t(H_t) = Cooperate$.

Is the grim-trigger a subgame perfect Nash equilibrium? Consider the case where both players are playing the grim-trigger. Payoffs are as listed in the previous section.

1. Grim-trigger: $\frac{3}{1-r}$
2. In period $t$ play Defect: $5 + r\left(\frac{2}{1-r}\right)$

So against the grim-trigger in which the other player is playing Cooperate, if you play Cooperate each period you get 3 each period. If you play Defect in one period, you get 5 for that period but the grim strategy kicks in and you get 2 for every period. For the grim-trigger to be an equilibrium choice (1) must give a higher payoff than choice (2).

The player would prefer to play grim-trigger if the following inequality holds.

$$\begin{aligned}
&\tfrac{3}{1-r} > 5 + r\left(\tfrac{2}{1-r}\right)\\
&3 > 5(1-r) + r2\\
&3 > 5 - 5r + 2r\\
&0 > 2 - 3r\\
&3r > 2\\
&r > \tfrac{2}{3}
\end{aligned} \tag{7.3}$$

As long as the players care about the future enough, $r$ is high enough, they will play Cooperate against the grim-trigger. If they are very myopic, $r$ is low, then they will cheat and take the high payoff today.

Just to tie things up we should also check that both players will keep playing Defect after the trigger has been pulled. Looking back at the discussion above we see that this must be the case.

## 7.3 Bertrand Competition

This section revisits the Bertrand pricing game presented in Chapter 3. It presents this model assuming logit demand, which is the more common assumption in the literature. It uses this model to estimate parameters of demand using data for retail gasoline from Perth Australia.

### 7.3.1 Two Firm Model

Consider a model where we have two firms that choose prices. The two firms sell similar products but are not exactly the same. For example, the products may be stores that are located in different places. The differentiation between the products is enough to induce some market power for each firm. That is for a small price increase, the firm is not going to lose all its customers. Below we will use this model to analyze retail gasolone. The gasoline itself is identical as it literally comes from the same pipe.[1] What is different between stations is there location and their brand.

In this model, Firm 1 chooses their price ($p_1$) to maximize the margin which is price less cost ($c_1$) multiplied by Firm 1's share ($s_1(p_1, p_2)$)

$$\max_{p_1} \quad (p_1 - c_1)s_1(p_1, p_2) \tag{7.4}$$

The logit model assumes demand has the following form.

$$s_1(p_1, p_2) = \frac{\exp(\delta_1)}{\exp(\delta_1) + \exp(\delta_2)} \tag{7.5}$$

where $\delta_1 = -\alpha_1 p_1 + \xi_1$ and $\delta_2 = -\alpha_2 p_2 + \xi_2$. In this model, demand for Firm 1 is determined by firm specific value $\xi_1$ and by Firm 1's price $p_1$. Sensitivity to price is determined by the parameter $\alpha_1$.

### 7.3.2 *J* Firm Model

Assume that we have $J$ firms and logit demand.

$$\delta_j = -\alpha_j p_j + \xi_j \tag{7.6}$$

$$s_j(p_j, p_{-j}) = \frac{\exp(\delta_j)}{1 + \sum_{j'=1}^{J} \exp(\delta_{j'})} \tag{7.7}$$

[1] Some stations may add in additives at the end.

where $p_j$ is the price charged by firm $j$, $\alpha_j$ represents the price sensitivity of firm $j$'s customers, and $\xi_j$ is a set of characteristics other than price that determine demand. The notation $-j$ means all the other firms that are not $j$.

### 7.3.3 Nash Equilibrium in R

The following are the functions for determining the Nash equilibrium using the logit demand system. The function `f_share()` determines the shares from the logit model given the vectors of $\xi$s, $\alpha$s and prices. The function `pi_f()` determines the vector of firm profits, while `pi_i()` determines the profits for firm $i$ given a vector of prices for the other firms. The function `pi_i_opt()` determines firm $i$'s best response to a set of prices, while `pi_opt()` determines the vector of best response. The function `ne_f()` determines the Nash equilibrium vector of prices.

```
> f_share = function(xi, price, alpha) {
+   delta = xi - alpha*price
+   exp_delta = c(1,exp(delta))
+   return((exp_delta/sum(exp_delta))[-1])
+ }
> pi_f = function(xi, price, alpha, cost) {
+   return((price - cost)*f_share(xi, price, alpha))
+ }
> pi_i = function(i, p_i, xi, price, alpha, cost) {
+   price[i] = p_i
+   return(pi_f(xi, price, alpha, cost)[i])
+ }
> pi_i_opt = function(i, xi, price, alpha, cost) {
+   a_i = optimize(f=pi_i, interval=c(0, 400),
+                  i = i,
+                  xi = xi,
+                  price = price,
+                  cost = cost,
+                  alpha = alpha,
+                  maximum = TRUE)
+   return(a_i$maximum)
+ }
> pi_opt = function(xi, price, alpha, cost) {
+   N = length(xi)
+   price_new = rep(NA, N)
+   for(i in 1:N) {
+     price_new[i] = pi_i_opt(i,xi,price,alpha,cost)
+   }
+   return(price_new)
+ }
```

Nash equilibrium is determined using a similar iterative approach that we used to determine equilibrium of the Cournot model in Chapter 3. It uses a `while()` to determine when the sequence of price choices converges.

```
> ne_f = function(xi, price, alpha, cost,
+                   tol=1e-10, maxiter=10000,
+                   trace=0) {
+   price_0 = price
+   diff = sum(abs(price_0))
+   iter = 1
+   converge = FALSE
+   while(diff > tol & iter < maxiter) {
+     price_1 = pi_opt(xi, price_0, alpha, cost)
+     diff = sum(abs(price_1 - price_0))
+     price_0 = price_1
+     if(trace > 0) {
+       print(diff)
+       print(iter)
+     }
+     iter = iter + 1
+   }
+   if(diff < tol) {converge=TRUE}
+   return(list(price=price_1,converge=converge))
+ }
```

## 7.4 Empirical Analysis: Retail Gasoline Pricing using R

This section estimates the parameters of the model assuming the Perth retail gas oline market is priced competitively, i.e., consistent with static Bertrand Nash equilibrium.

### 7.4.1 Perth Gas Price Data

Figure 7.1 presents the average margins by week for the gas stations. While there is a lot of variation in margins in 2008 and 2009, the average stays pretty steady. Things seems to change in 2010, with margins steadily increasing. Why would that be?

### 7.4.2 Estimating Parameters

In the analysis below, we aggregate up to the brand level and estimate the pricing game between the brands. We assume that margins and shares are the

```
> file = paste0(dir, "perth_gas_data.csv")
> dt = fread(file)
> dt[,.(
+   margin = mean(margin),
+   date = mean(date)
+   ),
+          by = c("week", "year")] |>
+   ggplot(aes(x=date, y=margin)) +
+   geom_point(color = "gray") +
+   geom_smooth(se = FALSE) +
+   labs(title = "Margin (cents/liter)",
+        x = "",
+        y = "")
```

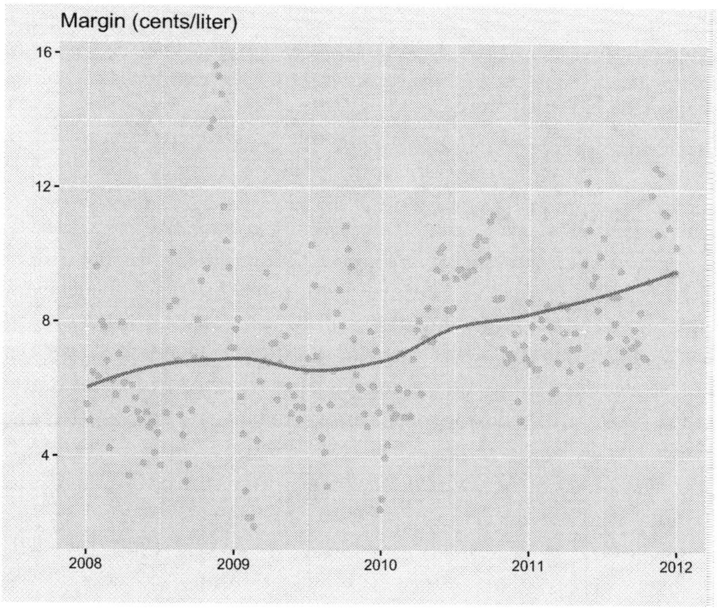


**FIGURE 7.1**
Plot of weekly margins from 2008, 2009, 2010, and 2011. There is a lot of variation from 2006 to 2010 but prices are moving around the same average. After 2010, the average price starts to increase.

result of Nash equilibrium of the static pricing game. The data provide daily prices, daily wholesale price, the location of the stations and the number of stations for each brand.

To estimate the firm's marginal costs, we regress prices on the distance between the station and the Kwinana terminal, which is located south of Perth.

The assumption is that variation in prices due to distance is determined by the trucking costs of the fuel. We use a quadratic on distance, brand dummies and week dummies to estimate marginal costs for each station. These values are added to the terminal price to get the estimated cost for each station. As we don't have access to quantity information, the brand share is assumed to be equal to the proportion of stations that the brand has.

The model requires estimates of two parameters for each firm. The price sensitivity parameter, $\alpha_j$, and the unobserved quality parameter, $\xi_j$. The first is found from the firm's first order condition when demand is determined by a logit model.

$$\alpha_j = \frac{1}{(p_j - c_j)(1 - s_j)} \tag{7.8}$$

The second value is from inverting the logit demand to get the unobserved characteristic as a function of the observed prices, shares, and the parameter $\alpha_j$.

$$\xi_j = \alpha_j p_j + log(s_j) - log(s_0) \tag{7.9}$$

where $s_0$ is the outside share.

### 7.4.3 Parameter Estimates

From Equation (7.8) and (7.9), we determine $\alpha$ and $\xi$ for each firm from the observed prices and shares. We are assuming that the observed prices and shares are determined as the outcome of a static Bertrand Nash equilibrium. We are also assuming that the outside good are the independent stations and the smaller brands, Mobil, Wesco and Better Choice. Finally, we are assuming that Caltex and Caltex Woolworths make pricing decisions as if they are the same firm.

To do this analysis we will aggregate up to annual average prices, costs, margins, and shares.

The first step is to select the variables and the year (2008) we will use. Also redefine some of the brand names to make the analysis easier.

```
> dt2 = dt |>
+   filter(year == 2008) |>
+   select(
+     date,
+     store = TRADING_NAME,
+     brand = BRAND_DESCRIPTION,
+     price = PRODUCT_PRICE,
+     margin
+     )
> dt2$brand[grep("Caltex", dt2$brand)] = "Caltex"
> dt2$brand[which(dt2$brand %in% c("Independent",
+                                  "Mobil",
```

```
+                                  "Wesco",
+                                  "Better Choice"))] = "Independent"
```

The next step is to calculate the shares by determining the number of stations for each brand and then calculating the share for each brand.

```
> stores = dt2[, .N, by = brand]
> stores$shares = stores$N/sum(stores$N)
> dt2 = merge(dt2, stores, by = c("brand"))
```

The next step creates a data set with prices, margins, shares, and costs averaged up to the brand level for 2008. This also calculates $\alpha$ for each brand.

```
> dt3 = dt2[, .(
+   price = mean(price, na.rm = TRUE),
+   margin = mean(margin, na.rm = TRUE),
+   share = mean(shares, na.rm = TRUE),
+   cost = mean(-margin + price, na.rm = TRUE),
+   alpha = 1/(mean(margin, na.rm = TRUE)*(1 - mean(shares,
+                                            na.rm = TRUE)))
+   ),
+           by = brand]
```

Next, is calculating $\xi$ for each brand. This calculation assumes that the independent stores are the outside option.

```
> index_ind = grep("Independent", dt3$brand)
> dt3$xi = NA
> dt3$xi[-index_ind] =
+   dt3$alpha[-index_ind]*dt3$price[-index_ind] +
+   log(dt3$share[-index_ind]) -
+   log(dt3$share[index_ind])
```

The final step is to use the function `ne_f()` to determine the Nash equilibrium.

```
> a_f = ne_f(dt3$xi[-index_ind],
+            dt3$price[-index_ind],
+            dt3$alpha[-index_ind],
+            dt3$cost[-index_ind])
```

Table 7.2 presents the estimates for $\alpha$ and $\xi$ for each brand. These values reconcile the observed equilibrium margins and the observed equilibrium shares. BP is able to have higher margins and only slightly lower share because its customers are less price sensitive than Caltex.

Can you compare the parameter estimates using the algebraic approach to the numeric approach to determining the Nash equilibrium? Are they exactly the same? Would you expect them to be?

**TABLE 7.2**
The table presents the prices, margins, market share, and estimates for $\alpha$ and $\xi$. BP is able to have higher margins and only slightly lower share because its customers are less price sensitive than for Caltex.

| | Brand | Price | Cost | Share | $\alpha$ | $\xi$ |
|---|---|---|---|---|---|---|
| 1 | Ampol | 153.70 | 142.53 | 0.02 | 0.09 | 12.93 |
| 2 | BP | 148.53 | 140.77 | 0.23 | 0.17 | 26.16 |
| 3 | Caltex | 146.14 | 139.46 | 0.30 | 0.21 | 32.56 |
| 4 | Coles Express | 144.65 | 139.92 | 0.10 | 0.23 | 34.28 |
| 5 | Eagle | 152.32 | 147.32 | 0.00 | 0.20 | 26.47 |
| 6 | Gull | 144.00 | 138.06 | 0.11 | 0.19 | 27.70 |
| 7 | Liberty | 145.23 | 138.06 | 0.03 | 0.14 | 19.70 |
| 8 | Peak | 139.98 | 136.33 | 0.04 | 0.29 | 39.42 |
| 9 | Shell | 152.11 | 142.57 | 0.07 | 0.11 | 17.20 |
| 10 | United | 138.25 | 135.42 | 0.02 | 0.36 | 48.59 |

## 7.5 Repeated Oligopoly

The standard differentiated goods Bertrand model of price competition suggests that the outcome we would expect, while higher than perfect competition, it is not collusion. In the static game, it is always better for the firms to "cheat" and lower their prices in order to increase profits. The question then is whether we would expect to see collusion when firms interact repeatedly.

The section adapts the static Bertrand game to an infinitely repeated setting and determines the optimal pricing under collusion.

### 7.5.1 Collusive Equilibrium

Choosing the static equilibrium Bertrand prices in each period can be supported as a subgame perfect Nash equilibrium of the infinitely repeated game. Like with prisoner's dilemma, it is always fine to play Defect every period.

Given this, we ask whether or under what circumstances can a trigger strategy support collusion. Let $\pi_N$ denote the per-period profits in a Nash equilibrium of the static game, $\pi_C$ denote the collusive profits and $\pi_D$ the profits from defecting and choosing an optimal price when the other firm is offering the collusive price. We have $\pi_D > \pi_C > \pi_N$. This is exactly the same as for a prisoner's dilemma.

Like the prisoner's dilemma, collusion can be supported by a trigger strategy if the following inequality holds.

$$\begin{aligned}
&\frac{\pi_C}{1-r} > \pi_D + r\left(\frac{\pi_N}{1-r}\right) \\
&\pi_C > (1-r)\pi_D + r\pi_N \\
&r(\pi_D - \pi_N) > \pi_D - \pi_C \\
&r > \frac{\pi_D - \pi_C}{\pi_D - \pi_N}
\end{aligned} \tag{7.10}$$

The number on the bottom is larger than the number on the top, and so for a large enough $r$ collusion is a Nash equilibrium of the infinitely repeated oligopoly game.

### 7.5.2 Identifying Collusion

Identifying collusion for policy or academic purposes is quite different from identifying collusion for criminal prosecution. A criminal case requires hard evidence, not some cool econometric specification. The best evidence includes credible witnesses, audio recordings, video recordings, and written documents. You may be surprised to learn that there isn't that much economics involved prosecuting a criminal collusion case. There is undercover work, wire tapping, etc, but no economics. Once the criminal case is proven, economists are called in to estimate damages.[2] Here again, identifying the collusion is not that difficult. At least it is not that difficult given that the FBI has already completed the task. The case record includes the dates when the collusion occurred and (hopefully) dates when the collusion did not occur. The hard part for the econometrician is working out which part of the difference in prices is due to the collusion and which is due to other changes.

Identifying collusion without the assistance of the FBI is difficult. To observe prices and quantities from a market, there is enough exogenous variation such that we can estimate the elasticity of demand. If the products are differentiated, we can use Bertrand Nash equilibrium to identify marginal costs and mark-ups. While assuming a static equilibrium or a dynamic collusive equilibrium gives different estimates of the mark ups and marginal cost, without some other information we can't tell the difference from the data. If we had data on marginal costs, then suddenly things get a lot easier. We can match the implied marginal costs from our proposed behavioral assumptions and see which ones fits better. Alternatively, if the courts tell us that there was a period where the static equilibrium determined prices, then we can use that period to identify marginal costs. We can then compare the implied markups from the static equilibrium to the observed markups to estimate the **super markups** associated with collusion.[3]

[2]This is usually the amount equal to the consumer surplus lost from the collusion.

[3]These are markups that are larger than we would expect from firms choosing price consistent with equilibrium of a static Bertrand game.

### 7.5.3 Choosing Super Markups

Choosing the collusive price seems like a simple enough problem for the firms. The collusive price should be equal to the price that a monopoly would charge. Not so fast Sonny Jim! Sure, the monopoly price would be optimal if all the firms behaved as one firm. They are not one firm and they do not behave as such. In each period, each firm may prefer to renege on the deal and choose a lower price, gaining share while everyone else is charging the collusive price.

The collusive price is the solution to the following maximization problem. The super markup (smu) is the difference between the collusive price and the static Nash equilibrium price where the marginal cost is normalized to zero.

$$\begin{array}{ll} \max_{p_c} & \pi_c(p_c) \\ s.t. & \pi_d(p_c) + \frac{r\pi_{NE}}{1-r} \leq \frac{\pi_c(p_c)}{1-r} \end{array} \tag{7.11}$$

where $\pi_c(p_c)$ are the per-period profits that one firm makes when everyone chargers $p_c$, $\pi_d(p_c)$ is the per-period profit when one firm is able to deviate and charge a lower price knowing that everyone else is charging $p_c$, and $\pi_{NE}$ is the per-period profit in a static Bertrand Nash equilibrium. The firm's preferences over profits in future periods is determined by the parameter $r$.

The problem states that firms can charge any collusive price they like as long as all the firms are unwilling to cheat on the agreement. In general, we find that the higher the price, the greater the value in cheating on the agreement.

In general, the colluding firms cannot charge anything they want, in fact they can't even charge the monopoly price. They are constrained both by the demand system and by the incentive of firms to cheat on the agreement.

### 7.5.4 Estimating Collusive Prices in R

The following functions are used to determine the collusive price when demand is determined by the logit model. The collusive price is assumed to be equal to the Nash equilibrium price plus a **super markup**. The function `f_smu_share()` determines the vector shares at a particular smu. The function `pi_d()` determines the optimal deviation profits given a particular smu. This places a constraint on what smu can be chosen.

```
> f_smu_share = function(smu, xi, price, alpha) {
+   f_share(xi, price + smu, alpha)
+ }
> pi_d = function(p_d, i, smu, xi, price, alpha, cost) {
+   price = price + smu
+   price[i] = p_d
+   share = f_share(xi, price, alpha)
+   return(share[i]*(p_d - cost[i]))
+ }
```

The function pi_smu() determines the vector of profits for a particular smu and pi_smu_int() is an intermediate function for optim().

```
> pi_smu = function(smu, xi, price, alpha, cost,
+                   r, lambda, pi_ne) {
+   N = length(xi)
+   pi_c = (price + smu - cost)*f_smu_share(smu, xi, price,
+                                           alpha)
+   PI_d = matrix(NA, N, 2)
+   for(i in 1:N) {
+     ai = optimize(f = pi_d,
+                   interval = c(0,2000),
+                   i = i,
+                   smu = smu,
+                   xi = xi,
+                   price = price,
+                   alpha = alpha,
+                   cost = cost,
+                   maximum = TRUE)
+     PI_d[i,1] = i
+     PI_d[i,2] = ai$objective
+     #print(i)
+   }
+   return(pi_c -
+          lambda*(((1 - r)*PI_d[,2] +
+          r*pi_ne - pi_c)^2))
+ }
> pi_smu_int = function(par, xi, price, alpha,
+                       cost, r, pi_ne) {
+   smu = par[1]
+   lambda = 1
+   return(sum(pi_smu(smu, xi, price,
+                     alpha, cost, r,
+                     lambda, pi_ne)))
+ }
```

## 7.6 Empirical Analysis: Collusion in Perth Gas Stations using R

This section estimates the extent of collusion in the the Perth retail gasoline market.

### 7.6.1 Super Markups

With all the parameters of the model estimated from the data under the static Bertrand Nash assumption, we use the parameters to determine the super markup.

As above, to determine the super markup, we need to determine the static Nash profits, the profits from cheating, and we need to make an assumption about discounting. We assume `r` is equal to 0.9. What happens at different values?

```
> margin_ne = (a_f$price - dt3$cost[-index_ind])
> share_ne = f_smu_share(0, dt3$xi[-index_ind],
+                        a_f$price, dt3$alpha[-index_ind])
> pi_ne = margin_ne*share_ne
> a1 = optimize(pi_smu_int,
+               c(0,20),
+               xi = dt3$xi[-index_ind],
+               price = a_f$price,
+               alpha = dt3$alpha[-index_ind],
+               cost = dt3$cost[-index_ind],
+               r = 0.9,
+               pi_ne = pi_ne,
+               maximum = TRUE)
> a1$maximum
  [1] 4.589335
> a1$maximum/mean(margin_ne)
  [1] 0.7117553
```

In this set up, the super markup is 4.59 cents a liter or 71 percent of the average static Nash margins. Not bad! If we look at Figure 7.1 we see that margins increased about 4 cents a liter between 2008 and 2012. This suggests that the collusive model is a more accurate representation of pricing behavior than the static pricing model.

### 7.6.2 Analyzing Mergers with Collusion

Previously we have discussed the effect of mergers when firm competition and pricing can be modeled as a static Bertrand Nash equilibrium. What if the firms prices are really being determined as a collusive agreement? Will the merger affect prices? You may think that the collusive price is as high as prices could get. What impact could the merger have?

Mergers can cause prices to increase when firms are colluding. With collusion, the merger leads to two changes in the market. First, it has an effect on the static Bertrand Nash equilibrium price and thus the punishment that can be imposed. Second, it changes the set of constraints that are placed on the problem. Post merger, there is one less firm that is trying to cheat on the deal.

In theory, these two effects work in different directions. The merger increases profits from the punishment phase, which makes incentivizing firms to collude harder. On the other hand, the number of firms that must be kept in line has been reduced.

Consider a merger between BP and Peak. Peak is very small and so is unlikely to be of concern using standard static Bertrand Nash pricing. To model the effect of the merger, we need to calculate the observed characteristics of the merged firm.

```
> index_m = 9
> dt3$brand[index_m]
  [1] "Peak"
> bp_s = dt3$share[2]/(dt3$share[2] + dt3$share[index_m])
```

The variable bp_s is the share of the merged firm that is BP. The following code calculates the new characteristics of the merged firm. The new firm's costs are the weighted average of each firms average costs. Similarly, the new firm's price sensitivity parameter and unobserved characteristics are weighted averages of the two firms. Do these assumptions make sense? What happens under alternative assumptions?

```
> brand_m = dt3$brand[-c(index_ind, index_m)]
> brand_m[2] = paste0(dt3$brand[index_m], " and ", dt3$brand[2])
> xi_m = dt3$xi[-c(index_ind, index_m)]
> alpha_m = dt3$alpha[-c(index_ind, index_m)]
> alpha_m[2] = bp_s*dt3$alpha[2] + (1  - bp_s)*dt3$alpha[index_m]
> price_m = dt3$price[-c(index_ind, index_m)]
> price_m[2] = bp_s*dt3$price[2] + (1 - bp_s)*dt3$price[index_m]
> cost_m = dt3$cost[-c(index_ind, index_m)]
> cost_m[2] = bp_s*dt3$cost[2] + (1 - bp_s)*dt3$cost[index_m]
```

Given the change caused by the merger, we can calculate the new Bertrand Nash equilibrium as well as the new equilibrium collusive price.

```
> a_f_m = ne_f(xi_m, price_m, alpha_m,
+              cost_m, maxiter=1000000)
> margin_m = (a_f_m$price - cost_m)
> share_m = f_smu_share(0, xi_m, a_f_m$price, alpha_m)
> pi_ne_m = margin_m*share_m
> a2 = optimize(pi_smu_int,
+                 c(0,20),
+                 xi = xi_m,
+                 price = a_f_m$price,
+                 alpha = alpha_m,
+                 cost = cost_m,
+                 r = 0.9,
```

```
+                  pi_ne = pi_ne_m,
+                  maximum = TRUE)
```

If the firms are playing Bertrand static game, then merger has the following impact on prices. Prices will increase 0.4 percent.

```
> (mean(a_f_m$price) - mean(a_f$price))/mean(a_f$price)
  [1] 0.004122352
```

If the firms are playing a collusive game, then the merger has the following impact on prices. Prices will increase 1.2 percent.

```
> (mean(a_f_m$price+a2$maximum) -
+      mean(a_f$price+a1$maximum))/mean(a_f$price+a1$maximum)
  [1] 0.01238215
```

```
> sum_tab_m = cbind(
+    as.numeric(a_f$price),
+    as.numeric(c(a_f_m$price[1:7], NA, a_f_m$price[8:9])),
+    as.numeric(a_f$price) + a1$maximum,
+    as.numeric(c(a_f_m$price[1:7], NA, a_f_m$price[8:9]))
+ a2$maximum
+ )
> rownames(sum_tab_m) = dt3$brand[-index_ind]
> colnames(sum_tab_m) = c("Price", "Price Merge",
+                         "Collusive", "Collusive Merge")
```

Table 7.3 presents the impact of the merger on prices under the two different models. Note that while BP's price goes down with the merger, the average price of BP and Peak increases. Remember the new firm is a weighted average of BP and Peak's observed and unobserved characteristics.[4] The impact of the merger is quite different under the two models of pricing behavior. The average price effect of the merger is small. It is 0.4% increase in prices in the static Bertrand case and 1.2% increase in the collusive case. The difference between the price increase assuming static Bertrand and collusion is over 100%. The assumed method by which firm's determine prices may have a large effect on our prediction of the merger.

## 7.7 Discussion and Further Reading

In the 1990s, industrial organization economists made great gains in understanding markets and competition. But there was something missing. The

[4]What may be other assumptions that could be made about the new post-merger firm?

**TABLE 7.3**
The table presents impact of the merger under the two pricing models. The first and third column are the pre-merger prices for the two models. The second and fourth columns are the new prices after the simulated merger between BP and Peak. Note that the new prices are given for BP. The impact of the merger is quite different under the two pricing models. For example, Shell's prices don't really change with the merger in the Bertrand static pricing game, while they increase substantially in the repeated game.

| | Price | Price Merge | Collusive | Collusive Merge |
|---|---|---|---|---|
| Ampol | 153.70 | 153.81 | 158.29 | 159.67 |
| BP | 148.53 | 145.77 | 153.12 | 151.62 |
| Caltex | 146.14 | 146.84 | 150.72 | 152.70 |
| Coles Express | 144.65 | 144.85 | 149.24 | 150.70 |
| Eagle | 152.32 | 152.33 | 156.91 | 153.18 |
| Gull | 144.00 | 144.27 | 148.59 | 150.13 |
| Liberty | 145.23 | 145.31 | 149.82 | 151.17 |
| Peak | 139.98 | | 144.57 | |
| Shell | 152.11 | 152.41 | 156.70 | 153.26 |
| United | 138.25 | 138.28 | 142.84 | 144.14 |

standard models didn't always capture the pricing behavior. A number of recent papers have shown that we need better models in order to capture the possibility of collusion in markets. Repeated games allow us to use much richer models of pricing behavior and show how collusion can be a more natural feature of how firms determine prices.

The paper by David Byrne and Nic de Roos does not provide us with any model techniques. Rather, the paper takes a very close look at an actual market and shows how firms in that market actually behave (Byrne and de Roos, 2019). Nathan Miller and Matt Weinberg review the joint venture between Miller and Coors. The authors use information on the premerger market to estimate parameters of the model. They show that the observed price increase post merger cannot be explained by our standard static Nash model (Miller and Weinberg, 2017). Nathan Miller, Gloria Sheu, and Matt Weinberg argue that a model that explicitly accounts for collusion is necessary to predict the effect of mergers in some industries (Miller et al., 2021).

A number of papers formally compare the predictions of static pricing models to observed pricing behavior. Miller and Weinberg (2017) use pre-merger pricing to estimate the parameters, then compare simulated pricing to actual pricing after the merger. Nevo (2001) and Backus et al (2021) compare the predictions of collusive pricing models to actual pricing in the ready-to-eat cereal market.

The pricing patterns seen in retail gasoline is very strange (Lewis, 2012). In the Perth data, we see the firms move to a very ordered pattern of pricing on a weekly basis. If you zoom in even closer you see that the brands are using one or two stations to signal which price they will move to for that week. Our analysis in this chapter is based on the work of DOJ Economist, Zhongmin Wang. Wang (2009) uses a mixed strategy dynamic model of to analyze pricing patterns in the Perth data. Nobel prizing winning economists, Eric Maskin and Jean Tirole, show that short term commitment to a price is necessary to get the type of pricing dynamics we see in the data (Maskin and Tirole, 1988). The West Australian state government introduced price regulation that allowed for this type of equilibrium. We generally call it "post and hold" regulation. The post means that the price is publicly displayed and the hold means that the price cannot be changed for some period of time, say 24 hours. By making the price publicly available, it allowed for strategies that are a function of each other's prices. Hold means that the firms can commit to the price which means that the equilibrium cannot devolve into a competitive pricing process. That is, each firm will choose a price to under cut its competitor, like in the original Bertrand game presented in Chapter 3.

# 8

# *Bargaining*

## 8.1 Introduction

When you take your first economics class you learn about two price setting mechanisms, perfect competition and monopoly. You learn about the case where no seller in the market has any power to determine the price and the case where the only seller has the power to determine the price. You may have also been exposed to the idea that the buyer has the power to determine the price, monopsony. What about the case when both the buyer and the seller have the power to determine the price? What happens in that case?

In the case where both sides of the transaction have the ability to determine the price, we need a bargaining model. We need a model that can help determine what the outcome will be. This chapter considers three models of bargaining, the **ultimatum game**, the **alternating offer game** and the **Nash bargaining model**.

While the **ultimatum game** is very simple, our predictions of what will happen in the game are both unsatisfactory and do not seem to agree with what happens when actual people play the game. The chapter goes through the various predictions for this game and then looks at what happens in experiments where real people play the game. In order to develop a game that predicts outcomes that seem more reasonable the chapter presents the **alternating offers game**. The chapter shows that while this game is quite complicated, its predictions are quite intuitive. The third model of bargaining is not technically a game. That said, it can be shown that the Nash bargaining model is equivalent to certain alternating offers games. So, although the model itself is not game theoretic, it does have a game theoretic foundation. More importantly, the Nash bargaining model is much simpler than the alternating offers game. This simplicity has allowed the model to become an important tool in empirical analysis of situations where bargaining is used to determine prices.

The chapter uses the Nash bargaining model to understand competition between hospitals in Florida. The empirical estimates are used to predict the effect on hospital prices of a merger between hospitals located in Palm Beach County, the home county for this book's publisher.

DOI: 10.1201/9781003351603-8

## 8.2 Ultimatum Game

The simplest way to model a bargaining game is with a "take-it-or-leave-it" offer. A TIOLI offer if you will. The game has two periods. In the first period, one player makes an offer on how to split the pie between the two players.[1] In the second period, the other player observes the offer and decides to accept or reject. If the second player accepts, then the two players split the pie as determined by the first player's offer. If the second player rejects the offer, then the pie vanishes and neither players gets anything.

What would you do if you are the first player? How much would you give to the second player? How much would you keep for yourself? What if the pie is \$5.00? Would you give the other player \$2.50? \$2.00? \$1.00? Whoa. \$0.00? How about if the pie is \$5,000,000.00? What if you are the second player? Would you reject an offer that you believe is unreasonably low? Would you reject an offer of \$2.00? What if the pie is \$5,000,000.00, would you reject the \$2.00 offer in that case?

This section considers the prediction when we use **Nash equilibrium** and when we use **subgame perfect Nash equilibrium**. It then brings in some data from experiments where actual people play the game and compares the observed outcomes to the predicted outcomes.

### 8.2.1 The Game

We can write down the ultimatum game using a formal representation. Odd's strategy is simple. She offers some number between zero and one, including zero or one. This number represents the share of the pie that Odd keeps if the offer is accepted. Even's strategy is more complicated. Even will accept certain offers and will reject other offers. The notation below is completely general. It includes strategies like reject all offers and accept all offers and accept all offers that are multiples of 0.1345888, etc.

- Players: Odd, Even
- Strategies:
  - Odd: Offer $x \in [0, 1]$
  - Even: If $x \in R$ reject, otherwise accept, where $R \subset [0, 1]$.
- Payoffs:
  - Odd offers $x$, Even accepts: $\{x, 1 - x\}$
  - Odd offers $x$, Even rejects: $\{0, 0\}$

[1]The term "pie" is used to refer to some outcome that can be divided up between the players. In many cases, this is some fixed amount of money.

where Odd gets the payoff associated with the first element of the set. Odd's strategy is denoted as $x$, which is the share of the pie that Odd receives. The remainder, $1 - x$, is the share received by Even. Even rejects Odd's offer if $x$ lies in a particular subset of the offers called $R$, which stands for reject. To make things simpler, we will generally assume that Even plays a "cutoff" strategy. That is $R$ is equivalent to the set $x$ where $x \geq y$, where $y \in [0, 1]$. If $x$ is large, then Even's share is small and he will reject the offer.

### 8.2.2 Game Tree

While this game is not that complicated, it turns out that presenting it in extensive form is very useful.

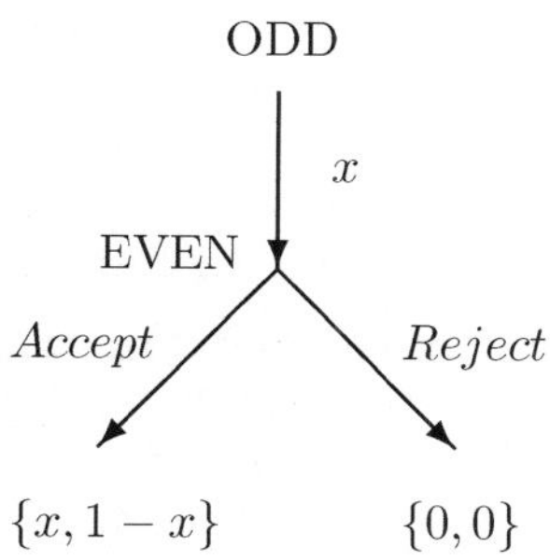


**FIGURE 8.1**
An extensive form representation of an ultimatum game. Odd makes an offer $x$ which is the share of the pie that they will keep. Even observes the offer and decides whether or not to accept or reject the offer.

Figure 8.1 presents the extensive form representation of the ultimatum game. In the first period, Odd makes an offer of $x$ which is some value between 0 and 1. It is the share of the pie that Odd will keep if Even accepts the offer. After observing Odd's offer, Even decides whether to accept or reject the offer. If Even accepts, then he gets $1 - x$. If he rejects, then no one gets nothing.

### 8.2.3 Normal Form

As mentioned above, while we generally use extensive form representations for dynamic games, there is nothing to stop us from using a normal form representation. In order to think about some of the predicted outcomes of this game, it is useful to think about the normal form representation.

We will simplify and just consider three strategies by Odd, $x \in \{0, 0.5, 1\}$. We will also limit Even to three strategies. We will further limit Even to cutoff strategies. That is, Even will choose some number such that he will accept any offer less than or equal to that number and reject any offer above that number. We will denote the cutoff value $y$. So $y = 0$ means that Even rejects any offer that gives him less than the full share. While $y = 0.5$ means that Even will accept an offer that gives him at least half. Lastly, $y = 1$ means that Even accepts any offer that Odd makes.

**TABLE 8.1**
Normal form representation of simplified ultimatum game. Odd's offer is $x$ which is the share Odd receives, with $1 - x$ going to Even. Even's strategy is denoted $y$, and it gives the level of the offer that she will accept. The payoffs are in the cells with Odd first.

| Odd, Even | $y = 0$ | $y = 0.5$ | $y = 1$ |
|---|---|---|---|
| $x = 0$ | 0, 1 | 0, 1 | 0, 1 |
| $x = 0.5$ | 0, 0 | 0.5, 0.5 | 0.5, 0.5 |
| $x = 1$ | 0, 0 | 0, 0 | 1, 0 |

Table 8.1 presents a normal form representation of a simplified version of the ultimatum game. What is the Nash equilibrium of the game? Is $x = 0.5, y = 0.5$ a Nash equilibrium? What about $x = 0, y = 0$ or $x = 1, y = 1$?

### 8.2.4 Nash Equilibrium

This is a pretty simple game. Odd makes an offer, Even observes the offer and decides to reject or accept the offer. What do you think is the likely outcome of the game? Do you think $x = 0.5$ in equilibrium? Do you think there will be an even split? Do you think $x = 1$? Do you think Odd will make an offer where she gets to keep the whole pie? What would Even do if he saw such an offer?

Is $x = 0.5$ a Nash equilibrium? It is if we are a bit more careful in what it is. Remember Even's strategy needs to state what will happen in every possible case, that is, every possible offer Odd could make. Consider the following set of strategies. Odd offers $x = 0.5$ and Even accepts any offer below 0.5 (or equal) and rejects any offer above 0.5. This is a Nash equilibrium.

To see that it is a Nash equilibrium, let's go back to our algorithm. In the algorithm, we first assume Odd plays her strategy. We then determine if it is optimal for Even to play the strategy stated in the proposed equilibrium. If it is, we reverse things and assume Even plays the strategy stated in the proposed equilibrium. Given that strategy, we determine if Odd's optimal strategy is the same as in the proposed equilibrium. If it is, then we have a Nash equilibrium!

The proposed equilibrium is $\{x = 0.5, y = 0.5\}$

- Assume Odd offers $x = 0.5$, what is Even's optimal response?
  - Even's Payoffs for his different strategies:
    * $y = 0.5$: 0.5
    * $y < 0.5$: 0
    * $y > 0.5$: 0.5
  - Even can't be made better off with $y \neq 0.5$, so $y = 0.5$ is optimal.
- Assume Even's cutoff is $y = 0.5$, what is Odd's optimal response?
  - Odd's Payoffs from her strategies:
    * $x = 0.5$: 0.5
    * $x < 0.5$: $x$ (which is less than 0.5)
    * $x > 0.5$: 0
  - Odd is not better off choosing some other offer, so Odd's strategy is optimal and the proposed set of strategies is a Nash equilibrium.

The offer that splits the difference can be part of a Nash equilibrium. Can any split be supported as part of a Nash equilibrium? Yes. All of them.

Any offer by Odd can be supported as a Nash equilibrium of the game. Consider the case where Odd offers $a$ where $a$ is between 0 and 1. For this case also assume that Even's strategy is for $y = a$. That is, Even accepts any offer less than $a$ (or equal to) and rejects any offer above $a$.

- Assume Odd Offers $x = a$, what is Even's optimal response?
  - Even's payoffs from his strategies:
    * $y = a$: $1 - a$
    * $y < a$: 0
    * $y > a$: $1 - a$
  - Even can't be made better off with $y \neq a$.
- Assume Even will accept any offer below $a$ ($y = a$), what is Odd's optimal response?
  - Odd's payoffs from her strategies:
    * $x = a$: a
    * $x < a$: $x$
    * $x > a$: 0
  - Odd is not better off choosing some other offer.

There is a set of Nash equilibrium of the ultimatum game such that $\{x = a, y = a\}$, where $x$ represents Odd's offer and $y$ represents Even's cutoff. Even will accept any offer below $y$ and reject any offer above $y$.

So Nash equilibrium predicts any outcome. That doesn't seem particularly useful nor does it seem intuitive. Do you think Even would really have a strategy that says he will accept any offer below a 10 percent share of the pie?

### 8.2.5 Subgame Perfection

So Nash equilibrium predicts any thing can happen. What about **subgame perfection**? Remember the definition. An outcome is subgame perfect if it is a Nash equilibrium of every subgame. Our ultimatum game has two (sort of) subgames. There is a subgame after Odd makes the offer $x$ and Even observes the offer. This subgame has one player, Even, and the strategy is just the action accept or reject. The other subgame is the whole game.

What is the Nash equilibrium of the first subgame? Let Odd offer $x$. Even is best off accepting that offer. If Even accepts he gets $1 - x$, while if he rejects he gets 0. Even is always at least weakly better off accepting Odd's offer.

So going back to the whole game the only strategy of Even's that is subgame perfect is for $y = 1$. That is Even accepts any offer Odd makes. Given that Even will accept any offer, Odd is best off offering $x = 1$.

So subgame perfection substantially reduces the number of outcomes that are an equilibrium. Now we just have one. It is for Odd to get the whole pie! Does that seem reasonable?

## 8.3 Empirical Analysis: Ultimatum Game in India using R

The ultimatum game may be one of the most studied games in economic experiments. How people play the ultimatum game give us an interesting view into different cultures. What do people think is a fair split? Will people reject splits that they do not think is fair even if it means giving up real money? Do the stakes matter? In particular, do the so called "fair" offers go away when the players are playing with real money? Is subgame perfection actually a better prediction than you thought?

The section analyzes data from an experiment conducted in northern India where the experimenters vary the stakes.

### 8.3.1 Data

These data are replication data for Andersen et al. (2011). The authors went into a very poor northern Indian village and ran an experiment where villagers played for real money. In terms of hours worked equivalent, the size of the pie varied from \$30 to \$48,000 (assuming \$20 per hour). So yes, real money.

Figure 8.2 presents the density of the offer percentages. These are the amounts that Even receives if he accepts Odd's offer. The modal offer is the 50-50 split and it is actually very rare to offer more than that! The proposer seems to be unwilling to give the responder more than half the pie. That said, there is quite a lot of density less than 50–50 and even some that is pretty low.

```
> file = paste0(dir, "20100982_DATA.dta")
> read.dta(file) |>
+   ggplot(aes(x = percent_offer)) +
+   geom_density(fill = "gray", alpha = 0.5) +
+   geom_vline(xintercept = 0.5, col = "gray") +
+   labs(title = "Density of Offer Shares",
+        x = "Offer Share Responder Receives",
+        y = "") +
+   # remove y-axis
+   theme(axis.text.y=element_blank(),
+         axis.ticks.y=element_blank())
```

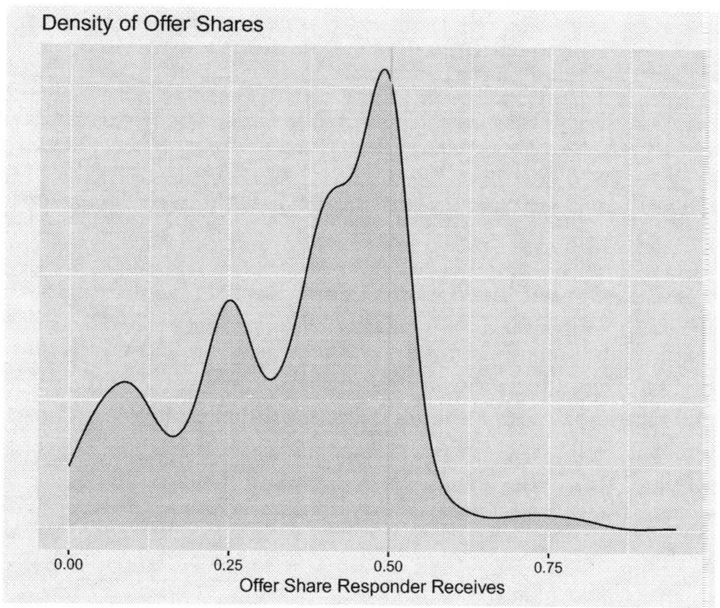


**FIGURE 8.2**
Density plot of share of the pie offered by the proposer (Odd) to the responder (Even). The bulk of the offers are just below half of the pie going to the responder.

Some responders make pretty lower offers.

### 8.3.2 Equilibrium Play or Fairness?

We see in Figure 8.2 that a large number of offers are around the 50% mark. Is this just the proposers playing fair or is it part of the equilibrium behavior?

That is, do the proposers believe that if they make an offer below the 50–50 split, it will be rejected?

While we observe the proposer's strategy, we don't observe the strategy for the responder. We could estimate it. If we assume all responders are the same, then we can estimate a model of their strategy.

Assume that we can represent the responder's strategy using a logit. We use `glm()` to estimate the logit. We can determine the probability of accepting the offer as a function of the percent of the pie offered. In order to plot it out, we can determine the predicted probability of accepting the offer for each actual offer we observe in the data.

```
> file = paste0(dir, "20100982_DATA.dta")
> data = read.dta(file)
> glm1 = glm(accept ~ percent_offer,
+            data = data,
+            family = binomial(link = "logit"))
> data = data |>
+   mutate(
+     accept_pred = predict.glm(glm1, type = c("response"))
+   )
```

Figure 8.3 looks nothing like what we would expect an equilibrium strategy to look like. If the strategy is Reject if $x < 0.5$ and Accept if $x \geq 0.5$, then we would expect an S-shape, going to 0 when offers are close to 0 and close to 1 when offers are close to 1, with a cross around 0.50. These players are accepting offers that are much lower than the equilibrium predicts. Similarly, it is not consistent with subgame perfection. In that case, we would expect something like a straight horizontal line at 1.00. There are many more rejections than we would expect if the strategies were part of a subgame perfect Nash equilibrium.

### 8.3.3 Do Stakes Matter?

What do things look like for a subset of the data where there are very large stakes? Let's restrict the data to when the stakes are 20,000 rupees. In this case, the average offer is just 12 percent of the pie and for the average person the pie is worth 24 days of work.

We can filter the data to only include the large stakes games and then estimate the predicted acceptance rate for these games.

```
> data = data |>
+   filter(
+     stakes_4 == 1
+   )
> glm2 = glm(accept ~ percent_offer,
+            data = data,
+            family = binomial(link = "logit"))
```

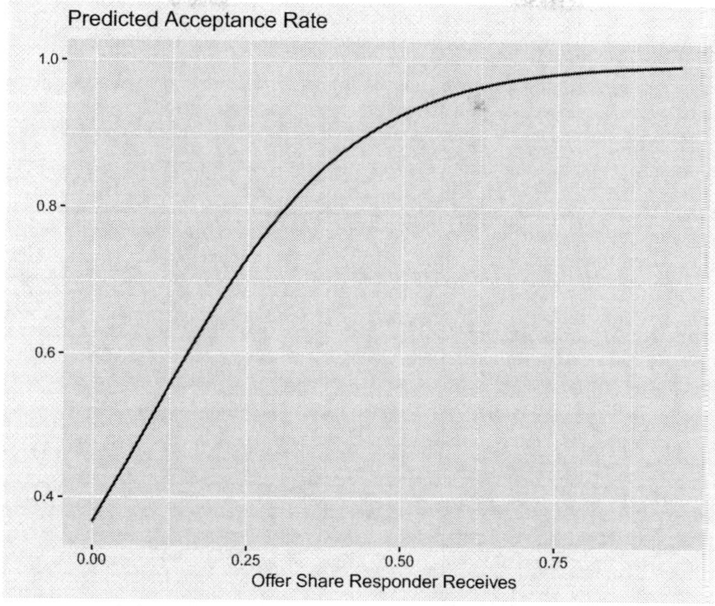


**FIGURE 8.3**
Plot of predicted acceptance rate as a function of the actual offer shares. The predicted acceptance percentage grows to approach 100 as the offer gets closer to 50 percent.

```
> data = data |>
+     mutate(
+       accept_pred = 7.5*predict.glm(glm2, type = c("response"))
+     )
```

Here is the code to create a `ggplot()` object that shows the density of offers and the predicted acceptance rate for the subset of experiments with large stakes.

```
> ggplot_pred_accept_hs = data |>
+   ggplot() +
+   geom_density(aes(x = percent_offer),
+                fill = "gray",
+                alpha = 0.5) +
+   geom_smooth(aes(x = percent_offer,
+                   y = accept_pred),
+               se = FALSE) +
+   scale_y_continuous(breaks = seq(0, 7.5, by = 7.5/2),
+                      labels = seq(0, 100, by = 50)) +
+   geom_text(aes(x = 0.2, y = 9,
```

```
+                  label = "Predicted Acceptance Rate (percent)"),
+              color = "gray") +
+   geom_text(aes(x = 0.4, y = 2,
+                  label = "Density of Offer Shares"),
+              color = "gray") +
+   labs(title = "",
+        x = "Offer Share to Responder",
+        y = "")

> ggplot_pred_accept_hs
```

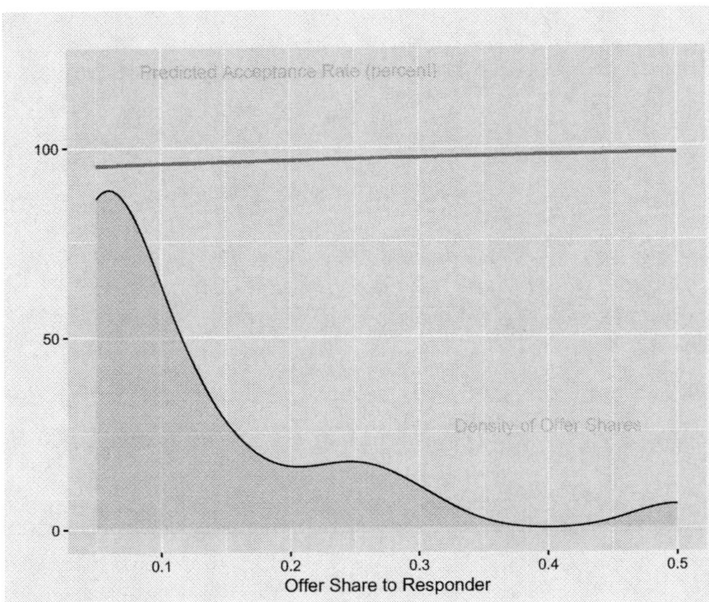


**FIGURE 8.4**
Density plot of offer shares and predicted acceptance rate for the subset of experiments with large stakes. The proposers makes offers that tend to be close to zero. The receiver's predicted acceptance rate is close to 100 percent.

Figure 8.4 presents the density of offers and the predicted acceptance rates. This graph is consistent with the subgame perfect Nash equilibrium. The weight of offers is 0, and the predicted acceptance rate is consistent with a strategy of accepting any offer.

Do stakes matter? Yes. Apparently they do. When they do, subgame perfection predicts the likely outcome of the game. Is that what you would have thought?

## 8.4 Two Period Alternating Offers Game

It is nice that the very strong prediction of subgame perfection in the ultimatum game can be born out in some actual games. But it is not that satisfying. Most of the results of the experiment are not consistent with either subgame perfection or Nash equilibrium. Nor are the predictions of the game consistent with our intuition. Why is the 50–50 split so dominant? It is only one of many predictions of the game.

If the simple model is doing a poor job of predicting outcomes of interest, then one solution is a more complicated slash realistic model. This section considers what happens when the model is made more realistic slash more complicated. An alternating offers game is one in which the two players take it in turns to make the offer, where the game only ends if the offer is accepted.

The section presents the game, the extensive form representation and finds the subgame perfect Nash equilibrium.

### 8.4.1 The Game

We formally represent the game with some compact notation. Compact is code for confusing. As before $x_t$ refers to the share that Odd gets, while Even gets $1 - x_t$. The $t \in \{1, 2\}$ refers to which period we are in, the first or second. The $y_t$ refers to the cutoff strategy. It's exact meaning depends on who the responder is. If the responder is Even ($y_1$) means that Even will accept any offer where Odd gets less than $y_1$.

In the second period, Odd and Even switch roles. The $x_2$ is still the amount that Odd gets, but it is Even that is making the proposal. This may be a function of the offer made by Odd in the first period. It is also a function of whether the offer was accepted or rejected but given that we only see this strategy if it was rejected we can ignore that part of the history. The $y_2$ refer to Odd's cutoff strategy. This means that Odd will REJECT any offer less than $y_2$. Remember we are always talking about the share that Odd gets. This is also a function of Odd's offer in period 1.

- Players: Odd, Even
- Strategies:
  - Odd: $\{x_1, y_2(x_1)\}$
  - Even: $\{y_1, x_2(x_1)\}$
- Payoffs
  - $x_1 < y_1$: $\{x_1, 1 - x_1\}$
  - $x_1 > y_1$, $x_2 > y_2$: $\{rx_2, r(1 - x_2)\}$

- $x_1 > y_1$, $x_2 < y_2$: $\{0, 0\}$

where $x_t$ is the offer in period $t$ and the amount received by Odd, $y_t$ is the cutoff amount in period $t$.

The probability $r$ is meant to capture the possibility that the parties risk failure by extending the negotiations. We will see that the size of this risk bears heavily on the negotiated outcome.

### 8.4.2 Game Tree

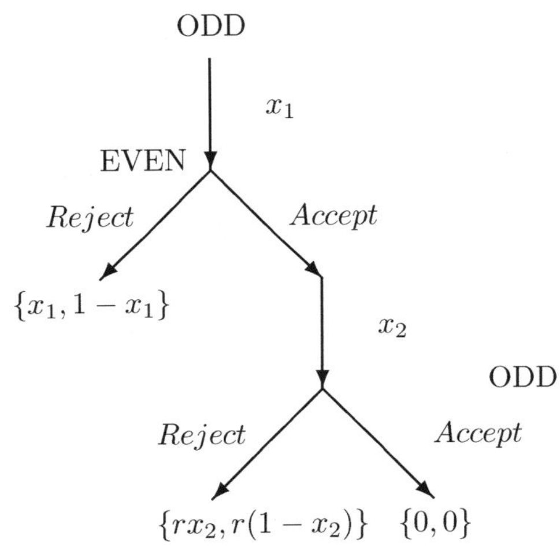


**FIGURE 8.5**
Two period alternating offers game. Odd makes first offer of $x_1$. If rejected, Even makes an offer of $x_2$. With probability $r$ the game continues. Odd's payoff is listed first in brackets.

Figure 8.5 presents the extensive form representation of the two-period alternating offers game. The size of the pie decreases from the first period to the second period. The amount of the decrease is determined by $r$.

### 8.4.3 Subgame Perfection

The second period is an ultimatum game with Even as the proposer and Odd as the responder. Given this, $y_2 = 0$ and $x_2 = 0$. That is, Odd will accept any offer and Even will offer 0.

Moving back to the first period, we can take the second period equilibrium as given. If Even accepts, then he gets $1 - x_1$ and if he rejects he gets $r(1 - x_2) = r$, where $r$ is the probability that the second period occurs. Given this, Odd will make an offer such that $1 - x_1 = r$ or $x_1 = 1 - r$.

In this case, the subgame perfect equilibrium is

$$x_1 = 1 - r, y_2 = 0, y_1 = 1 - r, x_2 = 0 \tag{8.1}$$

The outcome we actually see is an offer of $x_1 = 1 - r$. So having this probability that the game ends prior to the second period changes the outcome of the game. Instead of Odd getting the whole pie, they get the pie less a portion equal to the probability that the game continues to the next period

## 8.5 Infinite Alternating Offers Game

While we see some evidence that the subgame perfect Nash equilibrium does occur when real people play the ultimatum game, it is not clear that real people actually play the ultimatum game. Maybe there is some other game that more accurately represents what is happening when people bargain.

We saw in the previous section that adding both a second period and the probability that the game ends prior to the second period changes the outcomes. What happens if we add even more periods? What happens if we add an infinite number of periods?

In acknowledgement of the seminal contribution by Israeli economist, Ariel Rubinstein, we generally refer to this as the **Rubinstein bargaining model**.

The section presents the game, the extensive form representation, finds the subgame perfect Nash equilibrium and presents an algorithm for finding that outcome.

### 8.5.1 The Game

The game is as before, but now with an infinite number of periods. In each period, there is a proposer and a responder. If the period is odd, then the proposer is Odd, while the proposer is Even if the period is even. As before, the responder can either accept or reject the offer. If the responder accepts, the game ends and the players get the payoffs $\{x_t, 1 - x_t\}$. If the responder rejects the offer, then the game goes to the next period and the proposer and responder swap roles.

Similar to the previous game, the size of the pie changes over time. The parameter $r \in (0, 1)$ represents the change in the size of the pie from period to period. We can think of it as standard financial discount rate. Alternatively, we could think about it as representing the probability that the game will continue.

- Players: Odd, Even
- Strategies:

- Odd
  * If the time period $t$ is odd, then given the history of offers up to time period $t-1$, offer $x_t$ to Even.
  * If the time period $t$ is even, then accept or reject Even's offer based on $x_t$ and the history of offers up to $t-1$.
- Even
  * If the time period $t$ is even, then given the history of offers up to time period $t-1$, offer $x_t$ to Odd.
  * If the time period $t$ is odd, then accept or reject Odd's offer based on $x_t$ and the history of offers up to $t-1$.

- Payoffs
  - If at time $t$ the an offer $x_t$ is accepted:
    * Odd: $0, 0, 0, ..., r^t x_t$
    * Even: $0, 0, 0, ..., r^t(1-x_t)$
  - If at time $t$ the offer $x_t$ is rejected and the offer $x_s$ is accepted in periods $s > t$.
    * Odd: $0, 0, 0, ..., 0, ..., r^s x_s$
    * Even: $0, 0, 0, ..., 0, ..., r^s(1-x_s)$

### 8.5.2 Game Tree

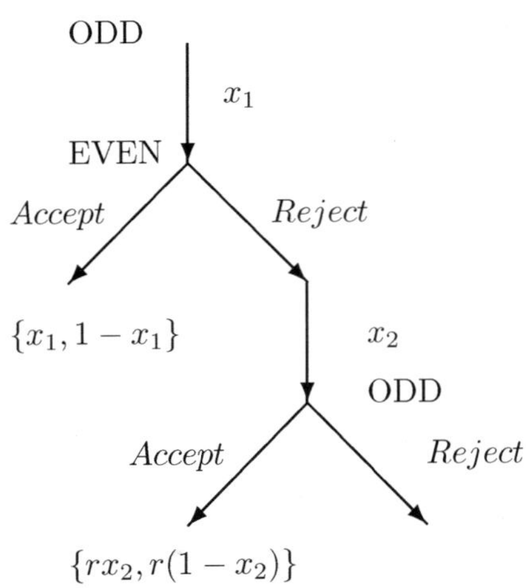


**FIGURE 8.6**
The first two periods of an infinite period alternating offers game.

Figure 8.6 presents the first two periods of an infinite period alternating offers game. In each period, the size of the pie decreases in proportion to the discount rate $r$.

### 8.5.3 Subgame Perfection

So we know how to solve for the subgame perfect Nash equilibrium. Simply go to the last period, work out the equilibrium for that game and then work your way backwards. OK, but what if there is no last period? What if the game has an infinite number of periods?

The standard solution is to approximate our infinite period game with a finite period game. Consider a game that ends at period $T$ (assume odd).

The last period is an ultimatum game where Odd offers $x_T$ and Even observes the offer and chooses whether to accept or reject. In the subgame where Odd has made the offer of $x_T$, Even's payoffs are

- Accept: $1 - x_T$
- Reject: 0

Even will accept any offer where $1 - x_T \geq 0$ (indifference assume Even accepts). Working backwards, Odd will choose the $x_T$ that is as small as possible, Odd offers $x_T = 1$. If the game gets to period $T$ the payoffs are $\{1, 0\}$.

Now let's do $T - 1$. This is an ultimatum game where Even makes the offer and Odd chooses whether or not to accept or reject. Even offers Odd $x_{T-1}$. Odd's payoffs are:

- Accept: $x_{T-1}$
- Reject: $r \times 1$

where $r$ is how much Odd discounts the future. If Odd rejects, she gets nothing immediately, but in one period she knows that she will get the whole pie of 1. But that pie gets discounted in proportion to $r$.

Odd's best response is to accept any offer such that $x_{T-1} \geq r$ and reject otherwise. Even knows this and wants to make $x_{T-1}$ as small as possible but still have Odd accept the offer. That is where $x_{T-1} = r$. If the game gets to $T - 1$, then the payoffs are $\{r, 1 - r\}$.

Now consider $T - 2$. This an ultimatum game where Odd makes an offer of $x_{T-2}$ to Even and Even decides to accept or reject. Even's payoffs are

- Accept: $1 - x_{T-2}$
- Reject: $r \times (1 - r) = r - r^2$.

Therefore, Even will accept any offer $1 - x_{T-2} \geq r - r^2$. Odd wants to make $x_{T-2}$ as large as possible, so they will choose $x_{T-2} = 1 - r + r^2$. If the game gets to $T - 2$ then the payoffs are $\{1 - r + r^2, r - r^2\}$.

If we let $r = 0.9$, then the payoffs if the game gets to $T$ are $\{1, 0\}$, if it gets to $T - 1$ they are $\{0.9, 0.1\}$ and if it gets to $T - 2$ the payoffs are $\{0.91, 0.09\}$

What happens in $T - 3$?

### 8.5.4 Game Ends in Period 1

Working all the way back to Period 1, Odd makes an offer $x_1$ and Even decides to Accept or Reject. In the game with $T$ periods, the offer of $x_1$ will be the amazingly complicated thing with lots of $r$s. However, as $T$ gets very large $x_1$ converges to 0.5. That is, the subgame perfect Nash equilibrium of the game is a 50-50 split! It is a super complicated game that makes a very simple and intuitive prediction.

### 8.5.5 Infinite Alternating Offers Game in R

We can use the computer to analyze more complicated games than what we looked at above. We can allow the two players to have different beliefs about when the game is going to end and different payoffs if the parties fail to reach a bargain.

Odd makes an offer at time $t$, assume that the next period the game ends in agreement and Even gets a payoff of $(1 - x_{t+1})V_A$, where $V_A$ is the size of the pie if the offer is accepted. Also assume that Even discounts the future by $r_E$. In addition, assume that if there is no agreement, this period Even gets $v_{EN}$.

By having different discount rates and non-agreement values, we can get different bargaining outcomes. You can see that what looks to be small differences lead to quite large differences in the bargaining outcomes.

What offer should Odd make?

Even will accept Odd's offer if and only if

$$1 - x_t \geq v_{EN} + r_E(1 - x_{t+1})V_A \tag{8.2}$$

Assume that the offer is such that the payoff for the responder makes them indifferent between accepting or rejecting the offer. In the function below `x` is the proportion of the pie received by the responder, `V_A` is the size of the pie, `v_N` is the period amount the responder gets if they do not accept the offer and `r` is the discount rate.

```
> offer = function(x, V_A, v_N, r) {
+    v_N + r*x*V_A
+ }
> T = 250
> r_odd = 0.99
> r_even = 0.98
> V_A = 1
> v_EN = 0
> v_ON = 0.002
```

Given the parameter values above the following loop determines the equilibrium of the game.

```
> odd_offers = rep(NA, T)
> even_offers = rep(NA, T)
> odd_offer_old = 0
> for (i in 1:T) {
+   even_offer_old = offer(1 - odd_offer_old, V_A, v_ON, r_odd)
+   odd_offer_old = offer(1 - even_offer_old, V_A, v_EN, r_even)
+   odd_offers[i] = odd_offer_old
+   even_offers[i] = even_offer_old
+   #print(odd_offer_old)
+ }
```

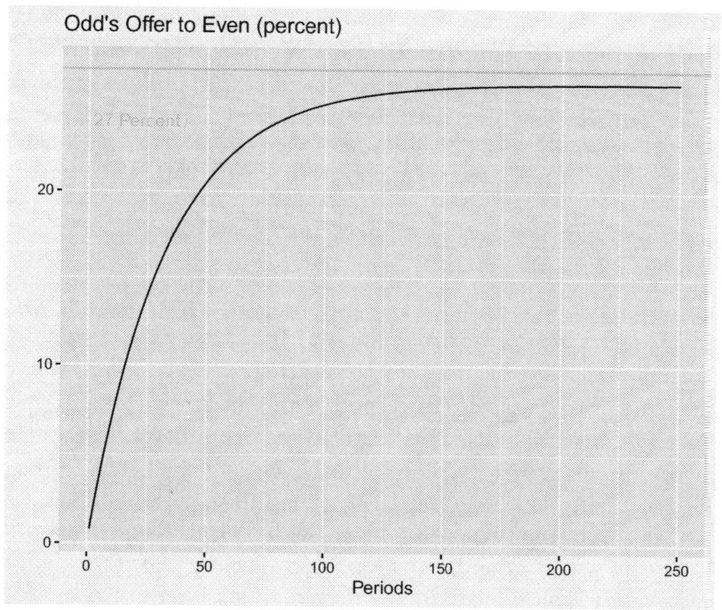


**FIGURE 8.7**
Line chart of offers to Even as the number of periods gets large. It shows that in equilibrium the offer to Even approaches 27 percent.

Figure 8.7 shows that the equilibrium offer to Even converges to 0.27, with Odd receiving 0.73 of the pie. Why is this split not even? Why is it in favor of Odd? What change could you make to get it more of an even split?

## 8.6 Nash Bargaining Model

John Nash developed one of the most important ideas in modern game theory - every (finite) game has what we now call a Nash equilibrium. It has an outcome which is "stable" in the sense that no player would want to deviate from that strategy if they knew which strategies all the other players were playing.

Nash was interested in another problem, how to determine the outcome when entities bargain. While the Nash equilibrium became the keystone concept in non-cooperative game theory, Nash himself was not able to work out how to model bargaining as a non-cooperative game. Instead he developed an alternative framework for analyzing bargaining problems known as the Nash bargaining model. The parameterization of the model presented below gives same split of the pie as the infinite period alternating offers game above. This suggests an equivalence between the two models.

The section presents the Nash bargaining model and shows how it is used to analyze competition between hospitals.

### 8.6.1 The Model

Consider a game where we have two players, say a hospital and an insurance company. The two firms are bargaining over how to pay the hospital for various services that the hospital provides to the insurance companies beneficiaries. The price that the hospital will receive depends on two sets of things. First, it depends upon what both the hospital and the insurance company get if negotiations break down. If there is only one hospital in an area, then the insurance company is not going to be able offer its beneficiaries much of a product if it can't come to a deal with the hospital. If most of the people in the area work for the same firm and are covered by the same insurance then demand for the hospital will drop dramatically if the hospital can't come to a deal with the insurance company. Second, it depends on how good each side is at bargaining, which we will conceptualize as the relative "bargaining weights." These are somewhat amorphous. Practically, these weights are often set to be equal. Below we look at how these weights relate to the $r$s used in the alternating offers game presented earlier.

$$\max_x \quad (x-a)^{\lambda}(1-x-b)^{1-\lambda} \tag{8.3}$$

where $a$ and $b$ are the alternative outcomes (the outcome if bargaining fails) for Odd and Even, respectively, $x$ is the share of the pie that goes to Odd and $\lambda$ is the bargaining weight.

Taking first-order conditions and simplifying.

$$\begin{aligned}
&\lambda(x-a)^{\lambda-1}(1-x-b)^{1-\lambda} - (1-\lambda)(x-a)^{\lambda}(1-x-b)^{-\lambda} = 0 \\
&\lambda(1-x-b) - (x-a)(1-\lambda) = 0 \\
&x = \lambda(1-b) + (1-\lambda)a
\end{aligned} \tag{8.4}$$

We see that Odd's share is increasing in the size of the alternative and decreasing in the size of Even's alternative. The size of Odd's share depends on how much Odd has to lose. If Odd's alternative is good, $a$ is large, then Even will need to offer Odd more to have her accept the bargain.

**Definition 14.** *A* Nash bargaining model *is an algorithm for determining the outcome from bargaining based on the player's payoffs from agreement and disagreement and from their relative bargaining weights.*

### 8.6.2 Nash Solution using R

Let's set up the problem so that it is equivalent to the infinite alternating offer model analyzed above.

$$max_x \quad (xV_A - v_{ON})^{\lambda}((1-x)V_A - v_{EN})^{(1-\lambda)} \tag{8.5}$$

where $V_A$ is the value of the agreement, $v_{ON}$ is the value to Odd if there is no agreement, and $v_{EN}$ is the same for Even, $x$ is the proportion of the pie that Odd receives and $\lambda$ is Odd's bargaining weight.

We can create a little numerical version of the model. You can see how things change when you change various parameters.

```
> lambda = 0.74
> nash_value = function(x) {
+    ((x*V_A - v_ON)^lambda)*(((1 - x)*V_A - v_EN)^(1-lambda))
+ }
> optimize(nash_value, c(0, 1), maximum = TRUE)
  $maximum
  [1] 0.7405195

  $objective
  [1] 0.5626717
```

The resulting share is similar to the results of the alternating offers game analyzed above. There is an equivalence between the bargaining weights in this model and the relative difference in the $r$s in the infinite period alternating offers model.

## 8.7 Modeling Hospital Competition and Pricing

How do we work out the effect of hospital mergers on prices when most of the customers don't pay anything or just a small fraction of the actual cost of the services? Insurers pay numbers closers to the actual costs but insurers don't really use the services. A solution is to use the Nash bargaining model to estimate how much the insurer will pay given the choices made by the insurer's beneficiaries.

In the early 2000s, economists of the FTC and in academia began rethinking how pricing worked in the hospital market. They realized that while hospitals provided services to patients, patients were not the ones that determined prices. Prices for hospital services are determined by the interaction of large hospitals bargaining with large insurers.

The section shows how the Nash bargaining model can be used to analyze hospital competition.

### 8.7.1 Bargaining Model

Consider a hospital and an insurer bargaining over the price of services ($p$). We have simplified things by assuming that the hospital only has one price and the insurer only has one set of beneficiaries. The insurer's payoff from a successful negotiation is just the value of the hospital to the insurer's beneficiaries ($v(h)$) minus the price paid to the hospital ($p$). The hospital's payoff is the price paid by the insurer ($p$) times the number of beneficiaries ($q$). If the bargaining fails, then the hospital gets no revenue from the insurer and the insurer gets the value of the alternative hospital to insurer's beneficiaries is $v(h')$ and the insurer pays $p'$.

- Successful:
  - Hospital: $pq$
  - Insurer: $(v(h) - p)q$
- Failure:
  - Hospital: 0
  - Insurer: $(v(h') - p')q$

The solution to the Nash bargaining model is as follows.

$$\max_p \quad (pq)^\lambda((v(h) - v(h') - (p - p'))q)^{1-\lambda} \tag{8.6}$$

For the hospital, the value of agreement is seeing the insurer's beneficiaries. For the insurer, it is the value to their beneficiaries of going to that hospital relative to the alternative less the relative price.

The first-order condition gives the following result. Let $A = pq$ and $B = (v(h) - v(h') - (p - p'))q$.

$$\begin{aligned} &\lambda A^{\lambda-1} q B^{1-\lambda} - (1-\lambda) A^\lambda B^{-\lambda} q = 0 \\ &\lambda A^{-1} B - (1-\lambda) = 0 \\ &\lambda B - (1-\lambda) A = 0 \end{aligned} \tag{8.7}$$

Substituting in the definitions of $A$ and $B$, we get the following.

$$\begin{aligned}&\lambda(v(h) - v(h') - (p - p'))q - (1 - \lambda)pq = 0\\ &p = \lambda(v(h) - v(h') + p')\end{aligned} \tag{8.8}$$

The price depends on the incremental value of the hospital to the insurer's beneficiaries, $(v(h) - v(h'))$, the price of the alternative hospital $(p')$ and the bargaining weight of the hospital $(\lambda)$.

To determine the market price, we need to estimate the incremental value of the hospital.

### 8.7.2 Demand for Hospitals

We don't know a beneficiary's value for a hospital but we can know their revealed preference. We used the same idea in Chapter 4 when looking at the choice of bookstores to enter a market.

A particular person will choose the hospital if the following inequality holds.

$$v_i(h) - v_i(h') > 0 \tag{8.9}$$

Our standard demand model replaces the inequality with the following. Again, this is what we did in Chapter 4. The hospital and the alternative hospital have observed characteristics $\mathbf{X}_h$ and $\mathbf{X}_{h'}$ respectively. The individual values characteristics according to weighting vector $\beta$. The unobserved characteristics are $\xi_h$ and $\xi_{h'}$ respectively.

$$(\mathbf{X}_h - \mathbf{X}_{h'})'\beta + \xi_h - \xi_{h'} > 0 \tag{8.10}$$

Under certain assumptions on $\xi$, we get the logit form for the probability that an individual will choose hospital $h$.

$$s_h = \frac{\exp(\delta_h)}{1 + \exp(\delta_h)} \tag{8.11}$$

where $\delta_h = (\mathbf{X}_h - \mathbf{X}_{h'})\beta + \xi_h - \xi_{h'}$.

The beneficiary often pays close to nothing for hospital services, so we generally ignore the beneficiaries out of pocket expenses to simplify the problem.

### 8.7.3 Willingness-to-Pay

From our demand set up and what we know from the bargaining model, we can calculate what an insurance company would be willing to pay to keep a hospital in network. That is, keep the hospital available to its beneficiaries at the discount prices. A common measure is called willingness-to-pay (WTP), equal to $-\log(1 - s_h)$. That is, an insurance company is willing to pay a lot more for the hospital when its beneficiaries are not willing to go to any other hospital.

Why such a weird formula? Remember above we have Equation (8.7), which states the price should be equal to the difference in the relative value of the hospital and the next best alternative plus the price of the next best alternative. Let's not worry about the last part. What is $v(h) - v(h')$? Let's make things simpler and assume that $v(h')$ is just the outside option.

The first thing to note is in our logit world with our assumptions on the distribution of the unobserved values, $v(h) = \log(\exp(\delta_h) + 1)$. Why? Another excellent question. This formula is the expected value of optimal choice.[2] As $h'$ is the outside option, $v(h') = \log(1) = 0$. It is the expected value when hospital $h$ is removed as an option. So $v(h) - v(h') = \log(\exp(\delta_h) + 1)$.

Now, Equation (8.11) tells us what $\exp(\delta_h)$ is. Rearranging that equation, we have the following relationship between it and the **diversion ratio** to the hospital.[3]

$$\begin{aligned} s_h(1 + \exp(\delta_h)) &= \exp(\delta_h) \\ \exp(\delta_h)(1 - s_h) &= s_h \\ \exp(\delta_h) &= \tfrac{s_h}{1-s_h} \end{aligned} \tag{8.12}$$

Plugging this back in we get our WTP formula.

$$\begin{aligned} v(h) - v(h') &= \log(\tfrac{s_h}{1-s_h} + 1) = \log(\tfrac{s_h+1-s_h}{1-s_h}) = \log(\tfrac{1}{1-s_h}) \\ &= -\log(1 - s_h) \end{aligned} \tag{8.13}$$

Now we have a measure of how much the insurer is willing to pay for the hospital as a function of stuff that we observe in the data.

This simple measure has turned out to be an extremely good predictor of what actually happens in hospital markets. We generally find that a 10% increase in WTP is associated with a 2% increase in hospital prices.[4] Below we will estimate this relationship on data from Florida hospitals.

## 8.8 Empirical Analysis: Hospital Competition with R

By the early 2000s, the two federal antitrust agencies, the FTC and DOJ, had an impressive string of losses in hospital merger enforcement. The FTC's Republican Chairman, Tim Muris, decided to put a large amount of the commission's resources in turning the record around. While most of this work was legal analysis, the FTC's Bureau of Economics became heavily involved in the effort. A lot of the important work in modeling hospital mergers and measuring their potential impact has been done by economists who have been in the Bureau.

[2]See Capps et al. (2003).

[3]The diversion ratio is the share of people who go to the hospital over the share of people who go to all the other options in the market.

[4]See Bob Town's expert report in a Virginia hospital merger case, https://www.vdh.virginia.gov/content/uploads/sites/96/2017/10/Expert-Report-of-Robert-Town.pdf.

The section uses publicly available data from Florida, the US Census Bureau and the Centers for Medicaid and Medicare Services (CMS) to analyze demand for hospitals and uses the parameter estimates to simulate a merger in Palm Beach County.

### 8.8.1 Data

The section uses publicly available discharge from Florida for 2018. These data provide detailed information on discharges from Florida's hospitals at various demographics and conditions. The analysis uses the demographic data and assume demand is the same across conditions. A more standard analysis would group by condition ("DRG") as well. In addition to this, it is usual in hospital merger cases to have zip code level data. Here we only know the county where the hospital is. We assume that everyone discharged from the hospital lives in the same county and people in the county only choose between hospitals in the county.

Added to this information, we will use the American Community Survey data to determine counts of people in the various demographic groups at the county level. We assume that people who do not visit one of the county's hospitals in a particular year choose the outside option.

WTP is calculated using the observed shares at the demographic levels by county.

To construct these measures, we use weights. These weights are the importance of the group to the hospital. That is if a hospital specializes in a particular disease then that will be captured by the weighting and may lead to a high price even if the hospital doesn't have a large share of the market more generally. For example, the hospital may have a low market share overall but high market share for child birth. In that case, the market share for women aged 25 to 54 will be a lot more important than the same hospital's market share for men 55 or older. In the data we have 6 demographic groups. We label things `_fw` for firm weight and `_cw` for county weight. As we go through each case we find the importance of that demographic to the hospital then we find the share of that demographic in the market for the hospital. Last we calculate the WTP and the hospital share where the weights are the importance of the demographic to the hospital.

```
> ages = c("0_24", "25_54", "55")
> genders = c("female", "male")
> file = paste0(dir, "hospitals.csv")
> df = read.csv(file)[,-1]
> df$WTP = 0
> df$share = 0
> for(i in 1:length(ages)) {
+   for(j in 1:length(genders)) {
+     col = colnames(df)==paste("Discharges_",
```

```
+                                  ages[i],"_",
+                                  genders[j],"_fw",sep="")
+     weight = df[,col]
+     col = colnames(df)==paste0("Discharges_",
+                                  ages[i],"_",
+                                  genders[j],"_cw")
+     share = df[,col]
+     df$WTP = ifelse(share==1, df$WTP, df$WTP -
+                            weight*log(1 - share))
+     df$share = df$share + weight*share
+   }
+ }
```

Lastly, we match the data above to hospital cost and pricing reports from the Centers for Medicare and Medicaid (CMS). Not all the hospitals match and so the analysis is limited to the cases where we have matches across the discharge data and the cost reporting data.

### 8.8.2 Pricing and WTP

Our analysis is limited to estimating the effect of mergers on WTP. The measure could be used in its own right for merger review, similar to the way Herfindahl-Hirschman Index (HHI) or Upward Pricing Pressure (UPP) is used.[5]

Usually in merger cases some sort of relationship between prices and WTP is presented. We combine estimates of WTP and share above with information about prices and costs from CMS for Florida hospitals in 2018. The code brings in the data which combines information on Florida hospitals with prices, demographic and competition measures. It then runs two linear regressions, price on WTP and price on share of market.

```
> require(data.table)
> file = paste0(dir, "hospital3.csv")
> dt = fread(file)


> lm1 = lm(Price ~ Wages + Beds + WTP, data = dt)
> lm2 = lm(Price ~ Wages + Beds + share, data = dt)
```

Table 8.2 presents the linear regressions of price on measures of competition, WTP and share. Both regressions show that there is a positive relationship, although there is a lot of uncertainty. The estimated elasticity of 0.2 between WTP and price seems to be consistent with other estimates.

[5]See the 2010 Horizontal Merger Guidelines, https://www.justice.gov/atr/horizontal-merger-guidelines-08192010 accessed August 5 2023.

**TABLE 8.2**
Linear regression estimates of the relationship of price on WTP and share using data from Florida hospitals for 2018.

| | *Dependent variable:* | |
|---|---|---|
| | Price | |
| | (1) | (2) |
| Wages | 0.048 | 0.047 |
| | (0.058) | (0.058) |
| Beds | 0.010 | 0.007 |
| | (0.067) | (0.067) |
| WTP | 0.223 | |
| | (0.141) | |
| share | | 0.357 |
| | | (0.221) |
| Constant | 11.288*** | 11.301*** |
| | (0.468) | (0.468) |
| Observations | 106 | 106 |
| $R^2$ | 0.034 | 0.035 |
| *Note:* | *p<0.1; **p<0.05; ***p<0.01 | |

### 8.8.3 Mergers and Willingness To Pay

Consider a hospital merger in the home county for this publisher, Palm Beach County. The hospitals are Bethesda East, Bethesda West, and Boca Raton Regional. To estimate the effect of the merger, we recalculate the WTP for the combined hospital. The code is the same as above but recalculated just for the new merged firm.

```
> df2 = df
> merger = c("BETHESDA HOSPITAL EAST", "BETHESDA HOSPITAL WEST",
+            "BOCA RATON REGIONAL HOSPITAL")
> df$merge = ifelse(df$Hospital.Name %in% merger, 1, 0)
> df2$merge = ifelse(df2$Hospital.Name %in% merger, 1, 0)
> df2[df2$merge==1,]$Discharges =
+   sum(df2[df2$merge==1,]$Discharges)
> for(i in 1:length(ages)) {
+   for(j in 1:length(genders)) {
+     col = which(colnames(df)==paste0("Discharges_",
+                                      ages[i],"_",
+                                      genders[j],"_fw"))
+     df2[df2$merge==1,col] = sum(df2[df2$merge==1,col])
+     col = which(colnames(df)==paste0("Discharges_",
+                                      ages[i],"_",
+                                      genders[j],"_cw"))
+     df2[df2$merge==1,col] = sum(df2[df2$merge==1,col])
+   }
+ }
```

It then creates a new data where the shares are passed from the merging hospitals to the new hospital. We drop the merged hospitals from the data.

```
> df3 = df2[-which(df2$Hospital.Name %in%
+                  c("BETHESDA HOSPITAL EAST",
+                    "BETHESDA HOSPITAL WEST")), ]
```

Given the post-merger data, the WTP and share can be recalculated.

```
> df3$WTP = 0
> df3$share = 0
> l = 1
> for(i in 1:length(ages)) {
+   for(j in 1:length(genders)) {
+     col = which(colnames(df3)==paste0("Discharges_",
+                                       ages[i],"_",
+                                       genders[j],"_fw"))
+     weight = df3[,col]
+     col = which(colnames(df3)==paste0("Discharges_",
```

```
+                                    ages[i],"_",
+                                    genders[j],"_cw"))
+      share = df3[,col]
+      df3$WTP = ifelse(share==1, df3$WTP,
+                          df3$WTP - weight*log(1 - share))
+      df3$share = df3$share + weight*share
+      l = l + 1
+    }
+ }
```

Now we can determine the impact of the merger on prices using the WTP change caused by the merger on the price in Palm Beach County.

```
> index_merge = which(df$merge==1)
> index3_merge = which(df3$merge==1)
> a = sum(df$WTP[index_merge], na.rm = TRUE)
> b = sum(df3$WTP[index3_merge], na.rm = TRUE)
> c = (b - a)/a
> lm1$coefficients[4]*c*
+   mean(exp(dt$Price[dt$county.x == "palm beach"]),
+                            na.rm = TRUE)
      WTP
 53583.48
> lm1$coefficients[4]*c
       WTP
 0.4429545
```

This analysis suggests that a merger between these hospitals in Palm Beach County will have a substantive effect on price, a 44 percent increase or $53,583.48 per discharge.[6] Across the Florida hospitals in the sample, the price goes up $333 per discharge or 0.3 of a percent.

## 8.9 Discussion and Further Reading

The ultimatum game is one of the most common games used in experiments. The results used here suggest that very high stakes games do provide support for subgame perfection as a predictor of the outcome. However, Cameron (2007) suggests that may not always occur.

Bargaining models have become very important in industrial organization, particularly analysis of mergers. They form the heart of antitrust analysis

[6]This should not be consider a legal analysis, but rather an illustration of how these methods are used in legal analysis.

of hospital mergers but have also been used in other mergers where similar dynamics is at play.

Capps et al. (2003) came along at just the right time for US antitrust authorities. The agencies had been on an impressive losing streak with trying to prevent hospital mergers. To the credit of the antitrust agencies, the learnings presented there, in Gaynor and Vogt (2003), Gowrisankaran and Town (2003) and others, were incorporated into antitrust enforcement. The FTC brought a retrospective case against a hospital merger in Chicago and used these methods to prove that the merger was anticompetitive. That is, the FTC showed that the observed post-merger price increases were caused by the merger.[7]

Economists at Federal Trade Commission have made major contributions to this literature. The analysis presented in this chapter is heavily influenced by Raval et al. (2017). Raval et al. (2022) provide a very interesting test of the modeling approach. Garmon (2017) and Balan and Brand (2018) are among other great papers testing and using these methods to analyze the effects of hospital mergers.

[7] `https://www.ftc.gov/sites/default/files/documents/cases/2005/10/051020initialdecision.pdf`.

# Part III

# Static Games of Incomplete Information

# 9

# *Bayes Nash Equilibrium*

## 9.1 Introduction

In the first two parts of the book, the games involved cases where all the players knew everything about what had happened or what was happening. We say that the players have complete information. In Parts III and IV of the book, we drop this assumption. The next two parts of the book consider games where players don't necessarily know who they are playing against, what their opponents payoffs are or what actions other players have played. In the first half the book, we could model chess. In this half of the book, we can model Stratego! If you don't know it, Stratego is a board game with pieces that have different attributes and abilities just like chess. The difference is that the pieces are all the exact same shape and the picture denoting the piece is only printed on one side. This means your opponent knows only that you have a piece in a particular square. They don't observe which piece. Stratego can be modeled as a dynamic game of incomplete information.

The problem with modeling games of incomplete information is that it is not clear how to do it. That we can model these situations is really due to some amazing work by an Hungarian mathematician, John Harsayni, who immigrated to Australia to escape the communists and then immigrated to the United States to escape the Australians.

Harsayni's insight was to think of games of incomplete information as games of complete information. Brilliant! What? Harsanyi realized that we could think of the problem of not knowing the other player's payoffs as a problem of not knowing the other player.

This chapter introduces the idea of **beliefs**. If something is unknown by the players, we need a way to quantify what is unknown. The assumption is that each player of the game places probability weights on the unknown events of the game. It is assumed that these probability weights (beliefs) are known to everyone in the game.

This chapter introduces the idea of a Bayesian game and the Nash equilibrium for such a game. It illustrates the idea using entry games. We saw entry games in Chapters 4, 5, and 6. This time it is assumed that the firms contemplating entering a market do not know about the characteristics of the other firms contemplating entering the market. Interestingly, when players

DOI: 10.1201/9781003351603-9

know about the characteristics of the other firms, the equilibrium is a lot more complicated than when players don't know about the characteristics of the other firms. Making the game more complicated makes the game easier to use for empirical analysis. Yale economist, Katja Seim makes this point in her 2006 *RAND Journal of Economics* paper on the video retail industry (Seim, 2006).

## 9.2 A Bayesian Game

We call a game with uncertainty over the types, a **Bayesian game** because we require players to use **Bayes rule** to update their **beliefs** (the probability weights). Although we are restricting the information available to players, we still require the players to process a lot of information. Newer game theory models relax some of these assumptions and explore the implications.

The section introduces a game based on an actual situation than can happen in undergraduate courses, it then formally defines the game and the equilibrium concept, **Bayes Nash equilibrium**.

### 9.2.1 A Grading Game

Let's analyze a game in which the players are students in a game theory class. Grading is done on a curve. Everyone who gets below the mean score for the class gets a B and everyone above the class mean gets an A. Assume that the students in the class are grade focused. They would prefer to only get As in their courses if they can help it. Also, students can drop the class at any time without penalty. This last bit is unrealistic, but it makes the analysis simpler.

For each student the problem is to determine whether or not to drop the class. Assume that each student observes their own raw score in the class. For example they may know that their grades add up to a grand total of 33 out of 100. While they don't know raw scores of the other students, they have been told the distribution of raw scores. They can see that the mean is around 50 and scores range from close to 0 up to close to 90.

Assume that each student will stay in the class if they believe that they will get an A but will drop the class if they believe that they will get a B. To simplify the problem, we can assume that each student plays a **cutoff strategy**. A strategy here is a mapping from the player's raw score, to either stay or drop. A **cutoff strategy** states that the player will stay in the course if their score is above some cutoff level and will drop the class if it is below.

### 9.2.2 Grading Game Simulation using R

It is easier to look at a simulation. Assume we have 100 students in the class and their raw scores are determined by a normal distribution with a mean of

50 and a standard deviation of 20.

```
> set.seed(123456789)
> N = 100
> score = rnorm(N, mean=50, sd=20)
> summary(score)
   Min. 1st Qu.  Median    Mean 3rd Qu.    Max.
   4.41   36.61   52.73   50.50   64.98   89.80
```

What will happen in this game? Let's assume that everyone plays a strategy where they never drop. In this case a student in the top half will get an A and a student in the bottom half will get a B.

```
> # Initial grades
> grade = ifelse(score > mean(score), "A", "B")
> table(grade)
grade
 A  B
54 46
```

Assume that students play the strategy that if their scores is above 50 they will stay and students will drop if their score is below 50. That is, if the student thinks they will get an A, they stay in the class. If the student thinks that they will get a B, they will drop the class.

Now, if this is the strategy played by all the players, the grades will be different. Remember the grades are curved based on students who are in the class. The score distribution for those students will be a truncated normal distribution (the top half).

```
> s = score > 50
> score_1 = score[s]
> summary(score_1)
   Min. 1st Qu.  Median    Mean 3rd Qu.    Max.
  50.18   56.89   64.37   64.95   73.29   89.80
> length(score_1)
[1] 55
```

If students play this strategy, then the mean jumps from 50 to 65 and 45 students drop the class. Given the new reality, the grades will be adjusted.

```
> # New grades
> grade_1 = ifelse(score_1 > mean(score_1), "A", "B")
> table(grade_1)
grade_1
 A  B
25 30
```

```
> mean(score_1)
  [1] 64.9546
```

The number of students who get an A drops to 25 and they have to have a grade above 65.

Assume that the students adjust to this new reality, so their strategy changes. They will stay if their grade is above 65 as this guarantees them an A for the class.

```
> s_1 = score > 65
> score_2 = score[s_1]
> summary(score_2)
     Min. 1st Qu.  Median    Mean 3rd Qu.    Max.
    65.36   71.05   73.49   74.06   78.25   89.80
```

Now the remaining students need to update their beliefs about their grade given the strategies of the other students. Given this update, another 30 students drop the class and the mean increases to 74.

```
> grade_2 = ifelse(score_2 > mean(score_2), "A", "B")
> table(grade_2)
  grade_2
   A  B
  10 15
> s_2 = score > 74
> score_3 = score[s_2]
> summary(score_3)
     Min. 1st Qu.  Median    Mean 3rd Qu.    Max.
    74.39   76.74   78.58   79.56   79.56   89.80
```

Updating again, another 15 students drop the class and the average grade increases to 80.

Will the class have any students in it?

Dartmouth College has a provision to stop this unraveling. While Dartmouth's econ department uses a curve, professors are allowed to include students who dropped the class in calculating the grade distribution, where such students are given the lowest score for the purposes of doing the calculation. A policy like this may still lead students to drop based on beliefs of their grade in the class, but it doesn't lead to the unraveling. What is the outcome of the game used by Dartmouth?

### 9.2.3 Definitions

Now we have a taste for how these games work, let's get more formal. What do we mean by a **game of incomplete information**?

**Definition 15.** *In a* game of incomplete information, *players don't necessarily observe the actions of other players or know the payoffs of the other players.*

In Parts I and II of the book, players know exactly what is happening or has happened at every moment. In Parts III and IV of the book, they don't.

**Assumption 2.** *There are a set of player* types *determining the payoffs the player will get in each outcome. This set of types are known to all the players.*

While players don't know exactly what is going on, we will make the assumption that they do know player **types**. Assumption 2 is a super important idea, it allows us use all the machinery we have developed for analyzing complete information games to analyze incomplete information games. The player's type captures all the information relevant to the game. If each player's type is observed by every other player of the game, then the game is one of complete information.

**Assumption 3.** *Each player knows their own type and the distribution of types (the probability that the other player is a particular type).*

What makes games of incomplete information hard to think about is the implications of Assumption 3. When analyzing the game, we must be careful to remember what exactly the players know and what they do not know.

### 9.2.4 The Game

A Bayesian game has the following form.

- Players: Set of players and a set of types.
- Strategies: A function mapping from type to actions.
- Payoffs: For each type and each outcome a payoff.
- Beliefs: A probability distribution over types that is known by each player.

Our strategies are more complicated than for static games of complete information. Again, a strategy is complete plan. In this case, the complete plan is determined prior to the player knowing their type. That is, the plan states what the player will do for each possible type that they could be.

We have added one more piece to the basic game description. We generally refer to this known probability distribution over types as **beliefs**. Players do not know the types of the other players in the game, but they do know probability that another player is a particular type.

### 9.2.5 Course Grading Game

For the game introduced above, we have the following formal description. In this game, the player's type is the raw score of the player in the class which is denoted $\theta_i$. The raw score is between 0 and 100. The player's strategy states whether the player will stay or leave based on the raw score they observe.

- Players: $N = 100$ students where each observes their score in the class, $\theta_i \in [0, 100]$.
- Strategies: Each player $i \in N$ chooses $s(\theta_i) \in \{0, 1\}$, where 1 means stay.
- Payoffs:
  - Stay and $\theta_i > m$: $A$
  - Stay and $\theta_i \leq m$: $B$

  Leave: 0, where $B < 0 < A$

  where $m$ is the mean grade of the students remaining in the class.
- Beliefs: $\theta_i \sim F$

When we write down the game we don't worry about whether there exists an equilibrium of the game. It is also going to turn out that the initial beliefs don't matter that much, so let's just call it $F$.

### 9.2.6 Equilibrium

Given our new set up, we need a new equilibrium concept.

**Definition 16.** *A* Bayes Nash equilibrium *is an outcome where for each type, the outcome cannot be improved upon given the strategies of the other players and beliefs about the distribution of types. Where beliefs are consistent with equilibrium strategies.*

First, we are still assuming Nash equilibrium. Strategies must be optimal given the strategies of the other players. What is new is this idea of beliefs.

**Definition 17.** *A player's* belief *is what the player knows about the distribution of types playing each strategy.*

The equilibrium concept requires that the beliefs of the players be consistent with the equilibrium strategies. We are assuming that players are choosing their optimal strategies given expected payoffs. The players may not know exactly what payoff will occur because they don't know exactly which type of player they are playing against.

Here it is important to point out the difference between assumptions about the game and assumptions about the predictions of the game (the equilibrium concept). In the game, the players only know their own type and the distribution of types for the other players. In a Bayes Nash equilibrium, it may be that players know exactly the type of the other players because their beliefs must be updated to be consistent with the equilibrium strategies. Hopefully, this distinction will become clearer as we work through examples.

### 9.2.7 Equilibrium of Course Grade Game

Is there an equilibrium where each student only stays in the class if their grade is an $A$? No.

We are going to show this using a proof by contradiction. Let $c$ be the cutoff such that if $\theta_i > c$, then student $i$ stays; otherwise, they drop. If $c$ is the same for every student in the class then $c$ is the minimum grade for the students that stay in the class. Given the letter grades are determined by the mean grade of students that stay in the class, then $m = \mathbb{E}(\theta_i|\theta_i > c)$. The curve requires that $A$ is given if $\theta_i > m$ and $B$ is given otherwise.

The proposed equilibrium requires that $m \leq c$. If $c$ and $m$ exists, then $m = \mathbb{E}(\theta_i|\theta_i > c) > c$, a contradiction.

## 9.3 Empirical Entry Game

We first looked at entry games in Chapter 4. Those were static games of complete information. We revisited entry games in Chapter 5 and again in Chapter 6. The last time we modeled them as dynamic games of complete information. Now we are going to revisit them again. This section models entry games as static games of incomplete information. Comparing our analysis in this chapter to the analysis in Chapters 4 and 5, the assumptions in this chapter make the game itself more complicated, but the equilibrium easier to analyze.

It is worth contemplating the difference between the assumption made here and in Chapter 5. In both cases in equilibrium the firms don't know exactly if the other firm will enter the market. In both cases, they know the probability that the other firm will enter. These models are different. In the model used in Chapter 5, each firm knows the value of the unobserved entry costs, $\xi$. The econometrician does not know this value, but the firms playing the game do. Here, the value of $\xi$ is unknown to both the econometrician and some of the players of the game. This is a subtle distinction but it substantially changes how we estimate the model.

### 9.3.1 The Game

- Players: Barnes & Noble (and $\xi_{1i}$), Borders (and $\xi_{2i}$)
- Strategies:
  - Given $\xi_{1i}$ (and $\mathbf{X}_i$) Barnes & Noble chooses enter or not enter.
  - Given $\xi_{2i}$ (and $\mathbf{X}_i$) Borders chooses to enter or not enter.
- Payoffs:

- Barnes & Noble
  - Enter: $\mathbf{X}_i'\beta_1 - \alpha_1 \Pr(D_{2i} = 1|\xi_{1i}) + \xi_{1i}$
  - Not Enter: 0
- Borders
  - Enter: $\mathbf{X}_i'\beta_2 - \alpha_2 \Pr(D_{1i} = 1|\xi_{2i}) + \xi_{2i}$
  - Not Enter: 0
- Beliefs: $\{\xi_{1i}, \xi_{2i}\} \sim \Phi_2(0, \Sigma)$, where $\Sigma = \begin{bmatrix} 1 & \rho \\ \rho & 1 \end{bmatrix}$.

where $D_{1i}$ indicates whether or not Barnes & Noble enters, while $D_{2i}$ indicates Borders entry.

This game is similar to the one we used in Chapter 4. There are unobserved characteristics of the firms and the market, $\xi$, that are now unobserved by the other player. These are the player types. Entry is determined by observed characteristics $\mathbf{X}_i$ of the market $i$ and the parameter $\beta$ which is the same across markets. The parameter $\alpha$ determines the extent to which competition from the other firm reduces the benefits of entry into a particular market.

### 9.3.2 Equilibrium

The difference between this game and what we saw in Chapter 4 is the assumption about what Barnes & Noble knows about $D_2$ in equilibrium. In Chapter 4 we assumed that in equilibrium, Barnes & Noble knew the exact value of $D_2$. They knew whether or not Borders was also entering the market. In this chapter we assume that in equilibrium Barnes & Noble only knows the probability that $D_2 = 1$. More precisely, in equilibrium Barnes & Noble accounts for Border's equilibrium strategy in Barnes & Noble's beliefs about $D_2$. It's value $D_2$ is determined by the following inequality.

$$\mathbf{X}_i'\beta_2 - \alpha_2 D_{1i} + \xi_{2i} > 0 \tag{9.1}$$

Assume that there exists an equilibrium in cutoff strategies, where the cutoff values are $\{c_{1i}, c_{2i}\}$, respectively. Barnes & Noble's expectation about Border's entry into market $i$ is as follows.

$$\begin{aligned} \mathbb{E}(D_{2i}|\mathbf{X}_i, \xi_{1i}) &= \Pr(\xi_{2i} > c_{2i}|\mathbf{X}_i, \xi_{1i}) \\ &= \Pr(\xi_{2i} > -\mathbf{X}_i'\beta_2 + \alpha \Pr(\xi_{1i} > c_{1i})|\xi_{1i}) \end{aligned} \tag{9.2}$$

In determining the probability that Borders will enter, Barnes & Noble must account for Border's beliefs that Barnes & Noble will enter. Things are a lot simpler if we assume that the unobserved term, the $\xi$'s are independent of each other. That seems pretty unrealistic. We would expect a lot of things about the market to be correlated. While Barnes & Noble doesn't know Border's costs of entering exactly, it does observe its own costs and can make an inference about Border's costs.[1]

Even still, there is a distinct advantage of using games of incomplete information to model entry. In Chapter 4, there were multiple equilibria at certain values of the $\xi$s. That is not the case here. The equilibrium is unique in the cutoff strategies of the firms. The only issue is determining what that equilibrium is!

### 9.3.3 Estimating Entry Games

Can we back out the distribution of entry costs by looking at the distribution firms in different markets? We can. However, we must solve for the equilibrium in order to do so. It is assumed that the entry costs are distributed bivariate normal. The question is whether we can estimate the correlation coefficient ($\rho$). In a standard bivariate probit model, it is assumed that the actions are correlated but the optimality of the actions only depend on each other through the correlation. Here the actions themselves are interdependent.

The seminal work of Guerre et al. (2000) suggests that it is not necessary to actually solve for the equilibrium. Instead it is simpler to do a "two-step" procedure and the correlation parameter.

In the first step, we use maximum likelihood to estimate the cutoff value. In this step, we are not making any claims about the parameter values we are estimating. We are assuming that the observed data is being generated by some sort of stationary process. As with any discrete choice type data, our identification is heavily reliant on parametric assumptions. The likelihood function is determined by the standard bivariate normal. The probabilities of the four states are as follows. Note the notation. There is a squiggly line over all the parameters to remind us that these are not the structural parameters but the parameter values coming from the **reduced form estimation** in the first step.[2]

Maximizing the log-likelihood function provides estimates of $\tilde{\beta}_1$, $\tilde{\beta}_2$, and $\tilde{\rho}$. The model for estimating these parameters is presented in Chapter 4.

With this information, we can determine probability of entry for each firm. These probabilities are estimated because they are based on the parameters estimates in the first step.

$$\begin{aligned} \Pr(D_{1i} = 1|\mathbf{X}_i, \xi_{2i}) &= 1 - \Phi\left(\frac{-\mathbf{X}_i\hat{\tilde{\beta}}_1 - \hat{\tilde{\rho}}\xi_{2i}}{\sqrt{(1-\hat{\tilde{\rho}}^2)}}\right) \\ \Pr(D_{2i} = 1|\mathbf{X}_i, \xi_{1i}) &= 1 - \Phi\left(\frac{-\mathbf{X}_i\hat{\tilde{\beta}}_2 - \hat{\tilde{\rho}}\xi_{1i}}{\sqrt{(1-\hat{\tilde{\rho}}^2)}}\right) \end{aligned} \tag{9.3}$$

Given the estimated probabilities of entry, we can estimate our structural

[1] To be clear, we assume that the firms each know all the observed characteristics, $\mathbf{X}_i$ as well as all the parameter values.

[2] We use the term reduced form to refer to standard empirical estimation techniques where we are not relying on assumptions about actors generating the data are behaving.

parameters.

The probabilities are now without the squiggly lines. This is a reminder that these probabilities are structural estimates. They come from the game theory model. The probabilities with the squiggly lines are observed in the data and don't depend upon any assumptions about how the data are generated.

We also have two extra parameters, $\alpha_1$ and $\alpha_2$. Both firms will enter if the following inequality holds. The $\xi$s appears in two different places in the inequalities.

$$\begin{array}{c} \mathbf{X}_i\beta_1 - \alpha_1 \hat{\Pr}(D_{2i} = 1|\mathbf{X}_i, \xi_{1i}) + \xi_{1i} > 0 \\ \mathbf{X}_i\beta_2 - \alpha_2 \hat{\Pr}(D_{1i} = 1|\mathbf{X}_i, \xi_{2i}) + \xi_{2i} > 0 \end{array} \tag{9.4}$$

where $\{\xi_{1i}, \xi_{2i}\} \sim \mathcal{N}_2(0, \Sigma)$ and $\Sigma = \begin{bmatrix} 1 & \rho \\ \rho & 1 \end{bmatrix}$.

### 9.3.4 Estimating Entry Games in R

Similarly, we can write out a function to represent the entry inequalities for the second stage of the estimation (Equation (9.4)). In the code, the extra `t` denotes the parameters estimated from the reduced form model (the squiggly line parameters). Chapter 4 estimates these parameters using `f_entry2()`.

```
> f_biprobit = function(X, beta_1, beta_2,
+                       alpha_1, alpha_2, rho,
+                       beta_1t, beta_2t, rhot) {
+   N = dim(X)[1]
+   xi_1 = Z_1
+   xi_2 = Z_2*sqrt(1 - rho^2) + rho*Z_1
+   Xb_1 = X%*%beta_1
+   Xb_2 = X%*%beta_2
+   Xb_1t = X%*%beta_1t
+   Xb_2t = X%*%beta_2t
+   p_00 = p_01 = p_11 = rep(0, N)
+   for(k in 1:K) {
+     D_1k = 1 -
+       pnorm(0, Xb_1t + rhot*xi_1[k], sqrt(1 - rhot^2))
+     D_2k = 1 -
+       pnorm(0, Xb_2t + rhot*xi_2[k], sqrt(1 - rhot^2))
+     pi_1k = Xb_1 - alpha_1*D_2k + xi_1[k]
+     pi_2k = Xb_2 - alpha_2*D_1k + xi_2[k]
+     p_00 = p_00 + (pi_1k < 0 & pi_2k < 0)
+     p_01 = p_01 + (pi_1k < 0 & pi_2k > 0)
+     p_11 = p_11 + (pi_1k > 0 & pi_2k > 0)
+   }
+   return(list(p_00 = p_00/K,
+               p_01 = p_01/K,
```

```
+                    p_11 = p_11/K))
+ }
```

The second step estimator, `f_biprobit`, uses simulation to calculate the probabilities. The code uses `pnorm()` to calculate the probability of entry given the estimated parameters from the first step.

## 9.4 Empirical Analysis: Mega Bookstore Entry (Again) using R

This section re-estimates the entry of Barnes & Noble and Borders under the assumption that each bookstore chain does not know the exact costs of entry of their competitor. This assumption seems more realistic than the assumption made in Chapter 4. While the assumption makes the estimator somewhat more complicated, it has the nice property that there aren't multiple equilibria.

### 9.4.1 Estimates

Table 9.1 presents the mean and standard deviation of the coefficient estimates from the first and second stages of the two-step estimator. The estimates show that population, college education and the number of bookstores in 1990 are all important determinants of entry into a county by these two mega bookstore chains. It also shows that Barnes & Noble is much less likely to enter than Borders and that the two firms do not like to compete. It is interesting to compare these results to the results presented in Chapter 4, 5 and 6.

Table **??** presents the model fit excercise comparing the two models with the actual data. The Bayes Nash equilibrium model does a better job of predicting the case where there is only one firm in the market, but does a worse job of predicting the case where there are two firms in the market.

### 9.4.2 Policy

Given our new estimates, we can reconsider the policy question analyzed in Chapter 4. In that analysis, the merger resolved a coordination problem for the firms. In general, the two firms prefer to have only one of them in the market, but when the firms are independent they cannot coordinate on which one. In Chapter 4, the firms knew whether or not the other firm was going to enter, under these model assumptions they do not. The merger resolves the information problem inherent in the modeling assumptions.

Table 9.2 presents the results of simulating the merger. We see fewer cases where the two firms compete as well as fewer markets with entry. Theoretically this model can have ambiguous predictions about the impact of the merger on consumer welfare, but the empirical estimates suggest that the merger would

**TABLE 9.1**
Results from estimates of the first and second stage of the estimates assuming a Bayes Nash equilibrium. The first set of columns labeled "First Stage" are estimates assuming that the two firms are making entry decisions that are strategically independent but statistically independent. These are the same as the results presented in Chapter 4 for the same model. The second set of columns labeled "Second Stage" assumes the firm entry decisions are strategically dependent, but each firm does not know exactly if the other firm will enter, but assumes their strategy is consistent with observed entry decisions.

| | First Stage | SD | BNE | SD |
|---:|---:|---:|---:|---:|
| const_1 | −15.03 | 0.09 | −15.03 | 0.23 |
| Pop_1 | 1.09 | 0.01 | 1.06 | 0.02 |
| Income_1 | −1.04 | 0.25 | −0.75 | 0.36 |
| College_1 | 5.51 | 0.34 | 5.85 | 0.40 |
| Stores_1990_1 | 0.28 | 0.07 | 0.44 | 0.09 |
| const_2 | −11.54 | 0.11 | −11.39 | 0.17 |
| Pop_2 | 0.66 | 0.01 | 0.63 | 0.02 |
| Income_2 | 1.74 | 0.25 | 1.80 | 0.39 |
| College_2 | 2.59 | 0.33 | 2.75 | 0.54 |
| Stores_1990_2 | 0.49 | 0.06 | 0.70 | 0.10 |
| alpha_1 | | | 0.87 | 0.22 |
| alpha_2 | | | 0.64 | 0.17 |
| rho | −0.08 | 0.10 | −0.01 | 0.11 |

**TABLE 9.2**
Comparison of actual entry in 2000 compared to simulated entry and simulated entry under a merger assuming a Bayes Nash equilibrium of the entry game.

| | Actual | Sim | Merger |
|---:|---:|---:|---:|
| none | 2919 | 2643 | 2857 |
| BN | 155 | 284 | 222 |
| Borders | 15 | 148 | 71 |
| both | 128 | 104 | 29 |

have lowered consumer welfare.

Compare these predictions to the predictions in Chapter 4. It is the same policy analysis. What is different is the assumption about the information that the competitors have.

## 9.5 Discussion and Further Reading

Modeling entry games as games of complete information leads to all sorts of weirdness (Bresnahan and Reiss, 1990; Tamer, 2003). However, when the game is more realistic by reducing the information available to the players, the game gets a whole lot simpler to analyze. Seim (2006) uses these ideas to analyze entry of retail video stores.

The estimator here is based on Guerre et al. (2000). The argument is that we can assume that the observed data are the result of equilibrium behavior. We can estimate a reduced form model in the first step and then impose equilibrium behavioral assumptions to back out the parameters of the underlying game theory model. We come back to this idea in Chapter 10.

This chapter uses data from Adams and Basker (2025) and their analysis of entry of the mega bookstores.

# 10

# *Auctions*

## 10.1 Introduction

According to the travel presenter, Rick Steves, the Aalsmeer auction house is one the largest commercial buildings in the world. Royal Flora Holland, the owner of Aalsmeer, sold 12.5 billion plants and flowers in 2016 through its auction houses. But with $5.2 billion in auction sales, Royal Flora Holland is nowhere near the biggest auction house in the world.[1] That honor may go to Google. Google sold $47.6 billion in search ads using what the economist, Hal Varian, called the biggest auction in the world (Varian, 2007).[2] While that is impressive, a single auction in 2015 almost beat Google's annual number. The US Federal Communication Commission's auction number 97 (AWS-3) raised $44.9 billion dollars for US taxpayers.[3] That pails in comparison to the fact that every week the US Federal government offers billions of dollars in securities auctions. A single 4-week T-bill auction for July 6 2023 was for $70 billion.

Auctions are used to sell and buy a large number of products. Governments use auctions to purchase everything from paper to police body cameras. The US Federal government uses auctions to sell oil drilling rights, FCC spectrum, 10 year bonds and timber access. You can sell and buy items from eBay.com using auctions.

The auctions at Aalsmeer are unique. The auction runs for a short amount of time with a "clock" clicking the price down as the auction continues. As the price falls, the first bidder to hit the button, wins, at whatever price the clock is at. A spokesperson for Aalsmeer stated that because the price falls, it is called a **Dutch auction**. Actually they got the causality backwards. Because the Dutch popularized these types of auctions for selling flowers, we call them **Dutch auctions**.

The auction style you may be most familiar with is called an **English auction**. In this auction, there is an auctioneer who often speaks very very fast and does a lot of pointing while bidders hold up paddles or make hand gestures. In English auctions, the last bidder wins and pays the price at which the bidding stops.

[1] `https://www.royalfloraholland.com`

[2] eMarketer.com, 7/26/16

[3] `https://www.fcc.gov/auction/97/factsheet`

DOI: 10.1201/9781003351603-10

Economic analysis of auctions began with William Vickrey's seminal 1961 paper, *Counterspeculation, Auctions, and Competitive Sealed Bid Tenders*. Vickrey pointed out that Dutch auctions and **sealed bid auctions** are **strategically equivalent**. In a standard sealed bid auction each bidder submits a secret written bid. The auctioneer chooses the highest bid, and the bidder pays the number written down in her bid.

Vickrey characterized what a bidder should optimally bid in such an auction. He then showed that the same bidder should bid exactly the same amount in a Dutch auction. That is, in a Dutch auction, the bidder should wait until the price falls to the number written down, and then hit the button. Vickrey showed that while these two auctions formats are strategically equivalent, they are not strategically equivalent to an English auction.

Vickrey wondered if there was a sealed bid auction that is strategically equivalent to an English auction. Vickrey invented a new auction. In a **Vickrey auction**, each bidder writes down a bid like in a standard sealed bid auction and the winner is the person who writes down the highest bid. However, the winner pays the amount written down by the second highest bidder. Vickrey showed that his auction is strategically equivalent to an English auction.

This chapter discusses two of the most important auction formats, sealed bid auctions and English auctions. It presents estimators for both. The sealed bid auction estimation is based on Guerre et al. (2000). The English auction analysis uses the order statistic approach of Athey and Haile (2002). In both cases, it presents analysis of timber auctions. The chapter tests whether loggers are bidding rationally in sealed bid auctions and whether loggers colluded in English auctions.

## 10.2 Sealed Bid Auctions

Sealed bid auctions are one of the most commonly used auction formats. These auctions are very prominent in procurement, both in government and in the private sector. In a sealed bid auction, each bidder writes down her bid and secretly submits it to the auctioneer. The auctioneer sorts the bids from highest to lowest (or lowest to highest if they are buying instead of selling). The winner is the highest bidder and she pays the amount she wrote down. This is called a **first price auction** because the price is determined by the highest bid or first price.

Vickrey pointed out that sealed bid auctions are strategically complicated. To see this, assume that a bidder's utility for an item is equal to their intrinsic value for the item less the price they pay for the item. For example, a logger bidding in a timber auction will earn profits from the logs less the price paid to the US Forestry service for access to the trees. If a logger bids an amount equal to her expected profits, then if she wins she will earn nothing from the logging. It is optimal for the logger to shade her bid down. The problem is that the more she shades down, the lower her chance of winning the auction. The bidder must calculate the trade off between the probability of winning the auction and the value of winning the auction.

The section presents the model of a sealed bid auction, it then simulates data from such an auction. The section develops an estimator for determining each bidder's type, or valuation for the item.

### 10.2.1 Sealed Bid Model

The sealed bid game has $N$ bidders and each bidder $i$ knows their own type $v_i$.

- Players: $N$ bidders each with valuation $v_i$ (type)
- Strategies: For each valuation (type) $v_i$, for bidder $i$, she chooses a bid $b_i(v_i)$.
- Payoffs:
  - $b_i > b_j \forall j \neq i$: $v_i - b_i$
  - $b_i < b_j \forall j \neq i$: 0
- Beliefs: $v_i \sim F$

We will ignore ties.[4]

If the bidder has the highest bid she wins and has a payoff $v_i - b_i$, which is her intrinsic value less the amount of the bid. If she loses she gets nothing.

**Assumption 4.** *Independent Private Values (IPV). Let $v_i \overset{iid}{\sim} F$, where $v_i$ is the value of bidder $i$ and $F$ is the distribution function.*

Assumption 4 makes the exposition a lot simpler. It also seems to be a reasonable approximation for the problems considered in the chapter. It states that a bidder's value for the item is unrelated to the values of the other bidders in the auction, except that they draw their valuation from the same distribution. The next chapter considers an alternative assumption where valuations are associated with each other.

### 10.2.2 Bayes Nash Equilibrium

In equilibrium, the bidder is assumed to maximize her expected returns from the auction. Assume that the bidder gets 0 if she loses. If she wins, assume she gets her intrinsic value ($v_i$) for the item less her bid ($b_i$).

$$\max_{b_i} \quad \Pr(\text{win}|b_i)(v_i - b_i) \tag{10.1}$$

[4] If the bid increments are small and the valuations are continuously distributed, then the probability of a tie is small.

If we take first-order conditions of Equation (10.1), then we get the following expression.

$$g(b_i|N)(v_i - b_i) - G(b_i|N) = 0 \tag{10.2}$$

Let $G(b_i|N)$ denote the probability that bidder $i$ is the highest bidder with a bid of $b_i$, conditional on there being $N$ bidders in the auction, and $g(b_i|N)$ is the derivative. $G(b_i|N)$ is the probability that she wins the auction.

We can rearrange this formula to show how much the bidder should shade her bid.

$$b_i = v_i - \frac{G(b_i|N)}{g(b_i|N)} \tag{10.3}$$

The formula states that the bidder should bid her value, less a shading factor which is determined by how much a decrease in her bid reduces her probability of winning the auction.

It will be useful for our code to write the probability of winning the auction as a function of the bid distribution as this distribution is observed in the data. Let $H(b)$ denote the distribution of bids in the auctions. Given Assumption 4, the probability of a particular bidder winning the auction is given by the following equation.

$$G(b_i|N) = H(b_i)^{N-1} \tag{10.4}$$

If there are two bidders in the auction, then the probability of winning is simply the probability that her bid is higher than the other bidder. If there are more than two bidders, it is the probability that her bid is higher than *all* the other bidders. The independent private values assumption, Assumption 4, implies that this is the probability that each of the other bidders makes a bid less than hers, all multiplied together.

We can also determine the derivative of this function in terms of the bid distribution observed in the data.

$$g(b_i|N) = (N-1)h(b_i)H(b_i)^{N-2} \tag{10.5}$$

where $h$ is the derivative of the bid distribution $H$.

### 10.2.3 Sealed Bid Simulation in R

In the simulated data, each bidder draws their value from a uniform distribution. Vickrey shows that the optimal bid in this auction is calculated using the following formula.

$$b_i = \frac{(N-1)v_i}{N} \tag{10.6}$$

In Vickrey's version of the game, bidders know the function represented by Equation (10.6). The uniform distribution simplifies the problem, which is why it is used. In each simulated auction, there are different numbers of simulated bidders.

```
> set.seed(123456789)
> M = 1000  # number of simulated auctions.
> data1 = matrix(NA,M,12)
> for (i in 1:M) {
+   N = round(runif(1, min=2,max=10)) # number of bidders.
+   v = runif(N) # valuations, uniform distribution.
+   b = (N - 1)*v/N # bid function
+   p = max(b)  # auction price
+   x = rep(NA,10)
+   x[1:N] = b  # bid data
+   data1[i,1] = N
+   data1[i,2] = p
+   data1[i,3:12] = x
+ }
> colnames(data1) = c("Num","Price","Bid1",
+                     "Bid2","Bid3","Bid4",
+                     "Bid5","Bid6","Bid7",
+                     "Bid8","Bid9","Bid10")
> data1 = as.data.frame(data1)
```

The simulation creates a data set with 1,000 auctions. In each auction, there is between 2 and 10 bidders. The bidders are not listed in order.

### 10.2.4 Sealed Bid Estimator

The estimator uses Equation (10.3) to back out values from observed bids. To do this, we calculate the probability of winning the auction conditional on the number of bidders. It should be straightforward to determine this from the data. Once we have this function, we use the formula to determine the bidder's valuation from their bid.

The first step is to estimate the bid distribution.

$$\hat{H}(b) = \frac{1}{N}\sum_{i=1}^{N} \mathbb{1}(b_i < b) \tag{10.7}$$

The **non-parametric estimate** of the distribution function, $H(b)$, is the fraction of bids that are below some value $b$.[5]

[5] A non-parametric estimator makes no parametric assumptions about how the bids are distributed in the data.

The second step is to estimate the derivative of the bid distribution. This can be calculated numerically for some given "small" number, $\epsilon$.[6]

$$\hat{h}(b) = \frac{\hat{H}(b+\epsilon) - \hat{H}(b-\epsilon)}{2\epsilon} \tag{10.8}$$

If there are two bidders, Equation (10.3) determines the valuation for each bidder.

$$\hat{v}_i = b_i + \frac{\hat{H}(b_i)}{\hat{h}(b_i)} \tag{10.9}$$

where $i \in \{1, 2\}$.

### 10.2.5 Sealed Bid Estimator in R

The estimator limits the data to only those auctions with two bidders. In this special case, the probability of winning is just given by the distribution of bids.[7] In the code, the `epsilon` stands for the Greek letter, $\epsilon$, and refers to a "small" number. See Equation (10.8).

```
> f_sealed_2bid = function(bids, epsilon=0.5) {
+   # epsilon for "small" number for finite difference method
+   # of taking numerical derivatives.
+   values = rep(NA,length(bids))
+   for (i in 1:length(bids)) {
+     H_hat = mean(bids < bids[i])
+     # bid probability distribution
+     h_hat = (mean(bids < bids[i] + epsilon) -
+       mean(bids < bids[i] - epsilon))/(2*epsilon)
+     # bid density
+     values[i] = bids[i] + H_hat/h_hat
+   }
+   return(values)
+ }
```

It is straightforward to calculate the probability of winning, as this is the probability the other bidder bids less. Given IPV (Assumption 4), this is the cumulative probability for a particular bid. Calculating the density is slightly more complicated. We can approximate this derivative numerically by looking at the change in the probability for a "small" change in the bids.[8] The value is calculated using Equation (10.3).

---

[6]This is a **finite difference estimator**.

[7]The probability of winning is the probability that your bid is higher than the other bidders in the auction.

[8]This is an example of using finite differences to calculate numerical derivatives. What happens with different values of `epsilon`?

The code creates a `ggplot()` object that shows the density of bids and the derived valuations from the two-person auctions.

```
> data2 = data1 |>
+   filter(
+     Num == 2
+   )
> ggplot_sealed = data.frame(
+   "bids" = c(data2$Bid1, data2$Bid2),
+   "values" = f_sealed_2bid(c(data2$Bid1, data2$Bid2))
+ ) |>
+   ggplot() +
+   geom_density(aes(values),
+                fill = "gray",
+                alpha = 0.5) +
+   geom_density(aes(bids),
+                fill="black",
+                alpha = 0.5) +
+   scale_x_continuous(limits = c(-0.5,1.5)) +
+   labs(
+     x = "value",
+     y = "",
+     title = "Density of Bids and Values"
+   ) +
+   geom_text(aes(x = 1.2, y = 0.8, label = "Bids"),
+             color = "gray") +
+   geom_text(aes(x = -0.2, y = 1.2, label = "Values"),
+             color = "gray") +
+   theme(axis.text.y=element_blank(),
+         axis.ticks.y=element_blank())


> ggplot_sealed
```

Figure 10.1 shows that the bids are significantly shaded from the true values, particularly for very high valuations. The figure presents the density functions for bids and derived valuations from the two-person auctions. The true density of valuations lies at 0.5 and goes from 0 to 1. Here the estimated density is a little higher and goes over its bounds. However, part of the reason may be the method we are using to represent the density in the figure.[9] You should try different values of epsilon to see how that changes things.

[9]The kernel density method assumes the distribution can be approximated as a mixture of normal distributions.

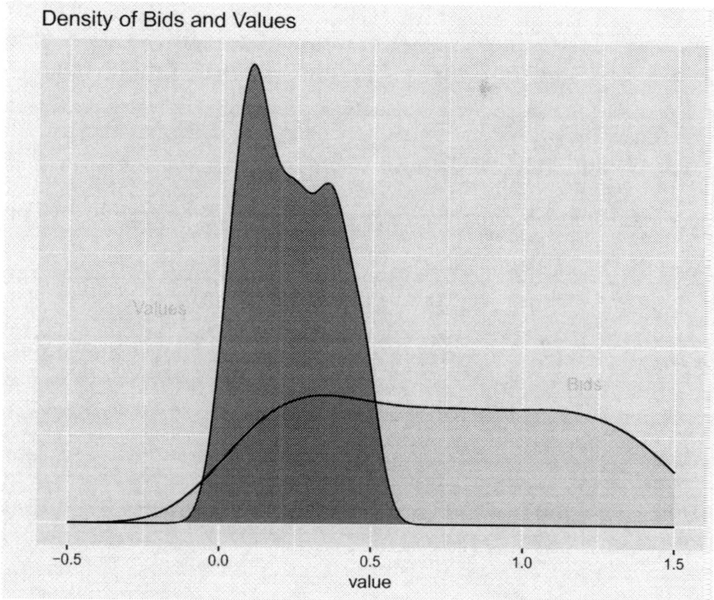


**FIGURE 10.1**
Plot of the density function for bids and values from 2 person auctions. The true distribution of valuations has a density of 0.5 from 0 to 1.

## 10.3 English Auctions

The auction format that people are most familiar with is the English auction. These auctions are used to sell cattle, antiques, collector stamps, and houses (in Australia). In the 1970s, they were also the standard format used by the US Forestry Service to sell timber access (Aryal et al., 2018).

Because bidders can observe each other's bid as the auction progresses, it is a dynamic game. To make our life a lot simpler, we will lean heavily on Vickrey's analysis and model these auctions as second price sealed bid auctions.

The section presents the game under the assumption of a second price sealed bid auction and determines the Bayes Nash equilibrium. It then switches focus to estimating valuations from English auctions. Often in these auctions, we only observe the winning bid, or the price, and possibly the number of bidders. Because of this limitation on the data we will use an idea of an order statistic and a result from Athey and Haile (2002).

### 10.3.1 Second Price Auction Game

The game is a second price sealed bid auction with $N$ bidders and valuations $v_i$.

- Players: $N$ bidders with valuation $v_i$
- Strategies: Each bidder $i$ chooses a bid given their valuation for the item, $b_i(v_i)$.
- Payoffs:
  - $b_i > b_j \forall j \neq i$: $v_i - b_2$, where $b_2$ is the bid of the second highest bidder.
  - $b_i < b_j \forall j \neq i$: 0
- Beliefs: $v_i \sim F$

In addition to assuming away the dynamics, we assume away bid increments that may make our life complicated.

### 10.3.2 Bayes Nash Equilibrium

Vickrey showed that English auctions are strategically very simple. Imagine a bidder hires an expert auction consultant to help them bid in an English auction.

- Expert: “What is your value for the item?”
- Bidder: “$2,300”
- Expert: “Bid up to $2,300 and then stop.”

In sealed bid auctions, there is an optimal trade-off between winning and profiting from the auction. In second price auctions, there is no such trade-off.

The optimal bid for bidder $i$ is to bid her value.

$$b_i = v_i \tag{10.10}$$

Equation (10.10) suggests that empirical analysis of English auctions is a lot simpler than for sealed bid auctions. If only that were so! To be clear, the “bid” in Equation (10.10) means the strategy described by the expert. In the data, we do not necessarily observe this strategy.

If we could observe all the bid strategies in the auction, then we would have an estimate of the value distribution. But that tends to be the problem. Depending on the context, not all active bidders in the auction may actually be observed making a bid. In addition, if the price jumps during the auction we may not have a good idea when bidders stopped bidding (Haile and Tamer, 2003).

Athey and Haile (2002) provide a solution. They point out that the price in an English auction has a straightforward interpretation. When valuations follow Assumption 4, the price is the second highest valuation of the people who bid in the auction. Consider the case when the price is lower than the second highest valuation. How could that be? Why did one of the bidders exit the auction at a price lower than her valuation? What if the price is higher than the second highest valuation? How could that be? Why would a bidder bid more than her valuation?

If the price is equal to the second highest valuation, then it is a particular order statistic of the value distribution. Athey and Haile (2002) show how the observed distributions of an order statistic uniquely determine the value distribution.

### 10.3.3 Order Statistics

To understand how order statistics work, consider the problem of determining the distribution of heights of players in the WNBA. The obvious way to do it is to take a data set on player heights and calculate the distribution. A less obvious way is to use order statistics.

In this method, data are taken from a random sample of teams, where for each team, the height of the tallest player is measured. Assume each team has 10 players on the roster and you know the height of the tallest, say the center. This is enough information to estimate the distribution of heights in the WBNA. We can use the math of order statistics and the fact that we know both the height of the tallest and we know that 9 other players are shorter. In this case we are using the tallest, but you can do the same method with the shortest or the second tallest, etc.

The price is more or less equal to the second highest valuation of the bidders in the auction.[10] The probability of the second highest of $N$ valuations is equal to some value $b$ which is given by the following formula:

$$\Pr(b^{(N-1):N} = b) = N(N-1)F(b)^{N-2}f(b)(1-F(b)) \tag{10.11}$$

The order statistic notation for the second highest bid of $N$ is $b^{(N-1):N}$. We can parse this equation from right to left. It states that the probability of seeing a price equal to $b$ is the probability that one bidder has a value greater than $b$. This is the winner of the auction and this probability is given by $1 - F(b)$, where $F(b)$ is the cumulative probability of a bidder's valuation less than $b$. This probability is multiplied by the probability that there is exactly one bidder with a valuation of $b$. This is the second highest bidder who is assumed to bid her value. This is represented by the density function $f(b)$.[11] These two values are multiplied by the probability that the remaining bidders

[10]Officially, the price may be a small increment above the bid of the second highest bidder. We will ignore this possibility.

[11]This is the derivative of $F(b)$.

have valuations less than $b$. If there are $N$ bidders in the auction then $N-2$ of them have valuations less than the price. The probability of this occurring given Assumption 4 is $F(b)^{N-2}$. Lastly, the labeling of the bidders is irrelevant so there are $\frac{N!}{1!(N-2)!} = N(N-1)$ possible combinations. If the auction has two bidders, then the probability of observing a price $p$ is $2f(p)(1-F(p))$.

The question raised by Athey and Haile (2002) is whether we can use this formula to determine $F$. Can we use the order statistic formula of the distribution of prices to uncover the underlying distribution of valuations? Yes.

### 10.3.4 Identifying the Value Distribution

Let's say we observe a two bidder auction with a price equal to the lowest possible valuation for the item; call that $v_0$. Actually, it is a lot easier to think about the case where the price is slightly above the lowest possible value. Say that the price is less than $v_1 = v_0 + \epsilon$, where $\epsilon$ is a "small" number. What do we know? We know that one bidder has that very low valuation, which occurs with probability equal to $F(v_1)$. What about the other bidder? The other bidder may also have a value equal to the lowest valuation or they may have a higher valuation. That is, their value for the item could be anything. The probability of value lying between the highest and lowest possible value is 1. So $\Pr(p \leq v_1) = 2 \times 1 \times F(v_1)$ and either bidder could be the high bidder. There are 2 possibilities, so we must multiply by 2. As the probability of a price less than $v_1$ is observed in the data, we can rearrange things to get the initial probability, $F(v_1) = \Pr(p \leq v_1)/2$.

Now take another value, $v_2 = v_1 + \epsilon$. The probability of observing a price between $v_1$ and $v_2$ is as follows.

$$\Pr(p \in (v_1, v_2]) = 2(F(v_2) - F(v_1))(1 - F(v_1)) \tag{10.12}$$

It is the probability of seeing one bidder with a value between $v_2$ and $v_1$ and the second bidder with a value greater than $v_1$. Again, the two bidders can be ordered in two ways.

We can solve $F(v_2)$ using Equation (10.12). We observe the quantity on the left-hand side and we previously calculated $F(v_1)$.

For a finite subset of the valuations, we can use this iterative method to calculate the whole distribution. More generally, we would use differential equations. For this to work, each bidder's valuation is assumed to be independent of the other bidders and comes from the same distribution of valuations (Assumption 4).

### 10.3.5 English Auction Estimator

The non-parametric estimator of the distribution follows the logic above.

The initial step determines the probability at the minimum value,

$$\hat{F}(v_1) = \frac{\sum_{j=1}^{M} \mathbb{1}(p_j \le v_1)}{2M} \tag{10.13}$$

where there are $M$ auctions and $p_j$ is the price in auction $j$.

To this initial condition, we can add an iteration equation.

$$\hat{F}(v_k) = \frac{\sum_{j=1}^{M} \mathbb{1}(v_k < p_j \le v_{k+1})}{2M(1 - \hat{F}(v_{k-1}))} + \hat{F}(v_{k-1}) \tag{10.14}$$

These equations are then used to determine the distribution of the valuations.

### 10.3.6 English Auction Estimator in R

We can estimate the distribution function non-parametrically by approximating it at $K = 100$ points evenly distributed across the range of observed values. The estimator is based on Equations (10.13) and (10.14).

```
> f_English_2bid = function(price, K=100, epsilon=1e-8) {
+   # K number of finite values.
+   # epsilon small number for getting the probabilities
+   # calculated correctly.
+   min1 = min(price)
+   max1 = max(price)
+   diff1 = (max1 - min1)/K
+   Fv = matrix(NA,K,2)
+   min_temp = min1 - epsilon
+   max_temp = min_temp + diff1
+   # determines the boundaries of the cell.
+   Fv[1,1] = (min_temp + max_temp)/2
+   gp = mean(price > min_temp & price < max_temp)
+   # price probability
+   Fv[1,2] = gp/2  # initial probability
+   for (k in 2:K) {
+     min_temp = max_temp - epsilon
+     max_temp = min_temp + diff1
+     Fv[k,1] = (min_temp + max_temp)/2
+     gp = mean(price > min_temp & price < max_temp)
+     Fv[k,2] = gp/(2*(1 - Fv[k-1,2])) + Fv[k-1,2]
+     # cumulative probability
+   }
+   return(Fv)
+ }
```

## 10.4 Empirical Analysis: Testing the Rationality of Loggers using R

In the 1970s, the US Forest Service conducted an interesting experiment. It introduced sealed bid auctions in 1977. Previous to that, most US Forest Service auctions had been English auctions.[12] In 1977, the service mixed between auction formats. As discussed above, bidding in sealed bid auctions is strategically a lot more complicated than bidding in English auctions. In the latter, the bidder simply bids her value. In the former, she must trade off between bidding higher and increasing the likelihood of winning against paying more if she does win.

Because of the experiment, we can test whether the loggers in the sealed bid auctions bid consistently with their actions in the English auctions. Our test involves estimating the underlying value distribution using bid data from sealed bid auctions and comparing that to an estimate of the underlying value distribution using price data from English auctions. These two value distributions are the same under the assumptions of the game theory model.

### 10.4.1 Timber Data

The data used here are from the US Forest Service downloaded from Phil Haile's website.[13]

In order to estimate the distributions of bids and valuations it is helpful to "normalize" them so that we are comparing apples to apples. The standard method is to use a log function of the bid amount and run a linear regression on various characteristics of the auction including the number of acres bid on, the estimated value of the timber, access costs and characteristics of the forest and species (Haile et al., 2006).[14]

The code brings in the data and uses `lm()` to create the object `lm1`. The regression creates dummy variables for the different tree species, regions, forests and districts using `as.factor()`. It then creates a normalized bid using the residuals from the linear regression.

```
> file = paste0(dir, "auctions.csv")
> df = read.csv(file)
> lm1 = lm(log_amount ~ as.factor(Salvage) + Acres +
```

[12] You may think of this as just some academic question. But the US Senator for Idaho, Senator Church, was not happy with the decision. "In fact, there is a growing body of evidence that shows that serious economic dislocations may already be occurring as a result of the sealed bid requirement." See *Congressional Record* September 14 1977, p. 29223.

[13] `http://www.econ.yale.edu/~pah29/timber/timber.htm`. The version used here is available from here: `https://sites.google.com/view/microeconometricswithr/table-of-contents`.

[14] Baldwin et al. (1997) discuss the importance of various observable characteristics of timber auctions.

```
+               Sale.Size + log_value + Haul +
+               Road.Construction + as.factor(Species) +
+               as.factor(Region) + as.factor(Forest) +
+               as.factor(District), data=df)
> # as.factor creates a dummy variable for each entry under the
> # variable name.  For example, it will have a dummy for each
> # species in the data.
> df$norm_bid = NA
> df$norm_bid[-lm1$na.action] = lm1$residuals
> # lm object includes "residuals" term which is the difference
> # between the model estimate and the observed outcome.
> # na.action accounts for the fact that lm drops
> # missing variables (NAs)
```

In general, we are looking for a normal-like distribution. Figure 10.2 presents the histogram of the normalized log bids. It is not required that the distribution be normal, but if the distribution is quite different from normal, you should think about why that may be. Does this distribution look normal?[15]

### 10.4.2 Sealed Bid Auctions

In order to simplify things, we will limit the analysis to two-bidder auctions. In the data, sealed bid auctions are denoted “S”.

```
> df1 = df |>
+   filter(
+     num_bidders == 2 &
+       Method.of.Sale == "S"
+   ) |>
+   mutate(
+     bids = as.vector(norm_bid),
+     values = f_sealed_2bid(norm_bid)
+   )
> summary(df1$bids)
    Min. 1st Qu.  Median    Mean 3rd Qu.    Max.
 -4.0262 -0.7987 -0.1918 -0.2419  0.3437  5.0647
> summary(df1$values)
     Min.   1st Qu.    Median      Mean   3rd Qu.      Max.
  -4.0262   -0.0218    0.9185    4.8704    2.3108  860.0647
```

Using the same method that we used above, it is possible to back out an estimate of the value distribution from the bids in the data. We see that

[15] It is approximately normal, but it is skewed somewhat to lower values. This may be due to low bids in the English auction. How does the distribution look if only sealed bids are graphed?

```
> df |>
+   ggplot() +
+   geom_histogram(aes(x = norm_bid),
+                  fill = "gray",
+                  alpha = 0.5) +
+   scale_x_continuous(limits = c(-3,2.5)) +
+   labs(
+     x = "bid",
+     y = "",
+     title = "Histogram of Bids"
+   ) +
+   theme(axis.text.y = element_blank(),
+         axis.ticks.y = element_blank())
```

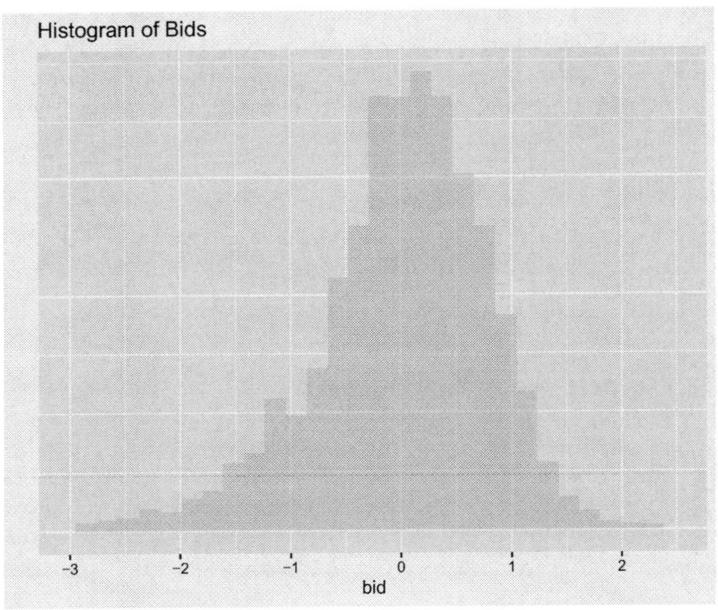


**FIGURE 10.2**
Histogram of normalized bid residual for US Forest Service auctions from 1977.

comparing the valuations to the bids, the bids are significantly shaded particularly for higher valuations. The negative numbers may seem odd but remember we have normalized the bids in the auction.

### 10.4.3 Comparing English Auctions to Sealed Bid Auctions

We can back out the value distribution from the English auctions by assuming that the price is the second highest bid, the second highest **order statistic**. The English auctions are denoted “A”.

The function `f_English_2bid` estimates the cumulative probability function for the value distribution.

```
> df2 = df |>
+   filter(
+     num_bidders == 2 &
+       Method.of.Sale == "A" &
+       Rank == 2
+   )
> Fv_english = f_English_2bid(df2$norm_bid)
```

To calculate the equivalent for the sealed bid auctions, we can use the `ecdf()` function.

```
> Fv_sealed = ecdf(df1$values)
```

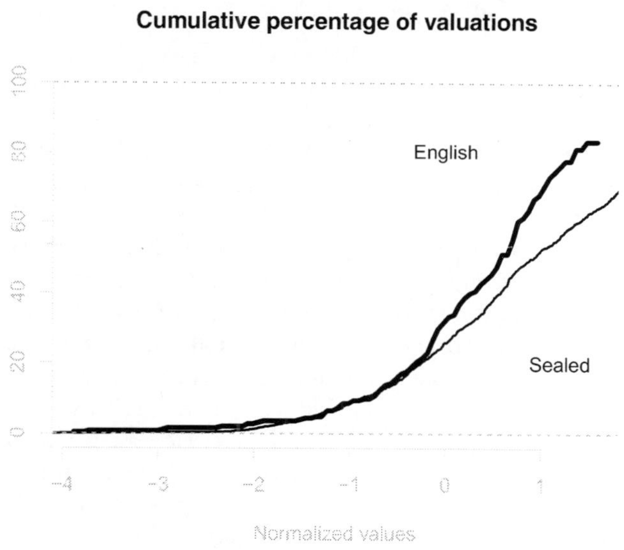


**FIGURE 10.3**
Comparison of estimated distributions from two bidder English and sealed bid auctions. The estimate from the English auction and the sealed bid auction are similar to about 0, then estimate from English auctions places more weight on lower valuations than the estimate from the sealed bid auctions.

Figure 10.3 shows that there is not a whole lot of difference between the estimate of the distribution of valuations from sealed bid auctions and English auctions. The two distributions of valuations from the sealed bid auctions and

English auctions lie fairly close to each other, particularly for lower values. This suggests loggers are bidding rationally. That said, at higher values, the two distributions diverge. The value distribution from the sealed bid auctions suggests that valuations are higher than the estimate from the English auctions. What else may explain this divergence?

## 10.5 Empirical Analysis: Testing for Collusion using R

Is there evidence that bidders in English auctions are colluding? This section presents a test of collusion based on using auction theory to back out the implied value distribution of the bidders in the larger English auctions. We can compare the implied distribution to the distribution we have estimated above.

### 10.5.1 A Test of Collusion

Consider the following test of collusion. Using large English auctions, we can estimate the distribution of valuations. Under the prevailing assumptions of the game theory model, this estimate should be the same as for two-bidder auctions. If the estimate from the large auctions suggests valuations are much lower than for two-bidder auctions, this suggests collusion.

Specifically, if the inferred valuations in these larger auctions look much like auctions with fewer bidders. That is, bidders may behave "as if" there are actually fewer bidders in the auction. For example, if there is an active **bid ring**, bidders may have a mechanism for determining who will win the auction and how the losers may be compensated for not bidding.[16] In an English auction, it is simple to enforce a collusive agreement because members of the bid ring can bid in the auction where their bids are observed.

Can we determine the size of the bid ring? How many people are bidding collusively? What if we have an auction with six bidders? If three of them are members of a bid ring, then those three will agree on who should bid from the ring. Only one of the members of the bid ring will bid their value. If we estimate the model under the assumption that there are four independent bidders, it will match the value distribution we estimated from the two bidder auction.[17]

[16] Asker (2010) presents a detailed account of a bid ring in stamp auctions.

[17] In the bid ring mechanism discussed in Asker (2010), the collusion actually leads to higher prices in the main auction.

### 10.5.2 "Large" English Auctions

Assume there are six bidders in the auction. From above, the order statistic formula for this case is as follows.

$$\Pr(b^{5:6} = b) = 30F(b)^4 f(b)(1 - F(b)) \tag{10.15}$$

As above, order statistics are used to determine the underlying value distribution ($F$); however, in this case, it is a little more complicated to determine the starting value.

Think about the situation where the price in a 6 bidder auction is observed at the minimum valuation. What do we know? As before, one bidder may have a value equal to the minimum or a value above the minimum. That is, their value could be anything. The probability of a valuation lying between the minimum and maximum value is 1. We also know that the five other bidders had valuations at the minimum. If not, one of them would have bid more and the price would have been higher. As there are six bidders, there are six different bidders that could have had the highest valuation. This reasoning gives the following formula for the starting value.

$$\Pr(b^{5:6} < v_1) = 6F(v_1)^5 \tag{10.16}$$

Rearranging, we have $F(v_1) = \left(\frac{\Pr(p<v_1)}{6}\right)^{\frac{1}{5}}$.

Given this formula we can use the same iterative method as for two-bidder auctions to solve for the distribution of valuations.

### 10.5.3 Large English Auction Estimator

Again we can estimate the value distribution by using an iterative process. In this case, we have the following estimators.

$$\hat{F}(v_1) = \left(\frac{\sum_{j=1}^{M} \mathbb{1}(p_j < v_1)}{6M}\right)^{\frac{1}{5}} \tag{10.17}$$

and

$$\hat{F}(v_k) = \frac{\sum_{j=1}^{M} \mathbb{1}(v_k < p_j < v_{k+1})}{30M\hat{F}(v_{k-1})^4(1 - \hat{F}(v_{k-1}))} + \hat{F}(v_{k-1}) \tag{10.18}$$

The other functions are as defined in the previous section.

We can also solve for the implied distribution under the assumption that there are three bidders and under the assumption that there are two bidders.[18] Note in each auction there are at least six bidders.[19]

[18] See Equation (10.11) for the other cases.

[19] For simplicity, it is assumed that all of these auctions have six bidders. Once there are a large enough number of bidders in the auction, prices do not really change with more bidders. In fact, these methods may not work as the number of bidders gets large (Deltas, 2004).

### 10.5.4 Large English Auction Estimator in R

We can adjust the estimator above to allow any number of bidders, $N$.

```
> f_English_Nbid = function(price, N, K=100, epsilon=1e-8) {
+   min1 = min(price)
+   max1 = max(price)
+   diff1 = (max1 - min1)/K
+   Fv = matrix(NA,K,2)
+   min_temp = min1 - epsilon
+   max_temp = min_temp + diff1
+   Fv[1,1] = (min_temp + max_temp)/2
+   gp = mean(price > min_temp & price < max_temp)
+   Fv[1,2] = (gp/N)^(1/(N-1))
+   for (k in 2:K) {
+     min_temp = max_temp - epsilon
+     max_temp = min_temp + diff1
+     Fv[k,1] = (min_temp + max_temp)/2
+     gp = mean(price > min_temp & price < max_temp)
+     Fv[k,2] =
+       gp/(N*(N-1)*(Fv[k-1,2]^(N-2))*(1 - Fv[k-1,2])) +
+       Fv[k-1,2]
+   }
+   return(Fv)
+ }
```

### 10.5.5 Evidence of Collusion

We limit the auctions to ones with more than five bidders.

```
> df3 = df |>
+   filter(
+     num_bidders > 5 &
+       Method.of.Sale == "A" &
+       Rank == 2
+   )
```

We can estimate the value distribution under the assumption that there are six bidders in the auction. We can also estimate the value distribution under the assumption that there are three bidders and two bidders. The results are presented in Figure 10.4.

```
> Fv_6 = f_English_Nbid(df3$norm_bid, N = 6)
> Fv_3 = f_English_Nbid(df3$norm_bid, N = 3)
> Fv_2 = f_English_2bid(df3$norm_bid)
```

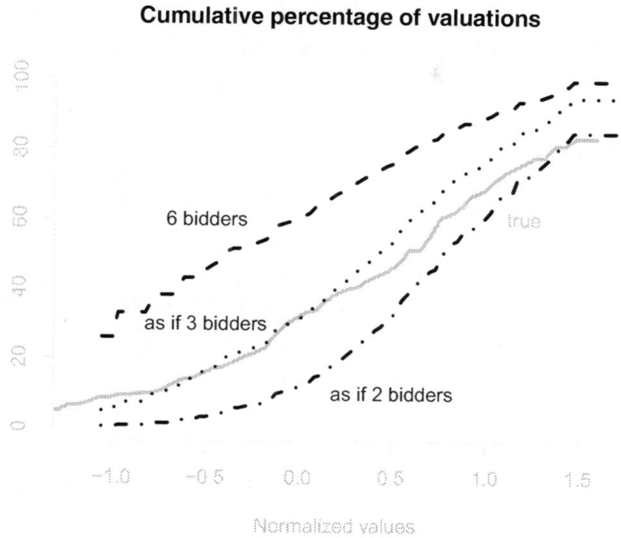


**FIGURE 10.4**
Comparison of estimated distribution of valuations from English auctions with at least 6 bidders. These estimates are compared to the estimate from 2 bidder auctions which is labeled "true." The estimate of the 6-bidder auctions suggests valuations are much lower than for the 2-bidder auctions.

Figure 10.4 suggests that there is in fact collusion in these auctions! Assuming there are 6 bidders in the auction implies that valuations are much lower than we estimated for 2 bidder auctions from both English auctions and sealed bid auctions. In the chart, the distribution function is shifted to the left, meaning there is greater probability of lower valuations.

In theory, these two estimates should be the same or very close. Remember we are estimating the underlying value distribution which is unrelated to the number of bidders in the auction. The estimate with 6 bidders suggests that bidders value the timber much lower than when there are 2 bidders. If we don't think there is any collusion in 2 bidder auctions, then these estimates provide the ground truth. These estimates tell us how the timber is valued. This implies that the reason the value distribution is lower is not that the values are lower, but that the bids are lower.

If there are 5 bidders in the ring, then bidders will bid as if they are in a 2 bidder auction. We can estimate the value distribution assuming that there are 2 bidders in the auction. The result is that the estimated value distribution is higher than the ground truth. Either the 2 bidders are bidding too much or there are more than 2 independent bidders in the auction.

If there are 4 bidders in the ring, then bidders will bid as if they are in a 3 bidder auction. Estimates assuming there are three bidders and two bidders lie above and below the true value, respectively. This suggests that bidders are behaving as if there are between two and three bidders in the auction. This implies that the bid ring has between four and five bidders in each auction.

These results are suggestive of an active bid ring in these auctions in 1977. It turns out that this was of real concern. In 1977, the United States Senate conducted hearings into collusion in these auctions. In fact, this may be why the US Forestry Service looked into changing to sealed bid auctions. The US Department of Justice also brought cases against loggers and millers (Baldwin et al., 1997). Alternative empirical approaches have also found evidence of collusion in these auctions, including Baldwin et al. (1997) and Athey et al. (2011).

## 10.6 Discussion and Further Reading

Economic analysis of auctions began with Vickrey's 1961 paper. Vickrey used game theory to analyze sealed bid auctions, Dutch auctions and English auctions. Vickrey also derived a new auction, the sealed bid second price auction.

The chapter considers two of the most important auction mechanisms, sealed bid auctions and English auctions. The sealed bid auctions are analyzed using the two-step procedure of Guerre et al. (2000). The first step uses non-parametric methods to estimate the bid distribution. The second step uses the Nash equilibrium to back out the value distribution.

While we observe all the bids in the sealed bid auctions, we generally only observe the high bids in English auctions. The chapter uses the order statistic approach of Athey and Haile (2002) to estimate the value distribution from these auctions.

Baldwin et al. (1997) and Athey et al. (2011) analyze collusion in US timber auctions. Aryal et al. (2018) uses US timber auctions to measure how decision makers account for risk and uncertainty.

# 11

# *Auctions with Affiliated Valuations*

## 11.1 Introduction

The previous chapter assumed that bidder's valuations are independent of each other. This substantially simplified both the game theory and the econometrics. In this chapter we are going to relax this assumption. Now we will allow for bidder valuations to be dependent. These types of auctions are called **affiliated private values auctions**.

The conical example is bidding on the right to drill for oil in the US's outer continental shelf (OCS). The amount of oil that can be pulled out of the ground has nothing to do with the bidder. Similarly, the price of that oil has little or nothing to do with the identity of the bidder.[1]

The chapter analyzes **common value auctions** and takes the game theory model to OCS auctions. The chapters asks whether bid rings should be allowed in these auctions.

## 11.2 Auctions with Common Values

Here we consider auctions where the bidder's valuations are interdependent. The simplest case of such interdependence is the pure **common value auction**. In this auction, bidder's don't know the exact value of the item they are bidding on. Each bidder draws a signal about the true value of the item such that if all the signals were aggregated, then that would provide a pretty close approximation of the true value of the item.

The classic example is bidders considering how much to bid for the rights to drill for oil in a particular area. The oil field has a particular amount of oil worth a particular value. While unknown, it has nothing to do with who is bidding. The value of the oil field is the same to all bidders. They just don't

[1]Of course, some bidders may have market power in the oil market.

DOI: 10.1201/9781003351603-11

know what it is. Oil companies, the bidders, will hire different geologists with different models, equipment, and expertise to estimate the value of the oil field. These geologists will come up with different guesses. We call these guesses, signals of the oil field's value.

The section presents the game and the **Bayes Nash equilibrium** of the auction. It discusses an idea called the **winner's curse**, which leads bidders to shade their bid down in equilibrium. The section presents an estimator for **common value auctions** and illustrates the estimator using simulated data.

### 11.2.1 Simple Model

The auction game has $N$ bidders with signals $s_i$. The signals are drawn from a normal distribution with mean $v$ and variance $\sigma^2$. The true value of the item is $v$. Each bidder's bid is a function of their signal, $b_i(s_i)$. The bidder with the highest bid wins the auction and pays their bid. If the bidder wins, they get the true value of the item, $v$. If the bidder loses, they get nothing.

- Players: $N$ bidders and signals (types) $s_i \in \Re$.
- Strategies: Each bidder $i$ observes a signal $(s_i)$ and chooses a bid $b_i(s_i)$.
- Payoffs:
  - $b_i > b_j \forall j \neq i$: $v - b_i(s_i)$
  - If $\exists j$ s.t. $b_i < b_j$: 0
- Beliefs: $s_i \sim \mathcal{N}(v, \sigma^2)$, where $v$ is the true value of the item.

If the player $i$ wins, they get $v$ which doesn't have a subscript. This is because it is the same for every bidder. What varies is between bidders, is their signal $s_i$. Bidders know that their signal is drawn from a normal distribution with mean $v$ but they do not know $v$.

### 11.2.2 Winner's Curse

The winner is the bidder with the highest bid. Consider what happens if all bidders bid their valuations. In that case, the winner bids $s^{N:N}$ (using order statistic notation) which is going to be substantially higher than $v$.

```
> set.seed(123456789)
> N = 10
> v = 5
> s = rnorm(N, v)


> max(s)
  [1] 6.415538
```

In the example with 10 bidders, the winner bids 6.42, which much higher than the true value 5. The winner really loses. The bidders should adjust their strategy by shading their bid down in order to account for the winner's curse.

### 11.2.3 Bayes Nash Equilibrium

We often see common value auctions that are also **first price auctions** Bidders are shading their bids for two different reasons. First, as in the previous chapter, they are shading their bid because it is a first price auction and they need to account for winning and paying what they bid. Second, these bidders have to account for the fact that if they win the auction it is because their signal is higher than everyone else's and thus higher than the true value of the item.

To make things simpler, let's assume that bidders bid their expected value for the item. Assume that bidders use **Bayes rule** and their expected value is conditional on both the signal that they observe and on the fact that they won the auction. Of course, they don't actually know whether they win the auction or not, but their bid is of no importance if they lose. The bidder is thinking through the cases. The only case of interest is the case where the bidder won the auction. They realize that if they won the auction, their bid must, by definition, have been the highest bid.

Again, trading off winning and the amount paid is left out for the moment. So how much information does the bidder have? You may think that they don't have much because all they have is one signal, but it turns out that they know a lot more than that. Because they won the auction, they know that their signal must be higher than everyone else's. So that tells them quite a lot about every body else's signal.

Chapter 10 illustrates how order statistics work. The probability that a particular signal $s$ is the highest order statistic ($s^{N:N}$) is given by the following equation assuming that all the signals are drawn independently from the distribution $F(s|\mu, \sigma)$. The probability that a particular $s$ is higher than all the other signals is then $G(s|N, \mu, \sigma) = F(s|\mu, \sigma)^{N-1}$ where $\mu$ represents the different possible values of the true value $v$ and there are $N - 1$ other signals. The derivative is then $g(s|N, \mu, \sigma) = (N - 1)f(s|\mu, \sigma)F(s|\mu, \sigma)^{N-2}$.

We can use this likelihood to determine the bidder's expected value for the item conditional on submitting the winning bid. We just need to use Bayes rule. To determine the expected value, we need the probability that a particular distribution is generating the signals we observe.

$$\gamma(\mu, \sigma|s = s^{N:N}) = \frac{g(s|N, \mu, \sigma)}{\sum_{\mu', \sigma'} g(s|N, \mu', \sigma')} \tag{11.1}$$

If we know all the possible $\mu$s and $\sigma$s and assume a normal distribution and assume that the prior over $\mu$s and $\sigma$s is uniform, then Equation (11.1) gives the probability that the observed signals are generated by a particular $\mu$ and $\sigma$.

Consider a simple version where $\mu = v$, the true value and $\sigma = 1$. In this case $g(s|N, v) = (N-1)\phi(s-v)\Phi(s-v)^{N-2}$ and $\phi()$ and $\Phi$ represent the standard normal's density and probability function, respectively.

The expected value of the item given that the bidder has the highest signal, $s = s^{N:N}$, is then given by following function.

$$\mathbb{E}(v|N, s) = \int_{v'} v'\gamma(v'|s = s^{N:N})d(v') \tag{11.2}$$

where $\gamma()$ is defined in Equation (11.1). Given $N$ bidders and a signal $s$ the expected value of the true bid is the integral over the possible true values weighted by the $\gamma()$ function assuming that $s$ is the highest signal. Remember $\sigma = 1$.

Now we have how much the bidder values the item, the next question is how much to bid. Remember this is a sealed bid auction, so it is still the case that the bidder is trading off the value of winning against the probability of winning.

$$b_i(s_i) = \mathbb{E}(v|N, s_i) - \frac{G(b_i|N, s_i)}{g(b_i|N, s_i)} \tag{11.3}$$

So we need to determine the probability of winning conditional on the bidder's signal. The simplest assumption to make is that equilibrium bids are monotonically increasing in the signal. This doesn't seem unreasonable.

Given our monotonicity assumption, the probability of winning is just the probability of having the highest signal. We have that $G(b_i|N, s_i) = F(s_i|v)^{N-1}$ and $g(b_i|N, s_i) = (N-1)f(s_i|v)F(s_i|v)^{N-2}$, which not coincidentally is similar function that we defined previously.

### 11.2.4 Common Value Auction in R

Below we create all the probability functions that we need for bidding in first price auctions with common values. It is a little confusing because we are using order statistics for two different things. First, there is the standard method discussed in the previous chapter where the bidder is determining the optimal bid for a first price auction. Second, the bidder is using order statistics to back out their expectation of the value for the item given the signal they observed and conditional upon winning the auction.

The signal distribution is a normal distribution denoted $F$, where $f$ is the density. We use `log_F()` and `log_f()`. Given `N` bidders, the distribution of the highest signal is denoted by $G$ and we use `log_G()` and `log_g()` for the density. The function `dnorm()` calculates the density of the normal distribution and `pnorm()` calculates the probability of the normal distribution.

```
> log_f = function(s, v, sigma = 1)
+   log(dnorm(s, v, sigma))
> log_F = function(s, v, sigma = 1)
+   log(pnorm(s, v, sigma))
```

```
> log_G = function(s, v, sigma = 1, N) (N-1)*log_F(s, v, sigma)
> log_g = function(s, v, sigma=1, N) log(N-1) +
+   log_f(s, v, sigma) +
+   (N-2)*log_F(s, v, sigma)
```

Given these probabilities, we can determine the bidder's expected value for the item and their bid. The expectation function `E_fun()` takes in two global variables `u` and `sig`. These are the possible parameters of the normal distribution determining the signals observed by the bidders. Lastly there is the bid function `b_fun()` based on Equation (11.3).

```
> E_fun = function(s, N) {
+   g_u = matrix(NA,length(u),length(sig))
+   u_mat = matrix(NA, length(u), length(sig))
+   for(j in 1:length(sig)) {
+     g_u[,j] = exp(log_g(s, u, sig[j], N))
+     u_mat[,j] = u
+   }
+   sum_g_u = sum(g_u)
+   gamma_u = g_u/sum_g_u
+   mu = sum(u_mat*gamma_u)
+   sigma = sqrt(sum(u_mat^2*gamma_u) - mu^2)
+   return(list(mu=mu, sigma=sigma))
+ }
> b_fun = function(s, N) {
+   v_bar = E_fun(s, N)
+   G = exp(log_G(s, v_bar$mu, v_bar$sigma, N))
+   g = exp(log_g(s, v_bar$mu, v_bar$sigma, N))
+   return(v_bar$mu - G/g)
+ }
```

### 11.2.5 Simulation of Common Value Auction using R

It helps to work through a simulation. There are 100 auctions and the number of bidders varies. The true value is 0 for each auction, and the signal is distributed standard normal. The function `seq()` calculates a sequence of numbers between the first and second values with the third value as the step. The function `rnorm()` generates random numbers from a normal distribution. The function `sample()` randomly samples from a set of numbers. The function `rep()` repeats a number a certain number of times.

The code uses `sapply()` to loop through the signals and calculate the expected value and bid for each signal.

```
> set.seed(123456789)
> M = 100
```

```
> N = NULL
> bids = NULL
> ids = NULL
> values = NULL
> u = seq(-10, 10, 0.15)
> sig = seq(0.1, 3, 0.15)
> Ns = sample(3:4, M, replace=TRUE)
> v = rep(0, M)
> sigma = 1
> for(i in 1:M) {
+   ids = c(ids, rep(i, Ns[i]))
+   N = c(N, rep(Ns[i], Ns[i]))
+   s_i = rnorm(Ns[i], v[i], sigma)
+   values = c(values,
+              sapply(1:length(s_i),
+                     function(j) E_fun(s_i[j],
+                                       Ns[i])$mu))
+   bids = c(bids,
+            sapply(1:length(s_i),
+                   function(j) b_fun(s_i[j],
+                                     Ns[i])))
+ }
```

The code below creates a density plot of the bids and the expected values. The bids are shifted down from the expected values. The expected values are shifted down from the signal distribution in equilibrium.

```
> ggplot_sim_cv_bids = data.frame(
+   bids = bids,
+   values = values
+ ) |>
+   ggplot(aes(bids)) +
+   geom_density(alpha = 0.5) +
+   geom_density(aes(values), linetype = 2, alpha = 0.5) +
+   labs(
+     x = "values/bids",
+     y = "",
+     title = "Density of bids and values"
+   ) +
+   geom_vline(xintercept = 0, linetype = 2,
+              color = "gray") +
+   geom_text(aes(x = -5, y = 0.2, label = "bids"),
+             color = "gray") +
+   geom_text(aes(x = 2, y = 0.2, label = "values"),
+             color = "gray") +
```

```
+    theme(axis.text.y=element_blank(),
+          axis.ticks.y=element_blank())
```

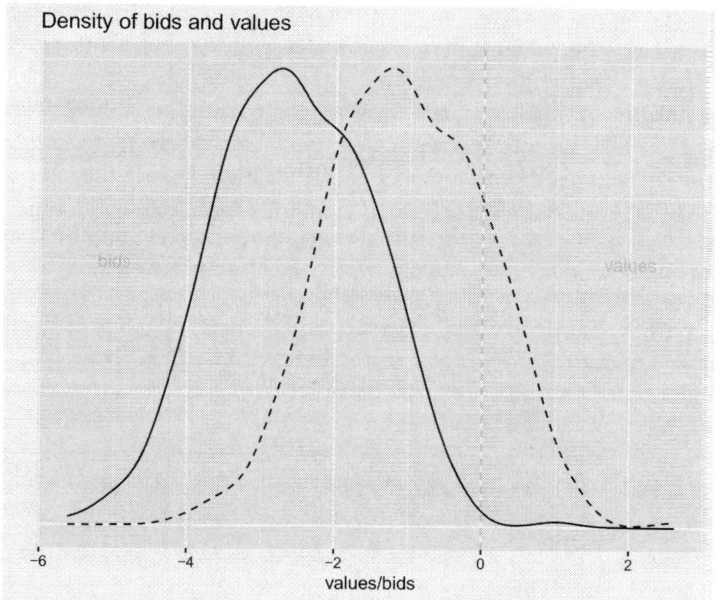


**FIGURE 11.1**
Plot of the density of the bids and expected values in a first price common values auction. Bids are shifted down from valuations because it is a first price auction.

Figure 11.1 presents the observed bids and the estimated values. As we saw in the IPV case, bids are significantly shaded down from values for first price auctions. Expected values are also significantly shaded down from the original signal observed by the bidder. These expected values are conditional on the signal assuming that signal is highest signal in the Bayes Nash equilibrium. The actual signal is distributed around zero.

### 11.2.6 Estimator for Common Values Auctions

The estimator reverse engineers the signal distribution from the bid distribution. We know from Equation (11.3) and the discussion in Chapter 10 that we can identify the distribution of the expected values conditional on the signal. Unfortunately, it not generally possible to uniquely determine the signal distribution from the expected value distribution. We will need to rely on parametric restrictions.

### 11.2.7 Common Values Estimator in R

To estimate the underlying signal distribution from the observed bids, we will combine maximum likelihood with simulation. The estimator chooses the $\mu$ and $\sigma$ of the signal distribution that maximizes the likelihood of the observed bids given the game theory assumption that bidders bid their expected value conditional on their signal being the highest.

The estimator works by taking a set of parameter values for the distribution of signals, `mu` and `sigma`, and simulating the resulting bids, `b_sim`. It simulates the signals, then loops through the simulated signals and creates the corresponding simulated bids. It then calculates the log likelihood of observing the observed bids (`bids_temp`) given the derived parameters from the bid distribution. It does this for each size of auction in the data.

```
> f_bid_ml = function(mu, sigma, bids_temp, N_temp, s) {
+   Ns = unique(N_temp)
+   log_lik = rep(NA, length(bids_temp))
+   for(i in 1:length(Ns)) {
+     N_i = Ns[i]
+     index = which(N_temp==Ns[i])
+     b_sim = sapply(1:length(s), function(i) b_fun(s[i], N_i))
+     mu_i = mean(b_sim, na.rm = TRUE)
+     sigma_i = sd(b_sim, na.rm = TRUE)
+     z_i = (bids_temp[index] - mu_i)/sigma_i
+     log_lik[index] = log(dnorm(z_i)) - log(sigma_i)
+   }
+   return(log_lik)
+ }
> f_bid_ml_int = function(par, bids_temp, N_temp) {
+   set.seed(123456789)
+   mu = par[1]
+   sigma = exp(par[2])
+   s = U*sigma + mu
+   return(-sum(f_bid_ml(mu, sigma, bids_temp, N_temp, s)))
+ }
```

This estimator requires three global variables `U`, `u`, and `sig`.

```
> U = rnorm(1000)
> a = optim(par = c(0, log(sigma)), f_bid_ml_int,
+              bids_temp = bids, N_temp = N,
+             control = list(trace=0, maxit=100000))
```

The code below estimates the signal distribution from the observed bids.

```
> ggplot_sim_cv_signals =
+   data.frame(
```

```
+     signals = rnorm(length(values),
+                     a$par[1],
+                     exp(a$par[2])),
+     values = values
+   ) |>
+     ggplot(aes(values)) +
+     geom_density(alpha = 0.5) +
+     geom_density(aes(signals), linetype = 2, alpha = 0.5) +
+     labs(
+       x = "values/signals",
+       y = "",
+       title = "Density of signals and values"
+     ) +
+     geom_vline(xintercept = 0, linetype = 2, color = "gray") +
+     geom_text(aes(x = -3.5, y = 0.2, label = "values"),
+               color = "gray") +
+     geom_text(aes(x = 2.5, y = 0.2, label = "signals (est.)"),
+               color = "gray") +
+     theme(axis.text.y=element_blank(),
+           axis.ticks.y=element_blank())
```

```
> ggplot_sim_cv_signals
```

Figure 11.2 presents the density of the expected values and the estimated signals in the simulated data. The estimated signals are pretty close to the true distribution, which is a standard normal distribution. The figure shows that the bidders substantially discount their bids from the observed signals. We see that this occurs for two reasons. First, their expected value (conditional on winning) is discounted from their signal. Second, because it is a first price auction, bidders discount their bid from the expected value of the item (see Figure 11.1).

## 11.3 Empirical Analysis: Signal Distribution from OCS Auctions using R

This section uses data on outer continental shelf (OCS) oil and gas tracts off Texas and Louisiana from 1954 to 1979.[2]

[2]These data are available from Penn State `https://capcp.la.psu.edu/data-and-software/outer-continental-shelf-ocs-auction-data/`.

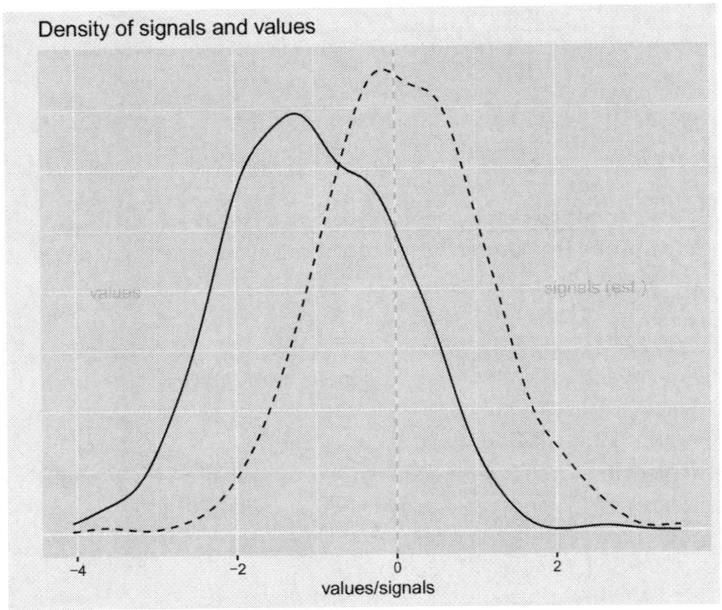


**FIGURE 11.2**
Plot of the density of expected values and signals (estimated) in a first price common values auction. The bidder's valuations are shifted down from the estimated signal distribution.

### 11.3.1 Data

The code brings in the data set, `book_ocs_ch11.csv`. Next we do the trick of creating a residual auction value by regressing bids on observed characteristics of the auctions. The function `as.factor()` is used to create dummy variables for the block code and date of the auction. The function `lm()` is used to estimate the linear regression. The residuals are then calculated, and the data set is converted to a data frame.

```
> file = paste0(dir, "book_ocs_ch11.csv")
> df = read.csv(file) |>
+   select(
+     lbid,
+     lvalue,
+     lcost,
+     BlockCode,
+     Date,
+     TractNumber,
+     nCompany
+   ) |>
```

```
+   na.omit()
> lm1 = lm(lbid ~ lvalue + as.factor(BlockCode) +
+             as.factor(Date) + lcost, data = df)
> df$res = lm1$residuals
> dt = setDT(df)

> ggplot_ocs_bids =
+   df |>
+   ggplot(aes(res)) +
+   geom_density(alpha = 0.5) +
+   geom_density(aes(rnorm(length(res),
+                          mean(res),
+                          sd(res))),
+                linetype = 2, alpha = 0.5) +
+   scale_x_continuous(limits = c(-4,4)) +
+   labs(
+     x = "residuals",
+     y = "",
+     title = "Density of residuals"
+   ) +
+   geom_vline(xintercept = 0, linetype = 2, color = "gray") +
+   geom_text(aes(x = 2, y = 0.2, label = "residuals"),
+             color = "gray") +
+   geom_text(aes(x = -3, y = 0.2, label = "normal"),
+             color = "gray") +
+   theme(axis.text.y=element_blank(),
+         axis.ticks.y=element_blank())

> ggplot_ocs_bids
```

Figure 11.5 presents the normalized bids for the OCS auctions. The figure also shows simulated bids from a normal distribution to suggest that a normal distribution is a reasonable approximation.

### 11.3.2 Estimating the Signal Distribution

We restrict the sample to just those auctions without coalitions in them.[3] We can use the estimator above to estimate the signal distribution for these auctions. In addition, the code creates an index of observations that have less than 3 bidders, more than 10 bidders and missing residuals. The code then calculates the initial values for the optimization routine and runs the routine. The code creates an object `index` that determines the auctions with less than 3 bidders and more than 10 bidders. It then uses `-index` to drop those auctions.

[3]Coalitions are discussed in detail in the next section. They are legal bid rings.

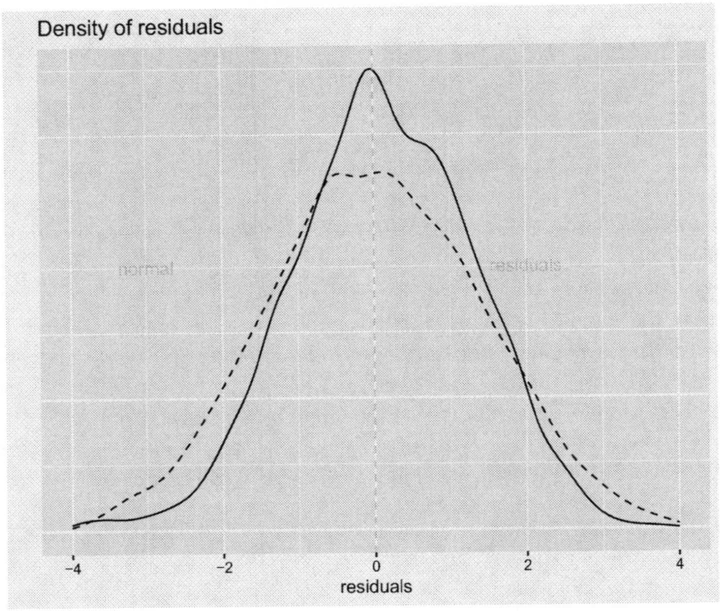


**FIGURE 11.3**
Plot of the density of the residual bids against a simulated data set drawn from a normal distribution with the mean and variance equal to the mean and variance for the normalized bids. This indicates that the normal distribution is a reasonable approximation of the bids.

```
> dt2 = dt[num_coy == N]
> index = c(which(dt2$num_coy < 3 | dt2$num_coy > 10),
+           which(is.na(dt2$res)))
> init = c(mean(dt2$res[-index]), sd(dt2$res[-index]))
> b1 = optim(par = init,
+            f_bid_ml_int,
+            bids_temp = dt2$res[-index],
+            N_temp = dt2$num_coy[-index],
+            control = list(trace = FALSE,
+                           maxit = 1000000))
```

The code then creates a plot of the estimated signal distribution and the observed bids.

```
> ggplot_est_cv_signals =
+   data.frame(
+     bids = dt$res,
+     signals = rnorm(length(dt$res),
+                     b1$par[1],
```

```
+                      exp(b1$par[2]))
+   ) |>
+     ggplot(aes(bids)) +
+     geom_density(alpha = 0.5) +
+     geom_density(aes(signals), linetype = 2, alpha = 0.5) +
+     scale_x_continuous(limits = c(-5,8)) +
+     labs(
+       x = "bids/signals",
+       y = "",
+       title = "Density of signals and bids"
+     ) +
+     geom_vline(xintercept = 0, linetype = 2,
+                color = "gray") +
+     geom_text(aes(x = -3.5, y = 0.2, label = "bids"),
+               color = "gray") +
+     geom_text(aes(x = 6.5, y = 0.2, label = "signals (est.)"),
+               color = "gray") +
+     theme(axis.text.y=element_blank(),
+           axis.ticks.y=element_blank())

> ggplot_est_cv_signals
```

Figure 11.4 presents the density of the normalized bids and the estimated signals in the OCS auctions. The figure shows that the bidders substantially discount their bids from the observed signals. We see that this occurs for two reasons. First, their expected value (conditional on winning) is discounted from their signal because of winner's curse. Second, because it is a first price auction, bidders discount their bid from the expected value of the item.

## 11.4 Auctions with Coalitions

The classic data set for considering common values auctions is the US federal government's off-shore drilling auctions. One surprising fact is that the auctions include "coalitions" of bidders. Basically, legal bid rings. Given the discussion about bid rings in the Chapter 10, it seems very odd that the government would allow collusion in these auctions.

An obvious policy question is whether the government should in fact allow bid rings in OCS auctions. You are probably thinking that the answer is obviously no. Actually it is not that obvious in the case of common value auctions.

The section works through how bidding in coalitions can be estimated. It then uses the estimated parameters from above and some characteristics of OCS auctions to simulate the policy.

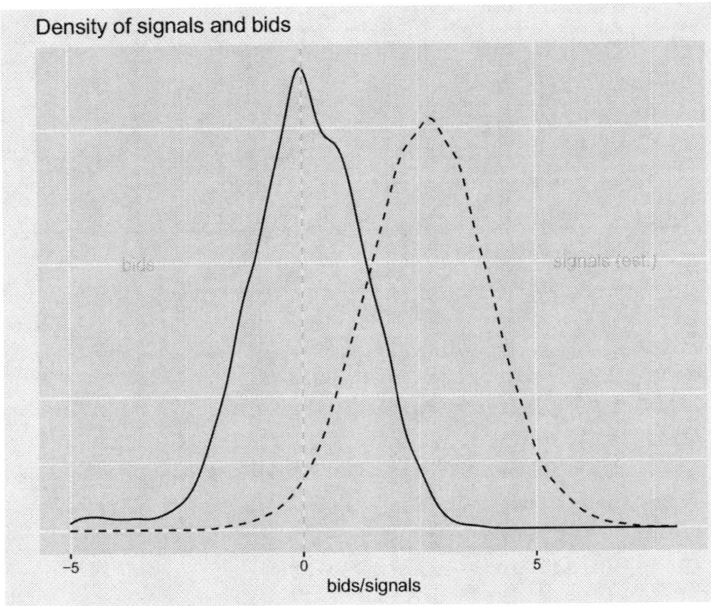


**FIGURE 11.4**
Plot of the density of normalized bids from all auctions and signals (estimated) in OCS auctions without coalitions that have between 3 and 10 bidders. The bidder's valuations are shifted down from the estimated signal distribution.

### 11.4.1 The Benefit of Coalitions

The reason for allowing coalitions is that coalitions allow bidders to pool their information about the item's value. By increasing the amount of information available to the bidders, the coalitions may lead to higher bids!

Remember we said that there are two reasons for bidders shading their bids. The first, discussed in Chapter 10, states that bidders shade in order to account for the probability of winning and trade off the probability of winning against how much they pay if they win. As the number of bidders increases, the probability of any particular bidder winning decreases, and so the smaller trade off leads to higher bids. As we said previously, a bid ring allows bidders to reduce their bids because their probability of winning is higher. The second reason for shading the bids is because of the information problem. Here the bid ring works in the opposite direction, by pooling information the bidders have a more precise signal of the value of the item which allows them to bid more.

How does a bidder's valuation change with more signals? Assume that the expected valuation for the ring will be the mean of the signals, conditional upon that mean being greater than all the other signals. From statistics, we know that we can approximate the distribution of this sample mean as a normal distribution with a mean equal to the true mean and the variance equal to true variance divided by the sample size.

For the $J$ members of the coalition, the probability of winning the auction with a particular average of signals ($\bar{s}$) is as follows.

$$\Phi\left(\frac{\bar{s}-\mu}{\sigma}\right)^{N-J-1} \tag{11.4}$$

where this gives the probability that the $J$ members of the coalition will observe a particular average of their signals multiplied by the probability that the other bidders outside the coalition will have signals below coalition s average.

What about for the bidders outside the coalition?

$$\Phi\left(\frac{s-\mu}{\sigma}\right)^{N-J-1} \Phi\left(\frac{s-\mu}{\frac{\sigma}{J}}\right) \tag{11.5}$$

The probability is the probability that a signal $s$ is observed and is greater than all the other bidders outside the coalition and greater than average of the signals that are in the coalition.

### 11.4.2 Estimating Coalitions in R

The probabilities with coalitions allowed are the following. For bidders in the coalition, their signal has less noise than for bidders outside the coalition. For all bidders, the number of independent bidders is lower. The code uses `_in` to refer to the bidders in the coalition and `_out` for the bidders outside the coalition.

```
> log_G_in = function(s, v, sigma=1, N, J)
+    (N-J-1)*log_F(s, v, sigma/sqrt(J))
> log_g_in = function(s, v, sigma=1, N, J) log(N-J-1) +
+    log_f(s, v, sigma/sqrt(J)) +
+    (N-J-2)*log_F(s, v, sigma/sqrt(J))
> log_G_out = function(s, v, sigma=1, N, J)
+    (N-J-1)*log_F(s, v, sigma)
> log_g_out = function(s, v, sigma=1, N, J) log(N-J-1) +
+    log_f(s, v, sigma) + (N-J-2)*log_F(s, v, sigma)
```

The expected values and bids for bidders in and out of the coalition are as you would expect.

```
> E_in = function(s, N_i, J_i) {
+    g_u = matrix(NA,length(u),length(sig))
+    u_mat = matrix(NA, length(u), length(sig))
+    for(j in 1:length(sig)) {
+      g_u[,j] = exp(log_g_in(s, u, sig[j], N_i, J_i))
+      u_mat[,j] = u
```

```
+    }
+    sum_g_u = sum(g_u)
+    gamma_u = g_u/sum_g_u
+    mu = sum(u_mat*gamma_u)
+    sigma = sqrt(sum(u_mat^2*gamma_u) - mu^2)
+    return(list(mu=mu, sigma=sigma))
+ }
> E_out = function(s, N_i, J_i) {
+    g_u = matrix(NA,length(u),length(sig))
+    u_mat = matrix(NA, length(u), length(sig))
+    for(j in 1:length(sig)) {
+       g_u[,j] = exp(log_g_out(s, u, sig[j], N_i, J_i))
+       u_mat[,j] = u
+    }
+    sum_g_u = sum(g_u)
+    gamma_u = g_u/sum_g_u
+    mu = sum(u_mat*gamma_u)
+    sigma = sqrt(sum(u_mat^2*gamma_u) - mu^2)
+    return(list(mu=mu, sigma=sigma))
+ }
> b_in = function(s, N, J) {
+    v_bar = E_in(s, N, J)
+    G = exp(log_G(s, v_bar$mu, v_bar$sigma, N-J+1))
+    g = exp(log_g(s, v_bar$mu, v_bar$sigma, N-J+1))
+    return(v_bar$mu - G/g)
+ }
> b_out = function(s, N, J) {
+    v_bar = E_out(s, N, J)
+    G = exp(log_G(s, v_bar$mu, v_bar$sigma, N-J+1))
+    g = exp(log_g(s, v_bar$mu, v_bar$sigma, N-J+1))
+    return(v_bar$mu - G/g)
+ }
```

### 11.4.3 Policy Simulation

In the mid-1970s, the Department changed the policy to make illegal for larger bidders to join forces, but that still allowed small bidders to join with big bidders or with other small bidders.

Assume that auctions with coalitions have the same signal distribution as auctions without coalitions. This allows us to use the estimates from the previous section above in the policy simulations.

This analysis is restricted to cases where there is just one coalition and there are more than two bidders. When the number of bids is smaller than the number of bidders, we have coalitions in the auction. This information is then merged back into the main data set. The code uses the estimates of `mu` and `sigma` from the previous section to simulate bids in and outside of the coalition. The number of bidders in the auction is from the data. It simulates the auctions allowing for a coalition of bidders and if the coalition is not allowed.

```
> dt1 = dt[, .(num = .N,
+               num_coy = sum(as.numeric(nCompany))),
+          by = TractNumber]
> dt = merge(dt, dt1, by="TractNumber")
```

```
> dt2 = dt[num_coy - num == 1 & num > 2]
> M = length(unique(dt2$TractNumber))
> N = dt2$num_coy
> mu = rep(b1$par[1], M)
> sigma = exp(b1$par[2])
> bids_sim = NULL
> bids_cf = NULL
> set.seed(123456789)
> for(i in 1:M) {
+   N_i = N[i]
+   s_i = rnorm(N_i, mu, sigma)
+   bids_i = sapply(1:N_i, function(j) b_fun(s_i[j], N_i))
+   bids_i_in = b_in(mean(s_i[1:2]), N_i, 2)
+   bids_i_out = sapply(1:(N_i-1), function(j)
+     b_out(s_i[j], N_i, 2))
+   bids_sim = c(bids_sim, bids_i_in, bids_i_out)
+   bids_cf = c(bids_cf, bids_i)
+   # print(i)
+ }
```

The code then creates a plot of the density of the bids in auctions with and without coalitions.

```
> ggplot_ocs_bids =
+   data.frame(
+     rings = bids_sim,
+     no_rings = bids_cf
+   ) |>
+     filter(
+       is.finite(rings) & is.finite(no_rings)
+     ) |>
+     ggplot(aes(rings)) +
+     geom_density(alpha = 0.5) +
+     geom_density(aes(no_rings), linetype = 2, alpha = 0.5) +
+     labs(
```

```
+       x = "Normalized bids",
+       y = "",
+       title = "Density of bids"
+     ) +
+     geom_text(aes(x = 3, y = 0.2, label = "rings"),
+               color = "gray") +
+     geom_text(aes(x = -3, y = 0.2, label = "no rings"),
+               color = "gray") +
+     theme(axis.text.y=element_blank(),
+           axis.ticks.y=element_blank())

> ggplot_ocs_bids
```

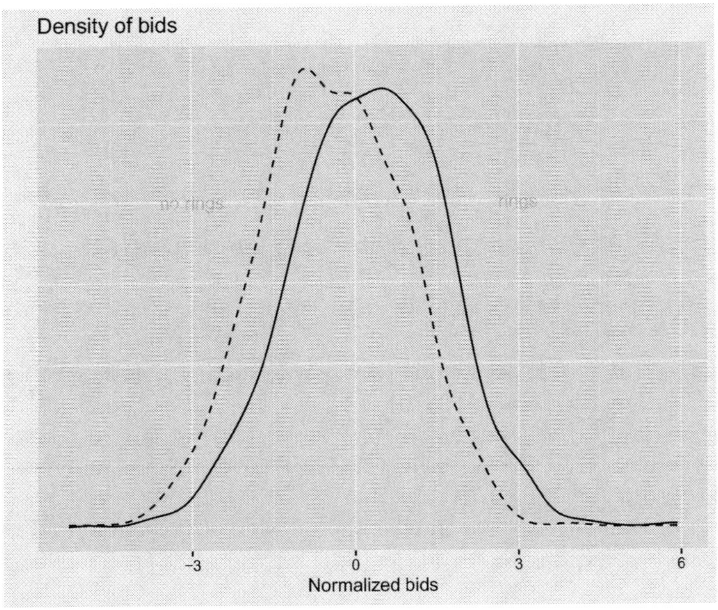


**FIGURE 11.5**
Plot of the density of the amount bid where coalitions are not allowed ("no rings") and coalitions are allowed ("rings"). The value of the bids have been normalized. The bids are higher when coalitions are allowed in these OCS auctions

Figure 11.5 shows that allowing bid rings (coalitions) tends to lead to higher bids! This analysis accounts for the fact that bidders will bid lower because the number of independent bidders is lower. Despite that, the bids are higher showing the advantage of aggregating signals in common values auctions.

## 11.5 Discussion and Further Reading

Chapter 10 made a simplifying assumption called independent private values (IPV). This assumption rules out the common values model. Laffont and Vuong (1996) present the main negative result of the common values literature. It states that without strong parametric assumptions it is not possible to identify the exact model generating the data in this setting. Despite this negative result, we could test for whether the auction is a common values auction (Haile et al., 2006).

The analysis in this chapter uses a parametric model. The chapter suggests that allowing cooperation among competitors may lead to better outcomes for the government in the sale of oil drilling rights. See Paarsch and Hong (2006) for more detailed analysis of the econometrics of auctions, including common value auctions. The OCS data used here have been analyzed in a number of papers, for example, Hendricks et al. (2003).

# Part IV

# Dynamic Games of Incomplete Information

# 12

# *Moral Hazard*

## 12.1 Introduction

The book has four parts covering complete information static games, complete information dynamic games, incomplete information static games, and incomplete information dynamic games. This part is the last.

This part considers two types of dynamic games of incomplete information. The first are games in which one player does not observe the other player's action before making their choice. What makes these games dynamic is that the first player's action is associated with a signal that is observed by the second player before making his choice. The second set of games are ones where the action of the first player is observed by the second player, but the second player does not know the first player's payoffs. The second set of games are discussed in the next chapter.

The first set of games are often given the moniker, moral hazard or principal-agent games. Consider a game between an employer and employee, where the principal is the employer and the agent is the employee. The employee takes actions like working hard and thinking carefully, but the employer doesn't get to observe this. The employer may observe signals of these things like the number of hours worked or profits of the firm. In these games, we are generally interested how the employer can get the employee to take actions that the employer prefers even when those actions are not directly observed.

This chapter presents the principal-agent game and applies it to the problem faced by financiers of whaling expeditions in New England in the 1830s. A new, at the time, form of company organization, the corporation, allowed many more people to invest in whaling. Ventures funded through corporations tended to give managers less of a stake in the outcome of the venture and tended to have worse outcomes.

DOI: 10.1201/9781003351603-12

## 12.2 Principal-Agent Game

In the 1830s, whaling was an important industry in the United States, particularly in communities in New England. The industry developed to hunt and kill sperm whales for oil and other products. The ships sailed literally around the world and the voyages would last years. To hunt and kill whales, you need a ship with the necessary equipment, supplies, a captain, and a crew of around 30. Obviously you need money for all of this. The investment will payoff when you sell the oil and other products rendered from the whales.

How do investors insure that they get a good return on their investment? How do they know that they have a good captain and crew? How do they know that the captain and crew are doing a good job when they are literally on the other side of the world and it is the 1830s?

These operations generally relied on a number of incentive mechanisms. The captain and crew were paid a small share of the profits. The investors hired an agent who was responsible for hiring the captain and crew, determining the voyage's route and communicating with the captain during the voyage. Under the law, the owner of the vessel controlled everything on the vessel including any equipment used. Because of this, investors generally bought an ownership share of the vessel itself.

Two organizational structures were used to finance these hunts. In one case a small number of investors hired an agent. The agent was generally given a large share of the ownership of the vessel and the investors may include family members or members of the local community. An alternative method for raising funds from investors was to create a corporation. These organizations provided some legal recourse for investors and made it clearer who owned what. Corporate whaling enterprises had a much larger set of owners investing smaller amounts of money. Like unincorporated ventures, they still hired an agent who was responsible for overseeing the operation. One big difference between the two organizations seems to be share of ownership given to the agent. For the corporations the evidence suggests that the agent earned a small fraction of the profits.

This section works through a formal game of the investor–agent relationship in 1830s whaling. It then goes through an example with simulated data.

### 12.2.1 Simple No Contract Game

Consider a simplified version of the game faced by whaling ventures. In this game, the agent chooses her effort level and the investor pays her based on what the outcome they observe.

- Players: Investor, Agent
- Strategies:

- Agent: $e \in \{e_L, e_H\}$
- Nature: $y \in \{y_L, y_H\}$, where $y_H > y_L$, as a function of $e$, $\Pr(y_H|e_H) = p_H > \Pr(y_H|e_L) = p_L$
- Investor: $w(y) \in \{w_L, w_H\}$

- Payoffs:
  - Agent: $w(y) - c(e)$, where $c(e_H) = c$ and $c(e_L) = 0$
  - Investor: $y - w(y)$
- Beliefs: $p = \Pr(e = e_H)$

In this game, the agent moves first and chooses how much "effort" to put into the venture. If they choose the high effort level $e_H$, then it costs them $c$, while the low effort level cost them nothing. Nature observes the effort level and chooses the outcome of the venture. The higher effort level increases the probability of the better outcome $y_H$. Lastly, the Investor observes the choice of Nature and pays the Agent $w$. The more she pays the agent, the worse off the investor is.

What is the Bayes Nash equilibrium of the game? Is there an equilibrium where the Agent chooses $e_H$? What is the subgame Perfect Nash equilibrium?

The last is a trick question. There is only one subgame, the whole game.[1]

The Agent's strategy is to choose an effort level $e_L$ or $e_H$. The Investor's strategy is to choose a payment level given the observed outcome $y$, $w(y)$ and the Investor's beliefs about the choice of the Agent.

### 12.2.2 Bayes Nash Equilibrium

One proposed equilibrium is for the Agent to choose the high effort level $e_H$ and for the Investor choose a payment that pays more if the outcome is $y_H$ than if the outcome is $y_L$. Such payments could make it worthwhile for the Agent to choose the higher cost effort level because that effort level increases the probability of getting a higher payment.

This is not an equilibrium. It is not optimal for the Investor to pay the Agent anything. What ever effort level the Agent chooses, the Investor prefer not to pay the Agent anything.

Assume the equilibrium is for the Agent to choose $e_H$ and the Investor pays the Agent $w(y_H) = w_H$ and $w(y_L) = w_L$ where the following inequality holds.

$$p_H w_H + (1 - p_H) w_L - c \geq p_L w_H + (1 - p_L) w_L \tag{12.1}$$

[1]A **perfect Bayesian Nash equilibrium** has a similar flavor to a subgame perfect Nash equilibrium. In a perfect Bayesian Nash equilibrium, the players' strategies must be optimal given their beliefs and the beliefs must be consistent with Bayes' rule at each information set along each equilibrium path that can be reached with positive probability.

The expected payoff from choosing the high effort level is higher than the expected payoff from choose the low effort level. We call this the incentive compatibility (IC) constraint.

If in the proposed equilibrium the payment $w(y)$ is such that the IC holds, then the Agent will choose the high effort level. What if Agent chooses $e_H$. What contract should the Investor offer? Let $w^*$ denote the Investor's payment if they observe $y_H$. In equilibrium, the Investor's expected profits are as follows.

$$p_H(y_H - w^*) + (1 - p_H)(y_L - w_L) \tag{12.2}$$

The Investor should offer $w^* = 0$.

In the proposed equilibrium, the Agent chooses $e_H$ and the Investor must believe that is true. Given all of that, why should the Investor pay the Agent anything?

### 12.2.3 Simple Contract Game

The problem in the previous game is that the Investor cannot commit to paying the Agent for their high effort level. The result is that the Agent is not willing to put in the effort. We need a contract. In general, we assume that such a contract can be enforced, for example in a court of law. Alternatively, it could be enforced in the court of public opinion or the court of the back alleys of the New England port town. If the Investor doesn't pay the Agent what is stated in the contract then the Investor gets punished. Making the contract enforceable limits what is contractable. We cannot contract on the effort level of the Agent because that is not observed by anyone but the Agent, certainly not in a court of law. Potentially, we can contract on the outcome ($y$) as that is observed. We will assume that the $y$ is both observable and verifiable by a court of law.

- Players: Investor, Agent
- Strategies:
  - Investor: Offer contract $w(y)$
  - Agent: Accept or Reject given offer $w(y)$
  - Agent: Given Accept of $w(y)$ choose $e \in \{e_L, e_H\}$
  - Nature: Choose $y \in \{y_L, y_H\}$ given $e$, $\Pr(y_H|e_H) = p_H > \Pr(y_H|e_L) = p_L$
- Payoffs: If the contract is accepted:
  - Investor: $\mathbb{E}(y(e) - w(y(e)))$
  - Agent $\mathbb{E}(w(y(e))) - c(e)$

  Assume that both get 0 if rejected.
- Beliefs: $p = \Pr(e = e_H)$

In the Bayes Nash equilibrium, the Investor offers a contract such that the following equality holds.

$$p_H w_H + (1 - p_H) w_H - c = 0 \tag{12.3}$$

The Agent is indifferent between their expected wage if they choose the high effort level and their outside option (which is assumed to be 0). We call this the individual rationality (IR) constraint. In this set up, $w_L$ will be negative, but you should just think of this as a value lower than Agent's outside option or alternative if they reject the contract.

Also, the Agent prefers the high effort level to the low effort level. The IC is as follows.

$$p_H w_H + (1 - p_H) w_L - c \geq p_L w_H + (1 - p_L) w_L \tag{12.4}$$

The difference between pay given the good outcome $w_H$ and pay given the bad outcome $w_L$ is large enough to induce the Agent to choose the high effort level despite the higher cost $c$.

The Investor then chooses $w_H$ and $w_L$ such that both the IR and IC constraints hold. Moreover, for it to be an equilibrium, it must be that $p_H(y_H - w_H) + (1 - p_H)(y_L - w_L) \geq 0$. We know from the IR constraint that it must be that the following in equality holds for the contract to be profitable.

$$p_H y_H + (1 - p_H) y_L - c \geq 0 \tag{12.5}$$

So the contract is profitable if the expected output from the project is greater than the cost to the Agent of completing the project.

### 12.2.4 More Complicated Contract Game

Let's model the principal-agent problem associated with investors contracting with an agent to run the whaling venture.

- Players: Investor, Agent
- Strategies:
  - Investor: Offer $w(y)$
  - Agent: Accept or Reject offer
  - Agent: Choose $e \geq 0$.
  - Nature: Choose $y$ given $e$, $y \sim F(e)$
- Payoffs
  - Agent: $\mathbb{E}(u(w(y))|e) - c(e)$
  - Investor: $\mathbb{E}(y(e) - w(y(e)))$
- Beliefs: $e \sim P$.

The Agent chooses an effort level, $e$, which incurs some cost $c(e)$. The contract is based on the outcome, $y$, where this is a random number drawn from a distribution that is determined by Agent's effort choice $F(e)$. Based on this outcome, the Agent is paid $w(y)$. One difference between this model and what we have seen previously is that the Agent's payoff is determined by their utility function $u(y)$. Here we are going to assume that the agent is risk-averse while the principal is risk-neutral. The principal would like to incentivize the agent by giving them a large share of the outcome. This implies that the agent is taking on a lot of risk. The principle will need to compensate the agent by paying them a very large amount that does not vary with the outcome of the venture. We will talk more about this in a bit.

### 12.2.5 Bayes Nash Equilibrium

One of the issues highlighted in this book is the difference between assumptions about the game and assumptions about equilibrium. This issue comes up again here. Look at the payoffs for the Investor. Her expected payoff is not conditional on the effort level of the Agent. In a Bayes Nash equilibrium, we assume that the Investor's beliefs are consistent with equilibrium behavior. In equilibrium, the Agent will choose a particular effort level $e^*$, so *in equilibrium* the Investor's expected payoff will be $\mathbb{E}(y(e) - w(y(e))|e = e^*)$.

The Agent will choose this effort level to optimize his expected payoff.

$$\max_e \quad \mathbb{E}(u(w(y(e)))) - c(e) \tag{12.6}$$

The Agent knows $e$ and so the cost function is not inside the expectation.

The Investor will choose $w(y)$ so as to maximize her profits.

$$\begin{array}{ll} \max_{w(y)} & \mathbb{E}(y(e) - w(y(e))|e = e^*) \\ s.t. & \mathbb{E}(u(w(e^*))) - c(e^*) \geq 0 \\ & e^* = \arg\max_e \mathbb{E}(u(w(y(e)))|e) - c(e) \end{array} \tag{12.7}$$

In words, the Investor will choose a payment as a function of the outcome $w(y)$, that maximizes her expected return subject to the Agent choosing to accept the contract and choosing their optimal effort level conditional on the payment offer.

### 12.2.6 Parameterized Model

Assume that the Agent's expected utility is a mean-variance utility. That is, the Agent's utility is increasing in the mean of his payoffs and decreasing in the variance of his payoffs. The agent's dislike of risk is represented by the parameter $r$.

$$\mathbb{E}(u(x)) = \mu_x - r\frac{\sigma_x^2}{2} \tag{12.8}$$

where $x \sim \mathcal{N}(\mu_x, \sigma_x^2)$. This utility function is a substantial simplification. It is often justified using a particular utility function and assuming the outcome is normally distributed.

The Agent doesn't like producing effort and the cost of the effort is governed by the parameter $k$ and the cost is increasing in effort at a geometric rate, $c(e) = -k\frac{e^2}{2}$. For a particular incentive rate, $b$, the Agent's expected utility is as follows.

$$b(\mu_x + e) - r\frac{b^2\sigma_x^2}{2} - k\frac{e^2}{2} \tag{12.9}$$

The variance of $bx$ is $b^2\sigma_x^2$. Also the dividing by 2 thing is useful once we get to the first-order condition.

$$\begin{aligned} b - ke &= 0 \\ e &= \tfrac{b}{k} \end{aligned} \tag{12.10}$$

The agent's optimal effort level is increasing in the power of the incentives and decreasing in the cost.

The production function is simply that $y(e) \sim \mathcal{N}(\mu + e, \sigma)$. The mean of output is increasing 1 to 1 with the effort level $e$.

The incentive contract used by the Investor is a linear function of output, $w(y) = a + by$, where $a$ is a constant and $b$ is the fraction of the output received by the Agent. When you look below you see there is no $a$. This is because we have implicitly solve the individual rationality constraint by choosing $a$ such that the Agent is indifferent between accepting the contract and rejecting the contract. We can also substitute in the Agent's optimal effort level.

$$\pi_I(b) = \mu + \frac{b}{k} - \frac{rb^2(\sigma^2)}{2} - \frac{b^2}{2k} \tag{12.11}$$

The first-order condition is then given by the following equation.

$$\begin{aligned} \tfrac{1}{k} - rb\sigma^2 - \tfrac{b}{k} &= 0 \\ 1 - krb\sigma^2 - b &= 0 \\ b &= \tfrac{1}{1+kr\sigma^2} \end{aligned} \tag{12.12}$$

The power of the incentive contract is decreasing in the effort cost of the Agent, the risk aversion of the Agent and the variance in the output.

### 12.2.7 Simulation with R

In the code, the function in Equation (12.11) is as follows. The parameters `mu` and `sigma` are global variables determined below.

```
> Pi_I = function(b, r, k) {
+    mu + b/k - r*(b^2)*(sigma^2)/2 - (b^2)/(2*k)
+ }
```

The parameters of the simulation have been calibrated such that the incentive contract is similar to the average for unincorporated ventures in the data.

```
> mu = 90000
> sigma = 40000
> k = 0.000005
> r = 0.0002
```

Given the set up we can solve for the optimal share of the output given to the agent and the optimal effort level.

```
> b1 = optimize(Pi_I, c(0, 1), maximum = TRUE, r=r, k=k)
> b1$maximum
  [1] 0.3846154
> b1$maximum/k
  [1] 76923.08
> Pi_I(b1$maximum, r, k)
  [1] 128461.5
```

The firm makes \$128,000 from the venture with the agent taking 39%. The agent's effort cost is \$77,000.

Now consider what happens if the power of the incentive contract is significantly reduced. Let $b = 0.05$, rather than the optimal level.

```
> 0.05/b1$maximum
  [1] 0.13
> Pi_I(0.05, r, k)/Pi_I(b1$maximum, r, k)
  [1] 0.7733832
```

The new level of optimal effort for the Agent is 13% of the optimal contract and the Investor's profits also falls but to just 77% of what they would be with the optimal contract.

## 12.3 Empirical Analysis: Whaling Corporations in the 19th Century using R

Wellsley College professor, Eric Hilt, documents the surprising failure of the corporate structure in whaling. In the first half of the 19th century, corporations were a relatively new type of institution in the United States. They provided a legal structure for people to raise money from investors where it was clear what rights investors did and did not have. Whaling in New England was generally financed by small groups of investors, many of whom knew each other or were from the same family. Corporations opened up whaling to a much broader range of investors. Given these advantages it is surprising that the corporate structure performed so poorly.

This section using data from whaling venture financial records to understand the contracts with agents used by whaling corporations. Were the contracts responsible for the poor performance of corporations?

### 12.3.1 Whaling Data

Bringing in the data called `whaling.csv`.

The code below reads in the data and plots average output by year and by how the venture was financed.

```
> file = paste0(dir, "whaling.csv")
> dt = fread(file)
> dt1 = dt[, .(lprod_wb = mean(lprod_wb,
+                                    na.rm = TRUE)),
+           by = .(ayear, corp)]
> ggplot_log_output = setDF(dt1) |>
+   mutate(
+     corp = as.factor(corp)
+   ) |>
+   ggplot(aes(ayear, lprod_wb, corp,
+                    linetype = corp)) +
+     geom_line() +
+     labs(
+       x = "Year",
+       y = "",
+       title = "Average log output by year"
+     ) +
+     theme(axis.text.y=element_blank(),
+           axis.ticks.y=element_blank())

> ggplot_log_output
```

Figure 12.1 presents a plot of the productivity of the whaling ventures. It is the ratio of the output of the venture in terms of the value of the oil and other products rendered from the whale to the size of the ship multiplied by the length of the journey.

The figure shows two things. First, productivity is falling dramatically over time. It is becoming harder and harder to find whales to kill. Second, the corporate ventures tend be less productive than closely held ventures.

Why are corporations doing so poorly?

### 12.3.2 Regressions

Does the pattern from Figure 12.1 hold when we are more careful about accounting for various factors determining the outcome. One of the big variables

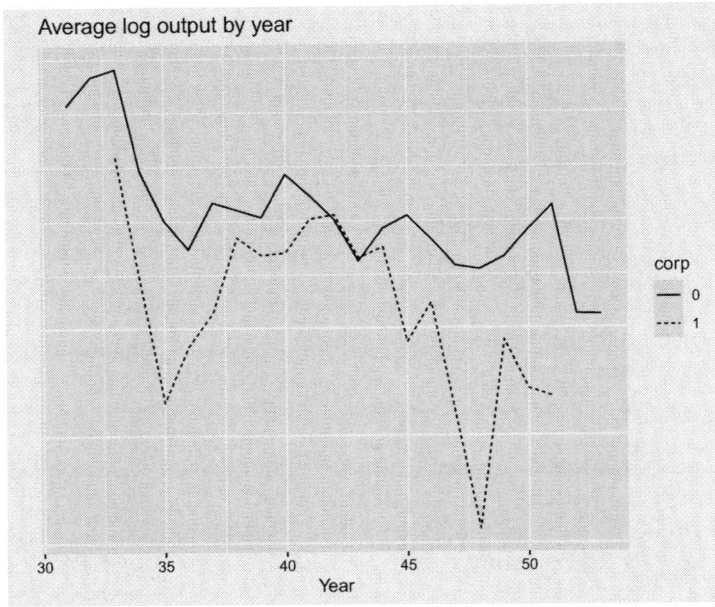


**FIGURE 12.1**
Line chart of average log of output by year and corporation. It shows output decreasing from 1835 to 1850, but corporate (corp = 1) output is generally lower.

is the agent themselves. The data include cases where the same agent is used by a closely held firm and by a corporation. Using fixed effects, we can account for differences across agents. We can hold the "agent-effect" fixed.

Table 12.1 presents the results from the **fixed effects analysis**. The empirical model is generally called two-way fixed effects. We have fixed effects for the agent (the first way) and fixed effects for the year (the second way) which are not presented in the table. The idea is that we can account for the agent and the year to isolate the effect of the financial structure. It shows that corporations have lower productivity even accounting for individual agent effects. Specifications (3) and (4) suggest that it may not be the corporate entity itself but due to the larger number of owners associated with the corporate structure. It also shows that having the captain die, is not good for the success of the hunt.

Why may this be happening? Corporations provided agents with a share of the output, but shares were substantially lower than what we see for the closely held ventures. What are the implications of this difference in compensation to the agent's effort level and profitability of the venture?

**TABLE 12.1**
OLS regressions of output on corporate form with agent and year fixed effects. Ownership structure is accounted for either through the dummy variable for corporation or by the number of investors.

| | *Dependent variable:* | | | |
|---|---|---|---|---|
| | lprod_wb | | | |
| | (1) | (2) | (3) | (4) |
| corp | −0.455** | −0.411** | | −0.279 |
| | (0.178) | (0.179) | | (0.442) |
| I(owners/10) | | | −0.126* | −0.126* |
| | | | (0.065) | (0.065) |
| I((owners/10)^2) | | | 0.014 | 0.014 |
| | | | (0.012) | (0.012) |
| atlantic | | 0.048 | 0.074 | 0.074 |
| | | (0.053) | (0.060) | (0.060) |
| pacific | | −0.107** | −0.102** | −0.102** |
| | | (0.044) | (0.049) | (0.049) |
| tons | | −0.001*** | −0.001** | −0.001** |
| | | (0.0003) | (0.0003) | (0.0003) |
| vesselage | | −0.003 | −0.004 | −0.004 |
| | | (0.002) | (0.002) | (0.002) |
| capexp | | 0.006 | 0.008 | 0.008 |
| | | (0.008) | (0.010) | (0.010) |
| capdied | | −0.338*** | −0.407*** | −0.407*** |
| | | (0.100) | (0.121) | (0.121) |
| Observations | 831 | 809 | 671 | 671 |
| $R^2$ | 0.300 | 0.339 | 0.343 | 0.343 |
| *Note:* | | $^{*}$p<0.1; $^{**}$p<0.05; $^{***}$p<0.01 | | |

### 12.3.3 Calibrating the Model

In order to get some sense of what happens when incentives of the agent are changed, we can use the observed data to calibrate the game presented above.

In order to get apples to apples, we can use the same trick that we used in Chapter 10. Regress the output measure on various observed characteristics and then use the residual as the normalized output.

```
> df1 = setDF(dt) |>
+   filter(
+     corp == 0
+   ) |>
+   na.omit()
> lm5 = lm(output_wb ~ atlantic + pacific + tons +
+              vesselage + dyear, data = df1)
> df1$res = lm5$residuals + lm5$coefficients[1]
> mu = mean(df1$res, na.rm = TRUE)
> sigma = sd(df1$res, na.rm = TRUE)
```

Given the normalized output, we can use the observed contracts from the closely held firms to back out the parameters on agent's costs ($k$) and risk-preferences ($r$).

```
> e_1 = mean(df1$res, na.rm = TRUE)
> sigma_1 = sd(df1$res, na.rm=TRUE)
> b_1 = mean(df1$agt_shr, na.rm = TRUE)
> k_1 = b_1/e_1
> r_1 = e_1*(1 - b_1)/((b_1^2)*sigma_1^2)
```

```
> Pi_I(b_1, r_1, k_1)
  [1] 269567.9
> e_1*(1 - b_1)
  [1] 73007.96
```

The closely held firm makes \$270,000, and the agent's effort costs are \$73,000. Looking at the corporations, we can again normalize output and estimate the effort level of the agent given the share received by the agent.

```
> df2 = setDF(dt) |>
+   filter(
+     corp == 1
+   ) |>
+   select(
+     output_wb,
+     atlantic,
+     pacific,
```

```
+      tons,
+      vesselage,
+      dyear,
+      agt_shr
+    ) |>
+    na.omit()
> df2$res = df2$output_wb -
+   predict.lm(lm5, df2) +  lm5$coefficients[1]
> e_2 = mean(df2$res, na.rm = TRUE)
> b_2 = mean(df2$agt_shr, na.rm = TRUE)
> e_2_pred = b_2/k_1
```

```
> Pi_I(b_2, r_1, k_1)
  [1] 183437.4
> e_2_pred*(1 - b_2)
  [1] 3718.121
```

In the case of the profit sharing under the corporate form, the venture's profits fall to $183,000 and the agent's effort would be only $3,700.

```
> e_2_pred/e_1
  [1] 0.02094994
> Pi_I(b_2, r_1, k_1)/Pi_I(b_1, r_1, k_1)
  [1] 0.680487
```

Given the share offered to agents working for corporations, the predicted effort level falls to just 2% of the unincorporated effort level. That said, the predicted profits only fall to 68% of the unincorporated profits. The reason is that while the output falls, the share of output received by the investors is higher. In addition, the investors don't have to compensate the agent for risk.

```
> e_2/e_1
  [1] 0.6883835
```

While our model predicts that the Agent's effort level will drop precipitously. The effort level does fall by a large amount to 69% of the unincorporated level. The predicted output is much lower than what we actually observe in the corporate ventures. This discrepancy suggests that corporations are providing incentives. They are just not providing incentives in the form of share of output. They must be using other mechanisms like direct supervision and the threat of firing, rather than giving high powered incentives. Those incentives may leading to significantly lower effort levels by the Agent, but the returns to the Investors are not necessarily that bad.

## 12.4 Discussion and Further Reading

Wellsley College professor, Eric Hilt, has created an amazing data set on how whaling firms actually worked. It is very cool to bring modern economic theory to bear on why closely held firms worked so well in the 1830s (Hilt, 2006).

Theoretical work on incentives and the principal-agent problem substantially improved our understanding of contracts and performance pay systems (Lazear and Oyer, 2013). The classic paper on the principal-agent problem is Holmström (1979).

# 13

# *Adverse Selection*

## 13.1 Introduction

This chapter considers **adverse selection** problems. These games are generally split between **signalling** and **screening** games. In a **signalling game**, the player with an unknown type chooses an action and that action choice may provide information to the other player. In a **screening game**, the uninformed player goes first and offers some choices to the player whose type is not known. The observed choice may provide information to the uninformed player.

In either case, the question is whether the observed action provides information about the player's type. We may have equilibria in which the action provides information about the player's type, we call these **separating equilibria**. We also may have equilibria where the action does not provide information about the player's type. We call these **pooling equilibria**.

The chapter presents the most famous game of unknown types, George Akerlof's model a used car market and the lemons problem. It then uses these ideas to think about the market for health insurance. It considers whether government subsidies or taxes can be used to solve the lemons problem in health insurance. The chapter analyzes these policies with a model calibrated using data from the Medical Expenditure Panel Survey (MEPS).

## 13.2 Akerlof's Lemons

Outside economics, there is a perception that economists are gungho believers in free market capitalism. For economists that study markets and how markets work or don't work, it is odd to be called "believers." George Ackerlof is an economist through and through, he even won the Nobel prize in economics. He is also married to one of the most influential economists in the world former Treasury Secretary Janet Yellen. Akerlof showed how a simple misallocation of information in a market, would cause that market to fail.

This section introduces Akerlof's lemons market and provides a formal game-theoretic version of the problem.

DOI: 10.1201/9781003351603-13

### 13.2.1 The Lemons Market

Consider a car market. In this market, there are two types of cars, good cars and lemons (bad cars). Unfortunately, consumers cannot observe the type of car prior to purchase. They can, however, resell the car on the used car market after observing the car's type.

Now consider a bunch of new cars become available. The key thing about new cars is that the seller of the cars will sell the car no matter if it is good or a lemon. Consumers buy the car and learn the car's type. After the consumers learn the car's type they can sell the car on the used market or keep it. If prices in the used car market are roughly similar to prices in the new car market, then those customers who purchased lemons will dump them on the used car market and go back to the new car market hoping to get a good car.

Customers on the used car market are going to notice that the proportion of lemons just got a lot higher and so the price in the used car market will fall. Where previously, people that owned good cars may have been willing to sell their car on the used car market, the price dropped because of all the lemons entering the market. So these sellers are going to keep their good cars. With owners of lemons dumping them and owners of good cars keeping them, the used car market becomes increasingly filled with lemons. In response, the price in the used car markets drops. This only exacerbates things. As the price falls more, the only cars that are profitable sell in the used car market are lemons. Finally, the used car market is just lemons.

The market fails.

### 13.2.2 Lemons Game

Akerlof's original argument was not game theory. We can reorient it as a game.

- Players: Seller of car type Bad (b) or Good (g), Buyer
- Strategies:
  - Buyer: Offer price, $p$.
  - Nature: Determine car type, $\{b, g\}$
  - Seller: Observe car type and offered price, then Sell or Keep car.
- Payoffs:
  - Seller: Sell: $\mathbb{E}(p|Sell)$, Keep: $V$
  - Buyer: $\mathbb{E}(V|Sell, p) - p$
  - $V \in \{b, g\}$
- Beliefs: $\Pr(V = g) = \pi$

The Buyer offers a price for the used car. The Seller observe her type, actually the car's type. Given that observation and the observed price offered by the Buyer, the seller chooses to sell or keep the car. Again it is important to separate out what happens in the game and what happens in equilibrium. In equilibrium, the Buyer's price needs to adjust for the strategy of the Seller.

### 13.2.3 Bayes Nash Equilibrium

There is no Bayes Nash equilibrium in which there is a transaction of good cars.

Remember when determining a Bayes Nash equilibrium, we need to update the player's beliefs about the other player's type. Another way to say it, is that beliefs have to be consistent with player strategies in equilibrium.

Consider the case where only $b$ type sellers sell. In this case, the Buyer updates his beliefs on the type of Seller that sells a car on the used car market. Given these updated beliefs, the Buyer will offer a price of $b$. Bad car sellers will sell, but good car sellers won't. This is an equilibrium. A Seller with car of type $b$ is willing to sell at price $b$. A Seller of car type $g$, is not willing to sell at price $b$. The Buyer believes that all cars being sold are type $b$. The buyer offers $b$. We have a **separating equilibrium**.

What if both types of Sellers sell on the market. Both types sell, so the Buyer does not update their beliefs. In this case, the price offered by the buyer will be $\pi g + (1 - \pi)b$ where $\pi$ is the probability that the Seller is has a good car. As this price is greater than $b$, sellers of cars of type $b$ will also sell. The price offered is $\pi g + (1 - \pi)b < g$. Because this low price is below the value of the good car, the good-type seller will not want to sell. That is to say, this is not an equilibrium. There is no **pooling equilibrium**.

In the Bayes Nash equilibrium of the lemon's market, all cars sold in the used car market are lemons. Is this an accurate representation of the actual used car market? What mechanisms exist that encourage sellers of good cars to sell in the used car market?

## 13.3 Insurance

The most well-known example of **adverse selection** is in the context of insurance. Insurance is the idea that two agents can trade because they have different risk preferences. In actual fact, the insurance company is able to be less risk-averse because it is able to diversify across a large number of bets.

This section presents a model of insurance where buyers of insurance are risk-averse. It then considers what happens when buyers are of different unobserved types. Some buyers tend to be sicker. These buyers will value insurance more. There will be a tendency for the market to behave like Akerlof's lemons market.

The sicker types buying insurance and the healthy types not buying insurance. The question is what types of policies could we implement to encourage healthier people to buy insurance.

### 13.3.1 Model

Assume there are two possible states of the world, sick and healthy. In the healthy state, an individual earns a certain income $y$, but in the sick state, the individual earns $y$ and pays $h$ in costs. These may be medical expenses or loss of income, etc. The individual's utility in the two states are $u(y)$ and $u(y-h)$. Importantly, the individual is better off receiving the expected outcome rather than expected utility over the two outcomes. That is the following inequality holds.

$$\begin{aligned} u(\pi y + (1-\pi)(y-h)) &= u(y-(1-\pi)h) \\ &> \pi u(y) + (1-\pi)u(y-h) \end{aligned} \tag{13.1}$$

where $(1-\pi)$ is the probability that the sick state occurs for this individual.

If the inequality holds, we say that the individual is risk-averse. Mathematically, their utility function is concave.

Because of the higher utility of the expected value than the separate states, the individual is willing to buy a product that pays less in the healthy state but more in the sick state.

Consider a product that pays $h$ in the sick state at a cost (premium) of $p$ that is paid in every state.

$$\pi u(y-p) + (1-\pi)u(y-h+h-p) = u(y-p) \tag{13.2}$$

If the premium is low enough, then the individual will prefer to be insured. If $p = (1-\pi)h + \epsilon$, then the individual strictly prefers insurance from Equation (13.1) (if $\epsilon$ is small enough).

For the insurance company, they would like to offer the product if $p - (1-\pi)h \geq 0$. So there exists a positive $\epsilon$ where the insurance company finds it profitable to offer insurance and the individual is willing to purchase the insurance.

### 13.3.2 Game with Unknown Types

So in a world with perfect information (albeit with uncertainty), it is profitable for insurance companies to offer insurance. What if the riskiness of the person is not observed by the insurer? We have two types $\pi \in \{\pi_l, \pi_h\}$, where $\pi_l$ has a much higher probability of being in the sick state. The individual gets to observe their type prior to buying insurance, while the insurance company does not get to observe the individual's type.

In this situation, while both types may like to buy insurance, it is difficult to offer insurance to people with a low probability of being sick. This is basically the same problem as Akerlof's lemons market.

- Players: Buyer of type (Low ($\pi_l$), High ($\pi_h$)), Insurer
- Strategies:
  - Insurer: Offer a contract that pays $h$ in the sick state at a premium $p$ (which paid in all states).
  - Nature: Choose Buyer type $\{\pi_l, \pi_h\}$.
  - Buyer: Buy (or not) insurance at the offered premium after observing their type.
- Payoffs:
  - Buyer: Buy: $u(y-p)$, Don't Buy: $\pi u(y) + (1-\pi)u(y-h)$
  - Insurer: Buy: $p - (1-\pi)h$, Don't Buy: 0
  - where $\pi \in \{\pi_l, \pi_h\}$
- Beliefs: $\Pr(\pi = \pi_l) = q$

### 13.3.3 Bayes Nash Equilibrium

Consider the case where both types are willing to buy insurance and let $q$ be the proportion of "sick" individuals. In this case, the insurer will want the premium to be such that $p-(1-q\pi_l-(1-q)\pi_h)h \geq 0$. For the "sick" individual, they will purchase if the following inequality holds.

$$u(y-p) \geq \pi_l u(y) + (1-\pi_l)u(y-h) \tag{13.3}$$

Similarly, for the healthy individual.

Is there a Bayes Nash equilibrium of this game? Let use simulation to see what happens.

### 13.3.4 Simulation using R

Consider the following simulated game. The buyers are risk averse, with a constant absolute risk-aversion utility function, with the parameter $r$ determining the extent of their aversion.

$$u(C) = \frac{C^{1-r}}{1-r} \tag{13.4}$$

This is a relatively simple utility function that is used a lot in the literature.

Healthy individuals only get sick 1% of the time, while sick individuals get sick 10% of the time. The economy has a population where 90% are the healthy type. If an individual gets sick, they lose half of their income.

One difference between the simulation and the set up of the game above is that premiums are assumed to be set at the actuarially fair rate. This is equivalent to saying that the Insurer makes zero profits which is equivalent to saying that there is perfect competition in the insurance market.

```
> u_f = function(C, r) {(C^(1 - r))/(1 - r)}
> Eu_f = function(r, x, p) sum(p*u_f(x, r))
> prem_f = function(pi, h) (1 - pi)*h
```

The function `u_f()` is the utility function with constant absolute risk aversion. The function `Eu_f()` is the Agent's expected utility and `prem_f()` calculates the actuarially fair premium.

The parameters for the simulation are as follows. There is no particular rhyme or reason for the choice. The sick type has a 10 percent probability of being in the sick state, while the health type has a 0.1 percent probability of being in the sick state.

```
> r = 0.7
> pi_l = 0.9
> pi_h = 0.999
> q = 0.9
> y = 1
> h = 0.5
```

Utility for the uninsured for each type of individual.

```
> Eu_f(r, c(y, y-h), c(pi_l, 1 - pi_l))
  [1] 3.270751
> Eu_f(r, c(y, y-h), c(pi_h, 1 - pi_h))
  [1] 3.332708
```

Assume we have a **pooling equilibrium**. That is, both types purchase insurance.

```
> prem = prem_f(q*pi_l + (1 - q)*pi_h, h)
```

Given this premium of half of one percent of income, would both types purchase insurance?

```
> Eu_f(r, c(y - prem, y - prem),
+        c(pi_l, 1 - pi_l)) >
+   Eu_f(r, c(y, y-h), c(pi_l, 1 - pi_l))
  [1] TRUE
> Eu_f(r, c(y - prem, y - prem),
+        c(pi_h, 1 - pi_h)) >
+   Eu_f(r, c(y, y-h), c(pi_h, 1 - pi_h))
  [1] FALSE
```

No. The healthy type is not willing to purchase insurance at that premium.

Is there an equilibrium where only the sick types are insured? Is there a **separating equilibrium**?

```
> prem = prem_f(pi_l, h)
> Eu_f(r, c(y - prem, y - prem),
+        c(pi_l, 1 - pi_l)) >
+   Eu_f(r, c(y, y-h), c(pi_l, 1 - pi_l))
  [1] TRUE
> prem
  [1] 0.05
```

Yes. The sick individual is willing to pay a premium of 5% of her income. But 90% of the population is uninsured.

### 13.3.5 Mandatory Insurance

In the simulation above, only 10% of the population purchases insurance and the premiums are very high.

One solution to this problem is to make insurance mandatory. Or more accurately have some sort of fine or tax for those that don't take up insurance. That is, the payoff to the individual is decreased by the amount of the tax when the individual chooses not to purchase health insurance.

```
> tax = 0.05
> prem = prem_f(q*pi_l + (1 - q)*pi_h, h)
> Eu_f(r, c(y - prem, y - prem),
+        c(pi_l, 1 - pi_l)) >
+   Eu_f(r, c(y - tax, y-h - tax),
+          c(pi_l, 1 - pi_l))
  [1] TRUE
> Eu_f(r, c(y - prem, y - prem),
+        c(pi_h, 1 - pi_h)) >
+   Eu_f(r, c(y - tax, y-h - tax),
+          c(pi_h, 1 - pi_h))
  [1] TRUE
> prem
  [1] 0.04505
```

By adding a tax of 5%, we make the non-insurance expected utility lower, so the high type is willing to pay a higher premium to be insured. This allows risks to be pooled and makes it profitable for the insurance company to offer insurance that the healthy individuals are willing to accept.

Under this policy, the sick types do very well. Their premiums drop from 5% of income to 4.5% of income.

## 13.4 Empirical Analysis: Health Insurance using R

One of the concerns policy makers have with the health insurance market is that many people do not carry insurance. People who are generally healthy don't carry insurance. This tendency means that Akerlof's lemons problem leads to a market failure. People who tend to be sicker will have insurance, while healthier people will not have insurance.

This section looks at the actual health insurance market in the United States using the Medical Expenditure Panel Survey (MEPS). This data set provides information on how much health costs people actually have, how much income they have, and whether or not the people actually buy insurance.

### 13.4.1 Willingness to Pay

For each subgroup, we can calculate the expected utility in the case where they are uninsured, the case where they are insured in separating equilibrium (the baseline case), and for the case when they are insured under a pooling equilibrium (counterfactual case). To be clear, the analysis assumes that the current observed data are from a separating equilibrium. In each subgroup, there are healthy types that are choosing not to get insurance. The insurance company is assumed to offer a premium that is actuarially fair given the types that purchases insurance.

The utility is a constant absolute risk aversion (CARA) function with risk-parameter `r`. To derive expected utility, we take the average income for the subgroup, the probability of having medical expenditure, the average expenditure, and the standard deviation of expenditure. In the code, the expected utility is calculated numerically. The code uses a trick of creating a global variable and using transformations of the uniform and the standard normal distributions.

```
> set.seed(123456789)
> K = 1000
> U = runif(K) #uniform distribution
> Z = qnorm(U) #transform to standard normal
> EU_exp = function(r, income, exp_pos, exp_mean, exp_sd) {
+   exp = income - (U < exp_pos)*(Z*exp_sd + exp_mean)
+   return(Eu_f(r, ifelse(exp > 0, exp, 0), rep(1/K,K)))
+ }
```

### 13.4.2 Premiums

There are three premiums we can calculate in the data. First, there are the actual premiums paid by beneficiaries in the data. These are called "out of pocket" premiums. Many Americans have their premiums subsidized. For many working Americans, the premiums are paid in part or full by the firm that they work for. These workers are accepting at least some amount of lower salary in order to get the subsidy on the insurance premium from their employer. There is also a substantial tax benefit to workers who get health insurance, which is paid for by the American tax payer. Second, we can calculate the actuarial fair premium for each subgroup that has insurance. Lastly, we can calculate what the actuarial fair premium would be if the uninsured became insured for each subgroup.

The actuarial fair premium is calculated as the expected expenditure for the subgroup conditioning on having insurance (`UNISURD == 2`). The out of pocket premium is read from the data.

```
> file = paste0(dir, "meps_full.csv")
> dt1 = fread(file)
> dt_ins = dt1[UNINSURD == 2,.(premium = premium,
+                               premium_alt = exp_pos*exp_mean),
+              by = c("age_group", "SEX", "edu_group")]
```

MEPS data are used to calculate average out of pocket premiums and health expenditures for each subgroup.

The pooled premium is calculated as the average expected expenditure for each subgroup where both insured and unsured individuals are included in the average.

```
> dt_pool = dt1[, .(premium_pool = mean(exp_pos*exp_mean)),
+                 by = c("age_group", "SEX", "edu_group")]
```

The two data sets are merged back into the original data.

```
> dt2 = merge(dt1, dt_ins,
+             by = c("age_group", "SEX", "edu_group"))
> dt2 = merge(dt2, dt_pool,
+             by = c("age_group", "SEX", "edu_group"))
> dt2$premium = dt2$premium.y
```

The code below calculates the premiums by age group and plots the results.

```
> ins = dt2[,.(premium = mean(premium),
+             premium_alt = mean(premium_alt),
+             premium_pool = mean(premium_pool)),
+                 by = age_group]
> ins = ins[order(age_group)]
> line_prems = setDF(ins) |>
+   ggplot(aes(x = age_group)) +
+   geom_line(aes(y = premium_alt), linetype = 2) +
+   geom_line(aes(y = premium)) +
```

```
+    geom_line(aes(y = premium_pool), linetype = 3) +
+    labs(x = "Age",
+         y = "",
+         title = "Premium ($)") +
+    scale_y_continuous(limits = c(0,5000)) +
+    annotate("text", x = 50, y = 4000, label = "Seperating") +
+    annotate("text", x = 62, y = 3000, label = "Pooling") +
+    annotate("text", x = 60, y = 500, label = "Actual")

> line_prems
```

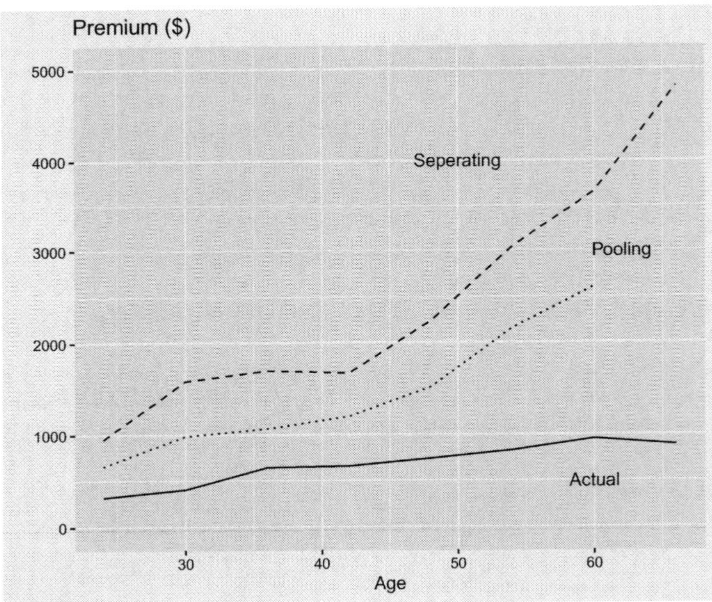


**FIGURE 13.1**
Line graph of average premiums for different age groups in the MEPS data. The lines show that premiums are increasing with age. The actual out of pocket premiums is substantially lower than the actuarial fair premiums for the insured population. The actual premiums increase from about \$200 to \$1,000 from 25 to 65. The actuarial fair premiums increase from \$1,000 to \$4,500 over the age-range. Pooled premiums increase from about \$750 to \$3,000.

Figure 13.1 presents line charts for average premiums across the different age groups. It presents the actual out of pocket premiums paid by insured beneficiaries, as well as the actuarially fair premium for the observed case and for the case where everyone is insured. It shows that the actual premium paid is substantially lower than the actuarially fair premium, particularly for the older beneficiaries. It also shows that premiums would come down in a pooling equilibrium.

### 13.4.3 Tax Policy

We may have too many people who are choosing to not have health insurance. The uninsured can be a burden on society because when they get sick they may not be able to cover the medical expenses, which moves those expenses to the hospitals or the government. There are a couple of policies that we could use to encourage a higher uptake of health insurance. We could either make choosing to have health insurance cheaper or we can make choosing not to have health insurance more expensive. The US uses the former. Taxes are not paid on the health insurance portion of income, giving a tax subsidy to employed people for purchasing insurance. An alternative policy of taxing those that don't choose to get insurance was originally planned for the insurance exchanges created as part of the Affordable Care Act.

The tax policy charges a tax on people who choose not to buy health insurance. If the tax is high enough, then most people will choose to have health insurance and the insurance premiums will fall.

To determine how behavior changes, calculate the expected value for being insured and being uninsured for the people in the data who are currently insured and for people who are currently uninsured. Below these are denoted `dt2_in` and `dt2_un`, respectively.

```
> r = 0.7
> dt2_in = dt2[UNINSURD == 2,
+              .(EU_uns = EU_exp(r,
+                                income + premium_alt - premium,
+                                exp_pos,
+                                exp_mean,
+                                exp_sd),
+                EU_ins = EU_exp(r,
+                                income - premium,
+                                0,
+                                0,
+                                0)),
+              by = "id"]
> dt2_un = dt2[UNINSURD == 1, .(EU_uns = EU_exp(r,
+                                               income,
+                                               exp_pos,
+                                               exp_mean,
+                                               exp_sd),
+                EU_ins_sep = EU_exp(r,
+                                    income - premium_alt,
```

```
+                                   0,
+                                   0,
+                                   0),
+                   EU_ins_pool = EU_exp(r,
+                                    income - premium_pool,
+                                    0,
+                                    0,
+                                    0)),
+              by="id"]
```

The function tax_policy() calculates the percentage of the population that is insured given a tax rate. The code is a bit ugly. It calculates the expected utility if the uninsured individual remains uninsured with the new tax. It then mergers, the results back into the original data. The function reports the percent of the population that choose to be insured under the policy. It sums the proportion of the population already insured and then for the population that is uninsured it determines whether the expected utility under a pooling equilibrium is greater then the tax.

```
> r = 0.7
> tax_policy = function(tax) {
+   dt21 = dt2[UNINSURD == 1,
+              .(EU_tax = EU_exp(r,
+                                (1 - tax)*income,
+                                exp_pos,
+                                exp_mean,
+                                exp_sd)),
+              by = "id"]
+   dt2_un1 = merge(dt2_un, dt21, by = "id")
+   dt2_un2 = merge(dt2_un1, dt2, by = "id")
+   return((sum(dt2$count[dt2$UNINSURD==2], na.rm = TRUE) +
+              sum((dt2_un2$EU_tax < dt2_un2$EU_ins_pool)*
+                   dt2_un2$count, na.rm = TRUE))/
+             sum(dt2$count, na.rm = TRUE))
+ }
```

The code below calculates the percentage of the population that is insured under different tax rates and plots the results.

```
> line_ins_prop = data.frame(tax = seq(0, 1, 0.01),
+             ins_prop = sapply(1:101, function(i)
+               tax_policy(i/100))) |>
+   ggplot(aes(x = tax, y = ins_prop)) +
+   geom_line() +
+   labs(x = "Tax rate",
+        y = "",
```

```
+          title = "Percent insured") +
+     scale_x_continuous(limits = c(0, 0.4)) +
+     scale_y_continuous(limits = c(0.75, 1)) +
+     geom_hline(yintercept = 1, linetype = 2)

> line_ins_prop
```

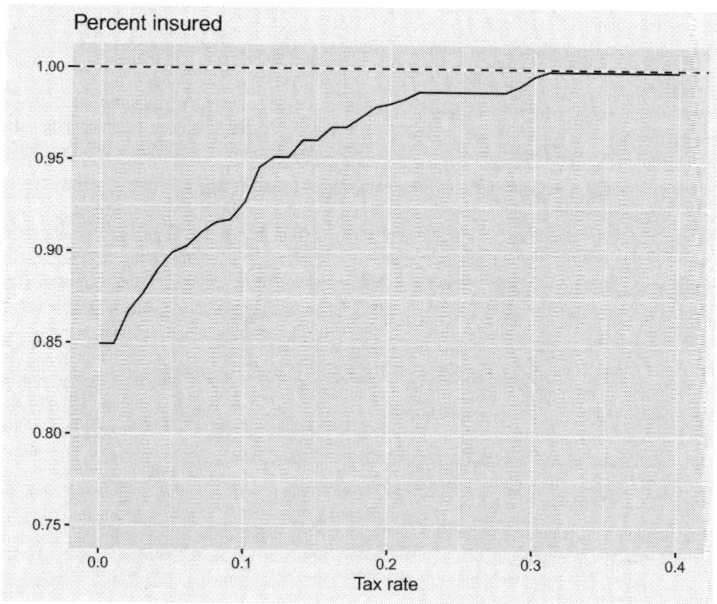


**FIGURE 13.2**
Line graph of the percentage of the population insured against the tax rate. Around 85% are insured without any tax, the proportion reaches 100% with a tax rate over 30%.

Figure 13.2 suggests that it is not that easy to encourage people to purchase insurance by imposing a tax penalty on the uninsured. The relationship is non-linear meaning that a small tax can get a large proportion of people onto insurance, but the relative effectiveness falls the higher the tax. The analysis suggests taxes 30 percent rage are needed to get everyone insured.

## 13.5 Discussion and Further Reading

Adverse selection is most famously illustrated by Akerlof (1970) and the idea of a lemons market. Adverse selection and moral hazard have become key to understanding health insurance markets. These ideas formed the basis for policies introduced in the Affordable Care Act that aimed to reduce the number of uninsured. See Pauly (1974). The tax on the uninsured, called the "individual mandate" was subject to lawsuits and was eventually repealed. So did it work? A recent survey by Brookings economist, Matt Fiedler finds that the tax *may* have reduced the uninsured rate. It turns out to be very tricky to determine what happened given all the changes that occurred with the introduction of the ACA and the complexity of the policy (Fiedler, 2020).

# Bibliography


Christopher P. Adams, *Learning Microeconometrics with R*, The R Series, CRC Press, December 2020.

Christopher P. Adams and Emek Basker. The Battle of the Big-Box Bookstores. January 2025.

Victor Aguirregabiria. Empirical industrial organization: Models, methods, and applications. University of Toronto, June 2021.

George A. Akerlof. The market for "lemons": Quality uncertainty and the market mechanism. *The Quarterly Journal of Economics*, 84(3) 488–500, August 1970.

Steffen Andersen, Seda Ertaç, Uri Gneezy, Moshe Hoffman, and John A. List. Stakes matter in ultimatum games. *American Economic Review*, 101(7): 3427–3439, December 2011. `https://doi.org/10.1257/aer.101.7.3427`.

Gaurab Aryal, Serafin Grundl, Dong-Hyuk Kim, and Yu Zhu. Empirical relevance of ambiguity in first-price auctions. *Journal of Econometrics*, 204(2):189–206, 2018.

Orley Ashenfelter, David Ashmore, Jonathan Baker, Suzanne Gleason, and Dan Hosken. Empirical Methods in Merger Analysis: Econometric analysis of pricing in FTC vs Staples. *International Journal of the Economics of Business*, 13(2):265–279, 2006.

John Asker. A study of the internal organization of a bidding cartel. *American Economic Review*, 100(3):724–762, 2010.

Susan Athey and Philip A. Haile. Identification in standard auction models. *Econometrica*, 70(6):2170–2140, 2002.

Susan Athey, Jonathan Levin, and Enrique Seira. Comparing open and sealed bid auctions: Evidence from timber auctions. *The Quarterly Journal of Economics*, 126:207–257, 2011.

Matthew Backus, Christopher Conlon, and Michael Sinkinson. Common ownership and competition in the ready-to-eat cereal industry. NBER Working Paper No. 28350, Janurary 2021.

David J. Balan and Keith Brand. Simulating Hospital Merger Simulations. *The Journal of Industrial Organization*, 71(1):47–123, 2018.

Laura H. Baldwin, Robert C. Marshall, and Jean-Francois Richard. Bidder collusion at forest service timber sales. *Journal of Political Economy*, 105 (5):657–699, 1997.

Steven Berry, James Levinsohn, and Ariel Pakes. Automobile prices in market equilibrium. *Econometrica*, 60(4):889–917, 1995.

Timothy F. Bresnahan and Peter C. Reiss. Entry in monopoly markets. *Review of Economic Studies*, 57(4):57–81, 1990.

Timothy F. Bresnahan and Peter C. Reiss. Empirical models of discrete games. *Journal of Econometrics*, 48:57–81, 1991a.

Timothy F. Bresnahan and Peter C. Reiss. Entry and competition in concentrated market. *The Journal of Political Economy*, 99(5):977–1009, 1991b.

David P. Byrne and Nicolas de Roos. Learning to Coordinate: A Study in Retail Gasoline. *The American Economic Review*, 109(2):591–619, 2019.

Colin F. Camerer. *Behavioral Game Theory: Experiments in Strategic Interaction*. Princeton University Press, 2003.

Lisa A. Cameron. Raising the Stakes in the Ultimatum Game: Evidence from Indonesia. *Economic Inquiry*, 37(1):47–59, 2007.

Cory Capps, David Dranove, and Mark Satterthwaite. Competition and Market Power in Option Demand Markets. *The RAND Journal of Economics*, 34 (4):737–763, Winter 2003.

George Deltas. Asymptotic and small sample analysis of the stochastic properties and certainty equivalents of winning bids in independent private values auctions. *Economic Theory*, 23:715–738, 2004.

Matthew Fiedler. The ACA's Individual Mandate In Retrospect: What Did It Do, And Where Do We Go From Here? *Health Affairs*, 39(3): 429–435, March 2020.

Drew Fudenberg and Jean Tirole. *Game Theory*. MIT Press, 1991.

Christopher Garmon. The accuracy of hospital merger screening methods. *The RAND Journal of Economics*, 48(4):1068–1102, 2017.

Martin Gaynor and William B. Vogt. Competition among hospitals. *The RAND Journal of Economics*, 34(4):764–785, 2003.

Robert Gibbons. *Game Theory for Applied Economists*. Princeton University Press, 1992.

Gautam Gowrisankaran and Robert J. Town. Competition, Payers, and Hospital Quality. *Health Services Research*, 38(6):1403–1422, 2003.

Emmanuel Guerre, Isabelle Perrigne, and Quang Vuong. Optimal nonparametric estimation of first-price auctions. *Econometrica*, 68(3):525–574, 2000.

Philip A. Haile and Elie T. Tamer. Inference with incomplete model of English auctions. *Journal of Political Economy*, 111(1), 2003.

Philip A. Haile, Han Hong, and Matt Shum. Nonparametric tests for common values in first-price sealed-bid auctions. Yale, November 2006.

Lars Peter Hansen. Large sample properties of generalized method of moments estimators. *Econometrica*, 50(4):1029–1054, 1982.

Kenneth Hendricks, Joris Pinkse, and Robert H. Porter. Empirical implications of equilibrium bidding in first-price, symmetric, common value auctions. *Review of Economic Studies*, 70:115–145, 2003.

Eric Hilt. Incentives in corporations: Evidence from the american whaling industry. *Journal of Law & Economics*, 49(1):197–227, 2006.

Bengt Holmström. Moral Hazard and Observability. *The Bell Journal of Economics*, 10(1): 74–91, Spring 1979.

Ali Hortaçsu and Joonhwi Joo. *Structural Econometric Modeling in Industrial Organization and Quantitative Marketing: Theory and Applications*. Princeton University Press, 2023.

Harold Hotelling. Stability in competition. *The Economic Journal*, 39(153): 41–57, March 1929. https://www.jstor.org/stable/2224214.

David E. Kalist. Data from the television game show friend or foe? *Journal of Statistics Education*, 12(3), 2004.

Kate Ho and Ariel Pakes. Hospital Choices, Hospital Prices, and Financial Incentives to Physicians. *The American Economic Review*, 104(12), December 2014.

Jean-Jacques Laffont and Quang Vuong. Structural analysis of auction data. *The American Economic Review*, 86(2):414–420, May 1996.

Edward P. Lazear and Paul Oyer. *Personnel Economics*, chapter 12, pages 479–519. The Handbook of Organizational Economics. Princeton University Press, Princeton, 2013. Robert Gibbons and John Roberts (eds).

Matthew S. Lewis. Price leadership and coordination in retail gasoline markets with price cycles. *International Journal of Industrial Organization*, 30(4): 342–351, July 2012.

John A. List. Friend or foe? a natural experiment of the prisoner's dilemma. *Review of Economics and Statistics*, 88(3):463–471, August 2006.

Mark D. Manuszak and Charles C. Moul. Prices and endogenous market structure in office supply superstores. *The Journal of Industrial Economics*, 56(1):94–112, March 2008.

Eric Maskin and Jean Tirole. A theory of dynamic oligopoly, ii: Price competition, kinked demand curves and edgeworth cycles. *Econometrica*, 56(3): 571–599, May 1988.

Nathan H. Miller and Matthew C. Weinberg. Understanding the price effects of the millercoors joint venture. *Econometrica*, 85(6):1763–1791, November 2017.

Nathan H. Miller, Gloria Sheu, and Matthew C. Weinberg. Oligopolistic price leadership and mergers: The united states beer industry. *The American Economic Review*, 111(10):3123–3159, October 2021.

Aviv Nevo. Measuring market power in the ready-to-eat cereal industry. Econometrica, 69(2):307–342, 2001.

Harry J. Paarsch and Han Hong (with M. Ryan Haley). *An Introduction to the Structural Econometrics of Auction Data*. MIT Press, 2006.

Mark V. Pauly. Overinsurance and public provision of insurance: The roles of moral hazard and adverse selection. *Quarterly Journal of Economics*, 88(1): 44–62, February 1974.

Martin Puterman. Markov Decision Processes: Discrete Stochastic Dynamic Programming. *John Wiley and Sons*, 1994.

Devesh Raval, Ted Rosenbaum, and Steve Tenn. Semiparametric discrete choice model: An application to hospital mergers. *Economic Inquiry*, 55(4): 1919–1944, 2017.

Devesh Raval, Ted Rosenbaum, and Nathan E. Wilson. Using disaster induced closures to evaluate discrete choice models of hospital demand. *The RAND Journal of Economics*, 53(3):561–589, 2022.

Elie T. Tamer. Incomplete simultaneous discrete response model with multiple equilibria. *The Review of Economic Studies*, 70(1):147–165, January 2003.

Katja Seim. An empirical model of firm entry with endogenous product-type choices. *The Rand Journal of Economics*; 37(3): 619-640, Autumn 2006.

Raphael Thomadsen. The effect of ownership structure on prices in geographically differentiated industries. *The RAND Journal of Economics*, 36(4): 908–929, 2005.

Jean Tirole. *The Theory of Industrial Organization.* MIT Press, 1988.

Hal Varian. Position auction. *International Journal of Industrial Organization*, 25(6):1163–1178, December 2007.

Zhongmin Wang. (Mixed) Strategy in Oligopoly Pricing: Evidence from Gasoline Price Cycles Before and Under a Timing Regulation. *Journal of Political Economy*, 117(6):987–1030, December 2009.

# Subject Index


**A**

- adverse selection xviii, 233, 235, 246
- affiliated values auctions 197; *see also* common, values auctions
- Akerlof's lemons market 234–236, 240, 246
- alternating offers bargaining game, *see* Rubinstein alternating offers bargaining game

**B**

- Bayes/Bayesian
  - game 164–165, 167
  - Nash equilibrium 163–164, 168–169, 173, 178, 183–184, 198–199, 203, 221, 223, 224, 235, 237
  - rule 164, 99
- beliefs 148, 163–171, 178, 184, 198, 221–224, 234–235, 237
- Bertrand
  - game xvii, 39–40, 45, 47, 62, 118, 120, 122, 124–125, 130–132
  - Nash equilibrium 120, 122, 125–126, 128–129
- best response function 40–44, 48–49, 51, 115, 119, 147
- bid ring 192, 196–197, 207, 209, 211, 214
- biprobit/bivariate probit 75, 77–78
- bootstrap 76–77, 93, 95

**C**

- common
  - knowledge 3
  - values auction 197, 200, 203–204, 206, 209, 214; *see also* affiliated values auctions
- coordination games xvii, 79
- Cournot game xvii, 27, 39–40, 42–45, 62, 120
- cutoff strategy 143, 164

**D**

- dominant strategy xvii, 11, 13, 18–22
  - equilibrium xvii, 19–20, 22
- Dutch auction 176–177, 196

**E**

- endogeneity 25
- English auction xviii, 176–177, 183–196; *see also* Vickrey auction
- entry game 24–27, 29–30, 38, 63, 67–68, 71, 76, 79, 100, 103–104, 163, 169, 171–172, 175
- equilibrium 19–20, 22, 24, 26, 28; *see also* Nash, equilibrium
  - refinement 24, 103, 107
- extensive form representation xv, 103–104, 106–107, 135, 143–145; *see also* game tree; normal form representation

**F**

- field experiment 17
- finite difference estimator 181
- first

price auction 177, 199–200, 203, 205, 209; *see also* sealed bid auctions
mover advantage 108
fixed effects, *see* two-way fixed effects

**G**
game tree 106–108, 135, 144, 146; *see also* extensive form representation
generalized method of moments (GMM) 108
Grim-trigger strategy 117

**H**
Hotelling game/line xvii, 39, 47–48, 61–62

**L**
lemons market, *see* Akerlof's lemons market
linear regression xv, 15, 17, 33, 51, 156, 157, 188, 206
logit xv, 15–17, 118–119, 122, 126, 140, 153–154; *see also* probit; softmax

**M**
marginal cost 27, 39, 40, 42, 44–45, 47–48, 52–54, 121–122, 125–126
maximum likelihood estimation xv, 68, 74, 171, 204
method of moments 53, 92; *see also* generalized method of moments
mixed strategy xv, 79–80, 82–86, 92, 95–98, 100, 132; *see also* pure strategy
moral hazard xviii, 133, 150–152

**N**
Nash
bargaining model/solution xviii, 133, 150–152
equilibrium xv, xvii, 18–19, 22–24, 26, 28–29, 38–41, 43, 47–49, 51, 58–59, 79–85, 92, 95–100, 105–108, 110–111, 114–117, 119–126, 128–129, 134, 136–138, 140, 142–143, 145, 147–148, 150, 163, 168, 196
non-parametric estimate 180, 186–187, 196
normal form representation xv, 10, 12, 20–21, 26, 81, 83, 85, 89, 96, 103, 115, 135–136; *see also* extensive form representation

**O**
oligopoly xvii–xviii, 39, 61, 113, 124–125
Order statistic 177, 190, 193, 196, 198–200
Ordinary least squares, *see* linear regression

**P**
Pareto optimal 11, 114
pipe operator 13
pooling equilibrium 233, 235, 240, 243–244; *see also* separating equilibrium
principal-agent problem xviii, 219–220, 223, 232
prisoner's dilemma game xvii, 3, 10, 12–13, 15, 20–23, 114–117, 124–125
probit 15–17, 65, 67, 74–75, 77–78, 171–173; *see also* logit
pure strategy 82–84; *see also* mixed strategy

**R**
reduced form estimation 171–172, 175; *see also* structural modeling
Rubinstein alternating offers bargaining game xviii, 133, 143–151

**S**

screening game 233; *see also* signaling game
sealed bid auctions xviii, 177–178, 183–184, 188–192, 195–196, 200; *see also* first, price auctions
second-price sealed bid auction, *see* Vickrey auction
signaling game 233; *see also* screening game
softmax 71; *see also* logit
strategically equivalent 177
structural modeling xiv, 31–33, 35, 38, 59, 100, 113, 171, 172; *see also* reduced form estimation
subgame 108–109, 138, 147, 221
  perfection xv, xviii, 103–104, 107–108, 110, 138, 140, 142–144, 147, 159
super markups (smu) 125–128

**T**

two-step estimation xv, 171, 173, 196
two-way fixed effects xv, 228
type 164, 167–170, 178, 193, 233–240

**U**

ultimatum game xviii, 133–138, 143–145, 147, 159

**V**

Vickrey auction 177; *see also* English auction

**W**

weakly dominant strategy 13, 19; *see also* dominant strategy
  equilibrium 19
Winner's curse 198–199, 209

**Z**

zero-sum game xvii, 80–82, 89, 96

# Author Index


**A**
Adams, Christopher P. 73, 79, 99, 112, 175
Akerlof, George A. 233–236, 240, 246
Andersen, Steffen 138
Aryal, Gaurab 183, 196
Ashenfelter, Orley 25, 38
Ashmore, David 25, 38
Asker, John 192
Athey, Susan 177, 183, 185, 186, 196

**B**
Backus, Matthew 131
Baker, Jonathan 25, 38
Balan, David J. 160
Baldwin, Laura H. 188, 196
Basker, Emek 73, 79, 112, 175
Berry, Steven 62
Bertrand, François 42, 45, 47
Brand, Keith 160
Bresnahan, Timothy F. 24, 30, 32, 38, 79, 175
Byrne, David P. 113, 131

**C**
Cameron, Lisa A. 159
Capps, Cory 154, 159
Conlon, Christopher 131
Cournot, Augustine 40

**D**
Deltas, George 193
de Roos, Nicolas 113, 131
Dranove, David 154, 159

**E**
Ertaç, Seda 138

**F**
Fiedler, Matthew 246

**G**
Garmon, Christopher 160
Gaynor, Martin 159
Gibbons, Robert xvii, xxi
Gleason, Suzanne 25, 38
Gneezy, Uri 138
Gowrisankaran, Gautam 159
Grundl, Serafin 183, 196
Guerre, Emmanuel 171, 175, 177, 196

**H**
Haile, Philip A. 177, 183–186, 188, 196, 214
Hansen, Lars Peter 100
Harsayni, John 163
Hendricks, Kenneth 215
Hilt, Eric 226, 232
Hoffman, Moshe 138
Holmström, Bengt 232
Hong, Han xxi, 188, 214–215
Hosken, Daniel 25, 38
Hotelling, Harold 47, 61

**K**
Kalist, David E. 13, 17
Kim, Dong-Hyuk 183, 196

**L**
Laffont, Jean-Jacques 214
Lazear, Edward P. 232
Levin, Jonathan 196
Levinsohn, James 62
Lewis, Matthew S. 131
List, John A. 17, 138

**M**
Marshall, Robert C. 188, 196
McFadden, Daniel 63
Miller, Nathan H. 113, 131
Morgenstern, Oskar 18

**N**
Nash, John C. 58
Nash, John F. 18, 22, 58, 61, 80, 82, 150
Nevo, Aviv 131

**O**
Oyer, Paul 232

**P**
Paarsch, Harry J. xxi, 215
Pakes, Ariel 62
Pauly, Mark V. 246
Perrigne, Isabelle 171, 175, 177, 196
Pinske, Joris 215
Porter, Robert H. 215

**R**
Raval, Devesh 160
Richard, Jean-Francois 188, 196
Reiss, Peter C. 24, 30, 32, 38, 79, 175
Rosenbaum, Ted 160
Rubinstein, Ariel 145

**S**
Satterthwaite, Mark 154, 159
Seim, Katja 164, 175
Seira, Enrique 196
Sheu, Gloria 113, 131
Shum, Matt 188, 214
Sinkinson, Michael 131

**T**
Tamer, Elie T. 79, 175, 184
Tenn, Steve 160
Thomadsen, Raphael xx, 49, 51
Town, Robert J. 154, 159

**V**
Vickrey, William 177, 179, 183, 184, 196
Varian, Hal 176
Vogt, William B. 159
von Neumann, John 18
Vuong, Quang 171, 175, 177, 196, 214

**W**
Weinberg, Matthew C. 113, 131
Wilson, Nathan E. 160

**Z**
Zhu, Yu 183, 196

FSC
www.fsc.org
MIX
Papier | Fördert
gute Waldnutzung
FSC® C083411

Kolibri 360 GmbH
Ferdinand-Jühlke-Straße 7
99095 Erfurt, Deutschland
produktsicherheit@kolibri360.de